The Concepts of Physics

The Concepts of Physics

Amit Goswami
University of Oregon

D. C. Heath and Company
Lexington, Massachusetts • Toronto

Cover photograph: M. C. Escher, "Waterfall," Escher Foundation—
Haags Gemeentemuseum—The Hague

To Maggie,
the first reader of this book,
and to all future readers,
with love.

Preface

An introductory physics textbook at the conceptual or descriptive level requires that the fundamental principles be introduced in an appealing manner with the use of very little mathematics. I have written this book in an entertaining and enthusiastic style in order to take advantage of students' natural curiosity about physics. When students feel the excitement inherent in its concepts their interest in physics is stimulated and their desire to learn intensified. Above all it is this sense of fascination that *The Concepts of Physics* attempts to convey.

The topics covered and the order of coverage are relatively traditional, but in order to motivate students and to set the tone for later chapters on Einstein's special and general relativity and cosmology, I have chosen to begin with a novel prologue on the basic topics of space and time. (Cover it in a cursory manner if you wish.) The next six chapters deal with Newtonian mechanics, culminating with Newton's law of gravitation. The approach is a blend of the historical, conceptual, and applied. I emphasize the processes that led Galileo, Kepler, and Newton to their discoveries—an approach that is particularly stimulating to the nonscientist. A chapter is then devoted to reference frames as a precursor to Einstein's relativity.

Chapters 8 to 11 are an introduction to the concepts of matter, energy, and waves, including (a) discussions of scaling, (b) a thorough treatment of entropy, (c) a discussion of black holes and an entertaining discourse on heat death of the universe, and (d) a detailed analysis of those properties of waves that distinguish them from matter. The important topic of energy and society is introduced here and further developed in Chapters 13, 14, 17, and 24, where both technological and environmental aspects of energy are discussed. The contemporary problems of energy and environment are covered more thoroughly than in most introductory physics texts.

Chapter 12 deals further with the classical aspects of sound and light, which were introduced in Chapter 11, and covers the intriguing topic of optical illusions. Chapters 15, 16, and 18, which deal with electromagnetism, emphasize the concept of field and Maxwell's discovery.

The final eight chapters of the book deal with modern physics. Chapters

19 to 21 are perhaps the most novel in the entire book. These chapters deal with special relativity and quantum mechanics, again blending historical and conceptual approaches. The next two chapters systematically present topics from all major fields of modern physics practiced today. Chapter 24 contains an objective discussion of the pros and cons of nuclear power. Finally, Chapters 25 and 26 deal with Einstein's general relativity and the structure of the universe, a culmination of the continuing theme of space and time.

The treatment of topics is basically nonmathematical, with the qualitative and conceptual aspects of physics stressed throughout. However, I believe that mathematical logic is occasionally simpler than verbal reasoning; one arithmetic problem can sometimes aid comprehension more than pages of descriptive physics. On such occasions I have presented numerical problems and their solutions as an option along with relevant motivational material (for example, see "Why Are Hindus Vegetarian?" in Chapter 14). This optional material is set off from the main text.

Throughout the text a special effort has been made to communicate the human aspects of physics. Besides using historical anecdotes I have delved into philosophy (see "Can Matter Be Replaced by Fields?" in Chapter 25) and poetry. However, the format of the book is such that it allows easy distinction between the motivational elements and the text itself. Original cartoons also emphasize this humanistic side of physics.

Other pedagogical aids include the following: a summary is given for each chapter; the exercises at the end of each chapter are divided into conceptual and arithmetic questions; three mathematical appendices are given at the end of the book along with a list of physical constants and a comprehensive conversion table. The answers to numerical exercises are also given. In addition, references to related articles in *Scientific American* and similar journals are provided.

This book has evolved from a full-year course, but it is quite suitable for use as a one-semester text. Chapters 1, 2, 9, 10, 13–15, 17, 22–24, and parts of Chapters 8, 16, and 18 stress energy and the environment; the Prologue and Chapters 1, 2, 4, 6, 7, 9–11, 15, 19, 20, 25, 26 and parts of Chapters 16 and 18 emphasize the theme of space and time.

Scientists and nonscientists are often portrayed as belonging to two different cultures. It is my intention that this book help bridge the gap between them.

As with the development of physical theories, the writing of a physics book is greatly facilitated by the existence of the "shoulders of giants." To begin with, I must acknowledge the help received from such great books as the *Feynman Lectures in Physics* and Kenneth Ford's *Classical and Modern Physics*. I have also been encouraged by the spirits expressed in three other books: Robert H. March's *Physics for Poets*, Fritjof Capra's *The Tao of Physics*, and Kim Malville's *A Feather for Daedalus*.

Many colleagues have helped me by offering their kind criticisms, among them R. L. Zimmerman, M. J. Moravcsik, P. L. Csonka, K. Park, D. Soper, N. G. Deshpande, R. J. Donnelly and J. W. McClure. Special thanks are due J. Wilson at Lander College and J. Overly at the University of Oregon for their reviews of the manuscript and their many suggestions for improvement.

Many friends, both scientists and nonscientists, have helped in terms both of facts and criticism and of encouragement. Alan Goodman attracted my attention to the works of the romantic poets in relation to determinism. Damon Knight edited and critically read the first few chapters from a nonscientist's point of view. Kate Wilhelm came through with crucial words of encouragement. Orhan Nalcioglu, Steve Texin, Fleetwood Bernstein, and Linda Wallmark gave valuable criticism of various chapters. I am also grateful to the staff at D. C. Heath for putting everything together.

Finally, this book could not have been written without the help and enthusiasm of my wife Maggie, who not only aided me with such things as editing, but whose spiritual help was instrumental in the writing of the book.

A. G.

Contents

To The Student

Physics deals with your environment. You look at the sky and the rivers, you hear the sounds of the ocean, you touch and are touched by the wind, and you wonder. The physics of these sights and sounds enables you to focus your awareness of the unity of these things. Physics gives you a unified view of all natural phenomena. It is exciting to look at nature from one unified, fundamental point of view, and I want you to experience that excitement.

Prologue: Questions about Space and Time

The most incomprehensible thing about the universe is that it is comprehensible.

ALBERT EINSTEIN

■ Questions about Space

People have asked questions about space and have attempted answers to some of these questions since the dawn of civilization. In fact, some of our current ideas about space have their origin in the thinking of the ancients.

The two major concerns of the ancients about space involved the questions of its absoluteness and whether it is finite or infinite. With the rise of modern science, further questions were added to the list. What is the geometry of space? Does space evolve with time or is it unchangeable? How many of these questions have you wondered about? It is very likely that you have thought about most of these aspects of space in one way or another.

This Prologue is about these questions. It will take the remainder of the book to see what answers we have found and how reliable those answers are.

Let us, then, begin at the beginning, with the first question.

Is space absolute?

What do we mean by the statement that space is **absolute?** We mean that space exists independently of matter; space comes first, and matter is placed in space as grains of wheat can be placed in a box. In this "box" model of space, we cannot think of matter existing without space, but of course space can exist without matter. Correspondingly, the developers of this concept thought of vast regions of empty space in which the stars or matter "float." The following quotation, from the writings of Ko Hung of the ancient Chinese school of thought, illustrates the point: "The sun, the moon and the company of stars float freely in the empty space, moving or standing still."*

There is, however, an alternative way of looking at space. Space can be regarded as a concept necessary for the ordering of material objects. From this viewpoint space is just a generalization of the concept of place assigned to matter. Matter comes first and the concept of space is secondary, necessary only in the relative description of material objects. Space exists because matter exists; empty space has no meaning. The ancient Greeks, led by the great philosopher Aristotle, thought about space in this fashion.

How do you picture space? As a container of material objects or as a collection of adjacent boundaries of material objects? Do you ever think of what we call outer space as a vast amount of nothingness? Or do you think that it is necessary to talk about space only in connection with matter and its positional quality?

In case you are curious about which concept has turned out to be correct, it may be pertinent to point out that the evolution of physics has encompassed both approaches in different periods. Isaac Newton, the great English physicist, built a theoretical framework for the study of motion that led him to believe in the concept of absolute space, as we will see in a later chapter. Much later, German-American physicist Albert Einstein was able to show that space is not completely independent of matter and thus cannot be regarded as absolute. The other important property of absolute space, namely, the existence of empty space, also has been challenged, first by the concept of force fields, now known to pervade such emptiness, and, more recently, by the experimental observation of background radiation in the universe.

Thus we will find that the ancient Chinese were wrong, as was Newton. However, we should not undervalue the important role the concept of absolute space played in each of their cultures. The concept of a vast empty space in which the earth is just a small grain gave the Chinese a profound humility that still is part of their philosophy. And Newton's work provided the foundation for the modern scientific era of civilization.

Is space finite or infinite? Bounded or unbounded?

There is a beautiful story in the Buddhist literature. A monk was interrogating the gods in order to find out where the universe ends but to no avail. Finally, the god of creation himself appeared. The monk asked him where the universe ends. To the monk's surprise the creator took him aside and said, "These gods, my subordinates, hold me in such respect that there is nothing I cannot see or understand. So I just couldn't answer you in their presence. You see, I do not know where the universe ends."

The story illustrates the view that no one, not even the creator himself, can imagine the end of the universe. On the other hand, many ancient mythologies have it the other way: a picture of space that is finite in extent. Perhaps the most famous of these pictorial representations of space is that given by Aristotle and

---

*J. Needham, *Science and Civilization in China*, vol. 3 (London: Cambridge University Press, 1959).

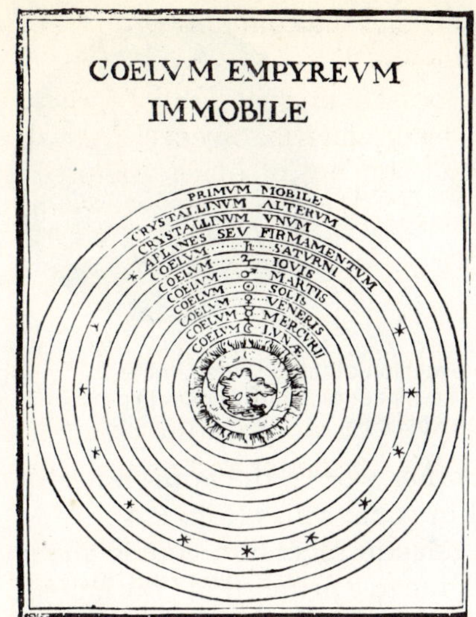

FIGURE P.1 A late medieval construction of Aristotle's universe. The terrestrial region, consisting of the four elements earth, water, air, and fire, is separated from the celestial region by the sphere of the moon (lune). Beyond are the concentric spheres, carrying the planets and the sun. The last sphere carries the fixed stars. (From a woodcut of 1508).

immortalized by the Italian poet Dante in his *Divine Comedy*. According to this representation, the earth is the innermost sphere in a series of concentric spheres, with the outermost sphere carrying the fixed stars and representing the limits or boundary of space (Fig. P.1). It is not surprising that Aristotle's space would be a finite one. Space in his view was a secondary phenomenon of matter, and since matter is finite, it is only logical that space would be finite too.

Isn't this idea of space having a boundary paradoxical? If space has a boundary, what happens if you look outside this boundary? What happens if you stand on such a boundary and raise your hand or shoot an arrow? Such nontrivial questions were asked by the Greeks and continued to be asked in the medieval era when Aristotle was rediscovered.

It is clear that a boundary separates two regions. Thus space, which is only one region, cannot have boundaries; space is unbounded. At the same time we cannot jump to the conclusion that space has to be infinite. Strangely, it is possible to think of space as finite yet unbounded. In an infinite space we think of straight lines as extending forever. But imagine instead a "straight" line that, when extended great distances, turns back onto itself instead of continuing indefinitely. This is what happens in a finite but unbounded space. A complete understanding of this idea requires a consideration of the geometry of space, which we will defer until Chapter 25. For now, consider Fig. P.2, which presents some analogies.

Steady state versus "big bang" universe

The questions mentioned so far are all ancient questions. Let's now turn our attention to a rather recent controversy that has arisen mainly from the advances of observational astronomy. Perhaps you know that our solar system, with the star we call our sun at its center, is a part of a collection of stars known as the Milky Way. Such a collection of stars is referred to as a **galaxy** (Fig. P.3). There are many, many galaxies in our universe. Now we come to the surprising thing. Observational evidence indicates that other galaxies are moving away from ours, and the further away a galaxy is from ours, the faster it is receding.

This unsettles a classical philosophical belief, namely, that space does not change or evolve with time. We are, of course, identifying space with the universe, which seems sensible since the universe is all we can observe.

The **steady state theory** of the universe, as developed by three physicists, Fred Hoyle and Hermann Bondi of England and Thomas Gold of the United States, shifts the burden of change from space to matter. In this theory, matter is continuously created out of nothing at a rate that precisely offsets the change created by the receding galaxies. Thus the overall distribution of matter in space

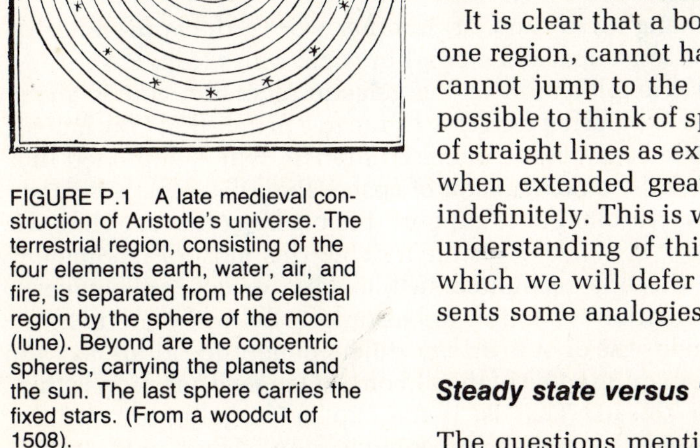

(a)

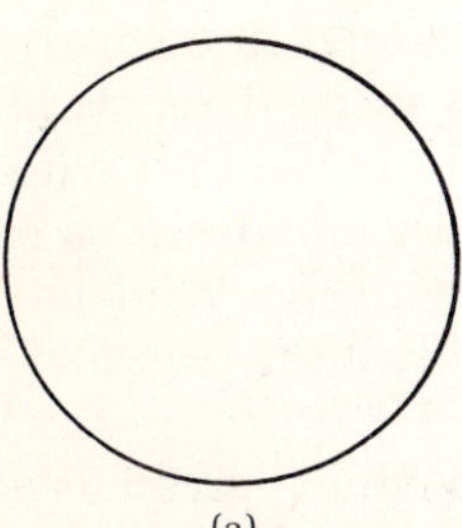

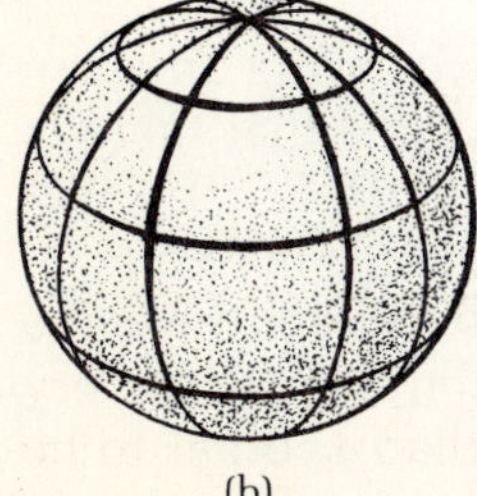

(b)

FIGURE P.2 (a) A circle, a one-dimensional curve, has no end or boundary. (b) The two-dimensional surface of a sphere has no surface boundary. We can theorize about a finite, boundless space as the three-dimensional generalization of the one-dimensional circle and the two-dimensional surface of the sphere, but we cannot draw it. You will have to imagine what we mean.

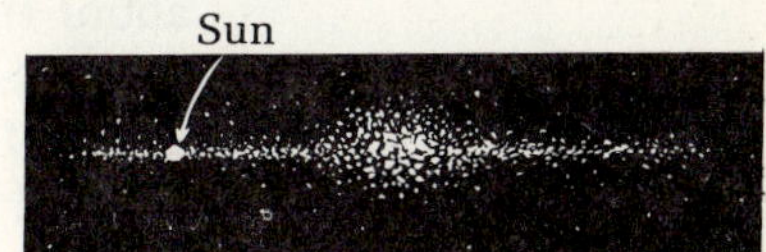

FIGURE P.3 A schematic picture of our galaxy, the Milky Way, and the position of our sun in it.

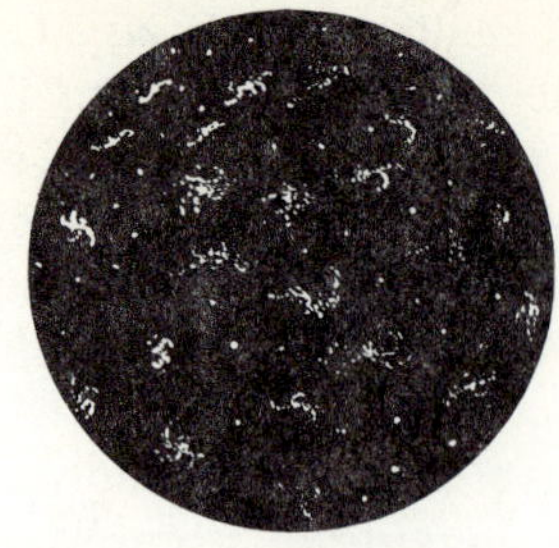

FIGURE P.4 The evolution of the steady state universe. As the universe expands, new matter is created so that the distribution of matter in any segment of the universe always looks the same.

FIGURE P.5 The evolution of the universe in the "big bang" model. The matter distribution gets thinner as the universe expands.

is considered to be unchanging in time, or, in other words, the appearance of space is not changing in time; space is eternal in some sense (Fig. P.4).

A more straightforward explanation of the galactic recession comes about when we assume that the expanding universe is due to a primordial **"big bang."** In this theory the universe began as a condensed fireball, which exploded in a big bang and was followed by an expansion of space (Fig. P.5). Most scientists now believe that the observational data support this big bang theory. Thus we see that the universe was "created." This theory confirms most of the ancient conceptions, including the one recounted in the Bible. In scientific theory, however, the date for "creation" is found to be some 10 billion years B.C.

The big bang theory opens the door to an assortment of new questions, as well as to the possibility of reshaping some of the old ones. One new question is this: Is the universe open or closed? That is, is the expansion going to continue forever (open universe), or is the expansion going to stop (closed universe)? If the expansion continues indefinitely, then our universe is infinite; if the expansion stops at some future date, then our universe is finite.

Many thousands of years ago, the philosophers of India imagined a model of the universe that contains the elements of both the steady state and big bang theories. In this model an infinite universe is created, maintained in a steady state for a long time, and then destroyed, only to be created again. This cycle goes on endlessly. Brahma is the God of the creation, Vishnu is the God who maintains the created universe, and Shiva, the King of the dancers, destroys it when the time comes. More recently, the American physicist John Wheeler, among others, has talked about a similar **cyclical universe** with alternate expansion and contraction cycles (Fig. P.6).

Which of these models is correct? The observations so far have not given us a final answer. We will discuss the current status of these models and the observational data in Chapter 26.

One question frequently brought up in the context of the expanding universe is about the center of the universe. Unfortunately, we must state that there is none. But, you may ask, if the galaxies are moving away from us, doesn't that imply that we are at the center of the universe? The answer again is no. You can see this if you consider that observers in another galaxy would feel the same way, since they too would see all other galaxies receding from them. So every galaxy recedes from every other, and none has more right than another to be called the center.

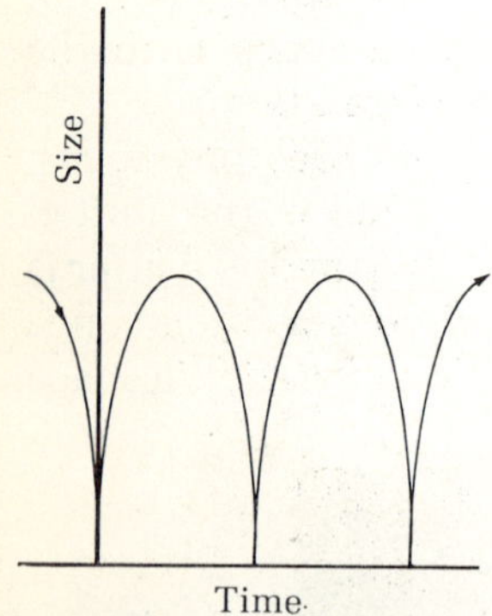

FIGURE P.6 In the cyclical model of the universe, its size alternately expands and contracts.

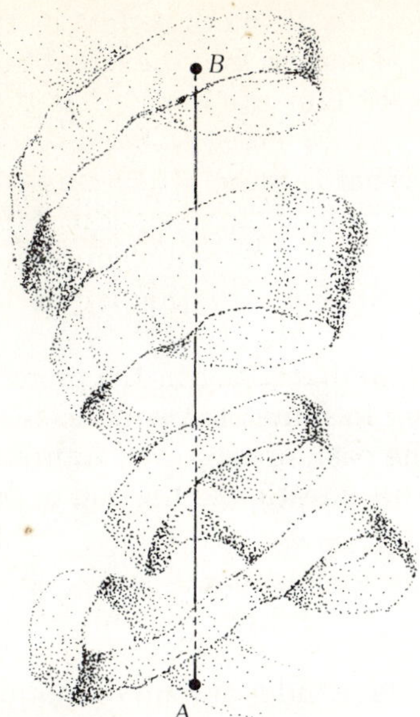

FIGURE P.7 Artist's conception of the hyperspace imagined by science fiction writers. There are two ways to go from *A* to *B*. One way is to follow the strip the long way, which is like regular space travel. The other way is to follow the straight line *AB* going through the strip and the hyperspace.

More exotic ideas about space

Then there are questions you may have about space for which we have no answers. For example, no one knows anything worthwhile about the science fiction writers' hyperspace. We can tell you what is meant by hyperspace. If space is curved, as in Fig. P.7, then we can imagine taking a shortcut, as shown, from *A* to *B*. The trouble is that so far nobody knows how to go about finding such a short route, even if one exists. However, one model that has been proposed assumes that the short trip involves a parallel universe. Some scientists recently have suggested that tunnels to such parallel universes may actually exist through holes in space called "black holes." In their model, mass disappears through the black holes to other parallel universes. In fact, some writers even imagine that the disappeared masses reappear into the other universe through "white holes." As evidence for "white holes," these people cite the phenomenon of **quasars** (an abbreviation for "quasi-stellar radio sources," which are very bright, move at very high speed, and perhaps are very far from us).

Much as we would like to know more about such things as hyperspace and parallel universes, at this time they must be regarded as purely imaginary. Why, then, mention them at all? The answer is best given in the words of the American poet Peter Nowell*:

> Quasars:
> the furthest "stars"
> with the presence of a hundred galaxies.
> What lies beyond
> the blizzard of lights
> we call the milky way;
> beyond the dogma
> of our sluggish imagination?
> Push the limits
> and space will yawn to receive us
> once more.

The answers will come as we "push the limits."

■ What Is Time?

G. J. Whitrow, in his book *The Nature of Time*, tells an interesting anecdote.† Samuel Marshak, a Russian poet, was on the streets of London without a watch when he was struck by an irresistible urge to know what time it was. So he went up to a man on the street and asked in his foreign English, "What's time?" The man looked very surprised and replied, "But that's a philosophical question. Why ask me?"

Although, like the man on the street, you probably would be reluctant to answer the question, "What's time?" you have probably given at least some thought to this concept. What questions emerge?

---

*"Pushing the Limits," by Peter Nowell. Reprinted by permission from the Christian Science Monitor. © 1976 The Christian Science Publishing Society. All rights reserved.
†G. J. Whitrow, *The Nature of Time* (London: Penguin Books, 1975), p. 1.

One thing is clear—time is associated with flow, with motion. Just as the concept of space is intimately connected with that of matter, the concept of time evolves around that of motion. Which comes first? Time or motion? Is time absolute?

In contrast to space, time is one-dimensional. What is more, it is a one-way street, always flowing toward the future. What determines the direction of this one-way street called time's arrow? Is time infinite? Does it have a beginning or an end? All these questions are scientific, as we will see. No, time is not just philosophy!

There is, however, one philosophical question, and this concerns the premise that time is linear. What is the basis of this premise? In some philosophies there is reference to nonlinear time, which means that the past, present, and future all exist together. Can this happen? We will discuss this question at the end of this section.

Time, absolute or relative?

The statement that time is **absolute** means that time is independent of motion, flowing irrespective of whether anything else changes. Time goes on even without motion, but motion or change must occur in time.

This view of time is found, for example, in the philosophy of ancient India:

> Time generated yonder sky, time also these earths; what is and what is to be stands out sent forth by time.
> Time created the earth; in time burns the sun, in time are all existences; in time the eye looks abroad.*

In modern times Newton made the following statement in his celebrated book, *The Principia*, proclaiming that time is **absolute:** "Absolute, true and mathematical time, of itself and from its own nature, flows equably without relation to anything external."

Clearly, according to Newton, moments of absolute time in sequence are like points of a straight line, and the rate at which the points on this straight line succeed each other is a constant. Newton was aware that there may be no ideal way to measure the ticking of absolute time, so he added: "Relative, apparent and vulgar time, is some sensible measure of absolute time, estimated by the motions of bodies, whether accurate or unequable, and is commonly used instead of true time; such as an hour, a day, a month or a week." So what Newton was really saying is that absolute time, for which no clocks may exist, is nevertheless a valid concept.

Most people, even today, think of time much as Newton did. Historically, Newton's ideas have been questioned for almost as long as they have been around. German mathematician G. W. Leibniz, Newton's contemporary, developed an alternative theory of time as an ordering of events. Events are fundamental and moments of time are relative in Leibniz's theory. Two events are simultaneous because they occur together, not because they occupy the same

---

*Swami Nikhilananda, trans., *The Upanishads* (New York: Harper & Row, 1963), p. 20.

moment of absolute time. The separateness of events from one another insures that time progresses. If there were no change, no separateness of events, the universe would exist as a single event and time would stop. Leibniz's theory is called the **relational theory of time** and is a precursor of Einstein's theory of relativity.

The difference in Einstein's approach to the problem of time is that he took the measurement process more seriously than did anybody before him. In showing that **time is relative,** depending on the observer, Einstein derived physical consequences of the relativity of time, consequences that could be checked experimentally. Before Einstein, time, for all practical purposes, was just a parameter for the description of motion. The discussions of the nature of time were basically philosophical. Einstein changed all this and truly established the scientific nature of time. Einstein's theory of relativity requires fuller development than can be given here and therefore will be deferred until a later chapter.

Einstein's relativity also led to a combination of the two separate concepts of space and time into one unified concept of space-time. Amazingly, the ancient Chinese coined a word, "yu-chou," which means space-time. Does this mean that the Chinese anticipated relativity by thousands of years?

The arrow of time

There are two primary aspects of natural events: events that repeat themselves and events that do not. For example, the rotation of the earth is a repetitive event; such events enable us to develop quantitative measures of time. On the other hand there also are events that are nonrepetitive, progressive, and irreversible: if a glass breaks, its fragments get scattered; if you open a bottle of perfume, the molecules of the perfume distribute themselves through the room*; as an object gets hotter, its molecules exhibit ever more disorderly motion. All these events are examples of **order-to-disorder** transition. Thus the glass, when unbroken, is an orderly juxtaposition (side-by-side placement) of its molecules, whereas the broken glass signifies a more disorderly state of the molecules (Fig. P.8).

Some primitive civilizations had very little awareness of events of a progressive nature. Life for the people of these civilizations was one of repetition of the same tradition over and over again from generation to generation. Their concept of time was essentially cyclical. In contrast, advanced civilizations have noticed the progressive nature of time and developed the concept of linear time.

If we knew of universal processes that record the progress of time at a uniform rate everywhere, we could even devise a master clock on the basis of the rate of these processes. As far as we know, no such universal process exists. On the other hand there exists a physical quantity associated with the order-to-disorder processes; its increase in magnitude can be associated with the passage of time from the past to the future. This quantity is called **entropy** and denotes the amount of disorder or randomness associated with a system (an assembly of

---

*The submicroscopic particles of which bulk matter is made are called molecules.

FIGURE P.8 Some examples of the order-to-disorder transition from everyday life.

related objects). It is a law of nature that for all large closed systems, the entropy always increases. And the direction of entropy increase establishes the direction of the **arrow of time.**

There is a story about filming a scene in a silent movie in a time-reversed fashion.* The heroine was supposed to be tied to the railroad tracks while a train was about to run her over. Of course in the story she would be saved—the train would stop just in the nick of time; but the actress was reluctant to risk her life. So they shot the whole sequence backward, starting with the actress tied to the tracks while the train was just ahead of her at a full stop. Then the train was run in reverse. But what do you suppose people saw when the film was run backwards? In those days trains ran by burning coal to make steam in a boiler. In the backward-running film, the smoke appeared to be flowing into the stack instead of flowing out, giving away the movie's secret. In real life smoke coming out of a stack or chimney (an ordered system) becomes more dispersed as it reaches the outside, and so order increasingly changes into disorder and entropy increases. Thus it was easy to tell the real direction of time in the sequence.

You may argue that if a film producer used an electric train, which could certainly be done today, then nobody would be able to tell how the film was made. While it is true that such a movie could fool most viewers, the truth could not escape scientific scrutiny. An expert could always tell the direction of time by taking careful notice of natural processes that must proceed in the direction of increasing entropy, thus giving adequate clues for determining the direction of time in the film.

---

*J. Bronowsky, *Insight* (New York: Harper & Row, 1964), p. 68.

Does time have a beginning or an end?

We have talked previously about the big bang model, which portrays the creation of the universe. Since the creation the universe has been expanding, but an interesting question is this: what existed before the creation? A story about theologian St. Augustine illustrates the point. One day St. Augustine was talking to a gathering on this kind of question, the beginning of time and so on, no doubt from a theological point of view. Somebody from his audience made a rather thought-provoking inquiry: "Now look here, Augustine. You have told us that in the beginning God created heaven and earth. You have also told us that God is immortal with no beginning and no end. What then was God doing before He created heaven and earth?"

What happened before the beginning of creation is a very legitimate question, then and in the present scientific framework. [Incidentally, St. Augustine gave a rather unsympathetic answer: "He (God) was creating hell for people who ask such questions." Unfortunately for our purposes, his answer is not scientifically useful.] Questions like "what existed before the beginning" presuppose that the creation of the universe occurs *in time:* this implies that time is absolute. We can counter this by asserting that the creation of the universe is also the creation of time itself.

As hard as it is to imagine that time has a beginning, perhaps it is even harder to imagine that time has an end. In some models of the universe that physicists study today, time does appear to have an end. On the other hand, remember the cyclical model of the universal expansion-contraction. In this model the universe itself is a cyclical event; there is no element of progressiveness, no arrow of time. Thus as a whole the universe is timeless, atemporal. Only within each cycle can we notice progressive events and mark an arrow of time. Thus time in each cycle exists with a finite beginning and a finite end, but as a whole the universe is timeless.

It is not an easy matter to choose between these models. Perhaps the most amazing thing about modern science is that today such questions are not philosophical ones. We almost have the necessary data to answer such questions; maybe we will find the answer in the years to come, perhaps even within your lifetime.

■ Scales and Ladders of Space and Time

The vistas covered by physics range in extremes. On one hand we talk about the size of the universe and on the other about the size of an atom (the submicroscopic component of matter). The first is a large number, and, written in long form, we would have to write 26 zeros after 1 before we could express the size of the universe in meters (abbreviated m). For the size of the atom, the problem is writing a very small number: in the usual decimal notation, it is approximately 0.0000000001 meter.

In talking about the very large and the very small, a mathematical notation called the **power-of-ten**, or **scientific**, **notation** is highly useful. In this notation we write a number as a product: the first number of the product is a number between 1 and 10 and the second is 10 raised to some power.

What do we mean by linear time? This is the idea that the course of time can be represented by a line, with the succession of instants of time depicted by points on the line. All the questions we have considered thus far implicitly assume that time is linear.

This concept of linear time seems to be the product of the evolution of our consciousness, to some extent, and it depends on our ability to distinguish between past, present, and future. There is some evidence that animals with very poorly developed consciousness cannot make such a distinction and live in a sort of eternal present. There is also some evidence that many primitive civilizations thought of time as cyclical, with events repeating periodically. This is one kind of nonlinear time. However, it is not very interesting since we can "disprove" its reality.

Before discussing another kind of nonlinear time, let us make a few observations on the psychology of linear time. How do we, for example, think of the past? The answer is, of course, through our memory, through records. Although memory does play some tricks sometimes, particularly in our perception of the duration of a particular time sequence, our memory is generally reliable in its recall of the past. Thus the past "exists" in our consciousness. The way we consider the future is more subtle. "Future" is a creative concept. We can think about the future because we have a limited capability of creating it in the sense of making reasonable plans for it. Actually, the concept of the present—now—is perhaps most baffling. Now is where the past ends and the future begins, but it is difficult to be more specific.

Civilized people more or less always have believed in the linearity of time, but the concept of nonlinear time nevertheless has existed. Witness the following quotation from the Bible: "That which hath been is now; and that which is to be hath already been. . . ."* The traditional linear view of events unfolding in time is changed in this nonlinear view into one of the unraveling of events that have already happened.

The following episode in the Bhagavad Gita, a classic of ancient Indian philosophy, elaborates the point. A fierce battle to reestablish the "right" is about to begin, but the battle is between two factions who are blood relatives. Just before the battle, Arjuna, the leader of the "right" faction, loses his appetite for fighting and killing his relatives. Now Krishna, the incarnation of God, must convince Arjuna that it is all right for him to kill the enemy, even though they are relatives. At the peak of the

argument Krishna says to Arjuna: "The enemy has already been killed, by me. Oh Arjuna, I urge you to carry out the causal connection." Krishna then gives Arjuna the power of spiritual vision so that Arjuna can see for himself the future that is ahead of them. And Arjuna sees array after array of enemy soldiers, all dead, in a glorious display of the entire past, present, and future.

The Krishna-Arjuna anecdote makes one thing very clear: the actual killing of the enemy by Arjuna is a causal connection—it is a human necessity. The live enemy and the dead enemy both exist as two separate realities.

A similar view of time as past and future coexisting with the present is found in the Zen way of looking at things. Consider an example cited by psychologist Robert Ornstein.* According to the linear view, we see a piece of wood burning into ashes as a cause-effect sequence. In the Zen view firewood and ashes are two independent stages; firewood stays firewood and ashes are ashes. Firewood doesn't become ashes; the Zen monk feels no necessity to make the causal connection.

German philosopher Friedrich Nietzsche has expounded views similar to those of Zen. In his *Vision and Enigma*, Neitzsche writes:

Look at this gateway! . . . it has two faces. Here two roads come together: these no one has yet gone on to the end. This long lane backwards—that takes one eternity. And that long lane forward—that is another eternity. They contradict each other, these lanes;—they hit each other at their foreheads;—and it is here at this gateway that they are upon each other. The name of the gateway is inscribed above: 'Moment.'†

This is the definition of Nietzsche's moment. Later he writes, and the similarity with Zen now becomes apparent, "And if everything has already existed, what do you . . . think of this Moment. Must not this gateway too—have already existed: And are not all things closely entangled in such a way that this Moment draws all coming things after it? Consequently—itself also?"‡

A final quotation to illustrate our point is from poet T. S. Eliot's "Burnt Norton":

---

*Eccles. 3:15

*Robert Ornstein, *The Psychology of Consciousness* (San Francisco: Freeman, 1972), p. 90.
†Quoted by D. W. Dauer, "Nietzsche and the Concept of Time," in *The Study of Time*, vol. II, ed. J. Fraser and N. Lawrence, (New York: Springer-Verlag, 1975), p. 85.
‡Ibid, p. 87.

And the end and the beginning were always there
Before the beginning and after the end
And all is always now . . .
Quick now, here, now, always—
Ridiculous the waste sad time
Stretching before and after.*

———
*T. S. Eliot, *Four Quartets* (New York: Harcourt Brace Jovanovich, 1943), p. 3. Courtesy Harcourt Brace Jovanovich and Faber and Faber Ltd., London.

The Bible, Bhagavad Gita, Zen, Nietzsche, and T. S. Eliot: they represent different cultures at different times. But they all express the same intriguing thought: a view of time other than linear time is possible. This other view could come from looking at space-time from the outside, from a different dimension. Perhaps consciousness is a different dimension of our space-time-matter-motion world.

Consider an example. The radius of the earth is 6,400,000 meters. In the power-of-ten notation, we write 6.4×10^6 meters. In this power-of-ten notation, the decimal point at the end of the number has been shifted to the left until it is just after the first digit. The decimal point has been moved through 6 places and 6 is the power of ten. So the exponent, or the power of ten, really tells how many places you moved the decimal point.

For small numbers we have to shift the decimal point the other way, to the right, in order to locate it immediately after a whole number. Now the number of places we move the decimal point appears as a negative exponent of ten. For example, we write 0.0000000001 as: 1×10^{-10}; the decimal point has been shifted to the right by 10 places, making 10 the negative power of ten.

Another way to visualize a negative exponent is to think of it as 1 divided by the corresponding, positive, power-of-ten number:

$$10^{-10} = \frac{1}{10^{10}}$$

Here are some more examples for practice. The number of atoms in a 12-kilogram (abbreviated kg) block of wood is roughly 600,000,000,000,000,000,000,000,000. In scientific notation we just write this number as 6×10^{26}. (Verify this.) The size of a typical virus, those molecular organisms that give us the flu, for example, is 0.00000002 meter. In the power-of-ten notation, this number is 2×10^{-8} meter; the decimal point has to be shifted 8 places to the right.

How do we multiply two numbers written in scientific notation? We could, of course, express them in the long form, multiply, and convert back. But there is an easier way.

The rule can be stated fairly simply: in multiplying numbers written in power-of-ten notation, multiply the numbers in front of the powers of ten and add the exponents to get the new power of ten. An example will make the rule clearer.

What is $(2.5 \times 10^5) \times (3 \times 10^8)$? It is

$$(2.5 \times 3) \times 10^{5+8} = 7.5 \times 10^{13}$$

Note that we can multiply large and small numbers following the same rule:

$$(3 \times 10^{15}) \times (1.5 \times 10^{-7}) = 3 \times 1.5 \times 10^{15-7} = 4.5 \times 10^8$$

or two small numbers:

$$(4 \times 10^{-3}) \times (3 \times 10^{-6}) = 12 \times 10^{-9}$$

In the last number, 12 is greater than 10, so we rewrite it as 1.2×10^1, multiply this with 10^{-9}, and finally get

$$1.2 \times 10^{-9+1} = 1.2 \times 10^{-8}$$

Since

$$\frac{1}{10^n} = 10^{-n}$$

for any value of n, we can also carry out division of large and small numbers using this same technique. As an example, consider

$$\frac{6 \times 10^{15}}{3 \times 10^8}$$

This is equal to

$$\frac{6}{3} \times 10^{15} \times 10^{-8} \qquad \text{or} \qquad \frac{6}{3} \times 10^{15-8}$$

or, finally, 2×10^7.

The concept of **order of magnitude** is also useful. The order of magnitude of a number is the closest power of ten to which it is equal. For example, the order of magnitude of 2.5×10^8 is 10^8. The order of magnitude of 6×10^8 is 10^9, however, because it is closer to 10^9 (6 is more than halfway between 0 and 10). We also adopt the convention that 5×10^8 has the order of magnitude of 10^9.

Often we indicate order of magnitude by a special symbol, $\sim$. The size of the atom is of the order of 10^{-10} meter, and we can write this as

$$\text{size of the atom} \sim 10^{-10} \text{ meter}$$

The scales and ladders of space

Distance in space is measured in two basic units. You are familiar with the foot (abbreviated ft)—a scale that originated through the age-old use of the human foot as a standard to measure distance. More recently we have introduced the meter scale. A meter was intended originally to be 1/40,000,000 of the circumference of the earth. We use the original meter scale even today, although very accurate measurements now reveal that the correspondence of our meter stick to the earth's circumference is not exactly what was intended. A meter is only slightly longer than a yard (abbreviated yd) and 1500 meters is a "lean" mile (abbreviated mi).

$$1 \text{ ft} = 0.30 \text{ m}, \qquad 1 \text{ m} = 3.3 \text{ ft}, \qquad 1 \text{ mi} = 1609 \text{ m}$$

Both the meter and the foot are distance scales suited to everyday phenomena. The diameter of a human hair is 1/10,000 of a meter (10^{-4} m), and 10,000 meters (10^4 m) is about the distance that we can survey with the naked eye. The ratio of the largest and smallest sizes that we can perceive with our senses is about

$$\frac{10^4}{10^{-4}} = 10^8$$

In addition to using the power-of-ten notation to express such distances as 10,000 meters, it has become customary to use some prefixes. For example, "kilo" is a prefix denoting "thousand," so a kilometer (km) means a 1,000 meters and 10,000 meters can be written, and often is, as 10 kilometers. Here are a few other prefixes:

$$
\begin{aligned}
\text{mega (M)} &= 10^6 \\
\text{centi (c)} &= 10^{-2} \\
\text{milli (m)} &= 10^{-3} \\
\text{micro } (\mu) &= 10^{-6} \\
\text{nano (n)} &= 10^{-9} \\
\text{pico (p)} &= 10^{-12}
\end{aligned}
$$

Thus a microsecond means 10^{-6} second, a millionth of a second. A centimeter (cm) is 10^{-2} meter or a hundredth of a meter, and so on.*

For very great distances we sometimes use the light-year as a unit. You probably know that it takes starlight at least a few years to reach us, even from the nearest star outside our solar system. Now light travels 3×10^8 meters (or 186,000 miles) in 1 second, and the distance it travels in a year is called a light-year. In terms of meters,†

$$1 \text{ light-year} = 9.6 \times 10^{15} \text{ meters} \approx 10^{16} \text{ meters}$$

Interstellar distances (within a galaxy) are usually of the order of a few light-years. Compared to this the size of an entire galaxy is of the order of 100,000 light-years, and the intergalactic distances run to a million light-years.

The largest size we can think of is, of course, that of the entire universe. We can determine a "size" for the universe in the context of the big bang theory. If we could establish the age of the universe—that is, the time that has elapsed since the big bang creation—we could define a radius of the universe as the distance that light has traveled from the big bang in all that time. Now it turns out that the "age" of the universe can be estimated to be some 10 billion years, probably a few billion years more. But let's work with the 10-billion-year figure, to keep things simple. In 10 billion years light travels 10 billion light-years. Thus the size of the universe is 10 billion (10^{10}) light-years. Since 1 light-year is approximately 10^{16} meters, the size of the universe is also given as $\approx 10^{10} \times 10^{16} \approx 10^{26}$ meters.

Now we want to talk about sizes in the microcosm, the submicroscopic world inside bulk matter. We have mentioned molecules before: a **molecule** is the

---

*If you are having difficulty memorizing such new words as the pico, maybe the following kind of word association will help. Following the given meaning of pico, what do you get for 10^{-12} boos? A picoboo, of course. A list of such word associations has been compiled by Philip A. Simpson (quoted in R. L. Weber, comp., *A Random Walk in Science* [New York: Crane, Russak, 1973], p. 61).

†The symbol $\approx$ stands for approximately equal.

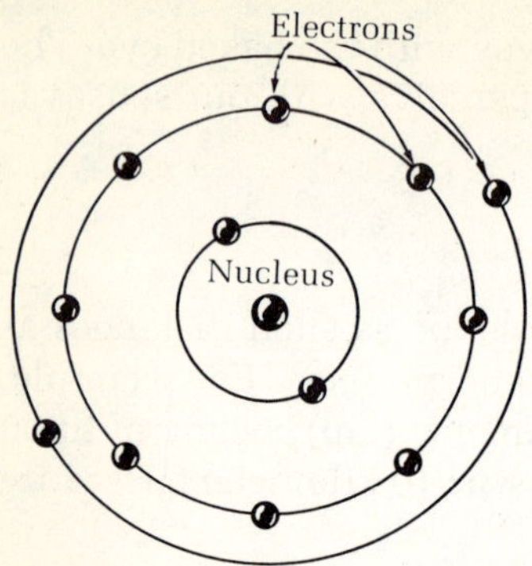

FIGURE P.9 The atom is pictured as a tiny solar system with the nucleus at its center and the electrons orbiting around it.

TABLE P.1   The ladder of spatial sizes.

	Meters (all numbers are orders of magnitude)
Size of the electron	no size to speak of
Size of subatomic particles	10^{-15}
Size of the atomic nuclei	10^{-14}
Size of the atom	10^{-10}
Size of a molecule of oil	10^{-9}
Size of a virus molecule	10^{-8}
Smallest size of a smoke particle	10^{-6}
Thickness of hair	10^{-4}
Diameter of a pencil	10^{-2}
Height of a human	1
Length of a football field	10^{2}
Distance our eye can see on earth	10^{4}
Diameter of earth	10^{7}
Distance from the earth to the sun	10^{11}
Distance to other nearby stars	10^{17}
Size of our galaxy	10^{21}
Distance between galaxies	10^{22}
Size of the universe	10^{26}

smallest particle of a substance that still retains the properties of the substance—it is the building block of all substances. The size of a molecule varies from 10^{-10} meter for small molecules to 10^{-8} meter for very large molecules.

It turns out that molecules themselves are made of still smaller particles called **atoms,** of a size on the order of 10^{-10} meter. The atom further consists of a core or **nucleus,** which carries most of its matter, and particles called **electrons,** which move around the nucleus forming something like a tiny solar system (Fig. P.9). The nucleus is the analogue of the sun, and the electrons are like the planets on a highly miniaturized scale. Now here is a surprising thing. Although the atom has a size of about 10^{-10} meter, the size of its nucleus is much smaller, only about 10^{-14} meter or so. Thus most of the atom is empty of matter.

But the atomic nucleus is not the end of the ladder. It too is found to be a composite; it is made up of subatomic particles called **protons** and **neutrons.** The sizes of these particles turn out to be about 10^{-15} meter. At the bottom line in the hierarchy of sizes are particles that have no size to speak of. The electron seems to be a particle of this kind, and there are perhaps a few others that also fall in the same category. (See Table P.1.)

The ladder of time

Let us now talk about a ladder of time (Table P.2). Human perception of time is limited to about one-tenth (10^{-1}) of a second in the lower limit and 100 years (3.2×10^{9} sec*) in the upper limit. Nature, however, shows little respect for our

---

*Although s is the standard abbreviation for second, this text uses sec for reasons of greater familiarity to students.

	Seconds *(all numbers are orders of magnitude)*
Shortest lifetime of a subatomic particle	10^{-24}
Shortest lifetime of an excited atomic nucleus	10^{-18}
Time it takes light to penetrate a windowpane	10^{-11}
Shortest time measurable by electronic clock	10^{-9}
Time of firecracker explosion	10^{-5}
Shortest time distinguishable by human ear	10^{-1}
Time between heartbeats	10^{0} (1)
One day	10^{5} (8.64×10^{4} exactly)
One year	10^{7} (3.2×10^{7} more exactly)
Human life span	10^{9}
Human civilization	10^{11}
Age of earth	10^{16}
Age of universe	10^{17}

limited perception. There are phenomena involving times as small as 10^{-24} second; on the other side of the coin, the longest time that makes sense is the age of the universe. This is estimated to be 10^{10} years or 3.2×10^{17} seconds, since 1 year $= 3.2 \times 10^{7}$ seconds.

OUTLOOK

It was English poet William Shakespeare who said:

> All the world's a stage
> And all the men and women merely players.

For the entire universe it seems that space and time set the stage; the players are matter of all kinds—bulk matter, stars and galaxies, and the submicroscopic constituents of bulk matter: atoms, molecules, and the like. We don't know exactly what the play is. The matter moves about exhibiting an astonishing variety of patterns and seems to follow a set of rules that we identify with physical laws. We do not know all the rules yet, but we are hopeful. By studying the motion of the players—matter—we also find out things about the stage— space and time. And maybe when we know enough about the stage and the players, the nature of the play will reveal itself.

Thus the study of motion is the key to opening the door to the mysteries of the universe. We make up frameworks or models by means of which we attempt to understand the seemingly complicated motion of the bodies around us. Isaac Newton was the first builder of such a model, or paradigm, the one we call classical mechanics today. For most of the macroscopic world at motion, Newton's paradigm is adequate. But when the arena is extended, when we study high-speed objects and the world of submicroscopic particles, Newton's model fails. When matter moves with high speed (close to the speed of light), a new model, Einstein's relativity, has to be used. And in the realm of the submicroscopic particles, which are invisible to us in any direct sense and which we can study only through our instruments, we must replace Newton's model by quantum mechanics (quantum mechanics involves the idea of quanta, or discrete bundles, of energy and other quantities as basic ingredients).

This book contains a great deal of physics. Certainly it contains many ideas, and some of them are not even fully developed, as you will see. How much can you expect to learn by reading these six hundred or so pages in the allotted time of a school year?

The following approach may help. Joshu, perhaps the most famous of all Zen masters was once asked by a monk, "Does a dog have Buddha nature or not?" To this Joshu replied, "Mu!" (Pronounced moo.) Many later masters have been asked by their disciples to explain this koan (Zen parables like this one are called koans). My favorite of these explanations goes something like this: If Joshu had said "no," he would not be telling the truth since, according to Zen Buddhism, every creature has Buddha nature. On the other hand, saying "yes" would be equally misleading, since that would leave out the experiential aspect of this truth.* You see, every creature has Buddha nature; but this is just a statement until the experience of oneness is felt. Thus the answer is "mu," which in this case transcends both yes and no. It recognizes the spiritual experience one can attain of the truth when liberated from a discriminating mind.

So if somebody asks you after you have read this book, "Do you understand physics?" perhaps you will be able to answer, "Mu." And when asked to explain your answer, perhaps you can respond: "To say yes would be untrue, since I do not know how to work out all the mathematical problems that a physicist can work out. But to say no would be equally untrue, since the questions and answers relating to physics are in my consciousness, because I have been able to put them together at least to some extent. The answer then has to be 'mu,' neither yes nor no, but transcending both."

---

*The word "experiential" means something based on subjective experience.

QUESTIONS

Review and reason

1. Explain the notion of absolute space. What is the alternative view?
2. Can space have boundaries? Explain.
3. One way to understand the idea of an infinite number is by way of examples. For illustration, only a little thought is needed to see that the set of even integers is infinite. Find a few such examples of infinity from the world of mathematics.
4. Explain the difference between the steady state and big bang models of the universe.
5. Explain the difference between the absolute and relational views of time.
6. Explain the phrase "time's arrow."
7. What is entropy? What is the connection between entropy and time's arrow?
8. Find a few examples of order-disorder transitions from your everyday experience. What happens to the entropy in each of these cases?
9. In Buddhist literature we find a reference to circular time as illustrated in Fig. P.10. Discuss circular time in the context of the cyclical model of the universe.

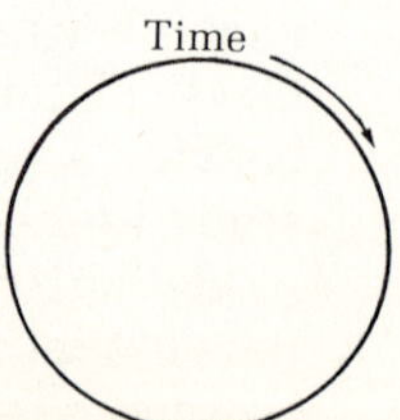

FIGURE P.10   Circular time.

10. Einstein was once invited to a party as guest of honor. Eleven o'clock came and Einstein was impatient to go to bed; however, it was against protocol for him to leave the party before the other guests. His host tried to comfort him by pointing out that the guests had started to leave, but it would take them a while because nobody "wants to leave an Einstein party." Einstein commented that "they remind me of time." Asked to explain what he meant he said, "Always going, but never gone." Comment on this comment. If there was no motion in the universe, would time still be "always going"? Does your answer depend on your model of time? Explain.

Arithmetic

1. There may be 10^{11} galaxies in the entire universe. Write this number in long form.
2. The size of the proton, one of the subatomic particles, is 0.0000000000000014 meter. Express this in power-of-ten notation.
3. Light takes 30,000 years to travel from the center of the Milky Way, our galaxy, to us. How far are we from the galactic center? In light-years? In meters?
4. How old is the universe, approximately?
5. Assuming the universe to be finite, what is the order of magnitude of its size? Show how you would estimate it.

Motion: From Zeno to Newton

Mechanics [the study of motion] is to physics what the skeleton is to the human figure
. . . after even a brief study of its functions one experiences with mounting excitement the discovery of an
astonishingly successful design, of a structure that is ingeniously complex, yet so simple as to be almost
inevitable.

GERALD HOLTON

■ 1.1   The Trials and Tribulations of the Ancient Greeks

Everything changes. And all changes can be traced to motion of some kind. Thus the study of motion is the key to finding the structure of the universe. Somehow the ancient Greeks knew about this and they tried and did manage to develop a theory of motion. The Greeks made a lot of mistakes in treating motion, but nevertheless they asked a few very important questions. So this is a good place to start our discussion of motion: the Greek ideas.

The paradoxes of Zeno the Greek

Suppose you are a man named Zeno who lives in ancient Greece, and you want to make the lives of fellow philosophers miserable. What do you do? You make up paradox after paradox to show that it is impossible to develop a theory of motion. These paradoxes can then be called **Zeno's paradoxes**. Here is a sample. This paradox goes by the name of "Achilles and the Tortoise."

Achilles runs ten times as fast as a tortoise. The two have a race, with Achilles given a handicap of a hundred yards. Can Achilles catch up with the tortoise? Not according to Zeno. By the time Achilles has gone the hundred yards to the tortoise's starting position, the tortoise has moved up another ten. As Achilles tries to make up the ten, the tortoise has gone one more yard, and so on. It seems that whenever Achilles comes to the point where the tortoise was, the tortoise

FIGURE 1.1   Can Achilles catch up with the tortoise? The answer is yes.

has moved from there, at least a little bit. Thus Achilles can never catch up with the tortoise.

Can you see what's wrong with Zeno's argument? Since we can divide a finite interval of time into an infinite number of infinitesimal (immeasurably small) segments, the reverse is also true. The infinite number of infinitesimal time segments needed by Achilles to catch up with the tortoise still adds up to a finite amount of time.

Here is a second paradox of Zeno, this one known as "The Stadium." Zeno "proves" that a runner can never reach the end of the track field. Suppose the runner first runs half the length of the field. He then will have to run half of the remaining half, which is one quarter of the total. Then he covers one half of what still remains, and so on ad infinitum. The point again is that a finite line is being divided into an infinite number of infinitesimal segments, and Zeno is claiming that it takes an infinite amount of time to cover this infinite number of segments.

The resolution of this paradox is essentially the same as for the previous one. The infinite number of time segments that it takes to cover the infinite number of line segments still adds up to a finite amount of time (Fig. 1.2).

Paradoxes like Zeno's show that the Greeks had great difficulty in comprehending the infinitesimal. Unfortunately, even the simplest concepts of motion

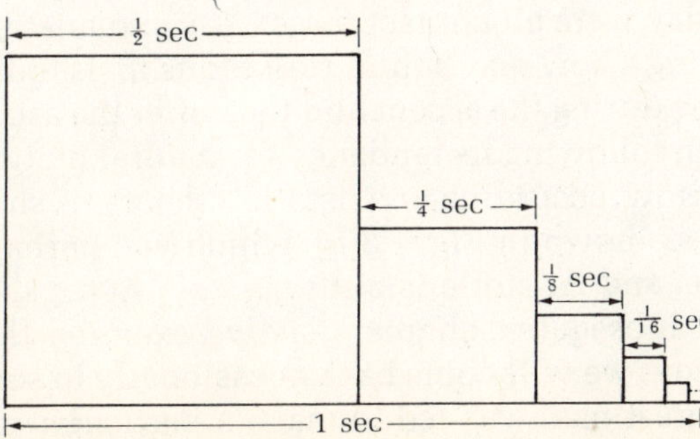

FIGURE 1.2   Zeno's paradox, "The Stadium." Suppose half the field takes $\frac{1}{2}$ sec to cover, half of the remaining half takes $\frac{1}{4}$ sec, half of the remaining half takes $\frac{1}{8}$ sec, and so forth. But as you can see, all these blocks of time, even though the number of blocks can be infinite, still add up to 1 sec.

need an appreciation of the infinitesimally small, as we will see in the next section. The Greeks never quite solved this problem. In fact, Newton was the first person to handle the math correctly, and that was some twenty-two centuries after Zeno.

The plights of Aristotle

Aristotle, born about 300 B.C., has been called by some the greatest thinker who ever lived in Western culture. His theories were destined to dominate much of medieval thought until the advent of modern science.

What is so interesting about Aristotle's theoretical framework is its completeness. He developed a comprehensive theory of the universe in which space, time, matter, and motion each play a very definite role. Much of the failure of his theory can be traced to the fact that he tried to develop so much based on so little.

Aristotle's ideas about the structure of the universe have been mentioned already. His concentric spheres (illustrated in the Prologue) were designed to be consistent with his model of the (apparent) motion of the planets and the stars around the earth. Heavenly motion was thought to be geocentric (earth-centered) in his day. Moreover, all heavenly motion was considered to be strictly circular, circles being recognized as the perfect curve and therefore the correct path for the motion of perfect objects. (Heavenly objects were regarded as perfect at that time.) Each of the concentric spheres was assumed to carry one or more heavenly objects, which were borne around the earth by the rotation of the spheres carrying them (Fig. P.1). Thus Aristotle's picture of the motion of heavenly objects is a thoroughly mechanical model, one that has been discarded by modern science. Strike one against Aristotle.

Aristotle had only slightly better luck in his attempt to explain terrestrial motion. He classified motion in two categories, natural and violent. Natural motion, according to Aristotle, was motion due to the tendency of objects to return to their places; and he assigned every terrestrial object a natural place. The falling motion of bodies was regarded as the natural motion of these bodies returning to their natural place. Italian scientist Galileo much later showed that Aristotle's ideas about the motion of falling bodies did not agree with experiment. Strike two.

Terrestrial motion that was not natural was classified by Aristotle as violent, needing a force as the agency for change in motion. Since the forces known in his day were all contact forces, his explanation of, say, the motion of an arrow leaving a bow may sound ridiculous to us today. He said that the air pushes the arrow during the ascent and that, after the ascent is over, the arrow falls straight down following its tendency for natural motion. The motion of a projectile like an arrow, according to Aristotle's theory, is shown in Fig. 1.3(a). The actual path is also shown in Fig. 1.3(b), which you perhaps recognize as a parabola. Strike three, and Aristotle is out!

In subsequent chapters, as we examine the currently accepted theories of motion, we will come back occasionally to some of Aristotle's ideas to see why the great man was led to make a particular mistake.

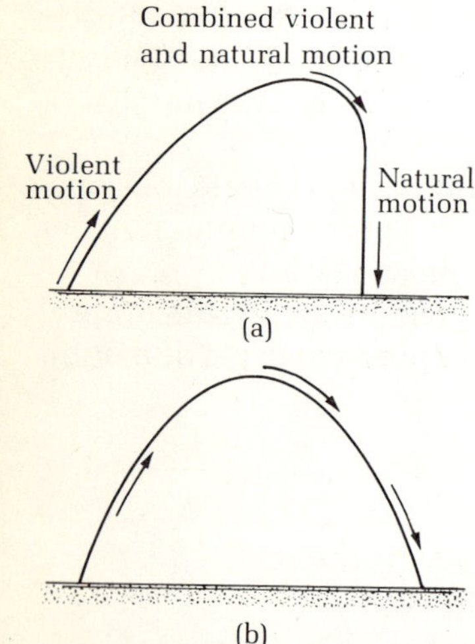

FIGURE 1.3   (a) The path of the arrow according to Artistotle's theory. The drawing follows a medieval construction. (b) The parabolic trajectory of a projectile.

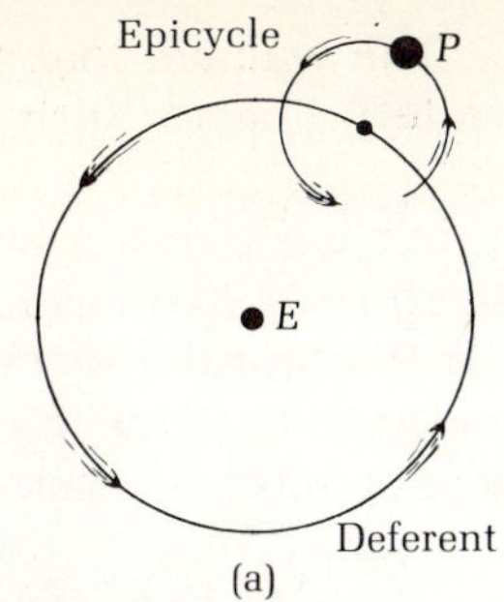

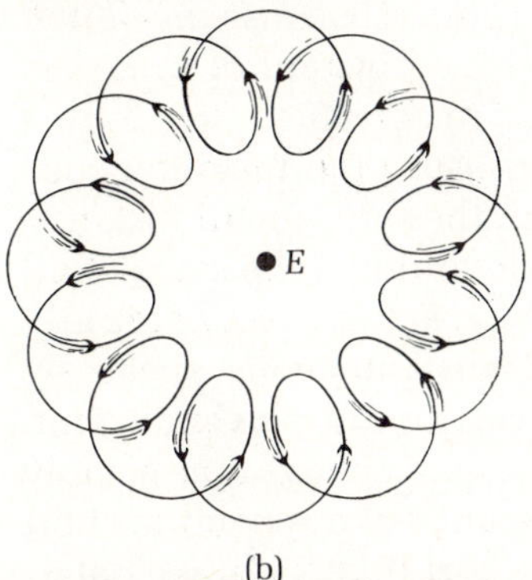

FIGURE 1.4   (a) Ptolemy's construction of the motion of heavenly objects in cycles and epicycles. The planet *P* moves in the epicycle. The center of the epicycle itself rotates around the earth *E* in a bigger circle called the deferent. (b) The retrograde loops of the apparent path of the planets are reproduced when we combine the two motions shown in (a).

The Ptolemaic system

Let's now consider the monumental work of Ptolemy, one of the greatest astronomers of ancient Greece. Ptolemy attempted to refine the geocentric theory of planetary motion so that precise predictions were possible. Of course, like all geocentric theories, Ptolemy's too ended up in oblivion, but only after centuries-long domination of much of scientific thought.

The Aristotelian system we have mentioned does not explain some obvious facts about planetary motion. For example, the distances of the planets from the earth are not in reality fixed, but in Aristotle's theory they are. Ptolemy found a way to modify Aristotle's theory so that the occasional retrograde motion (the apparent reversal of the direction of motion) of the planets could be explained. He retained the ideas of a stationary earth and circular motion (who would dare to give up circular motion in those days?) but abandoned the concentric spheres. In his scheme the motion of the moon, the planets, and the sun around the earth consisted of a superposition of motion in cycles and epicycles (Fig. 1.4). As you can see from the figure, each of the planets was assumed to have two simultaneous rotary motions. While the planet rotated in its epicyclic path, the center of the epicycle itself described the bigger circular path (the deferent) about the earth. By superimposing the two motions, one could reproduce the retrograde loops of the planets [Fig. 1.4(b)]. Ptolemy's system, though complex, was quite successful as a model and was used navigationally for centuries.

As you may know, Polish astronomer Nicolaus Copernicus established in A.D. 1543 the **heliocentric** (sun-centered) nature of our planetary system, which initiated an avalanche of new ideas. Actually, the idea of a heliocentric system wasn't entirely new. Even some Greeks, notably Aristarchus of Samos, had toyed with the idea. But somehow the early work never made any mark. Fortunately, Copernicus was followed by three great scientists, Galileo Galilei (1564–1642), German physicist Johannes Kepler (1571–1630), and Isaac Newton (1642–1727), whose works combined to establish the modern ideas of motion which today we call classical mechanics.

■ 1.2   Speed, Velocity, and Acceleration

Remember the words of the song in the motion picture *The Sound of Music*? "When you sing, you begin with do, re, mi." Well, in the study of motion we begin with the notion of speed. **Speed** is the ratio of distance traveled and the time of travel; it is the rate at which distance is covered.

$$\text{speed} = \frac{\text{distance traveled}}{\text{time of travel}} \tag{1.1}$$

For example, a speed of 60 miles per hour (mi/h) means that a distance of 60 miles is being traveled in the time of an hour. Yet there is also a subtlety here. To illustrate this, American physicist Richard Feynman tells a story of an elderly woman caught speeding by a policeman.* "Lady," the policeman snaps,

---

*R. P. Feynman, R. B. Leighton, and M. Sands, *The Feynman Lectures in Physics* (Reading, Mass.: Addison-Wesley, 1963), p. 8-3.

"You were doing 60 miles an hour in a 30-mile zone." The woman, looking straight at the policeman, says, "But officer, how can you tell? I haven't driven an hour yet."

Now this reply didn't save her from getting a ticket, but her point is well taken. She might go rather fast for a short while, but if she travels 30 miles in 1 hour, no more, she would be within the 30-mile-per-hour limit for the average speed as calculated from the formula (1.1). The formula is good only for calculating **average speeds**. Unfortunately, speed limits do not refer to average speeds.

Perhaps it is good to know from the outset that police officers and physicists usually do not operate on the notion of average speed. This is because they often are faced with the need to determine the speed at a given instant of time, the **instantaneous speed**. For example, the speed limit for road traffic refers to this instantaneous speed. The question is, how do we measure the instantaneous speed? Of course, if the speed were to remain constant, the same for all instants, there would be no problem. The instantaneous speed would be the same at all instants of time and would be equal to the average speed. For the case of rapidly changing speed, however, we must determine the instantaneous speed by taking our measurement using a very short time interval around a given instant. How short is short? Short enough so that the speed does not change appreciably within the period of the measurement. For a car one-tenth of a second is short enough. The speed of a car hardly changes in such a short time. So if we determine the distance traveled by a car in one-tenth of a second around the given instant, and divide it by the time interval (which is $\frac{1}{10}$ sec), we get its instantaneous speed, for all practical purposes. In fact, this is roughly what the speedometer of a car does.

If the speedometer measures that a car is going 8.8 feet in 0.1 second, which is the same thing as 88 feet/second or 60 miles/hour, it registers 60 on its dial. Incidentally, the equality of 60 miles/hour and 88 feet/second, which you can verify by converting miles into feet and hours into seconds and carrying out the arithmetic, is useful to memorize for quick reference.

Consider an example. Suppose you live off campus, 5 miles away. Every morning you have to make the trip to the college through heavy traffic, perhaps taking $\frac{1}{2}$ hour. Your average speed is given by Eq. (2.1):

$$\text{average speed} = \frac{\text{distance traveled}}{\text{time}}$$

$$= \frac{5 \text{ miles}}{\frac{1}{2} \text{ hour}} = 10 \text{ miles/hour}$$

Has your instantaneous speed been the same 10 miles/hour for the entire half hour? Unlikely. Sometimes you probably speeded up to 30 miles/hour and at other times slowed down to 5 miles/hour or even to a standstill, as would have been recorded on the speedometer.

Speedometers can tell us the instantaneous speed of an automobile, but how can we deal with speed changes at an even faster rate, say at an arbitrarily fast rate? Clearly we now have to choose arbitrarily small intervals of time in order to define the instantaneous speed. Mathematically, we call such time intervals

infinitesimal. Even though the time interval is infinitesimal, the ratio of infinitesimal distance (traveled during the infinitesimal interval of time) and the infinitesimal interval of time remains finite. It is this ratio that matters, and it is this ratio that defines the instantaneous speed mathematically.

The paradoxes of Zeno mentioned in Section 1.1 exemplify how much difficulty the ancient Greeks had in coping with even the addition of infinitesimals, let alone taking a ratio as is called for in the concept of instantaneous speed. In fact, nobody before Newton was able to construct a mathematically complete definition of the instantaneous speed.

One final remark on the subject of speeding and police officers. With this advanced knowledge, if you get yourself into the situation of the elderly woman and are interrogated by a police officer, your reaction may be quite different. If the police officer says, "You are going 60 miles per hour in a 30-mile zone," you might be tempted to give the following lecture: "I know what you are trying to say, officer, but your language is far from precise. You see, you really don't know how far I would have gone in an hour. You really should say that I was traveling at a speed of 88 feet per second or, even better, at 8.8 feet in 0.1 second. Of course, you could also say that my instantaneous speed was 60 miles per hour; but that involves the concept of infinitesimals and may be too complicated for you to go into if you were asked for a clarification. . . ."

You can guess what the results of such an encounter would be. Well, never mind about police officers: just remember that when physicists talk about speed, they almost always mean instantaneous speed.

Velocity and vectors

In describing motion we want to know where and when. But knowing the speed of an object is not enough for us to predict where the object will be after, say, an hour. If the object is moving at a speed of 60 miles/hour, it could be anywhere on a circle with a radius of 60 miles after an hour has elapsed. So what we need in addition to speed is a specification of the object's direction of motion.

We use the word **velocity** to signify both the speed and the direction of motion of an object. Although there is a tendency to use the two words "velocity" and "speed" interchangeably, speed really designates only the magnitude, or numerical value, of the velocity; the direction of the velocity is the direction of motion of the object. It is quite customary to indicate the direction of motion with an arrow, a familiar practice in road signs. So we use an arrow to show the direction of the velocity of an object (Fig. 1.5).

Quantities that have a direction as well as a magnitude are called **vectors** by mathematicians. So velocity is an example of a vector quantity. In contrast, quantities that do not have a direction are called **scalars**. The **mass** (quantity of matter in a body) and the **temperature** (degree of hotness or coldness) of a body are examples of scalar quantities; magnitude alone describes these quantities completely.

If you know the distance of an object from a given origin, can you tell where the object is? No, because you also need the direction, that is, whether the object is on the north or northwest or wherever. When distance from an origin is specified with a direction, we get a vector quantity that is called the **position**

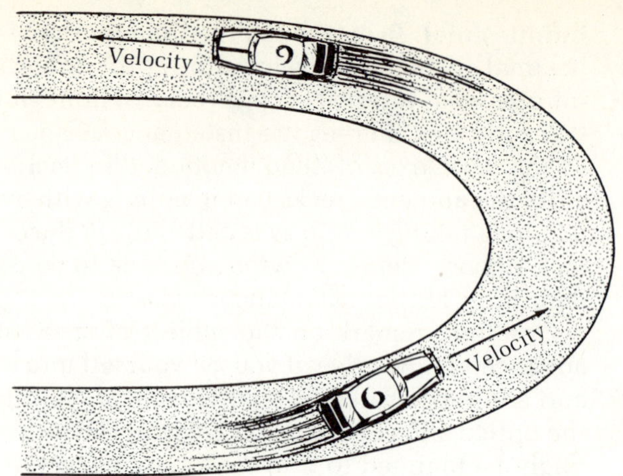

FIGURE 1.5 The velocity of the car is represented by an arrow in the direction of motion.

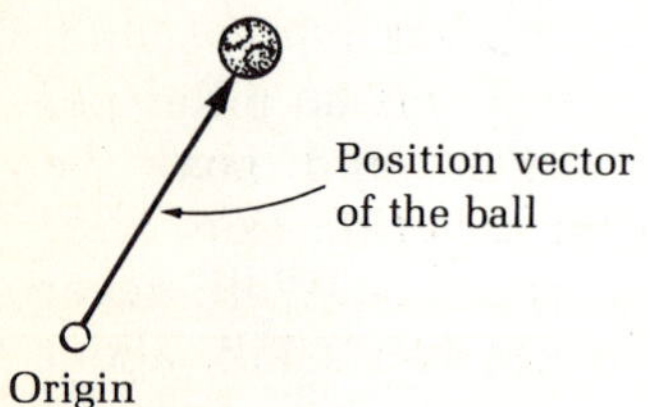

FIGURE 1.6 The position vector of a ball with respect to an origin.

vector (Fig 1.6). The velocity vector is the time rate of change of this position vector.

There is one main advantage in recognizing that physical quantities, like the position or the velocity of an object, are well-known mathematical quantities, namely, vectors. In dealing with the vector quantities we encounter in physics, we can use all the knowledge on vectors that has been amassed by mathematicians. For example, we often encounter situations where the total, or net, velocity of an object is the sum of two velocities: the velocity of a canoe on a river is the sum of the velocity due to paddling and that due to the river current. How do we find the sum of two vectors? Mathematics gives us the rule for the addition of vectors. But first, let us develop a simple geometrical picture for vectors.

A vector is represented by an arrow. The length of the arrow designates the magnitude of the vector quantity and the direction of the arrow designates the

MALCOLM'S LOGIC

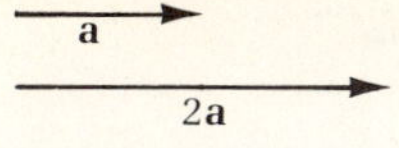

FIGURE 1.7   A vector twice the magnitude of another is represented by an arrow twice as long.

FIGURE 1.8   The vector $-\mathbf{v}$ is represented by an equal and opposite arrow.

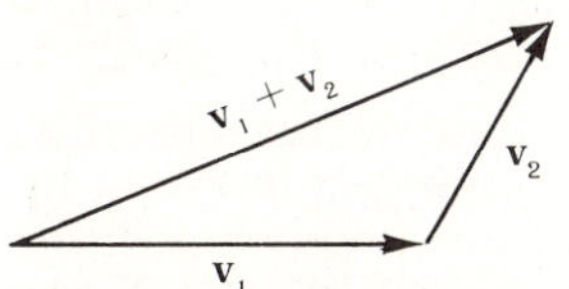

FIGURE 1.9   The head-to-foot method of addition of two vectors. Draw the two vectors with the head of one touching the foot of the other. The arrow from the foot of the first to the head of the second vector gives the sum.

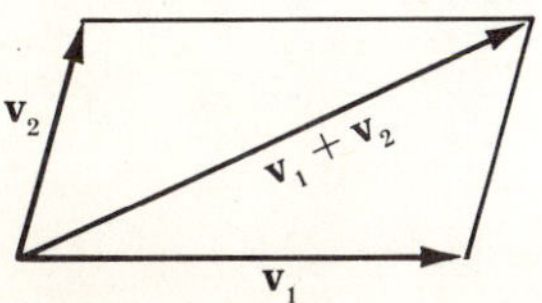

FIGURE 1.10   The parallelogram method of addition of two vectors. Draw the two vectors foot to foot, and complete the parallelogram. The long diagonal is the sum.

FIGURE 1.11   In vector addition, 2 + 2 can be (a) 4, (b) 0, (c) $2\sqrt{2}$, and also other values between 0 and 4 depending on the angle between the two vectors being added.

direction of the line of action of the vector. An arrow twice the length of another represents a vector that has twice the magnitude of the other (Fig. 1.7). We will, from now on, represent a vector by using boldface for the designated symbol (e.g., $\mathbf{v}$ for the velocity and $\mathbf{r}$ for the position vector) if the occasion calls for it.

Two vectors are equal only if they have the same magnitude and direction. Their physical locations are not important—that is, we can move them around (as long as we do not change their length and direction). A vector multiplied or divided by a scalar is a vector in the same direction as the original; only the magnitude is changed.

How do we represent a vector equal but opposite to a given vector? By using an arrow of equal length in the opposite direction. Symbolically, we represent an equal (in magnitude) and opposite (in direction) vector by the negative vector. Thus if $\mathbf{v}$ is a vector, $-\mathbf{v}$ is the equal and opposite vector (Fig. 1.8).

The addition rule for two vectors can be explained by referring to Fig. 1.9. Suppose we have a vector $\mathbf{v}_1$, represented by an arrow, and another vector $\mathbf{v}_2$, represented by a different arrow. To add them geometrically, we lay the arrows in such a way that they lie head to foot, the head of one at the foot of the other. Then the sum vector $\mathbf{v}_1 + \mathbf{v}_2$ is given by the arrow drawn from the foot of the first vector to the head of the second, as shown.

Another sometimes useful way to perform vector addition is to make a parallelogram with the two vectors, as shown in (Fig. 1.10), joined foot to foot. The diagonal from the foot is the vector sum.

So you see that the addition rule for vectors is different from the one you are familiar with for scalars. It should be clear that the magnitude and direction of the sum of two vectors depend not only on their individual magnitudes but also on the angle between them. For scalars if you add 2 and 2, you always get 4. When you add two vectors each of magnitude 2 units, they can add up to a vector having a magnitude anywhere between 0 and 4, depending on the angle between them (Fig. 1.11). So 2 and 2 do not necessarily make 4 for vectors.

Now we can consider the motion of a canoe being rowed across a river that has a current. The actual velocity of the canoe is the velocity imparted by paddling plus the velocity of the current. To be specific, suppose you paddle the canoe across the river toward the north, giving it a speed of 3 miles/hour; we represent it by a vector as shown in Fig. 1.12. Suppose the current velocity is 4 miles/hour in the easterly direction. Then your resultant velocity is given by the sum of these two velocities as determined by the rules of vector addition.

A slightly different composition of velocities occurs in the case of raindrops that fall on you as you run into the rain (when even an umbrella cannot save

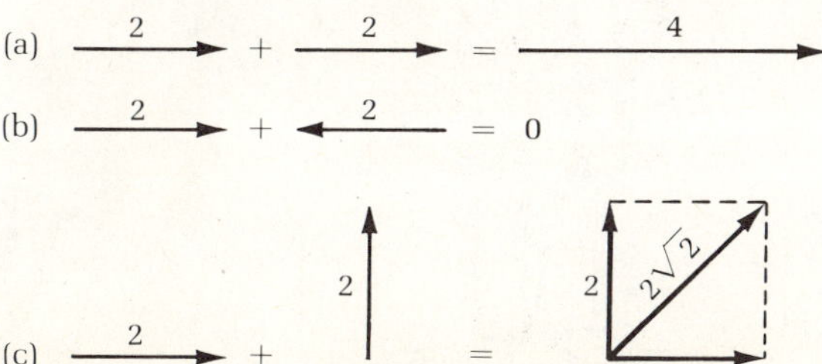

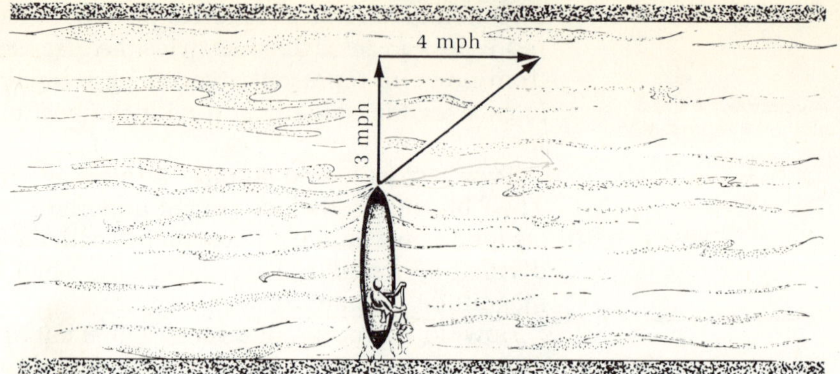

FIGURE 1.12 The resultant velocity of the canoe is the vector sum of the paddling velocity, 3 mi/h due north, and the velocity of the current, 4 mi/h due east.

you). If there is no wind, we know that rain falls vertically. But if you run, say, with a velocity represented by the vector $\mathbf{v}_M$ the rain acquires a horizontal velocity of $-\mathbf{v}_M$ relative to you in addition to its vertical velocity vector $\mathbf{v}_R$, as shown in Fig. 1.13. The total velocity of the raindrops is the vector sum of $\mathbf{v}_R$ and $-\mathbf{v}_M$ as shown; thus the rain comes from a slanted direction and you get wet.

Coming back to the distinction between speed and velocity, one of the most important things to notice is that motion at constant speed (uniform motion) is not necessarily motion at constant velocity. For example, a car on a zig-zag course may be traveling at a constant speed, but is its velocity constant? No, because as the direction of motion changes, so does the velocity. You can change the velocity of an object by changing its magnitude (the speed), or its direction, or both. Only if both the magnitude and the direction of the velocity remain the same is the motion one of constant velocity. When the velocity of the motion changes, the motion is called accelerated.

FIGURE 1.13 (a) A man standing in vertically falling rain. (b) When the man runs, the velocity of the rain relative to the man is composed of the sum of two velocity vectors. The two velocity vectors are the vertical velocity of the rain $\mathbf{v}_R$ and $-\mathbf{v}_M$, a vector equal and opposite to $\mathbf{v}_M$, the velocity of the man. The sum of the two vectors $\mathbf{v}_R - \mathbf{v}_M$ has a slanted direction as shown. This is why the rain falls slanted toward the man.

Acceleration

The foremost familiar example of acceleration is probably the speeding up of your car as you press down on the gas pedal. Perhaps you have noticed that it takes a certain amount of time to achieve a desired amount of change in the speed. The rate of change of speed or, more generally, the rate of change of velocity, is defined as the **acceleration** of an object.

$$\text{acceleration} = \frac{\text{change in velocity}}{\text{time}} \qquad (1.2)$$

Again, this definition really applies only to the **average acceleration**. For **instantaneous acceleration** we must use an infinitesimal time interval and take the ratio of the infinitesimal change in velocity (during the infinitesimal time interval) and the time interval. In everyday life, however, we most often find situations where the acceleration remains a constant. In this case the instantaneous acceleration is equal to the average acceleration and can be determined from Eq. (1.2). So we confine our discussion of acceleration to this equation.

When you press on the gas pedal, suppose your car accelerates at a uniform rate (another way of saying that the acceleration remains constant during the time under consideration) from zero to a speed of 60 miles/hour in 11 seconds. What is the acceleration? From Eq. (1.2), since the change in velocity is 60 miles/hour and the time interval is 11 seconds,

$$\text{acceleration} = \frac{60}{11} \text{ mi/h/sec}$$

Notice that the unit of acceleration (mi/h/sec) includes two units of time, one coming from the unit of velocity and one from the definition of acceleration. However, there is one thing that physicists do not like about this unit for acceleration: there are two different units for time, namely, hour and second. But a change is easy to accomplish. Since

$$60 \text{ mi/h} = 88 \text{ ft/sec}$$

we have

$$\text{acceleration} = \frac{60 \text{ mi/h}}{11 \text{ sec}} = \frac{88 \text{ ft/sec}}{11 \text{ sec}}$$

or 8 (ft/sec)/sec (read "8 feet per second per second"). Furthermore, it is customary to write (ft/sec)/sec in the shortened form ft/sec^2 (read "feet per second squared").

In other words, 8 ft/sec is the *amount* of change in velocity that takes place in the acceleration process. The other "per second" tells us that it takes 1 sec for this much change in velocity to occur. Thus 8 (ft/sec)/sec is the *rate* of change of velocity. In 2 sec the change in velocity is 16 ft/sec; in 3 sec it is 24 ft/sec; and so on (Table 1.1).

Suppose that initially the velocity is 22 ft/sec (15 mi/h). Then if we apply an acceleration of 8 (ft/sec)/sec, at the end of the first second the velocity will be 30 ft/sec; at the end of the second second it will be 38 ft/sec; and so on (Table 1.2). The increase of velocity is 8 ft/sec for every second of time of acceleration that elapses.

TABLE 1.1   The increase of velocity with time for accelerating motion. (Acceleration at the rate of 8 ft/sec².)

Time (sec)	Velocity (ft/sec)
0	0
1	8
2	16
3	24
4	32
5	40

TABLE 1.2   Another example of accelerating motion. (Acceleration = 8 ft/sec².)

Time (sec)	Velocity (ft/sec)
0	22
1	30
2	38
3	46
4	54
5	62

Another familiar example of accelerated motion is that of falling bodies. In the absence of air, all objects on earth are found to fall toward the ground with the same acceleration of 32 ft/sec². The acceleration of fall of an object on earth is referred to as g. Thus

$$g = 32 \text{ ft/sec}^2$$

If we drop a stone vertically from a tall building, and the stone initially is at rest, its velocity after 1 sec will be 32 ft/sec. After 2 sec its velocity is 64 ft/sec; and so forth.

The consideration of the motion of falling bodies has played an important role in the history of science. We will consider this history in Chapter 4. Later on we will further discuss this unexpected fact of the motion of falling bodies on earth—that all objects fall with the same acceleration g. In the meantime, file it away in your memory as one of the most important facts of nature.

Both the acceleration of a car and that of falling bodies are examples in which the speed of the object changes as a result of the acceleration. We mentioned before that acceleration also occurs with a change in the direction of motion or in the direction of the velocity vector. An example of this is circular motion. An object on a rotating turntable moves at an uniform speed; but its velocity, which at every instant is directed tangentially to the circle at the instantaneous position of the object (Fig. 1.14), changes direction. Thus it is accelerated motion.

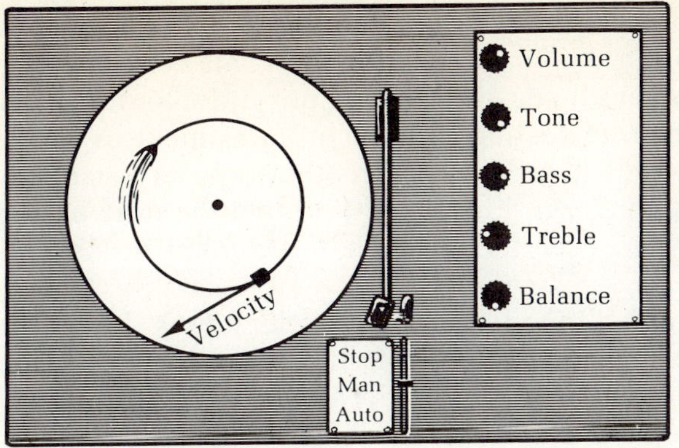

FIGURE 1.14 The velocity vector of an object in circular motion is along the tangent to the circle at the instantaneous position of the object, as shown. Obviously the velocity of the object moving in a circle continuously changes in direction, even though the speed may remain constant. The motion is thus accelerated.

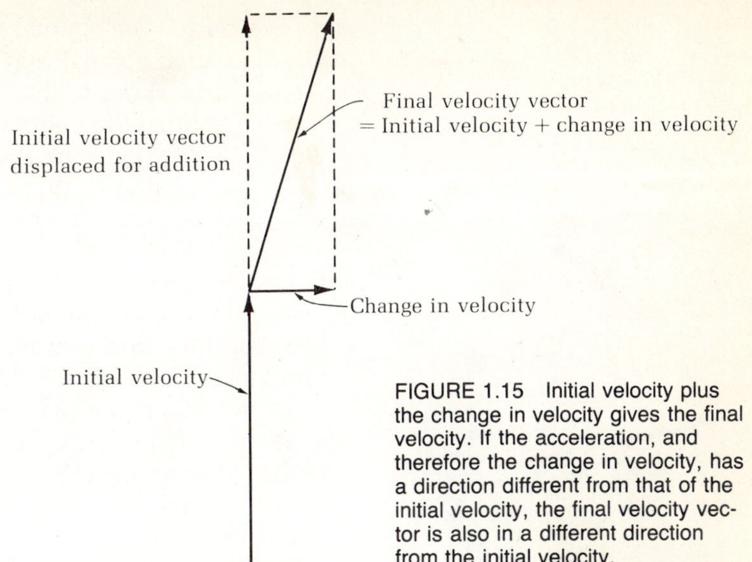

FIGURE 1.15 Initial velocity plus the change in velocity gives the final velocity. If the acceleration, and therefore the change in velocity, has a direction different from that of the initial velocity, the final velocity vector is also in a different direction from the initial velocity.

Now consider an example in which both the magnitude and the direction of the velocity change in accelerated motion: a car making a right turn. Every experienced driver knows that it is best to press the brake pedal (i.e., change the speed) while making the turn (changing the direction).

Like velocity, the acceleration of an object is also a vector quantity. If the direction of the acceleration vector is the same as that of the velocity, then the effect of acceleration is to increase the speed. But if the acceleration vector is in a direction opposite to that of the velocity vector, then its effect is to decrease the speed, a situation we usually refer to as deceleration. Notice that, once we recognize the vector nature of acceleration, there is no longer any necessity of distinguishing between acceleration and deceleration: the direction of the acceleration vector gives us the information automatically; deceleration is just a special case of acceleration.

When the direction of the acceleration vector is at an angle to the line of action of the velocity vector, the change in the velocity vector has a different direction (of acceleration) from that of the initial velocity. We now must add the original velocity vector to this change in order to get the final velocity vector. From the rules of vector addition, it should be clear that the total of two vectors having different directions must lie in a direction somewhere between them (Fig. 1.15). Thus the final velocity has a direction that is different from that of the initial velocity. In effect, then, we have accomplished a turn.

SUMMARY AND OUTLOOK

In this chapter we described motion and the precise meanings of words like speed, velocity, and acceleration. Although these words are familiar in your everyday life, they have subtleties in their use in physics. One example of this subtlety is the distinction between average and instantaneous speed. Another is the recognition of velocity and acceleration as vector quantities.

We started the chapter with a discussion of the pitfalls of the ancient Greeks in their attempts to describe motion. You have seen the problems Zeno encountered in working with infinitesimal quantities. It was not until hundreds of years later that the mathematical art of dealing with the infinitesimal was developed by Newton, who then was able to go ahead with other aspects of motion. The crucial step in this development was his recognition that the ratio of two infinitesimal quantities can be finite even though each of the quantities is arbitrarily close to zero.

In Section 1.1 we also talked about the plights of Aristotle. It took the genius of Galileo to find the right approach to the motion of falling bodies; in correcting Aristotle on this, Galileo also established the modern scientific approach to problems, namely, the use of logic and experimentation, hand in hand. Galileo's work is given in Chapter 4. Before going into that, we will examine in the next two chapters the Newtonian framework for describing motion in terms of forces that act on objects. Physics starts taking a coherent shape only after you have learned this framework, which is the fundamental basis for most of physics.

QUESTIONS

Review and reason

1. An object is in motion. What does this mean?
2. What is the source of confusion in the logic of the paradoxes of Zeno? How are the paradoxes resolved?
3. Draw the trajectory of a baseball after it leaves the batter's bat according to Aristotle's theory of motion. Compare this picture with one drawn from your own experience.
4. How does Ptolemy's system of cycles and epicycles explain the retrograde motion of the planets?
5. Explain the difference between the concepts of average and instantaneous speeds.
6. What is meant when we say that something is infinitesimally small? Can the ratio of two infinitesimal quantities be finite?
7. Explain the difference between speed and velocity.
8. Explain the difference between the concepts of a scalar and a vector. Give an example of each.
9. What are the two ways of adding two vectors? Draw diagrams.
10. You hit a baseball straight north with a velocity $\mathbf{v}_b$, but there is a wind toward the east with velocity $\mathbf{v}_w$. Draw a diagram showing how the resultant (net) velocity of the ball can be determined.
11. You are in a boat alongside another boat, both with identical velocities. What is your velocity relative to the other boat? What is the other boat's velocity relative to yours?
12. Your friend has left in a hurry and has forgotten his suitcase. You rush in your car onto the highway and there he is on the back of a pickup truck looking philosophical, no doubt trying to forget the matter of his luggage. You equalize speed and align your car with the truck. You are now ready to throw him the suitcase. Do you (a) throw the suitcase at him taking straight aim as if he were at rest or (b) make some special adjustments because he is in a moving truck? Explain your answer.
13. What is meant by saying that an object is in uniform motion?
14. When an object accelerates uniformly, which of the following quantities remains constant, velocity or acceleration?
15. Which units are the correct units of acceleration: m/sec, ft/sec^2, m/sec^2, ft/sec, (ft/sec)/sec, (m/sec)/sec? Explain.
16. Which of the following are examples of accelerated motion? (a) A car speeding up while remaining in a straight-line course; (b) a car slowing down while taking a right turn; (c) a car rounding a curve without any change in its speed; (d) a stunt car while in the air after being driven off a cliff. Explain your answer in each case.

1. Suppose you go on a hike and travel a mile in 5 long hours. What was your average speed during the journey?

2. Suppose you drive with a speed of 30 mi/h for a half hour and then at 60 mi/h for another half hour. What is your average speed for the entire hour?

3. You are driving along a straight highway at a uniform speed of 55 mi/h and approach another car traveling in the same direction at 50 mi/h. What is the velocity of the other car relative to you?

4. You are traveling in a train at a speed of 30 mi/h on a straight track when a car passes you going in the opposite direction on a parallel road. You estimate the velocity of the car relative to you to be 70 mi/h. Can you estimate the velocity of the car relative to the ground?

*5. In the canoe problem (Fig. 1.12), you can find the numerical value (magnitude) of the resultant velocity by using Pythagoras's theorem of geometry. According to this theorem, the square of the hypotenuse of a right-angled triangle is given by the sum of the squares of the other two sides. Find the magnitude of the net velocity of the canoe using this theorem.

6. A car accelerates at a uniform rate of 5 ft/sec^2 starting from rest. What is the velocity of the car after 1 sec? After 5 sec?

7. You throw an object upward with a speed of 32 ft/sec. The object decelerates at a rate of 32 ft/sec^2. How long does it take to reach the maximum height? (Note: Maximum height is reached when the upward speed has been reduced to zero.)

8. A car accelerates uniformly from 0 to 60 mi/h in 10 sec on one occasion and from 10 mi/h to 70 mi/h in the same time on another occasion. Is there a difference in the value of the acceleration between the two cases? Why or why not?

---

*Optional.

The Newtonian Framework

Nature and nature's laws
Lay hidden in night,
God said, "Let Newton be,"
And there was light.

ALEXANDER POPE

■ 2.1 Newton's Laws of Motion

Newton stated three laws that summarized his framework of the theory of motion. These laws mark the beginnings of dynamics, the branch of mechanics in which the main concern is why things move. As we will see in Chapter 6, this framework enabled Newton to explain aspects of falling motion discovered by Galileo and Kepler. As the implications and contents of these laws take shape in your mind, you will gain a new perspective on the whole subject of motion, the perspective of an insider. Also, armed with the tools of Newton's laws, you will be able to solve simple examples of motion yourself.

Newton's first law of motion: The concept of inertia and true natural motion

Here is the statement of the first of Newton's laws:

All bodies continue in their state of rest or of uniform linear motion (motion with constant velocity in a straight line) unless acted on by a net external force.

The term **inertia,** as used in physics, applies to the tendency of objects to behave according to Newton's first law of motion. The inertia of rest refers to the tendency of a body at rest to stay at rest, a fact that can hardly be considered unusual. Rest was considered to be an object's natural state even by the ancient Greeks. As our everyday experiences remind us, a force indeed is necessary to overcome a body's inertia of rest, just as the law states.

The second part of the law states that an object in uniform linear motion has a tendency to continue in this state; this tendency is referred to as the inertia of motion. This aspect of the law is not only novel, but at first sight it may seem to contradict your everyday experience. Perhaps you think that an object keeps moving only if a force continuously pushes on it. This is a mistaken notion (Don't be embarrassed, you are in pretty good company. Even Aristotle had a similar misconception.), one that can arise easily from a superficial observation of our everyday experiences of the behavior of moving objects. Aren't we always seeing objects in motion come to a stop unless there is a force to sustain the motion? While driving a car, you find that you have to keep pressing down on the accelerator pedal, otherwise the car comes to a stop in a matter of seconds. If you start a pencil rolling on your desktop, it eventually stops rolling. So where is the evidence for this inertia of motion?

Well, we notice several things in all cases such as those described above. First of all, there is a period when the object does keep moving (albeit slower and slower) without any force pushing on it. Second, it is not difficult to determine what slows down and eventually stops the motion. An object moving through air, such as an automobile, encounters resistance to its motion from the air; the force of air resistance is called **drag.** An object sliding on a solid surface is acted on by a force of resistance from the surface, a force commonly referred to as the **friction force.** The rougher the surface is, the larger the friction force is. For a rolling object the friction force of a surface is smaller than that for a sliding object, but it is still considerable. Thus a pencil rolling on a desktop is brought to rest by the friction force of the desk. A moving car comes to rest (unless you press the gas pedal) because of both the adversely acting air drag and the road friction on its tires. So in your attempt to understand the inertia of motion, ask not what keeps an object moving but what stops it. If there were no frictional or drag forces, there would be numerous examples of an object's inertia of motion around us.

Today, thanks to space-age technology, you can see occasional demonstrations of the law of inertia in space experiments. Newton, and Galileo before him, discovered this concept by the power of reasoning and idealization. You can do it too. Imagine a ball rolling in a uniform linear motion on a horizontal floor. You know the ball comes to a stop due to friction if the floor is rough. But imagine the floor becoming smoother and smoother and thus more and more frictionless. What happens to the motion of the rolling ball? Take advantage here of your experiences of motion on low-friction surfaces such as icy winter roads. You can see that the ball will be increasingly able to retain its motion. Now idealize this trend to the case where there is no friction—the floor is perfectly smooth. Now you can see that the ball will keep rolling forever. Congratulations, you have rediscovered Newton's first law of motion, for yourself!

The most important and novel aspect of Newton's first law of motion is the recognition of the true nature of **natural motion,** motion that does not need a force to be sustained. Natural motion is now understood to be motion with

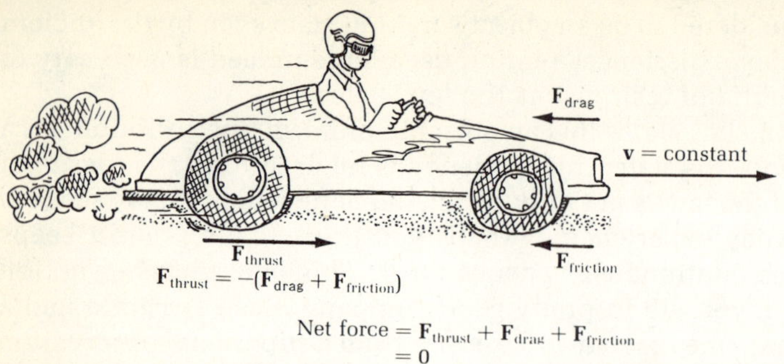

$$\mathbf{F}_{\text{thrust}} = -(\mathbf{F}_{\text{drag}} + \mathbf{F}_{\text{friction}})$$

$$\text{Net force} = \mathbf{F}_{\text{thrust}} + \mathbf{F}_{\text{drag}} + \mathbf{F}_{\text{friction}}$$
$$= 0$$

FIGURE 2.1   For a car moving at constant velocity, the sum of the drag force and the friction force balances the forward thrust.

constant velocity in a straight line. Let us compare this with Aristotle's notion of natural motion. According to Aristotle, the tendency of objects to come to rest at their natural place was what gave rise to natural motion. Thus he thought that the motion of a falling stone coming back to earth was natural motion. Additionally, Aristotle felt that the "perfect" circular motion of heavenly objects also must constitute natural motion. Perhaps Aristotle was misguided because of his lack of knowledge of the phenomenon of gravitation. Newton, armed with his understanding of the gravity force, could see that both the falling motion of earthly objects and the circular motion of heavenly bodies take place under the action of the gravity force. Therefore he had no difficulty in discerning them from natural motion. (Gravitation will be discussed in depth in Chapter 6.)

Newton's first law also gives us a qualitative definition of **force.** A force is recognized to be the agent that changes the state of rest or state of natural motion of an object. But only external forces can perform this feat; the law is very clear on this. Internal forces have no effect on an object's inertia. Importantly, only the *net*—the total—of all the external forces counts. There is no difference, in effect, between a situation where several external forces act on an object with their sum being zero and another situation where there is no force at all on the object. In both situations there is no net force, and therefore natural motion (or rest) will continue. Thus a car driven at a constant velocity in a straight line furnishes us with a perfect example of natural motion because there is no net force acting on the auto. The force of the forward thrust exactly balances the opposing forces of air drag and ground friction (Fig. 2.1).

Some examples of inertia

Most of us have had the experience of trying to push a heavy object like a stalled car or a boulder and not getting anywhere until we push extra hard. This is an

example of the object's inertia. Every object has this resistance against an attempt to change its state of rest.

It is also a common experience that the more massive the object, the more resistance there is to its being moved—it is easier to move a bicycle from rest than a car. So an object's inertia has something to do with its mass: the greater the mass, the greater the inertia. Thus it is more difficult to change the state of natural motion of a more massive object than that of one of lesser mass. We wouldn't think much of stopping a child's wagon that was coasting downhill, but can you imagine trying to stop an automobile that was rolling downhill?

It is interesting to note that long before Galileo and Newton farmers used the principle of inertia, perhaps with no idea of what the principle was. After they beat the wheat from the wheat stalks, the wheat kernels would be mixed with pieces of husk. To separate the kernels from the husk, the farmers used the technique of separation by inertia. They would toss the mixture up in the air on a windy day so that the force of the wind would blow the refuse away. The kernel, because of its greater inertia, would fall straight down.

Suppose we drop a stone from the mast of a moving ship. Where will it land? This question has historical significance. Aristotle believed that the earth did not move. His argument was that an object thrown vertically upward returns to the same place, proving that the earth does not move away from under the object. We can apply this argument to the case of the ship and say that the ship should move away from under the stone, which then should fall some distance behind the foot of the mast. Galileo, who invented the concept of inertia, said this did not happen. His explanation was that the stone, because of its inertia, would fall at the foot of the mast regardless of whether or not the ship was in motion. The stone, as it falls, would continue to move with the horizontal speed of the ship at the time it was dropped (Fig. 2.2). Thus Aristotle's proof that the earth does not move was no proof at all.

There is another incorrect idea based on the preceding argument of Aristotle. We now know that the earth does move. So if we went up in a helicopter and hovered overhead for a time while the earth rotated below, when we got down we would be at a different place. What a convenient way to travel. Unfortunately, this method of travel doesn't work, again because of inertia. The hovering helicopter, because of its inertia, continues to move horizontally with the earth below (Fig. 2.3).

FIGURE 2.2   (a) Aristotle's construction for the trajectory of a stone dropped from a moving ship as observed by somebody on the shore. (b) What an observer on the shore actually sees. The stone falls at the foot of the mast regardless of the ship's motion.

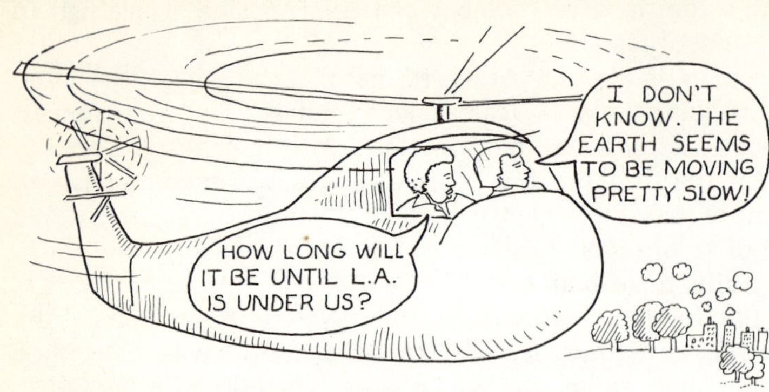

FIGURE 2.3   Can you travel in a hovering helicopter? What do you think?

Motion is relative

Another interesting feature of the first law is that the relationship between the state of rest and that of uniform linear motion is clearly recognized. Today almost everybody is aware of this relationship; it is usually expressed by the statement, "motion is relative." (A more complete statement is, "uniform linear motion is relative.")

For example, think about some time when you have traveled at constant velocity, such as in a car at night or in a night train, and per chance there were no bumps on the road or the track. As a result you might have gotten a feeling that you weren't moving at all. Even when you looked out the window, the feeling persisted for a few seconds. It seemed that you were at rest and the scenery was moving away from you. Of course, pretty soon your thinking mind took over. "Trees don't move," you reminded yourself, and concluded that it was you who was moving.

But suppose you are in a space capsule in uniform linear motion deep in space, with no window from which to look at your environment, the stars. Can you tell if you are moving? Absolutely not; at least, not from any experiments from inside the capsule. And even if you had a window, and you looked at the stars, you wouldn't be able to tell which is moving, you or the stars. This idea is called **relativity of motion.**

An important consequence of the relativity of motion is that the velocities of all moving objects that we measure are really relative velocities, that is, velocities relative to the observer. By comparing different sets of observations, we can compute the relative velocity of an object with respect to another, such as the earth or the sun, but we can never compute the absolute velocity of an object. So it does not make sense to talk about an absolute velocity or absolute (uniform linear) motion, since we can never observe or measure it.

Newton's second law: The connection of force and acceleration

Newton's first law generated the idea that a force changes a body's state of rest or state of natural motion; that is, a force changes the velocity of an object. The second law of motion quantifies this idea. It gives us the precise relationship between a force and the acceleration (the change in velocity per second) it produces.

How much acceleration is produced by a given amount of force? The answer to this question is the basis of quantitative calculations and predictions on the motion of objects. Newton made the assertion that in any situation,

> *the net external force always equals the product of mass and acceleration.*

This is the second law of motion.

$$\text{force} = \text{mass} \times \text{acceleration}$$

In symbols, writing F for force, m for mass, and a for acceleration, we get

$$F = ma \tag{2.1}$$

which is the mathematical statement of Newton's second law of motion. This equation is perhaps the most famous of all in physics. The quantity, **mass,** makes its formal debut in this equation. What is mass, exactly? Newton himself defined mass as the quantity of matter contained in an object. Additionally, mass can be looked upon as the inherent property of an object which determines its response to changes in its state of inertia.

Rewriting Eq. (2.1) in a slightly different form will make the last point clearer. We can transpose a to the other side of the equation, and write the equation as an equation for mass:

$$m = \frac{F}{a}$$

that is,

$$\text{mass} = \frac{\text{force}}{\text{acceleration}} \tag{2.1a}$$

This equation tells us that to produce a given change, or acceleration, in the motion of an object, it is the mass of the object that determines how much force must be applied. If an object has twice the mass of another, it will need twice the amount of force for gaining the same amount of acceleration. We say that the amount of force necessary is in direct proportion to the body's mass.*

To see how the acceleration depends on the mass of the object for a given amount of force, we rewrite Eq. (2.1) in the form

$$a = \frac{F}{m} \tag{2.1b}$$

This equation tells us that if the mass is doubled, the acceleration for a specific amount of force is reduced to half. This is called a relationship of inverse proportionality. If mass is increased by a certain factor, the resultant acceleration is decreased by the same factor, and vice versa.

Looking at the second law in these ways, we certainly get a good idea of what mass is. But now another question can be asked: Is the second law really just a

---

*See Appendix 2 for a review of the idea of proportions.

definition of mass? If you look at mass as the quantity of matter contained in an object, as Newton did, you can sidestep this question.

What is the right way to look at mass, or is there a right way? Actually, there is no right way. Newton's laws of motion represent a summary of human experiences in the form of what we might call "self-evident axioms." Mathematician-author Jacob Bronowsky has compared these laws with Euclid's self-evident axioms, which are the basis of Euclid's geometry (which you perhaps studied in high school). In Euclid's geometry some concepts like a point or a straight line are left largely undefined. It is the same with the concept of mass in Newton's system. We will have more to say about this later.

Notice also that the second law does not tell us how to identify or determine the forces acting in a given situation. We will have to supplement the second law with various force laws, that is, laws that tell us how to determine the forces acting on objects. Only then will we have a comprehensive theory of motion. As for what specific forces are acting in a given situation, these can be identified only by experience and experimentation.

Now let's consider the question of units. The unit of mass in the metric system is the kilogram.* Acceleration in this system is measured in units of meter/second² (abbreviated m/sec²). The metric unit of force, which is called the newton (abbreviated N), can now be defined in the following way:

> *A newton is the amount of force that accelerates 1 kg of mass through an acceleration of 1 m/sec². Thus 1 N is the same as 1 kg·m/sec².*

It is interesting to note that a newton as a unit of force is spelled with a lowercase n. There is a popular word game called Scrabble in which proper names are not allowed—unless of course the name happens to be that of a physicist who has a physical unit named after him. See, there is an advantage in learning the names of some of these units!

In the engineering system, force is measured in units of the pound (lb), and mass in units of slugs (no abbreviation). A pound of force can accelerate a slug of mass through 1 ft/sec². Probably the pound, still used as the unit of weight in the United States, is very familiar to you. But remember, *weight is a force*.

There is a simple formula for finding the mass in slugs, given the weight of an object in pounds:

$$\text{mass (in slugs)} = \frac{\text{weight (in pounds)}}{32} \tag{2.2}$$

The reasoning behind this formula is that the weight of an object on earth is given by the force of earth's gravity on it, and 32 ft/sec² is the acceleration due to earth's gravity, g [see Eq. (2.1a)]. The weight and mass of an object are directly proportional to each other.

---

*The metric system we use here is also called the MKS system or the meter-kilogram-second system of units. The engineering system commonly used in America is called the foot-pound-second (FPS) system. Note, however, that the pound is a unit of force. There is also another metric system (called the CGS system) which uses the centimeter as the unit of length, the gram as the unit of mass, and the second as the unit of time.

The metric equivalent of Eq. (2.2) is

$$\text{mass (in kilograms)} = \frac{\text{weight (in newtons)}}{9.8} \qquad (2.2a)$$

since the acceleration of gravity, g, is 9.8 m/sec² in MKS units.

The relationship between a newton and a pound of force is useful to know:

$$1 \text{ N} = 0.2248 \text{ lb} \simeq \tfrac{1}{5} \text{ lb}$$

The relationship between the kilogram, the MKS unit of mass, and the slug, the FPS unit, is given as

$$1 \text{ kg} \quad = 2.21 \text{ slugs}$$
$$1 \text{ slug} = 0.453 \text{ kg}$$

But the slug is used only in textbooks. You are probably more familiar with the pound when talking about masses—that is, you really talk about weights. The pound is a unit of force, so we cannot directly compare a kilogram and a pound. Fortunately, though, we can express the weight of a kilogram-mass in pounds and the mass of an object of weight 1 lb in kilograms:

A 1-kg mass weighs 2.21 lb.
A 1-lb weight has a mass of 0.453 kg.

For rough estimates remember that 1 kg is approximately *equivalent* to 2 lb of weight and 1 lb of weight is approximately equivalent to $\tfrac{1}{2}$ kg of mass.

A simple numerical illustration may be of assistance. Suppose a force of 25 N acts on an object of mass 5 kg. What is the acceleration? From Eq. (2. lb),

$$\text{acceleration} = \frac{\text{force}}{\text{mass}} = \frac{25}{5} = 5 \text{ m/sec}^2$$

One thing should be pointed out about solving numerical problems like this. If you use the standard unit for every given quantity in your formula (in any one system of units), the answer will always be in the standard unit also. In the calculation above we used the standard units of kilogram for mass and newton for force in the MKS system. Thus the number that results is in the standard unit for acceleration, namely, meter/second². This is one reason we encourage you to use standard units (in any system) every time you do a problem. Convert all units to standard units before you start solving a numerical problem (unless it is obviously not necessary to do so).

The vector nature of force

Force is a vector quantity; hence every force has a direction associated with it. The directions of the individual forces acting on a body must be taken into account when we calculate the net or total force on it. In other words, the net force must be determined by adding the individual force vectors, following the rules of vector addition.

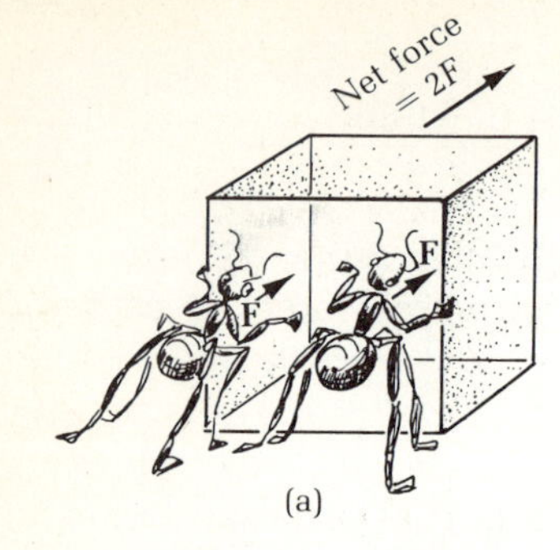

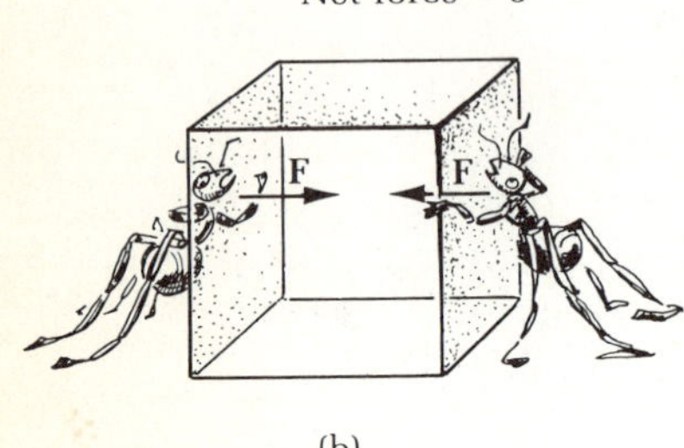

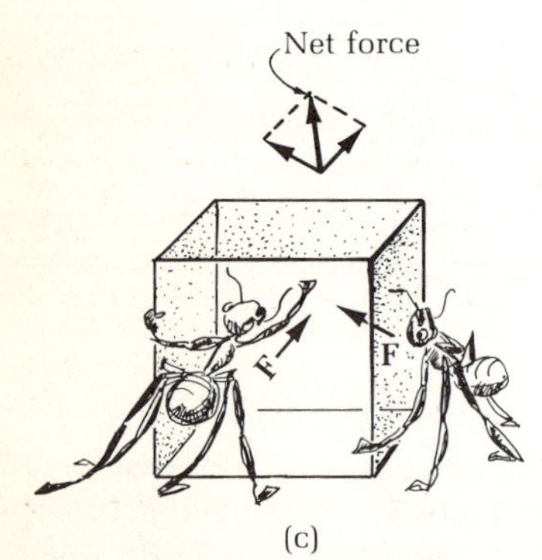

FIGURE 2.4   (a) When two ants cooperate to push a sugar cube, each exerting a force F as shown, the net force is $2F$. (b) But ants are not that clever, contrary to what you may have thought. Often they are found to push a sugar cube at opposite ends, in which case the net force is zero. (c) When two ants push from different directions not in the same line, the net force is given by the sum, as determined by the parallelogram rule.

Consider two ants each applying a force of magnitude F on a sugar cube. If the forces have identical directions (actually very unusual in the ant world), then the net force is $2F$ [Fig. 2.4(a)]. If, on the other hand, the forces act in opposite directions (this is often seen with ants, believe it or not), then they cancel each other out and the net force is zero [Fig. 2.4(b)]. If the individual forces act in some intermediate directions [Fig. 2.4(c)], we must determine the net force by the rule of vector addition, as shown in the figure.

For a given net force acting on a body, the direction of the body's acceleration is always along the direction of the net force. For a moving object this means the following: If the net force is directed along the direction of motion, there will be an acceleration in the same direction; the body will speed up. On the other hand, if the direction of the net force is opposite to the direction of motion, the acceleration vector is directed opposite to the velocity vector, and the speed will decrease. If the net force has a line of action different from the direction of motion, the direction of acceleration is different from the direction of the original velocity vector. In this case the direction of the motion must change. The object must take a turn.

Mass

Let us briefly discuss mass again. So far we have defined mass in two different ways. Mass can be defined from Newton's law. The second law says that forces are proportional to the acceleration and that the ratio of force and acceleration for an object acted upon by the force is its mass [Eq. (2.1a)]. If two objects have the same mass, they will be affected by a force in the same way in terms of acceleration. On the other hand, mass can also be defined as Newton defined it: mass is the quantity of matter contained in an object.

Neither of these definitions is completely satisfactory. The first definition makes the second law merely a definition of mass. The second definition is unsatisfactory since it suggests that the mass of an object should be the sum of the masses of its constituent submicroscopic particles. However, this is not strictly true because of the presence of forces among the constituent submicroscopic particles (as will be seen later). In this context the theory of relativity

TABLE 2.1   Some important masses (kg).

Our galaxy	4×10^{41}
Our sun	2×10^{30}
Earth	6×10^{24}
Moon	7×10^{22}
A mountain	$\sim 10^{15}$
A car	$\sim 10^{3}$
A book	~ 1
A postage stamp	2×10^{-5}
A speck of dust	$\sim 10^{-11}$
A red blood corpuscle	$\sim 10^{-13}$
Proton	1.67×10^{-27}
Neutron	1.67×10^{-27}
Electron	9.1×10^{-31}

accomplishes an important step: it shows that mass is equivalent to energy. This last fact has led to the suggestion that the mass of a particle is due to the forces themselves, which give rise to energy and therefore mass. But so far our efforts in calculating mass from the basic forces have not been successful.

At the opposite pole is a principle known as Mach's principle, stated by Austrian physicist-philosopher Ernst Mach, which suggests that the mass of an object is due to faraway stars. According to this principle, if the rest of the universe were empty, the mass of an object would be zero. In any case, mass is a very important property of matter. From time to time we will return to this subject and try to understand it better.

Table 2.1 gives the masses of some of the objects that you see around you and may be curious about.

Weight, mass, volume, and density

Swiss psychologist Jean Piaget has done extensive experiments with children of all ages. His studies suggest that it takes a while before children start to develop the distinction between weight and volume. For example, in one of Piaget's experiments, young children were asked if the level of water in a glass would rise if a pebble were put into the water. Almost every child forecast that the level of water would rise. But when asked to explain, children under 9 years of age tended to say that the rise was because of the weight of the pebble. Clearly, in the minds of young children, the concept of volume as the amount of space occupied by an object has not yet developed.

The confusion between weight and volume is even more apparent in another experiment. Children were asked: If a small pebble and a bigger (but lighter) block of wood are submerged in the water, which one will make the water rise higher? Their answer generally was that the pebble would make the water rise higher. This misconception, at least in part, is due to the children's identification of weight with volume. Since weight is proportional to volume in their minds, invariably it makes sense to them that the heavier pebble will raise the water more.

A similar misconception occurs when college students are asked: Which is more massive, a 1-pound ball of lead or a 1-pound ball of feathers? Quite a few students will answer that the lead ball is more massive. Of course, the weights being equal, the masses must be equal too, since weight and mass are directly proportional to each other. But somehow many people miss this point.

The missing link in these types of misconceptions is the concept of **density**. What is density? It is a measure of how tightly packed matter is inside the material. More precisely, density is defined as the mass per unit volume:

$$\text{density} = \frac{\text{mass}}{\text{volume}} \tag{2.3}$$

With the help of the concept of density, we can now see the source of confusion in the case of determining which is more massive, a lead ball or a feather ball, both of the same weight. Because lead has greater density than feathers, some students tend to feel that the lead ball should have more mass.

In the case of Piaget's experiments, many children felt that volume is proportional to weight, whereas in reality volume is directly proportional to weight but inversely proportional to density. The pebble has greater weight but less volume than the block of wood because matter inside a pebble is much more tightly packed than it is inside wood; hence the pebble's density is much greater. Of course, as we grow up we learn to see volume as the quantity of space occupied by an object, so this particular confusion tends to disappear.

Table 8.1 (in Chapter 8—take a look) lists the densities of a few important materials. The MKS unit for density is kilogram per cubic meter (kg/m³) and follows from the definition [Eq. (2.3)]. The density in MKS units tells us how many kilograms of a substance are contained in an amount that fills a space of 1 cubic meter. Density can also be defined as the weight per unit volume, and it is defined this way in the FPS system. Thus the FPS unit of density is pound per cubic foot (lb/ft³). When density is measured in this way, we refer to it as **weight density,** in contrast to **mass density** for the MKS system.

Newton's third law: Forces come in pairs

Newton's third law of motion states

> *To every action force there is an equal and opposite reaction force.*

Thus forces occur only in action-reaction pairs. The third law tells us that a force is a mutual interaction between two partners. Action by one partner must be accompanied by a reaction on it from the other partner. Without the reaction, the action cannot happen in the first place. Both partners play an equal role; there is no lack of democracy here (Fig. 2.5). Why, then, in many examples of motion, such as a gun firing a bullet or a horse pulling a cart, does one of the partners appear to be more active than the other? Again, to answer this question we need to look deeper, beyond what our first intuition might say.

In the case of a gun firing a bullet, we find on a closer look that the gun also

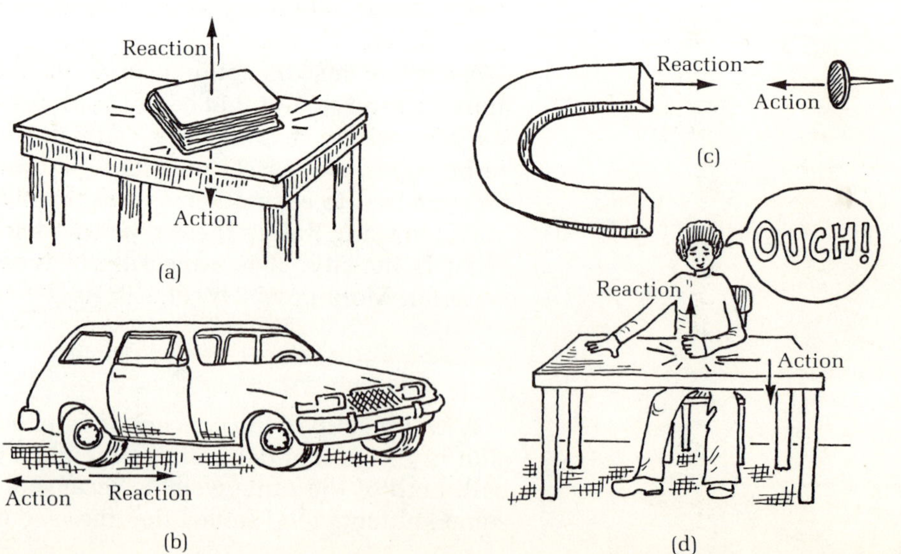

FIGURE 2.5 Some examples of action-reaction forces. (a) The book pushes on the table; the table exerts an equal and opposite reaction force on the book. (b) The car exerts a thrust on the road; the road exerts a reaction force (friction), which enables the car to accelerate forward. (c) A magnet attracts a thumb tack, but the thumb tack reacts back. (d) The boy hits the table, and the table hits back.

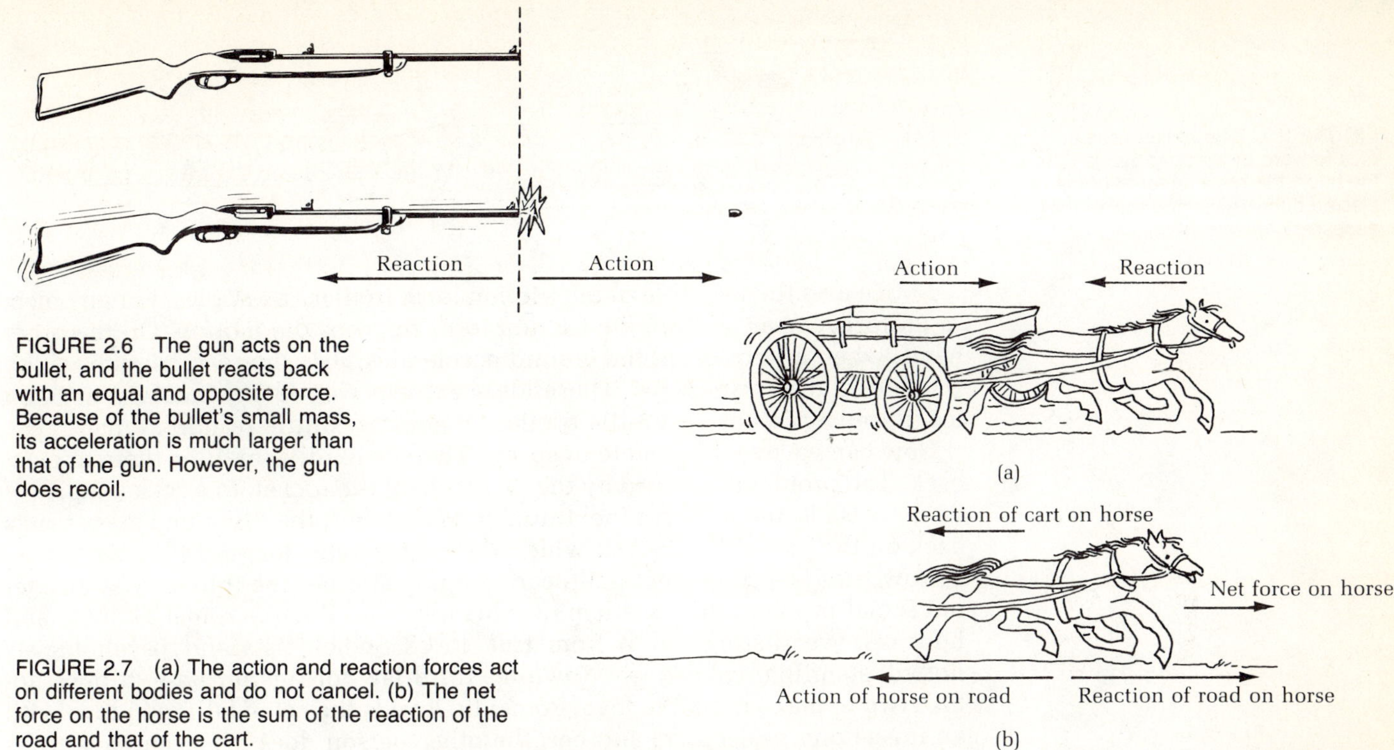

FIGURE 2.6   The gun acts on the bullet, and the bullet reacts back with an equal and opposite force. Because of the bullet's small mass, its acceleration is much larger than that of the gun. However, the gun does recoil.

FIGURE 2.7   (a) The action and reaction forces act on different bodies and do not cancel. (b) The net force on the horse is the sum of the reaction of the road and that of the cart.

moves—it recoils. Furthermore, when we realize that the gun has a much larger mass than the bullet, we can understand why the recoil velocity of the gun is so much smaller than the velocity of the bullet (Fig. 2.6). Both are acted on by the same amount of force. However, the second law tells us that the acceleration of the bullet is, because of its smaller mass, very much greater than that of the gun. This is what causes the enormous difference between the velocities.

The case of the horse pulling a cart is somewhat more complicated. At first sight it may even seem that there should be no motion, since, according to the third law, the cart also pulls the horse with an equal and opposite force. That is, the cart is accelerated forward but the horse is accelerated backward. The paradox is resolved when we recall that an object's acceleration depends on the net force acting on it. What are the forces on the horse? First there is the reaction force of the cart acting on it in the backward direction. Second there is the reaction force of the road acting on the horse; the direction of this reaction force is forward. And this is the key idea. The reaction force of the road is what enables the horse (or you or anyone) to walk or run in the first place. However, when the horse is hitched to the cart, the reaction force of the cart manages to cancel a part of the forward force on the horse; that's why a horse slows down somewhat when pulling a cart (Fig. 2.7).

Thus, according to Newton, a horse (or a person) walks or runs on the surface of the earth by kicking the earth backward so that the earth can provide a reaction force forward. You may ask: "Does the earth accelerate backward in the process? The answer is yes, although the amount of acceleration is so tiny that it is not detectable.

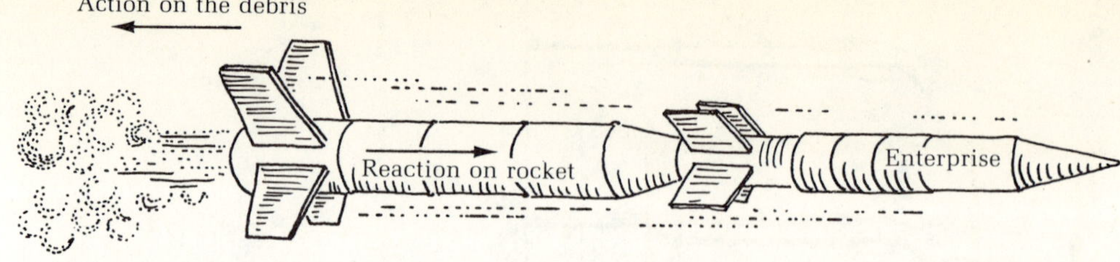

FIGURE 2.8 The rocket accelerates forward by ejecting debris from
the back; the reaction force of the
ejected debris gives the rocket the
necessary forward force.

Notice also the new role of the friction force in these examples. For an object in uniform linear motion, the friction force opposes the motion. On the other hand, when an object on the ground accelerates, it is the road friction which provides the forward force. These ideas explain why it is difficult to walk on ice. Because ice has very little friction, it generates little reaction force.

How can we ever accelerate in space? There is hardly anything there to react back. The problem is solved by the principle of the rocket. In a rocket a part of its mass is ejected as burnt fuel through its rear end; the discarded mass reacts back on the rest of the rocket, which then accelerates forward (Fig. 2.8).

Now here is a somewhat lighthearted application of the third law to an age-old social problem: the problem of romantic love. What is romantic love, and how can we distinguish it from true love? Sociologists and psychologists notwithstanding, we can use Newton's third law for this purpose. It does not take two to make romantic love. Romantic love is the act of falling in love with an image one projects on another; the other person does not even have to be aware of its existence. In contrast, true love demands that two people fully interact in the sense of Newton's third law with all its trimmings. For example, true love requires a perfect equality in the action-reaction pair.

■ 2.2 Examples and Applications of Newton's Laws

Let us first consider some examples of the first law. Take the case of a book lying on a table (Fig. 2.9). It is at rest, so from Newton's first law we conclude that the net force on it is zero. Let us analyze the forces that are acting on the book. There is, of course, the weight of the book in the downward direction. There must be a force that balances it. A little thought tells us that the balancing force must be the support of the table. Thus the **support force** must be equal and opposite to the weight force; the two forces add to zero. If the weight is represented by a vector **W**, the support of the table can be represented in this case by the vector $-\mathbf{W}$. Note especially that both forces, the weight and the support, act on the book; they are not an action-reaction pair.

Our next example is that of a downhill skier traveling at a constant velocity, an example of natural motion. Again the net force must be zero on the skier. What are the forces on her? Look at Fig. 2.10. There is, of course, the weight acting vertically downward. Another force is that of support; but our experience tells us that the direction of the support force is always normal, or perpendicular, to the base of support, as shown. Thus the support force itself is not opposite to the weight force and cannot balance it. Fortunately, there is also the force of friction that opposes the motion, acting in a direction opposite to the direction

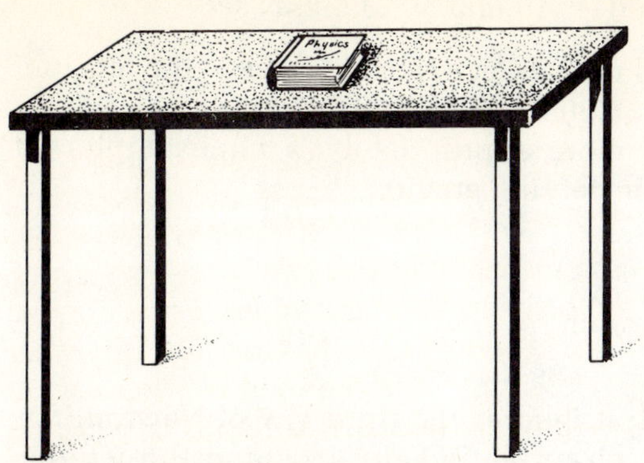

FIGURE 2.9 Book on a table. The forces on the book are equal and opposite, and hence they cancel.

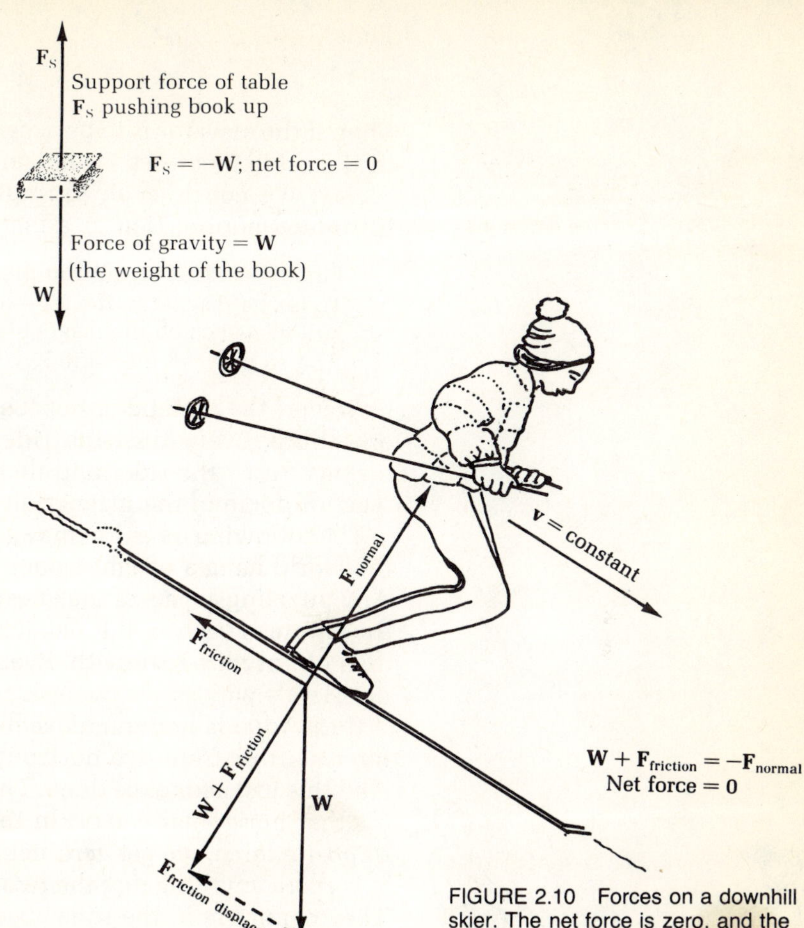

FIGURE 2.10 Forces on a downhill skier. The net force is zero, and the skier continues in natural motion.

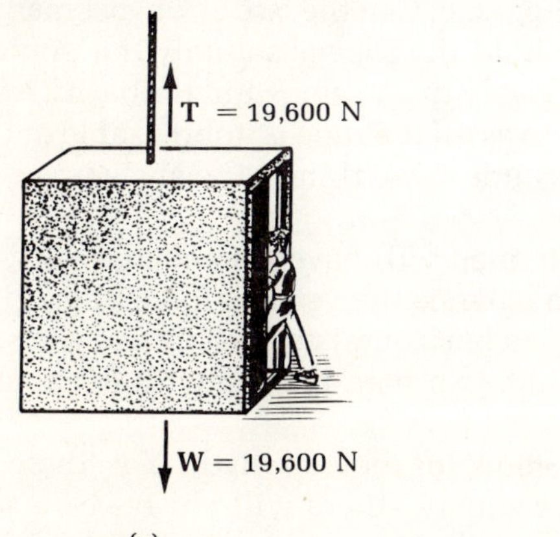

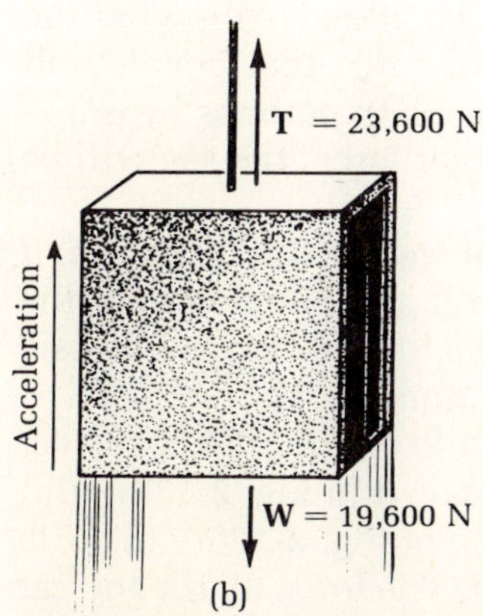

FIGURE 2.11 (a) Stationary elevator. The tension T in the cable exactly balances the weight force W of the elevator. (b) When the tension is larger, the elevator accelerates upward.

of motion. In this case the three vectors are in three different directions. But when we add them, following the rule of vector addition, we get zero.

Now let's consider the motion of an elevator. When it is at rest, the **tension force** (T) in the elevator cable balances the elevator's weight force (W). Suppose the mass of the elevator is 2,000 kg (2 metric tons). Then the elevator's weight can be determined from Eq. (2.2a):

$$\text{weight } W = 2{,}000 \times 9.8 = 19{,}600 \text{ N}$$

The numerical value of the tension force T in the cable is exactly this amount, 19,600 N, when the cable is at rest [Fig. 2.11(a)].

Now suppose the elevator is accelerated upward at a rate of 2 m/sec². What is the value of the tension now? Clearly the tension force upward must now be greater than the downward weight force, so that the net force ($T - W$) is able to give the elevator the required acceleration of 2 m/sec².

$$\text{net force} = T - W = \text{mass of the elevator} \times \text{acceleration}$$
$$= 2{,}000 \text{ kg} \times 2 \text{ m/sec}^2 = 4{,}000 \text{ kg·m/sec}^2 = 4{,}000 \text{ N}$$

Thus*

$$T = 4,000 + W = 4,000 + 19,600 = 23,600 \text{ N}$$

Then if the elevator is to be accelerated upward at a rate of 2 m/sec², the tension in the cable increases to the value of 23,600 N [Fig. 2.11(b)].

Next we consider an example (or, more appropriately, a nonexample) of third-law motion. Here is a method for defying gravity:

> Finally—seated on an iron plate,
> To hurl a magnet in the air—the iron
> Follows—I catch the magnet—throw again
> And so proceed indefinitely.†

However, the method is not foolproof; it ignores the third law of Newton, for one thing. Every time the rider of such an antigravity machine throws the magnet "up," the rider and the plate must recoil "downward" because of the reaction force of the magnet on them.

The following question may appear as a puzzle: Can the two strongest men in the world hang a weight from a rope and hold the rope absolutely horizontal? Any puzzling aspect of the question is probably due to some unfamiliarity with the nature of vectors. It is usual for people to ask if the rope is unbreakable or the men of unlimited strength. Even if you assume these things, the answer is still no. Here's why.

If the rope is horizontal, then the strong men will have generated a vertical force starting from two horizontal ones to balance the vertical force of gravity. And this just cannot be done. The sum of two horizontal vectors always gives us another horizontal vector. In the case of the two men pulling with equal and opposite force, we get zero net force.

Is there anything that the two men can show for their strength? Yes, there is. They cannot hold the rope absolutely horizontal—there will always be a sag. But by pulling hard they can keep the sag small. Look at Fig. 2.12. If there is a sag, the tension forces along the lines of the sag can be added, following the parallelogram rule. This gives us a vertically upward force that equals and cancels that of the weight. You can easily convince yourself, perhaps by drawing a few more diagrams, that if the men pull with larger force, the sag will be smaller.

And now a paradoxical problem. There is a toy car of weight 1 lb on a table. It is being accelerated with a weight of 3 lb attached to it and hanging over the edge of the table as shown in Fig. 2-13(a). What is the net force acting on the car? Don't worry about the pulley, it is assumed to be frictionless.

Perhaps common sense tells you that the answer is 3 lb of force. This answer is wrong. You are probably thinking of the situation shown in Fig. 2-13(b). But there is a difference between the two cases. In the case of Fig. 2-13(a), the 3-lb force must accelerate both weights, not just the toy car. The force on the toy car itself is less than 3 lb.

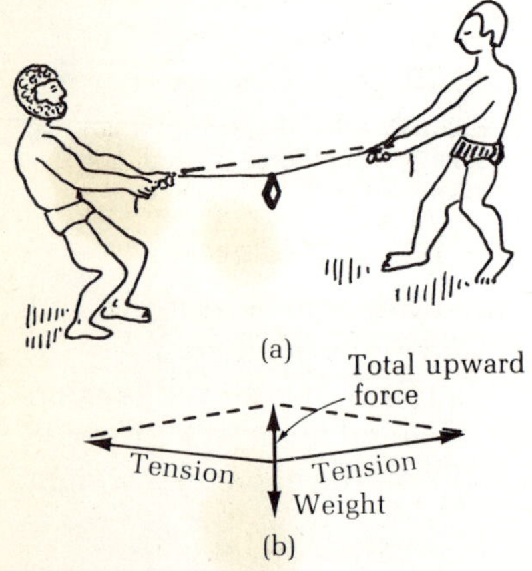

FIGURE 2.12   (a) There is always a sag since we can never get a vertical balancing force starting with horizontal forces. (b) When there is a sag, the tension forces along the two halves of the rope add, giving the vertical balancing force.

---

*Notice how the units check out. This is also a useful way of handling the units.

†From *Cyrano de Bergerac*—Brian Hooker translation by Edmund Rostand. Copyright 1923 by Holt, Rinehart and Winston. Copyright 1951 by Doris C. Hooker. Reprinted by permission of Holt, Rinehart and Winston, Publishers.

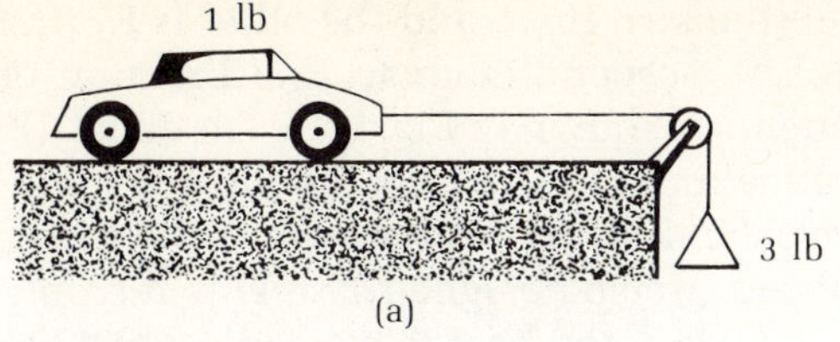

FIGURE 2.13 (a) What is the force on the toy car? If you think it is 3 lb, think again. (b) Is this the situation you are thinking of? A person can pull the string applying a 3-lb force on the car, but can the weight do that?

The mathematics of the problem works out this way. First, we find the common acceleration, which is the total force divided by the total mass. The total force is 3 lb; the mass has to be calculated from the weights using Eq. (2.2),

$$\text{total mass} = \frac{\text{total weight}}{32} = \frac{4}{32} = \frac{1}{8} \text{ slug}$$

Thus

$$\text{acceleration} = \frac{\text{total force}}{\text{total mass}} = \frac{3}{\frac{1}{8}} = 24 \text{ ft/sec}^2$$

Notice that since we are using standard FPS units for force and mass, the acceleration is given also in standard FPS units.

Next we find the force on the toy car by applying Newton's second law to the car alone. The force F is the mass ($\frac{1}{32}$ slug) times the acceleration (24 ft/sec^2) of the car:

$$F = \tfrac{1}{32} \times 24 = \tfrac{3}{4} \text{ lb}$$

Note that since we used standard FPS units for the mass and acceleration, the force also is in the standard FPS unit.

We have looked at several examples in which two force vectors add together to give a net force. Two vectors that add to give a net vector are called **components** of the final vector. Conversely, any vector can be **resolved** into two components.

Often the resolution of a vector into its components comes in handy to get insight into a problem. Usually the problem itself will suggest the desired direction of one of the components. The other component vector is usually sought in a direction perpendicular to the first direction. The magnitude of both components is then determined by using the parallelogram rule; when added according to this rule, their sum or **resultant** must be the original vector. The procedure is illustrated in Fig. 2-14.

As an application of the preceding ideas, let's consider a sailboat. In particular, let's see how a sailboat manages to sail against a prevailing wind (sailors call this "close-hauling"). First, though, do you know that a prevailing wind always pushes a sail in a direction perpendicular to the sail's plane and *not* in the direction of the wind? This is true of the interaction of all gases (and also liquids) with a smooth surface—the force of interaction is always perpendicular to the surface. Once you know this, the rest of the mystery of close-hauling can be cleared up by resolving the force (call it **F**) perpendicular to the plane of the sail into two components: one is **F**$_\parallel$ (the symbol $\parallel$ means parallel) which is

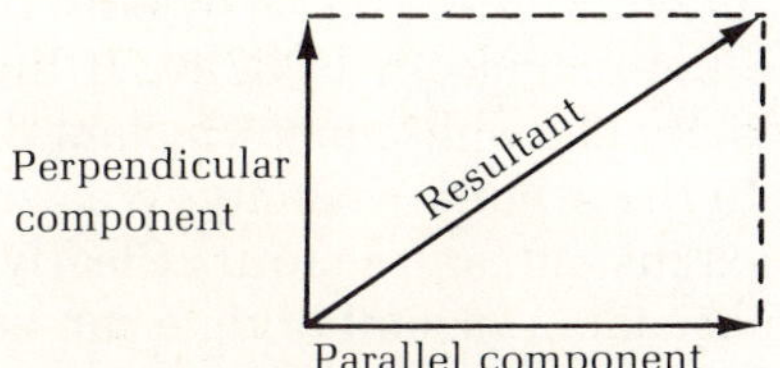

FIGURE 2.14 To find the components of a vector, first draw a line through the foot of the vector along the given direction. Next draw a line perpendicular to the first line. Finally, complete the parallelogram, whose long diagonal is the given vector. The component of the given vector along the desired direction is called the parallel component, and the component in the perpendicular direction is called the perpendicular component.

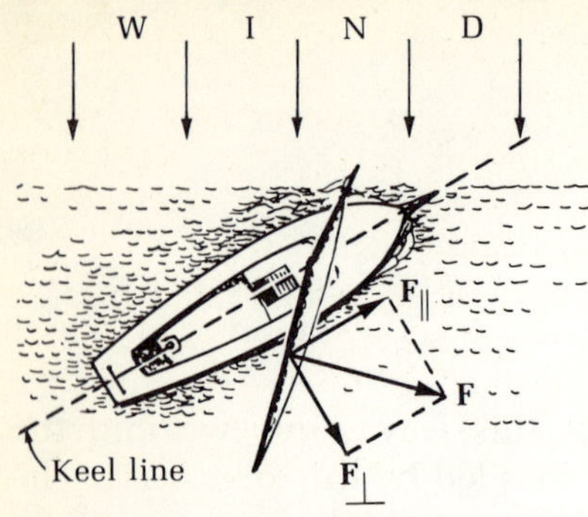

FIGURE 2.15 Sailing almost against the wind. The force due to the wind, **F,** is perpendicular to the plane of the sail. We can resolve it into a component $F_\parallel$ parallel to the keel line, and another $F_\perp$, perpendicular to the keel line. $F_\parallel$ propels the boat. The effect of $F_\perp$ is canceled by the frictional resistance of the water.

along the direction of motion of the boat (the keel line), and the other is $\mathbf{F}_\perp$ (the symbol $\perp$ means perpendicular), which is perpendicular to the direction of motion. The resolution of **F** into components is shown in Fig. 2.15. Obviously $F_\parallel$ is the desired component; $\mathbf{F}_\perp$ is a nuisance force trying to move the boat sideways. Fortunately, $\mathbf{F}_\perp$ is offset by the resistance of the water against the heavy deep keel (this is why the keels of sailboats are so designed). So $\mathbf{F}_\parallel$ is the only effective force and will haul the boat even when the boat is headed against the direction of the wind, as you can see in the figure. Unless the plane of the sail is parallel to the keel line or to the wind direction, there is always a finite force component $\mathbf{F}_\parallel$ available to accelerate the boat in any desired direction. We can, in fact, prove that the best result (maximum acceleration) is obtained when the plane of the sail bisects the angle between the wind direction and the keel line.

■ 2.3   The Newtonian Framework

In the study of motion, we try to find the answers to two questions: Where? When? The objective is to determine the position of a body at all times. How does the framework of Newton's laws work toward meeting this objective?

Given the net force of an object at all times, Newton's second law of motion enables us to calculate the acceleration of the object at all times according to the equation

$$F = ma$$

Now acceleration is the rate of change of velocity with time. The knowledge of acceleration, in turn, enables us to determine the change in velocity. If the velocity is known at some initial time, the velocity at a subsequent instant of time can be determined by adding the change to the original velocity. We can then proceed to determine the velocity at the next instant, and so forth. Clearly, if we are diligent enough, the velocity at any subsequent instant of time can be determined.

Velocity is the rate of change of position of an object with time. So the same procedure described above can be used to determine the subsequent positions of an object if we are given its initial position, by using the knowledge about its velocity. In effect, then, once the forces on an object are known, as well as the initial values of its velocity and position (together referred to as the initial conditions), we can determine the position of an object at any subsequent time using Eq. (2.1).

The starting point of the whole calculation is the forces. What do we know about the forces that act on objects under different situations? What are the rules for figuring out forces?

I remember a high school writing assignment: Write an article on an automobile. Following my scientific bent of mind, I learned aspects of automobile driving, all technical, of course, and wrote the article, carefully explaining how an auto accelerates and decelerates, how its performance is affected by road conditions, and so on. I entitled my article, "The Automobile and Newton's Laws." I had learned some mechanics in my science class and did not hesitate to show my advanced knowledge. It was a big surprise to me when the teacher selected a friend's article instead of mine as best. When I read my friend's article I understood why his was chosen. His article was named, "One Day in the Life of an Automobile." He wrote about the intimate relationship of a man and his auto; the man takes care of the auto and in the process learns about himself. The article did not have a single bit of technical information about automobiles or automobile driving, but somehow after I read it I felt that I had learned more.

All this came back to me recently when I was reading *Zen and the Art of Motorcycle Maintenance.** This book, very similar in theme to my friend's article, talks about a man who takes care of his motorcycle. However there was one important difference. This book does not underrate the technical aspects. I felt better. Suddenly I realized that my article was as important as my friend's and both of us committed the same mistake—ignoring the other side of the coin.

Following this preamble is the content, even if not in the original form, of my automobile story. I wish I had a copy of my friend's article, but you may feel his way about the auto, or know someone who does, and hence may be able to imagine its content.

Often you are faced with quickly accelerating your car from rest (zero speed) to a highway speed of some 60 mi/h in a matter of a few seconds. The acceleration is found to be 8 ft/sec² when a car is accelerated from 0 to 60 mi/h (88 ft/sec) in 11 sec. We know from experience that almost any car can handle this kind of acceleration, provided the road conditions are good. On the other hand, on wet roads this much acceleration can be dangerous. Why?

When we say good road conditions we mean a road capable of generating a certain amount of friction. Friction is reduced drastically when the road is wet. Since road friction is the key to generating the forward force required for acceleration, we see at once why it is trickier to accelerate on a wet road.

Let's calculate some numbers. The first thing we have

*R. M. Pirsig (New York: Morrow, 1974).

to find is the magnitude of the friction force on a car. There happens to be a simple rule for figuring this out. The friction force is a fraction (called the **coefficient of friction**) of the normal support force on an object. Now for a car on a level road, the normal support force of the road balances one-fourth the weight of the car at each of the four wheels (Fig. 2.16). Thus for a car with rear-wheel drive, the magnitude of the normal support force on the rear wheels is equal to half the weight of the auto. Then we get

force of road friction at back wheels

$$= \text{coefficient of friction} \times \text{normal support force}$$

$$= \text{coefficient of friction} \times \frac{\text{weight of car}}{2}$$

For a dry road the coefficient of friction is about 0.6. The weight of the car can be assumed to be 3200 lb. Thus

$$\text{force of road friction} = 0.6 \times \frac{3200}{2} = 960 \text{ lb}$$

$$\text{mass of car} = \frac{\text{weight (in pounds)}}{32}$$

$$= \frac{3200}{32} = 100 \text{ slug}$$

see Eq. (2.2)]. Thus the maximum acceleration attainable for a 3200-lb automobile with two-wheel drive on a road with a coefficient of friction of 0.6 is given as

$$\text{acceleration} = \frac{\text{force of road friction}}{\text{mass}}$$

$$= \frac{960}{100} = 9.6 \text{ ft/sec}^2$$

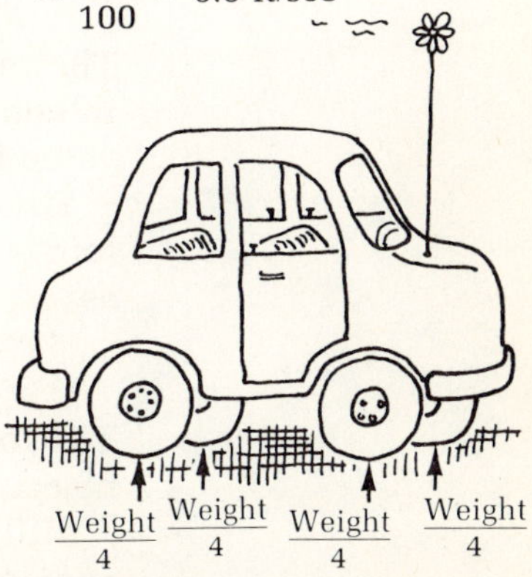

FIGURE 2.16   For an auto on the road, there is a support force equal to one-fourth the weight of car at each of the four wheels of the car.

This is clearly adequate compared to the 8 ft/sec² of acceleration that you need.

On a wet road the coefficient of friction is reduced to, say, 0.3. Then the force of friction is reduced to

$$0.3 \times \frac{3200}{2} = 480 \text{ lb}$$

and the maximum acceleration is reduced to

$$\frac{480}{100} = 4.8 \text{ ft/sec}^2$$

This is considerably smaller than the 8 ft/sec² needed to accelerate from 0 to 60 mi/h in 11 sec. Therefore, we must spend a longer time to accelerate on wet roads.

Actually, racers manage to accelerate faster than the 9.6 ft/sec² that we have calculated above. The reason is that racetracks and racing tires are made so that it is possible to get a coefficient of friction as high as 1 or even larger.*

The larger the acceleration, the greater the force that comes into play. Let us estimate the force of impact on a car that hits a wall at a speed of 30 mi/h. In a matter of a fraction of a second, the car is forced to decelerate from 30 mi/h to 0 speed. Suppose this happens in $\frac{1}{8}$ sec. Then

the magnitude of the deceleration is (since 30 mi/h = 44 ft/sec)

$$\text{deceleration} = \frac{\text{change in velocity}}{\text{time to stop}}$$

$$= \frac{44 \text{ ft/sec}}{\frac{1}{8} \text{ sec}} = 352 \text{ (ft/sec)/sec}$$

This is exactly 11 times the acceleration due to gravity, which is 32 ft/sec.²

The force of impact on the person in the car due to the collision is then given as

$$\text{force} = \text{mass} \times \text{deceleration}$$

$$= \frac{\text{weight (in pounds)}}{32} \times \text{deceleration}$$

using Eq. (2.2) again. If the person's weight is 150 lb, the force of impact is

$$\frac{150}{32} \times 352 = 1650 \text{ lb}$$

This is 11 times the person's weight, and human bodies do not usually accommodate such a large force.

---

*We can also think of using four-wheel drive to get a greater acceleration. But this poses other problems. So we find a very rare use of four-wheel drive on cars on racetracks.

Forces

We have already talked about one force in particular, the force of friction. For example, the rule for finding the friction of the road on an automobile tire is given by

force of friction = coefficient of friction × normal support force on the object

The value of the coefficient of friction depends on the nature of the two objects in contact. For example, its value for rubber on cement is different from its value for rubber on wood.

Similar rules have been established for the evaluation of some of the other forces we encounter. The force of air drag on a moving automobile is given as

$$\text{drag force} = \text{force constant of drag} \times (\text{velocity})^2 \qquad (2.4)$$

The force constant of drag depends on some of the physical properties of the object (under drag), such as shape.

Still another example is **spring force**, the force that acts on a mass at the end of an elongated spring. The force in this case is proportional to the elongation of the spring (Fig. 2.17), a law that was first established by English physicist Robert Hooke, a contemporary of Newton.

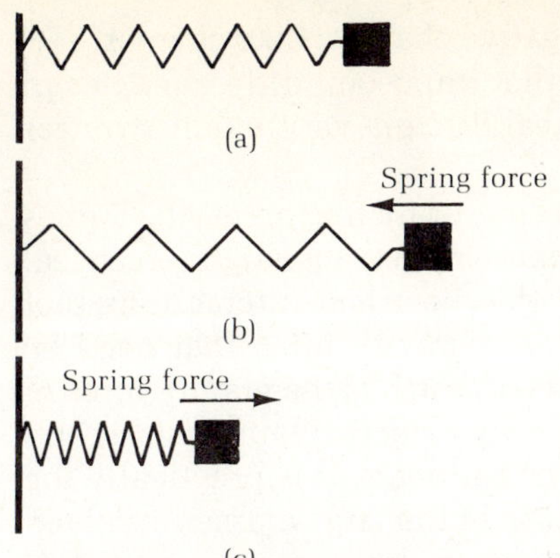

FIGURE 2.17 (a) Unstretched spring, no force on the mass. (b) The spring is stretched, and there is now a spring force acting on the mass. The spring force is proportional to the amount of extension of the spring and acts to restore the spring to its unstretched position. (c) The spring is compressed; again the mass is acted on by a restoring force proportional to the amount of compression of the spring.

$$\text{spring force} = \text{spring force constant} \times \text{elongation}$$

The spring force constant also depends on some physical properties of the spring, and hence it varies from one spring to another.

There is also **weight force**, which is really the force of earth's gravity on an object. Newton himself discovered the law that determines the **force of gravity** between any two objects:

$$\text{force of gravity} = G \times \frac{(\text{mass of object No. 1}) (\text{mass of object No. 2})}{(\text{distance between them})^2}$$

G is also a force constant (the force constant of gravity), but it has one special feature. G is the same for all objects interacting gravitationally; it is a *universal* constant.

There are some important differences between the gravity force and forces like friction or spring force. For one thing, the force laws for the latter are not universal, so the constants of proportionality depend on the characteristics of the interacting objects. Another difference is that these latter forces have a rather limited range of validity; for example, Eq. (2.4) for determining the air drag works only for objects at fairly high velocity. For objects at low velocity, the drag force varies linearly with velocity, and even then approximately. Actually all three "laws" of force mentioned—friction, drag, and spring—are found to be approximate laws on close examination. Any attempt to find a more accurate law with a greater range of validity makes the force law more and more complicated. The law for the gravity force, on the other hand, does not have these problems.

It is natural, then, to distinguish between forces such as friction and those like gravity. We call gravity a **fundamental force** because it is fundamentally simple. Forces like air drag, on the other hand, are regarded as **nonfundamental.**

With the advent of submicroscopic physics, it has been possible to show that the nonfundamental forces really arise from another fundamental force at the atomic level. This is the **force of electrical interaction.** This fundamental force results from the electrical interaction between charged particles. There are two kinds of **charges,** positive and negative. Of the three subatomic particles mentioned previously, the electron is negatively charged, the proton carries a positive charge, and the neutron has no charge. Atoms and molecules, the submicroscopic constituents of matter, have equal numbers of protons and electrons and therefore have no net charge. Thus matter in bulk is neutral—ordinarily, that is. If by some means we manage to cause an imbalance of the charges—for example, by removing a few electrons—the object will acquire a charge. This is what happens when we rub glass with silk. The glass loses a few electrons and acquires positive charge, while the silk, which gets the extra electrons, becomes negatively charged.

The force law for electrical charges is given by **Coulomb's law:**

$$\text{electrical force} = k \times \frac{(\text{charge on object No. 1}) \times (\text{charge on object No. 2})}{(\text{distance between them})^2}$$

where k is the force constant of electrical force. Thus the law is very similar to the gravity law, with two important differences. The force is attractive only for

unlike charges, that is, between positive and negative charges. Like charges, a +
and a + or a − and a −, repel each other. Another important difference lies in
the numerical values of the forces. The electrical force is very much stronger
than the gravity force.

Electrical forces dominate the everyday working of matter on earth. Forces
such as the friction force arise in a complex manner from the slight electrical
imbalances that exist at the atomic scale and the electrical interactions that
result from these imbalances. The only important gravity force that need be
taken into account for the motion of most objects on earth is the gravity force of
the earth itself. In discussing the motion of heavenly objects (planets, stars, and
the like), on the other hand, gravity completely dominates; it is practically the
only force that need be considered. This is because of the large masses involved
(so that the gravity force is very large) and also because heavenly objects maintain a moderate amount of charge neutrality as a whole (so that electrical forces
arising from any imbalances are small).

To finish this discussion, let's examine two other fundamental forces, both
with extremely short ranges (they operate over distances of the order of 10^{-15} m
or less). These forces vanish when the distance of the interacting objects is
larger than the size of the atomic nucleus, so they are important only in nuclear
phenomena. These are the **strong nuclear force** and the **weak nuclear force.**
The strong force is even stronger than the electrical force within the range of its
activity. The weak force, as the name suggests, is quite a bit weaker than the
electrical force but still a lot stronger than the gravitational force. These two
nuclear forces are also important in the supposed evolutionary processes of
stellar systems.

Table 2.2 summarizes all four fundamental interactions. And one more thing:
The magnetic phenomena that you are familiar with really arise from electrical
charges in motion. Electricity and magnetism are related and together are referred to as **electromagnetism.** The table, therefore, lists the electromagnetic
force rather than the electrical force.

TABLE 2.2   The fundamental interactions of matter.

Interaction	Strength Compared to Nuclear Force	Range
Strong nuclear force	1	Short range, about 10^{-15} m, the size of the atomic nucleus
Electromagnetic	10^{-2}	Infinity
Weak nuclear force	10^{-14}	Extremely short range, even less than nuclear size
Gravitational	10^{-40}	Infinity

Absolute space and absolute time

There is one problem associated with the first law of motion that bothered
Newton. When he asserts that a force-free object undergoes natural motion—
uniform motion in a straight line—a question naturally crops up: Motion with

respect to what? It could not be with respect to material objects like the earth or even the sun, because these bodies have accelerated motion of their own. True natural motion must be defined, according to Newton, with respect to absolute space because, as he wrote in the *Principia*, "Absolute space, in its own nature, without regard to anything external, remains always similar and immovable."

Newton felt that similar considerations also must apply to time. If time were defined with respect to the motion of material objects like the earth, the law of inertia would not be valid because of irregularities in the motion of the earth. This tempted Newton to make a similar statement defining the concept of absolute time: "Absolute true and mathematical time, of itself and from its own nature, flows equably without regard to anything external. . . ."

Newton's absolute space does not depend on matter and his absolute time does not depend on motion. We will find later that physical space and physical time do not have the Newtonian attributes of absoluteness.

But hasn't the question that we started with in this discussion come back to haunt us? Today we believe that the first law plays a dual role: it defines a special set of reference frames and at the same time remains a meaningful law of nature.* If a "free body" (a body free of external forces) is observed to be unaccelerated from a certain frame of reference, that frame of reference is one for which Newton's first law is valid. Such a frame of reference is now called an **inertial frame of reference.** The dual role of the first law can be appreciated by noting that if we have established an inertial reference frame by looking at the motion of one free object and finding that it is natural, the motion of all other free bodies also is guaranteed to be natural (unaccelerated) viewed from this reference frame.

SUMMARY

Now you have been introduced to the basic structure of dynamics: the interplay of force and motion. The structure is stated in terms of the three laws of motion formulated by Newton. The first law gives the correct description of force-free or natural motion; natural motion is motion with constant velocity in a straight line. An object's tendency to continue in natural motion, if no external force acts on it, is called inertia. The state of rest is recognized as a special case of natural motion. You cannot discern between rest and natural motion anyway, since natural motion is relative.

Among the three laws, the second law of motion is the most useful. It gives the quantitative relationship of the net external force acting on an object and the acceleration that results from it. We can state the law in words—force is the product of the object's mass and acceleration—but the law is more well-known in the form of the mathematical equation

$$F = ma$$

This is the one common thing about physical laws: they are precise mathemat-

---

*A reference frame is essentially a coordinate system. Three rods tied together at a point in a mutually perpendicular fashion define a reference frame. In other words, a reference frame is three coordinate axes at an origin. See Chapter 7 for a more detailed discussion.

ical statements between physical quantities, and they are most easily conceptualized as such.

There are some subtleties associated with the concept of force which you must have noticed. For example, force is a vector. Also, the third law of motion tells us that forces can occur only as action-reaction pairs. A force is an interaction between two partners. Another interesting aspect of some of the forces (like the friction force) is that they are not fundamental but are derived from the fundamental electrical forces at the submicroscopic level. Altogether there are only four different kinds of fundamental forces between matter of any kind; all other forces are derived.

QUESTIONS

Review and reason

1. State Newton's first law of motion and explain the concept of inertia.
2. An automobile is parked on the roadside. Draw a diagram showing the important forces acting on it. Is there any uncanceled force?
3. An object is under the action of several forces that all add up to zero. What can you say about the velocity of the object?
4. An ice skater glides along a straight path with constant speed on a frictionless level surface of ice because of which of the following? (a) The force of wind is pushing the skater along; (b) the ice exerts a pulling force on the skater; (c) the skater's inertia; (d) none of the above.
5. In some parts of South Asia, there is a simple machine called "dheki" (rhymes with Becki), which is used by farm women to separate the rice from the husks. When the beater comes down on the rice underneath, the husks move further away from the beater than do the grains. Why?

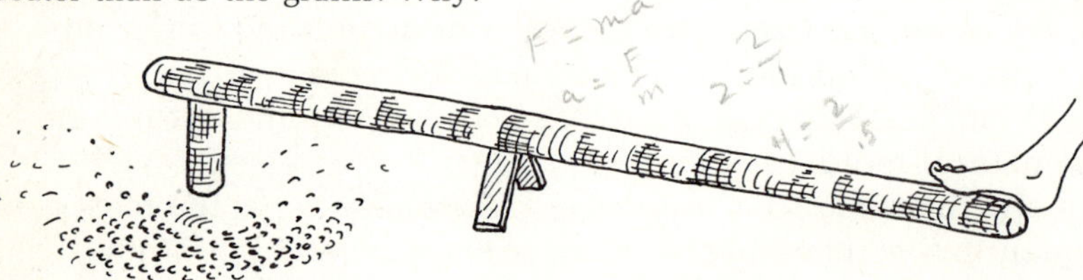

6. Two methods for dusting a jacket are to shake it or to sweep it off with a brush. Discuss the physical principle used in each case.
7. Have you ever tried to jump into a moving vehicle? It's easy if you first run and acquire a velocity equal to that of the moving vehicle and then jump in. Why?

8. In many parts of the world, people often get off a moving bus without waiting for it to stop. They face the same way as the direction of motion of the bus, jump off, and take a few steps backwards after they touch the ground. Explain why this should help them keep their balance.
9. You are riding in a moving helicopter. Upon seeing your friend's house below, you quickly drop a note wrapped around a piece of rock (you were prepared for such an eventuality of course!). Will the message fall on target? Why or why not?
10. State and explain Newton's second law of motion.
11. The same force is applied to two objects. One of them accelerates at a rate of twice the other. What conclusion can you reach about their relative masses?
12. Define a newton of force. How does it compare with a pound?
13. Distinguish between the concepts of weight, mass, volume, and density.
14. Two objects A and B have identical mass, but A occupies more volume than B. Which one has more density?

15. Fig. 2.18 shows two devices, both called balances. One has a spring. When an object is placed on the pan as shown, the spring is compressed, and the extent of the compression of the spring is recorded on a calibrated (standardized) scale. Does this give direct information about the object's weight or its mass? In the other balance an object can be placed on one pan and its weight matched by placing calibrated masses on the other pan. Is this a direct way of measuring weight or mass? Explain.

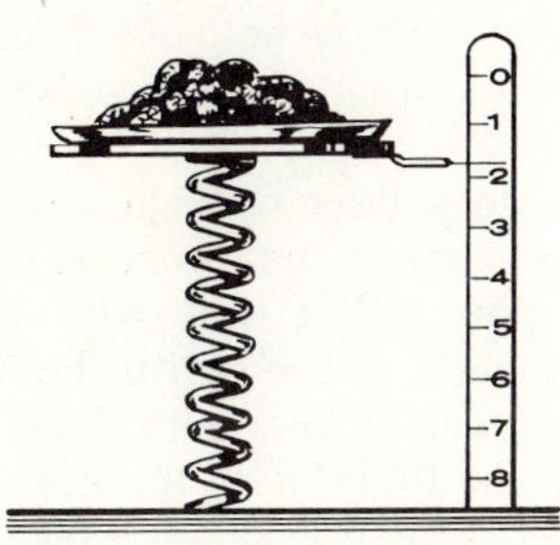

Spring Scale

FIGURE 2.18

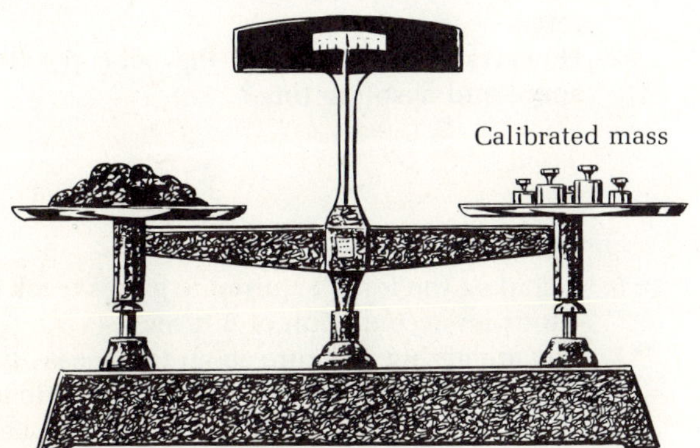

Two-Pan Balance

16. Discuss the application of the third law of Newton in rowing a boat.

17. Imagine that you are stranded on a frozen lake of completely frictionless ice. Can you come out of there by walking, running, or even crawling? Why or why not? Assuming you are fully dressed (why shouldn't you be?—we are just giving you a hint), can you suggest a way of getting off the ice?

18. There is a European folk tale in which a hero, after falling into a swamp and beginning to sink into the mud, saves himself by grabbing his own hair and pulling himself out slowly but with a calm assurance that comes only with experience. Is this possible or is the story in flagrant violation of an important physical law? Explain.

19. Fig. 2.19 shows a painting hanging from a nail in the wall. T is the force of tension along the two wires of suspension and W is the weight of the painting. Since the painting is in a state of rest, the forces must cancel. Show how they cancel by drawing a parallelogram of forces.

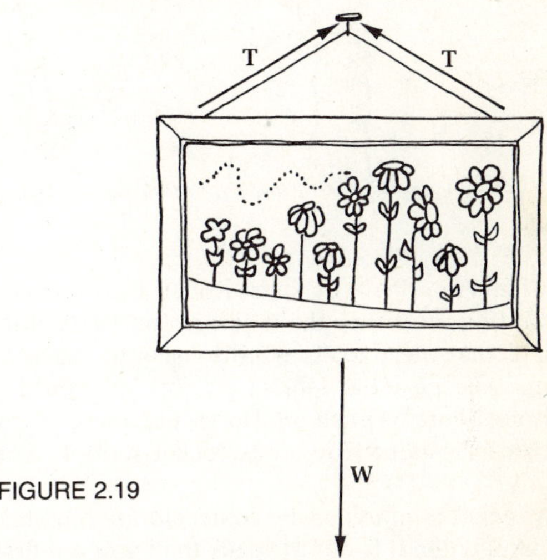

FIGURE 2.19

20. A flower pot hangs from the ceiling as shown in Fig. 2.20. If you give a quick jerk on the bottom string, which string is more likely to break first? Why? If you pull the bottom string slowly and firmly, which string is more likely to break first? Explain.

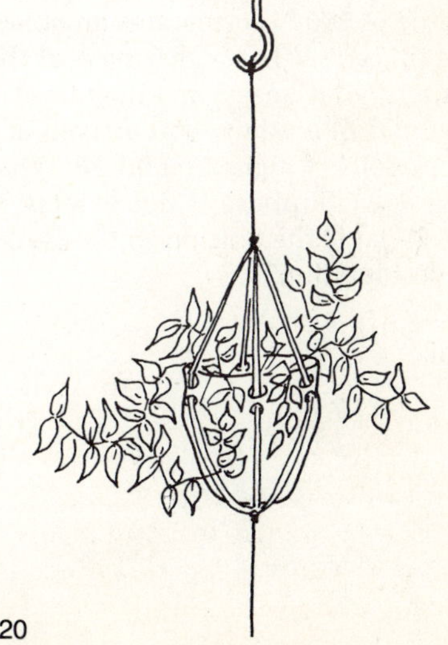

FIGURE 2.20

21. Fig. 2.21 shows a child on a swing. What is **T** and what is **W**? Draw the parallelogram of forces for finding the resultant force.

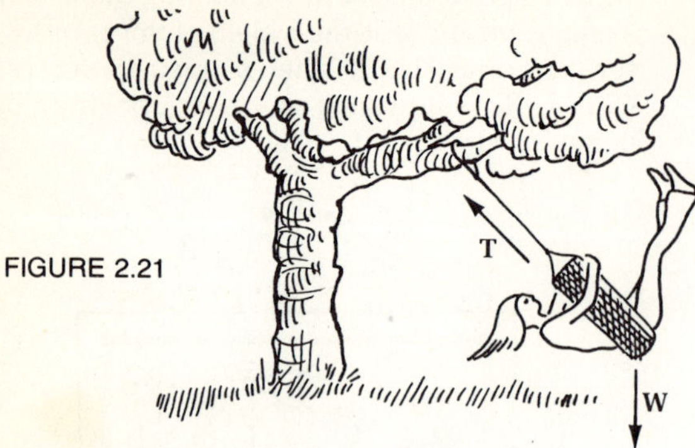

FIGURE 2.21

22. When American physicist Robert Goddard proposed sending rockets to the moon in the 1920s, his critics said that the rockets would cease to move in outer space because the debris they emitted would have no atmosphere to push on. Do you agree with this criticism? Explain. How does rocket motion work anyway?

23. A rocket is much easier to accelerate (i.e., takes less force) when it is out in space than when it first blasts off the ground. Why?

24. A man is approaching the shore in a boat. Being in a hurry and thinking that if he reduced the load on the boat, it will move faster toward the shore, he throws a heavy piece of his luggage forward onto the shore. Does this act accomplish his objective? Why or why not?

25. A woman in a boat has lost her oars. She ties a rope to the front end of the boat and continuously pulls the rope toward herself. Will she be able to get the boat moving this way? Will this method work for a spaceship stranded in space? Why or why not?

26. Explain with the help of a diagram how a force can be resolved along and perpendicular to a given direction.

27. How do sailors manage to close-haul, that is, sail almost against the direction of the wind?

28. Can you sail a boat with the plane of the sail completely aligned with the direction of the wind? Can you sail a boat with the plane of the sail aligned with the direction of the keel line? Explain.

29. A spring force is called a restoring force. Why?

30. Distinguish between fundamental and nonfundamental forces. Give an example of each.

31. Discuss two different methods of measuring a force.

32. How was Newton led to the concepts of absolute space and absolute time?

Arithmetic

1. An object having a mass of 20 kg is acted on by a force of 100 N. What is its acceleration?

2. What net force is needed to accelerate a body of mass 3 kg to an acceleration of 3 m/sec^2?

3. A force of 500 N accelerates an object by the amount of 10 m/sec^2. What is the mass of the object?

4. What is your mass in kilograms? Your weight in newtons? Show how you arrived at your answers.

5. An elevator weighs 20,000 N. What is its mass in kilograms? Suppose it decelerates at a rate of 2 m/sec^2. What is the tension in the elevator cable during the deceleration?

6. Calculate the force required to give a truck of weight 4 tons an acceleration of 8 ft/sec^2.

*7. You are giving a lecture about the necessity of wearing seat belts. Can you do some arithmetic to demonstrate dramatically how necessary seat belts really are? For example, estimate the force the seat belt counters in the case of a quick stop and discuss what could happen if the seat belts were not fastened.

*8. A 5-lb weight causes a 2-in elongation of a spring. Calculate the elongation when a weight of 12 lb acts on the spring.

---

*Optional.

3 Momentum and its Conservation

Discovering the laws of physics is like trying to put together the pieces of a jigsaw puzzle. We have all these different pieces, and today they are proliferating rapidly. Many of them are lying about and cannot be fitted with the other ones. How do we know that they belong together? How do we know that they are really all part of one as yet incomplete picture? We are not sure, and it worries us to some extent, but we get encouragement from the common characteristic of several pieces. They all show blue sky, or they are all made out of the same kind of wood. All the various physical laws obey the same conservation principles.

RICHARD P. FEYNMAN

■ 3.1 Momentum and a Conservation Law

René Descartes, who was perhaps more of a philosopher than a physicist, developed a concept in the theory of motion that has turned out to have great importance. This is the concept of **momentum,** which Descartes referred to as the quantity of motion. Momentum is the product of the mass and velocity of an object:

$$\text{momentum} = \text{mass} \times \text{velocity}$$

In symbols,

$$p = mv \tag{3.1}$$

where p stands for momentum, m for mass, and v for velocity. Since velocity is a vector, so is momentum. The direction of the momentum vector is the same as that of the velocity vector, the direction of motion.

Actually, momentum is an everyday word. When an object is hard to stop, we say it has great momentum. How is that compatible with the definition above?

Stopping a moving object involves a change in its momentum. From Eq. (3.1)

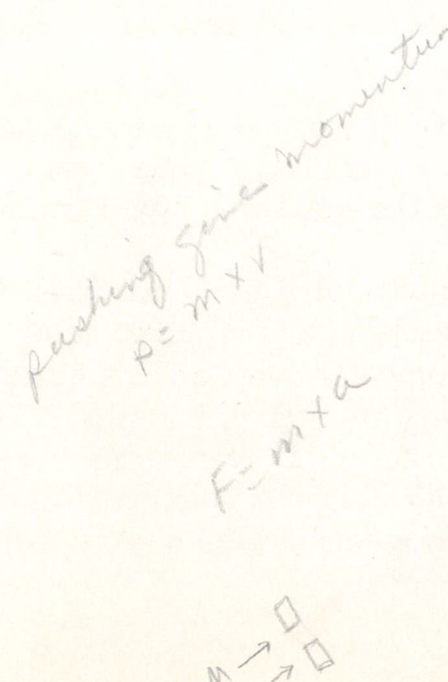

we find that the change in momentum of an object can only be accomplished through a change in velocity if the mass of the object is held constant, which is the usual case. We can express this as an equation:

$$\text{change in momentum} = \text{mass} \times \text{change in velocity}$$

If we divide both sides of this equation by the time in which the change takes place, we get

$$\frac{\text{change in momentum}}{\text{time}} = \frac{\text{mass} \times \text{change in velocity}}{\text{time}}$$

$$= \text{mass} \times \left(\frac{\text{change in velocity}}{\text{time}}\right)$$

But (change in velocity)/time is nothing but the acceleration. Thus the right-hand side of the equation above is equal to the product of mass and acceleration, a product that equals the applied force according to Newton's second law of motion. Thus we get the following expression:

$$\frac{\text{change in momentum}}{\text{time}} = \text{force} \tag{3.2}$$

There is one aspect of this relationship between the change in momentum and the force that is important to point out. It is a new expression for the force. Force is now given as the time rate of change of momentum. In fact, this new way of looking at the effect of a force on the motion of an object is more general than our previous $F = ma$. Newton actually recognized this and stated his second law of motion originally in the form of Eq. (3.2). If the mass of an object does not change, the two expressions for the force are entirely equivalent, as our logic indicates. However, there are occasions when the mass does change—for example, in the case of an accelerating rocket. In such a case Eq. (3.2), not Eq. (2.1), would be the appropriate equation to use. Later on we will see that the mass of an object also changes in the domain of relativity, and again Eq. (3.2) is the appropriate law of motion in relativity.
Eq. (3.2) can also be written as*

$$\text{change in momentum} = \text{force} \times \text{time} \tag{3.3}$$

Why is something with a large momentum hard to stop? Because a large change in momentum in a given amount of time involves a large force. Thus a heavy truck is harder to stop than a car even if both have the same velocity; it is the momentum that matters.

Equation (3.3) helps us to understand some common advice given in a variety of sports. When you start playing soccer, hitting a baseball, swinging a golf club, or rolling a bowling ball, one piece of advice that is repeated over and over again is to "follow through." Now what does following through accomplish? It increases the amount of time in which a given force acts on the target. Eq. (3.3) makes it clear that the longer the time of impact—for example, when you hit a baseball—the greater the change in momentum. So whenever the objective of

---

*The product force × time is called **impulse**.

the game is to impart the maximum change in momentum for a given force, maximizing the time, or the follow-through, helps.

Interestingly, there is one kind of sport where the objective is quite different. In karate, for example, the idea is to impart as much force as possible—in short, to maximize the force.* Remember,

$$\text{force} = \frac{\text{change in momentum}}{\text{time}}$$

Clearly, in this case we gain force by maximizing the change in momentum but minimizing the time of impact. Since time occurs in the denominator on the right-hand side of the equation above, the shorter the time of impact is, the larger the force is. This is why a karate chopper brings his or her hand down very quickly with no follow-through.

Why does a glass tumbler break when dropped on a cement floor but will often survive a fall to a carpeted floor? The answer is that in the latter case the same change in momentum occurs over a longer time since the carpet stretches the time of impact, and thus the force of the impact is smaller.

There is one important thing to note: since motion is relative, only relative velocities are considered. In fact, when you think about it, all velocities are relative. Thus on the surface of the earth, we always reckon velocity with respect to the ground—the velocity of the earth itself does not matter. There is a true story of a French pilot in World War II who stretched his hand out of his small airplane window and caught a bullet. How is this possible? The velocity of the bullet must have been about the same as the velocity of his plane. This can happen, because a bullet slows down to such a velocity toward the end of its flight path due to the action of air drag. Since only relative velocity matters, the momentum of the bullet was negligible as far as the pilot was concerned. Thus he would have had little problem in stopping the bullet.

---

*More exactly, the objective of karate is to maximize the pressure, which is force per unit area. So it is also important to apply force on as small an area as can be managed—for example, by using the edge of the hand rather than the palm.

A numerical example

Do you suppose you could ever run fast enough to acquire a momentum comparable to that of a small automobile, even though it is moving at a snail's speed of $\frac{1}{2}$ m a second (roughly 1 mi/h)? It is likely that you can. Suppose the car's mass is 1000 kg (1 metric ton). Then the momentum of the car is*

$$1000 \text{ kg} \times 0.5 \text{ m/sec} = 500 \text{ kg·m/sec}$$

You probably can run as fast as 6 m/sec (which roughly corresponds to a $4\frac{1}{2}$ minute mile), at least for a short while. If your mass is 85 kg, you can reach a momentum of

$$85 \text{ kg} \times 6 \text{ m/sec} = 510 \text{ kg·m/sec}$$

You have exceeded the momentum of the car. However, your friends with a lesser mass have to run faster to reach the same magnitude in momentum.

The law of momentum conservation

We have already mentioned that Descartes was a philosopher. There is a very important difference between the way a philosopher looks at things and the way a scientist looks at things. This is how Danish physicist Niels Bohr felt about philosophers: he said, that to be a philosopher one must know something about most things. If you generalize this trend (which is a characteristic of the philosopher), very soon you conclude that a better philosopher is one who knows less and less about more and more things; and of course the best philosopher would be one who knows practically nothing about practically everything.

Since Descartes was one of the best philosophers of his time, one might assume that he did not know much about physics.† How does one find answers for a subject without knowing much about it? In Descartes' case, he formulated a general principle and started to work his way from the top down. He didn't go very far in his downward journey, but the principle he started with proved to be very sound. Today we call it the **law of conservation of momentum.**

What is a conservation law? It is the simple statement that a quantity does not change in time. It says that if we start with a definite value for this quantity that is to be conserved, then checking it time and again, we will find that the value representing the quantity remains the same, while other things have changed. Momentum happens to be such a quantity that is conserved. *In all the turmoil of the universe, we believe that its total momentum never changes.*

What is true for the entire universe is also true for any isolated system. Incidentally, we use the word "system" whenever we want to focus our attention on a particular group of particles or objects; the rest of the universe is then the environment for our system. An isolated system means one without any inter-

---

*Since momentum = mass × velocity, the standard MKS unit for momentum is obtained as the product of the units for mass and velocity, giving us the kilogram-meter/second, abbreviated as kg·m/sec and read "kilogram meter per second."

†This comment is intended lightheartedly, of course. Descartes' discovery of momentum conservation alone testifies to his credibility as a physicist.

action with its environment—that is, there are no external forces acting on the system.

It is easy to see the validity of the momentum conservation law from the second law of motion of Newton expressed in the form of Eq. (3.2) or Eq. (3.3). These equations make it clear that when the net external force is zero, there is no change in the value of the momentum of a system, which is precisely what conservation of momentum means.

How about internal forces? Here Newton's third law comes into play and tells us that all internal forces are action-reaction pairs and therefore impart equal and opposite momenta to the two objects of the pair. Momentum is a vector quantity. For the whole body the momenta contributed by the reaction-action pairs cancel out and thus there is no contribution to the total momentum from them.

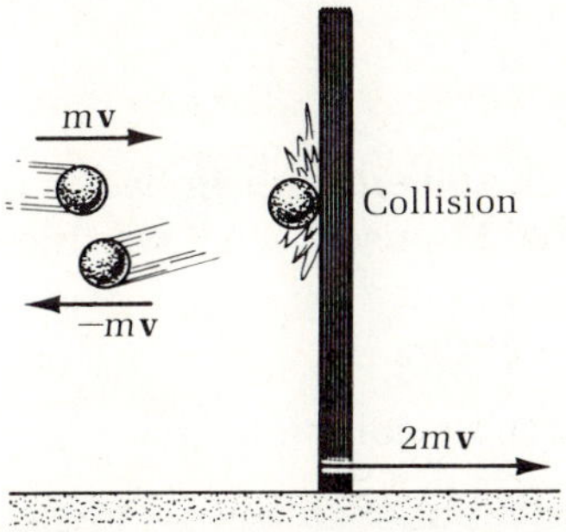

It is interesting to discuss again some of the examples of motion that we treated before with the help of the third law. This time, however, we will use the principle of conservation of momentum. When we walk we gain forward momentum for ourselves by giving the earth an equal and opposite amount of momentum. When we throw a ball at a wall, the ball rebounds. The change in momentum in the process is twice the original magnitude. Why? Remember, momentum is a vector. Suppose the forward momentum of the ball is $m\mathbf{v}$ (Fig. 3.1). When it rebounds its momentum is $-m\mathbf{v}$, a vector of the same magnitude but in the opposite direction. The change in momentum is given by the difference, final momentum − initial momentum:

$$-m\mathbf{v} - m\mathbf{v} = -2m\mathbf{v}$$

The minus sign simply indicates that the direction of momentum is now opposite to that of the original momentum: the change in magnitude is $2\,m\mathbf{v}$. How is momentum conserved then? According to the third law the wall receives an equal and opposite force from the ball. Thus the momentum of the wall, or, since the wall is fixed to the earth, the momentum of the whole earth, must change by the same amount in the opposite direction. But because the earth is immensely massive, we are not actually aware of its velocity change.

Consider rocket motion. The momentum of the rocket plus everything in it never changes. How then can the rocket accelerate forward? By ejecting debris through its rear with momentum in the backward direction, the rest of the rocket gets an increase in the forward momentum.

Clearly, everything that can be explained from the third law can also be explained with the principle of conservation of momentum. Today we believe that Descartes' principle of momentum conservation is even more fundamental than Newton's third law.

Two-body collisions on a pool table

The case of the two-body collision deserves special attention. You may be familiar with some of the results here if you are a pool player.

For simplicity we will keep our collisions simple, with no "English" or spin. First consider a head-on collision. What does momentum conservation tell us in this case? Just this simple thing:

FIGURE 3.1 When a ball rebounds from a wall, its momentum is reversed to $-m\mathbf{v}$, but the wall, and the earth with it, gains a momentum of $2m\mathbf{v}$. The total final momentum is $2m\mathbf{v} + (-m\mathbf{v}) = m\mathbf{v}$, the same as the initial momentum. Hence momentum is conserved.

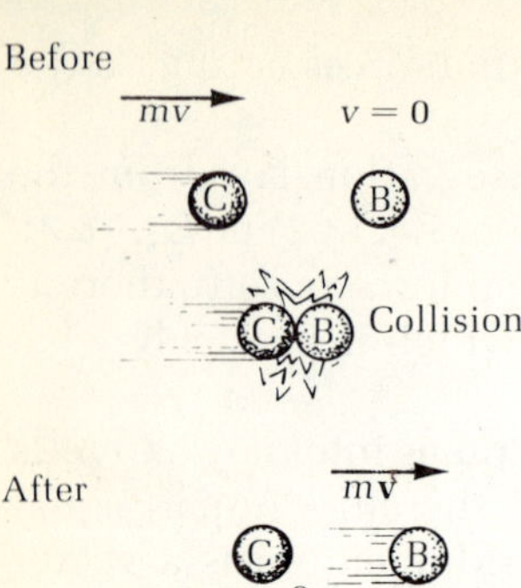

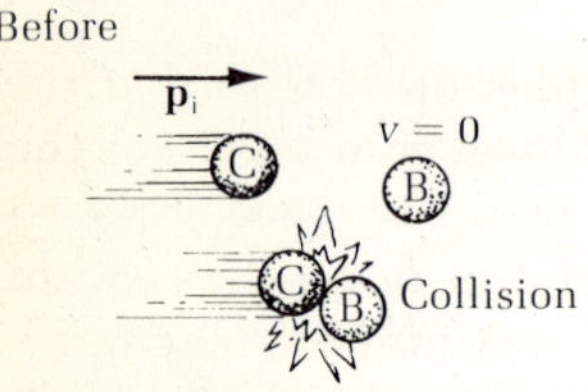

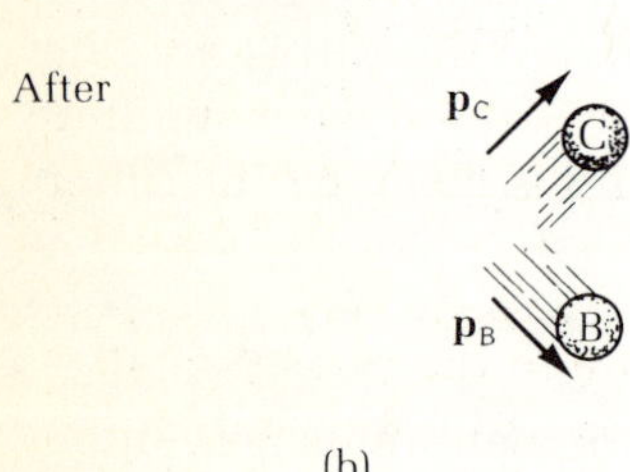

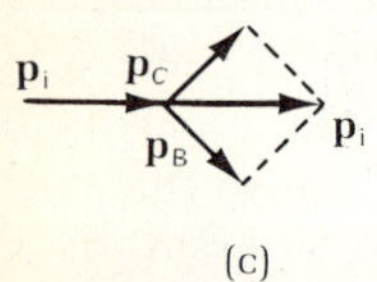

FIGURE 3.2 (a) A head-on collision between the cue ball C and another billiard ball B on the pool table. The cue ball comes to rest and the other ball picks up all of its momentum following the law of conservation of momentum. (b) Momentum is conserved also for glancing collisions. (c) The final momentum of the cue ball $\mathbf{p}_C$ and that of the struck ball $\mathbf{p}_B$ add up vectorially to the initial momentum p_i.

total momentum before collision = total momentum after collision $\qquad$ (3.4)

Thus if a cue ball starts with a certain momentum and comes to rest in a head-on collision with another ball initially at rest, the other ball must pick up the entire momentum of the cue ball—magnitude, direction, and all. Since the other ball has the same mass as the cue ball, it must move away with exactly the same velocity that the cue ball had before collision.

Even in more complicated collisions, if you look carefully you will find that the final momenta of the cue ball and the ball that it hits always add up vectorially to the initial momentum of the cue ball (Fig. 3.2). This is the way momentum conservation controls the situation, and a knowledgeable player takes advantage of it.

EXAMPLE

A bullet of mass 0.025 kg strikes a 1-kg wooden block and ends up embedded in it. The block (with the bullet in it) acquires a speed of 15 m/sec. What was the velocity of the bullet?

The key to the answer is the conservation of momentum:

$$\text{momentum before collision} = \text{momentum after collision}$$

The momentum before collision is the momentum of the bullet alone (the block is at rest initially). However, we cannot calculate the momentum of the bullet yet, because we don't know its velocity. This is one of those problems where we have to assume a symbol for the unknown quantity and do a little symbol manipulation to get what we want. Let v be the velocity of the bullet. Then its momentum is $0.025v$. After the collision the block and the bullet together have a total mass of 1.025 kg. Since the velocity is 15 m/sec, the final momentum is

$$1.025 \text{ kg} \times 15 \text{ m/sec} = 15.375 \text{ kg·m/sec}$$

Equating this to the initial momentum, we get

$$0.025v = 15.375$$

Solving for v we obtain

$$v = \frac{15.375}{0.025} = 615 \text{ m/sec}$$

This example illustrates a practical way for measuring the velocity of a high-speed bullet by making a much easier measurement of a smaller velocity.

■ 3.2 Center of Mass

FIGURE 3.3 When a body falls without any spin, all its constituent particles fall the same way. This is true of all translational motion.

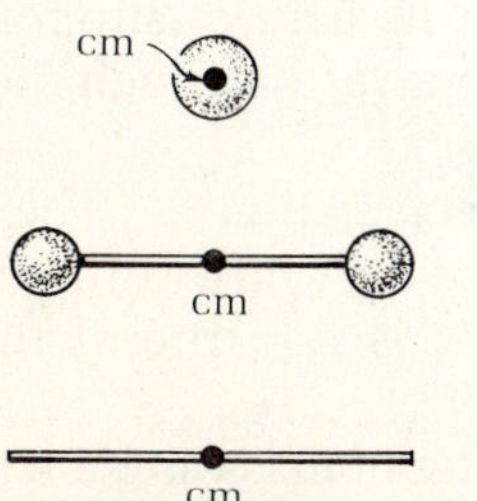

FIGURE 3.4 The center of mass (cm) of a symmetrical object of uniform density is at its geometrical center.

We have been applying the laws of Newton to large objects and small objects alike. Actually the laws are designed to apply to the motion of particles, that is, geometrical mass points. For extended objects (objects of finite size as opposed to mass points), we must carefully construct the equations of motion for the whole system of particles, starting with those of the components. Fortunately, it's not often as difficult as it sounds. And for translational motion (where every component of the body moves the same way, as in Fig. 3.3), which we have considered so far, a particle actually can represent a large object as a whole. This is because there is a point where we can assume all the mass of the object is concentrated. The location of the particle, this point, is called the **center of mass** of the body.

The center of mass of a system is the average position of the masses of all its components.* For a symmetrical object like a sphere—a ball, for example—the center of mass coincides with the geometric center. The same reasoning gives the center of mass of a dumbbell to be its midpoint (Fig. 3.4). On the other hand, for an object of an irregular shape—a hammer is a good example—there is more mass concentrated at one end and the center of mass lies nearer to the heavier end.

The center of mass is really the **center of gravity**, or the balance point of an object. Weight and mass are proportional to each other, so the average of the positions of the masses and the average of the positions of their weights is usually the same. However, we prefer to use the phrase "center of mass," because an object has mass even when it does not have weight (for example, in empty space away from all gravitating objects).

This other way of looking at the center of mass gives us an easy method to determine the center of mass even for an irregular object. All we have to do is to find the balance point. If we suspend an object—that is, balance the weight of the object by a support (Fig. 3.5)—the center of mass must lie vertically below (or above) the point of support. Why? Because the weight of the object is a vertical force acting through the center of gravity. In order to balance the object, the support force must be along the same vertical line. Now this still does not

---

*A more complete definition is this: the center of mass is the weighted average of the position vectors of all the component masses from any arbitrary origin. The phrase "weighted average" just means that when you add the position vectors for the purpose of averaging, you have to multiply each by a weighting factor, which is given by the ratio of the mass of the component and the total mass of the system. For further clarification, see the example at the end of the section.

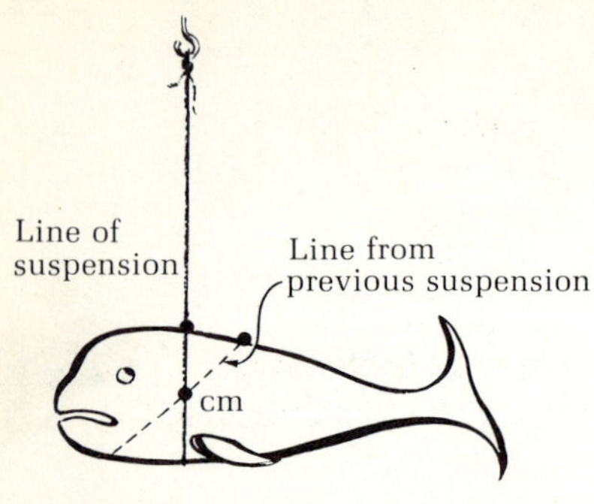

FIGURE 3.5   The point of intersection of two lines of suspension gives the center of mass of an object of irregular shape.

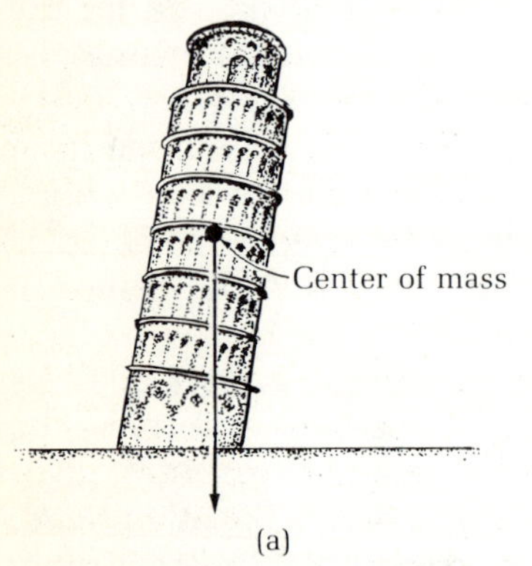

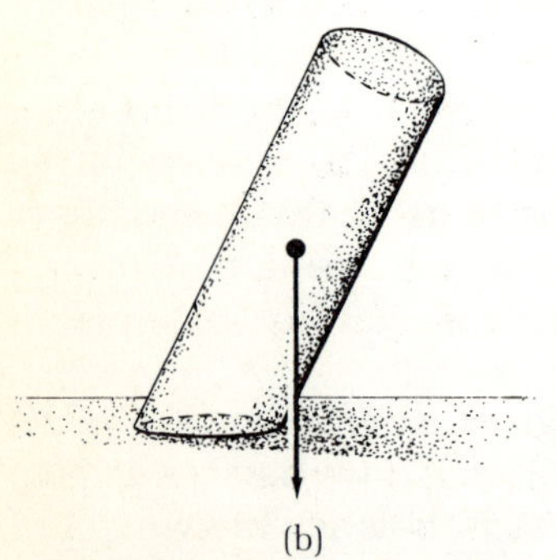

FIGURE 3.7   An object is stable only if the perpendicular dropped from its center of gravity passes through its base of support. This is true of the leaning tower of Pisa, accounting for its stability. But the other cylindrical structure shown here does not satisfy the criterion and hence is unstable.

tell us where on this line the center of mass is. However, we can suspend the body from another point, and again the center of mass must lie on the vertical line through the point of suspension. But now we have two lines and the assertion that the center of mass lies on both. This is possible only if the center of mass is at the point of intersection of the two lines. In effect, we have determined the center of mass.

The center of mass of a body does not necessarily have to be located within the body. If the distribution of mass warrants it—for example, in the case of a ring or a boomerang—the center of mass lies outside the body (Fig. 3.6).

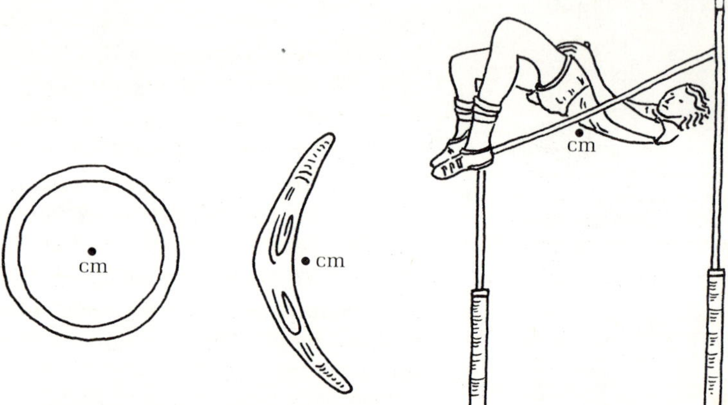

FIGURE 3.6   The center of mass can be outside an object's body, as for a ring, a boomerang, and a high jumper. The last example is of particular interest since this is how today's athletes manage to jump so high. Their center of mass never passes over the bar as they clear it.

Stability

The stable **equilibrium** of an object—the state of balance as a result of the cancellation of forces acting on it—crucially depends on the location of its center of mass.* This is the general rule: An object will topple unless the vertical line through its center of gravity passes through its base of support. Only then will the force of weight of the object, acting downward through the center of gravity, be canceled by an upward support force.

As an example consider the stability of the two objects in Fig. 3.7. The leaning tower of Pisa is stable, in spite of its leaning, because the vertical line from its center of gravity falls well within its base. But this is not true of the other cylindrical object, so that object topples.

It is advantageous to design objects with a low center of mass and a wide base because this better prevents them from toppling. Racing cars are designed with

---

*More generally, in order for an object to be in equilibrium, the **torques** (turning effects of forces) on it must also cancel. The concept of torque is discussed in Chapter 5.

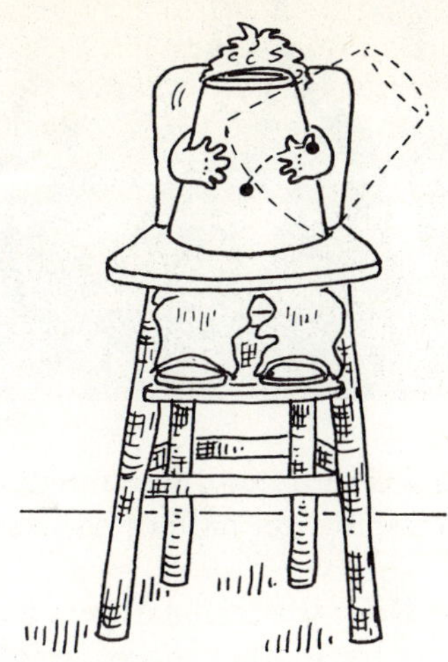

FIGURE 3.9   The child is unlikely to topple this glass! Note that tipping raises the center of mass.

FIGURE 3.10   Can you stand from this position without adjusting any part of your body? Try it and see.

FIGURE 3.8   A racing car has a low center of gravity to insure stability.

low centers of gravity, compared to a regular car, to assure them a greater stability (Fig. 3.8). Tipping an object raises its center of mass (Fig. 3.9), so eventually, if you tip sufficiently, all objects will topple. But if the center of gravity is low and the base is wide, you will have to raise the center of gravity quite a bit before the vertical line through the center of gravity falls outside the base (which is when the body must topple).

In South Asia and some parts of Africa, it is quite common for village women to carry vessels on their heads. It is a beautiful sight and a marvel that they are able to do it without tipping the vessel. Notice that putting a full vessel on her head actually raises a woman's center of gravity, yet she is able to keep her stability. To do this she has to travel quite upright and keep her head straight. Only these actions can guarantee that her center of gravity will always be above her support base, her feet. Since women have a lower center of gravity than men, women may be more adept in performing this feat than men.

Here is an everyday experience connected with the center of mass of our body. When we are seated on a chair with our back straight and our feet in front of the chair, as is customary (Fig. 3.10), we cannot get up unless we do one (or both) of two things. We can either shift our feet under the chair or we can bend forward. Both actions bring our center of gravity directly over our feet so that they are able to balance our body, which they must do when we are in the standing position. Remember, then, that the key to maintaining your balance is to be in a position such that your center of gravity is directly above your feet.

Newton's second law and the center of mass

For the translational motion of an extended body or a system of masses, we can now rewrite Newton's second law in terms of the acceleration of the center of mass, denoted as a_{cm}:

$$F = ma_{cm} \tag{3.5}$$

where F denotes the net external force and m the total mass of the body, as before. Equation (3.5) tells you that if F is zero, a_{cm} is zero. In the absence of any net external force, the center of mass of a body in motion exhibits natural motion. Parts of an object may wobble wildly, but if the center of mass still exhibits natural motion, we are assured that there is no net external force on the

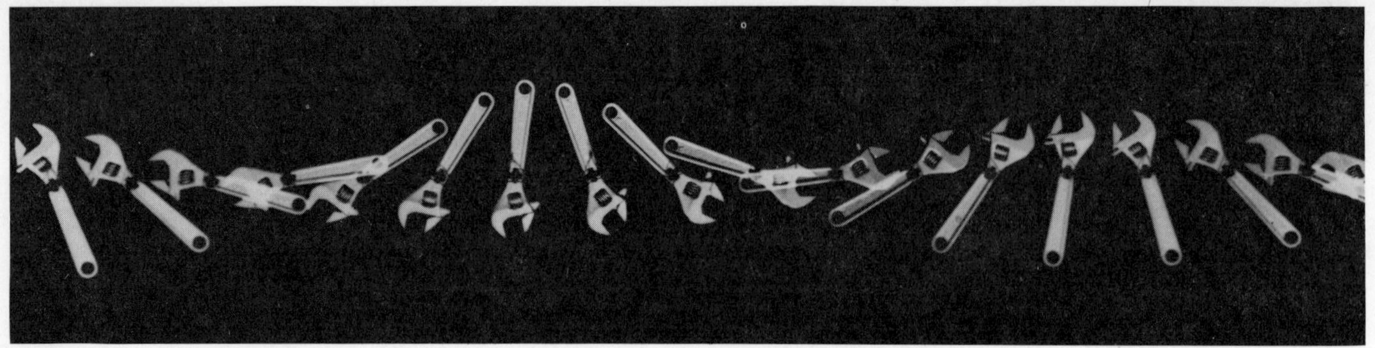

object. Figure 3.11 shows a series of flash photographs of a moving wrench in such a motion; the center of mass is moving in a straight-line path of constant velocity, although the wrench is spinning.

The center of mass concept thus enables us to consider the translational part of the motion of an object even when its components have other more complicated motion as well. If the center of mass is initially at rest, then in the absence of external forces, it must stay at rest forever. Suppose an object splits into two fragments, one twice as heavy as the other. To keep the center of mass in its original place, the heavier fragment must move at a speed of one-half that of the lighter one. Then only their weighted average, or the center of mass, will remain stationary at all times. In the case of an exploding bomb, the heavier pieces must move away with smaller speed, and the fragments must be distributed in such a way that the center of mass remains in the same place it was in before the explosion. Notice that these conclusions can also be reached by applying momentum conservation to the problem. If we say "momentum is conserved" and "velocity of the center of mass remains constant," we are saying the same thing.

A numerical example*

Imagine yourself and a friend in a skating rink of perfectly frictionless ice. You and she are joined by a weightless rope 9 ft long. Now the two of you start moving toward each other by pulling on the rope hand over hand. Where will you meet her?

The way the problem is set up, perfectly frictionless ice is the important component. The two of you and the rope form a system (an asymmetric dumbbell) with no external horizontal force. Thus the center of mass of the system does not move. So the two of you will meet at the center of mass of this system.

Thus the problem turns out to be one of locating the center of mass, the weighted average of your position and her position from any origin (the rope has been assumed to be weightless). Now we need your weights. Suppose you weigh 200 lb and your friend weighs 100 lb. Take your original position as the

---

*We closely follow the treatment by Kenneth W. Ford in *Classical and Modern Physics* (New York: Wiley, 1972), p. 293.

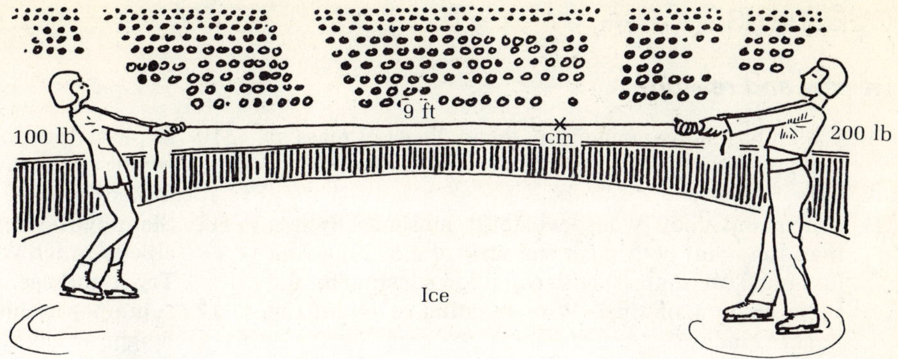

FIGURE 3.12 How do we know where the center of mass of this system is? The numerical calculation in the text gives the answer.

origin (Fig. 3.12). The weighting factor is the ratio of the individual weights to the total weight of the system.

Distance of the center of mass from the origin
(which is your original position)

$$= \frac{\text{friend's weight} \times \text{distance from origin} + \text{your weight} \times \text{distance from origin}}{\text{total weight of the system}}$$

$$= \frac{100 \times 9 + 200 \times 0}{300} \text{ ft} = \frac{900}{300} \text{ ft} = 3 \text{ ft}$$

Thus you will meet her at a distance of 3 ft from your original position.

We hope this gives you the idea of how to find the center of mass (or the center of gravity, which is the same thing), at least in simple cases like this. Remember, for two masses the center of mass is always on the line joining them and is closer to the heavier of the two. To determine exactly where it is, you have to use the procedure outlined in the example.

SUMMARY

The subject of Chapter 3 is really an extension of the Newtonian framework. It introduces the concept of momentum (product of mass and velocity) of an object, which is actually a natural vehicle for expressing the second law of motion (in addition to $F = ma$). Force is the rate of change of momentum. What follows from this way of looking at the second law is of utmost importance. The result of this viewpoint is the law of conservation of momentum. That is, if the net external force is zero, momentum stays constant in time; momentum is conserved.

The concept of center of mass rounds out all the fundamental concepts of the Newtonian framework as applied to translational motion (a motion in which all the components of a body exhibit the same motion). The center of mass is the average position of all the particles of a body. The important concept of equilibrium is also introduced in this context: an object can be in equilibrium only when all the forces acting on the body are canceled. Since the weight of an extended object acts through its center of mass, this means that the vertical line from the object's center must pass through its base of support in order for it to be stable.

Review and reason

1. Which has more momentum: (a) an object of mass m and speed $3v$ or (b) an object of mass $2m$ and speed $2v$?

2. A man and a boy (whose weight is substantially less than the man) both push the wall of a skating rink (assumed frictionless) with equal force lasting for the same amount of time. Who is going to attain the larger velocity?

3. A woman throws a basketball to the floor and catches it on the rebound. How many times does the momentum of the basketball change? When is the change largest?

4. Imagine that you have been selected as the target of a ball-throwing practice. You can choose either your stomach or your head as the target. What will your choice be? Explain.

5. Discuss the importance of follow-through in sports.

6. What does a conservation law mean? Make up a conservation law for a situation outside the physical sciences, even if it is an approximate or invalid one (example: some people think that the total amount of love that a person can bestow on others is conserved). Discuss some applications of your conservation law and point out how it could be tested experimentally.

7. Does a bow recoil when an arrow is shot from it? Discuss this from a momentum conservation point of view.

8. Observe and discuss a few examples of momentum conservation in pool shooting.

9. When two cars have a head-on collision, where is it most likely that the objects in the cars will end up, in the front of the cars, the backs, or the sides?

10. Explain the notion of center of mass. How is the center of mass of an object of irregular shape determined?

11. Stand with your heels against the wall. You won't be able to touch your toes without bending your knees. Try it and see. Now explain why.

12. A jumper is able to lift his center of gravity through a height of 2.8 ft. His center of gravity is 3.5 ft above-ground when he stands erect. Thus in a jump his center of gravity is raised only to a height of $2.8 + 3.5 = 6.3$ ft. Yet he clears 7 ft. How is this possible?

13. A bomb explodes from a position of rest somewhere in outer space. Where would the center of mass be after the explosion?

14. Imagine a rocket in the process of ejecting debris from its rear. Does the momentum of the rocket plus debris change? Does the momentum of the rocket itself change? What happens to the center of mass of the rocket plus debris?

15. A truck carrying gunpowder, and traveling with a velocity of 20 mi/h, explodes. What is the velocity of its center of mass after the explosion? Neglect interaction of pieces with earth.

16. A woman standing on a wooden plank resting on frictionless ice walks from one end of the plank to the other. Does the plank move, and, if so, in what direction? What happens to the center of mass of the woman plus plank?

17. A boxcar is at rest on a frictionless track with a woman standing at one end. If the woman walks from one end to the other, will the boxcar move? Why or why not?

Arithmetic

1. Which object has more momentum? (a) A mass of 100 kg moving with a velocity of 10 m/sec; (b) a mass of 1000 kg moving with a velocity of 1 m/sec; (c) a mass of 500 kg moving with a velocity of 5 m/sec.

2. Suppose you and your sister are standing on frictionless ice in a skating rink. You push her gently to get started. If your mass is twice her mass, how will your velocities compare?

3. A railroad car weighing 3 tons and traveling on a straight track at a speed of 5 mi/h collides with another car of identical weight that is at rest. After the collision the second car is found to be moving with a velocity of 5 mi/h. What is the velocity of the first car after the collision?

4. A mass of 10 kg hits an object of the same mass at rest and they stick together. Afterwards the combined masses are found to move with a velocity of 30 m/sec. What was the velocity of the first mass before the collision?

5. Here is a way to rescue your pet poodle from a frozen lake of frictionless ice. Chew some gum, say of a mass 10 grams, which is the same as 0.01 kg. Now throw the gum at your dog as hard as you can. The gum sticks to its body, and since momentum is conserved, the

poodle plus gum must now move with the original momentum of the gum. The poodle's mass is 3 kg, and the poodle plus gum is found to move with a speed of 0.1 m/sec. Calculate the speed of the gum before it hits the poodle.

6. Two weights of 20 lb and 50 lb are hung from opposite ends of a rod. To balance the system you have to locate the center of mass. Suppose the length of the rod is 7 ft. Where should you put a prop in order to achieve balance? (The weight of the rod is negligible and can be ignored.)

Falling Bodies on Earth and in the Heavens

The leaves fall, as if from far away,
like withered things from gardens deep in sky;
they fall with gestures of renunciation.

And through the night the heavy earth falls too,
down from the stars into the loneliness.

And we all fall. This hand must fall.
Look everywhere: it is the lot of all.*

RAINER MARIA RILKE

■ 4.1 Galileo Solves the Case of Falling Bodies

Does the title of this section sound like that of a Perry Mason mystery? As you probably know, Perry Mason is the fictional lawyer who, in his efforts to free his clients, perennially fights the establishment represented by the local district attorney's office. His opponents support the dogma of circumstantial evidence that Perry Mason has to tear down. His weapons are sharp logic (which refuses to compromise with even a single bit of unexplained evidence) combined with new experimentation and the collection of new evidence. Sometimes he even performs a razzle-dazzle in order to achieve his goal.

It may suprise you to know that Galileo had to do all this and more, some four hundred years ago, before he could establish the law that correctly describes the motion of falling bodies. His opponents consisted of the powerful Church with the eighteen-hundred-year-old dogma of Aristotelian science behind it. Galileo had to gather much new experimental evidence in order to prove his point. In

---

*From *The Duino Elegies and the Sonnets to Orpheus* by Rilke. Translated by A. Poulin. Copyright © 1975, 1976, 1977 by A. Poulin, Jr. Reprinted by permission of Houghton Mifflin Company and St. John's College, Oxford, and Hogarth Press.

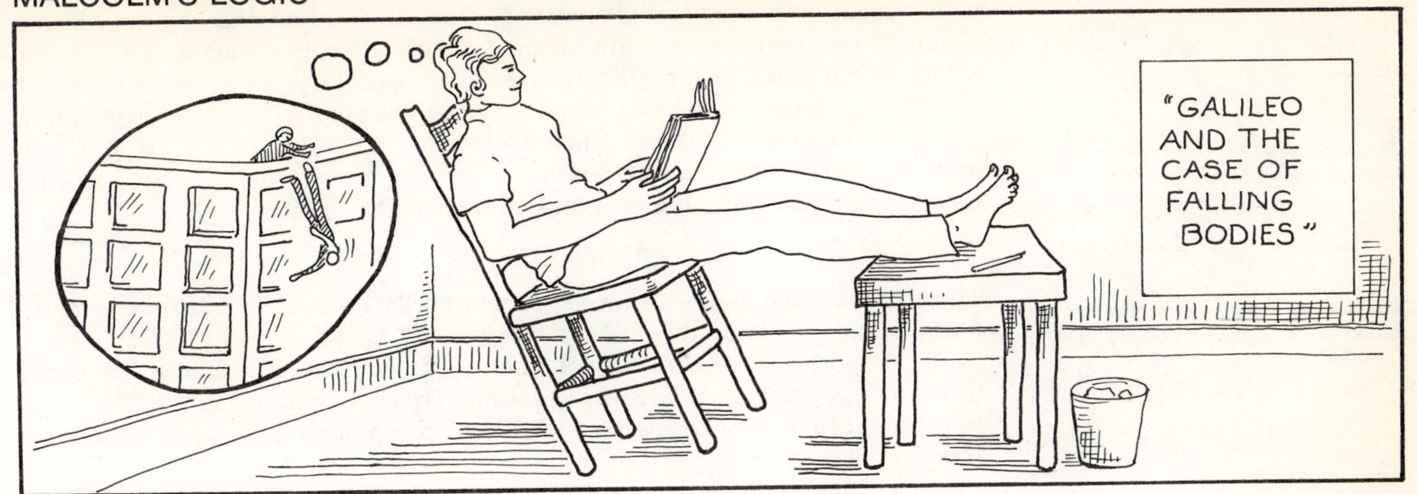

his attempts to convince the world that he, not Aristotle, had the right key to the understanding of falling body motion, Galileo not only presented his case in the best possible light but also, taking advantage of every weakness in Aristotle's theories, sought to expose him as the villain. Galileo was not satisifed, however, with just tearing down Aristotle's ideas. As we all know, tearing down is easier than building. Perhaps here is the most noteworthy point of similarity between Galileo and Perry Mason. The fictional Perry Mason is never satisfied with just demolishing the prosecuter's case against his client but must also always expose the real criminal.

Aristotle had the erroneous notion that falling was a natural motion of objects, representing the return of earthly objects to their natural place. Since heavier objects are more strongly attached to the earth, he thought that they should try harder to return. Thus he arrived at the following law: Objects fall in the same medium with a speed proportional to their weight. Thus according to Aristotle, heavier objects fall faster. A 10-pound boulder should fall ten times faster than a 1-pound stone. That sounds absurd to us today. Apparently Aristotle did not realize that his theory leads to such horrendous conclusions when applied to all cases of falling body motion. Yet, as we will see later in the chapter, taken in proper context the law turns out to have some merit.

As you can guess, Galileo would not give Aristotle the benefit of the doubt. In his experiment Galileo dropped simultaneously an iron ball and a much lighter wooden ball from the top of the leaning tower of Pisa.* He demonstrated to the public that the two objects reached the ground more or less simultaneously (in glaring contradiction to Aristotle). Thus an object's time of fall has no relation to its weight. However, as Galileo himself guessed, this is exactly true only in a vacuum.†

---

*Actually, historians are not sure that Galileo did his falling body experiments from the leaning tower of Pisa. The records kept by Galileo tell us only that he dropped objects from a tower. It is in the writings of his biographer, Viviani, that this tower is described as the famous leaning tower of Pisa, and the experiments are said to have taken place before a crowd of scholars and students. Whether Viviani purposely created a legend nobody knows.

†Vacuum essentially means space void of matter, in this case air.

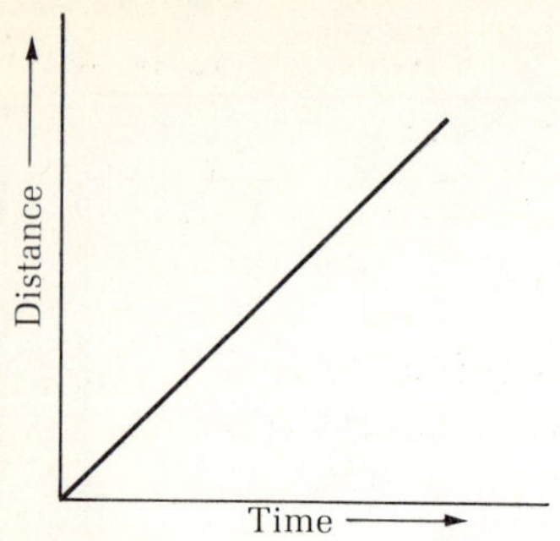

FIGURE 4.1   Distance versus time graph for an object moving with uniform speed.

The next question Galileo studied was this: Is falling motion one of constant velocity, or is it accelerated motion? Aristotle implied that falling objects acquire a constant velocity almost immediately after they start to fall.

It turns out that it is not difficult to prove or disprove whether or not an object is falling at a constant velocity. For example, if we are told that a car is traveling at a constant speed of 30 mi/h, then we conclude that the car will go 30 mi the first hour, then 30 mi in the next hour, and so forth. Equal distance will be traveled in equal time. For motion with constant velocity.

$$\text{distance traveled} \propto \text{time} \tag{4.1}$$

The symbol $\propto$ stands for "is proportional to." This means that as time is doubled, the distance is doubled, and so forth.

The proportional relationship of distance and time can be readily visualized by drawing a graph of distance against time. This is shown in Fig. 4.1. The graph is a straight line. For this reason a relationship expressed by Eq. (4.1) is called a **linear** relationship.

Is the relationship of distance traveled with time a linear one for falling objects? Galileo dropped balls from twice the original height and found that the time of fall was not doubled. The relationship was not a linear one. Thus falling objects do not descend with a constant velocity. The speed increases as the object descends: hence motion is accelerated.

For accelerated motion the relationship of distance and time is obviously more complicated. Galileo was able to show with the help of some mathematics that if the acceleration remains constant, then

$$\text{distance traveled} \propto (\text{time})^2 \tag{4.2}$$

Such a dependence is called **quadratic** and gives a parabolic curve when drawn in a graph, as shown in Fig. 4.2.

It remained for Galileo to show experimentally that the distance-time relation

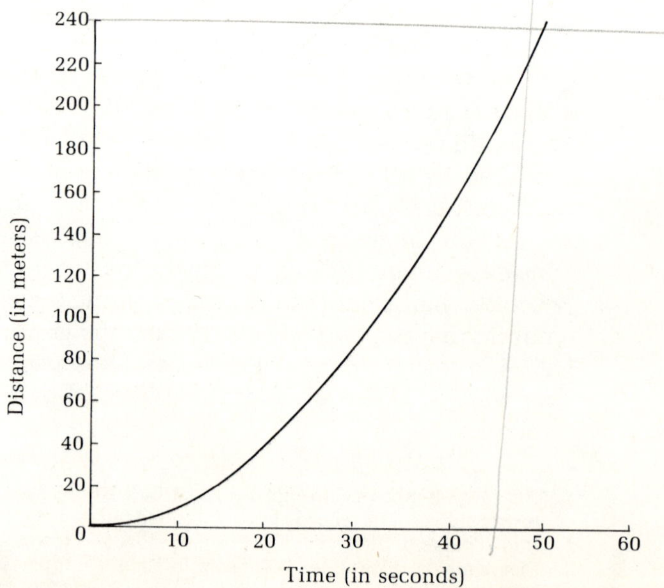

FIGURE 4.2   Distance versus time graph for an object moving with uniform acceleration. The graph is a parabola.

in falling motion is indeed given by Eq. (4.2). This turned out to be a very
difficult task. To appreciate his difficulties, we have to be aware of the fact that a
stone takes only a little over 3 sec to fall from a 150-ft tower to the ground. In
Galileo's time, however, the clocks were not accurate enough for the experimental study of distance-time relation when the time interval was so short.

Nevertheless, Galileo found a way. He discovered that the acceleration of a
falling object is reduced on inclined planes and the time of travel is thus
increased. He hoped that the essential aspects of the motion of free fall would
not change for an object rolled down an inclined plane. It turned out that he was
right.

Using inclined planes Galileo was finally able to verify that the relation of
distance traveled with time was indeed a quadratic one for falling objects.
However, even with inclined planes the experiment needed the utmost ingenuity. Here is Galileo's own account of his experiments:

> We took a piece of wooden scantling,* about 12 cubits long,† half a cubit wide,
> and three finger-breadths thick. In its top edge we cut a straight channel a little
> more than a finger in breadth; this groove was made smooth by lining it with
> parchment, polished as smooth as possible, to facilitate the rolling in it of a
> smooth and very round ball made of the hardest bronze. Having placed the
> scantling in a sloping position by raising one end some one or two cubits above
> the other, we let the ball roll down the channel, noting, in a manner presently to
> be described, the time required for the descent. We repeated this experiment
> more than once in order to be sure of the time of descent and found that the
> deviation between two observations never exceeded one-tenth of a pulse beat.
> Having performed this operation until assured of its reliability, we now let the
> ball roll down only one-quarter the length of the channel; and having measured
> the time of its descent, we found it to be precisely one-half of the former. Next
> we tried other distances, comparing the time for the whole length with that for
> the half, or for two-thirds, or for three-fourths, or indeed for any fraction.
> In such experiments, repeated a full hundred times, we always found that
> the distances traversed were to each other as the squares of the times, and this
> was true for any inclination of the . . . channel along which we rolled the
> ball. . . .
>
> For the measurement of time, we employed a large vessel of water placed in
> an elevated position; to the bottom of this vessel was soldered a pipe of small
> diameter giving a thin jet of water, which we collected in a small cup during the
> time of each descent, whether for the whole length of the channel or for a part of
> its length; the water thus collected was weighed on a very accurate balance; the
> differences and ratios of these weights gave us the differences and ratios of the
> time intervals, and this with such accuracy that, although the operation was
> repeated many, many times, there was no appreciable discrepency in the
> results.

And that is how Galileo discovered the following law, which to this day
remains one of the most important laws in physics:

> *All objects fall (in vacuum) with the same constant acceleration indepen-
> dent of their weight or any other physical property.*

---

*A small piece of lumber, such as a two-by-four.
†An ancient measure of length: the distance from the tip of the middle finger to the elbow; roughly
  equal to 18 to 22 inches.

■ 4.2 HOW OBJECTS FALL: SOME EXAMPLES

The relationship expressed by Eq. (4.2) can be written as one of equality by introducing a constant of proportionality C. Thus for motion with constant acceleration, we have

$$\text{distance traveled} = C\,(\text{time})^2$$

A little mathematics shows that the constant of proportionality is related to the acceleration (constant in this case) which we denote by a:

$$C = \frac{a}{2}$$

Thus the distance-time relationship for motion under constant acceleration is given by the equation

$$\text{distance traveled} = \tfrac{1}{2} \times \text{acceleration} \times (\text{time of travel})^2$$

or in symbols

$$d = \tfrac{1}{2}at^2 \tag{4.3}$$

using the symbols d for distance, a for acceleration, and t for time.

For falling objects on earth, the constant acceleration has been measured to be* 32 ft/sec², or 9.8 m/sec². This acceleration is referred to as one g (of acceleration). Notice how quickly a falling object accelerates (see Fig. 4.3). Starting from zero velocity, it speeds up to a velocity of 32 ft/sec at the end of 1 sec; at the end of 2 sec its velocity is 64 ft/sec; and at the end of just 3 sec it attains a speed of 96 ft/sec, which is more than 60 mi/h.

Substituting the numerical value of the acceleration of fall, namely, 32 ft/sec², in Eq. (4.3), we get the following distance-time relationship for the special case of falling objects on earth:

$$\text{distance of fall (in feet)} = 16\,(\text{time of fall, in seconds})^2 \tag{4.4}$$

Table 4.1 displays the numerical values of the velocity attained and the distance fallen for the first 5 seconds. Perhaps after a glance at the table you can appreciate Galileo's difficulty with the time measurements.

You can measure the approximate time taken by nerve impulses to travel from one part of your body to another via the brain by using Eq. (4.4). The experiment has a fun part and a more scientific part. Let's start with the fun part. Get a friend. Then hold a dollar bill above her hand, centered between her open thumb and forefinger, and challenge her to catch it as you release the bill without warning. No matter how hard she tries, she won't be able to catch it. Why? Let's figure out the time of fall for a distance equal to $\tfrac{1}{2}$ the length of a dollar bill. The length of a dollar bill is 6 in. Thus from Eq. (4.4)

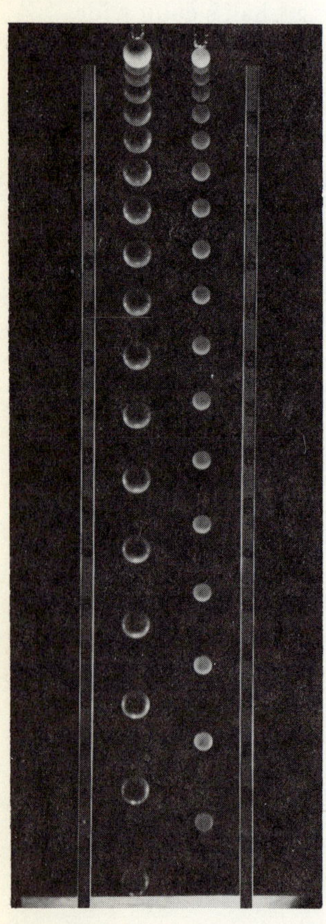

FIGURE 4.3 A series of flash photographs of a falling billiard ball. The ball on the right is a Ping-pong ball, which is affected by air resistance. See later. (Education Development Center, Inc., Newton, Mass.)

---

*More accurately, 32.2 ft/sec². We will, however, use 32 for convenience.

TABLE 4.1   Data for falling bodies.

Time of fall (sec)	Velocity (ft/sec)	Distance Traveled (ft)
0	0	0
1	32	16
2	64	64
3	96	144
4	128	256
5	160	400

$$(\text{time in seconds})^2 = \frac{\text{distance fallen (in feet)}}{16} = \frac{\frac{1}{4}}{16} = \frac{1}{64}$$

Therefore,*

$$\text{time of fall} = \sqrt{\tfrac{1}{64}} = \tfrac{1}{8} \text{ sec}$$

The time, as calculated here, is less than the time it takes for the nerve impulses to go from her eye to her brain and then to her fingers. So by the time her brain's instructions have reached her fingers, the bill is gone.

Now you can find the value of the time of travel of the nerve impulses. Increase the length of fall of the dollar bill between your friend's fingers, in small amounts, until she is able to catch the bill when released. She should be able to catch it when you let just about two-thirds of the bill fall through her fingers. Now find the travel time of her nerve signals using the procedure above and the equation.

$$\text{travel time for nerve impulses (in seconds)} = \sqrt{\frac{\text{distance (in feet)}}{16}}$$

Falling objects and the laws of Newton

Galileo ran into a roadblock after establishing that falling motion is a case of accelerated motion; he formulated a law governing such motion but couldn't, or at least didn't, explain it. For Newton, however, the accelerated nature of falling motion meant one simple thing: a force must be responsible for the acceleration. From Newton's second law we can deduce the fact that this force must be equal to

$$\text{mass} \times \text{acceleration of falling}$$

or, using the symbols W for this force of weight, m for mass, and g for the acceleration of fall,

$$W = mg \tag{4.5}$$

Actually Newton did not stop here. Remember the third law, which says that

---

*The square root, indicated by $\sqrt{}$, of a quantity is a quantity which, when multiplied by itself, is the original quantity. In this case $\tfrac{1}{8} \times \tfrac{1}{8} = \tfrac{1}{64}$, the original number.

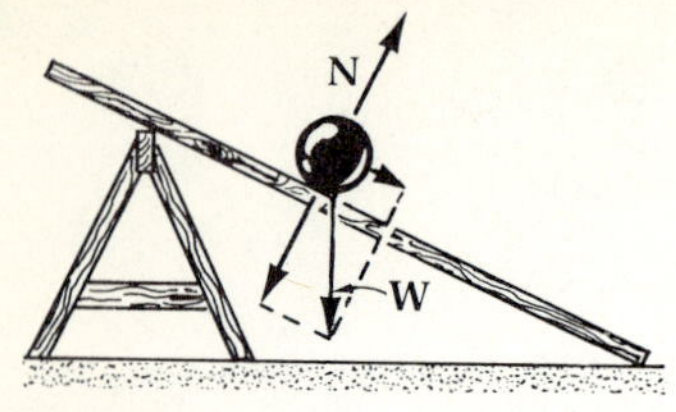

FIGURE 4.4 An object falling on an inclined plane. The component of the weight **W** in the direction perpendicular to the plane is canceled by the equal and opposite force of support **N**. But the component of **W** along the incline is unbalanced and is responsible for the somewhat reduced acceleration of the falling object along the plane.

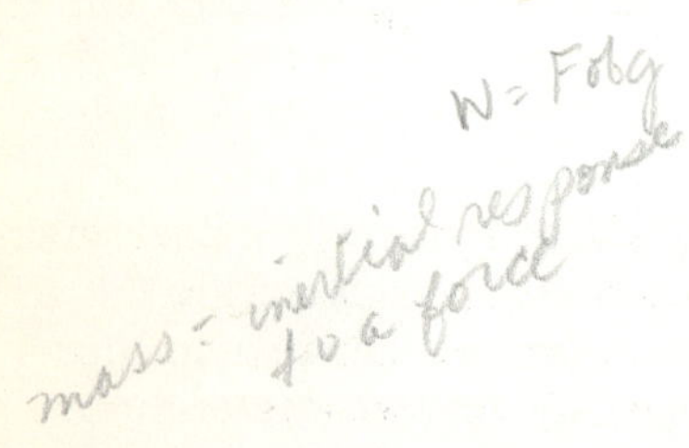

force is an interaction; it takes two objects for an interaction to occur. So Newton started looking for a second body and very soon discovered that it must be the earth. It is earth's gravity force that makes an object fall. The acceleration of fall is acceleration due to gravity. In case you are wondering if a falling body also pulls on the earth, the answer is yes. The effect, however, is quite imperceptible due to the very large inertia of the earth.

With the concept of the weight force acting downward on a body, we can now understand why the acceleration due to gravity is "diluted" along an inclined plane, as in Galileo's experiment. For an object on an incline, the force of the weight **W** still acts vertically downward (Fig. 4.4). But the normal support force **N** on the object acts in a direction perpendicular to the incline. Since **N** is not in the same direction as **W**, it cannot cancel the effect of **W** entirely; it compensates for only a part of **W**. The difference between the two acts along the incline and hence provides the acceleration along the incline. However, the magnitude of the difference is less than that of **W**; in effect, the pull of gravity has been reduced by using the incline.

There is a mathematical way of looking at what we have just discussed. The vector **W** can be looked upon as the sum of two mutually perpendicular vectors, called the components (see Section 2.2), one along the incline and one in the perpendicular direction, as shown in Fig. 4.4. Note that the components are such that they add up to the original vector **W** according to the parallelogram rule. Then we can say that the component of **W** that is perpendicular to the plane of the incline is canceled by the normal support force. But the component of **W** along the incline is unbalanced and therefore accelerates the falling body down the incline.

Mass, weight and apparent weightlessness

The weight of an object is equal to the force of gravity acting on it. It is a different quantity than the mass of an object, which measures its inertial response to a force. Clearly the weight of an object will be quite different when the object is taken to the moon, since the force of lunar gravity is only one-sixth that of earth's. So an object on the moon weighs only one-sixth of its earthly weight. On the other hand, the mass of an object remains unchanged when taken to the moon. So accelerating a cart is no easier on the moon than on the earth, but lifting a weight is.

There is another interesting thing about weight. Our sense of weight comes from the reaction force of the floor or other support that we stand on. When we stand on the floor, we are actually at rest—there is no net force on us; the force of our weight is exactly balanced by the force of the support of the floor.

What is **apparent weightlessness**? It is not the absence of the force of gravity, but the absence of a force or support. In free fall we feel weightless because of the lack of support under us. You may have experienced the feeling of partial weightlessness in a roller coaster ride, which makes many people dizzy and uncomfortable.

Motion of a sky diver

We have just noted that free fall makes some of us feel uncomfortable because of the weightlessness experienced. From this you may get the impression that

skydiving is a very unpleasant hobby. Far from it! The reason is that for most of the downward journey, the sky diver enjoys natural motion with no net force acting on him. How is this possible?

The reason can be found in the fact that the force of air drag on the sky diver increases with velocity [see Eq. (2.4)]. Thus as the sky diver accelerates downward under the influence of gravity, his downward velocity goes up, and so does the drag force that opposes the force of gravity. At some value of the velocity, the air drag becomes sufficiently large to reduce the acceleration of the sky diver to zero. The sky diver then rests on the air cushion with complete ease, as if supported by a floor, his feeling of weight restored. The velocity at which this happens is called **terminal velocity**. The sky diver reaches terminal velocity when the upward drag force acting on him balances the downward gravity force, and there is no longer a net force acting on him. Motion from then on is natural motion with constant velocity. For a 70-kg sky diver starting in a spread-eagle position in a typical case, the terminal velocity is approached in about 12 sec, after falling through a distance of about 400 m. The value of terminal velocity in this case is approximately 54 m/sec, or 120 mi/h. Toward the end of the fall, the sky diver has to slow down to a much smaller velocity, with the help of a parachute for safe landing. Typically a sky jump takes place from a height of 1 mi and the whole journey takes about 30 sec. So the sky diver spends a considerable portion of the total time falling at terminal speed, and this is where the fun of skydiving is.

Objects like feathers or even raindrops acquire their terminal velocity quickly. In fact, so do particles of air pollution from smokestacks. In this case the value of the terminal velocity is rather small, which means that the particles take long periods of time to settle down, a rather unfortunate situation. (See Fig. 4.5.)

There is one more interesting thing about falling motion at terminal velocity. For objects of similar size and shape, the value of the terminal velocity is very nearly proportional to the weight of the object. So if you were a researcher who studied only those cases of falling motion involving terminal velocity, what

FIGURE 4.5 Examples of objects falling in air with terminal speed (and therefore in accordance with Aristotle's law): (a) raindrops; (b) smoke particles; (c) parachutist; (d) a ball bearing falling through a viscous liquid.

would your conclusion be? You might be tempted to discover the following law of falling bodies: Objects fall with constant velocity in proportion to their weight. That sounds like Aristotle's law. So Aristotle's law has some validity after all, if not for all falling motion, at least for falling motion at terminal velocity.

American physicist Robert March, in his book *Physics for Poets*, speculates that perhaps Aristotle based his considerations on the motion of objects falling through a viscous liquid (because such objects reach terminal velocity very quickly).* The terminal velocity reached by a falling object in a dense liquid is low, and therefore the object falls rather slowly through it. Thus the motion would be much easier to time.

---

*Robert H. March, *Physics for Poets* (New York: McGraw-Hill, 1978), p. 18.

MALCOLM'S LOGIC

Unfortunately for Aristotle—assuming he did use this approach—falling motion through a dense medium is basically different from falling motion in vacuum or air. A stone falls thousands of feet before it attains terminal velocity in air. Except for a few objects like feathers, falling motion in air is mainly accelerated motion.

It is perhaps most interesting that both Aristotle and Galileo were unable to study the falling motion of objects like stones in air because of the inadequacy of the clocks available in their times. They both would have had to retard the falling motion in order to increase the time of fall so that they could measure it. Aristotle may have chosen to do this by considering objects falling through a dense liquid. Unfortunately this would have altered the nature of falling motion—from an accelerated motion to motion at constant velocity—leading Aristotle to his wrong law. Galileo, on the other hand, used the inclined plane to reduce the acceleration, and luckily for him this did not qualitatively alter the nature of the motion. Looking at the story this way vindicates Aristotle to a great extent. It would seem that the man who was the greatest scientist of his time would not describe a physical law without making some sort of observations. He probably just made an honest, if unlucky, mistake.

Superposition of motion

The motion of a projectile thrown horizontally from a rooftop follows a curved trajectory, a curve mathematicians call a parabola. A simple construction shows that this is due to the superposition of two motions: a horizontal motion of constant speed, with which the projectile is thrown, and a downward motion due to gravity, which takes place with an acceleration. Thus the horizontal distance traveled is proportional to the time. In contrast, the vertical distance fallen by the projectile during the same time is proportional to the square of the time. Therefore, if the time is the same, the vertical distance fallen is proportional to the square of the horizontal distance covered, starting from the origin. This relationship between the vertical and horizontal coordinates is unique to the parabola. The superposition of the two motions is shown in Fig. 4.6.

We also can construct the trajectory of a projectile thrown at an angle to the vertical (for example, the trajectory of a baseball after it leaves the hitter's bat), if we neglect the effect of air resistance. While the ball is rising, the vertical motion is decelerated. The velocity in the vertical direction decreases until it reaches zero. After this happens, the trajectory is the same as that shown in Fig. 4.6, since from that point on the ball can be regarded as one thrown horizontally

FIGURE 4.6 Constructing the parabolic trajectory of a projectile. (a) Simulated horizontal motion at constant velocity (if gravity were switched off). (b) Vertical motion alone, showing acceleration under gravity. (c) Combined motion. The trajectory is a parabola.

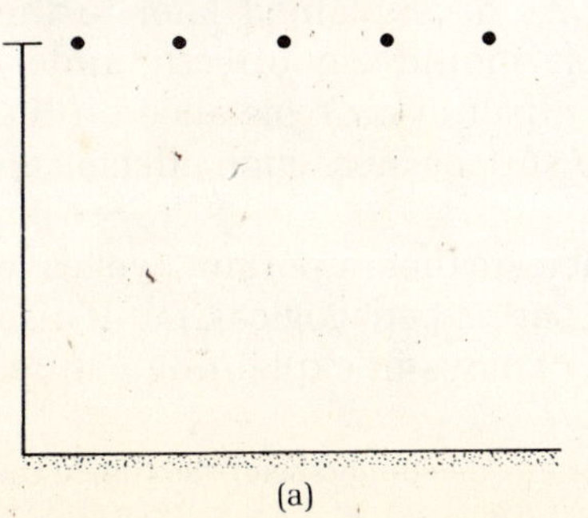

(a)

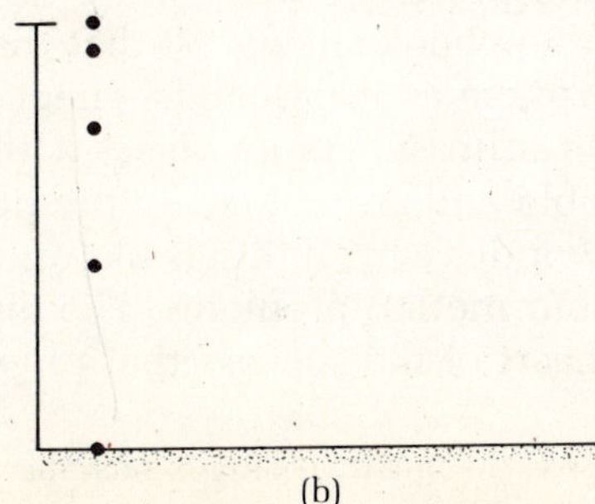

(b)

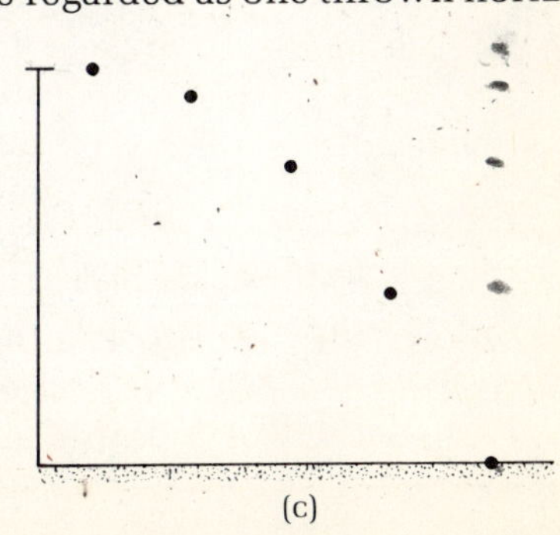

(c)

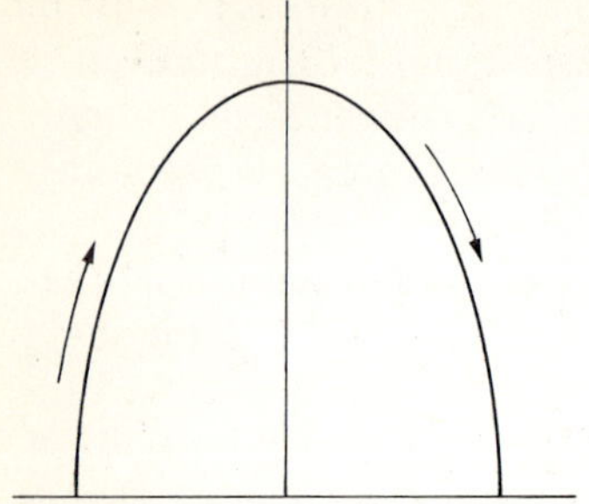

FIGURE 4.7 The rising part of a trajectory is a mirror reflection of the descending part.

FIGURE 4.8 The bullet will hit the can if the can is dropped at the same time that the bullet is fired.

from a height. Now suppose we make a motion picture of the whole trajectory as it takes place and run it backward. Now the ascent looks like the descent and vice versa. Clearly, then, the rising portion of the trajectory must also be a parabola, and the whole trajectory must appear as shown in Fig. 4.7.

The main lesson to be learned from the study of motion in two dimensions is that motion along either of two mutually perpendicular directions is independent of the presence of the other. There are many dramatic demonstrations of this. In one demonstration a gun is sighted at an elevated target, which is dropped by means of a trip mechanism at the instant the bullet leaves the muzzle of the gun. The bullet will always hit the target; it does not matter what the initial speed of the bullet is.* How is this possible? If there were no falling motion due to gravity, the bullet would not drop, and the released target would not fall from its line of sight. So there would be a hit. The effect of gravity accelerates each body downward at the same rate, and therefore they fall the same distance in any given time (Fig. 4.8). Thus whenever the bullet reaches the line of fall of the target, it will be the same distance below the initial position of the target as is the target.

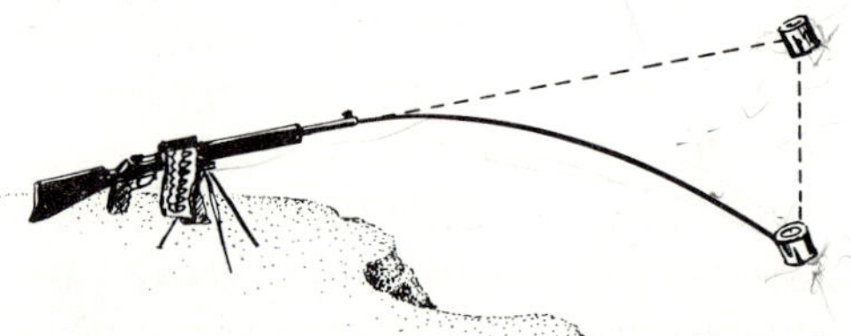

Effect of air resistance on the path of a projectile

A student once asked a professor to explain a simple problem. If a suitcase were dropped from an airplane, where would the suitcase be relative to the airplane in subsequent times until it reached the ground? The professor replied, "I don't know."

The student was stunned. He was testing the professor, you see; he himself knew the "answer." The inertia of the suitcase would carry it horizontally in pace with the airplane while it fell. The independence of the horizontal and vertical components of motion should guarantee that at all stages the suitcase would be vertically beneath the airplane.

The student gave the professor a hint. "Shouldn't it stay directly beneath the airplane as it falls?" The professor didn't seem interested in changing his position. "I don't know," he insisted.

Actually the professor was telling the truth. As he explained later to the student, the "textbook answer"—that the suitcase should stay directly underneath the airplane as it descends—neglects the effect of air resistance on the motion of the suitcase. For an object with a large surface area, such effects can be considerable and are not easily predicted.

You may wonder why, if things like air resistance are that important, we leave them out of our motion problems. The reason in part is pedagogical, but it also goes to the heart of the success that physics now enjoys in explaining nature.

---

*Unless, of course, the initial velocity is such that the bullet hits the ground before getting to the line of fall of the target.

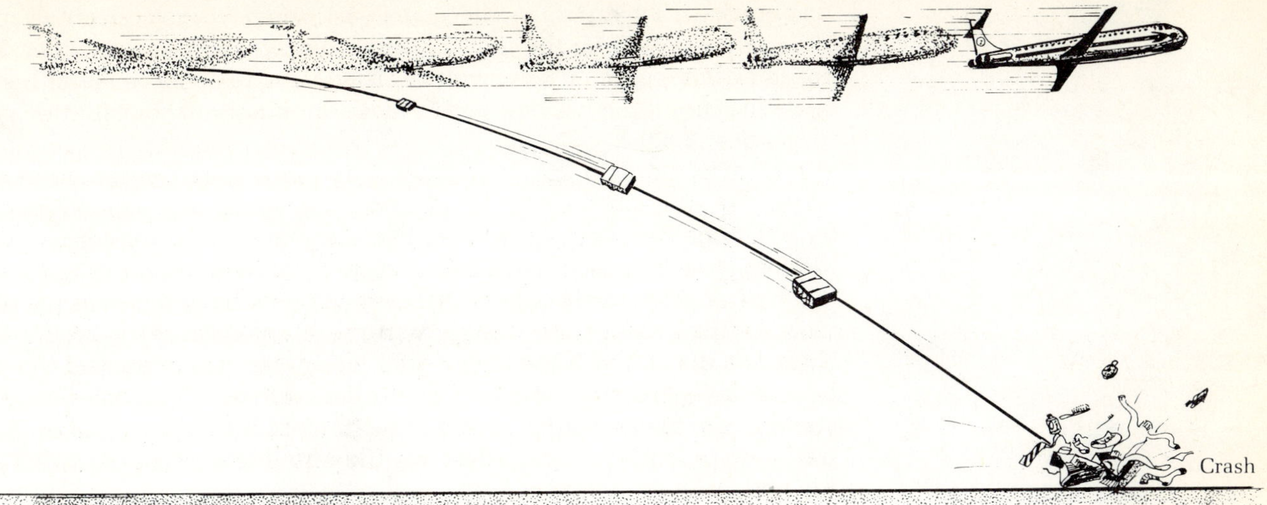

Effects such as air resistance play important but often nonessential roles in the overall nature of things. By leaving them out, the physicist isolates the real problem, the solution of which leads to a profound understanding. The nonessential things can be taken into account later. This is, of course, how Galileo made the breakthrough in the understanding of the motion of falling bodies.

So for the people who have the job of finding a suitcase after it is dropped from an airplane, life is a bit harder, as it is for those who deal with high-speed objects like bullets and missiles. Indeed, for missiles it is necessary to take into account the effect of air resistance, which requires quite detailed calculations. The general effect of air resistance on a missile is such that it does not quite reach the same height nor the same range as shown in Fig. 4.9. The actual path isn't quite a parabola either.

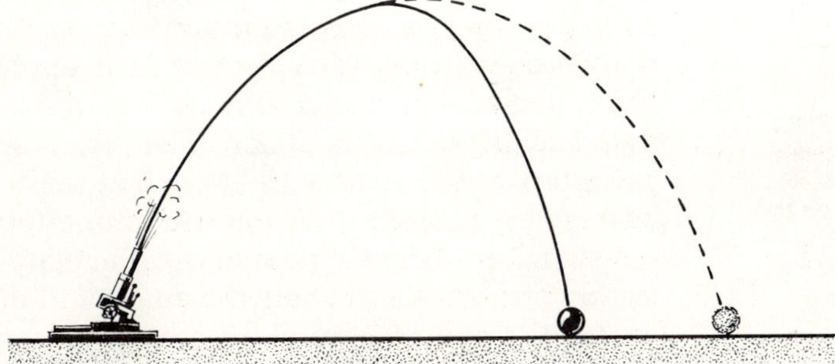

FIGURE 4.9 The ideal (dotted line) versus a common actual trajectory of a missile in the air.

■ 4.3 Kepler's Laws of Planetary Motion

A very good case can be made for the proposition that the breakthrough that occurred in Western science in the sixteenth century was largely due to young men connected with the Church and supported by the Church. Such young men would have had a very considerable amount of spare time on their hands in which to pursue scientific interests. Copernicus, who began the breakthrough with his suggestion of the heliocentric universe, was just such a young man, as was Galileo.

In contrast, Tycho Brahe, the Danish astronomer who collected very accurate observational data on the motion of the planets, was an aristocrat. It was also not too unusual for an aristocrat in those days to spend most of his time seriously looking at the heavens and to make an important contribution such as this to science.

Tycho Brahe also made another important contribution to posterity: he employed an assistant, a brilliant, eccentric young man named Johannes Kepler, who codified Brahe's data on planetary motion in the form of three laws. Kepler's laws represent such a departure from the then prevalent notion about the motion of heavenly objects that they can only be regarded as the result of the most daring imagination. As we will see, Kepler's ideas played a key role in Newton's discovery of the law of gravitation. Newton once said that the reason he was able to see so much of scientific truth was because he stood on the shoulders of giants. Rather than a reflection of his modesty, Newton's statement was simple truth; he did stand on the shoulders of giants like Galileo and Kepler.

Kepler, like his predecessors, started his model of the solar system by fitting the motion of all the planets in perfect circles. It was known that it was impossible to fit the planets into circular orbits with the sun at the center. So in one of his trial attempts, Kepler tried circles with the sun off center. Kepler almost succeeded with this model, but not quite. There was a slight discrepancy between his calculated positions of the planet Mars and Brahe's observations, a small discrepancy but clearly outside the margin of experimental error. So Kepler had to abandon this model and try another one.

After many, many trials Kepler reached a revolutionary conclusion: the planetary orbits were not circles at all. For two thousand years after Plato originated the circle idea, people had regarded heavenly motion as occurring in perfect circles. It turned out that this was not so.

Kepler found that the curve that described the orbits of the planets is an **ellipse,** a sort of elongated circle. The Greeks had studied this curve two thousand years before, and its properties were well known. You can draw an ellipse by making a loop with a piece of cord, anchoring the two ends of the cord with thumbtacks pinned to a drawing board, and then putting a pencil in the loop and drawing with the pencil. The pencil will draw an ellipse (Fig. 4.10). The positions of the two tacks are called the **foci** of the ellipse. The orbits of the planets are ellipses with the sun at one focus.

One noteworthy element of the planetary orbits is that most of them are very nearly circles. We indicate the amount of departure from a circle by a quantity called the **eccentricity** of an ellipse. Eccentricity is the ratio of the distance between the foci and the major diameter (the major axis as it is usually called), as shown in the figure. A circle can be looked upon as an ellipse of zero eccentricity. An ellipse of small eccentricity looks almost like a circle. In Fig. 4.11 notice how, as the eccentricity e becomes closer and closer to 1, the elliptical shape becomes more and more pronounced. When $e = 1$, the ellipse becomes a straight line. In Table 4.2 the eccentricities of the orbits of all the planets are shown.

Notice in Table 4.2 that Mars has the second largest eccentricity among the first six planets, those that were known in Kepler's time. It was pure luck that

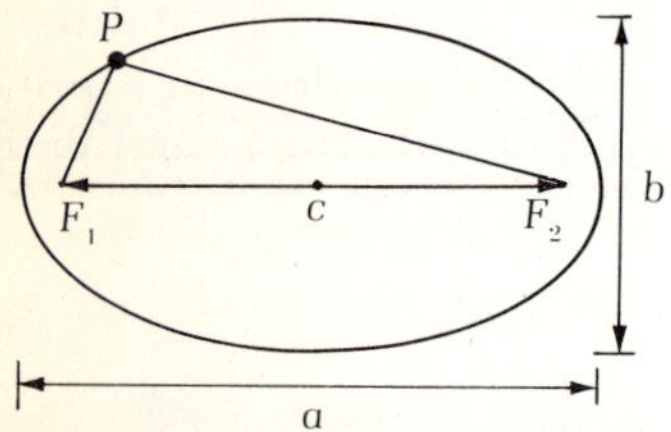

FIGURE 4.10 An ellipse with its major axis a and minor axis b defined as shown. F_1 and F_2 are its two foci. Its shape is described by its eccentricity e, which is the ratio of the distance c between the foci and the major axis; that is, $e = c/a$.

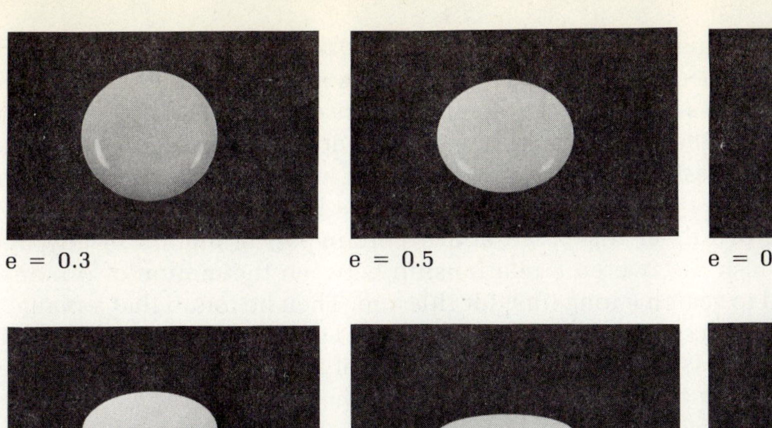

FIGURE 4.11 Ellipses of different eccentricities. The pictures were made by photographing a saucer at various angles. (Project Physics.)

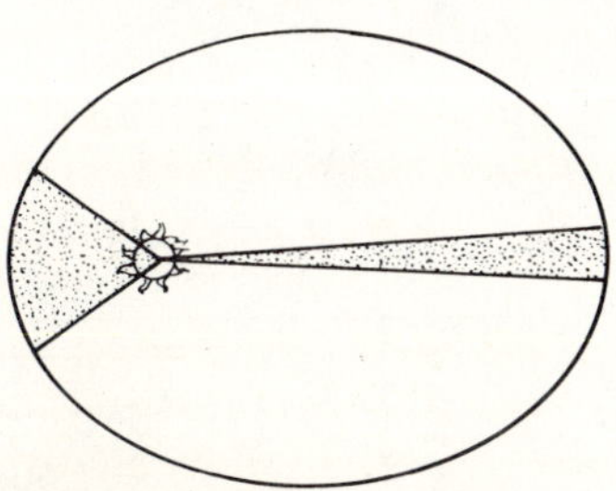

FIGURE 4.12 Illustration of Kepler's law of areas. The areas swept by a planet's position vector in equal times are equal.

Mars was the planet Kepler studied thoroughly; he might never have hit upon ellipses as the solution to the problem if he had studied one of the planets with smaller eccentricities.

At the same time that Kepler became convinced that the planetary orbits were not perfect circles, he also realized that the motion was not uniform either. The question now arises: what is the nature of the nonuniformity? If the speed varies, when do the planets speed up most and when do they travel most sluggishly? Kepler found the answers to these questions also.

Kepler found that the planets speed up when they are close to the sun and slow down when they are farther away. Further study revealed a remarkable thing. Suppose we take two positions of a planet on its orbit separated by a definite period of time, say a week, and then consider two other positions separated by the same period in another part of the orbit (Fig. 4.12). We now draw lines from the sun to the planet in each of the positions (these lines are the position vectors of the planet with the sun as the origin). The area enclosed by the two lines at one segment of the orbit turns out to be equal to the area generated at the other segment. The area swept by the position vector in the same

TABLE 4.2 Eccentricities of the planets.

Planet	Orbital Eccentricity
Mercury	0.206
Venus	0.007
Earth	0.17
Mars	0.093
Jupiter	0.048
Saturn	0.056
Uranus	0.047
Neptune	0.009
Pluto	0.249

period of time remains the same irrespective of the planet's position in the orbit. Thus the planet speeds up as it approaches the sun and slows down when it moves away in such a manner as to maintain precisely the same rate for sweeping the area. The speed of the planet is maximum at the **perihelion**, the point of its orbit that is closest to the sun, and minimum at the **aphelion**, the furthest point from the sun.

The findings of Kepler discussed above pertain to the motion of any one planet. Kepler also discovered a relationship between the motion of different planets. He had to search a long time for this one. Then he found that a planet's **period** of revolution, which is the time it takes to complete one revolution, is longer the further it is from the sun. The square of the time period (T) actually varies as the cube of the average distance from the sun (R). In symbols,

$$T^2 \propto R^3 \tag{4.6}$$

Does this sound too complicated? Perhaps an illustration will help. Suppose R is tripled. Then the cube of R increases by a factor of 3^3, or 27. Kepler's relation, Eq. (4.6), says that the square of the time period T increases by the same factor of 27. This means that the time period itself increases by a factor of $\sqrt{27}$, which is 5.2. You can verify the validity of the equation yourself with the help of Table 4.3, where R and T are given for each planet. You can show that the ratio of the square of T for any two planets is equal to the ratio of the cube of their average distance R from the sun.

TABLE 4.3 The parameters of planetary orbits.

Name of Planet	Average Distance from the Sun (10^6 km)	Period (years)
Mercury	57.9	0.24
Venus	108.1	0.62
Earth	149.5	1.00
Mars	227.8	1.88
Jupiter	778	11.86
Saturn	1426	29.46
Uranus	2868	84.01
Neptune	4494	164.79
Pluto	5896	246.69

Kepler's work is summarized in the form of three laws of planetary motion (now called Kepler's laws).

First law, or the law of elliptical orbits:

The planets revolve in orbits that are ellipses with the sun at one focus.

Second law, or the law of equal areas:

The line from the sun to the orbiting planet (the planet's position vector with the sun as origin) sweeps equal areas in equal times.

Third law, or the harmonic law:

> *The square of the time period to complete the orbit is proportional to the cube of the average distance from the sun.*

Kepler's work opened up, more forcefully than ever before, the possibility of another question: What causes the planetary movements to obey such simple laws? Kepler and others of his time were unable to find the right answer to this question. Interestingly Kepler, unlike some of his predecessors (including Brahe), believed that the answer to this question was to be sought in terms of dynamics or forces and not in geometry. Unfortunately, his notions of how forces caused motion were clouded by the ideas of Aristotle. Aristotle classified motions that were not natural as violent; and any kind of violent motion, according to Aristotle, needed a force to push the object as long as the violent motion lasted. So people in Kepler's time thought that angels must be pushing the planets (from behind them and around the orbits). Kepler's own thoughts were much more advanced than this, but they were not adequate.

Sixty years later Newton finally discovered the theory behind Kepler's laws.

Kepler's laws and Newton

One of the main reasons that Aristotle and his medieval followers had so many misconceptions about motion is that theology played a large role in their thinking. Thus the apparent circular motion of heavenly objects was perceived as the ideal of perfect motion, needing no explanation. In Aristotle's theory, motion of objects in the sky was regarded as natural motion.

The research of Johannes Kepler changed all that. Ellipses, although simple closed curves, were not regarded as perfect. So it was clear that planetary motion, since it was elliptical, needed an explanation after all. Kepler and his contemporaries did look for such an explanation; unfortunately Aristotle's influence still prevented their seeing clearly enough to find the right one.

Newton, having discovered the state of natural motion of objects to be uniform linear motion, suspected that motion in a curve must be accelerated motion. The immediate implication was that the nearly circular motion of the planets around the sun must be caused by a force. Newton was able to prove mathematically that the acceleration, and therefore the force, is directed toward the sun. Newton boldly asserted that this force is of the same kind as that of the earth on a falling object. The force responsible for the planetary motion is the pull of the sun's gravity. The gravity force is universal—all objects have it.

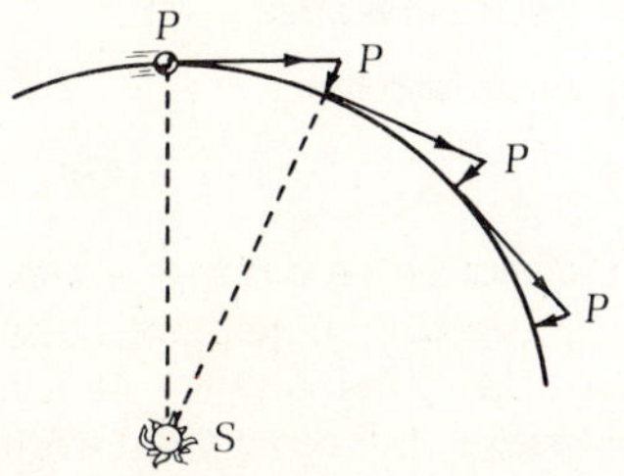

As we have already noted, Newton was led to postulate the gravity force in order to make Galileo's law of falling bodies consistent with his own laws of motion. The second step was to recognize that the motion of the planets around the sun is also a kind of falling motion: the planets continuously fall toward the sun (Fig. 4.13).* Once it is realized that the motion of the planets is also falling motion, it becomes clear that the force responsible for it can be gravity also.

FIGURE 4.13 *P* is a planet and *S* is the sun. This is how a planet "falls" as it orbits the sun.

---

*This point is further discussed in Chapter 6.

The subject of Kepler illustrates a very important point. Kepler is one of the few scientists to record in detail the agony and ecstasy involved in the pursuit of scientific truth. Sometimes he heard the music of the planets; at other times he tried to fit the planets into a symbolism based on the perfect solid shapes of the ancient Greeks. Imagine the intensity of the creative fervor of the person who wrote:

> I am free to give myself up to the sacred madness. I am free to taunt mortals with the confession that I am stealing the golden vessels of the Egyptians in order to build of them a temple for my god, far from the territory of Egypt. If you pardon me I shall rejoice, if you are not enraged, I shall bear up. The die is cast and I am writing the book—whether to be read by my contemporaries or by posterity matters not.

The quotation is from the preface of Kepler's *Harmonies of the World* in which he published his third law, the law that connects the period of a planetary orbit with its distance from the sun. It is also clear from this preface that he considered his solution of the problem of the motion of the planets "very excellent and very perfect."

The point here is that discovering a physical law is a spiritual experience, and Kepler's writings leave us abundant evidence that this man was indeed guided by spirituality and intuition rather than the cold logic that you may be accustomed to associate with a scientist. And why should it not be this way? What is spirituality? Is it not a tuning in to the music of the universe? It is only natural that finding the laws of the universe becomes easier if we can tune in to its music.

That Kepler was highly tuned in to the cosmic music becomes very clear when we realize another fact. Two more works based on his intuitive ideas have subsequently turned out to be extremely interesting in view of later developments, although their merit was far from clear at the time. The first is his work based on the idea that the planets move in unique special orbits determined by some basic underlying principle. When Kepler was a very young man, he was taken by the fact that there were six planets and five "perfect" solids: cube, tetrahedron, dodecahedron, isocahedron, and octahedron. An idea of fitting the mystery of the planetary distances from the sun with the five solids took shape in his consciousness and the result was as follows:

Start with the cube, the simplest of the solids. A cube can be enclosed by one sphere and likewise can itself enclose one sphere. He called the outer sphere the sphere

of Saturn and the inner one the sphere of Jupiter. The sphere of Jupiter was to contain the next regular solid, the tetrahedron, which in turn would contain the sphere of Mars. This sphere would contain the next solid, the dodecahedron, which would hold the sphere of earth, and so on (Fig. 4.14). Kepler discovered that the radii of these spheres corresponded fairly well with the mean distances of the respective planets from the sun, except for Jupiter; but that is only one disagreement in six items of comparison.

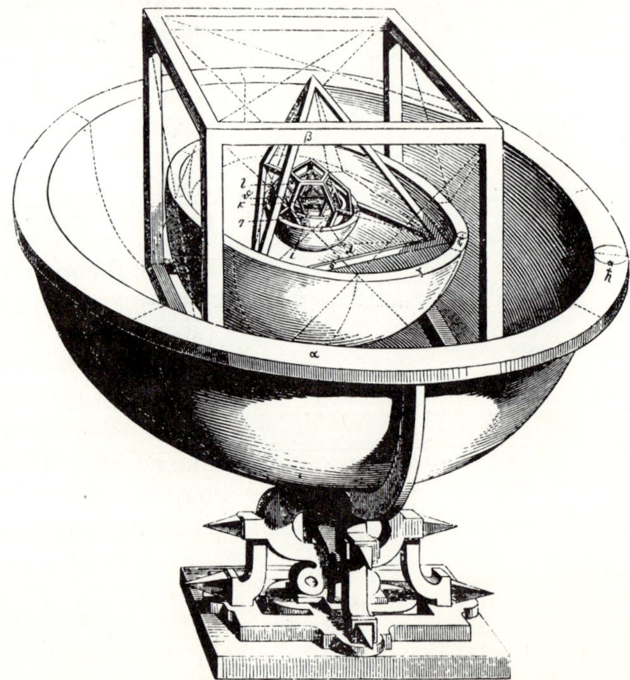

FIGURE 4.14   Kepler's planetary model with five perfect solids.

Kepler did not stop here. He ascribed a definite music to each of his spheres, which together constituted the harmony of the world, his world. This picture of the planetary system, which was dearer to Kepler than even his laws, was not taken very seriously. And of course we now know that it is not a good model for planetary motions. The radii and the periods of the planets almost certainly do not follow from a deep underlying principle. Strangely enough, however, the Keplerian picture of the harmony of the world has found its place in our understanding of nature—in the model of the atom based on quantum mechanics. In the atom it has been possible to understand the average distances of the electrons from the nucleus, and also the time periods associated with electronic motion, from fundamental underlying principles. So fundamental are these principles that electrons of a hydrogen atom vibrate with the same period anywhere in the universe. It is truly a harmony of the world.

The second neglected work of Kepler that has attracted some recent attention is his book called *The Dream*.* It is basically an exploration of lunar geography; Kepler was anticipating a voyage to the moon more than three hundred years before it became a reality. *The Dream* is an exceptional document in the history of science. The variety and depth of Kepler's vision is clearly marked in almost every page of this remarkable book. For example, Kepler anticipated quite a few of the properties of the gravity force that was later independently established by Newton.

Kepler has been called a sleepwalker, probably because he seemed to have fumbled into his discoveries of the planetary laws that made him famous.† Or perhaps he was a real sleepwalker. Who else could conceive of such a "dream"?

*See *Kepler's Dream*, ed. John Lear (Berkeley: University of California Press, 1965).

†A. Koesler, *The Sleepwalkers* (New York: Macmillan, 1959).

Another important thing: Newton did not stop at postulating gravity merely as a physical idea. Some of his contemporaries, notably Robert Hooke, had similar physical ideas. But Newton found a mathematical expression that enables us to calculate the gravity force between any two objects—and thus the motion of all objects under natural forces of gravity. His discovery of the mathematical expression for the gravity force must be regarded as one of the highest intellectual feats ever achieved. As we will see in Chapter 6, the main ingredients for Newton's work were the laws of planetary motion of Johannes Kepler and the law of falling bodies of Galileo.

Kepler's second law and the conservation of angular momentum

If we multiply the rate at which the area is swept by the position vector of a planet by its mass, we get a quantity called the **angular momentum** of the planet. So Kepler's second law, which says that the rate at which area is swept by the position vector remains a constant, can also be stated in the following form:

The angular momentum of a planet remains a constant.

Thus Kepler's second law implies the conservation of this new quantity, the angular momentum of a planet.

For motion in a circle, the angular momentum of an object takes on a simple form. In this case the rate at which area is swept by the position vector of the object is found to be the product of its velocity v and position r. Therefore, the angular momentum of an object in circular motion is the object's mass m times vr or mvr. Thus the angular momentum is also the product of the object's momentum mv and position r (Fig. 4.15). Further implications of the concept of the angular momentum and its conservation are discussed in the next chapter.

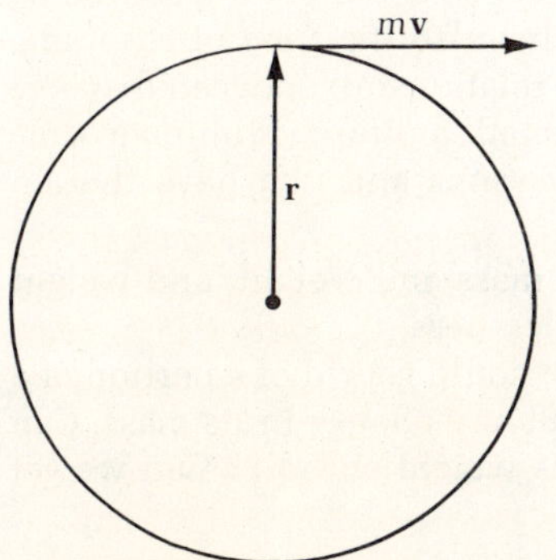

FIGURE 4.15 For circular motion, the angular momentum is always *mvr*.

SUMMARY

This chapter has several important themes. The law of falling bodies discovered by Galileo and the three laws of planetary motion discovered by Kepler act as introductions to the subject of gravitation, which is further discussed in Chapter 6. The discussion of falling bodies also introduces you to some aspects of accelerated motion, in addition to those discussed in Chapter 1. The distance-time relationship,

$$d = \tfrac{1}{2} at^2$$

is an important new feature. Also important are the fact that the path of a projectile is a parabola and the reason this is so. Kepler's second law introduces you to a very important concept, that of angular momentum. Angular momentum is the product of momentum and position (mvr).

In addition to examining the ideas in this chapter, we hope you have availed yourself of the opportunity of looking at the discovery processes of the two scientific giants, Galileo and Kepler. Science grows out of the experiences of people like Kepler. It is only by looking at them that we begin to get an inkling of what science is all about.

QUESTIONS

Review and reason

1. Two balls weighing 10 lb and 50 lb, are dropped from a 100-ft high tower. Which ball falls to the ground first (a) according to Galileo and (b) according to Aristotle? Who is right according to your own experience?

2. Discuss how a distance-time graph tells us whether a certain motion is natural or accelerated. Draw diagrams.

3. Fig. 4.16 shows a velocity versus time graph. From the graph can you tell at which times the object is undergoing accelerated motion and at which times it is moving at constant speed?

*4. In police work, investigators often determine the speed of a car from the length of the skid marks left by the braking. Explain why it is not correct to assume that the speed of the car is proportional to the length of the skid marks. Can you prove that the speed is actually proportional to the square root of the length of the skid? (Note: The proof involves a bit of symbol manipulation. You can start by asking, what is the relationship between the deceleration of the car and its initial speed, the final speed being zero? This will involve also the time of stopping. Now ask, what is the relationship between distance and time in accelerated motion? Eliminate time between the two equations and you have the answer.)

5. Distinguish between mass and weight, and weight and apparent weightlessness.

6. Describe a device that could get rid of a portion of a body's weight without any change in its mass. Can we get rid of a body's weight entirely? Can we get

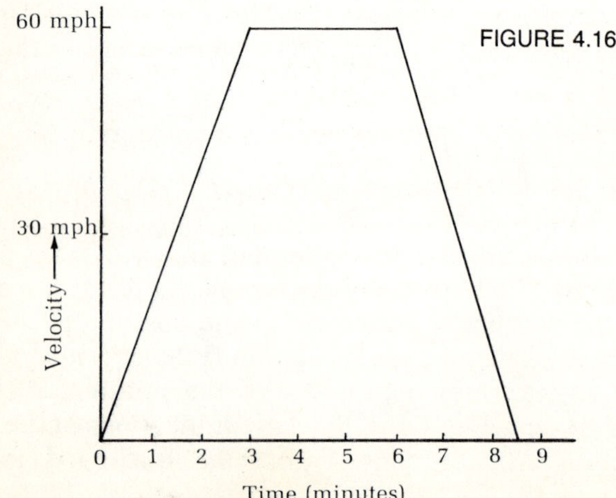

---

*Optional.

rid of a part or all of a body's mass without changing its weight? Explain.

7. Explain how a body falling in air can acquire terminal speed.

8. Suppose you observe, scientifically, little metal balls falling through liquid Prell, a hair shampoo. What law of falling bodies do you expect to rediscover upon analysis of your observations? Incidentally, what exactly do your scientific observations involve?

9. Which will reach the ground first, a 100-lb stone or a 10-lb stone, (a) when dropped simultaneously in vacuum from a height of 10 ft and (b) when dropped similarly in air from a height of 3 mi? Give your reasoning.

10. What is a parabola? Why is the path of a projectile a parabola?

11. Suppose a hunter takes direct aim at a snake hanging from a tree and shoots. The snake lets go of the tree and drops at exactly the same instant the bullet leaves the gun barrel. Will the bullet hit the snake?

12. If we neglect air resistance, the time taken by a ball thrown upward to reach its maximum height is exactly the same as the time it takes to fall to the level of throw. This follows from the symmetry of the upward and downward journeys—they are mirror reflections of each other (in the absence of air resistance). Now suppose we do not ignore the effect of air resistance. Would it take a longer time, a shorter time, or the same time to fall back compared to the time taken to reach the top?

13. Using the procedure described in the text, draw an ellipse.

14. State Kepler's laws of planetary motion.

15. Do some outside reading on the two contemporary geniuses Galileo and Kepler. As a result of your readings, how would you compare their styles as scientists?

16. What is meant by the eccentricity of an ellipse? The comets travel around the sun in elliptical orbits of large eccentricities. Draw such an orbit using your own imagination. Show the position of the sun and the aphelion and perihelion positions of the comet. Where does the comet have the largest orbital velocity? How do you know?

17. The moon revolves around the earth with a period of about 27 days. Assuming the validity of Kepler's laws for lunar motion about the earth, can you guess if the period would increase, remain the same, or decrease if the moon were to recede from the earth? Explain.

*18. The poem by Rilke in the beginning of the chapter has a line: "the heavy earth falls too." Do you agree? Explain your answer.

19. An object of mass m moves uniformly with speed v in a circle of radius r. What is its angular momentum?

Arithmetic

1. On the moon a free-falling object starting from rest acquires a speed of 5 ft/sec after 1 sec of fall. What is the acceleration of fall on the moon? Make a table of time, velocity, and distance fallen (in the style of Table 4.1) for falling objects on the moon.

2. A rocket is designed to accelerate at a uniform rate of 2 g's. How long will it take to reach a speed of 8 km/sec starting from rest?

3. There is a B.C. cartoon in which Peter asks Thor how deep the water level in his well is. "I don't know," says Thor. Peter throws a pebble in the well and starts counting seconds. He counts to sixteen when the splash is heard. "Your well is sixteen seconds deep," he declares. Can you convert Peter's "sixteen seconds deep" into a depth in feet?

*4. Galileo wrote: "So far as I know no one has pointed out that the distance traveled during equal intervals of time, by a falling body from rest, stand to one another in the same ratio as the odd numbers beginning with one." Explain.

5. A planet's position vector sweeps an area of 30 million square miles in 10 days. How many square miles will it sweep in 30 days? Explain how you arrived at your answer.

6. Imagine two planets of a star, one four times further away than the other from their sun. Assuming the validity of Kepler's laws, compute the ratio of their period of revolution.

---

*Optional.

Angular Momentum and Rotational Motion

I spin, you spin,
We all spin with earth-spin.

■ 5.1 Rotational Motion

In the last chapter we ignored one aspect of the motion of many falling objects. Many objects when they fall exhibit **rotational motion** as well as falling motion (Fig. 5.1).* Sometimes the properties of rotational motion introduce new elements of stability against falling—this is why a bike rider doesn't fall while moving. At other times spinning facilitates falling in a particularly desired fashion—diving is a case in point.

We will start with the role of angular momentum conservation in rotational motion and go on from there. One thing we will find is that the concepts needed to understand rotational motion are fairly analogous to those of translational motion. Thus instead of momentum we have angular momentum; instead of inertia we have rotational inertia. Also, the role of force is played by a quantity called torque; and so forth. In case it is not obvious, we should also point out that the theoretical framework we outline for rotational motion holds for all rotating objects, not just falling ones.

Angular momentum consideration in rotational motion

For a body rotating about a line, or "axis," the component masses of the body can be thought of as moving in circles with centers on the axis. So, as pointed out in Section 4.3, we can assign an angular momentum value mvr to each of these component masses. If we add up all these values for the different parts of

FIGURE 5.1 While the center of mass of the juggling pin describes a parabolic trajectory, the pin wobbles about its center of mass.

---

*In general, motion of an object around and around a center or an axis is called rotational motion.

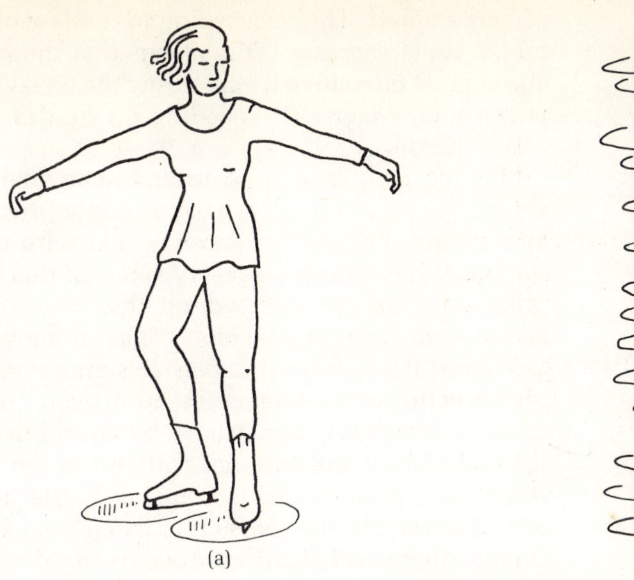

FIGURE 5.2 The figure skater uses the principle of angular momentum conservation.

the body, we get a total, which can be called the **total angular momentum** of the object. Admittedly, carrying this out may not be as simple as it sounds. But we can actually guess the results without having to carry out a complete calculation of the angular momentum, as you will see.

Consider the spinning motion of a figure skater (Fig. 5.2). We often see her start to spin at a slow rate with her arms extended. Then she draws in her arms and voila! The spin becomes faster. What's the mystery?

The explanation is to be found in the conservation of angular momentum. Initially, in the extended position, her arms rotate in a circle with a large r. When she draws in her arms, however, the radius r of the circle of rotation is small. Now if angular momentum is conserved, mvr must remain the same in both positions of her arms. This is the key to the explanation. When r is large, v, the speed of rotation, is small; when r is small, however, v must increase so that the product mvr retains the same value. The figure skater knows the importance of the conservation of angular momentum, and she puts it to good use.

Recently certain astronomical objects called **pulsars** have been discovered. Pulsars rotate with amazingly small periods of less than a second, which corresponds to very large speeds of rotation.* It turns out they are none other than **neutron stars,** which were predicted to exist as one of the end products of stellar evolution. Neutron stars are highly dense stars composed almost entirely of neutrons, the neutral component of the subatomic particles inside the atomic nucleus. Why do pulsars rotate so fast? The answer again comes from the conservation of angular momentum. The typical rotation period of a star like our sun is 30 days. This corresponds to a comparatively small value of the velocity v in the formula mvr for the angular momentum. Of course, for a regular star, r is large. But when the star collapses to become a neutron star, r

---

*The faster the speed, the smaller is the time it takes to make a complete rotation. This time is called the period of rotation. The speed and the period, then, are inversely proportional to each other.

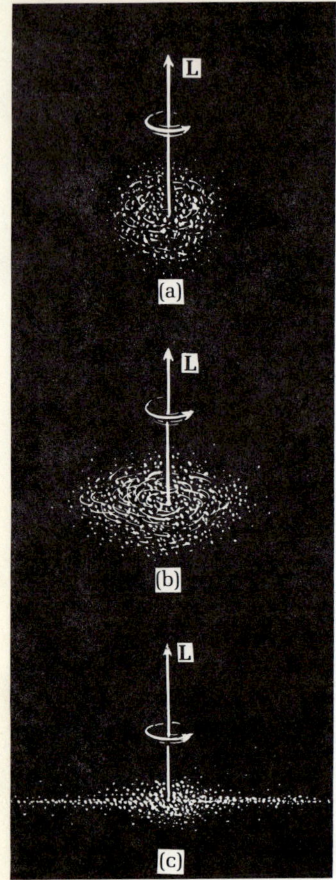

FIGURE 5.3 The evolution of the pancake shape of a galaxy. (a) The beginning: a cloud of gas of spherical shape. (b) The galaxy starts to flatten. (c) The final shape.

FIGURE 5.4 A gyrocompass.

becomes small. Then according to the law of conservation of angular momentum, v must increase. This means that the period of rotation of the collapsed object must be reduced, explaining the behavior of pulsars. It's like the behavior of the figure skater in speeding up by drawing in her arms, only now on a cosmic scale.

Like momentum and position, the angular momentum of an object is also a vector quantity. The following example illustrates the vector nature of angular momentum. Perhaps you are familiar with the flattened pancakelike shape of some galaxies. Our own galaxy is one of this kind. Galaxies get this shape if the pregalactic matter from which they were created has considerable angular momentum. The original cloud of gas is shown in Fig. 5.3(a). It is assumed that, as viewed from above, the particles are rotating in a counterclockwise pattern. The direction of the total momentum vector **L** in this case is as shown.* Because the angular momentum has to be conserved (magnitude, direction, and all), such a cloud of gas does not collapse in the direction perpendicular to **L**. The outer objects, instead of collapsing, settle down in orbits around the central part. Angular momentum acts as a deterrent against collapsing in this direction. On the other hand, if collapse occurs in a direction parallel to **L**, the value of the angular momentum is unaffected. Therefore, nothing prevents such a collapse. The galaxies thus assume the disc shape, as shown in Fig. 5.3(c).

Isn't it remarkable that the second law of Kepler, originally discovered in connection with the motion of solar planets, leads to a conservation principle that determines the shapes of entire galaxies? This is the way with all unifying principles.

It is also remarkable that we not only have uncovered such general principles as the conservation of angular momentum but also have put them to good use. A case in point is the gyrocompass, in which the unchangeability of the direction of the conserved angular momentum vector is utilized. The mechanical construction of the gyrocompass ensures that its angular momentum is unaffected by any motion of the ship; that is, its angular momentum is conserved (Fig. 5.4). Once the gyrocompass is set spinning with its angular momentum vector pointing toward the North Star, it will always point in that direction no matter where the ship is.

Rotational inertia

The angular momentum "hangup" of objects (to use the terminology of American physicist Freeman Dyson†) perhaps reminds you of inertia, the hangup of objects in natural motion. Rotating objects have a similar tendency to keep on rotating, a tendency that rightly can be called **rotational inertia.**

Let us find some examples of rotational inertia among objects around us. Remember the top you played with as a child? You couldn't balance the top on its point without spinning it; otherwise it fell over. Yet once you set it spinning,

---

*If the rotation were clockwise, as seen from above, the direction of **L** would be downward in the figure. This is the rule: If the rotation is seen to be counterclockwise from a certain direction, the angular momentum points toward the viewer. If the rotation is seen to be clockwise from a certain direction, it is pointed directly away from the viewer.

†Freeman J. Dyson, "Energy and the Universe," *Scientific American*, September 1971, p. 51.

the top would stay upright and spin for a long time. This is an example of rotational inertia.

Have you ever tried to sit on a bicycle at rest? It too tends to fall over. Yet once you get the bicycle going, the rotational inertia of the spinning wheels creates an extra resistance to change, and you are quite stable sitting on it. Thus the spinning motion—or rather the rotational inertia associated with spin—offers extra stability for moving objects (Fig. 5.5). Gun barrels are designed to have spiral bores to set the bullets spinning as they are ejected (the name "rifle," in fact, comes from the rifling, the spiral grooving inside the bore). The spinning motion guarantees better stability of the bullet, helping to keep it on its path against slight disturbances.

Many machines you are familiar with in your everyday activities have rotating parts—wheels, gears, shafts, and so on. The rotational inertia of such parts plays an important role in the performance of machines. Consider, as an example, the role of the flywheel in an automobile engine. The engine of your car does not burn fuel continuously but rather in short bursts. What turns the engine crankshaft between the bursts and smooths out the otherwise choppy ride? The flywheel (Fig. 5.6), which is connected to the engine, does this very important chore. With its large rotational inertia, the flywheel, once started, keeps rotating even between the fuel-burning phases of the engine. Since it is connected to the crankshaft, the crankshaft keeps turning smoothly throughout the driving operation.

As in the case of translational inertia, an object having a large rotational inertia not only is hard to stop when it is in rotation, but it is also hard to get started. Tightrope walkers in a circus take advantage of this particular idea from physics. Perhaps you have wondered about the beautiful umbrella that some rope-walkers carry while doing their act. An open umbrella has a large rotational inertia and, therefore, a considerable amount of resistance to rotation. If the tightrope walker starts to tumble, she and the umbrella start to rotate about the wire, but their rotational inertia resists spinning, giving her valuable time to bring her center of gravity back in line with her support.

Can you see anything in common between the flywheel and the umbrella, two objects with relatively large values of rotational inertia? For both the distribution of mass is such that there is plenty of mass extending away from the axis of rotation. For translational inertia it is the mass of the object that is the determining factor. For rotational inertia the determining factors are not only the mass but also how the mass is distributed in the body. The quantity that is analogous to mass, but which takes into account the distribution of mass in the object, is called the **moment of inertia** of the object.* The moment of inertia plays the same role in rotational motion as mass does in translational motion. The more the mass is distributed away from the rotation axis, the greater is the moment of inertia.

FIGURE 5.6   A flywheel.

---

*Some authors use the phrase "rotational inertia" to denote moment of inertia.

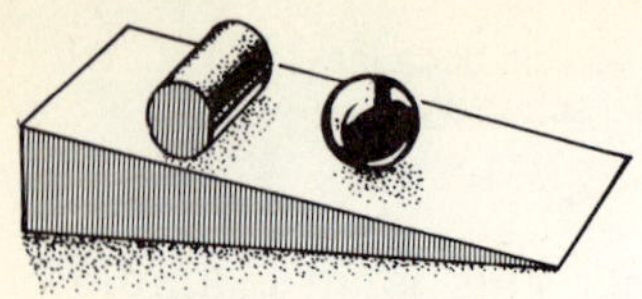

FIGURE 5.7 On an incline a sphere rolls faster than a cylinder.

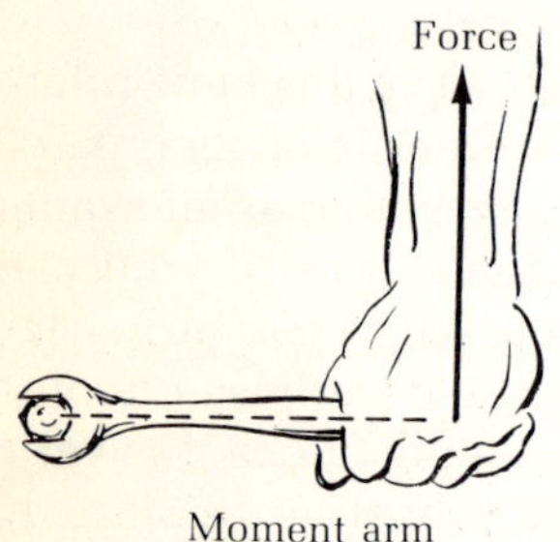

FIGURE 5.8 Generating a torque, the turning effect of a force.

Thus the moment of inertia depends on the shape of an object. For example, consider a sphere, a cylinder, and a hoop of the same mass and radius. The sphere has the smallest moment of inertia and the hoop has the largest, with the cylinder in between. I remember once getting a call from a student, quite a frantic call, on the evening of the day I taught Galileo's falling body experiments, including Galileo's use of the inclined plane. This enterprising student was doing some of Galileo's experiments for himself. Then he started rolling two objects, one a sphere and the other a cylinder, on an inclined plane, expecting that they would roll down the incline together. But they didn't and that was what his excitement was about (Fig. 5.7).

The explanation for the student's result has to do with the relative values of the moment of inertia of rolling spheres and cylinders. The cylinder has a greater moment of inertia, so it is harder to set it rolling—it resists rolling. The sphere, with a smaller moment of inertia, will invariably roll faster and reach the bottom first.

Galileo, of course, never said that any two rolling objects on an inclined plane should reach the bottom together. He said that was so only for objects falling in air (strictly, in a vacuum).

Torque

Most rotational motion we see around us does not continue unchanged forever. What is the agency that produces changes in the rotation of an object? In other words, what is the analogue of force in the case of rotational motion? This agent is **torque** (pronounced "tork"). Torque is the turning effect of a force. Let's consider a simple example to illustrate the concept: turning a bolt with a wrench. The force is applied on the wrench some distance away from the pivot point, or the axis of rotation (Fig. 5.8). In fact, the farther away you apply the force from the center of the bolt, the better is the turning effect. Thus the turning effect, or the torque, depends not only on the force but also on the distance from the center or axis of the turn. This distance is commonly called the **moment arm.** Formally, the torque is defined as the product of the applied force and the moment arm:

$$\text{torque} = \text{force} \times \text{moment arm} \tag{5.1}$$

Whenever you produce a torque, or a turning effect due to the force you apply, notice that you are applying the force off center, that is, away from the axis of rotation. Thus when you open a door, you want to turn it about the line through the hinges. Hence for best results you apply the force on the knob some distance from that axis and directed perpendicular to the plane of the door. You pedal your bike some distance away from the sprocket axle (directed perpendicular to it), which is the axis of rotation in this case. You apply force on the steering wheel of your car tangentially on the rim, separated by some distance from the steering column (Fig. 5.9).

You may already have determined that, like force, torque is a vector. It is actually a product of two vectors: (1) the position vector of the physical point of application of the force, which is the distance from the center of turn to the point, and (2) the applied force (Fig. 5.10). The quantity "moment arm" in Eq. (5.1) is the value of the component of the position vector in the direction

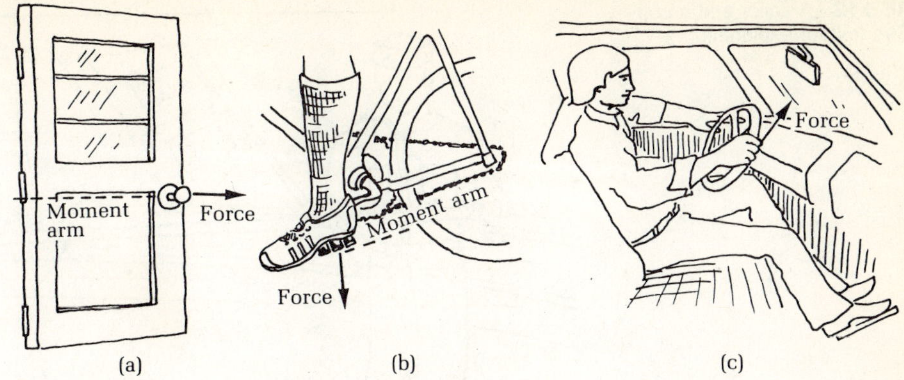

FIGURE 5.9 Examples of torque. In each case the force is applied away from the center or axis of rotation.

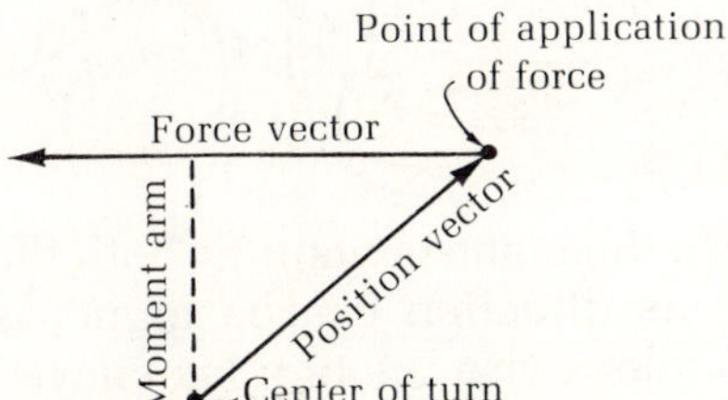

FIGURE 5.10 Torque is the product of the position vector and the force vector. The moment arm is the component of the position vector in a direction perpendicular to the force vector.

perpendicular to the force vector (Fig. 5.10). Now if we generate a vector as a product of two vectors, we get a nonzero vector only if the two vectors are not parallel to each other. If they are parallel, the product vector is zero. Furthermore, the magnitude of the product is maximum when the two vectors are perpendicular to each other.

Thus the torque is zero when the force is applied parallel to the position vector [Fig. 5.11(a)]; the moment arm is zero in this case. The torque is maximum, for a given magnitude of the force and the position vector, when the force is applied perpendicular to the position vector [Fig. 5.11(b)]. In this case the moment arm is equal to the full magnitude of the position vector, that is, the entire distance from the center of the turn to the point of application of the force. If the force is applied at any other angle to the position vector, the torque is reduced from its maximum value. The moment arm, which is equal to the component of the position vector in a direction perpendicular to the force [Fig. 5.11(c)], is smaller in this case than the magnitude of the position vector itself.

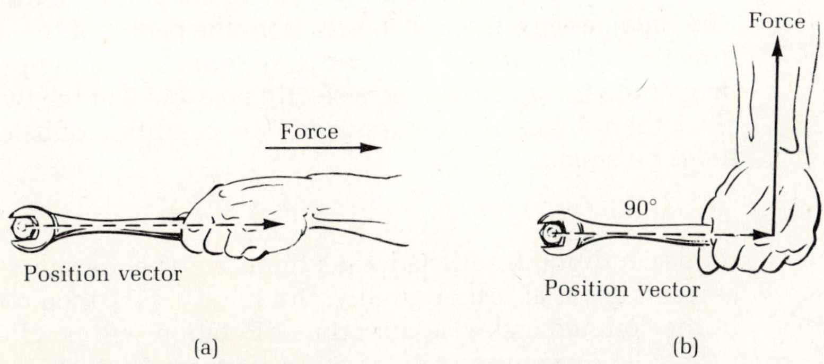

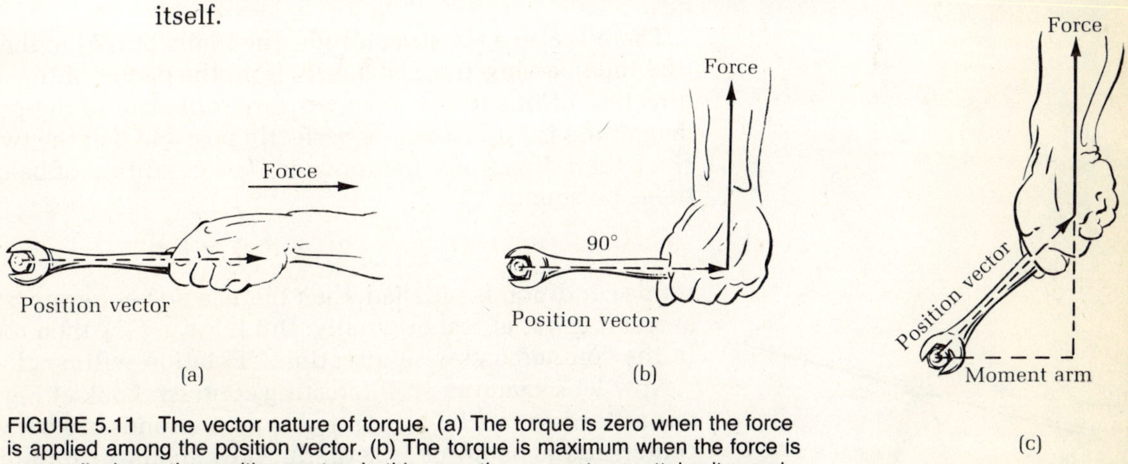

FIGURE 5.11 The vector nature of torque. (a) The torque is zero when the force is applied among the position vector. (b) The torque is maximum when the force is perpendicular to the position vector. In this case the moment arm attains its maximum value—the magnitude of the position vector. (c) When the force is applied at an angle other than 0° and 90°, the torque takes on a value intermediate between zero and the maximum. The moment arm is now the projection of the position vector in a direction perpendicular to the force vector. The moment arm can also be looked upon as the perpendicular distance of the force vector from the center.

FIGURE 5.12 A sailor and a capstan. See text for explanation.

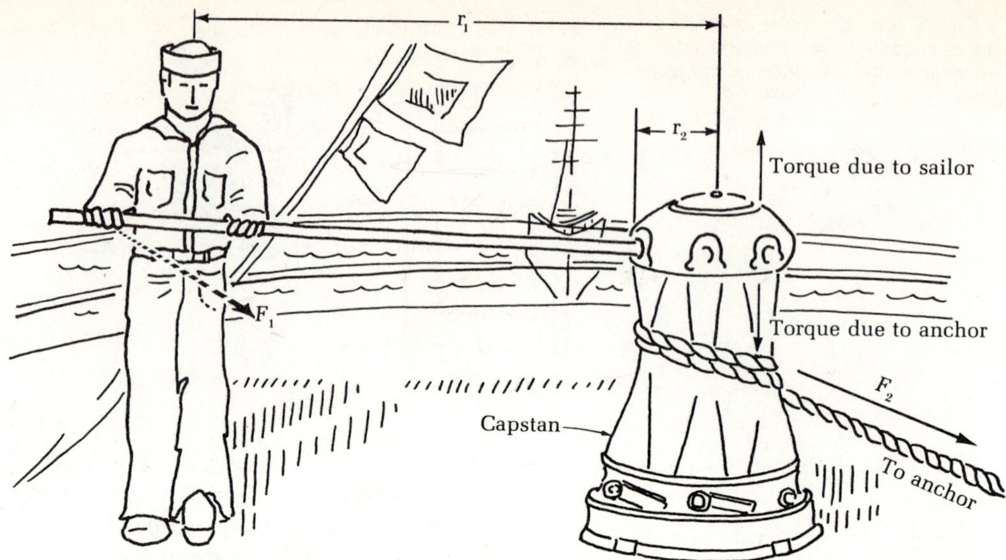

The line of action of torque is perpendicular to the plane containing both the position and the force vectors. The sense of its direction can be found as follows: If a torque is responsible for a counterclockwise rotation (as viewed from a certain direction) about an axis, its direction is toward the viewer. On the other hand, if the torque produces a clockwise rotation (viewed from the same direction), the torque is directed away from the viewer. In Fig. 5.12, a sailor is turning the capstan in a counterclockwise direction (as seen from above) by applying a horizontal force F_1 at a distance r_1 from the fulcrum (the center of turn). The direction of his applied torque is then upward, as shown. Since F_1 is applied perpendicular to the position vector, the moment arm is equal to the magnitude of the position vector, r_1. Thus the magnitude of this torque is $F_1 r_1$.

There is also a resisting torque due to the pull F_2 of the anchor on the hawser (the rope), acting at a distance r_2 from the center of the capstan to its rim. The direction of this torque is downward, opposite to the previous torque, and its magnitude is $F_2 r_2$. Now it is perfectly possible that the two torques will balance since their directions are opposite. The condition of balance is that the magnitudes be equal:

$$F_1 r_1 = F_2 r_2 \tag{5.2}$$

If this condition is satisfied, then there is no net torque, so the capstan will stay at rest if it was at rest originally. But if $F_1 r_1 > F_2 r_2$, then rotation will be induced in the counterclockwise direction.* Rotation will be clockwise if $F_2 r_2 > F_1 r_1$.

Now let's examine an interesting contrast. Look at Fig. 5.13. Here equal and opposite forces are being applied to the two ends of a bar with the fulcrum at the center, so the net force is zero. But how about the torque? Notice that the rotation due to each torque is counterclockwise (as seen from above). Both torques are directed upward and thus they add, producing a net torque that is double

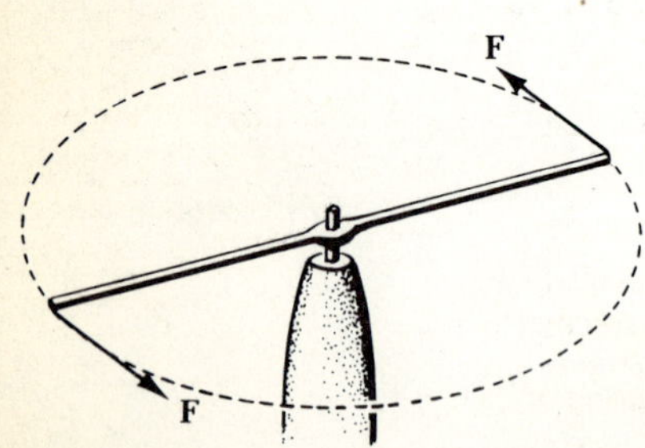

FIGURE 5.13 A couple. The total force is zero but the torque is not.

---

*The symbol $>$ means "greater than." The reversed symbol $<$ means "less than."

the torque had there been only an individual effort. Such a pair of equal and opposite forces not acting along the same straight line and producing a net torque is called a **couple.**

The relationship of torque and angular momentum

There is another way of saying that an unbalanced torque produces a change in the state of rotation of an object. We know that a rotating object has angular momentum. Thus we can express the effect of a torque on an object by saying that the torque changes the angular momentum of the object. In fact, to complete the analogy to the relationship of force and momentum for translational motion, torque is given by the rate of change of the angular momentum:

$$\text{torque} = \frac{\text{change of angular momentum}}{\text{time}} \tag{5.3}$$

Now we can state the law of conservation of angular momentum in the following general way:

> *The total angular momentum of a body is conserved if there is no net external torque acting on the body.*

This implies that all internal torques within the body must cancel one another. Internal torques have no effect on the angular momentum; only the net external torque counts.

Coming back to the Kepler orbits of the planets, what does angular momentum conservation imply there? It implies that the external torque is zero. So the gravity force responsible for the motion of the planets cannot give rise to any torque. This is possible only if the force is along the position vector, which it is (Fig. 5.14). Such a force in circular motion is called **centripetal,** meaning center-seeking. We will have more to say about this kind of force in the next chapter.

A numerical example

Suppose you take your little brother with you to play on the teeter-totter in the park. The difference in weight between the two of you can be a problem. Of course, instinctively everybody knows how to compensate for this: The heavier person has to sit closer to the center (Fig. 5.15). We can find out exactly where by using arithmetic and Eq. (5.2).

To be specific, suppose your weight is 140 pounds and your little brother's is 56 pounds. He sits at one end at a distance D from the fulcrum. Let's assume that you sit at a distance d such that the torque due to your weight about the center exactly balances the torque due to his. This condition is expressed by Eq. (5.2). Substituting in the equation the parameters of the problem, we get

$$56D = 140d$$

Solving for d we find

$$d = \frac{56}{140}D = \frac{2}{5}D$$

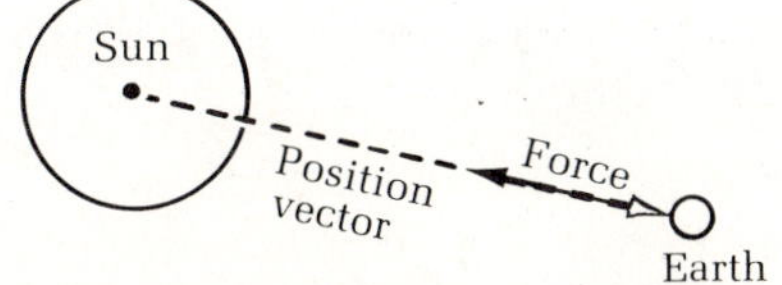

FIGURE 5.14 Angular momentum is conserved in planetary motion because the motion is torque-free. The force is directed along the position vector.

FIGURE 5.15 On a teeter-totter with a child. The big person must sit closer to the center.

So you have to sit at two-fifths of his distance from the center to get a good start.

Completing the story

We now have in our hands a large number of concepts for rotational motion, each of which has its own translational counterpart. However, there is one major omission. What is the rotational parallel of velocity?

Since rotational motion is angular motion (the body turns through an angle with respect to an initial position), we define **angular velocity** to describe the velocity of rotational motion. Angular velocity is the rate of change of the angle of the turn with time:

$$\text{angular velocity} = \frac{\text{amount of angle turned}}{\text{time}} \tag{5.4}$$

How is an angle measured? You may say in degrees. This is true, but there is another measure of angle that is often used instead: the radian. The radian is defined as an arc of a circle equal to its radius (Fig. 5.16). The relationship of a degree to a radian involves the number π (pronounced "pi"; it is a lowercase Greek letter). This number has been known since antiquity to be the ratio of the length of the circumference of a circle and its diameter:

$$\pi \text{ radians } = 180°$$

or

$$1 \text{ radian} = \frac{180}{\pi} \text{ degrees}$$

Numerically, the value of π is approximately 3.14. Thus 1 radian $= 180/3.14 = 57.32°$. Finally, we can define the unit of measure for angular velocity. It is the radian/second, abbreviated rad/sec.

Suppose an object in circular motion completes a turn. How much of an angle has it covered? A complete turn is 360 degrees. In radian measure, this is 2π radians. If T is the time period—the time taken by the object to complete one turn—then from its definition, the angular velocity, denoted by the lowercase Greek letter ω (pronounced "omega"), is given as

$$\omega = \frac{2\pi}{T} \tag{5.5}$$

The angular velocity and the period are inversely proportional to each other, which you may have guessed intuitively.

We can use this formula to find the angular velocity of the rotating earth. The period of rotation is 1 day, which is 86,400 sec. Thus

$$\omega = \frac{2\pi}{T} = \frac{2 \times 3.14}{86,400} = \frac{6.28}{8.64 \times 10^4} = \frac{6.28}{8.64} \times 10^{-4}$$

$$= 0.727 \times 10^{-4} = 7.27 \times 10^{-5} \text{ rad/sec}$$

In a similar vein, we can define an **angular acceleration** as the rate of change of angular velocity. Both angular velocity and angular acceleration are vectors, just as are their linear counterparts.

In terms of the angular velocity, we now have a slightly different way of

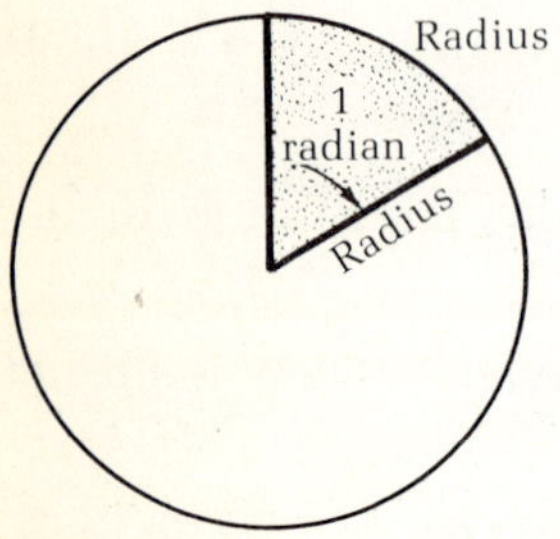

FIGURE 5.16   A radian is the arc of a circle equal to its radius.

TABLE 5.1 The parallels between translational and rotational motion.

Aspects	Translation	Rotation
Inertial property	Mass M	Moment of inertia I
Aspect of motion	Velocity of the center of mass V_c	Angular velocity ω
	Momentum $P = MV_c$	Angular momentum $L = I\omega$
Agent of change	Force	Torque
Law of motion	Force = rate of change of momentum	Torque = rate of change of angular momentum
Condition for a conservation law	Absence of external force	Absence of external torque

looking at the angular momentum of a rotating object, working in complete analogy with the case of translational motion. The translational analogue of angular momentum is momentum, which is the product of mass and velocity. The rotational analogue of mass is moment of inertia, and that of velocity is angular velocity. Thus angular momentum must be given as the product of moment of inertia and angular velocity. Indeed, it works out this way. Let us denote the moment of inertia by I and angular velocity by ω. Then the angular momentum L of a rotating object is given by the expression

$$L = I\omega \tag{5.6}$$

Let us use this definition of angular momentum to obtain a further understanding of the case of the spinning figure skater. We now can say that initially her moment of inertia is large, her arms being spread (remember that the moment of inertia is increased if there is more mass distributed away from the axis of rotation). She starts with a small angular velocity in this position, giving herself a certain amount of angular momentum. When she draws her arms in, she reduces her moment of inertia. Therefore, angular momentum conservation demands that her angular velocity increase, and she spins faster.

We will not go into further details of rotational motion in this text. Rotational motion is complicated, no denying that. Our effort here has been to assure you that it is possible to understand the basic aspects of rotational motion with a few concepts that happen to be analogous to the fundamental concepts of translational motion. Table 5.1 should serve as a quick reference for these parallels.

■ 5.2 Fun with Falling and Spinning

> And we, who have always thought of joy
> As rising, would feel the emotion
> That almost amazes us
> When a happy thing falls.*

---

*Rainer Maria Rilke, *The Sonnets of Orpheus*, trans. A. Poulin (Boston: Houghton Mifflin, 1977). Copyright © 1975, 1976, 1977 by A. Poulin, Jr. Reprinted by permission of Houghton Mifflin Company, J. B. Leishman and Hogarth Press.

Is falling fun? If you put spin into it, it can be—divers will vouch for that. Falling cats know about spin. Perhaps it is not a fair thing to say that a cat dropped upside down from a height is "a happy thing," but we do know that a cat never ends up unhappy—it always lands on its feet. We are going to see how. And we will start with an improvement on an idea of Christopher Columbus that has to do with standing an egg on its end (yes, it can be done).

Columbus's egg

Can you stand an egg on its end? Perhaps you have heard about Columbus's solution of this problem. If you haven't, here is the story.

After his discovery of America, Columbus went back to Spain, where he was greeted as a hero. But, as usual, not everyone felt he deserved all that credit. "Anybody could have done it," some of them started saying. So one day Columbus assembled his antagonists together and challenged them with this problem: stand an egg on its end!

"How can we do that?" they protested. Some tried anyway, but in vain. "I can do it," said Columbus. He broke the shell of the egg (he chose a boiled egg, of course), peeled off the broken end, and smoothly stood the egg on its flattened end. "But anybody can do that!" was the anguished outcry. Columbus calmly responded, "That's what you have been saying about my discovery of America, too."

Columbus's antagonists got the message, of course. However, if there had been a good logician among them, he could have pointed out that the cracked egg is not the same as the unbroken egg. Columbus's challenge was to stand an egg on its end, as is, and that problem still hadn't been solved.

If you have played with tops, you probably can think of a more acceptable solution to standing an egg on end than breaking the egg. You can use a property of a top, rotational inertia. Set the egg spinning on its end (Fig. 5.17). The rotational inertia of the spin will keep the egg standing on its broad end (or even its narrow end), at least for a time. Do you think anybody can do that? Of course they can.

Incidentally, in a public demonstration of this method of standing an egg on its end, always be sure to use a boiled egg, as did Columbus, but for a different reason. A raw egg, with all that liquid inside, does not respond well to your spinning effort. Interestingly, if you do manage to get a raw egg spinning, you will notice that it is harder to stop than a boiled one, because the liquid inside goes on moving even after you have stopped the outer shell. In fact, this gives you a handy way to tell if an egg is raw.

Diving: the art of piking and tucking

Diving from a springboard into a swimming pool is a spectacular exhibition of skill, yet its basics are easy to comprehend and learn. We will discuss only the very basic aspects of dives. You jump, giving yourself a rotational motion and you want a vertical entry into the water with your head down.

A couple of things you know already. The center of mass of the diver describes the usual parabolic path of a projectile. And after you leave the board,

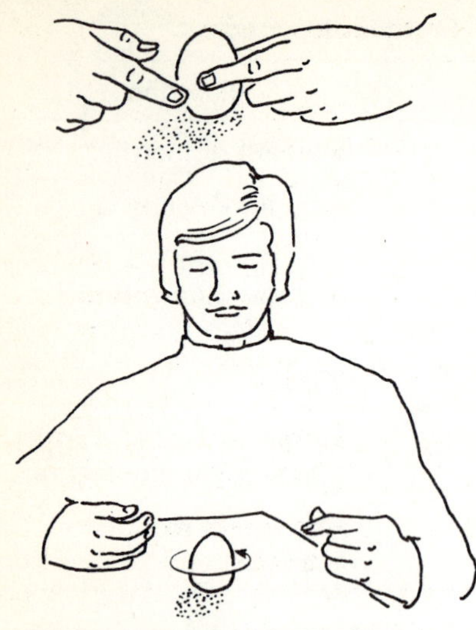

FIGURE 5.17 See text for how to tell a boiled egg from a raw one.

there is no further change in the angular momentum; it is conserved during the fall.

Basically the problem of diving is to find the right entry angle. It is very difficult to give yourself just the right rotation in the beginning so that you end up with just the right vertical entry. The problem is solved by the "piking" technique.

In piking, the diver draws his limbs in toward his body (Fig. 5.18), a process that causes his moment of inertia to be greatly decreased (by something like a factor of 3). Because angular momentum, the product of moment of inertia and angular speed, remains constant, his angular speed must increase to compensate for the decrease in moment of inertia. The diver must then unpike to adjust to the position of correct entry. Thus piking on the whole gives him more control on his angle of entry. Compared to this, an unpiked dive (Fig. 5.19) is almost always liable to end up with the wrong entry into the water.

How does a diver manage those somersaults we often see during a dive? This is done by "tucking," shaping the body almost into a bundle (Fig. 5.20). The moment of inertia is now reduced even more than in piking and the angular velocity is proportionately increased. With the increased angular velocity, the

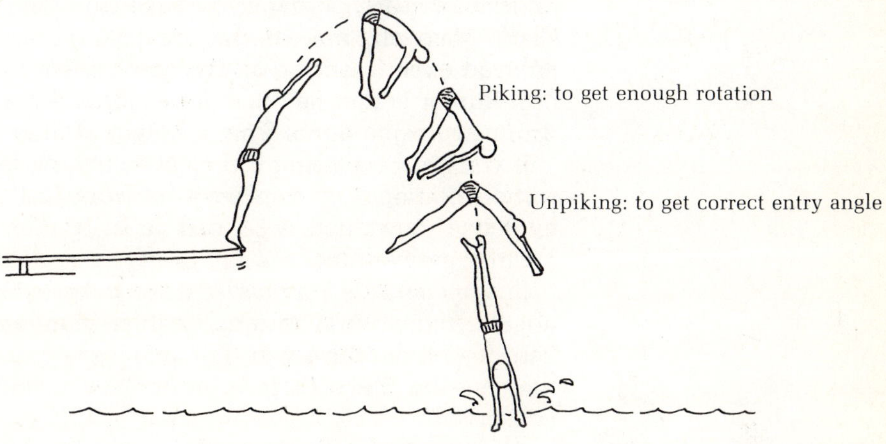

FIGURE 5.18   A piked dive.

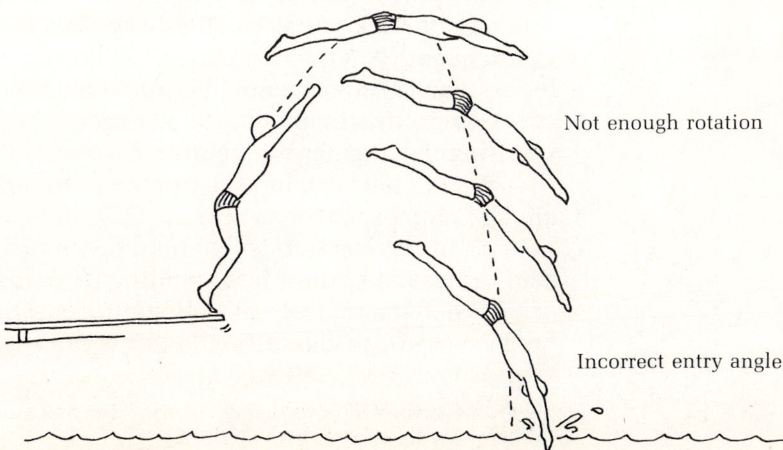

FIGURE 5.19   An unpiked dive. It is difficult to get the correct entry angle.

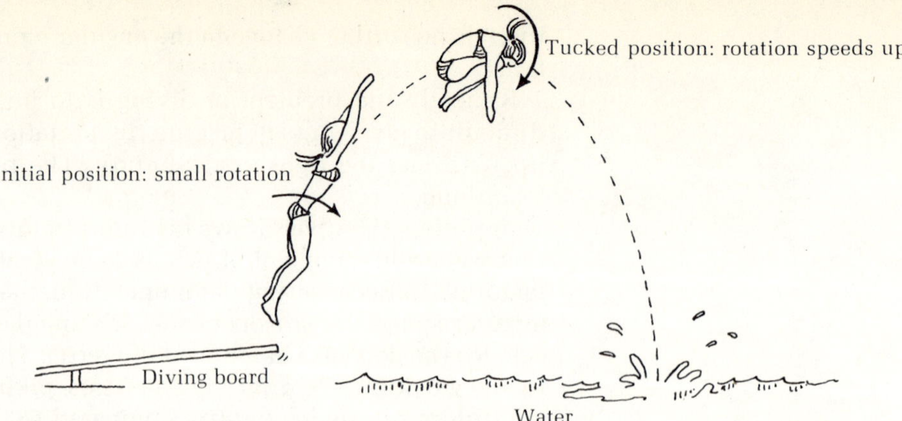

FIGURE 5.20 Tucking is the trick for performing airborne somersaults.

diver has time to complete a few somersaults in the air before entering the water.*

The case of the falling feline

When a cat falls, it always lands on its feet. If you were fond of cats during your childhood years, perhaps you have wondered about this. There are stories that Clark Maxwell, one of the greatest theoretical physicists who ever lived, enjoyed experimenting on the mechanism of a cat's ability to straighten itself out. Rumor is that he once demonstrated that a cat can right itself even when dropped upside down from a height of only a couple of inches.

If you are wondering if the cat somehow manages to turn by giving itself an initial rotational motion with its front feet just before release, observational evidence shows not. A cat can do its trick even when released with no initial angular momentum.

Imagine an axis lengthwise through the center of mass of the cat. If the initial angular momentum is zero, angular momentum about this axis must remain zero at all times during the fall; this is the demand of the conservation of angular momentum. Since there is no torque in the problem, angular momentum is conserved.

The sequential pictures of Fig. 5.21 show you what the cat does to right itself. The key thing to notice in the pictures is that the cat rotates its front parts one way and its back parts the other way. The falling cat's angular momentum is a vector quantity. If the two parts of its body turn in opposite ways—thus acquiring equal and opposite angular momentum—the total angular momentum will still be zero. Instinctively, the cat seems to know how to get around without violating the angular momentum conservation.

Further details can be constructed with the help of Eq. (5.6), considering the angular momentum of each of the parts of the cat's body. The important thing to note in this connection is that both the front and the back parts of its body have some adjustable limbs: by extending its limbs the cat can increase the moment of inertia of the relevant part. By pulling its limbs in, the moment of inertia can be decreased. Incidentally, the tail is not essential in the mechanism. Indeed,

---

*If you are interested in further detail, see "The Mechanics of Swimming and Diving," R. L. Page, *Physics Teacher*, February 1976, p. 72.

cats without tails can also perform this feat. Remember, angular momentum conservation requires the following:

1. The front and back must turn in opposite ways at any particular time.
2. $I_f\omega_f = I_h\omega_h$, where the subscripts f and h denote front and hind, respectively.

Now look at the pictures again, and let's construct a frame-by-frame analysis of what's going on. The cat, after being dropped from an upside-down position initially, pulls in its front limbs, making I_f small, whereas the hind legs are extended, so I_h is large. Thus the angular velocities behave as follows: the front part turns faster than the back (the back, of course, goes the opposite way at this stage). After the front part has turned quite a way, the cat reverses its strategy. Now it draws in its hind legs, extends its front ones, and reverses the direction of turn of the back part. Of course, the front part will now reverse its direction (toward its original configuration), but more slowly than before because the front legs are extended, and I_f is large. After the back part has turned some, the cat once more speeds up the front part, again by drawing in the front limbs; and so on. A few repetitions seem to achieve the objective: the cat has turned through 180° in midair all by itself. Recently the Austrians have revolutionized the efficiency of ski turns by using the cat's technique.*

FIGURE 5.21 Sequential pictures of a cat upending itself in midair.

------

*See J. I. Shonle and D. L. Nordick, "The Physics of Ski Turns," *Physics Teacher*, December 1972, p. 491.

SUMMARY

The subject of rotational motion is introduced in Section 5.1. The important concepts—rotational inertia, angular momentum, torque, angular velocity, and so on, which arise in almost any discussion of rotational motion—are most understandable in analogy to their translational counterparts—inertia, momentum, force, and velocity, respectively. In translational motion momentum is conserved when the external force is zero. Similarly, in rotational motion angular momentum is conserved when the net external torque (which is the turning effect of a force, or force times moment arm) is zero.

In contrast to mass, the moment of inertia of an object depends on the distribution of mass around the center or axis of rotation of the object. The angular momentum L of the object is given as the product of the moment of inertia I and the angular velocity of rotation

$$L = I\omega$$

QUESTIONS

Review and reason

1. Fig. 5.22 shows a girl rotating on a supposedly frictionless turntable with outstretched hands. Suppose she pulls in her arms. What will happen to the speed of rotation? What will happen to her angular momentum?

FIGURE 5.22

2. A car is moving in a circular ramp in a counterclockwise direction as seen from above. What is the direction of its angular momentum?

3. When seen from above, the main rotor of a helicopter rotates counterclockwise. What is the direction of its angular momentum? A helicopter also has a small tail rotor (Fig. 5.23) mounted on a horizontal axis, which rotates counterclockwise when seen from the right of the pilot. What is the direction of the angular momentum vector of this rotor? Assuming that the two angular momenta have the same magnitude, draw the parallelogram of vectors to find the direction of the total angular momentum. Explain the purpose of the tail rotor.

4. What is rotational inertia? Give a few examples.

5. Tightrope walkers often use a long stick. Why?

6. When you walk on a narrow curb, with the slightest feeling of imbalance you probably will extend your arms instinctively. Do your arms know something you don't? Or do you know the cause of this instinctive behavior? Explain.

7. Which will roll down an incline fastest, a sphere, a ring, or a disk?

8. When you push open a heavy door, what gives you a better torque, pushing close to the hinges or as far away from them as possible? Where would you put a wedge to keep a door open, close to the hinges or away from them? Explain.

*9. The scales in a doctor's office usually balance a large weight with a small one. Can you explain their operation? (Think in terms of torques. For example, consider a teeter-totter with asymmetric weights on the two sides.)

FIGURE 5.23

---

*Optional.

10. Can we have a torque without a net force? Can we have a force without a torque? Give examples.

11. A river starts in the north and joins the ocean near the equator. On its way it picks up a huge amount of sediment, which it deposits in the ocean. How does this affect the moment of inertia of the rotating earth? How does it affect the angular momentum? How does it affect the length of the day?

12. A few billion years from now, our sun may expand into a red giant (red giants are stars of much greater size than normal stars). What will then happen to the speed of rotation of the sun? Will the angular momentum of the sun undergo any change?

13. Why do trapeze artists roll up in a ball as they dismount with aerial somersaults?

14. There are legends that say a person can escape injury from the impact of a fall to the ground from as high as a five-story building by means of executing somersaults in the air. Do you believe this? Why or why not?

Arithmetic

1. The unit of torque is the meter newton (m·N). Explain. If a force of 5 newtons is exerted at the end of a $\frac{1}{5}$-meter wrench perpendicular to its length (see Fig. 5.8), what is the magnitude of the torque on the bolt?

2. Here is a teeter-totter problem. Suppose you (with a weight of 150 lb) take your kid brother (with a weight of 75 lb) to play on the teeter-totter with you. He sits at the very end of the 10-ft teeter-totter. Where will you sit to ensure balance?

*3. This is the same problem as exercise 2, with one addition. Your brother has brought his pet cat along (with a weight of 5 lb), which chooses to sit on the teeter-totter on the same side as your brother but at a distance of 2 ft from the center. Where will you sit this time and still maintain balance?

4. As you know, the earth revolves around the sun in 1 year (3.2×10^7 sec). Compute its angular velocity of revolution.

---

*Optional.

6 Newton and Gravitation

"What is gravity?" my 10-year-old daughter asked suddenly. If she were an Australian aborigine, I thought, her language would make it clear exactly what she wanted to know: the characteristics or the underlying reasons? The mechanics or the mechanism?

"The sun attracts the earth and the moon and is attracted by them. In fact, all masses attract each other in this way. This is gravity." I made a feeble attempt. How does one explain such a difficult subject to a child? My daughter was not satisfied.

"Gravity is Newton's idea. My teacher said that Newton discovered gravity. You didn't mention Newton."

I should have, I agreed. Gravity is Newton's way of putting together the harmony of the universe. The creator of an idea is as important as the idea because the idea isn't nature; it is the basis of our human models to describe nature.

■ 6.1 The Law of Universal Gravitation

Newton communicated his idea of universal gravity in the form of a mathematical law, which is usually referred to as the law of universal gravitation. It is a very important requirement of physics to be able to write down your perception of nature in the form of a mathematical equation. This is not because physics is mathematical, but because nature is precise and expressible in this simple and elegant way. Mathematics makes your ideas available for scrutiny by anybody else who wants to examine them in an objective manner. Also, detailed experimental tests of a law are possible only when it is given an exact mathematical form.

The purpose of this section is to give you some idea of how Newton arrived at his celebrated law. The process of a discovery is as interesting as the discovery itself. So this section traces some of the different bits and pieces of information that were available in Newton's day, and what one *could* derive from them.

The first clues must have come from Galileo's experimental results. What can we derive about the gravity force from these results? Planets move around the sun in nearly circular orbits. If one object moves around another in a circle, what can we say about the force between the two? Assuming it is the gravity force that keeps the planets moving around the sun, what can we learn about the force from the laws that we know hold for such motion, namely, Kepler's well-defined laws? These are the questions to which we will address ourselves.

The clue from Galileo

Remember Galileo's law? The acceleration toward the earth is the same for all falling objects on earth. What does this tell us about the force that causes the acceleration? Let's find out.

From Newton's second law of motion, every force is related to the acceleration it produces by the following simple rule:

$$\text{force} = \text{mass} \times \text{acceleration}$$

Thus the force of gravity must be equal to the product of the mass m of the object and the acceleration g of fall:

$$\text{force of earth's gravity} = m \times g$$

This enables us to write the following expression for g:

$$g = \frac{\text{force of earth's gravity}}{m} \tag{6.1}$$

How can g be independent of m? Equation (6.1) suggests a way that this can happen. If the force of gravity itself contains m as a factor, then m cancels out from both the numerator and the denominator, being a common factor. And then the expression for g will no longer contain m; in other words, g is independent of m.

Thus we see that the force of gravity must be proportional to m, the mass of the falling object. But a force cannot exist all by itself—remember the third law of Newton? It takes two to make a force. What's the other body involved? The earth, of course. But then the reaction force on the earth must be proportional to the mass of the earth, M_e, because that's the way it is with the gravity force. The gravity force is proportional to the mass of the object on which it acts. Furthermore, the third law tells us that the action and reaction forces are equal in magnitude. But how can this happen?

We can argue the following way. The force on m is proportional to m; this is the action force. The force on the earth is proportional to the mass of the earth; this is the reaction force. The equality of action-reaction pairs can be preserved if we assert that both forces are proportional to the product of both of the masses m and M_e.

Now we can generalize. What is true for the earth and an object falling to the earth must be true for any two gravitating objects (that's what universality of gravity is all about). Thus we conclude that the force of mutual gravity between any two objects is proportional to the product of their masses, m_1 and m_2:

$$\text{force of gravity between } m_1 \text{ and } m_2 \propto m_1 \times m_2 \tag{6.2}$$

FIGURE 6.1   Newton. (Crown copyright. Victoria and Albert Museum.)

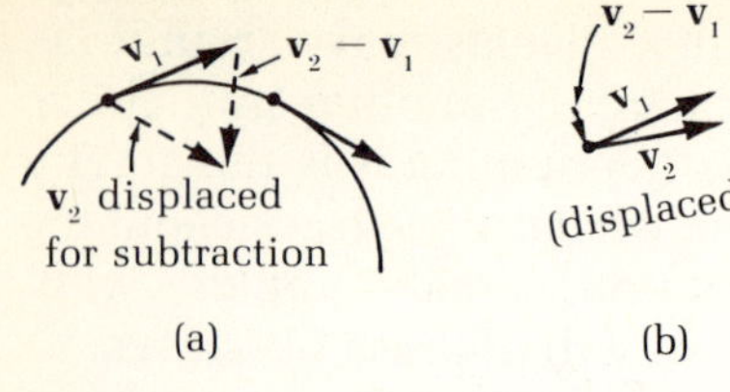

FIGURE 6.2   (a) Subtracting the vector $\mathbf{v}_1$ from $\mathbf{v}_2$. (b) As $\mathbf{v}_2$ becomes closer to being parallel to $\mathbf{v}_1$, the difference $\mathbf{v}_2 - \mathbf{v}_1$ is close to being perpendicular to both of them.

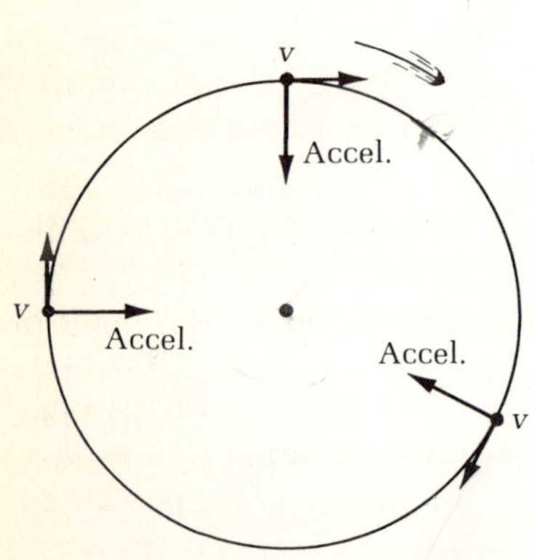

FIGURE 6.3   For circular motion of an object, the magnitude of the acceleration remains constant, the direction of the acceleration vector keeps changing, but the vector is always pointed toward the center of the circle. It is called centripetal acceleration.

Uniform circular motion

It is a matter of common experience that for motion in a circle, the direction of the instantaneous velocity at any position on the circle is along the tangent to the circle at that point. In Fig. 6.2, $\mathbf{v}_1$ denotes the velocity vector at a certain time t_1, and $\mathbf{v}_2$ denotes the velocity vector at a slightly later time t_2. Since the motion is uniform, the magnitude of the velocity (call it v) remains constant. Thus the two arrows are drawn with equal length (recall that the length of the arrow denotes the magnitude of the vector it represents). The change in velocity is given by the difference of the two vectors $\mathbf{v}_2$ and $\mathbf{v}_1$, namely, $\mathbf{v}_2 - \mathbf{v}_1$. This change takes place during the time interval $t_2 - t_1$. Thus we can find the average acceleration during this period as

$$\text{acceleration} = \frac{\mathbf{v}_2 - \mathbf{v}_1}{t_2 - t_1}$$

To proceed any further we have to know the rules for taking the difference of two vectors. Mathematics gives us the following rule:

> Draw the vectors $\mathbf{v}_2$ and $\mathbf{v}_1$ foot to foot. Then the difference vector $\mathbf{v}_2 - \mathbf{v}_1$ is represented by the arrow that can be drawn from the head of $\mathbf{v}_1$ to the head of $\mathbf{v}_2$.

In Fig. 6.2 we carry this procedure out for the two vectors $\mathbf{v}_1$ and $\mathbf{v}_2$ as shown, redrawing the vector $\mathbf{v}_2$ so that it is parallel to itself in order to bring it foot to foot with $\mathbf{v}_1$. The average acceleration is along the dotted line as shown. To obtain the instantaneous acceleration, we have to visualize what happens when the time interval $(t_2 - t_1)$ becomes very small, say, infinitesimal—that is, when t_2 is very close to t_1. As shown in Fig. 6.2(b), the two vectors $\mathbf{v}_2$ and $\mathbf{v}_1$ become almost parallel to each other when the time t_2 is close to the time t_1. The angle between the two vectors is much smaller now, and look at the difference vector. It appears to be almost at right angles to both $\mathbf{v}_2$ and $\mathbf{v}_1$. Stretch your imagination a little bit further. Then you will be able to see that in the limit of t_2 being arbitrarily close to t_1, the difference vector $(\mathbf{v}_2 - \mathbf{v}_1)$ becomes perpendicular to the tangent. The direction of $(\mathbf{v}_2 - \mathbf{v}_1)$ in this limit is therefore along the direction of the position vector for an object on a circle. Thus the acceleration vector for circular motion is always directed along the position vector toward the center (Fig. 6.3). For this reason it is called the **centripetal acceleration,** centripetal meaning center-seeking.

It is slightly more complicated to derive an expression for the magnitude of the centripetal acceleration. A little mathematics tells us that the magnitude is given as

$$\text{centripetal acceleration} = \frac{v^2}{R} \tag{6.3}$$

where v is the magnitude of the velocity and R is the radius of the circle.

Thus circular motion is accelerated motion, even though the speed is constant. The acceleration is directed toward the center of the circle. From Newton's laws of motion, we know that a force must provide this acceleration. Such a force must also be directed toward the center, and hence it is called a **centripetal force.**

You can easily verify that an object needs a centripetal force in order to go around in a circle. Try this simple experiment: whirl a stone tied to the end of a string in a circular path. You will find that you have to keep exerting a force on the string in order for the stone to continue moving in a circle (Fig. 6.4). From this experiment you can also verify the significance of Eq. (6.3) for the centripetal acceleration. The faster you whirl the stone—that is, the greater you make v—the greater is the centripetal acceleration. In fact, centripetal acceleration increases in proportion to the square of the speed. That is, if v increases by a factor of 3, the centripetal acceleration increases by a factor of 9. And you must exert a proportionately greater force on the string to provide the stone with the centripetal acceleration it needs to stay in circular motion. Eq. (6.3) also tells you that a greater force is needed to move the stone in a circle of smaller radius than in one of a larger radius.

When you take a curve in your car—for example, on a sharply curved highway entrance or exit ramp—have you noticed that the road is banked? This also has to do with centripetal force. You need a centripetal force to keep your car going in a circle. If the car is on a horizontal road, then the only source of centripetal force is the sideways friction on the tires [Fig. 6.5(a)]. Tires can handle a small amount of curve, but not a sharp one like the highway entrance ramp. The banking of the road makes part of the force of the road on the car available to provide additional centripetal force. If you are wondering how this is so, take a look at Fig. 6.5(b). The weight force on the car still acts vertically downward, but the force exerted on it by the road is in a direction perpendicular to the road. Part of the force of the road, its vertical component, balances the weight force. But the horizontal component of the force is in the radial direction toward the center and acts as the centripetal force.

The planets revolve around the sun in nearly circular orbits. Their motion is thus accelerated, with the acceleration directed along the radius toward the sun. Newton recognized this to be the attractive gravity force exerted by the sun on a planet. Since all forces involve interactions, a planet also attracts the sun toward it. Generalizing to the case of any two objects, we can say the following:

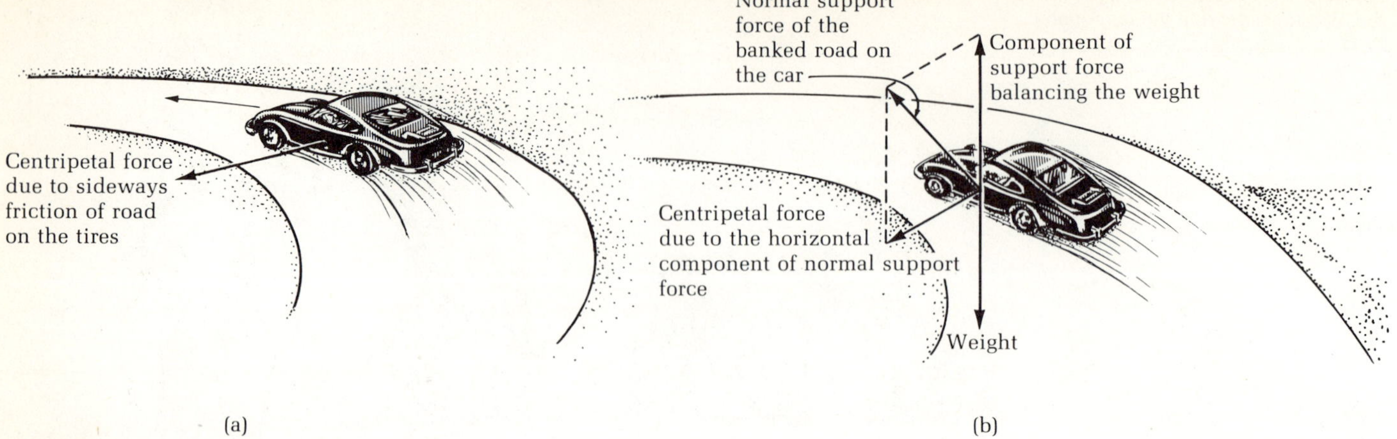

FIGURE 6.5   (a) Car taking a curve on a level road. Sideways friction provides the centripetal force. (b) Car taking a curve on a banked road. The horizontal component of the normal support force of the bank provides the centripetal force.

When two objects attract each other gravitationally, the force exerted on each is directed toward the other along the line joining them. The next question is: How does the magnitude of the force vary with the distance between the two objects?

The hint from Kepler's laws

How does the force of attraction of the sun on the planets depend upon the distance? It is reasonable to assume that the force should decrease as the distance from the sun increases. (All familiar interactions behave this way: the further away you go, the less is the attraction.) Now we have to express this mathematically.

We could start with the assumption that the force varies inversely as the distance R:

$$\text{force} \propto \frac{1}{R}$$

This means that if the distance increases by a factor of 2, the force will decrease by the same factor.

Or we could start instead with the assumption of an inverse square law:

$$\text{force} \propto \frac{1}{R^2}$$

In this case, if the distance increases by a factor of 2, the force decreases by a factor of the square of 2, or 4. Or we could assume other kinds of power dependence—inverse cube, inverse fourth power, and so on.

But which of these is the right assumption? The one that agrees with the observational data, the laws of Kepler. Newton found that only the inverse square dependence of the force on distance explains the first and the third laws of Kepler, although Kepler's second law could be explained by any of these power law dependencies. This was a crucial result, requiring some high-powered mathematics from Newton. Actually, some of Newton's contemporaries, notably Robert Hooke, had had some inkling of the answer but not the mathematics to prove it.

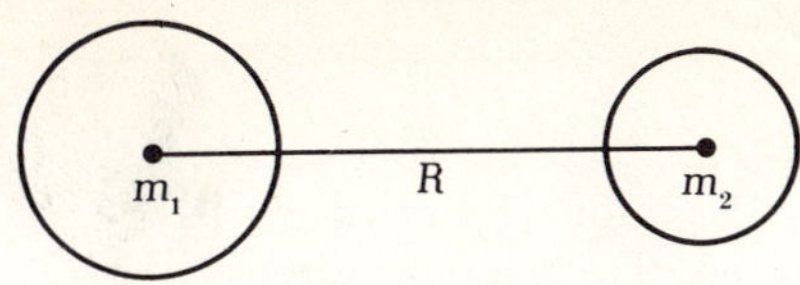

FIGURE 6.6 For extended objects the distance is measured from the center of mass of one object to the center of mass of the other.

Here is another consideration. In the theory of gravitation, very frequently we deal with huge objects like the planets and the stars. How do we measure the distance between such large objects since it can be done in more than one way? The distance is to be measured from the center of mass of one object to the center of mass of the other. For spherical objects the center of mass is at the center of the sphere. Thus for two spheres, we measure the distance from the center of one to the center of the other (Fig. 6.6).

The statement of the law

Now we have all the pieces of the puzzle. The analysis of circular motion has told us that the gravity force is directed along the line joining the two interacting objects. Each of the two objects is acted on by an attractive force directed toward the other. For the magnitude of the force we find that

1. The force is proportional to the product of the masses of the two interacting objects. Calling the force F, and the masses m_1 and m_2, we have

$$F \propto m_1 m_2$$

2. The force is inversely proportional to the distance between the two objects. Calling this distance R, we have

$$F \propto \frac{1}{R^2}$$

If F is proportional to both $m_1 m_2$ and $1/R^2$, it must be proportional to the product $m_1 m_2/R^2$, or

$$F \propto \frac{m_1 m_2}{R^2}$$

Introducing a constant of proportionality G, the force constant of gravity, we can write an equation for F:

$$F = G \left(\frac{m_1 m_2}{R^2} \right) \tag{6.4}$$

Eq. (6.4) is the mathematical expression for Newton's law of universal gravitation. In words, we can state the law as follows:

Every object attracts every other object gravitationally along the line joining the two objects. The magnitude of the force of gravitation between two objects is directly proportional to the product of their masses and inversely proportional to the square of the distance between them.

This concludes the logical analysis that could have led Newton to his gravity law. Well, did it? We don't know. On the other hand, there is a legend that Newton discovered the gravity law in a sudden flash of inspiration when he saw an apple fall to the ground. Can there be any truth to that? See the box for an analysis of the apple incident in which this other aspect (inspiration) of a creative discovery is discussed. In any case, it is not an either-or proposition. A great discovery like Newton's needs both ingredients: perspiration (the cold logical analysis) and inspiration.

You must have heard the apple story. It was 1665, and a plague had broken out in Cambridge, England. The University of Cambridge had to close, and Newton, who was teaching at the university, had to move to his mother's farm in Lincolnshire. There one day in the garden, Newton watched an apple fall to earth. This triggered in his consciousness the idea of universal gravity: Every object attracts every other with the force of gravity.

What does the apple story mean to you? Apples fall every day; hardly anyone notices. This was just as true in Newton's time as it is in ours. Yet the apple story tells us that Newton was enlightened by this trivial event. How can this happen?

In psychology there is a word, a very revealing word: gestalt. Basically it means the whole, the perception of an entire pattern instead of its scattered fragments separately. Suddenly the pattern clicks in the mind of the beholder. That's when truths are discovered. Gestalt isn't logical; it is not putting the fragments together in a logical way. In fact, gestalt is the antithesis of logic; it is totally experiential.

The perception of the gestalt is in the realization of the harmony of a musical pattern of notes. It is the sudden burst of pleasure you may get by looking at some of M. C. Escher's drawings when you recognize the "wholeness" of the pattern.

Let me quote Rabindranath Tagore, one of the most gifted poets to come out of the East in the last century. He became a poet after the following experience at a very young age. In his own words:

> I still remember the day in my childhood when I was made to struggle across my lessons in a first primer, strewn with isolated words smothered under the burden of spelling. The morning hour appeared to me like a once-illumined page, grown dusty and faded, discoloured into irrelevant marks, smudges and gaps, wearisome in its moth-eaten meaninglessness. Suddenly, I came to a rhymed sentence of combined words, which may be translated thus—"It rains, the leaves tremble." At once I came to a world wherein I recovered my full meaning. My mind touched the creative realm of expression, and at that moment I was no longer a mere student with his mind muffled by spelling lessons, enclosed by classroom. The rhythmic picture of the tremulous leaves beaten by the rain opened before my mind the world which does not merely carry information, but a harmony with my being. The unmeaning fragments lost their individual isolation and my mind revelled in the unity of a vision. In a similar manner, on that morning in the village, the facts of my life suddenly appeared to me in a luminous unity of truth. All things that had seemed like vagrant waves were revealed to my mind in relation to a boundless sea. I felt sure that some Being who comprehended me and my world was seeking his best expression in all my experiences, uniting them into an ever-widening individuality which is a spiritual work of art.*

So Tagore's revelation of the gestalt of the universe was triggered by a simple rhyme about the simple image of a leaf trembling in the rain. Newton's realization came when he saw an apple rushing toward earth. Newton didn't record his feelings after the apple incident, but my guess is that his account would not be different in essence from that of Tagore.

The understanding of the whole from pieces is a spiritual experience. Psychologists may call it an "aha" experience; the Zen Buddhists may call it "enlightenment"; scientists may refer to it as "intuition." By whatever name you call it, it clearly touches the same place in the consciousness of the observer. In the realm of creativity, the physicist is not different from the poet or the Zen monk. It is the same shadow of the sky that falls on the rivers and the lakes. Scientists express their enlightenment in the form of laws that seek to describe nature. The same enlightenment leads a Zen master to write a haiku or T. S. Eliot to write *Four Quartets*. The difference lies only in the expression; the realization and the ecstasy are the same. Science, at its base, is as experiential as poetry or Zen.

Profound realization comes only to a few. For the contemporaries of Newton, the fragmental evidence was all there. The constancy of acceleration of falling objects on earth, Kepler's planetary orbits, the relationship of the planetary orbital period to the distance from the sun, all these different bits and pieces were there for somebody to gather together into a coherent new force. Even the ideas that the earth attracts apples and that the sun attracts the planets were already around. Only the gestalt was missing. It was for Newton to experience it and tell the world about it.

The chronology of the publication of Newton's discovery has baffled many experts on the history of science. Everybody agrees that Newton figured out most of the mathematics and logic of the law of gravity within a short

---

*Rabindranath Tagore, *The Religion of Man* (New York: Macmillan Publishing Company, 1931), p. 93. Copyright 1931 by Macmillan Publishing Co., Inc. Courtesy Macmillan Publishing Company and George Allen & Unwin, Ltd. London.

time after the apple incident. But he didn't publish anything until almost twenty years later. This time lag has had different explanations from different people. I have my own point of view on this, which I would like to share in the form of the following story:

> A young monk approached his Zen master and asked him what it was he had done before gaining enlightenment. The Zen master's reply? "Before I was enlightened, I chopped wood and fetched water. Now that I am enlightened, I chop wood and fetch water."

Newton started his scientific career with the study of optics, the branch of physics dealing with light. Shortly after the discovery of gravity, he went back to studying optics.

■ 6.2   The Tests and Triumphs of Newton's Theory of Gravitation

Newton himself tested his gravity law by applying it to the motion of the moon around the earth. He also found some beautiful applications of the gravity law to explain some age-old mysteries of nature; one example is his explanation of the tides. Other scientists, particularly the French physicists and astronomers, started doing calculations for astronomical objects of the solar system, including the effect of gravity not only of the sun but also of other planets. The triumph of this kind of detailed work was demonstrated spectacularly when two new planets (Neptune and Pluto) were discovered based on the resulting predictions. The attraction of gravity also seems to control the motions of the stars themselves, as revealed in the revolution of the stars around the center of a galaxy and, even more dramatically, in the motions of double-star systems. At the level of the galaxies, we find evidence of the working of the gravity force in the shaping of these objects and also in their clustering tendency. Finally, the entire universe in its expansion seems to be slowing down by the braking action of the gravity force.

These are all examples of large-scale phenomena. This is not surprising, since the gravity force is large only on a large scale. Nevertheless, it is important to note that it has been directly verified that even relatively small masses attract each other gravitationally, following Newton's law. In fact, measurement of the gravitational force constant G was a result of this kind of experiment. Another beautiful application of Newton's gravity law on a relatively small scale is in the area of the artificial satellites.

The apple and the moon

The fall of the apple triggered in Newton's mind the idea that the force of earth's gravity must reach as far as the moon (Fig. 6.7), which falls under the same force, except that it is reduced because of the larger distance. Now for the apple in England, the distance was 4,000 mi from the center of the earth; for the moon the distance is 240,000 mi, or 60 times further than the apple. According to the gravity law, the force of gravity at the moon must decrease by a factor 60^2, or 3,600. The acceleration of fall of the moon must also be slower by the same

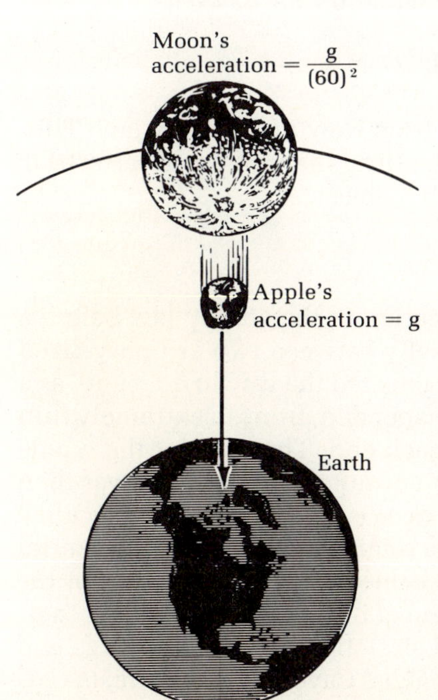

FIGURE 6.7   Newton's gestalt: the apple and the moon.

factor when compared to g. Thus Newton calculated that if the gravity law was correct, then the acceleration for the moon was

$$a = \frac{g}{3,600} = \frac{9.8 \text{ m/sec}^2}{3,600} = 0.0027 \text{ m/sec}^2$$

Fortunately, the acceleration of the moon in its nearly circular orbit was already known to Newton from another consideration. From (Eq. 6.3) for the centripetal acceleration,

$$\text{acceleration of the moon} = \frac{v^2}{R}$$

where v is the velocity of the moon's revolution and R is the radius of its orbit. From measurements, $R = 3.8 \times 10^8$ m. And v can be calculated as follows: The moon travels the circumference of its orbit (which is $\pi \times$ diameter $= \pi \times 2R = 2\pi R$) in one time period T of 27.32 days. Thus its orbital velocity v is given as

$$v = \frac{\text{distance traveled}}{\text{time of travel}} = \frac{2\pi R}{T} \tag{6.5}$$

Substituting $R = 3.8 \times 10^8$ m and $T = 27.32$ days $= 27.32 \times 8.64 \times 10^4$ sec $= 236 \times 10^4 = 2.36 \times 10^6$ sec, we get

$$v = \frac{2\pi R}{T} = \frac{2 \times 3.14 \times 3.8 \times 10^8}{2.36 \times 10^6} = 1011 \text{ m/sec}$$

Substituting these numbers for v and R in the equation for a, we get

$$a = \frac{v^2}{R} = \frac{(1011)^2}{3.8 \times 10^8} = 0.0027 \text{ m/sec}^2$$

This agrees with the preceding value obtained from Newton's gravity law. This was the very first time that a law for understanding the motion of terrestrial objects was found to work also for a "heavenly" object.

The determination of G

It was almost a hundred years after the publication of the gravity law before a method was found to measure the force of gravity between two ordinary-sized objects. English physicist Henry Cavendish discovered the way to do it and, as a result, came up with a value of G. Cavendish suspended, from an extremely thin thread, a light bar carrying a small sphere at each end. Then he put the whole thing inside a glass box as a protection from air currents. The glass box was then placed between two massive spheres, which were suspended in a balancelike fashion (Fig. 6.8). These large spheres could be rotated at will about the central axis. After the bar carrying the small spheres came to rest, the position of the bigger spheres was changed, causing a deflection of the bar inside the glass cage, which could only come from the gravitational attraction between the large and the small spheres. The amount of deflection enabled Cavendish to estimate G. G is found to be 6.67×10^{-11} MKS units.

Knowing the value of G, we now can verify how small the gravity force is

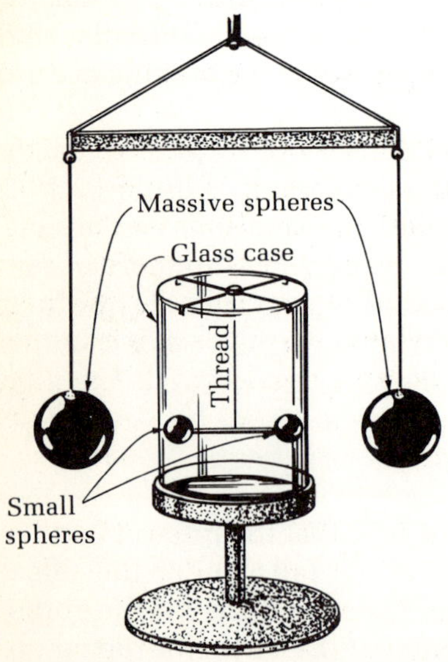

FIGURE 6.8 Cavendish's setup (see text for details).

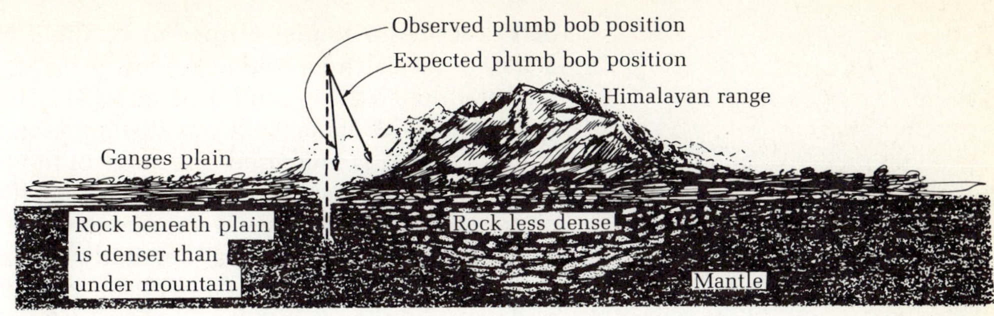

FIGURE 6.9 Mountains have their roots. Rock underneath the Himalayas is less dense than the rock beneath the plain on the left. Thus the attraction of the plain is greater compared to that of the Himalayan root on the plumb bob; this compensates for much of the attraction of the bob due to the mountain itself.

between two ordinary objects. Imagine two people 70 kg each standing 1 m apart. The value of the force of gravity between them is

$$\frac{Gm_1m_2}{R^2} = 6.67 \times 10^{-11} \times \frac{70 \times 70}{1^2}$$
$$= 6.67 \times 10^{-11} \times 7 \times 7 \times 10^2$$
$$= 328.3 \times 10^{-9} = 3.283 \times 10^{-7} \text{ N}$$

This is a very small force, quite imperceptible and ineffective for an object with as large an inertia as a human body. So if one of the two persons says to the other, "I am attracted to you," that person is not talking about gravity.

Everest experiment: application of the gravity law in geology

Gravitational measurements have played a very important role in geology. In one of the first major breakthroughs using this technique, British geologist George Everest (of Mount Everest fame) was engaged in the measurement of the gravitational attraction of a plumb bob toward the massive Mount Everest (Fig. 6.9). To his surprise he found that the attraction was much less than that predicted by Newton's theory. Since the validity of the gravity law was beyond question, geologists started looking for an explanation of the Everest experiment. Soon it was established that the attraction (on the plumb bob) was less than expected because the material beneath a mountain is less dense (and hence lighter) than underground material elsewhere. The density of matter beneath a mountain is more like that of the mountain itself. It is as if each mountain extends a root that goes considerably below the surface, forming a sort of "negative" mountain underneath. We will have more to say about this later.

Artificial satellites

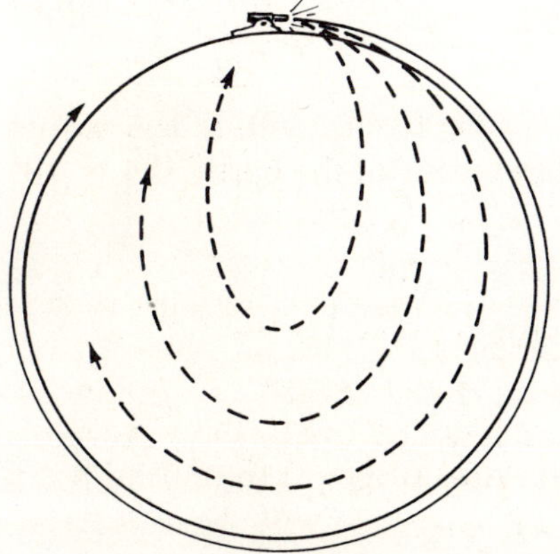

FIGURE 6.10 Newton's cannon on the mountain. When the velocity of the cannonball is 5 miles per second, it goes into a circular orbit around the earth.

Newton himself considered the case of a cannon fired from a high mountain in order to discuss when a cannonball becomes an artificial satellite (Fig. 6.10). Suppose the mountain is so high that we can neglect air drag. We have previously noted that the path of a projectile is a parabola. Actually the cannonball describes a Kepler ellipse in its flight forward and downward, with the center of the earth located at the lower focus. For the small flight path from the mountaintop to the ground, a parabola is indistinguishable from the near end of an ellipse, with its far focus as distant as the center of the earth. As the horizontal velocity is increased, the range of the projectile becomes larger. In terms of the Kepler ellipse, this means that the ellipse becomes rounder, closer and closer to being a circle. So if a cannonball is fired from the mountaintop with a high enough velocity, its path actually becomes a circle. The cannonball will fall downward at the same rate as the surface of the earth curves away, and it will never hit the ground. It will come back to its original point, though. So if we want the cannonball to continue indefinitely in its orbit, we had better move the cannon out of the way before the cannonball comes skimming over the mountain top. The cannonball now has become an artificial satellite.

Basically, this is the principle of satellite launching. The satellite is shot up by a rocket to an altitude of a few hundred miles and then turned to move horizontally, parallel to the earth's surface. Finally, it is given a forward thrust. If the forward thrust is sufficient, it will describe an orbit around the earth. For circular orbits the required forward velocity can be calculated easily. The force of earth's gravity on the object, $G M_e m/R^2$, must equal the mass of the object m times its centripetal acceleration v^2/R, where M_e is the mass of the earth, v is the desired velocity, and R is the radius of the circular orbit. We thus get the equation

$$\frac{mv^2}{R} = \frac{GM_e m}{R^2}$$

or, canceling common factors from both sides,

$$v^2 = \frac{GM_e}{R}$$

Taking the square root of both sides, we have

$$v = \sqrt{\frac{GM_e}{R}} \tag{6.6}$$

Numerically, $G = 6.67 \times 10^{-11}$ MKS units; $M_e = 6 \times 10^{24}$ kg; and R, the radius of the orbit, can be taken approximately to be the radius of the earth, 6.4×10^6 m. Substituting these numbers we get

$$v = \sqrt{\frac{GM_e}{R}} = \sqrt{\frac{6.67 \times 10^{-11} \times 6 \times 10^{24}}{6.4 \times 10^6}}$$

$$= \sqrt{6.4 \times 10^7} = \sqrt{64 \times 10^6} = \sqrt{64} \times \sqrt{10^6}$$

$$= 8 \times 10^3 \text{ m/sec} = 8 \text{ km/sec} = 5 \text{ mi/sec} = 18,000 \text{ mi/h}$$

A velocity higher than 8 km/sec will put the satellite in an elliptical path. If the velocity is as high as or higher than 11.2 km/sec, the satellite will escape earth's gravity altogether (more on this in Chapter 10). Today some of the most spectacular demonstrations of the validity of Newton's theory are provided by the many artificial satellites that orbit the earth, following his gravity law.

Discovery of new planets

English poet John Keats, in a poem called "on first looking into Chapman's Homer," compared his joy at getting a new insight into the realm of Homer's poetry with the feelings that an astronomer might have upon the discovery of a new planet. This is what Keats wrote:

> Oft of one wide expanse had I been told
> That deep brow'd Homer ruled as his demesne;
> Yet never did I breathe its pure serene
> Til I heard Chapman speak out loud and bold;
> Then felt I like some watcher of the skies
> When a new planet swims into his ken.

Keats was a romantic poet and had put himself on record that he did not like science much. But clearly even he perceived the discovery of a new planet as a worthwhile experience. Accordingly, the discovery of the planet Neptune, based on calculations using Newton's theory, must be regarded as one of the theory's most spectacular successes.

Before going into the story of Neptune, we should make one note. About sixty years before its discovery, the last previous solar planet to be discovered was found by an amateur astronomer named William Herschel, who had in his possession one of the better telescopes of the day. This planet was named Uranus. Uranus was actually visible to the naked eye, albeit barely, and previously had been always mistaken for a star. It is twice as far from the sun as Saturn.

The discovery of Uranus opened up the possibility of other solar planets. However, the sky is a very big arena for observation, and a faraway planet would be rather dim, making it difficult to observe. Fortunately, Newton's theory came to the aid of the observational astronomer just at the right time.

It happened like this. When the motion of Uranus was analyzed in terms of the theory, it was found that the planet did not behave as expected. This was around 1830. Newton's theory had achieved such respectability that it was unthinkable that it could be wrong. The logical conclusion reached by two young physicists was that the unusual behavior of Uranus must be due to the gravitational force of a hitherto unknown planet. These men calculated the exact location this new planet must have in order to make the motion of Uranus fit into Newton's theory. And a new planet, later named Neptune, was indeed discovered at the predicted place. The same reasoning led to the prediction and observation of still another planet, Pluto, as late as 1930.

A similar achievement of the Newtonian theory was the explanation of the mystery of comets. Throughout antiquity comets had always been regarded as the bearers of disaster. One reason for this perhaps was that the phenomenon was looked upon as a nonrepeating one. Now it was shown that comets orbit the sun with the same regularity as planets, although their orbital periods are very large. Through a study of the paths of the comets, it was discovered that they follow ellipses of rather large eccentricity. They are visible only when they come close to the sun, near their perihelion position. When they are farther away, these passing clouds of matter do not reflect enough sunlight to be visible (Fig. 6.11). Now that we know they are regular inhabitants of the solar system, returning to our skies with remarkable regularity, they have become objects of good omen. You may remember that some good things were predicted for the coming of the comet Kohoutek a few years ago (1973).

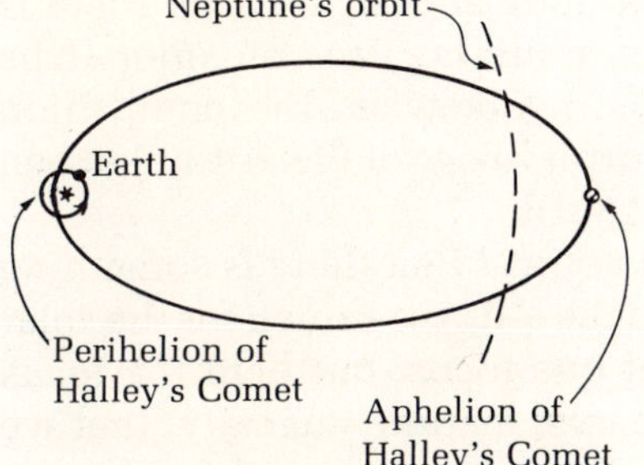

FIGURE 6.11 The elliptical orbit of Halley's comet.

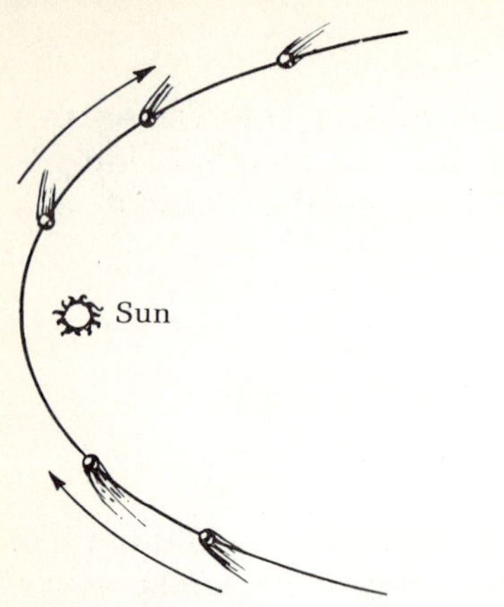

FIGURE 6.12   This comet is on a path of no return.

Let's consider one more thing about comets. They probably originate in interstellar space, to which some of them return after visiting the solar system. The comets in this case move in an open parabolic or hyperbolic path around the sun (Fig. 6.12). Only if a comet approaches the sun within the right range of velocity can it be captured by the sun and forced into a recurrent orbit.

Double stars

Ancient Egyptians knew about the star Sirius. When it appeared as a morning star in their skies, they knew it was the advent of the rainy season and that a flooding of the Nile was not far off. Sirius is the brightest star in our sky, and it was studied by Egypt and numerous other cultures and civilizations. Yet none of these ancient peoples would have guessed that Sirius does not follow a regular path in the sky. Who knows how they would have interpreted this fact if they had known it?

The irregularities in the motion of Sirius were discovered by astronomer Fredrich Bessel in 1834. By this time Newton's gravity law and its theoretical framework were firmly established in astronomical study. According to Newton's laws of motion, an isolated starry object like Sirius should move in the uniform linear path characteristic of natural motion, unless it is not isolated after all. Bessel suggested that the zigzagging of the path of Sirius was due to the existence of a companion star. This companion star was probably too dim to be seen with the naked eye, but perhaps it could be found if searched for carefully with the best available telescopes. The companion star was discovered soon after its prediction, providing more brilliant support for Newton's theory. (Since then many other double-star systems have been found.)

The two stars form one system. They are like two masses tied to a string, one at each end, the string being the invisible gravitational attraction between them. The two masses themselves wobble about their common center of mass. Only their center of mass exhibits natural motion, as shown in Fig. 6.13(a). Throughout all the years traced in the figure, Sirius and its companion have continuously changed their positions relative to their center of mass. The positions of each individual star about the common center of mass can be seen to trace an ellipse, as shown in Fig. 6.13(b). Notice that the distances (magnitude and direction) between the two stars are the same in parts (a) and (b), which have been drawn to scale. But this is just one way to look at their motion; there is another way. If we plotted the position of Sirius, assuming its companion to be stationary, again we get an ellipse [Fig. 6.13(c)]. Likewise, the companion describes an ellipse about Sirius, which is the mirror image of Fig. 6.13(c), when Sirius is assumed to remain stationary [Fig. 6.13(d)].

Being perceptive, you probably already have noticed that there is something peculiar about these pictures when compared to the ellipses shown for the solar planets. In the latter the sun is always seen at one focus, but here the focus seems to be misplaced. This, too, has a simple explanation—namely, that we are looking at the plane of revolution of the two stars not perpendicularly to the plane but from the angle at which we view them from earth. An ellipse viewed from a slanted direction still appears as an ellipse, but with a displaced focus. Viewed from a right angle, the appropriate body would indeed be seen as the focus.

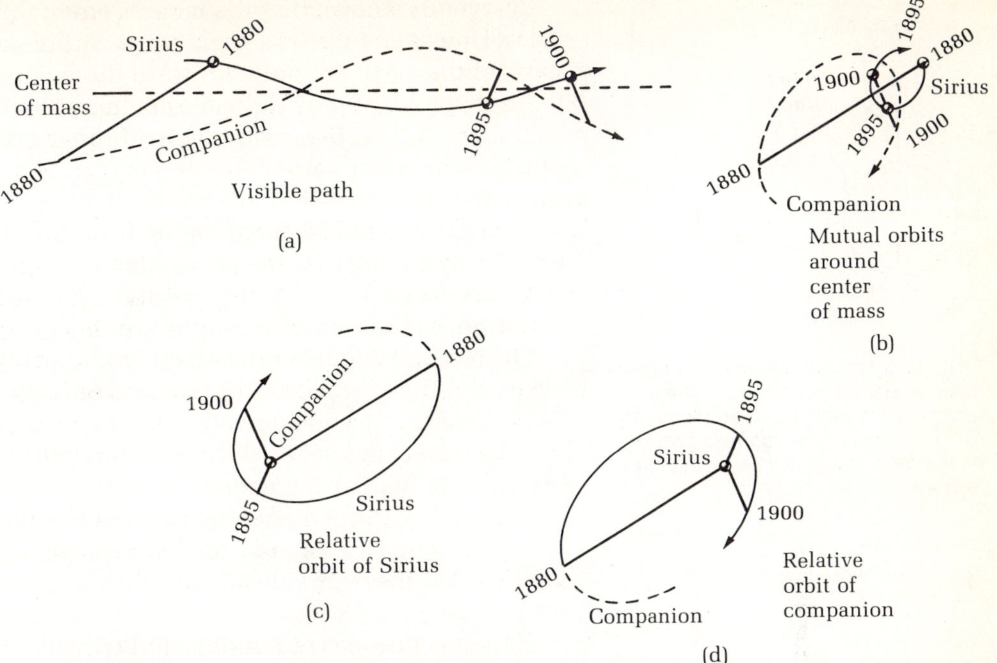

FIGURE 6.13 Paths of double star analyzed. See text for explanation. (By permission of William Collins Publishing Co., Inc., from *The Attractive Universe,* by E. G. Valens; photographs Bernice Abbott.)

Interestingly, the earth and the moon form a system similar to a double star; for this reason they are sometimes referred to as the double planet. They too are joined by the invisible string of mutual gravity. From the earth we get used to talking about how the moon describes an ellipse about the earth, but from the point of view of an observer on the moon, it would be perfectly sensible to say that it is the earth that goes around the moon. Actually, they both go around their center of mass, which is *not* at the earth's center.

Let us find the center of mass of the earth-moon double planet. The procedure is exactly the same as illustrated in Fig. 3.11; only the subjects have changed. Instead of a man and a woman, we now have the earth and the moon, and the rope is now the force of gravity. If we take the earth's center as the origin, then the location of the center of mass of the earth-moon system from this origin is (see Fig. 6.14)

$$\frac{\text{moon's mass}}{\text{total mass}} \times \begin{pmatrix} \text{distance from earth's center} \\ \text{to moon's center} \end{pmatrix} = \frac{1}{82} \times 240{,}000 \text{ miles}$$

$$= 2{,}926 \text{ miles}$$

Thus the earth revolves about a point inside its body about 2,900 miles from its center. The moon revolves about the same point. And this center of mass of the earth-moon double planet revolves in an ellipse around the sun.

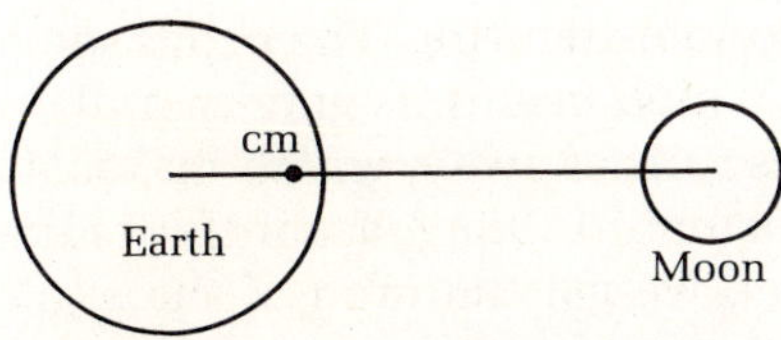

FIGURE 6.14 The center of mass of the earth-moon double planet is located on the line joining them at a distance of about 2900 miles from the earth's center.

Tides

The phenomenon of the tides needs special mention. We discuss this phenomenon not only because it is another application of the gravity law but also because it brings to light a fundamental aspect of the gravity force.

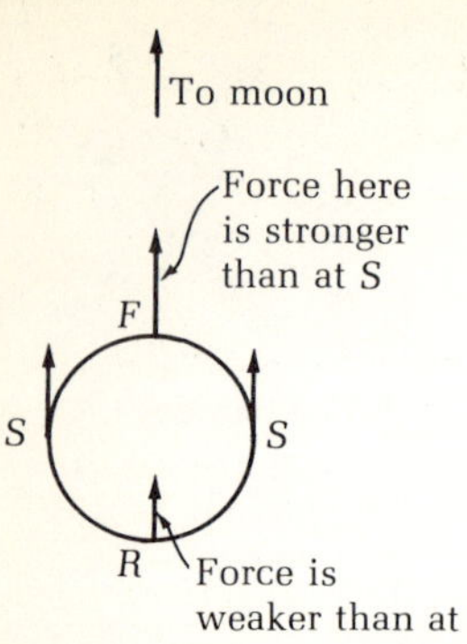

FIGURE 6.15 The attraction of the moon is stronger on the side of earth facing it *(F)* than the average force at *(S)*. The attraction at the rear side *(R)*, on the other hand, is weaker than the average.

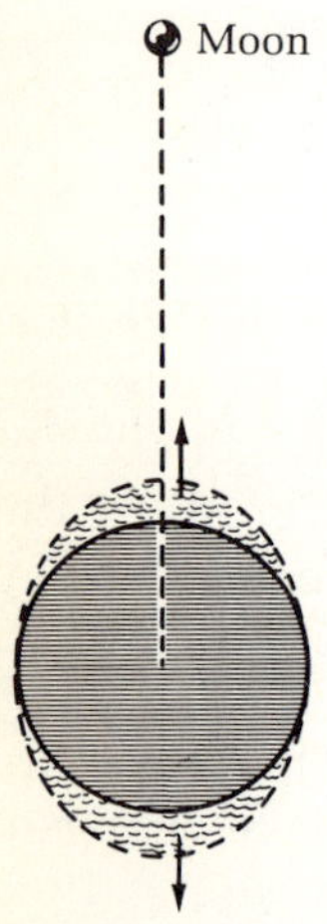

FIGURE 6.16 The explanation of the two tides during a full twenty-four hour day. The ocean water facing the moon has a bulge due to a positive gradient of the gravity force. At the same time, the water on the opposite side also shows a bulge due to the action of a negative gravity gradient; the earth underneath is attracted away from it.

Everybody knows that the moon's gravity pull has something to do with the tides. However, there is a slight twist. Suppose you are asked this question: If all parts of the earth are pulled toward the moon with the same force, and thereby displaced toward it by the same amount, how can you tell? The important point in understanding the relationship of lunar gravity and the tides is that all parts of the earth are not attracted the same amount by the moon's gravity force. The force is different at different points on earth. This is true of the gravity force on any large object. The force varies from one point to another, a fact we often express by saying that the gravity force in such a case has a **gradient.** Thus the tides are caused not by the gravity force of the moon but by its gradient or variation. Some numbers may make this clearer.

The moon is 240,000 miles away from earth's center; that is, it is 60 earth radii away. But the near side of the earth is only 59 earth radii away and the far side, 61 earth radii. As you know, the gravity force decreases with distance, so clearly it is stronger at the near side than at the center; on the other hand, it is weaker on the far side than at the center.

The force at the center of the earth can be thought of as the average pull of the moon's gravity. Compared to this average, there is an excess pull, a positive gradient, on the near side and a deficit, a negative gradient, on the far side (Fig. 6.15).

Thus it is now easy to understand why there should be two tides in a twenty-four-hour period. When the moon is drawing the near side of the earth away from the center, this part of the earth undergoes a tide. But at the same time, the opposite side also experiences a negative gravity gradient, meaning that the rest of the earth is pulled away from it. So there is a tidal bulge also at the far side (Fig. 6.16).

The effect of the sun's gravity on tides

Here is something to think about. The force of gravity on earth due to the sun is much larger than that due to the moon. The larger distance of the sun is more than compensated for by its huge mass. Let us do some figuring. The sun is some 27 million times more massive than the moon, which makes its gravity pull on the earth that much stronger. It is also 388 times further away, which makes its gravity pull weaker than that of the moon by a factor of $1/(388)^2$, according to the gravity law. When we multiply the two factors, we get the ratio of the sun's attraction on earth to the moon's:

$$27{,}000{,}000 \times \frac{1}{(388)^2} = \frac{27{,}000{,}000}{150{,}544} = 180$$

Now two questions arise. If the sun's force is that much stronger, why is the moon said to cause the tides? (Indeed, we know this to be true; the tides occur approximately at the same time the moon is overhead, although some lag exists because of earth's rotation.) What happens to the effect of the sun?

The answers are not hard to find, particularly if you understand the previous section on tides. Tides are not due to the average gravity force on a large body (which we compared above); they are caused by the gravity gradient, the variation of the gravity force over the volume of the attracted body. Since the sun is much further away, the difference of its gravity force on earth between the near side (or the far side) and the center is much smaller than in the case of the moon. This is roughly the explanation.

We can be quantitative, if you like. The gradient of gravity is found to exhibit an inverse cube law: it decreases as the cube of the distance. So the ratio of the gravity gradient due to the sun and the moon is

$$\frac{27,000,000}{(388)^3}$$

instead of the previous ratio in which we had 388^2. When we work out the arithmetic, we find the ratio in this case is 1/2.16 in the moon's favor. The effect of the moon's tide-producing force is 2.16 times stronger than that of the sun's.

Actually, although smaller than the moon's, the sun's tidal effect on earth is by no means negligible. Twice in a month, when the sun, earth, and the moon are lined up during the new and the full moon, the tidal bulge increases because the moon and the sun act in unison. The tides are then called spring tides. In contrast, when the sun and the moon pull at right angles to each other, the two effects subtract from each other, and the result is called the neap tide.

■ 6.3 Beyond Newton

Have you begun to develop a feeling of awe about the theory that explains so much with so little? In Chapter 2, we saw that the Newtonian framework consists of three ingredients: (1) Newton's second law of motion, (2) a law (or laws) to calculate the net force, and (3) a knowledge of the position and velocity of the moving object at some given instant of time (the so-called initial conditions). Here we have seen that framework in action. By combining

$$F = ma$$

with the gravitational force law

$$F = \frac{GMm}{R^2}$$

Newton and his followers were able to explain the motions of many of the objects in the heavens. The initial conditions were determined from present observations. Once this was done, information about the motions, both in the past and in the future, was completely determined. The success of Newton's gravitational theory led to the idea of determinism, which is held very strongly by many people.

A most interesting fact about the gravity force is that it acts through empty space—it is an action at a distance. But this is a very unusual attribute of a force. Most of the forces we encounter in everyday life are contact forces. Moreover, the action of the gravity force through any distance is assumed to be instantaneous. Is this reasonable?

The rise of determinism

It was the talent and prerogative of the French mathematicians to reduce the study of motion of objects—mechanics—to a branch of mathematics. Once they were convinced that the Newtonian theory of gravitation was right, there was no stopping them. They did not worry about action at a distance or about mechanisms for the gravity force, and they were quite satisfied with solving the calculable problems of celestial mechanics. As already indicated to some extent, their success was very impressive—so impressive that any nonsuccess was simply attributed to the limitations of their calculating machines. This is what Pierre Simon de Laplace, one of the most prominent physicists of that era, wrote:

> An intelligence that, at a given instant, was acquainted with all the forces by which nature is animated and with the state of the bodies of which it is composed, would—if it were vast enough to submit this data to analysis—embrace in the same formula the movements of the largest bodies of the universe and those of the lightest atoms: nothing would be uncertain to such an intelligence, and the future like the past would be present to its eyes.

Clearly Laplace believed that not only could the motions in the heavens be solved with the Newtonian world machine but also the motions of the particles in the microcosm. All that was necessary was to know the forces, plug them into the Newtonian framework programmed in the machine, and out would come the answer. This philosophy is called determinism, or, since it is based on a mechanical view of nature, mechanical determinism.

Before assessing this philosophy, it is interesting to dwell on the reaction of the romantic poets of the eighteenth century to this proclamation of total determinism by science. Perhaps you will sympathize with these poets. Some of their fears seem very relevant today, considering the current, somewhat-threatening status of science and technology in our society. The following is a quote from William Wordsworth, once the poet laureate of England:

> Thanks for the lessons of this spot-fit school
> For the presumptuous thoughts that would assign
> Mechanical laws to agency divine;
> And, measuring heaven by earth would overrule
> infinite power.*

It is also interesting to note that this same Wordsworth, when he was younger, wrote the following beautiful poem on Newton:

> I could behold
> The antechapel where the statue stood
> of Newton with his prism and his silent face,
> The marbel index of a mind forever
> voyaging through strange seas of thought, alone.†

---

*"Cave of Staffa," in *Complete Poetical Works of Wordsworth* (London: Macmillan, 1930).
†From "The Prelude," in *Complete Poetical Works of Wordsworth*.

What caused the poets' disillusionment? Was it the threat that determinism and the mechanical view of nature presented to the appreciation of the "poetic truth"? Or was it the threat to the future of human civilization brought about by the industrial revolution? Perhaps both. A poet can appreciate the experiential nature of science; a poet can accept even a coexistence of scientific truth and poetic truth. But when science threatens to take over and deprive the human soul of all its mystery, the poet has to speak up, as in these lines from British poet Thomas Campbell:

> When Science from creation's face
> Enchantment's veil withdraws,
> What lovely visions yield their place
> to cold material laws.*

Fortunately it turns out that mechanical determinism does not seem to provide the ultimate answer. Modern physics discovered that strict determinism does not hold for motion in the world of the atom. Even in the world of astronomy, the Laplacian doctrine can be questioned. Perhaps the answer lies in the assertion that there is never enough data available even to the ultimate computer for it to be able completely to determine the future. Thus there is always some amount of novelty and uncertainty.

Action at a distance versus fields

Gravity in Newtonian theory is force acting at a distance, propagating instantly from one interacting object to the other. This is very different from our everyday experience of forces, and indeed Newton himself was very perturbed by this. It was tempting to explain gravity by means of mechanical carriers. To Newton's credit, although he toyed with such ideas, he stuck to his theory. With the enormous success the theory achieved subsequently, the action-at-a-distance aspect became somewhat of a moot question.

Some developments in the nineteenth century forced physicists to look into this matter again. It was found that the phenomena connected with electricity and magnetism were described by laws that necessitated the introduction of a new concept, the propagation of a force via a field. Perhaps you have an intuitive idea of a force field, since we use this language quite often. A field is a condition in space that is created (around itself) by an object. A second object placed in the field "feels" the force. The field concept is a natural one for handling the question of the finite propagation time of interactions. It is an established fact that even interactions take a finite time to develop; no interaction is instantaneous. We will say more about this later.

The success of the field theory for electromagnetism makes it reasonable to assume that gravitational phenomena also can be understood in terms of such fields, gravitational fields. Einstein's general theory of relativity is such a field theory for gravitation, which we will take up in Chapter 25.

---

*From the poem, "To the Rainbow."

SUMMARY

With Newton's law of universal gravitation, you have reached the high point of your study of classical mechanics. The law is best expressed by the mathematical statement

$$F = \frac{Gm_1m_2}{R^2}$$

The force of gravity is an attractive force by which every object attracts every other with a force (whose magnitude is given by the expression above) directed along the line joining the two. Most of the rest of the chapter was devoted to various applications of Newton's gravity law. Among these is the experiment that measured the gravity force between two ordinary masses (Cavendish's experiment). The subject of artificial satellites and the equation for the orbital velocity of an earth satellite,

$$v = \sqrt{\frac{GM_e}{R}}$$

were presented. Another application of Newton's law involves the true cause of tides. They are not due to the average gravity force of the moon on the earth but are due to the variation of the gravity force over the volume of the earth (the gravity gradient).

Among gravity-related concepts introduced in the chapter, the concept of centripetal acceleration is important. The formula for centripetal acceleration is given as

$$\frac{v^2}{R}$$

Newton's discovery of the gravity law was a momentous event in the history of human civilization. In its application to celestial mechanics (the study of motion of heavenly objects), it has proven to be almost infallible, leading to the deterministic view of the world. Whether or not you agree with this view, perhaps you still can see how awesome it is.

QUESTIONS

Review and reason

1. The gravity force is greater on a large mass than it is on a small mass. Isn't it reasonable, then, to expect that a larger mass should fall to earth under gravity with a greater acceleration? Why doesn't this happen?
2. What is centripetal acceleration? Centripetal force?
3. Since earth is attracted by the sun's gravity, why doesn't it fall into the sun?
4. A bucket of water is swung in a vertical circle but the water does not spill out, even when the bucket is upside down. Explain.
5. Why does a car tend to skid on a curve on a level road if the speed is too high?
6. Why are the roadbeds for trains banked on a curve?
7. Why do the tires squeal when you take a turn with higher than usual speed?
8. What is meant by universal gravitation? State Newton's law of universal gravitation in the mathemati-

cal form. How is the distance from one object to the other reckoned when both objects are extended bodies?

*9. Explain "gestalt" in your own way. What constituted Newton's gestalt?

10. Show that one unit for G can be given as a meter³/(kilogram second²). (It may help to remember that a newton is equivalent to a kilogram meter/sec².)

11. Earth is known to slightly deviate from a perfect sphere, its equatorial radius being a little bit larger than the polar radius. Should the value of g be different between the equator and the pole?

12. A friend of yours claims that he lost weight when he went to the top of the Rocky Mountains. Do you agree? Explain your answer.

*13. Suppose there was a pole-to-pole tunnel through the center of the earth. Assuming that the earth is of uniform density, the force on an object in this tunnel at a distance r from the center can be shown to be proportional to r. Show that an object dropped at one mouth of the tunnel will oscillate back and forth between the two ends (assuming there is no friction).

14. Describe Cavendish's experiment for the measurement of G.

15. Northern India, because of the proximity of the Himalayas, is difficult to survey because a plumb bob does not hang quite vertically. Explain why. Is the deviation of the bob from the vertical as large as expected? If not, why not?

*16. Do you weigh more at night than during the day because of sun's gravity? In the cartoon Malcolm provides one rationale for giving an affirmative answer. Here is another reason to say yes. At noon, the sun is directly overhead and pulls all objects toward it against the pull of earth's gravity. But at midnight, the sun is on the other side, and therefore its pull has the same direction as the earth's gravity pull. Doesn't it make sense then to expect that our weight at midnight is greater than at midday? Actually, the answer is still no. Can you explain why?

17. The sun shows some wobbles in its motion through space. Can you give a reason?

18. Should Kepler's laws apply to the artificial satellites of the earth? Does an earth satellite necessarily travel with uniform speed? Explain.

19. Explain why there are two daily tides. Why is the tidal effect of the sun smaller than that of the moon?

20. What are the possible phases of the moon at neap tide, new, first quarter, or full? Explain your answer.

21. Do we have the highest tide during neap or spring? Explain.

*22. We have a high tide every 12 hours 20 min and 25 sec. Give one cause of the extra 20 min and 25 sec.

23. What does mechanical determinism mean to you?

Arithmetic

1. A car traveling at a speed of 20 m/sec takes a curve of radius 15 m. Calculate the car's centripetal acceleration.

---

*Optional.

2. In Cavendish's experiment, suppose the small mass is 0.04 kg and the large mass is 100 kg. They are placed so that the centers of mass are 0.1 m apart. What is the force of gravity between the two masses, given that $G = 6.67 \times 10^{-11}$ MKS units?

3. Calculate the gravity force between the earth and the sun. Find all the data for the problem from the text.

4. Suppose the size of a planet is decreased by a factor of 2 without any changes in its average density. How much does the mass of the planet change? How much does the force of gravity of the planet on an object on its surface change? How much does the acceleration of fall change due to the gravity change of the planet?

5. What is the gravity force of a 10 lb boulder on the earth? Does the force increase, decrease, or remain the same if you take the boulder to the North Pole?

6. An artificial satellite is put in a circular orbit around the earth at a distance 4000 mi from earth's surface. Calculate the orbital velocity *of the satellite*.

7. Science fiction writer Damon Knight has a story in which his hero takes a free-falling elevator ride through a tunnel through the center of the earth.* In a hurry, the hero figures out how long it will take to reach the other side: "Call the radius of the earth 4000 miles—about 20 million feet, for convenience. [Acceleration of] Gravity at the surface of the earth is 32 feet per second². The square root of 20 million over 32—would be 250 times the square root of ten . . . times pi. . . ." Well, how much is it? Can you write down the formula (using symbols) that our hero used? (Hint: The formula is not in the text; you have to figure it out from the above.)

8. The mass of Jupiter is roughly 1/1000 the mass of the sun, and it is approximately 775 million km away from the sun. How far away is the center of mass of the sun-Jupiter system from the center of the sun? Do you think the sun wobbles about this point?

---

*Damon Knight, *Beyond the Barrier* (New York: Doubleday, 1964), p. 128.

Inertial and Accelerated Reference Frames and Pseudoforces

My first train ride was as a little boy accompanying an older brother to his wedding. It was a night train. I woke up several times during the night, and, between periods of sleep, I thoroughly enjoyed the sounds and the feelings of the train. When morning came my brother asked me how the journey was. I said it was fine, great. And then I asked a little bashfully something that had been bothering me. How come there were two kinds of trains? The kind that we had been in just seemed to shake all night but didn't seem to go anywhere. I told him that the experience was very different from what I anticipated. And I would much rather ride on the other kind of train—the kind that moved, the kind that I had seen before.

Suddenly I became aware of the amused look on my brother's face; not only he but everybody within earshot was laughing. I started feeling embarassed; I must have said something funny without knowing it. Fortunately they were all adults.

Eventually someone explained that trains were all of the same kind, the moving kind, as I had called it. Our train *had* moved during the night, except that from the inside I couldn't tell without looking outside. Somebody else mentioned the phrase "reference frame" and something about things looking the same from the reference frame of a moving train as from the ground. I sat there, perhaps not completely understanding what was being said but experiencing every word.

■ 7.1 Frames of Reference

To determine the spatial location of an object in three-dimensional space, we need a set of three axes at an origin—for example, three sticks tied together at a point in a mutually perpendicular fashion. This is called a **frame of reference.** Given a frame of reference, the location of a point in space can be specified in many different ways. Each of these ways defines a specially named coordinate system. The most common coordinate system, which is also the easiest to use, is the Cartesian coordinate system, in which the distance of a point in space from

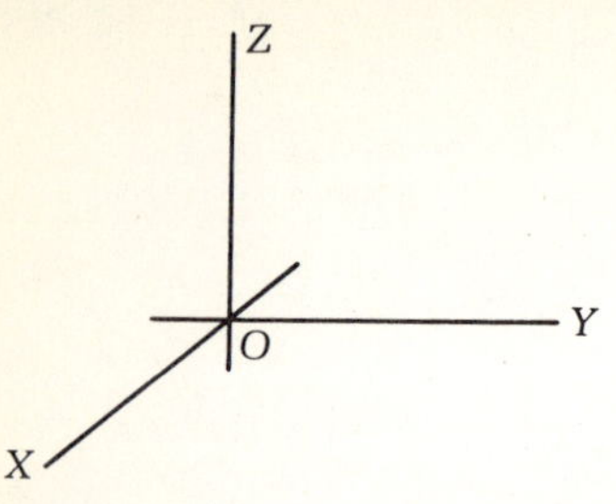

FIGURE 7.1   A reference frame: three mutully perpendicular axes *X*, *Y*, and *Z* at origin *O*.

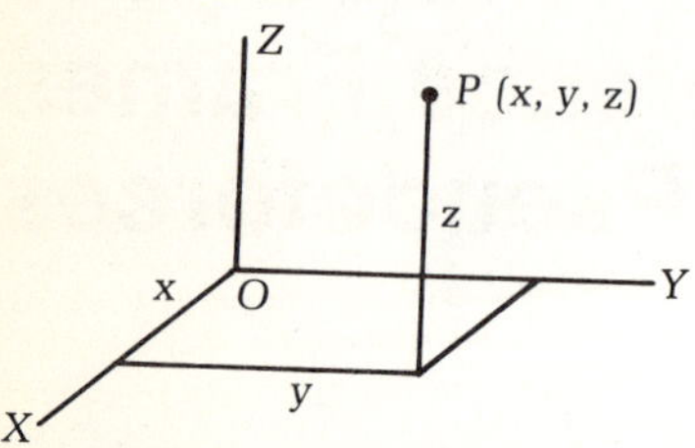

FIGURE 7.2   To reach the point *P* from the origin *O*, we have to travel a distance *x* parallel to the *X*-axis, a distance *y* parallel to the *Y*-axis, and a distance *z* parallel to the *Z*-axis. Thus *x*, *y*, and *z* are the coordinates of *P*.

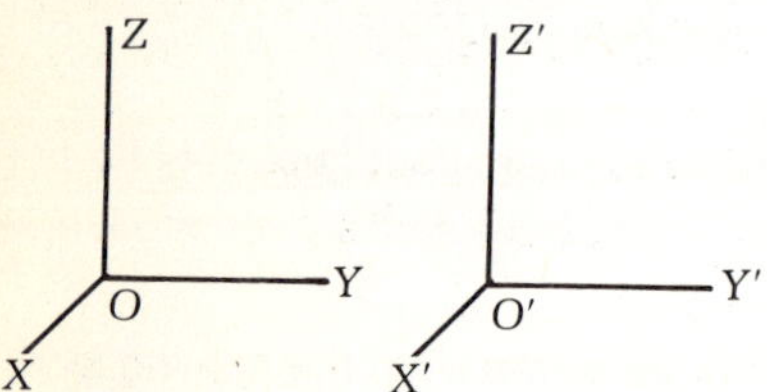

FIGURE 7.3   Two reference frames *OXYZ* and *O'X'Y'Z'* related to each other by spatial translation.

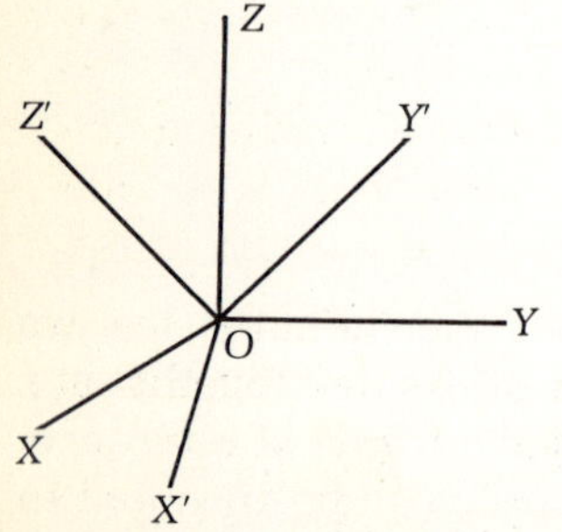

FIGURE 7.4   Two reference frames *OXYZ* and *O'X'Y'Z'* with a different orientation of axes.

each of the three axes is employed to specify the location of the point.

Figure 7.1 illustrates a frame of reference, a set of three perpendicular "rods," *X*, *Y*, and *Z*, attached at the origin *O*. A point *P* in space can be specified in this frame of reference by its coordinates *x*, *y*, and *z*, which are the distances we have to move from the origin *O* parallel to the axes *X*, *Y*, and *Z*, respectively, in order to reach the point *P* (Fig. 7.2).

Two different frames of reference may differ in a variety of ways. They may differ by a translation in space. That is, one reference frame is situated in a different location from the other but the orientation of the three axes of the translated reference frame is unchanged, as shown in Fig. 7.3. Instead of a different location, we can also think of a reference frame with the same origin as another but having a different orientation of the axes (Fig. 7.4). Two reference frames can also differ from each other in a combined way: a different location in space and also a different alignment of the axes.

A new element enters the picture when we consider moving reference frames. For example, an object on a moving train can be looked at from a reference frame mounted on the train or from one on the ground—but now the two reference frames are moving with respect to each other. For frames of reference moving rapidly with respect to each other, Einstein showed that the measure of time is different (see Chapter 19). Thus it is important that a frame of reference should also include a clock in addition to the three sticks.

Physical laws and frames of reference

Can physical laws depend on your frame of reference? For example, suppose you live in Oregon; in this case you look at things from a reference frame located in Oregon. From your observations you reach conclusions about the laws of nature. Are your laws going to be different from somebody else's whose reference frame is at another place?

We can ask a similar question about rotated reference frames that are oriented differently. Suppose you have a friend who looks at things from a reference frame that is somewhat oblique with respect to yours. Do you have to worry that your friend is going to find a different set of natural laws?

What does your experience have to say about this? Physical laws do not depend on frames of reference, so far as translations and orientations of the frames are involved (Fig. 7.5). Imagine the additional amount of international tension that could occur if the Russians were governed by a different set of physical laws just because their frame of reference is located in a different place on earth than ours.

This consistency of physical laws with respect to translations or orientations in space is an example of symmetry. What is symmetry? Suppose you perform an operation on a system and everything remains the same. This is an example of a symmetry principle. Thus the statement that physical laws do not change when looked at from a translated or a differently oriented frame of reference implies two symmetry principles. It is the existence of the underlying symmetry principles that gives us the unchangeability of the laws of nature under certain operations. In the case of translation, the underlying symmetry princi-

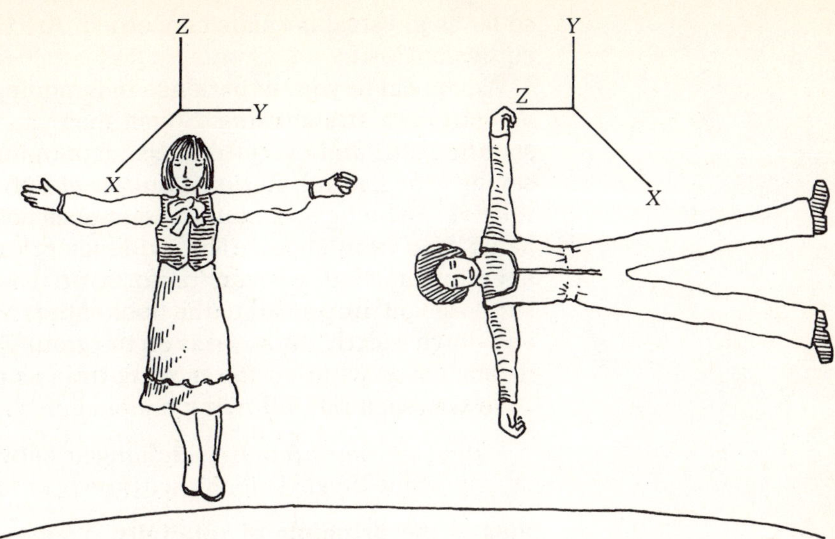

FIGURE 7.5 Viewing the world from a translated and a differently oriented reference frame. Does this change the perception of physical laws?

ple is **homogeneity** of space, which means that space is the same in one location as in another. Similarly, for orientation, the underlying symmetry is **isotropy** of space, which means that space is the same in one direction as another.

So we are asserting that physical laws have translational and orientational symmetry.* What do we really mean by these statements? How do we know for sure? Suppose we construct a system, a machine, with different interacting parts of various degrees of complication. If we transfer this machine from one place to another and watch its performance, it would be like looking at the laws that go into the performance of the machine from a translated frame of reference. What we find in every case is that the performance of a machine is not changed by translation.

Actually, in practice, verifying what we have just said can be tricky. It is easy to verify the validity of our statement for a machine that is totally self-contained, depending on nothing external to it. However, very few machines are like that. In fact, most machines depend on external factors in a major way—for instance, the operation of an automobile depends on road conditions, weather, and other external features. When we say that the performance of a car is the same in New York as in Los Angeles, what we mean is that it would be the same under identical road, weather, and other relevant conditions.

One important thing. The fact that physical laws do not depend on the choice of origin of our reference frame tells us that there is no preferred origin. One point of the universe is as good as any other point to serve as the origin or the center.

Inertial reference frames

What we have discussed so far points out that reference frames that differ from each other only in respect to position and orientation are completely equivalent

---

*Orientational symmetry is usually referred to as rotational symmetry. We have avoided the use of the word "rotation" to avoid confusion with rotating reference frames.

so far as physical laws are concerned. Are there other reference frames that are equivalent?

We appeal to your experience in a moving train, a train moving at a constant velocity in a straight line. Sometimes you have the feeling that your train is standing still and everything else is moving away from it. We express this by saying that natural motion (motion at constant velocity in a straight line) is relative. So being on a moving train does not seem to us to be any different from being on a train at rest. The evidence of our senses is really borne out by all experiments that we can perform on board a moving train. For example, suppose you drop a ball to the floor of the train. The ball will bounce straight up and down exactly the same as on the ground. You can test Galileo's law and find it remains as valid on the moving train as on the ground.

So we reach the following conclusion:

> *Physical laws remain unchanged between reference frames moving with constant linear velocity with respect to each other.*

This is the **principle of relativity.** Frames of reference that differ from each other by virtue of uniform relative (natural) motion are equivalent to each other. Combining this with a conclusion of the previous subsection, we can say that a reference frame is equivalent to any other reference frame that differs from it only in uniform relative (natural) motion, position, orientation, or any combination of these differences.

Previously we introduced the concept of an inertial reference frame. This is defined as a reference frame in which the motion of isolated objects—that is, objects without any force acting on them—is seen as a natural motion. We see now that any other reference frame that is related to an inertial reference frame by virtue of relative motion at constant velocity in a straight line, position, or orientation is itself an inertial reference frame, and all physical laws are the same for all such reference frames.

Suppose we consider a reference frame that is accelerated with respect to an inertial reference frame. Do physical laws or results of physical experiments remain unchanged when viewed from an accelerated reference frame? To answer this question, we will refer again to a common experience. When a car accelerates or an airplane takes off, you feel as though you are pushed against the back of the seat momentarily. In an airplane taking off, you are momentarily in an accelerating frame of reference. Somehow in such a frame, a mysterious force appears. Thus we can tell if we are in an accelerating reference frame because of the appearance of such a force. In contrast, we cannot tell the difference between two inertial reference frames.

Our home, the planet earth, is fairly close to being an inertial reference frame. However, it does rotate around its own axis and also revolves around the sun. Both of these motions are accelerated motions, so the earth is really an accelerated reference frame. And, to be sure, mysterious forces make their appearance on earth.

How do we find an inertial reference frame? Well, we can take off in a rocket with our sticks and clock. As we move away, every once in a while we can let go of an object and see if it exhibits natural motion. When and if it does, we are in an inertial reference frame.

Once we have found ourselves in an inertial reference frame, and we stop and look through our rocket window, we will see something remarkable. The stars will appear to be fixed, at rest. It is an experimental fact that inertial reference frames do not appear to rotate with respect to the "fixed" stars. Actually, many scientists believe that the fixed stars have a small amount of acceleration relative to the inertial reference frames. In any case, the fixed stars, at least approximately, form an inertial reference frame.

■ 7.2 Accelerated Reference Frames and Pseudoforces

The mysterious forces that make their appearance in an accelerated reference frame are called **pseudoforces** (false forces) to distinguish them from "real" forces arising from interactions between objects. Pseudoforces do not arise from the interaction of one body with another as Newton depicted in his third law. What causes the pseudoforces, then? We don't exactly know, except that somehow they are connected with the property of matter we call inertia. For this reason it is customary to refer to pseudoforces also as inertial forces.

We have already talked about the pseudoforce that appears in an accelerating car. The pseudoforce is in the direction opposite to the acceleration—it pins you to the seat. On the other hand, when a car is suddenly braked, we experience a pseudoforce in the forward direction, which is again the direction opposite to that of the acceleration. (Remember that deceleration, caused by braking, is acceleration in a direction opposite to the direction of motion). From the point of view of the accelerated reference frame of the car, it is this pseudoforce that makes us fall forward when a car undergoes a sudden braking.

Notice that an observer on the ground will account for your falling forward in an abruptly stopping car as simply due to your inertia; she will not say that there is any additional force. This is the point of view we adopted in Chapter 2, the point of view of the inertial observer. We now acknowledge that an observer in the accelerated frame has a very different experience, namely, that of being

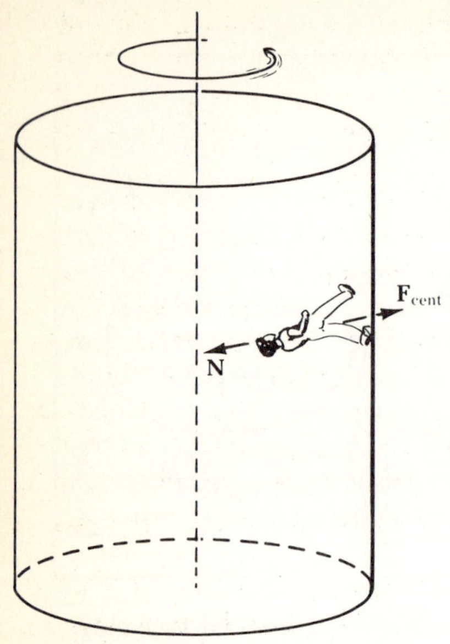

FIGURE 7.6 Artificial gravity in a rotating cylinder. The normal support force now arises from the reaction of the inner wall of the cylinder in response to the centrifugal force with which an inhabitant presses on the cylinder wall.

acted on by an additional force. For him this "pseudoforce" is as real as can be.

We can get a new perspective on the concept of weightlessness discussed earlier by reviewing the concept from the point of view of an accelerated reference frame. Imagine yourself in a freely falling elevator, an elevator whose cable has broken. From the point of view of the accelerating reference frame of the falling elevator, there is a pseudoforce acting on you in the upward direction. This pseudoforce is exactly equal and opposite to your weight force downward. So from within the accelerating frame, the forces acting on you cancel out; that is, you are apparently weightless. If you stood on a spring scale in the free-falling elevator, the scale would not register any weight.

Pseudoforces appear also in rotating or revolving frames of reference, since these too are accelerated frames. For example, there is the familiar **centrifugal force.** If you have ridden a merry-go-round you know how centrifugal force acts on you, trying to throw you outward. (Centrifugal means away from the center.) Consider an astronaut in an orbiting satellite. He is at rest relative to the satellite, within the accelerating reference frame of the satellite. In this frame the centrifugal force comes into play and balances the gravity force of the earth on him. Thus the astronaut is apparently weightless, this time because of the action of centrifugal force. Again, from an inertial observer's point of view, there is no centrifugal force. The astronaut and the satellite fall together under the action of the earth's gravity as they move around in their orbit.

Centrifugal force can be used to generate artificial gravity in space vehicles. Just let the vehicles rotate around an axis. Thus in a rotating cylinder there will be a centrifugal force acting outward. People living on the inner wall of the cylinder will be pushed against the wall by the centrifugal force. Thus the walls become the floor, and the centrifugal force will act like gravity (Fig. 7.6). This artificial gravity has an additional feature. It decreases as one climbs toward the axis of rotation. A very entertaining discussion of the experiences a person could have in such a rotating cylinder in descending (or ascending) a staircase toward the "floor" from the axis is given in Arthur C. Clarke's science fiction novel *Rendezvous with Rama.* * Centrifugal force in a rotating cylinder has been suggested as the provider of artificial gravity in the space colonies proposed by Princeton physicist Gerard O'Neill.†

We note one important thing here. It is known that all pseudoforces that come about in accelerating reference frames are proportional to the mass of the object on which they act. A more massive object is acted on by a proportionately larger pseudoforce. What other force behaves this way? Gravity, of course. This is why centrifugal force works so well as artificial gravity.

We mentioned before that the earth, by virtue of its rotation and revolution, is really an accelerated reference frame. What role do pseudoforces play on earth?

First, let us consider the centrifugal force arising from earth's revolution around the sun. In Chapter 6 there was a riddle (question 16). Since the direction of the sun's gravity force on a person is opposite to that of the earth during the day and the same at night, would a person's weight differ between day and

---

*Arthur C. Clarke, *Rendezvous with Rama* (New York: Balantine Books, 1973).
†Gerard K. O'Neill, *The High Frontier* (New York: Morrow, 1977).

night? One way to answer this is to look at the question from the point of view of the revolving reference frame of the earth. There is a centrifugal force due to the earth's revolution about the sun acting in a direction away from the sun and canceling the force of the sun's gravity exactly. Thus the sun's gravity has no effect on the weight of earthly objects.

The same cannot be said, however, of the centrifugal force arising from the earth's rotation around its own axis. In this case the centrifugal force, acting outward and opposite to the earth's gravity, actually manages to reduce our weight by about three parts in a thousand at the equator, where the full effect of the rotation is felt. At the poles, where there is no rotation (since the poles are on the axis), there is no weight reduction due to the centrifugal force. At all in-between points there is an intermediate reduction in the weight of an object. Note that the reduction of weight due to centrifugal force is in addition to that caused by earth's slight equatorial bulge, which causes an effect of about the same magnitude as the rotation. Thus the total reduction of weight from the poles to the equator is 0.5%.

There is another pseudoforce that comes into play whenever an object *is in motion* in a rotating reference frame. This is known as the **Coriolis force.** On earth the Coriolis force bends the path of all *moving* objects toward their right in the northern hemisphere and toward their left in the southern hemisphere. The magnitude of the Coriolis force increases in direct proportion to the velocity of the object.

The Coriolis force acting on local winds can cause phenomena like cyclones. Fluids (liquids and gases) have a tendency to move from points of high pressure to points of low pressure. Suppose there is a low-pressure point created somewhere in the atmosphere. Air will then rush toward it from all directions, as shown in Fig. 7.7(a). The Coriolis force, however, will bend the path of the approaching air, as indicated, to the right (in the northern hemisphere) and create a counterclockwise whirlwind. This is called a cyclone by weather scientists. If the air movement is violent, they call it a hurricane or perhaps a tornado.

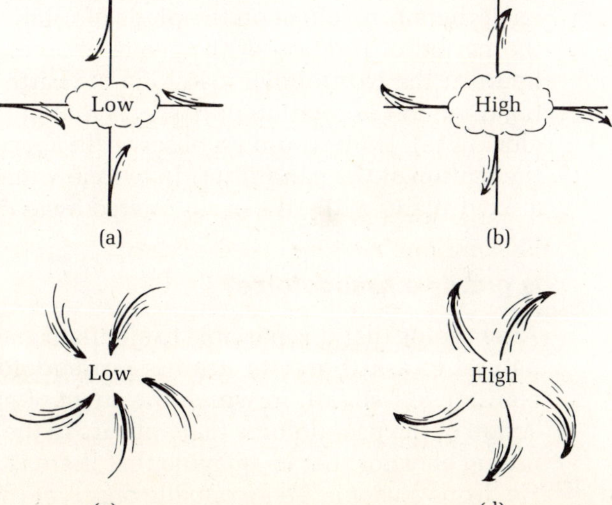

FIGURE 7.7   Cyclones and anticyclones. (a) and (b) The basic patterns of the movement of air for a cyclone and an anticyclone respectively. For the cyclone the air movement is inward and counterclockwise; for the anticyclone it is outward and clockwise. (c) and (d) Air movements in fully developed cyclone and anticyclone, respectively.

The reverse phenomenon, where air rushing out of a high-pressure zone is deflected clockwise (in the northern hemisphere), is called an anticyclone.

How do we know that the earth rotates?

In this day and age, with the astronauts able to take a look at the rotating earth from "up there," this may seem like a moot question. Historically, of course, whether or not the earth rotates (about its own axis) or stands still was an important question, debated in the public arena. At least it was until an engineer named Foucault conducted his pendulum experiment. This experiment converted most nonbelievers into believers (except for a few diehards).

What is Foucault's pendulum and how does it work? Almost every museum of science and industry has one, and you will enjoy taking a look at the operation if you haven't already. As the name suggests, it is a pendulum, consisting of a heavy bob hanging from a long cord from a hook fixed to a high ceiling. When at rest it is centered exactly on a graduated circle. When the pendulum is set to swing across the center along one of the diameters, the weight will swing from 0° to 180° and back. But this state of affairs will not last very long. Soon the pendulum will start its swing at 1° and end at 181°. And soon this too will change. The plane of vibration of the pendulum will rotate continuously in the clockwise direction (in the northern hemisphere). At the North Pole the pendulum will proceed all the way around in 24 hours, at which time it will be back at its original position of swing, only to start the whole motion over again. The time it takes the pendulum to return to its original configuration increases steadily as we go toward the equator, becoming 34 hours at 45°. On the equator there is no change in the plane of rotation.

The rotation of the plane of the swing of the pendulum is due to the Coriolis force arising from the earth's rotation. There is no other demonstration as dramatic as this of the Coriolis force and therefore of the rotation of the earth. It is uncanny to watch the pendulum changing its plane of swing for no apparent reason.

The Coriolis force is zero at the equator, which explains why at the equator itself there is no effect on the plane of rotation. It is no coincidence, either, that the period of rotation of the plane of swing of the pendulum is precisely 24 hours at the North Pole, matching the earth's rate of rotation exactly but with the direction of rotation opposite to that of the earth (the earth rotates counterclockwise). How would an observer in space in an inertial reference frame see the motion of the pendulum? In her view the pendulum would be oscillating in a fixed plane while the earth rotated away beneath the pendulum.

Is gravity a pseudoforce?

After seeing that it is possible to simulate gravity by means of pseudoforces, it is natural to ask if gravity itself is a pseudoforce. Previously we considered an elevator in free fall, in which the effect of gravity is canceled by the opposing action of the pseudoforce that appears in the accelerating reference frame of the falling elevator. Let us imagine that there is an observer in the falling elevator. He drops a watch. To an outside observer (perhaps someone looking through a

FIGURE 7.8 An elevator in space accelerating as shown with an acceleration of *g* simulates gravity toward the floor of the elevator. Its inhabitants fall with acceleration *g* toward the floor, just as a falling object would fall on earth. But in this elevator the floorward force causing the acceleration is a pseudoforce. Thus gravity and a pseudoforce are equivalent.

window) the elevator and the watch both seem to fall toward the earth with the same acceleration, following Galileo's law.

To the inside observer a strange thing happens. The watch does not fall. It appears to remain exactly where it was released. So the inside observer can conclude that there is no force of gravity on the watch; it is obeying the law of inertia. The effect of the force of gravity is not evident to the man in the elevator. If he knows about the earth and its force of gravity beforehand, of course, he will argue that there is a *pseudoforce* on the watch acting *upward*, arising from the fact that the reference frame is accelerating downward. The pseudoforce cancels the real force of gravity on the watch acting in the opposite direction.

But now suppose there is no earth and an elevator is accelerating upward in outer space with acceleration g. Then there will be a pseudoforce, directed downward this time, on any object in the elevator reference frame (Fig. 7.8). The *magnitude* of this pseudoforce is exactly equal to the object's weight on earth. Thus the earth's gravity is completely simulated. So is gravity a pseudoforce? Einstein put forward this famous **equivalence principle:**

> *Pseudoforces cannot be distinguished from the forces of gravity. Gravity is equivalent to a pseudoforce.*

Suppose we assert that the earth is a spaceship accelerating upward. There is no gravity, but only a downward pseudoforce that keeps us tied to the ground. One obvious problem with this assertion is that the earth is round. What happens to the people on the other side? They would be accelerating downward and the pseudoforce would be upward for them. Thus the assertion cannot be right. So to avoid this problem, Einstein proposed that the principle of equivalence holds *locally*. Gravity can be considered to be a pseudoforce only in a *small region of space at a time.*

Another way to see that gravity over a large region of space cannot be completely regarded as a pseudoforce is by considering the tides. Tidal effects arise from the local difference of gravity from one point to another, whereas in the accelerating elevator the simulated gravity is the same everywhere. The tidal effects of gravity cannot be simulated in an accelerated reference frame. Thus over a large region of space, the tidal effect of gravity must be regarded as "true," even if the average gravitational phenomenon is "pseudo."

Einstein was once invited to dinner with the king of Belgium. At the table he was seated next to the famous Polish-French physicist Marie Curie. Marie Curie took kindly to this young man. She had heard some things about his achievements, too. She asked Einstein what he was doing these days. He replied that he was thinking a lot about a person in a free-falling elevator performing experiments. Not surprisingly, Marie Curie was not impressed. "He's wasting his time," she probably thought.

Einstein worried about the problem for years. Finally, in 1915, he proposed a new theory of gravity (general relativity) that incorporates the ideas discussed above. One morning while his wife was making breakfast, Einstein came downstairs, still in his pajamas. He seemed all excited. "Darling, I've got a wonderful idea," he said to his wife, whereupon he disappeared into his study to spend the next fourteen days. It turned out that this wonderful idea was his new

FIGURE 7.9   A rotor. What's the mystery of this? (Courtesy Frank Hrubetz Company.)

theory of gravity using the idea of curved space. Space becomes warped in the presence of gravitation. We will have more to say on this in Chapter 25.

■ 7.3   Living with Pseudoforces

So far most of the examples of pseudoforces we have discussed are somewhat esoteric. In this section we are going to elaborate on one example from everyday life we already mentioned, accelerating elevators. We will also present two new ones that quite probably fall within your realm of experience. One example involves a rotor, a device one can ride in many amusement parks. When the device rotates, the platform is lowered away from under your feet (Fig. 7.9). But don't worry, you won't fall. It is a strange sensation, though, and the explanation is equally entertaining. The final example is from the bathroom. Accumulated water in the tub or the washbasin, when allowed to drain, often shows a tendency to form a counterclockwise whirlpool in our part of the world. This too is due to the action of a pseudoforce.

The accelerating elevator

Let us calculate some numbers this time. When an elevator accelerates downward, you know that there is a pseudoforce acting upward on all inside objects. The magnitude of the pseudoforce is proportional to the mass of the object, which is a very important thing to remember. It also is proportional to the acceleration of the elevator. The complete expression for the magnitude of the pseudoforce on an object of mass m inside an elevator of acceleration a is given by the product of m and a, ma.

Suppose there is a woman of weight 128 lb taking a ride in an elevator accelerating downward at 2 ft/sec². The woman's mass is then $128/32 = 4$ slugs. In the accelerating elevator, there is a pseudoforce acting upward on her of magnitude $4 \times 2 = 8$ lb. Thus the woman's *apparent* weight (during the period of acceleration of the elevator) will be

$$128 - 8 = 120 \text{ lb}$$

Indeed, if she took a bathroom scale along and stood on it during the downward acceleration, she could verify for herself that her weight is reduced to this value.

For an upward acceleration the effect would be the opposite. Now the pseudoforce acts downward, the same direction as that of gravity, and thus adds to her weight. For the same upward acceleration of 2 ft/sec², her apparent weight will show an increased reading on the scale of

$$128 + 8 = 136 \text{ lb}$$

If the elevator cable breaks what will the scale show? Zero, of course. A freely falling object apparently is weightless, as we have stated before.

Talking about freely falling elevators, let us briefly discuss a question that once appeared in a popular advice column. Suppose an unfortunate person was stuck in an elevator when the cable broke. Could he or she keep from being harmed as the elevator crashes by taking a jump into the air or holding onto a rod across the elevator? Although Einstein and other physicists have recorded their pleasure in thinking about the physics of the freely falling elevator, they apparently never worried about this particular question. You, however, may have a profound interest in it.

Taking a jump should help, and here's why. The bodily damage at the impact of the crash depends on the person's net downward velocity with respect to the ground at the moment of crash. Any way a person could reduce the net downward velocity should be of help. Taking a jump certainly should accomplish this. There are two problems with this solution, though. It may be very difficult to jump without a support, and a freely falling elevator does not render any support. Actually, however, because of air drag and other friction, there should be some support and taking a jump may not be impossible. The second problem is one of timing. If the person jumps too early he or she might hit the ceiling. And jumping too late, obviously, is the same as not jumping at all.

Even with a perfect timing, how much can a person's downward velocity be reduced? This depends on the height of the fall. For a 50-ft fall, perhaps a reduction of 10 to 15% of the velocity is possible. It is still unlikely that our unfortunate rider will be able to survive the crash.

What is the effect of holding on to a rail across the elevator? Apart from distributing the effect of the impact between the arms and the legs, there seem to be no other consequences.

The rotor

Usually the rotor comes in the form of a hollow cylindrical room that can be rotated about the central vertical axis of the cylinder. The riders stand against the wall and the rotor is set to rotate, its rotational speed gradually picking up. The excitement begins when, at a predetermined speed, the floor below the

FIGURE 7.10 The mystery of the rotor explained. See text.

passengers is opened, often revealing a deep pit. Children cry out—as might also a timid adult—but of course there is no danger of anybody falling in the pit.

The centrifugal force that comes into play in the rotating reference frame of the rotor is ultimately responsible for the thrill, but it is by no means obvious. Since the centrifugal force in this case is a horizontal force, how can it have anything to do with canceling the downward gravity force? The cancellation of the downward gravity is actually accomplished by the friction force, the friction of the cylinder wall with the passenger's bodies. The reason the friction is large enough in this case has to do with the centrifugal force. Here's why. The centrifugal force that acts outward presses everybody against the wall—indeed, if the wall were not there everyone would be thrown out along a tangent to the rotor wall. The wall exerts a normal force that is equal and opposite to the centrifugal force (Fig. 7.10). The friction force F_f that opposes a person's sliding down due to gravity is related to this normal force N by

$$\text{friction force } F_f = \text{coefficient of friction} \times N$$

What is the velocity v of the rotor? From the equilibrium of the passengers, we have two relations that are satisfied in a rotor ride for each individual passenger (Fig. 7.10):

$$\text{weight } W = \text{friction force } F_f \tag{7.1a}$$

$$\text{normal force } N = \text{centrifugal force } F_{\text{cent}} \tag{7.1b}$$

The magnitude of the centrifugal force is given by

$$F_{\text{cent}} = \frac{mv^2}{R} \tag{7.2}$$

where m is the mass of the passenger and R is the radius of the cylinder. Therefore,

$$N = F_{\text{cent}} = \frac{mv^2}{R}$$

The coefficient of friction in this case is about 0.4, so the friction force is $0.4N$, or $0.4mv^2/R$. Finally, this must equal the weight mg in Eq. (7.1a):

$$\frac{0.4\ mv^2}{R} = mg$$

As you see, m occurs on both sides of the equation and cancels out. Thus the result does not depend on m, the mass of any particular passenger. Canceling m and solving for v^2, we get

$$v^2 = \frac{Rg}{0.4} \qquad \text{or} \qquad v = \sqrt{\frac{Rg}{0.4}}$$

Typically R is 6 ft. Since g is 32 ft/sec², we get

$$v = \sqrt{\frac{6 \times 32}{0.4}} = \sqrt{6 \times 80} = \sqrt{480} \approx 22 \text{ ft/sec}$$

Vortices in the bathtub

Perhaps once in a while after the pleasure of a bath, you have removed the plug from the tub and watched the water rush down, twisting as in a whirlpool. More often than not, the whirlpool will be in the counterclockwise direction.

You have learned about the Coriolis force and are aware that this force causes the counterclockwise whirlwinds of the cyclones when air rushes in toward a low-pressure region from the outside. The similarity of these two patterns should arouse your sense of scientific reasoning. As water falls downward it is acted on by the Coriolis force; so it deviates to the right (we are in the northern hemisphere) and forms a counterclockwise pattern. Thus the Coriolis force is a part of your daily experience, after all. You do not have to see a tornado to see the Coriolis force in action.

But there is a snag. There is a chance that the very next time you go to a washbasin to do the experiment, you will find that the water whirls the wrong way. What's happening?

There is a simple explanation. The Coriolis force on the rushing water in a washbasin or a bathtub is a small force and other small forces that are usually present can make a difference. Seemingly inconsequential things, such as uneven heating of the water or a slight draft, can affect the outcome of this particular experiment. So it is safer to proclaim that there is a tendency of bathtub water to drain out with a counterclockwise whirl, a tendency attributed to the Coriolis force. If you watched the water drain from the tub 100 times, you would be sure to find it to whirl counterclockwise more than clockwise.

Still, this is not very scientific. If the effect is really due to the Coriolis force, we should perform the experiment with careful elimination of all these other counterforces and examine the results. The first such experiment was carried out by A. H. Shapiro of Massachusetts Institute of Technology and confirmed that the whirl is counterclockwise *always* when these other factors known to affect the flow are carefully removed. No draft, no uneven heating, careful settlement of the water for as long as twenty-four hours to eradicate any previous bias the water might have had before the plug was removed—Professor Shapiro saw to all these, and his efforts were rewarded.

The scientific verification of the whirling direction of bathtub vortices did not stop there. If the phenomenon is really due to the Coriolis force, then the effect should be reversed in the southern hemisphere. In a bathtub in Sydney, Australia, accumulated water should gurgle away down the drain in a clockwise vortex pattern. So another scientist whose curiosity was aroused, A. M. Binnie, went ahead and verified this.

To top it off, a team of scientists performed the experiment in a hotel in Nanuki, Kenya, a town located right on the equator.* As mentioned previously, the Coriolis force is zero on the equator itself. So at Nanuki, bathtub water will whirl this way or that, if at all, with no preference. And under the scientist's carefully controlled conditions, as above, the tub water should not whirl at all. And, indeed, when these scientists removed the plug, the water went straight down the drain in the bathtub.

---

*A member of this team, Magnus Pyke, has written an interesting account of the entire phenomenon in *Butterside Up* (New York: Sterling, 1976).

SUMMARY

This chapter serves three purposes. First, it clarifies some basic concepts regarding reference frames, symmetry principles, and the relativity of natural motion. Second, it points out that accelerated reference frames are special because of the existence of pseudoforces. Two such pseudoforces, namely, the Coriolis force and the centrifugal force, are discussed at length. The Coriolis force acts on all moving objects in the rotating reference frame of the earth. Its tendency is to deflect moving objects toward their right (in the northern hemisphere). The centrifugal force is the pseudoforce that tends to throw you outward when you are on a rotating merry-go-round. The magnitude of the centrifugal force is given as

$$F_{\text{cent}} = \frac{mv^2}{R}$$

Third, this chapter is also a prelude to Einstein's theory of gravitation, his celebrated general relativity theory. It is a fact that Einstein started his work by asking a simple question like this: How does an observer in a freely falling elevator perceive things? This led him to the discovery of the equivalence principle: One cannot distinguish between the gravity force and a pseudoforce that arises from being on an accelerated reference frame in a small region of space.

QUESTIONS

Review and reason

1. What is a reference frame? Distinguish between a reference frame and a coordinate system.
2. What do we mean when we say "physical laws have orientational symmetry"? How do we know?
3. What is an inertial reference frame? How can you tell if you are in an inertial reference frame?
4. What is a pseudoforce? You push a glass of water sliding down a table and have someone catch it at the other end. What happens to the surface of the water both initially and in the final phase of the experiment? Try it and see. Explain.
5. Why does the centrifugal force reduce our weight at the equator but not at the poles?
6. A needle is placed on a turntable. As the turntable starts rotating, picking up speed, the needle flies off even before the turntable reaches its final speed. Explain.
7. Mud flies off the tire when the car is in motion. Explain.
8. When a car makes a left turn, a passenger in the front seat slides toward the door. Explain this common experience (a) from an inertial reference frame point of view and (b) from the point of view of the passenger, his reference frame.
9. What is Coriolis force? An object on a turntable is at rest with regard to the turning turntable. Is there any Coriolis force acting on the object?
10. Explain the phenomena of cyclones and anticyclones, pointing out the role of the Coriolis force.
11. How does the Foucault pendulum prove that the earth rotates?
12. If you tie a stone to a rope and hang it from a three-story window and then put the stone into a swing, will the plane of swing of this pendulum remain fixed or will it change? Why?
13. Fast-moving artillery shells deflect a little bit to their right (with respect to the ground) in the northern hemisphere. Why?
14. Is gravity a pseudoforce? State the principle of equivalence.
15. The magnitude of the centrifugal force on an object resting on the edge of a rotating merry-go-round is given as mv^2/R. What are m, v, and R?
16. Bathtub water shows a tendency to whirl counterclockwise (in the northern hemisphere) as it drains. Explain.

1. You are in an elevator accelerating downward at a rate of 3 ft/sec². By how much does your weight change? Which way?

2. If your elevator accelerates upward at a rate of 3 ft/sec², by how much and which way does your weight change?

3. You are at the rim of a merry-go-round of radius 5 m rotating at a rim speed of 5 m/sec. Calculate the magnitude of the centrifugal force that acts on you.

4. Suppose you are in a space colony, a cylinder of radius 1 km. Calculate the speed of rotation of the cylinder at the wall if your entire earth weight is simulated.

8 Some Properties of Matter

The third basic entity of the world of classical physics was *matter.* (The first two are space and time.) The concept was hardly changed from the times of Leucippus to the beginning of the twentieth century: an impenetrable something which fills completely regions of space and which persists through time even when it changes its location.

MILIČ ČAPEK

■ 8.1 The Three States of Matter

This is perhaps the most elementary fact about matter: matter exists in three basic states, solid, liquid, and gas. Given a sample of matter, the state it assumes depends on its physical environment and also on its chemical composition. Thus, saying, as we usually do, that water is a liquid is not quite right; we all know situations in which water becomes solid ice or gaseous steam. The "waterness" of water is more fundamental than the fact that it ordinarily exists in the liquid form. We will see later that the waterness is the result of the submicroscopic (molecular) composition of water.

The first question we ask is this: how do we differentiate, at least qualitatively, between the three states? That is, what are the most obvious attributes of a solid, a liquid, or a gas? A solid has a fixed shape, which it tends to retain, a property that is called rigidity. Solids are rigid bodies. A solid also maintains a more or less fixed volume; it is almost incompressible. In comparison, a liquid is not rigid, but it too is relatively incompressible. A liquid sample does not have a fixed shape; it takes on the shape of its container. This is what we mean when we say that it is not rigid. But a liquid does tend to keep a constant volume—a property that distinguishes it from a gas, because a gas neither keeps a fixed volume nor a fixed shape. If you pour a liquid into a container, it will take on the shape of the container but only fill it to the extent of the liquid's volume. But put a gas in a container and it will not only adjust to the shape but also fill the container, taking on the container's volume as well.

To summarize, a solid is rigid and almost incompressible; a liquid is nonrigid but almost incompressible; and a gas is nonrigid and very compressible.

It turns out that because of their common property of nonrigidity, liquids and gases share many properties. An important one, for example, is their ability to flow. Because of this similarity, we give them a common name, **fluid.** The word "fluid," then, refers to both liquids and gases.

The atomic picture

Richard Feynman has said something to the effect that if most of human civilization was destroyed by some cataclysm and we were allowed to leave one piece of knowledge behind for the survivors, that piece would have to be the knowledge of the atom—that matter is made up of tiny constituents called atoms. It is possible to argue with this statement. After all, the idea has been around since Democritus of ancient Greece, but nobody seemed to be able to do anything with it until some other developments of science took place. But one thing we cannot argue with: the atomic picture is one of the most unifying pictures that people have ever constructed.

Previously we mentioned the waterness of water. According to the atomic picture, the waterness of water comes from the atomic constitution of the ultimate particles of water, the molecules of water. The idea is that if we could look at a drop of water with a microscope of sufficiently large magnifying power, we would "see" that this drop consisted of molecules of a very definite character. Molecules of no other substance would look the same. If you don't mind a rather idealized picture of the water molecule, it probably has the appearance of Fig. 8.1. The big piece is an atom of the element oxygen and the little pieces are the atoms of the element hydrogen.

The size of the atoms and molecules is of the order of 10^{-10} meter. So the magnification necessary for seeing molecules is about a billion, which we do not have at our disposal. Thus if you hold onto the motto "seeing is believing," you may have some difficulty in appreciating the power of the atomic theory. The point is this: the atomic picture is a model, no denying that. But so far, in whatever situation we have applied this model, the predictions always match the observations. Thus it is a very powerful model, and therein lies the usefulness of learning about it in spite of the fact that nobody has yet "seen" a single atom.

Previously we discussed the macroscopic differences of the states of matter that we call solid, liquid, and gas. Now equipped with the atomic picture, we can discuss the microscopic differences between these states.

It may have occurred to you already that there must be forces between atoms, forces that bind them together to form molecules. The same thing must be true of molecules. A solid piece of matter does not break apart because of these forces between its molecules.

The molecules of matter are also in constant rapid motion. In a solid the atoms and molecules form a very regular structure in the form of a crystal. Perhaps you can remember the last time you enjoyed the sight of the falling of hexagonal crystals of snow. These hexagonal snowflakes are built from tiny

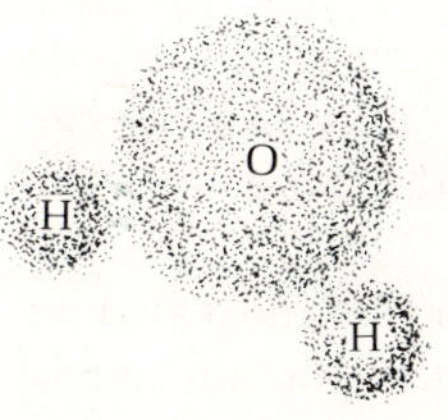

FIGURE 8.1   Idealized picture of a water molecule.

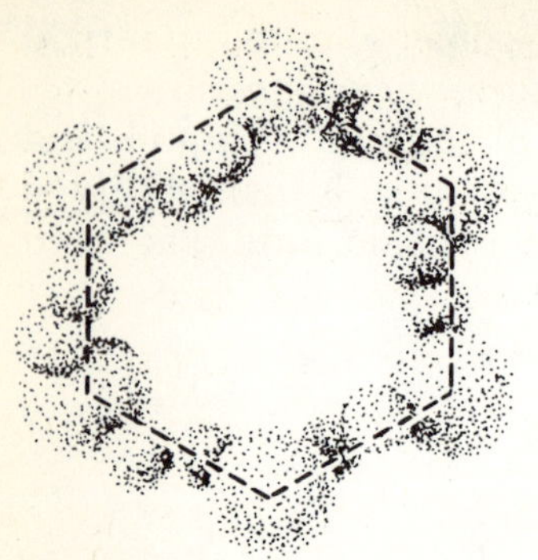

FIGURE 8.2   A hexagonal elementary crystal of ice.

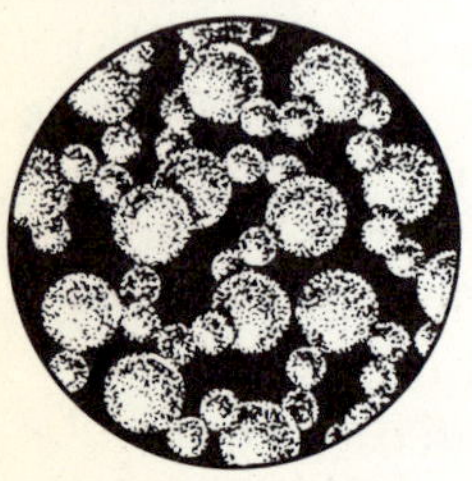

FIGURE 8.3   If a drop of liquid water was magnified a billion times, a portion of it might look like this.

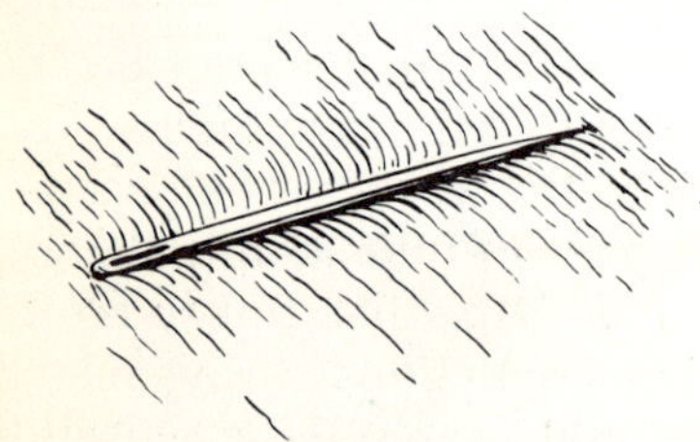

FIGURE 8.4   A magnified view of the depression created on the surface of water by a floating needle.

ones like that of Fig. 8.2, where, as you see, each of the corners is occupied by a molecule of water. The molecules can move, but only a little bit, about their positions in the crystals. In contrast, in liquid water the molecules are more free to move; they can slide by each other (Fig. 8.3). Finally, in a gas the molecules are almost completely free to go wherever they please. Clearly the forces between the molecules are at their strongest in the solid state, having weaker and weaker manifestations in the liquid and gaseous states.

We could compare the solid, liquid, and gaseous states of matter with the states of an audience during various times of watching a play. When the play is on, everybody is glued to his or her chair. People will occasionally move a little but they always will maintain some contact with their chair. During intermission their state is more like that of a liquid; people are seen to mingle and slide by each other, but they still clearly have a common pursuit. Finally, when the play ends, the crowd disperses and the individuals are now basically free to go wherever they wish. This is the analogue of the gaseous state.

As we mentioned previously the crystal structure of solids is the result of the orderly arrangement of the atoms (or molecules) of the solid. Now we will discuss one more interesting example of a phenomenon that finds a simple explanation in the atomic picture.

Have you heard the term **surface tension?** When placed carefully, a steel needle can be supported on the surface of water. Free, or almost free, drops of water take on a spherical shape; raindrops are a familiar example. These phenomena are caused by the large surface tension of water. There is a considerable force on the surface of a liquid, especially water. This force is called surface tension.

Surface tension originates from the tendency of the liquid to assume a surface of smallest area. Thus when a needle is creating a depression in the surface of the water, the water tries to readjust to a minimum area, in the process buoying up the needle (Fig. 8.4). And raindrops are spherical because for a given volume, a sphere has the minimum surface area.

How do we explain the contracting force of surface tension? In terms of molecules the explanation is almost obvious. A molecule inside the bulk of the liquid is acted on by forces from all sides due to its neighboring molecules; so all these forces cancel out. This is not so for a molecule on the surface. For it there are no neighboring molecules of the liquid on the top. Thus there is a net unbalanced attraction on the surface molecules that pulls them toward the inside of the liquid. As a result the surface contracts. Incidentally, if you are wondering whether a solid has similar peculiarities connected with its surface, the answer is yes. However, since solids don't flow, surface tension effects are much more difficult to observe.

One interesting thing about atoms around us on earth is that most of the physical and chemical processes they go through do not alter them at all but only rearrange them into different molecules or different physical states or both. Thus in all the turmoil about us here on earth, the atoms are something that are almost indestructible and perpetual.

Heraclitus of Greece, born about 500 B.C., thought that fire is the primary element of nature. He was quite taken by the ever-changeable character of fire and its forever restless flames. He saw the universe reflecting this restlessness.

Parmenides, a contemporary of Heraclitus, had just the opposite view. He thought that reality is forever; it does not change. The changes we see are only illusions.

You can see that both views are quite attractive as philosophies and each possesses a grain of truth. In the atomic picture, atoms are unchangeable in the course of everyday life; they represent the permanent aspect of our reality. Yet they are in incessant motion; they are forever restless.

Density

Density is the quality of matter that tells us how closely the submicroscopic particles of the given sample are packed. Intuitively, it is clear that solids will have higher density in general than liquids (there are some exceptions), and liquids have higher density than gases.

As you recall, the density (also called mass density) of a substance is defined as its mass per unit volume:

$$\text{density} = \frac{\text{mass}}{\text{volume}}$$

Now a problem arises in determining the density of gases because they are so compressible. The volume of a gas is rather sensitive to conditions of the environment. Since for a given sample of gas, the mass remains fixed, we must conclude that the density of a gas depends in a major way on the conditions of its environment. Any density value for a gas thus must include conditions of the environment.

In Table 8.1 we compare the density of some solids, liquids, and gases, measured under atmospheric conditions as specified. The weight density (weight per unit volume) is also given.

■ 8.2 Elasticity: Stresses and Strains in Solids

We mentioned before that a spring when elongated responds with a force that tends to bring it back to its original size once the external force is removed. Actually this property is shared by all solid matter to a varied degree. It is one aspect of the **elasticity** of solids. Aside from this tensile elasticity, solids also exhibit elasticity with respect to a bending force applied sideways (Fig. 8.5).

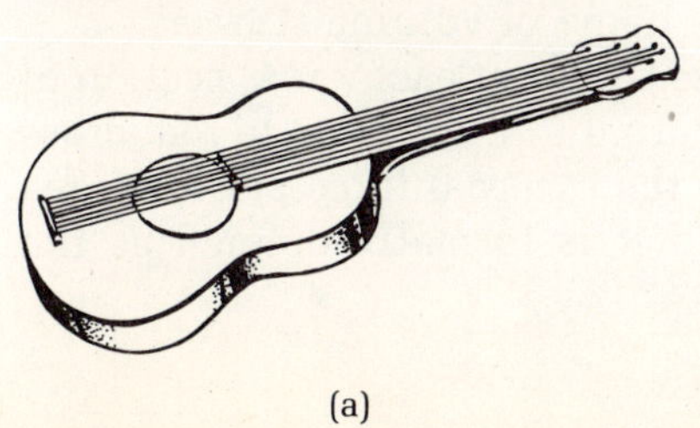

FIGURE 8.5   Different kinds of elastic deformation of solids. (a) Guitar strings under tension. (b) A rod being bent; an example of shearing strain. (c) a rectangular block undergoing volume strain.

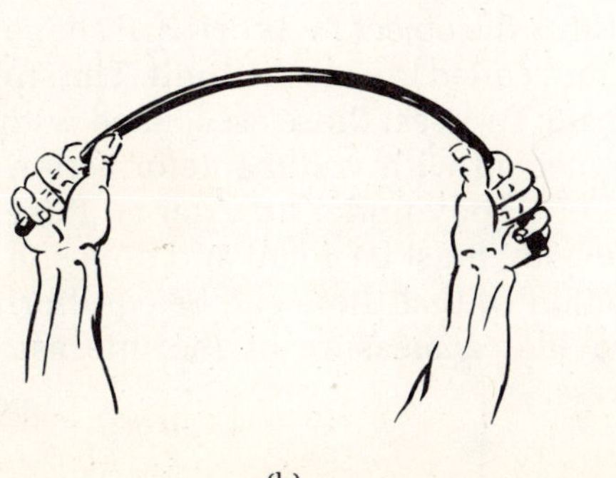

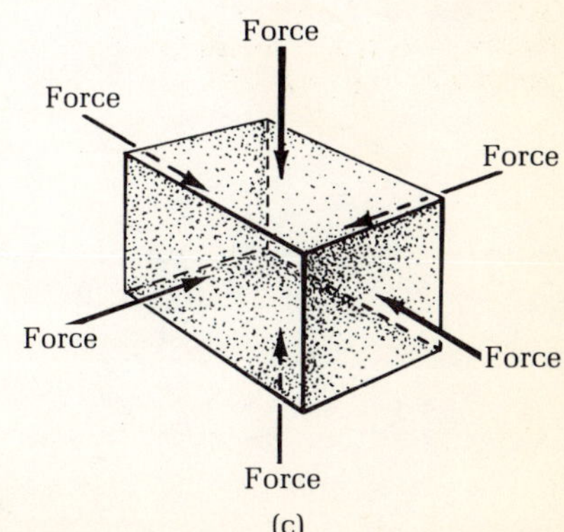

(a)

(b)

(c)

Substance	Density (10^3 kg/m^3)	Weight Density (lb/ft^3)
Solids		
Aluminum	2.7	170
Gold	19.0	1200
Human fat	0.92	58
Ice	0.92	58
Iron	7.8	480
Lead	11	700
Silver	10.5	650
Wood, oak (approx.)	0.8	50
Wood, pine (approx.)	0.4	25
Liquids		
Alcohol (ethyl)	0.8	50
Gasoline	0.68	42
Kerosene	0.8	50
Mercury	13.6	830
Oil	0.9	55
Water, pure	1	62.4
Water, sea	1.03	64
Gases		
Air	0.0013	0.08
Helium	0.0002	0.012
Hydrogen	0.00009	0.0054

This kind of elasticity usually comes under the name **shear.** Also, we can squeeze a whole solid from all sides and deform it. The tendency to recover from this kind of "bulk" deformation is also found to be considerable for solids. In general, we define elasticity as the tendency of solid objects to maintain their shapes and their ability (within limits) to return to their original shape after being deformed, once the deforming force ceases to act.

Consider a wire of iron placed under tension by a weight at one end. The wire is hanging from the ceiling. What do we mean when we say that the wire is under tension? Actually we are referring to the internal **stresses** set up in the wire. These stresses come into play whenever an object is deformed, and they balance the external force. When the external force is removed, the stress returns the object to its original shape, provided the deformation did not exceed a limit called the **elastic limit.** The stresses associated with elongation are called tensile stresses; those associated with shear are called shear stresses; and those connected with volume deformation go by the name of volume stresses.

For a body under an external force, the stress set up at each cross section of the body must be equal and opposite to the external force applied to that cross section so that there can be equilibrium. Thus the external force per unit area provides a measure of the internal force, and it is formally defined as the stress:

$$\text{stress} = \frac{\text{external force}}{\text{area}} \tag{8.1}$$

Shear is a force that causes sliding, bending or twisting strains on solid bodies. This word and its physical meaning neatly parallel certain personality types, which could be dubbed "shearist." Svengali could be considered a prime shearer; Charlie Brown and Walter Mitty are perfect shearees. Next time you see someone verbally twisting another's arm, you will be observing a shearer in action.

For a given force, a thin rod will stretch more than a thick one. It is the force *per unit area* that counts, which is larger for the thin rod. This is the reason for defining stress the way we did. The deformations of solids are also evaluated in a slightly different fashion. We are interested in the ratio of any change and the original, a quantity that is referred to as **strain.** For tensile deformation, for example, we are interested in the change of length divided by the original length.

Hooke's law of elasticity—that the restoring force is proportional to the deformation—can now be stated in a slightly different form:

The stress is proportional to the strain.
$$stress \propto strain$$

The coefficient of proportionality that appears when we write this relationship as an equality is called the coefficient or modulus of elasticity.

An illustration of the behavior of a metallic wire under different stresses according to Hooke's law is shown in Fig. 8.6, with greatly exaggerated elongations. In actuality, you need a magnifying glass to measure the elongation accurately enough to verify Hooke's law, but the point of the diagram is valid. The elongation is indeed proportional to the stress.

Hooke's law breaks down when the elastic limit of a material is reached. Beyond the elastic limit, some materials (called ductile materials) continue deforming without breaking, but the deformation is no longer reversible. The sample does not come back to its original shape even after the external force is removed. Copper is a good example of a ductile material. In contrast, some solids break at or near their elastic limits. Ice is a good example of this. Such materials are called brittle.

In everyday life there are many examples of elasticity. When we bend a bow, stretch a rubber band, or literally twist somebody's arm, we are counting on the elastic properties of the material under deformation. We expect the materials to go back to their original shape when we remove the deforming force. The physics of elasticity is also very important to material scientists or structural engineers, as you might guess.

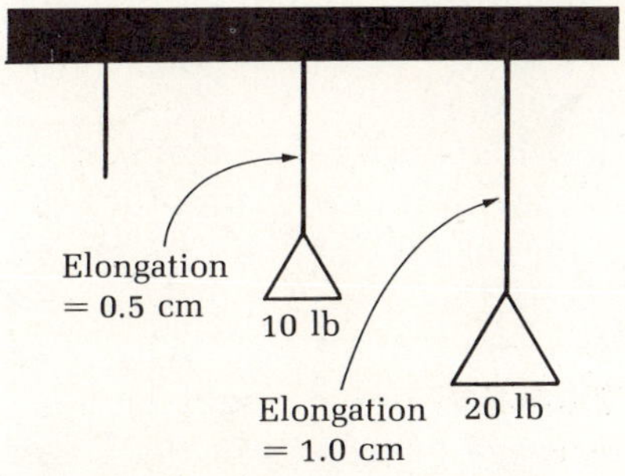

Strength of materials and scaling

What is **scaling?*** It means to increase or decrease all linear dimensions of a given structure in the same ratio without altering its shape. A most important question is this: do all physical properties of the structure remain unchanged?

FIGURE 8.6 Illustration of Hooke's law. The unstretched wire is shown on the extreme left.

---

*See Appendix 1 for an elementary discussion of scaling.

Novelist Jonathan Swift, with his very fertile imagination, gave us the creatures of Brobdingnag, who are scaled-up versions of humans. Are such giant humans really possible?

Unfortunately, there is a stability problem with the scaled-up structure. Galileo worried about this question and understood the basic physics involved.

The weight of a body can be given as

$$\text{weight} = \text{mass} \times g$$

Since mass is volume times density [see Eq. (2.3)], we can write

$$\text{weight} = \text{volume} \times \text{density} \times g \tag{8.2}$$

When we talk about scaling up a living organism, it is understood that the scaled-up model will have to be built of the same material, the same tissues. Thus the density remains the same and obviously so does g. From Eq. (8.2) the weight scales up by the same amount as the volume. Now the volume is proportional to the cube of the linear dimension L, and so the volume scales as the cube of L. This means that if we change the linear dimensions by a factor of 3, the volume will change by a factor of 3^3, or 27. Eq. (8.2) for the weight tells us that the weight also scales as L^3 and changes by a factor of 27 in this example.

The bone structure of the scaled-up body has to support this scaled-up weight. How does the strength of the bone structure scale? Presumably, the bones of the larger model must be made of the same material as the smaller version; thus the stress tolerance of the bone should be the same. Now the force a given structure can support is the product of stress and area [see Eq. (8.1), the definition of stress]. Since the maximum stress doesn't change, the maximum force the bones can support must scale as the area, as L^2. So the weight is supported by a bone structure that has a strength which scales as L^2 rather than L^3. As we scale up a body, the strength of its bone structure does not increase in the same proportion as its weight, and thus the bones will be quite ineffective in supporting the increased weight. What this means is that the creatures of Brobdingnag will be structurally unstable. There is a good reason that giants don't exist in nature.

Perhaps a numerical example is called for. Suppose we think of a human giant whose linear dimensions are twice those of a regular human, two times taller, two times wider, and so forth. His weight has to be 2^3 or 8 times that of a regular human, but, alas, his bone structure is only 2^2 or 4 times stronger. So his bones are just not strong enough to support his weight.

Of course, nature knows about this problem. So when it makes an elephant, it doesn't just scale up a smaller animal—a mouse, for example. It provides the larger animals with support bones that are right for supporting their increased weight. For example, a horse is about 3 times as tall as a wolf, but look at a horse's legs. They are not just 3 times thicker but more like 5 times. Why? The square of 5 and the cube of 3 are roughly equal. Thus the bone strength has been increased by nature in the same ratio as the increase of weight in the larger animal.

The word **pressure** is probably quite familiar to you on several accounts. First of all, who hasn't heard about atmospheric pressure, the pressure exerted by the atmosphere on every object in and under it? Second, in this very automobile-centered world, it is hard not to be conscious of phrases like "tire pressure." From medical checkups you must know about blood pressure. And this probably does not exhaust all the "pressure" language that you are aware of.

What does pressure mean? It is the force per unit area. In many situations around us, such as the stresses just encountered, it is this force per unit area that counts, not the total force. In this section we will consider some further examples.

In fluids the absence of rigidity gives a new twist to the problem, so much so that in the study of pressure in fluids we encounter novel and sometimes unexpected aspects. For example, we will see that the pressure is the same in all directions at any point of the fluid, although the magnitude varies with the depth of the fluid. Another interesting feature is that an external pressure is transmitted uniformly throughout the fluid. This has some spectacular industrial applications.

Pressure is measured by means of such instruments as the barometer, the pressure gauge, and the manometer. A barometer measures the pressure of the atmosphere and is important for weather forecasting, among other things. The auto mechanic measures tire pressure with a pressure gauge. But perhaps the pressure gauge you may be most curious about is the manometer, the contraption the office nurse puts on your arm when you go for a medical checkup. You know that she measures your blood pressure. But how? And what do her numbers mean exactly? It's all explained below. Read on!

Force and pressure

Let's start with a silly question: would you prefer to sleep on a bed of nails rather than in a regular bed? Of course not. But have you wondered why? It is not as comfortable, you would say. But why not? In both cases you act on the bed with the total force of your weight; the bed reacts back on you with the same force. What's the difference?

There is a difference—the surface area of contact between you and the bed. In the case of a regular bed, this area is roughly one-fourth the area of your body, because the bed fits your contour pretty well. In the case of the bed of nails, the area of contact is very small; perhaps only one-thousandth of the total area of your body is in contact with the support.

Why does the area matter? Because the larger the area of contact, the less is the pressure. Therefore, the regular bed feels comfortable. It is the pressure that matters. Pressure is force divided by area. The less the area, the more intense is the pressure. So you see now why you wouldn't be comfortable on a bed of nails.

Of course, the pressure area relationship can be used to advantage on occasion. For example, in digging a hole in the ground, you would rather use a sharp

instrument than a blunt one. The sharp instrument, by virtue of the small area of its pointed blade, can exert more pressure for the same total force.

Let us dwell on the fact that pressure is the force per unit area within the context of automobile tire pressure. As you may know, each of the tires bears the load of one-fourth of the weight of the car. So the total force on a tire is constant, namely, one-fourth the weight. Since

$$\text{pressure} = \frac{\text{force}}{\text{area}} \tag{8.3}$$

we have as well the following relation:

$$\text{pressure} \times \text{area} = \text{force} \tag{8.4}$$

So the product of air pressure and the area of the tire in contact with the ground must remain a constant, equal to the constant force in this case, one-fourth the car's weight.

Now you can understand why a tire becomes flat when it loses air pressure due to a leak. If the air pressure is reduced, the contact area must increase so that the product, pressure × area, remains constant. When the contact area of the tire with the ground increases, it looks flat (Fig. 8.7).

Some special things about pressure in fluids

There are several special features of the behavior of pressure in connection with fluids. When we talk about the pressure on a solid object as the force per unit area, we always think of a direction. The force, and therefore the pressure, has a definite direction. Strangely, for a fluid, this directional property doesn't exist; the pressure of a fluid is the same in all directions. Thus when you are under water, the water is exerting pressure on your body from all directions with identical magnitude. Try it and see. Stay at a certain depth of water and turn your head in several different directions. You can feel the water pressure on

FIGURE 8.7 A tire becomes flat when the air pressure is reduced. The reduced air pressure times the increased area of tire still equals one fourth the weight of the car.

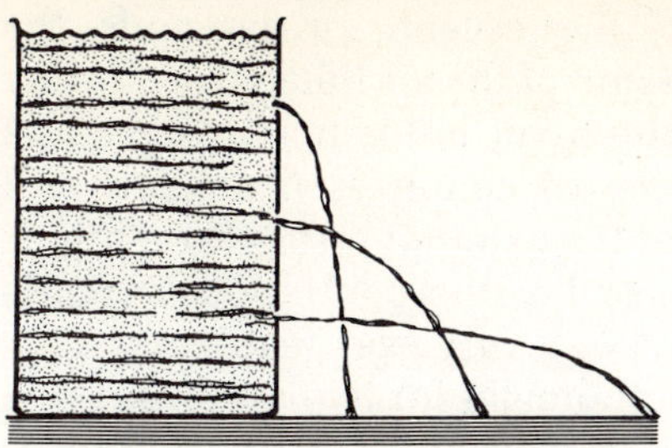

FIGURE 8.8 Pressure is highest at the bottom hole and exerted equally in all directions. Thus the water is forced out farthest from this hole.

your ears, but that pressure will not be affected at all by your changing the position of your head.

Notice that the directions for the simple experiment above said to stay at a certain depth. This is because pressure in a fluid is the same for all points at the same level but varies from level to level. You may have noticed that the pressure of water goes up in proportion to the depth when you dive underwater.

The complete formula for the pressure at a point in a fluid is

$$\text{pressure} = \text{weight density} \times \text{depth} \tag{8.5}$$

or, since weight density is mass density $\times$ g,

$$p = \rho g d \tag{8.6}$$

where ρ (pronounced "rho," a lowercase Greek letter) is the mass density of the fluid, d is the depth of the point below the surface, g is the acceleration of gravity, and p denotes pressure.

There is an interesting demonstration of this aspect of fluid pressure. Look at Fig. 8.8. Of the three holes in the side of the vessel, the water rushes furthest from the hole that is the lowest under the level of water (maximizing the depth), indicating that the fluid pressure increases with depth. Also important to note is that the size of the vessel and the volume of fluid it contains are quite inconsequential; only the depth matters.

Let us consider a practical problem that you may have wondered about, the problem of underwater diving. As the diver descends deeper and deeper into the water, the pressure the water exerts on the diver from all sides increases in direct proportion to the depth. The increased pressure of the water is transmitted to all the fluids in the body—the blood, the fluid in the cells, and so on. To maintain a balance of pressure in all parts of the body, the diver must inhale air at about the same pressure as the surrounding liquid. This equalizes the internal and external pressures, and no difference of pressure develops in the body, which otherwise could break a blood vessel or cause some other tissue damage.

■ DERIVATION OF THE PRESSURE FORMULA

Consider a liquid column of area A and depth d. The volume of the fluid is area times depth, or Ad, and the weight is volume times density times g, or, in symbols, $Ad\rho g$. The pressure at a point at the base of this column is the force per unit area (Fig. 8.9):

$$\text{pressure} = \frac{Ad\rho g}{A} = \rho g d$$

Since the weight density is equal to the product of mass density and g, we can also write this as

$$\text{pressure} = \text{weight density} \times \text{depth}$$

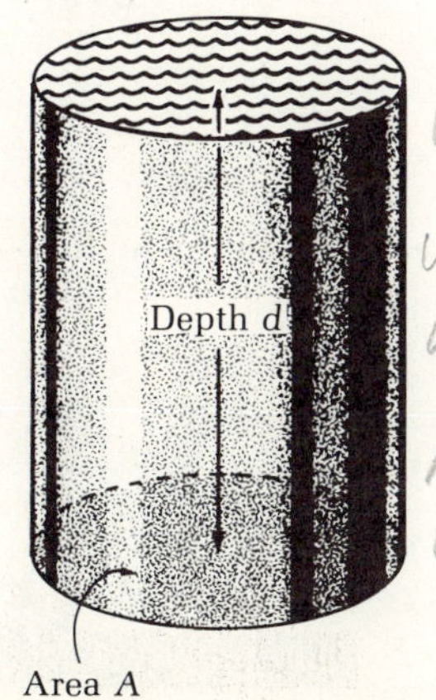

FIGURE 8.9 Pressure at the base is ρgd.

But now you can see the problem. As the diver ascends or descends, the depth varies and the diver must adjust the pressure of the air inhaled according to the depth. Many accidents occur because the diver holds his breath while ascending. The air inhaled at a greater depth was compressed air and will invariably damage his lungs as he comes up and the external pressure on him is reduced. So he must exhale as he comes up or breathe continuously while adjusting the pressure of his air tanks. This is not a very easy feat.

The pressure of the atmosphere decreases with height for two reasons. First, the depth of the air column above us decreases as we go up. But an even more important factor is that the density of the atmosphere decreases rapidly with height due to the effect of decreasing gravity. The direct result of this is that at high altitudes we suffer from lack of oxygen. Thus for high-altitude travel—the inside of a jet plane, for example—the vehicle must be pressurized to supply us with enough oxygen.

There is another interesting property of a fluid. *An external pressure applied to a fluid is transmitted in a uniform manner throughout the fluid, acting in all directions.* This principle is called **Pascal's principle** (named after French natural philosopher Blaise Pascal). It is the basis of the operation of the hydraulic lift, which is often used to lift very heavy objects like an auto. You can appreciate how it works by looking at Fig. 8.10. There is a piston of smaller area on one side and one of larger area on the other. Because pressure applied on the small piston must transmit undiminished through the fluid, we get the following interesting result. Since force is proportional to the area, if pressure is constant [Eq. (8.4)], the force on the other side must be larger in proportion to the area. For example, we can apply a 1-lb force on an area of 1 in² and magnify it 100 times by transmitting it through a fluid to act on a piston of area 100 in². Indeed, the larger piston will support a load of 100 lb. A hydraulic press and a hydraulic jack work on this principle.

It turns out that all these properties of a fluid really originate from just this one thing—the fact that fluids have no rigidity. You cannot apply a sideways force on a fluid. One of the first consequences of this absence of a sideways force is that the pressure of a fluid on the walls of its container is always perpendicular to the walls. Otherwise, according to Newton's third law, the container could also exert a sideways force back on the fluid, a force that a fluid cannot sustain. Without a reacting partner, the action cannot exist in the first place.

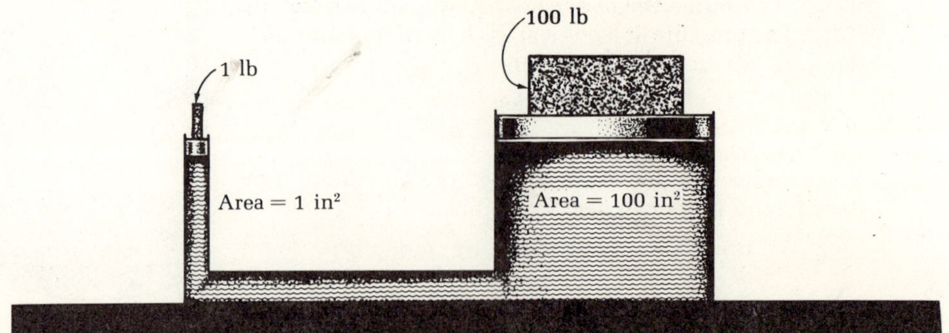

FIGURE 8.10 The pressure of 1 lb applied on an area of 1 in² is transmitted through the liquid unchanged. Thus it can support the weight of 100 lb acting on an area 100 times the original.

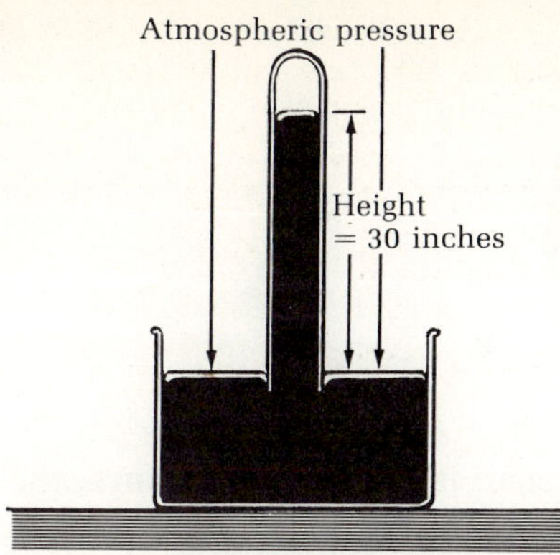

FIGURE 8.11 The principle of a mercury barometer. The atmospheric pressure holds the column of mercury up.

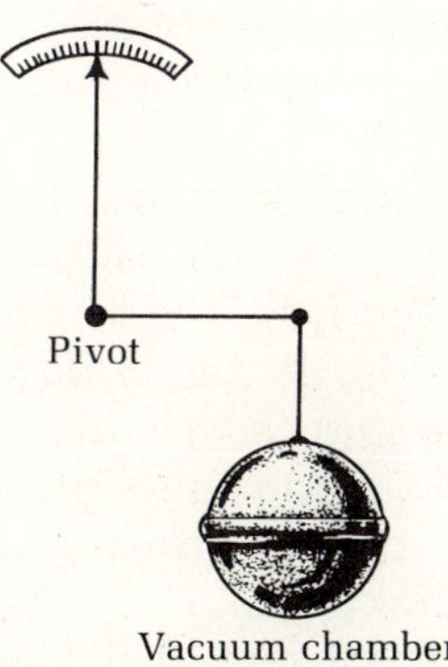

FIGURE 8.12 An aneroid barometer. See text for further explanation.

Units and measurement of pressure

Let us quickly settle the question of units. The MKS unit of pressure* is the newton per meter² abbreviated as N/m², and the standard FPS unit is pound per foot². Unfortunately, neither of these units is used much. The unit that you hear in connection with auto tires is pounds per square inch (abbreviated psi or lb/in²), a standard pressure for the tire of a standard-sized car being 28 lb/in². Another standard unit is the atmosphere (abbreviated atm), the pressure of our atmosphere at sea level. Sometimes this is also referred to as a bar (no abbreviation). Some useful conversion factors are as follows:

$$1 \text{ atm} = 1.013 \times 10^5 \text{ N/m}^2 = 14.7 \text{ lb/in}^2 = 1.013 \text{ bar}$$
$$1 \text{ lb/in}^2 = 144 \text{ lb/ft}^2$$

Still another important and often used unit of pressure is inch or millimeter of mercury. If you fill up a tube that is over 30 in long with mercury and invert it (Fig. 8.11) in a bowl of mercury, an interesting thing happens. Some of the mercury runs out, but at sea level a column measuring 30 in or so remains.

Consider the pressure at a point at the base of the column. The pressure at this point is due wholly to the weight of the column of mercury, since at the empty top of the tube, there is a vacuum (the air has been forced out). But the pressure at other points (outside the base of the column) at the same level on the surface of the mercury in the bowl is the atmospheric pressure. Since the pressure must be equal at all points of a liquid at the same level, it follows that the weight per unit area of a column of mercury 30 in (or 760 mm) high is the same as that of the atmosphere. Since mercury has a very high density, the weight per unit area of a 30 in column of mercury exactly equals the weight per unit area of the entire column of air above your head.

Mercury barometers are still popular for measuring pressure, and obviously they give the pressure in inches or millimeters of mercury. For conversion purposes,

$$30 \text{ in of mercury} = 760 \text{ mm of mercury} = 1 \text{ atmosphere}$$

The reason that mercury barometers work is that the atmospheric pressure holds up the column of mercury in the tube. Most household barometers today use another principle for measuring pressure. In the aneroid barometer, pressure is measured in terms of the extent to which the lid of a partially evacuated metal box is pushed in or out. The movement of the lid is transmitted to a pointer, which is calibrated in pressure units (Fig. 8.12).

Actually, even if you don't have a regular aneroid barometer in your household, if you look in your kitchen, you probably can find one without much difficulty. Any vacuum-packed can is an aneroid barometer (unfortunately without a dial), since it is a partially evacuated metal box. When the atmospheric pressure changes—for example, when there are rapid changes of weather—you can hear popping sounds from a vacuum-packed can as its lid moves up or down.

Now we consider one more important thing. When we say that the pressure of

---

*The unit newton per meter² is also called a pascal, abbreviated Pa.

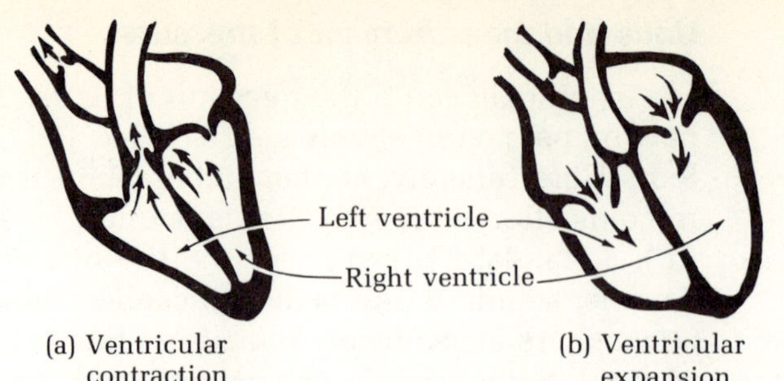

an automobile tire is 26 lb/in², what we really mean is the **gauge pressure,** the pressure read by the gauge. The gauge pressure is the difference between the unknown pressure and the atmospheric pressure. The real pressure, called the **absolute pressure,** of the air in the tube is the gauge pressure plus the atmospheric pressure. In our example, this turns out to be $26 + 14.7 = 40.7$ lb/in².

The human heart is a pump operated by the all-important heart muscles. On maximum contraction this pump can exert a pressure (called systolic pressure) of some 120 mm of mercury on the blood within it (Fig. 8.13). On relaxation the residual pressure (diastolic) comes down to about 80 mm of mercury. Thus the average blood pressure is the average of these two numbers:

$$\frac{120 + 80}{2} = 100 \text{ mm of mercury}$$

Under the action of pressure, blood is forced out of the ventricle of the heart into the aorta, through the branches of the arterial system onto the capillary beds, to be returned by the veins to the right auricle of the heart at just about zero pressure. The maximum difference of pressure between the arterial and venous blood is something like 120 mm of mercury.

Both the figures quoted for the systolic and the diastolic pressure, respectively, are gauge pressures as measured by a mercury manometer. You probably have personal experience of having your blood pressure being measured with this instrument, so the details may be of interest. The manometer, by the way, is a U-tube containing mercury, one end of which is connected to the external pressure to be measured while the other end is open to the atmosphere. The difference of the two levels of mercury (Fig. 8.14) gives us the gauge pressure.

In measuring the blood pressure, the manometer is connected to a closed bag, which is wrapped around the upper arm. Air is pumped into the bag by squeezing the rubber bulb until the air pressure within the bag is well above the systolic blood pressure within the artery. This cuts off the blood flow in the arteries of the forearm, and, as can be detected with a stethoscope, the sound of the pulse is cut off. Then air is released until the sound of the pulse returns. The return of the pulse sound signals the equalization of the bag pressure with the systolic pressure as a restricted amount of blood flow is resumed. The difference of the mercury levels in the manometer at this point gives the value of the systolic pressure. The restricted blood flow is associated with a characteristic tapping sound, which can be heard with the stethoscope. As more and more air

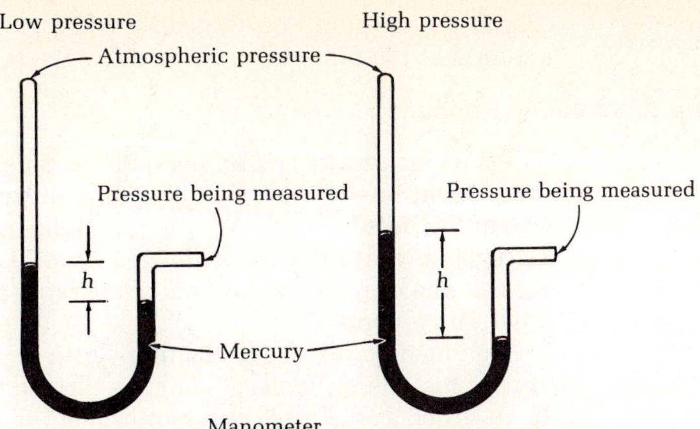

FIGURE 8.14   A manometer. The difference of the two levels of mercury is proportional to the difference of the pressure being measured and the atmospheric pressure.

is released, the tapping sound will stop as the pressure within the bag is reduced to the diastolic pressure and blood at low pressure is able to pass through. Now the diastolic pressure can be read from the gauge and the job is done. One more important thing: the bag must be attached to the arm at the same height as the heart since pressure depends on the height.

The fact that pressure depends on height sometimes causes problems. Occasionally when we are lying down and happen to rise suddenly, we feel dizzy. In the prone position the heart and the brain are at the same elevation, so the pressures in both are equal. In the standing position the head is somewhat above the heart. Pressure in a fluid increases with depth, so the heart pressure must be greater than the pressure at the brain (by about 35 mm of mercury). Now the average pressure of the heart is 100 mm, which is the pressure in the brain in the prone position. But when a person gets up suddenly the brain pressure has to drop from 100 mm to $(100 - 35) = 65$ mm. To maintain a constant flow of blood, the arteries in the brain region must expand to compensate for the drop in pressure. But such adjustments usually take some time. So, if you get up too quickly, you feel dizzy.

A numerical example

As a numerical example let us calculate the water pressure at a depth of 34 feet. For this purpose we can use either of the formulas for the relationship of pressure and depth, but it is a little more convenient to use Eq. (9.5), since the weight densities can be taken directly from Table 8.1. The weight density of water is 62.4 lb/ft^3 and the depth is given as 34 ft. So the pressure is given as

$$\text{pressure} = 62.4 \text{ lb/ft}^3 \times 34 \text{ ft} = 2121.6 \text{ lb/ft}^2$$

$$= \frac{2121.6 \text{ lb}}{144 \text{ in}^2} = 14.7 \text{ lb/in}^2 = 1 \text{ atm}$$

In the calculation we used the fact that 1 ft^2 = 144 in^2. Thus for every 34 ft the diver plunges, the pressure of the overhead water increases by 1 atm. It is also clear from this example that the atmospheric pressure can sustain a column of water only as high as 34 ft but no higher. This is the reason that a tube well, which uses a pump on the earth's surface to pump up the water, creating a vacuum at the top end, is limited to this 34 ft.

■ 8.4 Buoyancy: The Archimedes Principle

The ancient Greeks had an aversion to putting knowledge to work. There is a story about Euclid that illustrates their attitude. One day a man came to attend one of Euclid's lectures. After it ended the man asked: "That was all very interesting, but what's the use of all this?" Euclid looked at the man disdainfully and said to one of his slaves, "This man wants to make a profit out of knowledge. Give him a penny."

Archimedes was one of the first Greek scientists to break with the tradition. He put his science to use; in fact, he thrived on it. In this aspect perhaps he can be regarded as the true precursor of modern Western science—in particular, of American science, where the credo has always been to make practical use of knowledge.

The king of Syracuse (Syracuse was the city where Archimedes lived) knew only too well this uncanny ability of Archimedes to solve practical problems. Once the king wanted to find out if a certain crown was made of gold, as purported, and who but Archimedes could determine this without mutilating the crown?

Archimedes, needless to say, found the solution to the problem. There is a legend that he hit upon the right idea while in his bathtub and excitedly ran through the streets of Syracuse shouting, "Eureka! Eureka! I found it! I found it!" What he had discovered was a force that a fluid exerts on all objects immersed in it. We call this the **"buoyant force."**

Before discussing how the buoyant force solved Archimedes' problem, let's discuss a common example of this force. Perhaps you have attempted at one time or another to lift an object like a rock while it is underwater. Doesn't it feel lighter underwater than it would on the ground? This is due to the upward buoyant force canceling a part of the gravity force acting downward on the object (Fig. 8.15).

FIGURE 8.15　A buoyant force in action.

Archimedes not only discovered the existence of this force, he also determined the magnitude of the force. His finding is stated in a form that is now referred to as **Archimedes' principle:**

The buoyant force on an object immersed in a fluid is equal in magnitude to the weight of the fluid displaced by the object.

How do we find the weight of the displaced fluid? Clearly the volume of the displaced fluid is equal to the volume of the part of the solid that is immersed. (If the entire body is submerged, we take the total volume of the solid as the volume of the fluid displaced.) The mass of the displaced fluid is the product of this volume and the density of the fluid. Finally, the weight is mass times g.

Now we can go into Archimedes' solution. Archimedes determined the weight of the crown when suspended by a string in water and compared it with its normal weight. The ratio came out as 0.948.

$$\frac{\text{weight of crown when immersed in water}}{\text{weight of crown}} = 0.948$$

But the weight of the crown when suspended in water is the difference between its real weight and the buoyancy, or the weight of the displaced water. This gives the following relation:

$$\frac{\text{weight of crown} - \text{weight of displaced water}}{\text{weight of crown}} = 0.948$$

or

$$1 - \frac{\text{weight of displaced water}}{\text{weight of crown}} = 0.948$$

Rearranging, we have

$$\frac{\text{weight of displaced water}}{\text{weight of crown}} = 1 - 0.948 = 0.052$$

Inverting,

$$\frac{\text{weight of crown}}{\text{weight of displaced water}} = \frac{1}{0.052} = 19.3$$

You can see that the ratio of the two weights is really the ratio of two densities. (Since weight = volume × density × g, and the volume and g are the same for either of the two weights.) So the final result of Archimedes' analysis can be stated as follows:

$$\frac{\text{density of the material of the crown}}{\text{density of water}} = 19.3$$

Was the crown made of gold? Look at the densities given in Table 8.1. The ratio of the density of gold to that of water is found to be 19 according to the table. Archimedes' value was within experimental error of this value. Thus the crown was made of gold after all.

Archimedes' principle can tell us all about floating and sinking of solid objects submerged in liquids, which often is very important to know. In general, an object floats in a liquid if its density is less than that of the fluid; but if the object has a greater density, it will sink.

Let us see how this works, through a few examples. Consider ice floating in water. We know ice floats in water because ice has a smaller density than water. We can now figure out what proportion of the volume of a block of ice stays under water when it floats.

The ice block will reach an equilibrium in the water when the downward weight force is equalized by the opposite buoyancy force—that is, when the weight of the displaced water equals the weight of the block of ice. To be specific, suppose we consider a block of glacier of volume V floating in seawater (Fig. 8.16). The weight of the ice is equal to its total volume V times the weight density (it is more convenient to work with weight density for this kind of a problem). The weight of the displaced seawater is the weight density of seawater times the volume of the submerged portion of the ice, call it V_{sm}. The weight densities of ice and seawater are 58 and 64 lb/ft^3, respectively. So we get the relation

$$58V = 64V_{sm}$$

Therefore,

$$\frac{V_{sm}}{V} = \frac{58}{64} = 0.9$$

So icebergs are always found to float with only 10% of their volume above the sea; 90% of it stays below the surface of the water.

You can discover one interesting thing from Table 8.1. Ice, although less dense than water, is actually more dense than the alcohol in liquors. When alcohol is mixed with water in a mixed drink, you can tell how potent the drink is by looking at the ice cubes. Are they still floating normally or are they almost totally submerged? If ice tends to sink in a cocktail, you can be sure that the drink is spiked.

Let's consider some more examples. Iron is a dense material, as we all know. Is there a liquid in which iron can float? Yes; as you see from the table, mercury is such a liquid. It is considerably more dense than iron.

Of course, iron should sink in water, yet we know that ships made of steel (iron) do float. How come? The mystery is solved when we realize that the ship is built in such a way that it has a large amount of space filled with air. The

FIGURE 8.16   A floating iceberg.

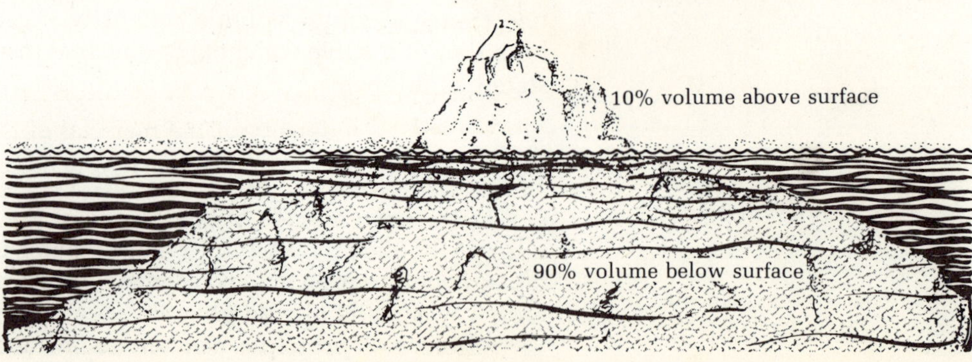

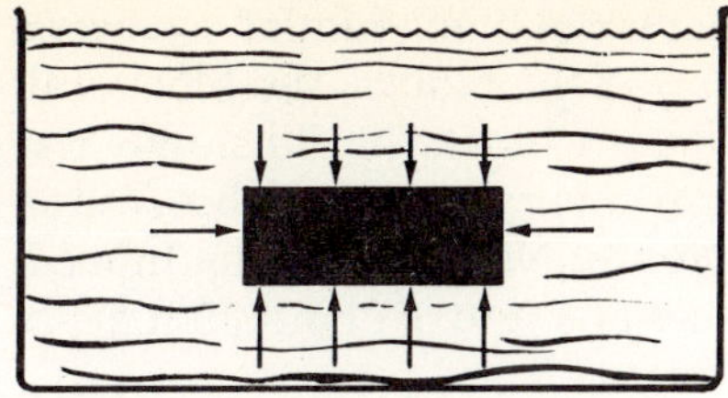

FIGURE 8.17 Pressure on the bottom of an object is larger than that on top, giving a net force upward. This is the buoyant force. The buoyant force balances the gravity force (not shown).

average density of the ship, which is the ratio of its total mass to the total volume, is less than that of water, so the ship floats.

We should mention that the buoyancy force on a submerged object can be understood in terms of the pressures acting on the object. Fig. 8.17 shows the pressures on an object submerged in a liquid; the pressures come from all directions. The liquid pressure on the sides of the body cancel one another, because they occur in equal and opposite pairs. (Of course, these are *not* action-reaction pairs.) But the same is not true for the pressure that acts on the top of the immersed body versus that which acts on the bottom. As you know, pressure of a liquid increases with depth. So the upward pressure on the bottom is greater than the downward pressure on the top. The net effect of this difference in pressure is that there is a net force upward on the object. This is the buoyant force.

Floating plates of the earth's crust

Why are mountains so high? Italian painter Leonardo da Vinci concerned himself with this problem to the extent of finding a solution. He decided that the mountains must be composed of rocks of smaller density than the usual crustal material and this somehow is responsible for the mountain's relatively higher elevation. Later research not only proved that he had a point but also raised new questions. We have mentioned the Everest experiment (see Section 6.2) and the question it raised: why do mountains extend a root underneath?

In general, we can ask: how do the continents manage to be supported so high above the ocean floor level? It is, of course, possible that the layer of the earth below the crust (called the mantle) is solid and rigid, and the crust simply rests on it, built the way it is with mountains and plains and ocean floors. But there are many problems with such a picture. We will mention one interesting consequence of a model like this. You know that the effect of erosion levels everything down to the same size given sufficient time. So, eventually, everything will become level and the ocean water will then flood the continents. This has been projected in many mythologies as the great flood. Does this reasoning have any validity?

There is an alternative way of thinking. Perhaps the mantle is not rigid but acts like a liquid in processes that involve a long time. Such a mantle would exert a buoyant force on the overlying crustal layer. With this reasoning we make up models that explain not only why the continental levels are higher than the ocean crust but also why mountains have roots.

One model of this kind is shown in Fig. 8.18. The crust is visualized as consisting of blocks of various heights floating on the denser mantle liquid underneath. The blocks are in equilibrium because the buoyant force of the liquid mantle cancels the downward gravity force. As you can see, in this particular model the densities of the crustal blocks are assumed to be identical. Later models allowed for different densities for different blocks and also for variation of density with depth in the same block (since density should increase with depth).

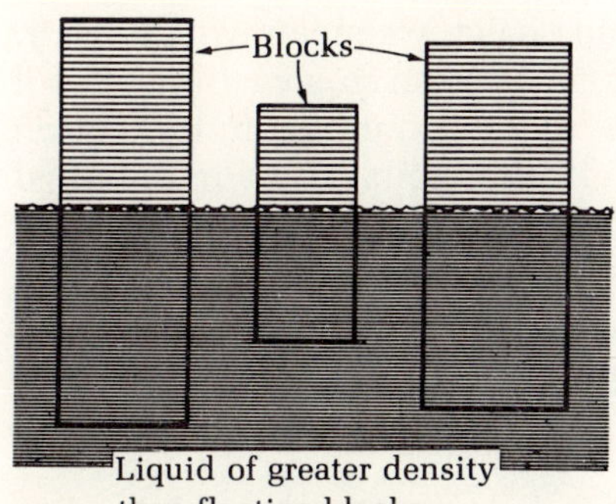

FIGURE 8.18 The earth's crust can be looked upon as blocks of various sizes floating in the liquid mantle, which is denser.

What evidence do we have for such an equilibrium—or **isostasy,** as it is called by geologists—other than what we have mentioned so far? One spectacular piece of evidence comes from the remnants of the last ice age, known to have occurred a few thousand years ago. It is clear that if the block model is correct, then if an extra weight of ice accumulates on one of the blocks, the block will sink further into the mantle. Later as the ice melts, the block should slowly rise until isostasy is recovered. This kind of isostatic recovery has been observed in Scandinavia as well as in the Hudson River region of New York state. Indeed, the land level is known to be rising in these places at a rate of about 30 ft every thousand years.

If the block model is basically right, then we don't have to worry about the great flood. The point is that the effect of erosion on one block manages to transfer matter from a higher one to one of lesser height, but no equalization of height can take place because of isostatic readjustments. The block that becomes lighter as a result of mass loss is now pushed up somewhat by the buoyancy of the underlying liquid mantle. On the other hand, the block that gains weight sinks a little more. Thus there will always be a difference in height and the great flood will never come.

Today geologists believe in the validity of the basic ideas of isostasy: the idea of floating blocks in an underlying liquid mantle. Today the concept of these blocks has been replaced by the concept of tectonic plates—about sixteen rigid plates making up the crust of the earth. In this plate tectonics model, the plates are dynamic (usually in motion of some sort) and their motion causes much of the intriguing geological mass movements, such as the building of mountains or the drift of the continents. However, isostasy remains one of the factors to be reckoned with in the consideration of the motion of the plates.

Buoyancy of air

Gases share the property of buoyancy with liquids. Of course, the weight of, say, an amount of air displaced by a solid object immersed in air is not much, so the buoyant force is usually small. Nevertheless, it is definitely not zero.

A reminder of the buoyancy of air is contained in the following exercise. Suppose we ask which is heavier, a ton of steel or a ton of cotton? If you are tempted to answer a ton of steel, stop. You are thinking about density and not weight. One ton is one ton; the weights are equal by the very statement of the problem. But now we can raise a very subtle issue.

Ordinarily when we talk about the weight of an object, we mean weight in air. Who cares to weigh a ton of steel or, worse, a ton of cotton in a vacuum? But if the weight of a certain amount of steel is equal to that of an amount of cotton when the measurement is carried out in air, will that equality be maintained when they are weighed in a vacuum? No, because the load of cotton has more volume. Thus it displaces more air and hence loses more weight when submerged in air than does the load of steel. So, in a vacuum, the one-ton load of cotton would actually be heavier. The difference of their weights in this case can be as much as a couple of pounds.

■ 8.5 Fluids in Motion and Bernoulli's Principle

In Eugene, Oregon, there is a river named Willamette, which is a pretty good place to swim in the summer except for one thing—it has whirls, or what is more commonly called rapids. Every once in a while a daredevil swimmer

FIGURE 8.19 The streamlines of motion of a (nonviscous) fluid through a pipe.

succumbs to the attraction of these whirls and loses control. Obviously rapids exert a large force on a swimmer, but where does the force come from?

If you don't relate to the plight of the swimmer, perhaps the following situation will ring a bell. You are traveling in your car on the freeway and there is a relatively slow-moving truck ahead that you must overtake. But as you go in the left-hand lane, for the short time that you travel beside the truck at high speed, there is a definite force that tries to sway your car toward the truck. What causes this?

Obviously the motion of fluids has something to do with this kind of phenomena. However, fluid motion is quite complicated, so we won't be dealing with all the details. Fortunately, a Swiss physicist named Daniel Bernoulli (1700–1782) formulated a principle that makes it rather easy to understand the passing truck phenomenon (and a few others).

To discuss the principle we will make a simple model of streamline motion, in which every particle of the fluid is assumed to move in the same path as every other particle preceding it and always in the same general direction as the overall direction of motion of the fluid (Fig. 8.19). Not considered in this model are things like air resistance or the viscosity (gumminess) of liquids and also things like eddies, which fall under the category of turbulent motion.

If a liquid follows such streamline motion through a pipe, its rate of flow (the volume that flows through in unit time) is the product of the cross-sectional area and the velocity of the fluid. And this rate remains a constant. If the cross section of the pipe varies, the velocity of the liquid changes in such a fashion that their product stays the same.

This may sound unfamiliar, but the fact that water flows faster through a constricted pipe is no secret—probably you are quite familiar with the idea. Now we ask: what speeds up the fluid? What causes its acceleration? There has to be a force—or, since we are talking about fluids, a difference of pressure. (Remember, for fluids it is more convenient to think in terms of pressure.) Thus it follows that if a fluid moves faster in a region, the pressure must be lower in that region compared to the surrounding regions where the velocity is less. Thus the difference in pressure causes the acceleration (Fig. 8.20).

The following is a simplified version of **Bernoulli's principle:**

The pressure of a fluid is smallest where the speed is the greatest and vice versa.

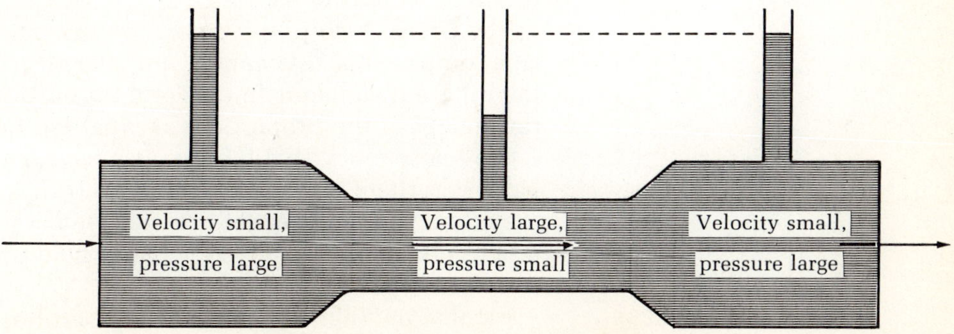

FIGURE 8.20 The pressure is lower in the narrow portion of the tube, because the velocity of flow of the fluid is faster there.

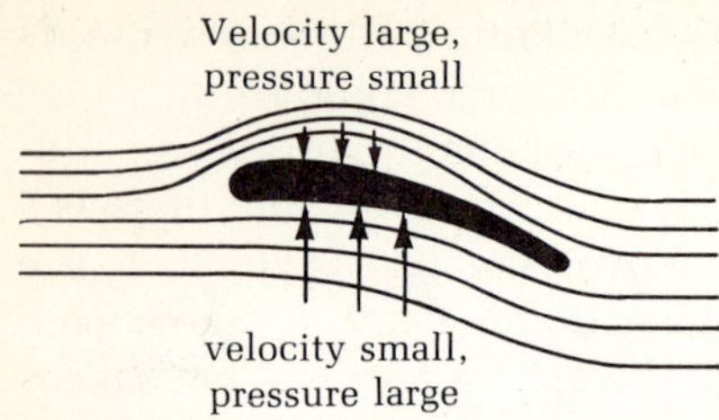

FIGURE 8.21 For an airplane wing the construction guarantees a faster flow and therefore a smaller pressure on the top of the wing. The higher pressure at the bottom is responsible for the lift force.

Thus near rapids the water speed is high, and therefore the pressure is low. Thus a swimmer is pushed by the higher pressure of the adjacent water toward the low-pressure region of the rapids. The same thing happens when you overtake a truck. The space between the car and truck can be thought of as a constricted pipe when compared to the surrounding space. The air flowing through it corresponds to higher speed and smaller pressure when compared to the other side of the car. It is the higher pressure from the other side of your car that pushes it toward the truck.

You can see the devastating effect of a tornado in a new light once you understand the Bernoulli principle. The still air inside buildings corresponds to higher pressure when compared to the pressure in the region of the high-speed wind of the tornado. So it is really the higher air pressure inside the buildings that pushes the walls of the building, causing its destruction. So if you are in a car (or inside a house) and encounter a tornado, keep your windows open. This will minimize the destructive effect by equalizing the pressure to some extent, contrary to what common sense might dictate.

Bernoulli's principle makes possible one of the marvels of modern technology: airplanes. The wings of the airplane are made in such a way that the air has to travel a longer distance past the upper surface, and therefore at a higher speed, when compared to the lower surface. As a result of the higher speed at the upper surface, Bernoulli's principle dictates that the pressure there be lower than that below the wing. Since the pressure up on the bottom face of the wing is higher than the pressure down on the upper surface of the wing, there is a net upward push on the wing (Fig. 8.21). This upward push is often called a **lift force.** The greater the difference between the speed of air flow between the upper and the lower faces of its wings, the greater is the lift force on the plane.

■ 8.6　Atoms and Chemistry

We end this chapter with a brief discussion of chemistry from the atomic point of view. The modern ideas of the atom were nurtured in the cradle of chemistry, and without the atomic ideas modern chemistry could not have developed.

As mentioned already, the basic idea of the atom originated in ancient Greece with Democritus, who called these elementary constituents of matt "atomos," meaning "indivisible." Although we now know that atoms are not indivisible, in most natural processes around us they are, and so it seems appropriate to retain the name.

The Greeks also propounded another interesting idea. They said that matter was composed of "elements," and that all substances were mixtures of atoms of the different elements in different proportions. The Greeks thought that a rearrangement of the proportions is what led to changes in substances.

We know today that these views have a suprising amount of truth in them. However, there is also some gross untruth in the Greek theories. For example, they thought there were four basic elements: earth, air, fire, and water. This and other details of their model actually held up rather than helped further development.

Finally, an Englishman named Robert Boyle, in 1661, put forth a useful new definition of what an element is. In his monumental work *The Sceptical Chymist* he wrote,

> I now mean by elements . . . certain primitive and simple and perfectly unmingled bodies, which not being made of any other bodies or of one another, are the ingredients of which all these perfectly mixt bodies are immediately compounded and into which they are ultimately resolved.

Thus the elements are perfectly "unmingled" and the compounds are "mixt." Never mind the spelling; the idea is basically correct. In our atomic picture, the elementary constituents of the elements are the atoms characteristic of the particular element. When two or more elements mingle to form a compound, it is because their atoms attach to each other forming the molecules. Consider an example. The macroscopic substances oxygen gas and hydrogen gas are elements because they cannot be broken up any further into simpler substances. But water can be decomposed into hydrogen and oxygen, and therefore water is not an element but a compound. When hydrogen is burned in oxygen, its atoms combine with those of oxygen—two of hydrogen to one of oxygen—to form the water molecule, the ulimate particle of water.

It took more than a hundred years after Boyle before all these ideas came together in the hands of such great chemists as Lavoisier, Dalton, and Avogadro. Through their work 90 naturally occurring elements have been identified (the table has now been extended to 106, counting those elements that have been made in the laboratory). A list of some of the most important elements and the chemical symbols used for their identification is given in Table 8.2. Also given for each element is a number called the **atomic weight,** which roughly tells us how much heavier that element's atom is when compared to the lightest atom, that of hydrogen. The atomic weight is really a ratio of the weight of the respective atom to the hydrogen atom. Don't mistake an atomic weight for the actual weight of the atom.

The actual mass of the hydrogen atom is very small,

$$m_\mathrm{H} = 1.67 \times 10^{-27} \text{ kg}$$

TABLE 8.2 Some Important Elements.

Element	Symbol	Atomic Weight (approximate)
Hydrogen	H	1
Helium	He	4
Carbon	C	12
Nitrogen	N	14
Oxygen	O	16
Aluminum	Al	27
Sulfur	S	32
Chlorine	Cl	35.5
Calcium	Ca	40
Iron	Fe	56
Copper	Cu	63.5
Silver	Ag	108
Uranium	U	238

Since the atomic weight of oxygen is 16, the mass of an oxygen atom is

$$(16) \times (1.67 \times 10^{-27}) = 26.7 \times 10^{-27} = 2.67 \times 10^{-26} \text{ kg}$$

which is also very small.

In many elements in their natural form, the atoms join together to form a molecule of the element. Of special importance are the diatomic gases: oxygen, nitrogen, and hydrogen, denoted as O_2, N_2, and H_2. The subscript of an atomic symbol in the formula of a molecule indicates the number of atoms of the element in that molecule. There are also elements whose atoms do not form molecules with each other. Such substances are called monatomic. Helium gas (He) and solid iron (Fe) are examples. However, atoms of most monatomic elements do form molecules with atoms of suitable other elements under suitable conditions.

Modern chemists have determined the atomic composition of thousands of molecules of different substances and have learned the art of making new ones in the laboratory (for example, plastics). They are even getting into the business of making complicated molecules like the proteins that are basic to life.

Once the atomic composition is known, the substance is denoted by writing down the symbols of each of the constituent atoms with each symbol subscripted by the number of that particular atom that the molecule contains. For a familiar example, water is denoted as H_2O. Octane, a component of gasoline, is written symbolically as C_8H_{18}; the notation makes it clear that a molecule of octane contains 8 atoms of carbon and 18 atoms of hydrogen.

If we know the composition of a substance, we can determine its **molecular weight.** The value obtained tells us the relative heaviness of its molecule compared with the hydrogen atom (the ratio of the weight of the molecule to that of the hydrogen atom). To calculate the molecular weight, we add the atomic weight times the number of atoms for all the elements in the molecule. As an example, the molecular weight of water is 18; 2 for the two hydrogen atoms and 16 for one oxygen atom added together to give 18. Carbon dioxide, denoted by CO_2, has a molecular weight of 44, which is 12 for the carbon atom plus 2×16 (for the two oxygen atoms).

Now a very important fact. The famous Italian chemist Amedeo Avogadro discovered, in 1811, a very important law, referred to as Avogadro's hypothesis for historical reasons. In essence, the law states that the amount of a substance whose mass in grams is numerically equal to its molecular weight (hereinafter referred to as a **gram-mole** of the substance) contains 6.02×10^{23} molecules. Thus a gram-mole of carbon, which is 12 grams, contains 6.02×10^{23} molecules of carbon (the molecules of carbon are really carbon atoms, since carbon is a monatomic element). By the same law, 18 grams of water contain 6.02×10^{23} molecules of water. The number 6.02×10^{23} is thus a very important number and is customarily called **Avogadro's number.**

Since we mostly use the MKS system in this book, where the unit of mass is the kilogram, let us define the kilogram-mole as the amount of a substance whose mass in kilograms is equal to its molecular weight. Thus a kilogram-mole is a thousand times more mass than the gram-mole and contains $1000 \times 6.02 \times 10^{23}$ molecules, which is 6.02×10^{26} molecules.

The knowledge of Avogadro's number enables us to determine the number of molecules in a glass of water, for example. A glass of water contains 8 ounces (oz), which is roughly 226 g; 18 g of water makes one gram-mole, so 226 g is equivalent to $226/18 = 12.56$ gram-moles. This many gram-moles contain $12.56 \times 6.02 \times 10^{23} = 75.6 \times 10^{23} = 7.56 \times 10^{24}$ molecules.

The chemists also discovered some striking regularities among the elements. Groups of elements were found to have almost identical chemical behaviors. The explanation of this had to come from the physics of the structure of the atom itself; this idea will be discussed in Chapter 22.

We can ask another question here: Why are the masses of many of the atoms of the heavier elements so close to integer multiples of the mass of the hydrogen atom? Does this mean that the higher elements are made up of hydrogen atoms in some sense? Today we know that the answer to this question is yes, in a way. The hydrogen atom is made up of a centrally located subatomic particle called proton, with a much lighter particle, the electron, orbiting around it. Heavier elements are composed also of a heavy central core (called the nucleus) and orbiting electrons. The protons, actually hydrogen nuclei, are constituents of the nuclei of heavier elements, which also contain particles called neutrons (with approximately the same mass as the proton). But now we are getting into nuclear physics, which is a story to be told in Chapter 23.

SUMMARY

This chapter is an introduction to matter. Matter around us exists in three states: solid, liquid, and gas. These states can be differentiated by their basic physical differences: solids are incompressible and rigid; liquids are incompressible but nonrigid; gases are neither incompressible nor rigid. A further understanding of the difference of these three states is achieved with the help of molecular theory. In a gas the molecules are free to move; in a liquid this freedom is somewhat impeded—the molecules can slide by each other but cannot fly away; in a solid the molecules can move only about fixed centers.

Elasticity concerns the deformation of solids. The important concepts of stress and strain are introduced here, along with the statement of Hooke's law: stress is proportional to strain. In connection with fluids (the common name of both liquids and gases), the concept of pressure plays a large role: pressure is force per unit area. Pressure in fluids has a variety of interesting and, to some extent, surprising aspects. For example, the pressure at a certain depth of a fluid depends on the height of the fluid above but not on its volume or shape. The exact relationship of pressure and depth is given as

$$P = \rho g d$$

Bodies submerged in a fluid are acted on by an upward force called the buoyant force. This was discovered by Archimedes in connection with his celebrated principle: A body submerged in a fluid loses weight in the amount of the weight of the fluid displaced.

In understanding some simple aspects of fluids in motion, one principle that is a great help was provided by Daniel Bernoulli: In regions of a fluid where the velocity of the fluid is large, the pressure is small, and vice versa. Bernoulli's principle helps us to understand how an airplane is able to float in midair as it travels. There is a lift force on the plane that results from an application of Bernoulli's principle in the design of the airplane's wings.

The final section of this chapter deals with some ideas of the atomic theory and some associated chemistry. Three concepts among those that were introduced will be recurrently used: atomic weight, molecular weight, and mole.

QUESTIONS

Review and reason

1. How do we distinguish among the solid, liquid, and gaseous states of matter?
2. What is the difference among a solid, a liquid, and a gas from an atomic point of view?
3. What is crystal structure? How is the crystal structure of solids explained from an atomic point of view?
4. What is surface tension? Find an example of surface tension phenomenon from your everyday experience. Explain surface tension from a molecular point of view.
5. Explain how a small insect can walk on water without drowning.
6. Explain the terms stress and strain. Name the different kinds of stresses and strains that exist in nature and give an example of each.
7. State Hooke's law. What is meant by the elastic limit?
8. What is meant by scaling? If the linear dimensions of a scale model are scaled by a factor x, by what factor do the perimeter, area, and the volume increase?
9. Why is it that in a Boeing 747 a smaller fraction of passengers get to use window seats in comparison to the window seats available in a smaller Boeing 727 plane? (Hint: How does the number of window seats scale? How does the total number of seats scale?)
10. Explain why the creatures of Brobdingnag, as envisioned by Swift, would be structurally unstable unless their bones are made of iron or something similar.
11. Why is the water bed more comfortable (for most people) than a regular bed?
12. Why is it that when you ski on fresh snow you don't sink, but the same snow will collapse if you try to walk on it?
13. How can a heavy tank go through a swamp when you can't walk on it without sinking through?
14. Why does a karate chopper use the side of the hand rather than the palm?

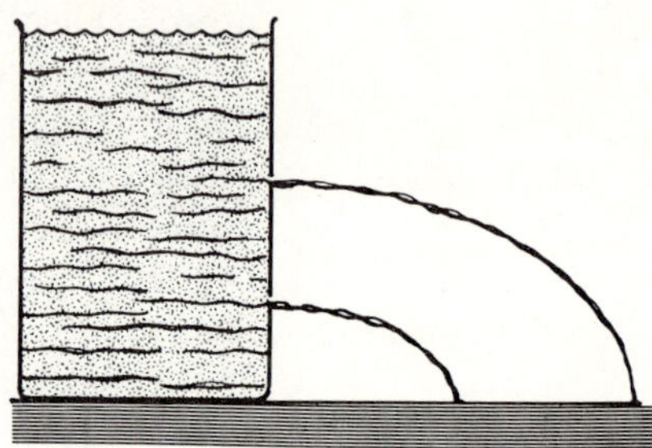

FIGURE 8.22

15. What's wrong with Fig. 8.22? Correct it.
16. Explain the principle of a hydraulic lift.
17. Can an underwater diver breath through a straw that has one end protruding above the surface of the water? Explain.
18. The direction of the pressure of a fluid is always perpendicular to the walls of its container. Explain.
19. What is meant by atmospheric pressure? If so much pressure is exerted upon us every moment of our existence, why don't we feel it?
20. Explain the principle of a mercury barometer.
21. Explain how we manage to drink a liquid with a straw. (Is the liquid pushed to our mouth or pulled?)
22. If you put a straw in a glass of soda pop and close the top of the straw with a finger, when you lift the straw out of the pop, a certain amount of pop will stay inside the straw (Fig. 8.23). What keeps the pop from flowing out of the straw?

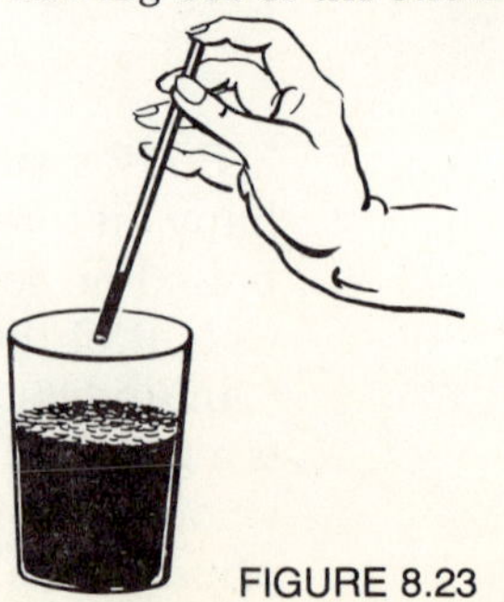

FIGURE 8.23

23. State and explain Archimedes' principle.
24. Football players are weighed underwater to determine the percentage of fat in their bodies. Explain.
25. Hold a sheet of paper as shown in Fig. 8.24 and blow on its top surface. The paper will rise (try it and see). Would you believe that airplane wings rise by using the same principle? Explain.

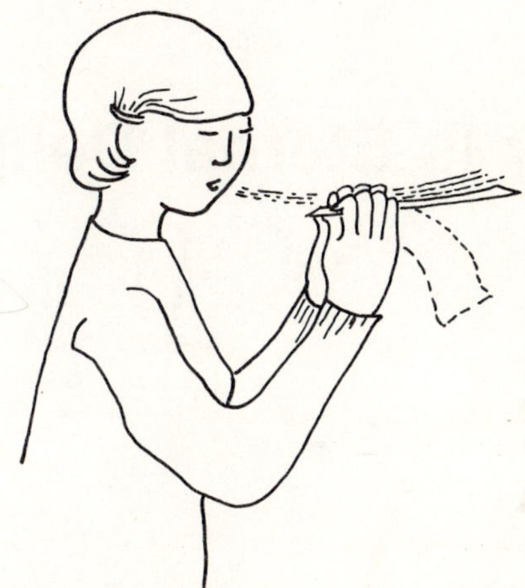

FIGURE 8.24

26. An airstream (as from a vacuum cleaner), when directed on a Ping-Pong ball as shown in Fig. 8.25, holds the ball in position. Explain. (Hint: Ask yourself why the ball doesn't move sideways and out of the stream.)

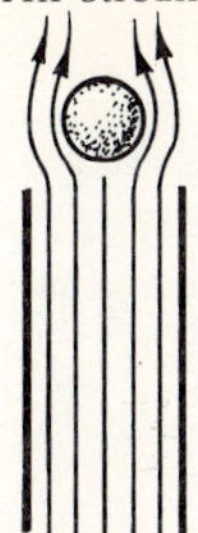

FIGURE 8.25

27. Is there any mechanical advantage in staying close behind a car or a truck when riding a bicycle? Explain.
28. What is meant by streamline motion of a fluid?
29. It is wise to advise young people not to stand too close to a railroad track when a fast-moving train is passing by, because of the risk of being pushed into the train. What does the pushing?
30. What is meant by the atomic and molecular weights of a substance? Explain the concepts of gram-mole and kilogram-mole.
31. What is the difference between an element and a compound?
32. A particular carbohydrate is represented by the chemical formula $C_6H_{12}O_6$. Explain the notation.

Arithmetic

1. You know about the Lilliputians of *Gulliver's Travels.* According to Swift's imagination, they are human beings scaled down in size by a factor of 12. As you may remember, one of the first things that the king did for Gulliver was to order his tailor to make some new clothes for this "giant." Assuming that the clothes the tailors made for Gulliver had the same thickness as their own, how many times more material did they need for a single suit of Gulliver than for one of their own suits?
2. There is a hole in a dam 20 ft below the water surface. What is the water pressure down there? Like the legendary Dutch boy who plugged the hole in a dike with his finger, you attempt to plug the hole in the dam with your thumb. How much force will the water exert on your thumb? (Hint: Start with an estimate of the surface area of your thumb.)
3. Calculate the pressure of water at the bottom of a fish tank filled with water 1 ft deep.
4. What is the absolute pressure on a diver 20 m below the surface of ocean water?
5. If the average gauge pressure of blood is 100 mm, what is its absolute pressure?
6. If a block of iron floats in mercury, what percentage of its volume will be submerged?
7. Assuming that firewood is 100% carbon, how many kilogram moles of carbon are contained in 24 kg of firewood? Also calculate the number of carbon atoms contained in the firewood.

9

Energy Is Eternal Delight

Man has no Body distinct from his Soul!
for that called Body is a portion of Soul
discerned by the five Senses,
the chief inlets of Soul in this age.
Energy is the only life and is from the Body;
and Reason is the bound or outward
circumference of energy.
Energy is eternal delight.

WILLIAM BLAKE

■ 9.1 What Is Energy?

Intuitively everybody knows what energy is. But try to explain what you know. You will find that this is not an easy task after all. Even physicists have difficulty finding an exact and valid definition of energy that is also brief and concise. So don't be surprised if it takes most of this chapter to explain what energy is.

Let us quickly try to find a working definition of energy. An interesting hint is found in the meaning of the Greek word "energos," from which the word "energy" originated. "Energos" means "container of work." We will change the words slightly and define **energy** as the capacity for doing work. But we have to be careful when applying this definition, because the word "work" is used colloquially with many different connotations. We will associate with **work** only those situations where a physical displacement of an object has been caused by a force acting on it.

Consider an example. Suppose you push a car and the car moves. This is an example of work under our reckoning, because a force (push) has caused physical displacement of an object (the car). On the other hand, imagine a situation where the car is stuck in the mud so badly that in spite of all your pushing, it

just refuses to move. Is any work done on the car in this case? The answer has to be no, according to physicists. You may object at this point. After all, you may be exhausted from the exertion of your push. The work you did in this case is connected with the movement of your muscles and is called *physiological work*, in contrast to the physical work we have defined. In a similar vein, the thinking you may have done while reading this paragraph is also classified as physiological work and cannot count as physical work.

Thus physical work always involves a force and a displacement. The amount of work done by a force in a particular situation is given by the product of the applied force and the displacement of the object on which the force acts.

$$\text{work} = \text{force} \times \text{displacement} \tag{9.1}$$

An object has energy if it can do physical work. A moving hammer has the ability to do work. This is easy to verify: we can let the hammer fall on a nail, thus exerting a force on the nail, which is physically displaced. This comprises work. Thus a moving hammer has energy.

From this kind of argument it is clear that any object in motion has energy. The energy of motion is called **kinetic energy** (Fig. 9.1). The faster the motion is, the greater the kinetic energy is. Since motion of an object is characterized by its speed, we can guess that the object's kinetic energy must increase as its speed increases. There is a simple quantitative relationship between the kinetic energy and the speed:

$$\text{kinetic energy} = \tfrac{1}{2} \times (\text{mass of the object}) \times (\text{speed of the object})^2$$

or, in symbols,

$$K = \tfrac{1}{2}mv^2 \tag{9.2}$$

where K stands for kinetic energy, m for mass, and v for speed. Notice that the kinetic energy is proportional to the square of the speed. This means that if the speed is increased by a factor of two, the kinetic energy increases by a factor of 2^2, or 4. Also, because the mass and the square of the speed are always positive, so is the kinetic energy.

Before going any further, some subtleties in the definition of work, Eq. (9.1), must be pointed out. Work is a scalar quantity; magnitude alone is all that is needed to specify it. But it is defined as the product of two vector quantities:

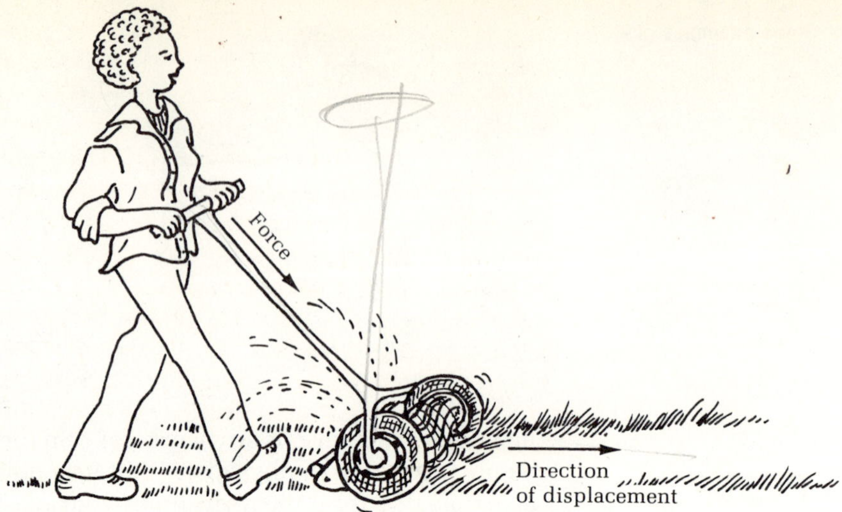

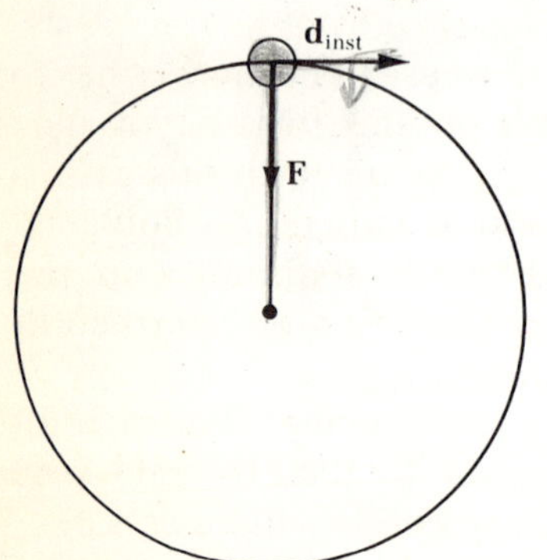

FIGURE 9.3   For circular motion the direction of the displacement at a given instant, $\mathbf{d}_{inst}$, is along the tangent to the circle. The force acts along the radius, which is perpendicular to the tangent. Therefore, the work done is zero, since the component of a force in a direction perpendicular to itself is zero.

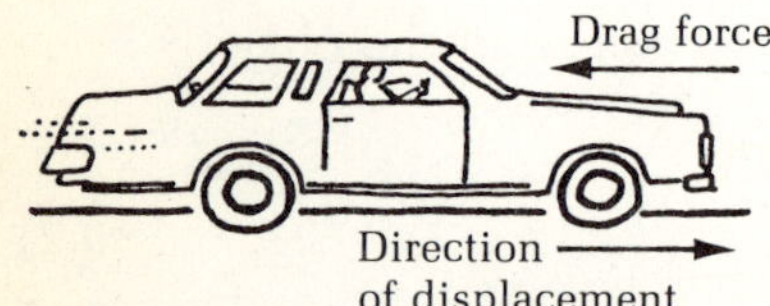

FIGURE 9.4   Since the direction of the air drag is opposite to that of the displacement of the car, the work done by it on the car is negative. The effect of the negative work is a reduction of the kinetic energy of the car. To maintain constant speed the car has to replace the depleted energy by burning gasoline.

both force and displacement are vectors. This is perfectly legitimate. Mathematicians call such a product (of two vectors forming a scalar) a scalar product. So work is the scalar product of two vectors, force and displacement. Now what does this imply? This is where some fine points enter in.

One fine point is that for a given magnitude of the force, the amount of work done by the force on a system depends on the direction of the force with respect to the displacement vector. We get *maximum* work if the force is in the *same* direction as the displacement. The work in this case is calculated as the product of the magnitude of the force and that of the displacement. However, if the force is not in the same direction as the displacement, not all of its magnitude is available in the direction of the displacement. In this case we must be careful to multiply only the component of the force along the direction of the displacement by the magnitude of the displacement in order to calculate the work (Fig. 9.2). To determine the component of a vector in a given direction involves some additional mathematics, which we won't go into, but it is important that you be aware of this subtlety.

Two special cases of the direction of force relative to the displacement, however, must be discussed. First, what if the force acts in a direction perpendicular to the direction of the displacement? An example is the circular motion of an object. The centripetal force is always perpendicular to the direction of the displacement at any given instant, the instantaneous displacement being along the tangent (Fig. 9.3). The component of a vector in a direction perpendicular to its own is zero. Thus the centripetal force directed along the radius has no component whatsoever along the tangential direction of the displacement. So the work done by the centripetal force on an object in circular motion is zero, which is why there is no change in the kinetic energy of such an object. The object moves with constant speed.

The second special case involves this question: What if a force works against the direction of the displacement, in the opposite direction (Fig. 9.4)? The work done by such a force must be regarded as negative work. For example, consider the effect of the drag force on moving automobiles. The effect is to reduce the kinetic energy of the car, to slow it down. The drag force acts against the direction of the displacement of the car and thus performs negative work on it, which leads to the reduction of the kinetic energy of the car.

Let us explore some aspects of energy spent by automobiles, using these concepts of work and kinetic energy. In a car, the chemical energy of gasoline is used to provide the kinetic energy of the car ($\frac{1}{2}mv^2$) every time it is accelerated from zero to the desired speed (unless you are driving downhill). The amount of energy involved in accelerating the car is thus proportional to (a) the mass of the car and (b) the square of the speed. Thus the more massive the car is, the more energy is needed each time the car is stopped and reaccelerated back to its original speed. The expense of providing kinetic energy of the car is even more pronounced as a function of the speed: it increases as the square of the speed. Since city driving involves many stops, it is clear that it is energetically advantageous to drive a less massive car and, more particularly, to drive at a lower speed.

How about driving on the highway? Here few stops are involved, so the energy expenditure in the initial acceleration does not make much difference in the overall situation. The energy of the gasoline is spent to keep the car running against the force of air resistance (drag) and also road friction (which we ignore in the following discussion) to a lesser extent. Forces like the drag force dissipate the kinetic energy by converting it into heat energy. The gasoline energy is used to replenish the loss of kinetic energy. You can think of the drag force as doing negative work on the car, while the positive work provided by the burning gasoline offsets the negative work; so the kinetic energy of the car remains a constant. Since

$$\text{work done} = \text{force} \times \text{distance}$$

for traveling a fixed distance, the amount of negative work done by the drag force, and therefore the amount of gasoline energy required, is directly proportional to the drag force. Since the drag force is proportional to the square of the speed, the gasoline energy spent is again proportional to the square of the speed of travel. Thus even for driving at a constant speed without stopping, it is advantageous to drive at a low speed.

Let's perform a little calculation. How much more energy is spent by driving a car at 70 mi/h in comparison to 55 mi/h, the current speed limit? The ratio of energy spent is roughly the same as the ratio of the squares of the respective speeds:

$$\frac{\text{energy spent at 70 mi/h}}{\text{energy spent at 55 mi/h}} = \left(\frac{70}{55}\right)^2 \approx 1.6$$

Thus the energy spent at 70 mi/h is 1.6 times that spent at 55 mi/h, or 60% greater. Thus there seems to be a 60% saving of energy by driving at the new 55 mi/h speed limit.

This, however, turns out to be an overestimate. First of all, when we drive a car, not all the gasoline energy is used to drive the wheels. The amount of the gasoline energy used elsewhere in the car (e.g., lights, water pump, generator) remains unchanged when we change speeds. Most importantly, in the estimate above we have ignored road friction, which is responsible for a significant portion of the energy dissipated. And the energy dissipated against road friction

does not change with speed. When we take all these other factors into account, the energy saved by driving at the lower speed amounts to some 20%, which is still quite considerable.

Next let's talk about a different kind of energy. Water stored at an elevation, as in a dam, has a potential capacity for doing work (Fig. 9.5). When released, this water falls on the massive turbines below it and displaces the turbine blades in a rotary motion, which is work. (It may be a little difficult to see how setting an object in rotary motion comprises work, since there is no translation of the body as a whole. The point is that each component particle of the rotor does undergo displacement in the process.) Thus water stored at an elevation has **potential energy,** which is a potential capacity for doing work. The potential energy arises from the position of the water with respect to the ground. The greater the height of the stored water, the greater is its potential energy.

Let us derive an expression for the potential energy of water on a hill. We can ask, how does the water get there? We can imagine that somebody hoisted it up in a bucket, and this involves doing work on the water. That is, we have to apply an external force that balances the weight of the water to lift it. The amount of physical displacement of the water is given by the height of the hill. Thus from Eq. (9.1) we get an expression for the work done by the applied force on the water:

$$\text{work done} = \text{applied force} \times \text{displacement}$$
$$= \text{weight of water} \times \text{height of the hill}$$

This amount of work now is stored in the water (due to its elevated position) as added potential energy. Thus the change in potential energy is given by the same formula:

$$\text{change in potential energy} = \text{weight} \times \text{height} \qquad (9.3)$$

Since weight = mg [see Eq. (4.5)], we can write the formula for potential energy of an object with respect to the ground as

$$\text{potential energy} = mgh \qquad (9.3)$$

where h denotes the height with respect to the ground.

This kind of potential energy is related to an object's position with respect to earth's gravity field; we call this the **gravitational potential energy.** There are other forms of potential energy (Fig. 9.6) connected with the other kinds of forces (e.g., electrical and nuclear) we mentioned previously (see Section 2.3).*

Energy exists in these two basic forms: kinetic and potential. All the different forms of energy you are aware of can be traced to these two basic forms.

Now it's time to introduce one more related concept. Often we are interested in the rate at which work is done or energy is expended rather than the total amount of work involved. This rate is called **power.** The faster work is done, the greater the power. This definition of power no doubt agrees with what you know from your experience. A 200-horsepower (abbreviated hp) car takes less time to travel uphill than a 90-horsepower car because the former does the same amount of work faster.

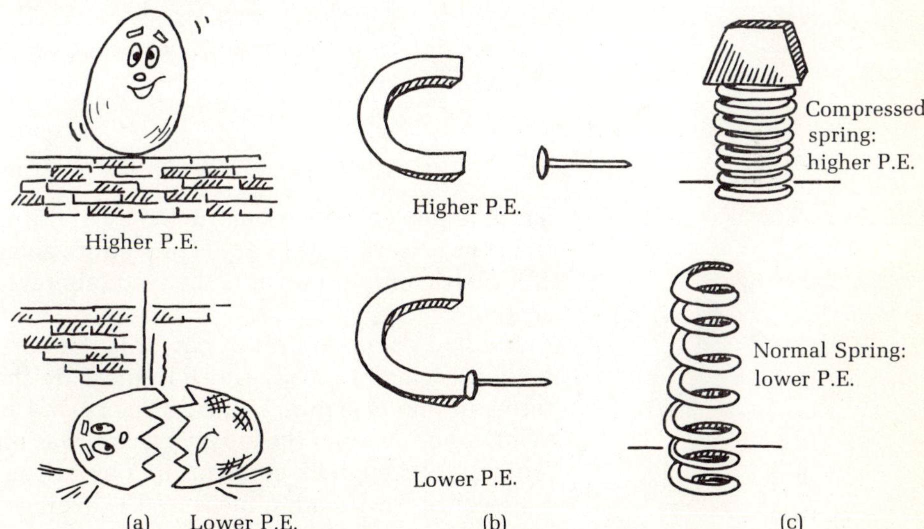

FIGURE 9.6 Some examples of potential energy: (a) gravitational potential energy; (b) magnetic potential energy; (c) potential energy of a spring.

*Forces that give rise to potential energy are called **conservative forces.** When calculating the work done on a body, we do not directly calculate the work done by conservative forces because this work is included as potential energy. We would be counting work twice if we added the work done by the conservative forces. In contrast, all frictional forces are nonconservative.

Mathematically, we have

$$\text{power} = \frac{\text{energy}}{\text{time}} \qquad (9.4)$$

Why are mountain roads winding? In overcoming the gravitational force, the same amount of energy is spent on a winding road as is spent in straight uphill travel. The work done against gravity is always weight times the vertical height of the hill. However, if the road is winding, it takes longer to make it to the top traveling at a certain speed when compared with a straight uphill trip. Thus the formula for power tells us that for a winding road less power is needed. This is the reason mountain roads are winding.

Units of energy and power

The description of a physical quantity is not complete without a unit for its measure. Since

$$\text{work} = \text{force} \times \text{distance}$$

the unit of work is the product of the unit of force and the unit of distance. In the engineering system, work is thus measured in the foot pound (ft·lb), which is the amount of work done when 1 pound of force displaces an object through a distance of 1 foot. In the metric system the unit of work is called the joule (abbreviated J). The unit could be called a newton meter (N·m) too: it is the work done by 1 newton of force displacing an object through 1 meter. Also,

$$1 \text{ ft·lb} = 1.356 \text{ J}$$
$$1 \text{ J} = 0.737 \text{ ft·lb}$$

Since energy is the capacity for doing work, the unit of energy is the same as the unit of work.

Since

$$\text{power} = \frac{\text{energy}}{\text{time}}$$

a unit of power can be constructed by dividing a unit of energy by a unit of time. In the engineering system this procedure gives us the unit of foot pound/second. The metric unit of power is the watt (abbreviated W), which is equal to 1 joule/second.

Another common unit of power is the horsepower (abbreviated hp). Historically it seemed to make sense to measure the power output of locomotives in terms of one of nature's most diligent and mobile workers—the horse. James Watt, who invented the steam engine, was often challenged with the question: How did his engines perform in comparison with the farmer's horse it was to replace? So Watt had to measure the average power of a horse. The number he obtained is only an approximate indicator of the average power of a horse. Nevertheless, his number has been the basis of the unit of horsepower. In terms of the usual units,

$$1 \text{ hp} = 550 \text{ ft·lb/sec} = 746 \text{ W}$$

How do you compare with a horse in the power department? You can do a

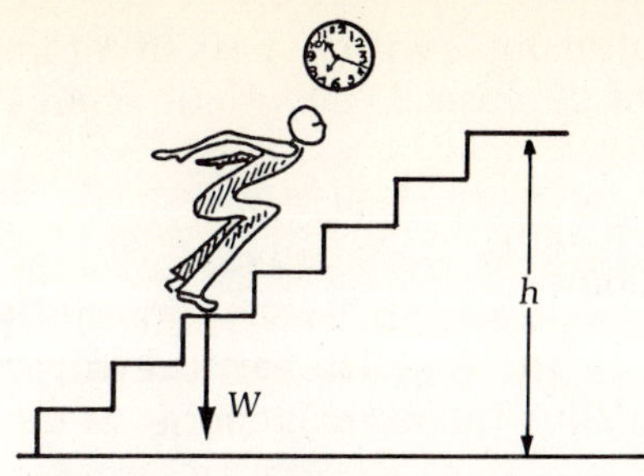

FIGURE 9.7   See text to figure out your horsepower rating.

simple experiment to find out. All you need is a watch that measures time in seconds and a rule to measure the height of a step on the stairs. Climb a flight of stairs and see how long it takes you. Suppose the time is t (in seconds). The work you perform in climbing the stairs is your weight times the (vertical) height of the stairs (number of steps $\times$ height of one step): $W \times h$ (Fig. 9.7). Then your power is given as $(W \times h)/t$ ft·lb/sec, if h is measured in feet and W in pounds. Since 1 ft·lb/sec = 1/550 hp, your power rating in horsepower is given as

$$\frac{Wh}{550t} \text{ hp}$$

where W is in pounds, h is in feet, and t is in seconds.

This writer once climbed 40 ft of stairs in 30 sec, both rough estimates. With a weight of about 150 lb, the horsepower rating was

$$\frac{150 \times 40}{550 \times 30} \approx \frac{1}{3} \text{ hp}$$

Needless to say the author was not particularly enthralled to find that he was no better than $\frac{1}{3}$ of a horse!

Interestingly enough, now we can obtain a unit of energy by multiplying a unit of power with a unit of time. This is how the commonly used energy unit called *kilowatthour* (kWh) originated. It is the energy expended when a machine having 1 kilowatt of power runs for an hour. For conversion purposes,

$$1 \text{ kWh} = 3.6 \times 10^6 \text{ J}$$

EXAMPLE 1   *Calculating work and power*

Suppose a crane lifts 2000 lb through 4 ft in 2 sec. How much total work is done? How much power is involved?

To figure out the total amount of work done, we have to use Eq. (9.1):

$$\text{work} = \text{force} \times \text{displacement}$$
$$= 2000 \times 4 = 8000 \text{ ft·lb}$$

For the calculation of power, we have to use the defining equation for power, Eq. (9.4). Thus

$$\text{power} = \frac{\text{work}}{\text{time}} = \frac{8000}{2} \frac{\text{ft·lb}}{\text{sec}} = 4000 \text{ ft·lb/sec}$$

Since 1 ft·lb is 1.356 J, 4000 ft·lb/sec is 5424 J/sec or 5424 W.

EXAMPLE 2   *Calculation of kinetic energy*

Suppose we calculate the kinetic energy of each of the three moving objects shown in Fig. 9.1. First we'll take the kinetic energy of the tennis ball. The mass of a typical tennis ball is about 58 g, which is 0.058 kg. Suppose its speed is given as 30 m/sec. Then the kinetic energy of the ball is

$$K = \tfrac{1}{2}mv^2 = \tfrac{1}{2} \times 0.058 \times (30)^2$$
$$= 0.029 \times 900 = 26.1 \text{ J}$$

Next we consider the kinetic energy of an automobile with a mass of 1 ton, which is roughly 1,000 kg, moving at a speed of 20 m/sec. The kinetic energy is

$$K = \tfrac{1}{2}mv^2 = \tfrac{1}{2} \times 1{,}000 \times (20)^2$$
$$= 500 \times 400 = 200{,}000 \text{ J}$$

Finally, the mass of the earth is given as 6×10^{24} kg. The earth revolves around the sun with a speed of 3×10^4 m/sec. Thus the kinetic energy of the earth's orbital motion is

$$K = \tfrac{1}{2}mv^2 = \tfrac{1}{2} \times 6 \times 10^{24} \times (3 \times 10^4)^2$$
$$= 3 \times 10^{24} \times 9 \times 10^8 = 27 \times 10^{32}$$
$$= 2.7 \times 10^{33} \text{ J}$$

EXAMPLE 3 *Calculation of gravitational potential energy*

A book weighing 5 lb is resting on a table (Fig. 9.8). The height of the table with respect to the floor is 3 ft. What is the potential energy of the book with respect to the floor?

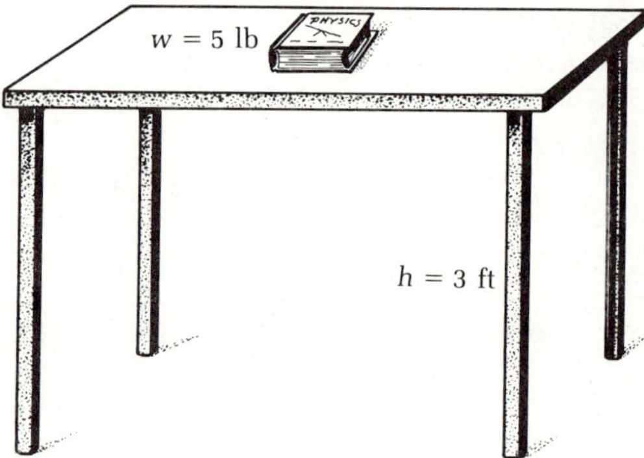

FIGURE 9.8 Book on a table. What is its potential energy?

From Eq. (9.3) the potential energy of the book with respect to the floor is given by

$$\text{weight} \times \text{height} = 5 \text{ lb} \times 3 \text{ ft} = 15 \text{ ft·lb}$$

EXAMPLE 4 *If we could only jump like a flea*

When you were a kid did you ever wish you could jump like a flea when you found your capability of jumping too limited for all the adventures you could invent? As an adult, if you happen to ponder the amazing jumping ability of the flea relative to its size, you might wonder whether fleas, if they were human size, could clear the Empire State Building.

But more thought reveals strange twists in these matters, things that certainly deceive the eye initially. The considerations that enter are those of scaling. The physical characteristics of a body do not scale the same way if you scale the size.

The work done in a jump is the product of the weight and the height. Thus we can express the height of a jump as the following ratio:

$$\text{height} = \frac{\text{work done by the muscles}}{\text{weight}}$$

Now weights of living matter scale as the volume, that is, as the cube of the linear dimension L; so weight scales as L^3. In size we are scaled up about 2000 times compared to the flea, so our weight is increased by a factor of 2000^3.

How about the work done by the muscles? Work is force times distance in this case; it is the force that relevant muscles exert times the extension of the muscles. The latter is a linear dimension, so it scales as L. Since the material of the muscle tissues in both the flea and the human is the same, the stresses that can be set up are also the same. Since the muscle force equals stress $\times$ area— and stress doesn't change—the force must scale as the area, L^2. So the work done scales as L times L^2, the first factor coming from the scaling of the extension and the second from that of the force of the muscles. So the work also scales as L^3.

Thus the height of a jump, which is the ratio of work and weight, does not scale up at all from a flea to a human. The weight of a human is 2000^3 times greater than that of a flea, but so is the work done. We now see that there is nothing to be gained by scaling up a flea to the size of a human. The flea's jump would not improve a bit. Of course, as we discussed earlier, such a giant flea would have other problems structurally.

EXAMPLE 5 *Energy expended in jogging**

In running, work is done by the muscle's force in accelerating and decelerating the legs with each stride of the runner, while the rest of the body maintains a more or less constant speed. Suppose the speed attained by a runner's body is v. Then with the lift off, as the runner accelerates one of his legs (of mass m) to the speed v, the kinetic energy of his leg increases to $\frac{1}{2}mv^2$, due to the work done by his muscles. Then the leg is brought to rest, and the opposing muscles have to perform an equal amount of work, $\frac{1}{2}mv^2$, to stop the leg. The total work done by the muscles in each stride is then mv^2. The mass of a leg is about 10 kg for a typical 70-kg man, and a typical velocity of jogging is 3 m/sec, which is roughly a 9-minute mile. Then energy expended *per stride per* leg is given as

$$mv^2 = 10 \times 3^2 = 90 \text{ J}$$

The length of a typical stride is 2 m, so that the runner takes one and a half (1.5) strides per second. We can then figure out the energy spent in 1 sec per leg as follows:

$$\text{energy per second} = \text{energy spent per stride} \times \text{number of strides per second}$$
$$= 90 \times 1.5 = 135 \text{ J/sec} = 135 \text{ W}$$

Multiply by 2 to account for two legs. Thus the power output of the two legs in jogging is 270 W. This power has to be provided by the leg muscles.

---

*We closely follow the treatment given in A. H. Cromer, *Physics for the Life Sciences* (New York: McGraw-Hill, 1976), p. 115.

However, leg muscles—or any machine for that matter—do not give a 100% return for the input energy. To take this into account, let us define a quantity called the **efficiency** of a machine:

$$\text{efficiency of a machine} = \frac{\text{work output}}{\text{energy input}} = \frac{\text{power output}}{\text{power input}} \qquad (9.5)$$

Now leg muscles operate at a 25% (or $\frac{1}{4}$) efficiency. This means that in order to provide 270 W of output power for the leg muscles, the body has to operate at four times as much power. This gives the power at which the typical male body works while jogging as

$$270 \times 4 = 1,080 \text{ W}$$

At this rate the energy expended in 1 h (3,600 sec) of jogging is

$$\text{energy} = \text{power} \times \text{time} = 1,080 \text{ W} \times 3,600 \text{ sec}$$
$$= 1,080 \times 3,600 \times \text{J/sec} \times \text{sec} = 3,888,000 \text{ J}$$

This is equivalent to burning about $\frac{1}{4}$ lb of body fat.

EXAMPLE 6   *Considerations for a tidal power plant*

In a tidal power plant we would like to raise the running water of the tide to a higher level, store it in a dam (or, if lucky, between two natural reefs), and then release the potential energy of the water by letting it fall on a turbine below. How much potential energy can we get for each tide? The difference in height between the high and the low tides (often called the height of a tide) is about 20 ft. Let us assume an area of 15 mi² for the water reservoir.

The volume of the water can be calculated as

$$\text{volume} = \text{area} \times \text{height} = 15 \times (5280)^2 \times 20 \text{ ft}^3$$

The factor $(5280)^2$ has arisen from the conversion of square miles into square feet.

What is the weight of this amount of water? The weight of one cubic foot of water (the weight per unit volume of a substance is called its weight density) is 62.4 lb. Thus the weight of the water in the reservoir for each tide is

$$\text{weight of water} = 15 \times (5280)^2 \times 20 \times 62.4$$
$$= 5.2 \times 10^{11} \text{ lb}$$

The average distance that the running water of the tide can be allowed to fall is half the height of the tide, in this case, 10 ft. So the total amount of potential energy of the tidal water is given as

$$\text{potential energy} = \text{weight} \times \text{average height of fall}$$
$$= 5.2 \times 10^{11} \times 10 = 5.2 \times 10^{12} \text{ ft·lb}$$

We can now calculate the amount of power available by dividing the potential energy by the time period of tides, namely, 12 h (12 × 3600 sec). Thus

$$\text{power} = \frac{\text{total potential energy}}{\text{period}} = \frac{5.2 \times 10^{12} \text{ ft·lb}}{12 \text{ h}}$$

$$= \frac{5.2 \times 10^{12}}{12 \times 3600} \text{ ft·lb/sec} = 1.2 \times 10^{8} \text{ ft·lb/sec}$$

$$= 1.2 \times 10^{8} \times 1.356 \text{ J/sec} = 1.53 \times 10^{8} \text{ W}$$

$$= 153{,}000 \text{ kW}$$

This is only a moderate amount of power for an electric plant by today's standards.

■ 9.2 Different Forms of Energy

Most forms of energy are quite familiar from our everyday experience. If you were asked to make a list, you probably would come up with one like this:

> Heat
> Sound
> Light
> Electricity
> Chemical energy
> Nuclear energy

along with the two that we have already discussed, mechanical kinetic energy and gravitational potential energy.

How do we know that all these are different forms of the same thing? We know that heat, electricity, chemical energy, and so on are forms of energy because in the course of our daily life we often see them being converted to mechanical work or mechanical energy. An automobile engine converts the chemical energy of gasoline into heat, a part of which is then converted into the mechanical energy of the car. Electric motors convert electricity into energy of mechanical motion. The examples are many.

It turns out that each of these different forms of energy is either kinetic or potential or a combination of these two basic forms. Take heat, for example; it is really a form of kinetic energy. At the submicroscopic level, matter is composed of atoms and molecules (see Section 8.1). These atoms and molecules in a body are always moving in a completely random fashion. The energy of this random motion is heat. The faster the molecules move, the hotter an object will get. We express this by saying that the temperature of an object increases with the average kinetic energy of its constituent molecules. **Temperature** is the macroscopic measure of how hot (or cold) an object is. The total heat energy of an object is given more or less by the total of all the kinetic energy of its constituent molecules.

Now let's consider an example. Fill a large pot with regular hot water from the tap. Also heat some water to boiling in another but much smaller pot. Now compare the total heat content and the temperatures of the water in the two pots. Obviously, the smaller pot has water at a much higher temperature, as you

can test by dipping a finger in both and comparing the degree of hotness. In contrast, the total heat energy of the water in each of the pots is determined by the sum of the energy of random motion of all the constituent molecules. Since there are many more molecules in the larger pot, it may very well have more heat energy.

Incidentally, this new way of looking at heat energy tells us that even a cold object has some heat energy, since the molecules of cold objects are still in some random motion, although the motion is slowed down, on the average.

The random thermal motion of molecules in a liquid like water can be readily witnessed by looking under a microscope at grains of pollen suspended in the water. The pollen will be found to move in a zigzag random fashion. This is known as **Brownian motion.** The phenomenon is explained by the collisions of the pollen with the water molecules. The collisions impart some of the random heat motion of the water molecules to the pollen.

Let us now turn to electricity. What is an **electric current?** We have previously mentioned electrons, which are the lighter constituents of atoms. Some of the atomic electrons in a so-called conducting material (for example, a metal) are quite free to move around. If the free electrons in a conducting wire are kept in a state of drifting motion (superimposed on their random heat motion) by applying a force, an electric current is the result. Thus the energy of an electric current is nothing but the kinetic energy of orderly motion of electrical charges.

At a close look, the phenomena of sound and light are recognized as the motion of waves; we will have more to say about this in later chapters. Note that these phenomena also represent energy of motion of a kind.

Chemical and nuclear energy have to do with potential energy, the first arising from attractive electrical forces within molecules and the second resulting from attractive nuclear forces within the atomic nucleus. Actually, there is some kinetic energy involved also, but the energy that is released in chemical and nuclear reactions comes from potential energy.

Thus what was so different at the outset doesn't look so different anymore.

A Russian book published in 1908 under the title *Physics for Fools* has some great experiments. Here is a sample that is ideally suited for a "foolish" demonstration that heat is a form of energy. Get one of your more gullible friends. Lay him down on a table and stack some books on the table against the top of his head. Now apply heat to his feet. Everybody knows that objects expand on heating. Your friend will, too, and the books at his head will start becoming displaced as he expands. But that's work. So you have just proven that heat has the capacity for doing work; it is a form of energy. Incidentally, it is quite possible that your friend will get up and run long before his expansion takes effect. And that's work too. Thus either way the demonstration works.

I can also give you a fool's proof that electricity is a form of energy. The idea is given in a story by English biologist-author J. B. S. Haldane. You have probably heard about electromagnets. If a rod of iron is placed inside a current-carrying coil of wire, the iron becomes magnetized; this is called an electromagnet. Now once upon a time, according to Haldane's story, a certain township became uncontrollably infected with rats. The rats began to eat up all the town's food supply, causing a near-panic situation. The town declared a reward for the solution of the rat problem. The winner came up with the following idea. Make a lot of cookies with little filings of iron in them and let the rats eat these cookies. Afterwards turn on the electromagnet, a huge one designed specially for the job. The magnetic force of the electromagnet would act on the iron in the belly of the rats and sweep them to a desired dump. Needless to say, the idea worked. Clearly it was the electrical energy in the coil of wire that performed the work.

All the different forms of energy are found to be made up of two fundamental forms, kinetic and potential. The atomic picture of matter enables us to appreciate this. We have gone through this picture in a very short space. It may take the whole book to become familiar with all facets of such a picture. We hope, though, that you see its beauty even after this short discussion. The following quotation will give you some idea of the ecstasy experienced by some. Physicist Fritjof Capra describes the following experience:

> I was sitting by the ocean one late summer afternoon, watching the waves rolling in, and feeling the rythm of my breathing, when I became suddenly aware of my whole environment as being engaged in a gigantic cosmic dance. Being a physicist, I knew that the sand, the rocks, water and air around me were made of vibrating molecules and atoms, and that these consisted of particles which interacted with one another by creating and destroying other particles. I knew also that the earth's atmosphere is continuously bombarded by showers of cosmic rays, particles of high energy undergoing multiple collisions as they penetrate the air. All this was familiar to me from my research in high energy physics, but until that moment I had only experienced it through graphs, diagrams and mathematical theories. As I sat on that beach my former experiences came to life. I "saw" the atoms of the elements and those of my body participating in this cosmic dance of energy; I "felt" its rhythm and I "heard" its sound, and at that moment I knew that this was the dance of Shiva, the lord of dancers worshipped by the Hindus.*

FIGURE 9.9 Shiva Nataraja, meaning the king of the dancers. In one hand he holds the fire for the destruction of the universe; in his other hand is the drum with which he announces the re-creation. (Courtesy of the Denver Art Museum, Dora Porter Mason Collection.)

One more comment. You may have heard about **mass energy.** Einstein showed that when motion is treated properly, with the right description of

---

*Reprinted by special arrangement with Shambhala Publications, Inc., 1123 Spruce Street, Boulder, Colorado 80302. From *The Tao of Physics*, by Fritjof Capra, 1975.

space-time, then it becomes clear that mass and energy are completely equivalent. So matter has energy by virtue of its mass alone; this is mass energy. We will say more about this later.

■ 9.3 Energy and Society

Perhaps today you hear about energy mostly in connection with the energy crisis. There may be a shortage of available energy resources, notably the cheapest of them, oil. Two questions come up immediately. First, what caused the energy crisis? And second, what can we do about it? The second question is hard to answer; actually, there is no short answer. We will not attempt even a brief discussion here, but we will return to this question in later chapters. Here we will discuss some of the aspects of the first question.

As you know, ours is a technological society, and modern technology has been growing at a very fast rate. To get an idea of how fast the growth is, let's consider transportation. The chariot was invented in 1600 B.C.; it set a speed record of about 25 mi/h that lasted until midway through the last century. But since then we have seen that maximum speed quadruple (through the development of the automobile) to 100 mi/h by the 1880s and quadruple again (through the invention and improvement of the airplane) by the 1930s. The next increase was even more dramatic. The rocket planes of the early fifties were able to travel ten times faster, at 4,000 mi/h, than the 400 mi/h of the thirties. And within a decade, with the advent of the Russian *Sputnik* and the American *Friendship* flights, we reached speeds of 18,000 mi/h for orbiting satellites and 25,000 mi/h for spaceships bound for other parts of the solar system. So in scarcely 100 years, our maximum speed has increased some 250-fold, from a meager 100 mi/h to 25,000 mi/h.

We have to pay a price for this kind of fast growth. Progress in technology has at least two price tags attached to it. First, technology needs energy and thus forces a rapid use of the available energy resources. This has brought us today's energy crisis. The second price tag of technology is environmental pollution, which will be discussed further in later chapters.

M. K. Hubbert draws a graph of human consumption of the common fossil fuel resources—coal, oil, and natural gas—which goes back in history to the days of the Egyptian Pharaohs and also projects into the future (Fig. 9.10).* This figure dramatizes the fast pace of energy demand in the modern era. The result? We will run out of our cheapest and most practical energy resources rather quickly. As you may know, it takes nature millions of years to accumulate fossil

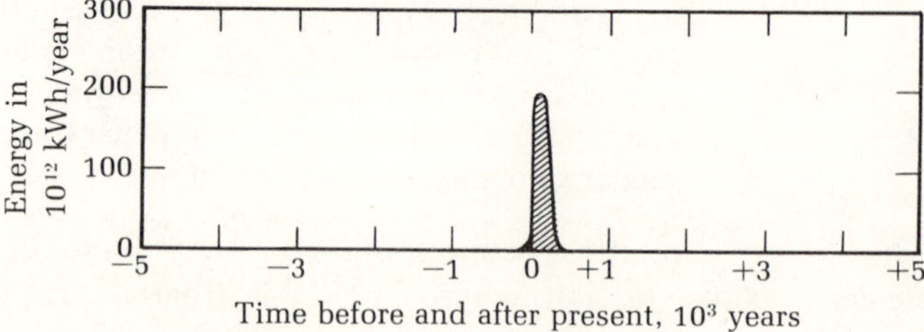

FIGURE 9.10   Hubert's graph. (Courtesy M. King Hubbert, "Man's Conquest of Energy: Its Ecological and Human Consequences," in A. Burt Klein, Jr., ed., *The Environmental and Ecological Forum 1970–1971,* U.S. Atomic Energy Commission, 1972, pp. 1–50.)

---

*M. K. Hubbert, *Resources and Man* (San Francisco: Freeman, 1969), p. 206.

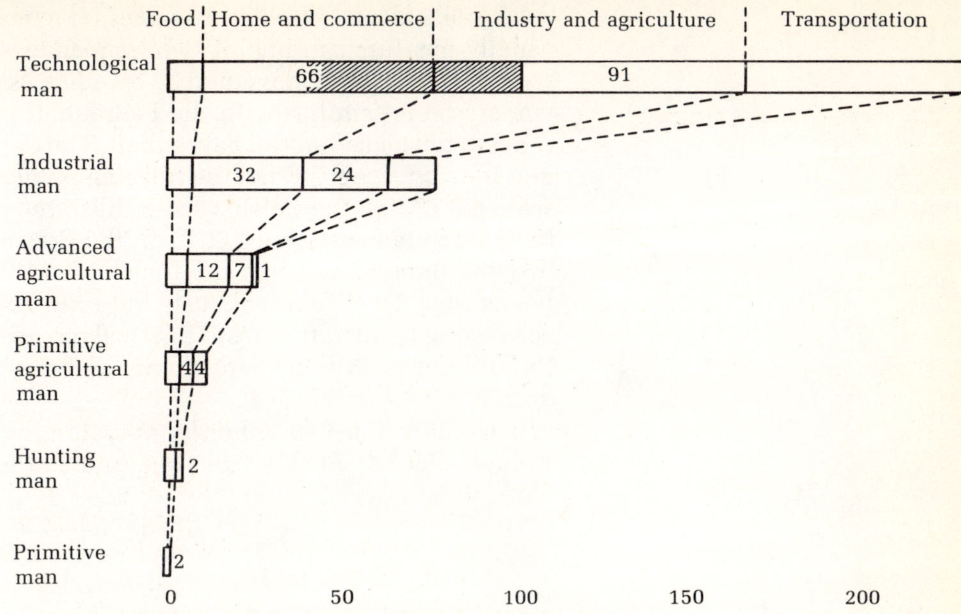

fuel energy from the sun, but it takes a technological civilization like ours only a few hundred years to exhaust it.

The September 1971 issue of *Scientific American* was completely devoted to the subject of energy and society. In it there is an article by geologist Earl Cook that displays the history of the growth of energy consumption in a very revealing graph (Fig. 9.11). Table 9.1, based on this graph, gives the total per capita daily energy consumption from primitive peoples through time to technological society.

TABLE 9.1   Power consumed at various stages of civilization.

Type of Society	Power Consumption per Person (Calories/day)
Primitive	2,000
Hunting	5,000
Primitive agricultural	12,000
Advanced agricultural	26,000
Industrial	77,000
Technological	230,000

Before we can discuss the table, we need to introduce you to another common unit of energy, the Calorie (abbreviated Cal), which is often used in connection with food and dieting.* A Calorie is the quantity of heat required to raise the temperature of 1 kilogram of water through 1 degree Celsius. But what is a Celsius degree?

---

*Not to be confused with calorie (cal) spelled with a lowercase c; 1 Cal = 1000 cal = 1 kcal.

Let's talk about temperature scales for a moment. Temperature of an object is usually measured in one of two temperature scales. Here we use the Celsius scale (°C), which was invented by Swedish astronomer Anders Celsius. Perhaps you are more familiar with the Fahrenheit scale (°F), which was invented by German physicist Gabriel Fahrenheit. The Celsius scale differs from the Fahrenheit in two respects. First, the freezing point of water is 32°F on the Fahrenheit scale but 0°C on the Celsius scale. Also, the boiling point of water is 212°F on the Fahrenheit scale but 100°C on the Celsius scale. The second difference is that the interval between the freezing and the boiling points of water is 180 degrees for the Fahrenheit scale but only 100 degrees on the Celsius scale. In converting temperature from one scale to another, you have to remember both the difference in their zero points and the difference in the magnitude of a degree.

If we denote a Fahrenheit temperature by T_F and a Celsius one by T_C, the formulas for conversion from one to the other are

$$T_C = \frac{(T_F - 32) \times 5}{9}$$

$$T_F = \tfrac{9}{5} T_C + 32$$

As an example we'll convert 86°F into a Celsius temperature. From the formula the corresponding Celsius temperature is

$$T_C = \frac{(86 - 32) \times 5}{9} = 54 \times \tfrac{5}{9} = 30°C$$

Now let's return to our discussion of the Calorie. The daily average per capita food consumption is something like 3000 Calories. It has been said that a good rounded kiss between a man and a woman generates approximately 1 Calorie of heat. These examples ought to give you an idea of how much energy 1 Calorie is. Also useful for reference is the relation*

$$1 \text{ Cal} = 4.2 \times 10^3 \text{ J}$$

Returning to Table 9.1 we notice that whereas primitive individuals each consumed only 2000 Cal of food and energy, which constituted their daily diet, the technological person in America demands more than a hundred times as much energy per day. This is a huge price to pay for technology; it represents the fantastic cost of modern living.

The power consumption in the table is expressed in the somewhat unusual units of Calories/day. Let us convert some of the numbers to the more usual unit of power, watts. Since

$$1 \text{ Cal} = 4.2 \times 10^3$$
$$1 \text{ day} = 8.64 \times 10^4 \text{ sec}$$
$$1 \text{ W} = 1 \text{ J/sec}$$

---

*It is customary to refer to the number 4.2×10^3 J/Cal, the conversion factor between the joule and the Calorie, as the **mechanical equivalent of heat.**

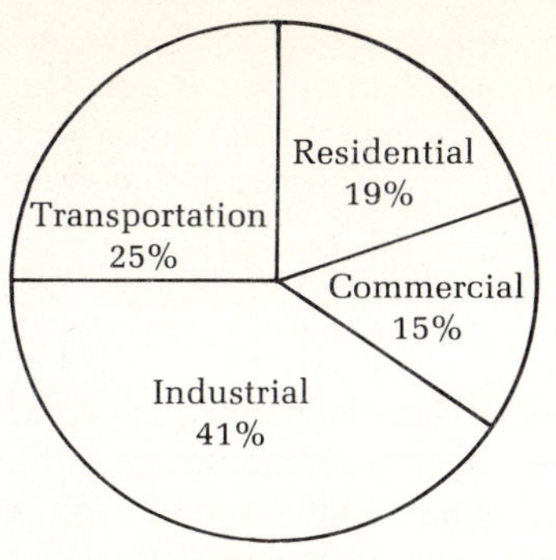

FIGURE 9.12 Splitting the U.S. energy budget into different sectors.

we carry out the conversion of units as follows: For the primitive society, the power of 2000 Cal/day is equivalent to

$$\frac{2000 \text{ Cal}}{1 \text{ day}} = \frac{2000 \times 4.2 \times 10^3 \text{ J/sec}}{86,400}$$

$$= \frac{8.4 \times 10^6}{8.64 \times 10^4} \approx 100 \text{ W}$$

Thus 100 W is the biological power of a human. In contrast, the power demand of a technological human is approximately 100×100 W, which is 10 kW.

What accounts for this huge increase in the energy demanded by the modern society? A look at the detailed graph of Fig. 9.12 will tell you the story. Transportation, of which the major chunk is your personal automobile, accounts for 20% of the total. The household and commercial sectors, which include the effect of consumerism in our society, account for some 25% or so of the total. The rest is largely the share of the individual in maintaining the industrial economy. It is also interesting that even our food energy bill has increased 5-fold from primitive to technological man. Can you guess why?

Author Alvin Toffler tells a story of an eleven-year-old boy dying of old age.* This is an unusual disease called pregoria—advanced aging. Toffler remarks that, metaphorically at least, the disease is commonplace in societies of high technologies. The energy crisis may be symptomatic of this disease in our society. We can only hope the outcome will not be the same.

Two natural laws are responsible for making it so difficult to solve the problem of the shortage of energy resources. One of these laws, the law of conservation of energy, is discussed next. The other, the entropy law, will be discussed in Chapter 10.

■ 9.4 The Law of Conservation of Energy

There are many different forms of energy. One form of energy, however, can be converted into another. What are the rules of energy conversion?

One of the rules is that in a conversion process, the total amount of energy always remains the same. Energy obeys a conservation law:

> *The sum of all the different forms of energy in the entire universe always adds up to the same amount.*

Mechanical kinetic energy + gravitational potential energy + heat + light + nuclear energy + mass energy + all other energy forms = a constant.

To make things more clear, we will take advantage of a parable told by physicist Richard Feynman. This is perhaps the best introduction to the subject of energy conservation that we can give.

> I want you to imagine that a mother has a child whom she leaves alone in a room with 28 absolutely indestructible blocks. The child plays with the blocks all day, and when the mother comes back she discovers that there are indeed 28 blocks; she checks all the time the conservation of blocks! This goes on for a few

*Alvin Toffler, *Future Shock* (New York: Random House, 1970), p. 20.

days, and then one day when she comes in there are only 27 blocks. However, she finds one block lying outside the window, the child had thrown it out. The first thing you must appreciate about conservation laws is that you must watch that the stuff you are trying to check does not go out through the wall. The same thing could happen the other way; if a boy came in to play with the child, bringing some blocks with him. Obviously these are matters you have to consider when you talk about conservation laws. Suppose one day when the mother comes to count the blocks she finds that there are only 25 blocks, but suspects that the child has hidden the other three blocks in a little toy box. So she says, "I am going to open the box." "No," he says, "you cannot open the box." Being a very clever mother she would say, "I know that when the box is empty it weighs 16 ounces, and each block weighs 3 ounces, so what I am going to do is to weigh the box." So, totaling up the number of blocks, she would get

$$\text{No. of blocks seen } + \frac{\text{weight of box } - \text{ 16 oz}}{\text{3 oz}}$$

and that adds up to 28. This works all right for a while, and then one day the sum does not check up properly. However, she notices that the dirty water in the sink is changing its level. She knows that the water is 6 inches deep when there is no block in it, and that it would rise $\frac{1}{4}$ inch if a block was in the water, so she adds another term, and now she has

$$\text{No. of blocks seen } + \frac{\text{weight of box } - \text{ 16 oz}}{\text{3 oz}} + \frac{\text{height of water } - \text{ 6 in}}{\frac{1}{4} \text{ in}}$$

and once again its adds up to 28. As the boy becomes more ingenious, and the mother continues to be equally ingenious, more and more terms must be added, all of which represent blocks, but from the mathematical standpoint are abstract calculations, because the blocks are not seen.

Now I would like to draw my analogy, and tell you what is common between this and the conservation of energy, and what is different. First suppose that in all of the situations you never saw any blocks. The term "No. of blocks seen" is never included. Then the mother would always be calculating the whole lot of terms like "blocks in the box," "blocks in the water," and so on. With energy there is this difference, that there are no blocks, so far as we can tell. Also, unlike the case of the blocks, for energy the numbers that come out are not integers.

What we have discovered about energy is that we have a scheme with a sequence of rules. From each different set of rules we can calculate a number for each different kind of energy, it always gives the same total. But as far as we know there are no real units, no little ballbearings. It is abstract, purely mathematical, that there is a number such that whenever you calculate it, it does not change. I cannot interpret it any better than that.*

Have you noticed the punch line: "There are no blocks?" Energy is an abstract mathematical concept, yet somehow the idea transcends mathematics. What makes the concept so useful is the conservation law.

In this section we will discuss some practical applications of the energy conservation principle. Energy conservation holds for any closed system—for

---

*Richard Feynman, *The Character of Physical Law* (Cambridge, Mass.: MIT Press, 1965), p. 69.

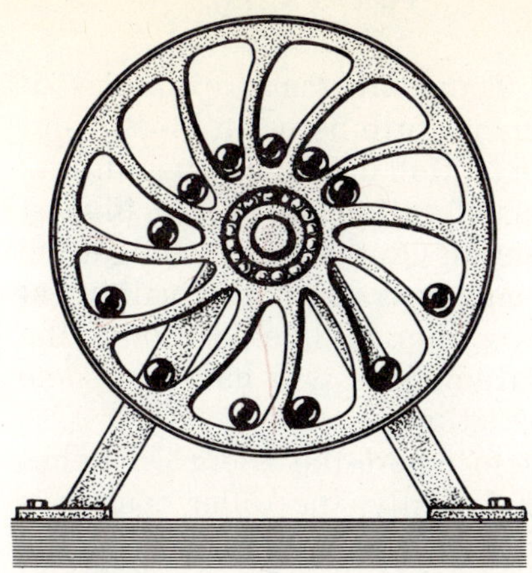

FIGURE 9.13 A perpetual motion machine. Of course, it doesn't work.

example, the biological system of the human body with its daily diet and exercise routines. Any machine, likewise, must accommodate the energy conservation law.

This brings us to the subject of *perpetual motion machines of the first kind* (yes, there is more than one kind of perpetual motion machine that people have tried to invent—in vain; the second kind will be discussed in the next chapter). Many people have tried to build wheels with various devices to keep them moving forever. No one has ever succeeded. The energy conservation law prevents them from performing this feat; that and the ever-present friction force.

Fig. 9.13 shows a perpetual motion wheel, one that even the famous Italian painter Leonardo da Vinci studied for awhile. This one is constructed with heavy balls rolling in compartments between the hub and the rim. The hope is that maybe there will always be enough balls close to the rim on one side (say the right) to generate an unbalanced torque that will keep the wheel rotating. It doesn't work. Why? Suppose initially there is such a situation. Someone starts the wheel rotating. But very soon there will be a slightly unbalanced torque on the other side (on the left). So the wheel swings back and forth. Does even this go on forever? No; the motion will continue only until the energy that the wheel was given initially is exhausted by frictional dissipation into heat.

There is a lot of talk today about the energy crisis. Wouldn't it be nice if we could make a power plant for which we pump some water up the hillside and then use this water to generate energy? Alas, this won't be profitable, since we'll have to spend at least as much energy to get the water up as we'll get from the water falling down. My favorite perpetual motion machine gets the water up the hill by imagining that space itself is warped in a very special way, as shown in Fig. 9.14. Unfortunately, as far as we know, physical space does not behave this way. So we just can't beat the energy conservation law. It is a "you can't win" law.

FIGURE 9.14 Another perpetual motion machine from the imagination of Escher. The waterfall can go on at least until all the water is evaporated away. Unfortunately, as far as we know, physical space is not warped like this. (M. C. Escher, "Waterfall," Escher Foundation—Haags Gemeentemuseum—The Hague.)

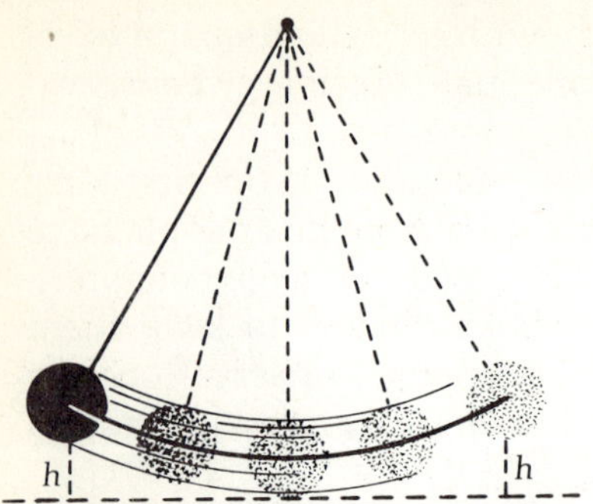

FIGURE 9.15 Conservation of energy for a simple pendulum. As the pendulum swings, the energy goes back and forth between kinetic and potential.

Examples of the energy conservation law

Interchange of kinetic and potential energy. There are many examples of motion where the form of energy changes back and forth between kinetic and potential energy. One of the simplest cases is the motion of a simple pendulum, a mass vibrating at the end of a string. When the pendulum bob is at its highest position (see Fig. 9.15), all of its energy is potential. On the other hand, at the vertical position of the pendulum, the kinetic energy is maximum, gaining at the expense of potential energy. Moreover, at any intermediate position of the pendulum, the sum of its kinetic and potential energies will have the same constant value (after allowing for the effect of friction).

The roller coaster (Fig. 9.16) is another system where the energy switches back and forth between kinetic and potential. Initially, the roller coaster is hoisted up to some height and then let loose. In all the subsequent motion, the total energy never can exceed this initial value. Thus none of the subsequent rises can exceed the starting elevation.

Application of the energy conservation law to human activities. Should the conservation of energy hold for the affairs of the living? The answer is yes, as far as we can tell. In fact, the law was first demonstrated by a medical doctor working with rats. Everybody more or less knows that there is a correlation between food intake, human metabolic activities, and gain in weight. If there is too much food intake and not enough expenditure of energy, then the body is going to have to store the left over in the form of fat, which is how we gain weight. On the other hand, the only way to lose weight is to let your activities get the upper hand over your diet.

One pound of human fat is equivalent to about 4000 Calories. If you do not immediately see the enormity of this number, compare the following figures. A moderate human activity like playing Ping-Pong can consume 400 Calories in an hour.* Heavy activities like playing soccer consume 500 Calories an hour. Even extremely heavy activities like rapid jogging or mountain climbing use up energy at a rate of only 1000 Calories/hour. Thus to get rid of just 1 pound of fat, you have to engage in 4 hours of mountain climbing or 8 hours of playing soccer or 10 hours of Ping-Pong.

The energy conservation law can act as a guideline for a diet and exercise routine. We have the rule

Energy of food intake − energy expended in metabolic activities = net gain or loss of energy which adds to (or subtracts from) the weight.

FIGURE 9.16 An ideal roller coaster. Subsequent rises cannot be higher than the initial one because of the constraint imposed by the law of conservation of energy. In a real roller coaster, the subsequent rises will have to be somewhat smaller to account for the inevitable friction that converts some of the mechanical energy into heat.

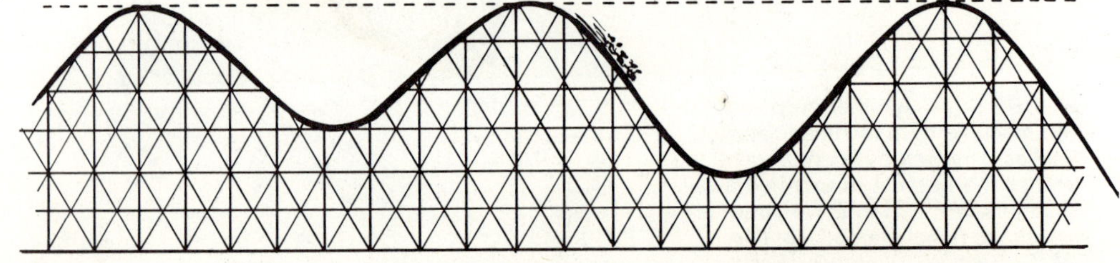

*Note that the energy is not actually "consumed" but changed into heat energy.

The Calorie value of various food items is available in any diet book. The net gain or loss of weight can be converted into energy units by using the equivalence of 1 pound of fat and 4000 Calories of energy. You can determine the energy that you expend in normal metabolic activities from the equation by carefully keeping track of your food intake and your weight gain (or loss) for a period of time. Once you know how much energy you expend in normal activities, you can shape your diet and exercise pattern according to your needs to lose or gain weight, using the same formula. Table 9.2 gives you a rough guideline for the energy expended in various activities.

TABLE 9.2   Metabolic rate for various human activities.

Activity Level	Metabolic Rate for a 140-lb Person (Cal/h)
Sleep	70
Light activity (listening to a lecture, some housework)	200
Moderate activity (walking rapidly)	400
Heavy activity (playing soccer, sawing wood)	500
Very heavy activity (bike racing, rapid jogging, mountain climbing)	1000

Energy conservation in simple machines. Simple machines such as the lever are designed to give us a mechanical advantage, which means that we can use less muscle force. We get more force from the machine than we put into it. This may give you the mistaken impression that a machine can help us out with energy, too. Can we perhaps get more energy out than we put in? No, this never happens; conservation of energy remains valid for all machines. In fact, if anything, by the presence of friction—inescapable in all machines—we tend to lose part of the energy input as heat. The useful output work always is less than the input energy. And thus the efficiency of a machine, which is the ratio of output work to input energy, is always less than 100%.

If we neglect this last fact (in a machine like a lever the effect of friction is small), we can use the energy conservation law to determine the mechanical advantage of a simple machine. Fig. 9.17 shows a simple lever. The idea is to apply a small force f through a large distance D at one end. But a large force F acting through a small distance d will now be exerted at the other end. The work that is put in is $f \times D$ and the work at the output end is $F \times d$. The two must be equal by the law of energy conservation. Thus

$$\frac{F}{f} = \frac{D}{d} \tag{9.6}$$

The mechanical advantage, which we define as the ratio of output force to input force, is equal to the inverse of the ratio of the distances through which the two forces act, respectively. Thus it is possible to use a rather small force, but through a large distance, to exert a large force through a small distance. The automobile jack, crowbars, and pliers all work this way.

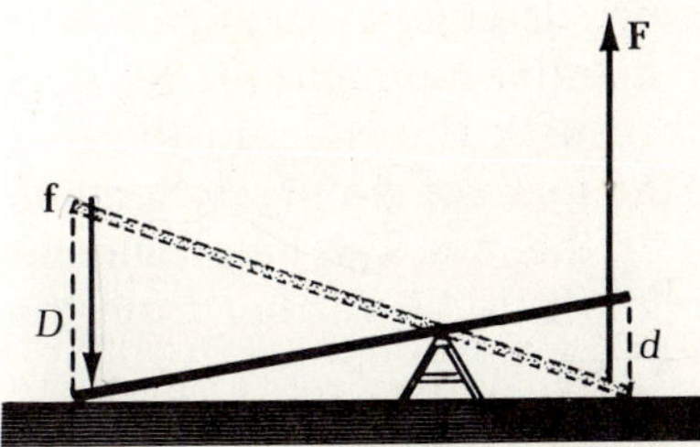

FIGURE 9.17   A lever. $f \times D = F \times d$. See text for explanation.

FIGURE 9.18 This apparatus not only tests the conservation of momentum but also gives a spectacular demonstration of the law of conservation of energy. See text for explanation.

The three conservation laws

We now have on hand the three most important conservation laws that nature follows: the conservation of momentum, the conservation of angular momentum, and the conservation of energy. We emphasize here that in many physical situations, not one, but two or even all three of the conservation laws can be important. Let's look at some examples.

There is a demonstration apparatus (Fig. 9.18) that shows that both momentum and energy conservation operate in a collision between objects. In this apparatus, if you take one ball at one end, separate it from the rest, and release it, the ball on the opposite end will come out with all the first ball's momentum. This is a demonstration of momentum conservation.

But now try this with two balls. In this case there are two ways to satisfy momentum conservation: two balls could come out of the other side with the same momentum, or one ball could come out with the sum of the momentum of each of the two original balls. However, when you do the experiment, you find that only one of these possibilities is realized. It is always two balls that come out at the other end. Why? The reason, of course, is the conservation of energy. If m and v are the mass and velocity of a ball, then the initial total energy for two balls is

$$\tfrac{1}{2}mv^2 + \tfrac{1}{2}mv^2 = mv^2$$

If two balls come out at the other end each with velocity v, obviously they have the same total energy as the original two balls. But if only one ball comes out with velocity $2v$ (as needed for momentum conservation), its kinetic energy is

$$\tfrac{1}{2}m(2v)^2 = \tfrac{1}{2}m(4v^2) = 2mv^2$$

Thus energy would not be conserved; the final energy would be greater than the initial energy. Since this is not allowed by nature, one ball will never emerge on the other side when you take two balls and release them.

We will now consider an example in which energy and angular momentum conservation become important simultaneously. Previously we talked about tides. The tides lead to the dissipation of considerable energy through friction. Where does this tidal energy come from? Tidal energy can come only from the rotational energy of the earth. The earth must slow down in its rotation due to the tides. And it does; our day becomes longer by something like a second in a thousand centuries.

But now here is an intriguing problem: There is no external torque from outside the earth-moon system, so the angular momentum of the earth-moon system cannot change. But if the earth slows down, its angular momentum is reduced. Is this a violation of the conservation of angular momentum? No, it is not. We must not forget about the moon, which causes the tides in the first place. The earth and the moon are connected by the force of gravity; when one loses angular momentum the other one must pick up the loss. The moon accommodates the extra angular momentum by shifting a little bit further from the earth.

Another thing we stress here about these great conservation principles is that they have been shown to be linked with some fundamental aspects of space and time. The conservation of momentum arises from the homogeneity of space (no part of space is different from any other). The conservation of angular momentum is the result of the isotropy of space (all directions are equivalent). And finally, the conservation of energy is the consequence of the homogeneity of time (time—past, present, and future—has identical properties). The "proofs" of these assertions are a little too involved for us here, but perhaps even without the proofs you can appreciate these links. According to the poet Rilke,

> Even if the world changes as fast
> as the shape of clouds,
> all perfected things at last
> fall back to the very old.*

The conservation principles are a few of the most "perfected things" in our physical model of the universe, and it is reassuring to find that indeed they "fall back" to very old concepts of some of the long-known symmetries of space and time.

SUMMARY

One of the most important concepts for our description of nature, namely energy, is introduced in this chapter. Energy is defined as the capacity for doing work, but that's just a starting point. Work is defined as the product of force and displacement. Another important concept is that of power, which is the rate of expending energy.

Energy exists in many different forms: heat, electricity, and so on. Historically, each of these forms of energy was studied somewhat independently. Only later were they understood to be one and the same thing. One of the effects of the historical development that still lingers on is the use of a different unit for each of the different forms of energy. For example, joule is used for mechanical energy, Calorie for heat and chemical energy, kilowatt hour for electricity, and so on. But remember, all these are units of energy, and therefore any form of energy can be expressed in terms of any of these units. Make use of the conversion table given in Appendix 4 whenever you need to convert one of the units into another.

It turns out on closer look using the atomic picture that all these forms of energy are really either kinetic, potential, or a combination of the two. Kinetic energy is energy of motion. Potential energy is a potential capacity for doing work; it is stored energy by virtue of mutual interaction between objects.

The formula for the kinetic energy of a moving object is

$$K = \tfrac{1}{2}mv^2$$

And the formula for calculating the gravitational potential energy of an object at a height h relative to the ground is

$$mgh$$

These are two useful formulas to remember.

---

*Rainer Maria Rilke, *The Sonnets of Orpheus*, trans. A. Poulin (Boston: Houghton Mifflin, 1977), p. 121. Copyright © 1975, 1976, 1977 by A. Poulin, Jr. Reprinted by permission of Houghton Mifflin Company and J. B. Leishman and The Hogarth Press.

Perhaps the most important unifying concept for the different forms of energy is the idea of the conservation of energy. The idea is expressed as the law of conservation of energy. Energy cannot be destroyed, nor can it be created. The total amount of energy of a closed system always remains the same. The form can change, but when you add up the contribution of each of the different forms, you always get the same number.

Some applications of the energy conservation law, for example to human metabolic activities, are given. It is also interesting that the energy conservation law (as well as the laws of conservation of momentum and angular momentum) discussed earlier is connected to a symmetry principle.

One important theme discussed in Chapter 9 is energy and society. It is generally agreed today that we are facing an energy crisis; the discussion on energy and society gives you a perspective on this issue. One important thing to remember is that subjects like the energy crisis are mostly quantitative. If you look at some of the numbers calculated and learn to calculate a few yourself, it will increase your understanding enormously. Become conversant with the units of energy, particularly power (which somehow many students find difficult). We will return to the theme of energy and society many times, particularly in Chapters 17 and 24.

QUESTIONS

Review and reason

1. What is work? For given magnitudes of force and displacement, what angle between them yields maximum work? For what angle do we get zero work? Can the work done by a force on an object be negative? Explain.

2. Is work done in any of the following cases? Explain. (a) You push a bike along the road. (b) You lift a chair. (c) You hold a chair up in the air for an hour.

3. Both torque and work are expressed in terms of identical units. Explain. If work and torque have the same units, do they express the same thing? Why or why not?

4. What is the equation for the kinetic energy of an object? Why is kinetic energy never negative?

*5. Which of the following has the most kinetic energy: (a) an object of mass $2m$ and speed $2v$; (b) an object of mass $4m$ and speed v; or (c) an object of mass m and speed $4v$? Explain.

6. What is potential energy? Give a few examples.

7. Which of the following has more potential energy: (a) an object of mass m at a height $3h$ aboveground; (b) an object of mass $3m$ at a height h aboveground?

8. Explain the concept of power. Why are mountain roads winding?

9. Compare the concepts of work, energy, and power.

10. The following good driving habits save gasoline. Explain why.
    *(a) Avoid so-called jackrabbit starts; accelerate gradually.
    (b) Maintain a steady speed; avoid unnecessary acceleration and braking.
    (c) Drive at a moderate speed.
    *(d) Avoid entering the highway in a hurry. Use the acceleration lane to accelerate gradually.
    *(e) When driving uphill, build up speed ahead of time so that you don't have to accelerate on the upgrade.

11. Consider each of the following units: joule, foot pound, horsepower, Calorie, and watt.
    (a) Which ones are units of work?
    (b) Which are units of energy?
    (c) Which are units of power?

12. Take a look at your electric bill. You will find that you have been charged for so many kilowatthours. Are you being charged for using power or energy?

13. The following statements have sneaky mistakes.

---

*Optional.

Correct them.

(a) Human biological power is rated to be 100 watts per day.

(b) A politician said the other day, "We need this river dam because it can bring us 100,000 kilowatts per year."

(c) A newspaper reports (obviously by a reporter who never took a physics course): "American electrical power need is 250 billion watts per year."

14. Take a look around you. Name some of the energy forms that you see.

15. Is it possible to have momentum without energy? Is it possible to have energy without momentum?

16. Distinguish between heat and temperature.

17. There are lots of stories about the legendary "Sasquatch." One story is that this elusive creature can jump as high as 15 ft. Show by the argument of scaling that such a feat may not be possible for the Sasquatch.

18. A hydroelectric power plant produces electricity, which is then used to operate a toaster. The forms of energy and its conversions are as follows: the water behind the dam has gravitational potential energy; the water turns the turbine, producing mechanical kinetic energy of rotation of the turbine blades; the turbine turns the electric generator, producing electrical energy; the toaster converts electrical energy into heat energy. This is a description of an **energy conversion pathway.** Construct a few energy conversion pathways from processes around you that involve at least three different forms of energy.

19. Rumor has it that Joule took a thermometer along on his honeymoon in the Swiss Alps. There he measured the temperature of the water of a fall after falling. Why should there be an increase of temperature?

20. Should the tires of a car get hot after braking to a halt? Explain.

21. What is the power requirement of the "biological person"? What is the power requirement of a "technological person"? Give a qualitative accounting for the huge difference.

22. Is there an energy crisis? If your answer is "yes," what do you think are some of the causes? If your answer is "no," give reasons for your answer.

23. Look for energy issues in your local newspaper for a week. Comment on the issues raised and also on the quality of the news coverage.

24. A child with a skateboard hitchhikes from the bottom to the top of a hill.

(a) The child now has more _________ than he or she had at the bottom of the hill.

(b) The child begins to coast down the hill on the skateboard so that some of his or her _________ is being converted into _________ .

(c) Eventually the child is back at the bottom of the hill. His or her extra energy at the top has been converted into _________ energy.

25. Physics professors often like to dramatize their confidence in the validity of nature's laws when they teach. In one such demonstration, an instructor makes a pendulum by suspending a cannonball to the end of a wire attached to the ceiling. He then stands at one side of the room, holds the ball against the tip of his nose, releases it and calmly awaits its return. Will he get hurt? Why or why not?

26. Julius Mayer, who is often credited with the discovery of the energy conservation law, is rumored to have hit upon the idea after noticing a correlation between the food intake of a sailor when working in the tropics and when in colder climates. He found that a sailor in the tropics eats less while performing the same kind of work. His conclusion: food energy (chemical potential energy), heat, and mechanical work are all equivalent. They are just different forms of the same thing, and their sum remains constant. Explain.

27.   Humpty Dumpty sat on a wall
        Humpty Dumpty had a great fall.
When did Humpty Dumpty have maximum kinetic energy? When did he have maximum potential energy? What happened to his energy after the fall?

28. A fluorescent light tube is hardly warm after a period of steady use, but an incandescent light bulb is too hot to touch even after brief use. Which gives off more light, a 40-watt fluorescent lamp or a 40-watt incandescent light bulb? How can you tell?

29. Give an example of the use of lever action from your everyday activities.

*30. The ancient Egyptians probably used the inclined plane as a machine to facilitate the building of great structures like the pyramids. An inclined plane enabled them to use less force (or power) for a given amount of work. Explain how.

31. What is the effect of tidal friction on earth's period of rotation? What happens to the angular momentum?

32. Each of the conservation laws of momentum, angular momentum, and energy is connected with a basic symmetry of space and time. Explain the symmetry principle corresponding to each of the conservation laws.

---

*Optional

1. Suppose you have to mow a 100-ft-long lawn. The force component you apply in the horizontal direction is 35 lb. What is the work done in mowing a strip of lawn the width of the mower and length 100 ft? Assuming there are 50 such strips in the entire lawn, what is the total amount of work involved in mowing it? Assuming an efficiency of 25% for the delivery of muscle energy, how much energy do you spend? If the job took an hour to finish, how much muscle power is involved?

2. Calculate the kinetic energy of the following:
   (a) a bullet of mass 30 g (0.03 kg) traveling with a speed of 700 m/sec;
   (b) a football player of mass 90 kg moving at a speed of 8 m/sec;
   (c) an automobile of mass 1000 kg traveling at a speed of 20 m/sec.

3. Which has the most potential energy?
   (a) a 50-lb object at a height of 100 ft
   (b) a 100-lb object at a height of 50 ft
   (c) a 75-lb object at a height of 75 ft

4. A mechanic is trying to decide what kind of hammer to use for a certain job. He finds that if he uses a light 1-lb hammer he can get twice the speed compared to one with twice its mass. With which hammer will he be able to generate more kinetic energy?

5. Suppose a person of weight 200 lb climbs a 100-ft flight of stairs in 30 sec. What is his or her horsepower rating?

6. Calculate the average power of a typical female while jogging. Assume a jogging speed of 3 m/sec, a length of 1.5 m for a stride (i.e., 2 strides/sec) and a mass of 8 kg for a leg. The efficiency of the leg muscles can be assumed to be 25%.

7. Assuming that the average power need per capita for an American is 10 kW, calculate the average power need for the entire nation. From this estimate calculate the energy demand per year for the United States.

8. A mass of 100 kg rests on a 100-m-high tower. What is the potential energy of the mass relative to the ground? If the 100-kg mass falls, what is its potential energy when it is halfway down? What is its kinetic energy at the halfway mark? Assuming that all its energy is converted into heat when it falls to the ground, how much heat is generated? In joules? In Calories?

9. You know that you maintain a constant weight on a diet of 3000 Cal/day. You want to reduce 2 pounds in the period of a week. How much should your Calorie intake be for the entire week?

10. Prying open the lid of a can using the handle of a spoon is an example of lever action. The rim of the can serves as the fulcrum. If a 3-lb force is applied at the end of the handle and the handle is lowered by a couple of inches in order to raise the lid $\frac{1}{10}$ in, how much force is applied on the lid?

Energy, Entropy, and the Universe

Humpty Dumpty sat on a wall;
  (a stable but precarious situation)
Humpty Dumpty had a great fall.
  (a natural event, gravity being what it is)
All the king's horses
And all the king's men
  (not a Maxwell's demon among them)
Couldn't put Humpty Dumpty together again.
  (the second law of thermodynamics still holds!)*

W. F. KIEFFER

■ 10.1   The Entropy Law

Entropy signifies the amount of disorder, or randomness, associated with a system. In a closed system all changes result in a net increase of entropy. Thus in the evolution of nature, *entropy always increases*. This is the entropy law.

How does one judge disorder as opposed to order? This is the question with which we will begin our discussion. We will continue with examples of order-disorder transitions and of the struggle between order and disorder that nature seems to display.

What does the entropy law have to do with energy? The point is that in ordered systems more energy tends to be available or useful; the opposite is true when the system is in disorder. Much of the energy in a disordered system is in the form of heat energy shared by zillions of molecules and tends to be unavailable for further utilization. Thus the entropy law, which tells us that the spon-

---

*Courtesy W. F. Kieffer, in *Chemistry: A Cultural Approach* (New York: Harper & Row, 1971), p. 388.

taneous direction of natural processes is from order to disorder, also tells us that energy goes from useful to unutilizable. So the entropy law is really an energy law—and a very important energy law, for that matter.

The energy conservation law can make you think that since energy cannot be destroyed, then everything is all right, even though you cannot create energy. There are such large reservoirs of energy in nature (for example, the heat energy of the oceans of air and water) that if we can tap these reservoirs, we never have to worry about running out of energy. However, the entropy law makes a shambles of this expectation. It tells us that the thermal energy of air molecules of the atmosphere or water molecules of the ocean is very likely beyond our reach.*

Tapping unutilizable energy is the quest of the perpetual motion machine of the second kind. The entropy law declares such machines to be utterly impossible.

One important consequence of the entropy law is that heat energy cannot be completely converted into useful mechanical work. In other words, heat engines—machines that convert heat into mechanical work—must operate at less than 100% efficiency. In practice, the efficiency of heat engines is quite severely limited through the considerations of the entropy law. We will have more to say about this later.

One interesting note. The entropy law should make us wonder if energy and the capacity for doing work are indeed the same thing. According to the entropy law, when energy is transformed from one form to another, some of the energy ends up in a form incapable of further utilization, for all practical purposes. This means that the capacity for doing work is not conserved, although energy is conserved in all such energy transformations. Thus it seems somewhat misleading to identify energy with the capacity for doing work.

Order, disorder, and order-disorder transitions

Previously we have defined entropy as a measure of the *disorder* of a system. Disorder is the lack of *order*. And what is order? To answer this question we must ask a second one: Order with respect to what? Suppose we have several large and small balls. The large balls are put in one corner of a flat container and the small balls in another corner (Fig. 10.1). Such a system is ordered with respect to the size of the balls. In contrast, suppose the balls are all mixed up, so that the neighbors of a particular ball can be a large ball or a small ball with equal likelihood. This system is disordered.

Suppose now that instead of balls we talk about molecules of two different kinds. Consider initially that we have a bottle of perfume. The molecules of the perfume are separated from those of the air by the bottle, and so we have an ordered state. If we open the bottle, the perfume molecules and the air molecules will be thoroughly mixed in a very short time. The system is now disordered—random—with respect to the air and perfume molecules. Thermal motion (with its continual incessant collisions) is the mechanism through which the disorder has come about. Also, this tendency toward disorder has come about spontaneously. On the other hand, once the system of mixed air and

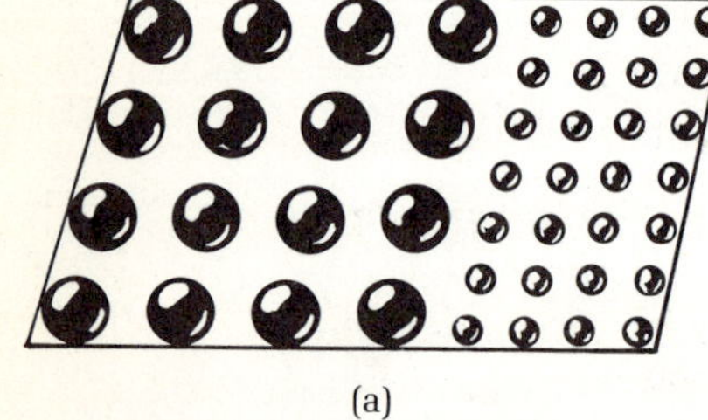

(a)

(b)

FIGURE 10.1   (a) The small balls and the large balls are separated. This is order. (b) A random assembly of large and small balls—disorder has taken over.

---

*However, see Section 13.3.

perfume molecules is created, we could wait a million years and never see the initial order reestablished. A system does not evolve spontaneously from disorder to order.

Let us consider another example. We mix some hot water with some cold water. In the hot water the molecules move faster, on the average, than in the cold. Upon mixing, the slower molecules will collide with the faster ones, resulting in energy sharing. Soon all the molecules will move with more or less the same velocity. Experimentally, this will result in a uniform temperature for the whole body of water instead of the cool and hot parts we started with initially. From the order-disorder point of view, the initial order here consists of a separation of the slow and fast molecules. There is an order with respect to speed. In the course of the natural process of mixing, this order is obliterated, and again we end up with disorder, or an increase of entropy.

In other situations it may be the spatial distribution of molecules that determines whether a given arrangement corresponds to order or disorder. A small volume of air at a given temperature is in a state of disorder with respect to the spatial distribution of the air molecules. We can examine any small portion of it and determine the number of air molecules contained in it. This number will be more or less the same for all portions.

Diamond or glass—which has more entropy, more disorder? In diamond the atoms are in a very definite crystalline arrangement; for this reason diamond has very special properties. In contrast, the arrangement of atoms in glass, which is amorphous and noncrystalline, is not so special. So glass has a greater entropy, corresponding to less order, order in this case referring to the arrangement of the atoms in space.

So in a discussion of order-disorder transition, always ask this question: Order with respect to what? Once you know which confirguration is order and which disorder, you can infer the natural direction of physical processes. You will find that it is always from order to disorder, in the direction of increasing entropy.

But of course not all systems end up in a state of maximum entropy. There is a rich variety of natural states exhibiting order of some kind. How is such order produced? For example, why do we find objects in a solid state or a liquid state—where the molecules have a rather orderly arrangement—instead of in the more disorderly gaseous state?

The riddle is solved when we realize that so far we have considered thermal motion only in the absence of any forces. But forces can produce order. For example, the order in a solid is caused by electrical forces between the molecules of the solid. Under the action of forces alone, a natural system tends to find the state of stable mechanical equilibrium. For example, a ball placed on the side of a bowl will come to rest at the bottom, which is the state of stable equilibrium (stable because it takes energy to displace the object from equilibrium).

When both forces and thermal motions are present in a system—which is the case in most of the physical world—there is a real struggle between the mechanism for order and that for disorder. How matters actually end up depends on the temperature. At low temperatures thermal motions are small, so the forces have the upper hand and we get solids and liquids. But as the temperature

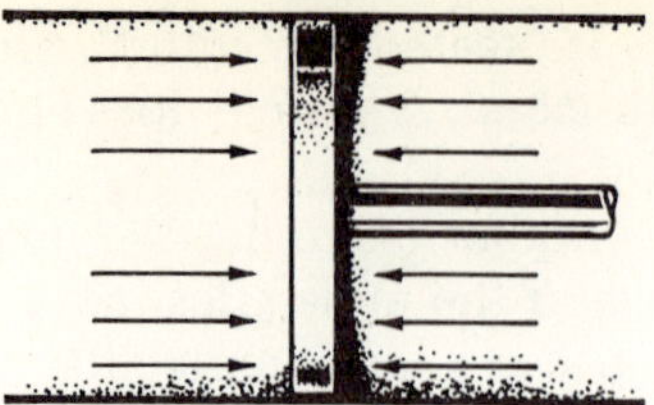

FIGURE 10.2 A piston placed in a system of uniform temperature. Any displacement of the piston arising from the collision of molecules on one face is canceled by the collisions on the other side.

increases, eventually thermal motion always dominates over forces and, at a sufficiently high temperature, all objects become gas.

It is also important to point out that when order is being created in a system, an equivalent or even greater amount of disorder is produced in that system's environment. There is no escape from the entropy law.

However, it is reassuring to know that it is possible to go uphill against entropy; life itself is such an uphill battle. We survive continuously in the direction of increasing order and decreasing entropy (in the molecular sense)— but at the expense of considerable increase of entropy in our environment.

Entropy and energy

Let us examine the entropy law from a slightly different angle. The entropy law tells us that, left to itself, every closed system tends to attain a state of **thermal equilibrium,** all parts of the system achieving the same temperature. This also means that there is no preference for a particular direction left in the system. If we were to count the molecules passing through a particular plane, we would find that about as many molecules pass in one direction as in another. Can we get work from such a system?

Suppose we put a piston in the system (Fig. 10.2). Any displacement of the piston achieved by the molecules moving one way will be canceled by the molecules moving the opposite way. There will be no net displacement or work. Thus we come to an extremely important statement of the entropy law:

> One cannot get work from a system in which all parts are at the same temperature.

This statement of the entropy law is also referred to, for historical reasons, as the second law of thermodynamics.*

We now see clearly that heat energy has a limited capacity for use. We can convert it into useful work or other forms of energy only when we introduce an object or an environment that is "colder" than our heat "source." Even so, only a portion of the heat of the source can be converted into useful work and the rest must be released to the environment. The latter heat is called "waste" heat or thermal pollution, and it is not recoverable.

Thus the second law of thermodynamics is really a law of energy. All the energy of a system in thermal equilibrium is unavailable for useful work. Thus the second law of thermodynamics can also be stated in this form:

---

*Thermodynamics is the subject of interchange of heat, mechanical energy, and other forms of energy. The law of conservation of energy is called the first law of thermodynamics. Remember that the second law of thermodynamics and the entropy law are one and the same thing.

In the course of development of nature, energy continuously goes from useful to unutilizable.

Whenever we convert energy from one form to another, some of the useful energy is converted into an unutilizable form, such as heat in the environment. Thus energy is not recyclable indefinitely. All the energy conversion pathways we encounter in nature ultimately end with heat energy in a form not further utilizable.

With each form of energy we can associate a certain amount of entropy, depending on how much disorder is involved. If there is much disorder—for example, when energy is present as heat in our environment—the entropy content is high. On the other hand, if there is little or no disorder, the entropy content is low. Gravitational energy, for example, has no random motion and therefore no entropy associated with it. Gravitational energy thus represents energy of the highest quality. Other forms of energy will be somewhere between these two extremes. Thus we can construct the following order of the various forms of energy according to their quality, starting from the ones of zero entropy (highest quality) and ending with the one of the highest entropy (lowest quality):*

> Gravitational potential energy, energy of rotational motion
> Nuclear energy
> Sunlight energy
> Chemical potential energy (fuels, food, etc.)
> Heat energy of our environment
> Energy of the cosmic microwave background†

Since the natural direction for the flow of energy is also that of increasing entropy, the value of such an ordering should be apparent. What we call quality, or low-entropy content, is really a statement of convertibility of that particular energy currency. The higher it is in the list, the more efficiently it can be converted. Thus gravitational energy is converted with close to 100% efficiency at the hydroelectric power plant.

Also, we note that energy of a high quality can be transformed spontaneously into one with a lesser quality, but the reverse is not true. The convertibility of the low-quality forms to higher ones is limited. Only a part of the low-quality form can be converted, and even then only if we can find an environment with a lower temperature than the operating temperature of the lower-quality form. For example, there is no way we can convert any of the cosmic microwave energy into a higher form, because there is no environment that has a lower temperature than that of this energy form. For the heat energy of our terrestrial environment, the situation is similar, although in principle we could run a heat engine between it as the source and outer space (which has a lower temperature) as the environment. In essence, then, the conversion of energy from a high-quality form to one lower on the list is really a degradation of energy.

Let us briefly mention the subject of the **perpetual motion machine of the**

---

*See "Energy in the Universe," F. J. Dyson, *Scientific American*, September 1971, p. 51.
†The waves that are produced and used to cook in a radar range are called microwaves. The cosmic microwave background is a residue from the big bang.

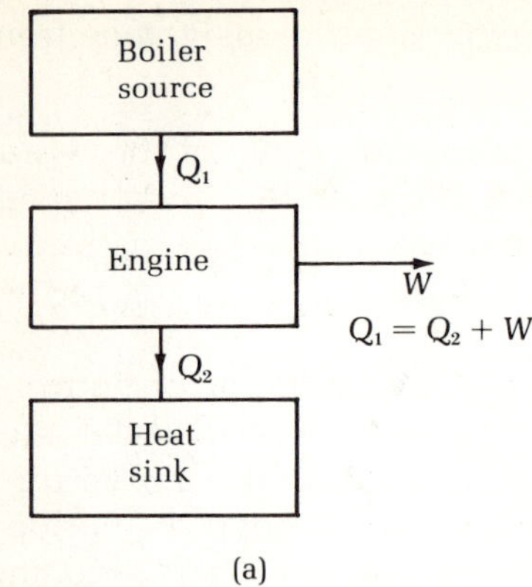

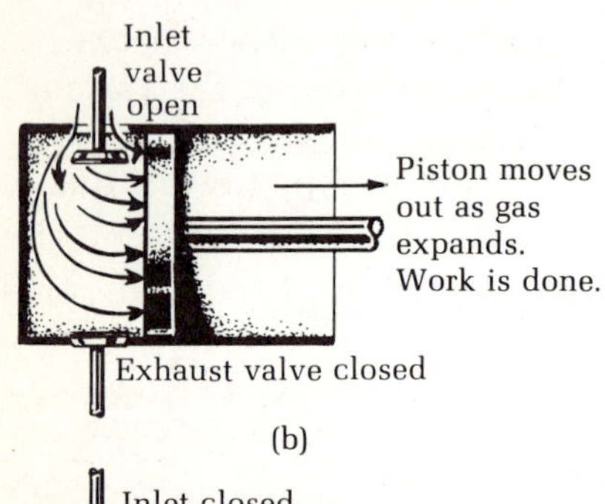

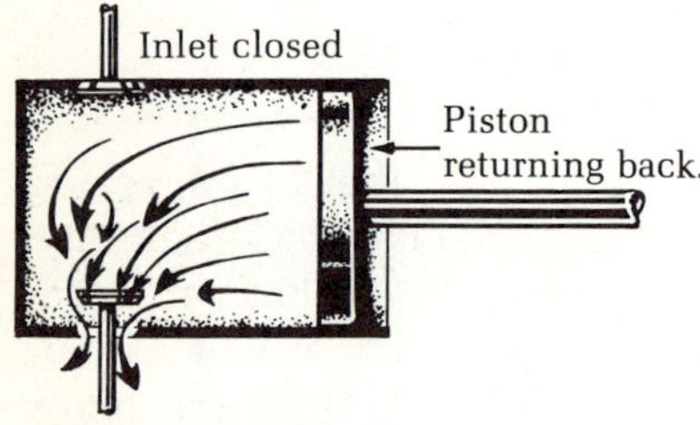

FIGURE 10.3   The performance of an actual heat engine. (a) A schematic diagram. Heat Q_1 is extracted from the source of high temperature and an amount Q_2 is dumped to the low-temperature sink. The difference $Q_1 - Q_2$ is the work W obtained by the engine. (b) The details of the performance. See text.

second kind, which purports to use unutilizable heat energy. Heat energy of the air molecules of the atmosphere is unutilizable since there is no appreciable difference of temperature in the atmosphere. The same statement can be made for most of the water of the ocean. Perpetual motion machinists have tried to build many ingenious devices that would tap these two vast energy sources—but to no avail, because of the entropy law.

How does a heat engine work?

The entropy law tells us that it is impossible to construct a perpetual motion machine in which, for example, the heat energy of the air is taken in by the machine and converted into useful work, which the machine then delivers to the outside world. Let us now consider an actual heat engine that *can* convert heat into mechanical work.

Fig. 10.3 shows a very simple heat engine. An energy fuel is burned in a boiler and the energy of the fuel is absorbed by a working fluid, like steam. The steam then can be allowed to enter into a cylinder through a valve. At this point the piston is as far to the left as possible. The expanding steam pushes the piston to the right. This is where external work is done by the engine—for example, a drive shaft can be attached to the piston for harnessing this work. When the piston reaches the other end of the cylinder, the exhaust valve opens and the steam goes out. This allows the piston to return to its original position, and the cycle begins again.

Notice several important points in this example. First, not all the thermal energy of the steam goes into moving the piston. Only those molecules of steam that happen to collide with it and have a momentum in the right direction will help move the piston. So the exhaust steam is bound to have some heat energy left, which has to be dumped into the outside environment. Second, notice that the outside has to have a lower temperature, otherwise there is no way the exhaust steam and feed it back into the cylinder to get more of its heat energy nal position. This is how the entropy law, the second law of thermodynamics, operates. We can convert the heat from a system into useful work only if there is a heat source *and* a heat sink into which we dump the waste heat.

It is inevitable that someone will ask the question: Why can't we take the exhaust steam and feed it back into the cylinder to get more of its heat energy converted into work? The second law tells us that we can't.* This would amount to running a perpetual motion machine that converts all of an object's heat energy into work.

Thus the net consequence of the entropy law for the operation of heat engines is that some of the energy will be dumped into the environment and wasted. The machine thus has a finite efficiency and always gives rise to thermal pollution of the environment. In today's power plants the efficiency factor is only about 35%. This means that when electricity is generated from heat (which itself comes from some sort of resource energy, but since this is a conversion from a high- to a low-quality form of energy, there is not much efficiency loss), only 35% is converted into electricity. The rest goes into the environment as thermal pollution.

---

*A more complete discussion of this point is given in Chapter 13.

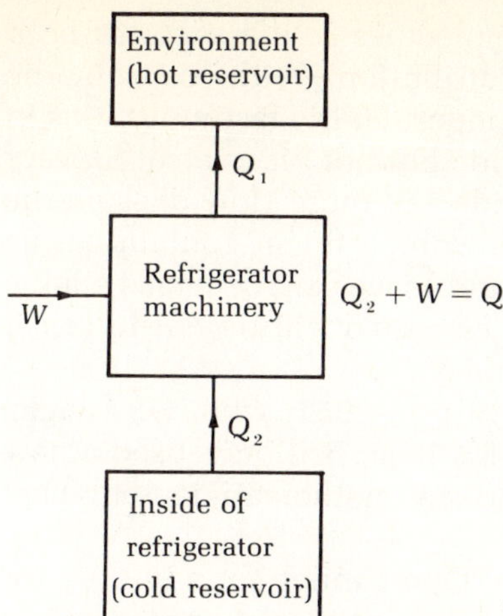

FIGURE 10.4 A schematic diagram of the working of a refrigerator.

The refrigerator

You may wonder if a refrigerator, which actually transfers heat from a cooler object to a hotter object, is not in violation of the second law of thermodynamics. Well, not really. The refrigerator has a motor, which puts work into the system. Thus it is nothing but a heat engine run in reverse (Fig. 10.4). The details of such a machine are actually quite complicated. But one thing should be clear even from our schematic diagram. The refrigerator takes out an amount of heat from its inside, and this, plus the heat equivalent of the energy needed to run the refrigerator, is dumped into the environment. Can you cool a room by keeping the refrigerator door open? No, this will actually heat up the room.

Entropy and information

Mathematician John Von Neumann once remarked, "No one knows what entropy really is." As you undoubtedly will agree from the discussion so far, such a point of view has some validity. It is our experience, though, that in such a case it is worthwhile to look at the subject of discussion from as many angles as we can. Maybe one of them will click and give us a flash of understanding. American author Gertrude Stein once complained to French painter Pablo Picasso after the maestro did her portrait, "It doesn't look like me." To this Picasso replied, "Never mind, it will." If you say that entropy doesn't make sense to you, perhaps you too can be reassured, "Never mind. It will." Take another look. Maybe the picture will be clearer this time.

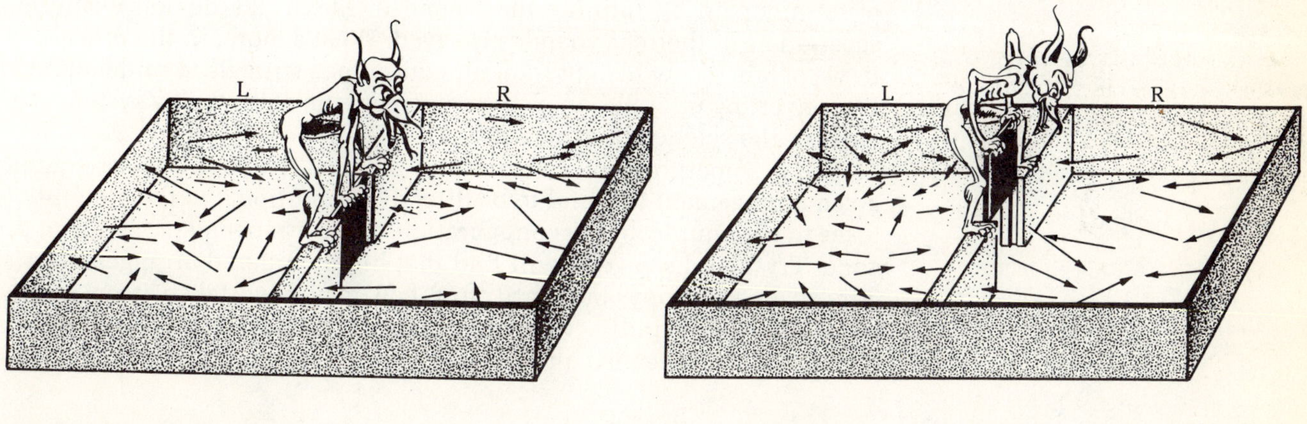

FIGURE 10.5 (a) Random assembly of slow (small arrows) and fast (large arrows) molecules in both compartments L and R. The arrows denote the velocity of the molecules. (b) The slow molecules have been separated from the fast ones by the demon and his trapdoor trick. Leon Brillouin showed that this is impossible to accomplish without increasing the entropy of the environment and expending energy. Information as to which molecules are slow and which are fast costs entropy as well as energy.

A new way of looking at entropy became necessary because, about a century ago, physicist Clark Maxwell invented a demonic paradox that shook the foundation of the second law of thermodynamics. Consider a vessel full of air at a uniform temperature. The vessel is divided into two compartments by means of a partition [Fig. 10.5 (a)]. Since the whole system is at the same temperature, we cannot run an engine between the two compartments, a consequence of the second law of thermodynamics. Now suppose, following Maxwell, that there is a trapdoor in the partition wall and a demon sits on this trapdoor. Although all the air is at the same temperature, this does not mean that all the air molecules have the same kinetic energy. Actually some move a little faster than others (of course, the distribution of the fast and slow molecules is the same in both compartments). The demon can change the distribution. With his demonic insight, he does, by opening and closing the trapdoor, let the fast molecules in on one side and the slow ones in on the other side but not *vice versa*. So very soon there will be more fast molecules on one side and more slow ones on the other [Fig. 10.5 (b)]. What this means is that the compartment containing the fast molecules is now hotter than the other side, and so we can run a heat engine between them and get work. This amounts to a decrease of entropy. And yes, it is a violation of the second law of thermodynamics.

"Not so!" said a physicist named Leon Brillouin in 1951, who, with logic, successfully exorcised Maxwell's "demon." In his logic Brillouin used a new interpretation of entropy developed by American mathematician-engineer Claude Shanon.

The basis of the new logic is the statement, "One cannot get anything for nothing, not even an observation."* The demon has to observe in order to find out which molecules are fast and which are slow. This observation for the purpose of gaining information has a price tag—an increase of entropy of the environment. Any learning transition, then, is really a process of buying information at the expense of increasing the entropy of the surroundings. The reciprocal process of converting the information for the purposes of decreasing entropy is also allowed, but the amount of the decrease is never as large as the increase in the learning process. So the net result is always an increase of entropy.

In terms of energy, it costs energy to gain information—enough energy to offset any gain by running the engine in Maxwell's demon example. So the second law of thermodynamics is saved. What's more, in the process of saving it, we obtain a new insight into our continuous struggle to understand entropy. By converting information, we can decrease the entropy of a system. This is the basis of the science of cybernetics.†

Here is something you can put into practice. The entropy associated with a given arrangement is found to be directly proportional to the amount of time it takes to communicate the configuration of the arrangement, using an optimum code. Try it and see. You will find that highly ordered arrangements (supposedly of low entropy, but until now you may have taken it on faith) can be communicated rapidly, whereas random arrangements take much longer to communicate. (See also question 12.)

■ 10.2 Energy, Gravitation, and Black Holes

We will now apply the concept of energy to the phenomenon of gravitation. So far we have talked about gravity mostly in connection with bound systems— terrestrial objects bound by earth's gravity or the planets bound by the gravity of the sun. How does an object free itself from the bonds of gravity? The concept of energy helps us answer this question.

---

*Dennis Gabor, discoverer of holography, is credited with this statement.
†Cybernetics is the comparative study of artificial intelligence (e.g., complex computers) and the human brain.

Intuitively, it is fairly obvious that an object can escape from the gravity field of its parent planet if it is given enough energy. Intuitively, it is also clear that the stronger the gravity field is, the harder it is to escape. But now we can ask this question: Are there any objects with such strong gravity that it is impossible to escape from them? Such objects, called "black holes," have been predicted theoretically and are now being sought.

These are the two topics we will discuss in this section: bound systems and black holes.

What makes an object bound?

To understand when an object is bound and when it is free, it is useful to take another look at the potential energy of an object when it is placed in the field of an attractive force like gravity. Previously we defined the gravitational potential energy of an object relative to earth as mgh. Actually, mgh represents the potential energy of the object relative to the earth's surface; h is the height of the object relative to the ground. For our present problem we will define the potential energy in a more general way, without making reference to a special place like the surface of the earth.

First, let us set the zero of the potential energy. When is the potential energy of an object arising from the force of earth's gravity zero? It is, of course, when the force is zero. In principle, the force becomes zero only when the object is infinitely far away. So we say that the potential energy is zero when the object is separated by an infinite distance from its interacting partner. We also assume that the object is at rest initially. Thus the kinetic energy is zero, and so is the total (kinetic and potential) energy of the object. Now suppose we bring the object close enough so that it is acted on by earth's gravity. Then it will be accelerated toward the earth and gain kinetic energy. But kinetic energy is always positive. Since the total energy has to remain zero—because it was zero initially (the law of energy conservation)—something negative must be added to the ever-positive kinetic energy to make the total energy remain constant. This something is the gravitational potential energy, which has to be negative.

The conclusion is correct for all objects under the influence of an attractive force. The potential energy in such a case is zero when the object is infinitely removed from the source of attraction. The potential energy takes on increasingly negative values as it is placed closer and closer to the source.

Is this confusing? The point to remember is that a negative number is actually larger when its absolute magnitude is smaller. Thus -5 is larger than -25. Zero is larger than any negative number. Thus as an object falls under the gravity force of the earth (or under any other attractive force), its potential energy becomes increasingly negative and thus continually decreases. Since energy is conserved, the amount of the decrease of potential energy goes into increasing the kinetic energy. When the object arrives at the point closest to the source (assuming it is not on a collision course), its potential energy is at its lowest point (as negative as it will get) and its kinetic energy has attained its maximum value. After this the object will recede away from the source, its kinetic energy will decrease, and its potential energy will increase (become less and less negative) until the object returns to infinity (Fig. 10.6). If the total energy is zero (as

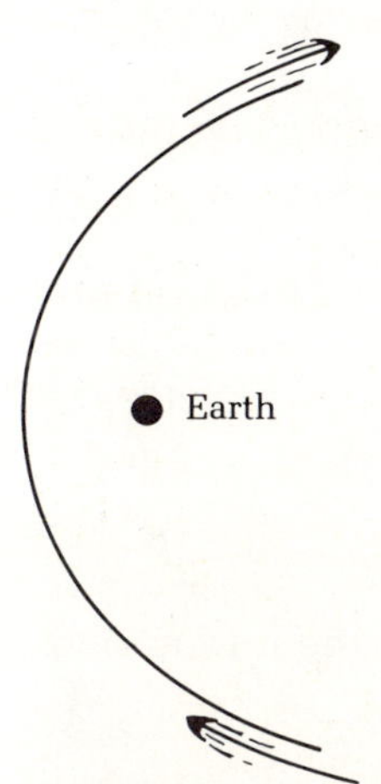

FIGURE 10.6   As an object of zero energy approaches the earth from infinity, its gravitational potential energy becomes more and more negative; the decrease of the potential energy goes into the increase of the kinetic. At the distance of the closest approach, the potential energy assumes its lowest value and the kinetic energy becomes maximum. After this the object recedes again. Its potential energy increases, tending to zero as the object approaches infinity.

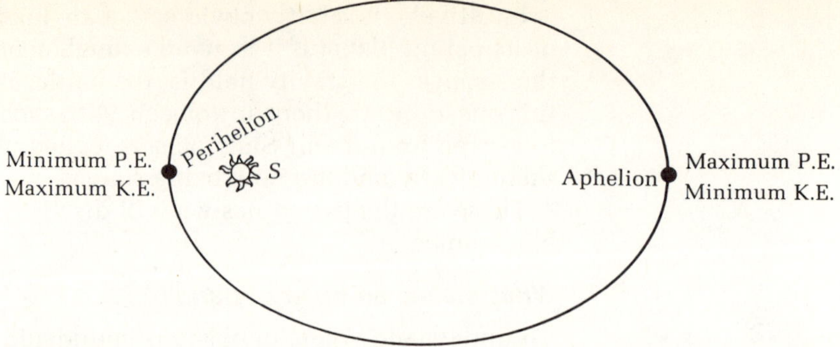

FIGURE 10.7 The planets are gravitationally bound to the sun, and their total energy is always negative. A planet can separate from the sun, but only within limits.

in this case) or positive, an object will not bind itself to an attractive source.

For bound objects the total energy is negative.* We can show with the following simple argument that if an object's total energy (kinetic plus potential) is negative, it can never escape from an attractive source such as the earth.

We note that as the object separates (from the earth, for example), its potential energy increases and becomes less negative. Initially, the value of the potential energy must have been more negative than that of the total energy since the total energy includes the contribution of the ever-positive kinetic energy. Thus as the object separates, its kinetic energy must decrease in order to conserve energy, so that we always get the same value for the total energy. But the kinetic energy cannot decrease beyond zero. When zero is reached the object is separated by the maximum allowable distance. It cannot go any further away or its total energy would change, which isn't allowed.

So we have proved than an object having negative total energy is bound—it cannot escape the field of gravity. The condition for escape, then, is that the total energy be zero or positive.

When the potential energy of an object is measured with respect to its value at infinity—where it is zero—we get a new expression for it. The gravitational potential energy of a mass m due to a planet or star of mass M at a distance R is now given as

$$- \frac{GMm}{R} \tag{10.1}$$

where G is the universal constant of the gravity force. Notice the minus sign.

Escape velocity

We now can calculate how much velocity an object must have in order to escape from the gravity of its parent planet or star system. Previously we have seen that the object is bound to the parent body only if the total energy is negative. If the total energy is zero or positive, the object is able to escape. The threshold for escape is obtained when the total energy is zero. Call the velocity of the object in this case v_{esc}; it is the **escape velocity.** Then the kinetic energy of the object is

$$\tfrac{1}{2}mv_{esc}^2$$

---

*The absolute value (numerical value without positive or negative sign) of the negative total energy of a bound system is called the **binding energy** of the system.

From the last subsection we know that if the object is at a distance R from the mass M, it has a potential energy of

$$-\frac{GMm}{R}$$

The total energy of the object is then given as

$$\text{total energy} = \text{kinetic energy} + \text{potential energy}$$

$$= \tfrac{1}{2}mv^2_{esc} - \frac{GMm}{R}$$

The minimum condition for the object's escape is that its total energy be equal to zero. This gives us

$$\tfrac{1}{2}mv^2_{esc} - \frac{GMm}{R} = 0$$

Simple transposition leads to the equations

$$\tfrac{1}{2}mv^2_{esc} = \frac{GMm}{R}$$

or,

$$v^2_{esc} = \frac{2GM}{R}$$

Taking the square root of both sides, we get the expression for v_{esc} as

$$v_{esc} = \sqrt{\frac{2GM}{R}} \qquad (10.2)$$

We can use this expression to calculate the escape velocity needed for a rocket to escape from the earth. Here the mass M is that of the earth, 6×10^{24} kg, and the radius R of the earth is 6.4×10^6 m. Since $G = 6.7 \times 10^{-11}$ MKS units, we have

$$v_{esc} = \sqrt{\frac{2 \times 6.7 \times 10^{-11} \times 6 \times 10^{24}}{6.4 \times 10^6}} = \sqrt{\frac{80.4 \times 10^{13}}{6.4 \times 10^6}} = \sqrt{12.5 \times 10^7}$$

$$= \sqrt{125} \times \sqrt{10^6} = 11.2 \times 10^3 \text{ m/sec} = 11.2 \text{ km/sec}$$

$$= 7 \text{ mi/sec} = 25,000 \text{ mi/h}$$

Thus a rocket destined for the moon must start from the top of the earth's atmosphere (to avoid the friction of the atmosphere) with a speed of roughly 25,000 mi/h.

We can calculate similar values for the escape velocity from the sun's gravity for an object on earth by using the solar mass for M and the earth-sun distance for R. This gives us

$$v_{esc} \text{ from sun} \approx 100,000 \text{ mi/h}$$

The value of the escape velocity is important not only for the discussion of space travel but also for an understanding of planetary atmospheres. Have you ever wondered why the moon does not have an atmosphere? The reason involves the low escape velocity needed to escape the moon. Let us figure out the escape velocity from the moon. The mass of the moon is 7×10^{22} kg and its radius is 1.7×10^6 m. Therefore, for objects on the moon,

$$v_{esc} = \sqrt{\frac{2 \times 6.7 \times 10^{-11} \times 7 \times 10^{22}}{1.7 \times 10^6}} = \sqrt{55.18 \times 10^5}$$

$$= \sqrt{5.52 \times 10^6} = \sqrt{5.52} \times \sqrt{10^6} = 2.35 \times 10^3 \text{ m/sec}$$

$$= 2.35 \text{ km/sec}$$

The molecules of a gas move continually in thermal motion. The speed of the primitive atmospheric molecules of the moon was greater than the necessary escape velocity, resulting in their escape into space. So the moon does not have an atmosphere. Even the earth, it turns out, has lost all the gaseous hydrogen in its atmosphere, since hydrogen's light molecules travel much faster than those of the heavier gases that cannot escape earth's gravitational pull. This has been a boon, of course, since the presence of explosive hydrogen in the atmosphere would be extremely hazardous to life on earth.

No escape from black holes

The escape velocity from an object of mass M and radius R has just been shown to be

$$v_{esc} = \sqrt{\frac{2GM}{R}} \qquad (10.2)$$

A very interesting situation arises if we use the following value of R in the expression for v_{esc}:

$$R = \frac{2GM}{c^2} \qquad (10.3)$$

where c denotes the velocity of light (numerically, 3×10^8 m/sec). Substituting this value of the radius R into Eq. (10.2) ($R = 2GM/c^2$ is called the **Schwarzschild radius**), we get for the escape velocity

$$v_{esc} = \sqrt{\frac{2GM}{2GM/c^2}} = \sqrt{2GM \times \frac{c^2}{2GM}}$$

$$= \sqrt{c^2} = c$$

The theory of relativity very convincingly tells us that no material object can surpass the speed of light, which is the maximum attainable speed. From the equations above it is clear that if

$$R < \frac{2GM}{c^2}$$

then

$$v_{esc} > c$$

According to the theory of relativity, such an escape velocity is impossible to attain. The conclusion has to be that from an object for which the size is the Schwarzschild radius R, no object, not even light, can escape.* The object is black because no light escapes from it, and hence it is called a **black hole.**

Before we tackle the million-dollar question of whether such things exist, let us study the kind of objects these black holes are expected to be. The most interesting black holes, for observational purposes, are predicted to be the end products of the collapse of massive stars under their own gravity (of mass at least three times that of the sun), the so-called gravitational collapse. Gravitational collapse is an important prediction of Einstein's general relativity; we'll say more on this later.

Suppose we assume that a collapsing star has a mass of 5 solar masses. Since the mass of the sun is 2×10^{30} kg, the mass of the star is $5 \times 2 \times 10^{30} = 10^{31}$ kg. We now ask this: What must the radius of the collapsed object be in order for it to qualify as a black hole? That is, what is its Schwarzschild radius?

From Eq. (10.3), since $G = 6.7 \times 10^{-11}$ MKS units,

$$R = \frac{2GM}{c^2} = \frac{2 \times 6.7 \times 10^{-11} \times 10^{31}}{(3 \times 10^8)^2}$$

$$= \frac{13.4 \times 10^{20}}{9 \times 10^{16}} \approx 1.5 \times 10^4$$

$$= 15,000 \text{ m} = 15 \text{ km}$$

Compare this value with that for the radius of the sun, 7×10^5 km, or even that of the earth, 6.4×10^3 km. These objects are over 400 times smaller than the earth but have a mass of 5 solar masses (almost 2 million earth masses) all packed into a very small volume.

Let us now calculate the density of matter inside collapsed stars that have become black holes. As before, we assume the mass of the black hole M to be 5 solar masses (10^{31} kg), in which case its Schwarzschild radius is 15 km (or 15×10^3 m), as shown above. The volume of the black hole, expected to be a sphere of this radius, is then given as

$$V = \tfrac{4}{3}\pi R^3 = \tfrac{4}{3}\pi (15 \times 10^3)^3 \text{ m}^3$$

Its density is given as

$$\frac{M}{V} = \frac{10^{31}}{\tfrac{4}{3}\pi (15 \times 10^3)^3} \text{ kg/m}^3 \approx 10^{18} \text{ kg/m}^3$$

This density is about 10^{15} times the density of water on earth.

The feature that's so interesting about such high density is that the force of

---

*The conclusion is fortuitous because Newton's theory of gravitation does not hold for such strong gravity objects. However, Einstein's theory of gravitation also predicts these objects.

FIGURE 10.8 An artist's conception of a black hole.

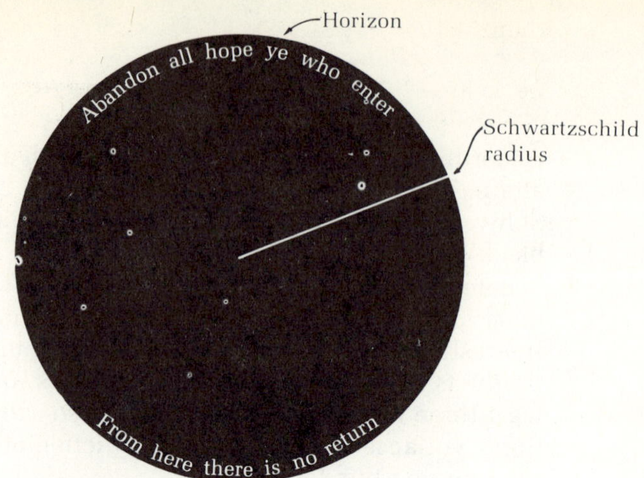

gravity inside such a black hole star is greater than that of any of the other interactions that exist between material particles. Thus in a black hole, gravity becomes supreme. The former star is now characterized only by its mass and its angular momentum, the two things that gravity respects. All other characteristics of the original matter that collapsed get lost in the shuffle; they make no difference as to the observable properties of a black hole.*

How can we observe a collapsed star that has become a black hole? Two properties of black holes are quite distinguishable. The mass of a black hole ought to be about three solar masses or greater. This mass continues to exert the usual Newtonian gravitational attraction on outside masses in its vicinity. And we can observe the effects of this gravitational attraction. For example, if a star were a component of a double-star system before becoming a black hole, it would continue to be a component of the system but now it would be invisible. So one way to look for a black hole is to study double-star systems in which one of the partners is invisible. Invisibility, of course, does not guarantee that the partner is indeed a black hole—it could be a dead star, for example, not giving out any appreciable light. But invisibility is a necessary criterion for any black hole candidate.

There is a second important property of a black hole that we can use for identification. A black hole has a clearly defined **horizon** into which everything can fall and from which nothing can escape (Fig. 10.8). The horizon defines a spherical region around the center of the black hole. The radius of this region is equal to the Schwarzschild radius. The consequence of this is that the black hole companion of a normal star continuously attracts matter from the normal star. Most importantly, this falling matter, under the strong gravity force of the black hole, gets so hot even before it is engulfed in the Schwarzschild radius that it emits very penetrating X rays, the same rays that doctors use for diagnostic purposes. Thus we could find evidence for a black hole partner of a double star by finding these X rays.

As you probably know, the earth's atmosphere protects us from all penetrat-

----

*See R. Penrose, "Black Holes," *Scientific American*, May 1972, p. 38, for a detailed discussion.

ing radiation like X rays. So the X rays from a black hole cannot be observed from an earth-based observatory. To overcome this limitation scientists from the United States and Italy jointly sent a satellite, named Uhuru, outside the atmosphere to search for prospective invisible, X-ray-emitting, massive partners of double-star systems. Six candidates were found, out of which one named Cygnus X-1 satisfied all the criteria for being a black hole.*

1. It is part of a double-star system.
2. The pattern of X-ray emission from it is as expected of a black hole source.
3. The energy characteristics of the X-ray source are simply explained as a conversion of gravitational energy into kinetic energy and heat during the process of mass accretion onto a black hole.
4. Conventional study of the visual partner has established the mass of the invisible compact star to be a minimum of 5 solar masses.

Even so, scientists are being properly cautious about proclaiming the discovery of a black hole star. Other models for the explanation of the observational data have not been completely ruled out.

In the meantime, there also have been some interesting speculations concerning other kinds of black hole matter. It has been suggested that small specks of black hole matter could have been formed during the big bang creation of the universe. Some people have even suggested that such a speck of black hole might have caused a huge unexplained catastrophe in Siberia in 1908, when trees burned, land was upturned, and an explosion was heard and seen for miles and for which no really satisfactory physical explanation has yet been found.

Some physicists have suggested that the specks of black holes could be a possible energy resource for use on earth. Still others have suggested that we use them in transportation. The possibilities are unlimited once these specks are found, *if* they are found.

It is also interesting to speculate if our entire galaxy is going to become something like a black hole. Since the mass of our galaxy is about 10^{41} kg, the Schwarzschild radius can be calculated to be about one-tenth of a light-year. If our galaxy were to become a black hole, it would shrink to this size from its present size of about 100,000 light-years. Even so, the black hole size would not be so small as to lead to an abnormally high density, as in a collapsed star. If you calculate the density, you will find it to be about 10^{-3} kg/m^3, which is a millionth of that of terrestrial matter. For such densities the local laws of matter are not expected to be affected. Therefore, if our galaxy becomes a black hole some day, as it probably will, it will do so not with a bang but with a whimper.

For the entire universe the mass is 10^{53} kg. The Schwarzschild radius turns out to be 10^{10} light-years, and the resulting density is given as 10^{-25} kg/m^3. The numbers fit the universe in which we reside just about right. So we may be living in an enormous black hole for all we know, although there is other evidence suggesting that we probably are not.

---

*For further details see K. S. Thorne "The Search for Black Holes," *Scientific American*, December 1974, p. 32.

After reading the discussion on black holes, questions may be swarming in your mind about universal destiny and about the fate of life on earth. Is it all going to end in black holes, all the matter in the universe? Earlier we mentioned the entropy law; in the processes of nature, entropy tends to increase forever. Gravitational energy, the purest form of energy, must, with the evolution of the universe, go into other disorderly forms of high entropy like light and heat. This is what happens in the collapse of matter into a black hole. On its way to becoming a mass concentrate, the original conglomeration of the mass that formed a star converts all its potential energy into light and heat, perhaps in a brilliant display of cosmic fireworks. In the process any planets and life thereon that might have been supported by the star vanish. If the star is not massive enough, the end product is a neutron star (if the mass is about 2 solar masses) or a white dwarf if the mass is less than 1.4 solar masses). On the way to these end products, entropy again is the big winner. Gravitational potential energy becomes other forms of energy of high-entropy content. White dwarfs and neutron stars, so far as we can tell, cannot support life on their planets—if indeed any planets should survive the processes of violent explosions that lead to these end products of stellar evolution.

Thus the future seems bleak, as bleak as the world portrayed by the English poet Algernon Swinburne*:

> Then star, nor sun shall awaken
> Nor any change of light
> Nor sound of water shaken
> Nor any sound or sight;
> Nor wintry leaves nor vernal
> Nor days nor things diurnal
> Only the sleep eternal
> In an eternal night.

Well maybe that's the way the universe is going to end; we don't know. There are, however, a few encouraging signs. First, the universe has a few hang-ups, to use the terminology of physicist Freeman Dyson.† As a result of these hang-ups, it will take the universe a very long time to run down, if it ever does so. Second, although entropy does seem to increase as the universe evolves, other principles also operate. Thus some stars die, yet other stars are born; order is created from the ashes of star death.

The hang-ups

The first question to answer is why the stars do not arrive at the end point of their evolution quickly, if that's what the physical laws demand. Well, if gravity were the only force operating in stellar evolution, this would be true; the stars would evolve to their end points quickly. It would not take more than 10 million years for a star to die. Ten million years may seem a long time to you,

---

*The idea of "heat death" described in the poem was already suggested by physicist Lord Kelvin. Obviously, Swinburne was quite taken by the idea.
†Freeman J. Dyson, "Energy in the Universe," *Scientific American*, September 1971, p. 51.

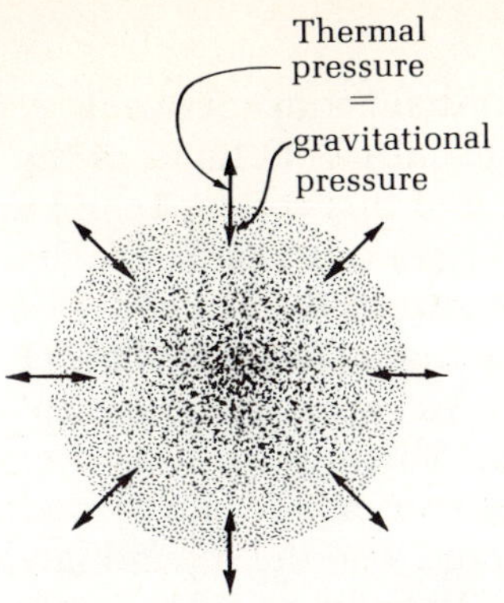

FIGURE 10.9   A star in equilibrium. The thermal pressure acting outward exactly cancels the inward pressure of gravity, establishing equilibrium.

but it actually is not enough time to complete the processes that create life on planets. So, naturally, other interactions of matter play the roles of coconspirators in creating the hang-up that keeps a star shiny and stable for a very long time. The principal role is played by the strong nuclear interaction.

In short, this is the picture of stellar evolution that scientists have put together over the past forty years. It is gravity indeed that brings the initial mass together. Contraction of the primitive mass releases energy, which becomes light and heat, half of one and half of the other. The light is radiated away; the thermal energy heats up the stellar matter to tens of millions of degrees of temperature. It is our common experience that the molecules and atoms of a hot gas confined in a vessel create considerable force on the wall of the container by their continued collisions, a situation often expressed by saying that hot gases generate an outward thermal pressure. Clearly, the outward thermal pressure opposes the inward pressure of gravity (Fig. 10.9). Can they balance out?

Yes, these forces will balance each other, but only temporarily. The problem is that a hot object has to radiate and the energy of radiation has to come from somewhere. If gravity is the only source, then further contraction is inevitable. Fortunately, there is another source of energy that comes into play after the initial contraction stage. In the very hot environment of temperatures of millions of degrees, atomic nuclei fuse and release energy. In essence the thermonuclear fusion process (as it is often called) is a conversion of mass energy into the heat and light of the star.

A physicist named Fritz Hautermans discovered this process of nuclear energy production in stars. Incidentally, it is the same energy that is released in hydrogen bombs, except that it is on a much larger scale in stars. Hydrogen fusion reaction gives a star enough energy to radiate, and it creates enough thermal pressure to balance gravity so that no further gravitational contraction occurs until the nuclear fuel is exhausted. And the fuel takes a long time to exhaust, because it is burning at a slow stable rate. We estimate that the thermonuclear fusion can go on for some ten billion years in a star like our sun before the fuel is gone. Stars have a long, long nuclear reaction hang-up.

How about galaxies? For them the hang-up is one we discussed in Chapter 5, angular momentum. The outer parts revolve around the galactic nucleus to ensure the conservation of angular momentum. Actually, the size of the galaxy also plays a role; the chance of a collision between intragalactic matter is slim because of the large sizes of the galaxies.

To be sure, size also is the hang-up that keeps the entire universe together for a very long time. Chances of collisions between galaxies are miniscule. Also, the universe is so huge that it takes about a hundred billion years for an object to fall through it. In short, then, size alone can keep the universe as it is for a very long time.

What is the conclusion from all this? For one thing, the heat death proposed by Kelvin does not seem to be impending. So there is no need to worry about it. Let us instead relax and let the universe take care of itself.

> Sitting quietly, doing nothing,
> Spring comes, and the grass grows by itself.*

---

*Quoted in Alan Watts, *The Way of Zen* (New York: Pantheon Books, 1957), p. 87. Copyright © 1957 by Pantheon Books, a Division of Random House, Inc.

SUMMARY AND OUTLOOK

These last two chapters have taken you through the important concept of energy and the two fundamental laws concerning it. Energy usually is defined as the capacity for doing work, but a complete understanding of this concept can be obtained only after you grasp the meaning of the two laws relating to energy. Energy exists in many different forms; these two laws govern the pathways of energy conversion from one form to another. The first of these laws is the law of conservation of energy: when we add up all the energies in various different forms, we find the same number for any closed system. This number does not change with time. The total energy of a closed system is conserved. The second energy law is most commonly called the entropy law (and also the second law of thermodynamics). The following two equivalent statements of the law are most revealing: (1) entropy always increases, and (2) it is impossible to get useful work from a system in which all parts are at the same temperature.

An important consequence of the energy laws is that perpetual motion machines are impossible to construct. Another consequence is that all machines have finite efficiencies. Efficiency of a machine is defined as the ratio of the work produced by the machine to the input energy.

Chapter 10 gives an in-depth review of the important and somewhat difficult concept of entropy introduced in the Prologue. Apart from the implications for energy, the entropy law relates to order-to-disorder transitions in nature. Another recent contribution to understanding entropy is the realization that even gathering information about a system costs us an increase of entropy in the system's environment.

Chapter 10 also gives you a first look at two fascinating topics: the black holes (to which we shall return again in Chapter 25) and heat death.

In Chapters 11 to 18 we will discuss several forms of energy: waves, thermal energy, and electricity. These chapters will add two more basic concepts of physics—waves and fields—to the list that includes (so far) space, time, motion, matter, and energy.

QUESTIONS

Review and reason

1. Explain the difference between order and disorder. In what respect are the following situations orderly or disorderly?
   (a) a volume of ordinary air looked upon as a mixture of oxygen and nitrogen molecules
   (b) The motion of grains of pollen suspended in water
   (c) an assembly of people listening to a lecture
2. If you leave a kettle of water on a table, will the water ever get hot? Suppose we argue like this. The table has heat energy, which it can deliver to the kettle, which gets hot, but the table gets cold at the same time; thus energy is conserved. So again, will the water ever get hot? Why or why not?
3. Are living organisms exempt from the entropy law? Explain your answer.
4. Discuss the validity of the entropy law in the following two cases: (a) you clean up your cluttered desk; (b) a child grows up to be an adult.
5. What is a perpetual motion machine of the second kind? Why doesn't it work?
6. How does an actual heat engine perform? How is the efficiency of a heat engine defined? Can the efficiency be 100%?
7. What is thermal pollution? Why is thermal pollution inevitable when resource energy is converted into useful work?
8. Can a kitchen be cooled by leaving the refrigerator

door open? Why or why not?

9. Grade the following forms of energy in the order of increasing entropy content (start from the one of lowest entropy): solar radiation, fossil fuel energy, nuclear energy, geothermal energy.

10. Does the Maxwell demon violate the second law of thermodynamics? Why or why not?

11. Consider a kettle of water placed on a table again. The table has molecules that are more energetic than the average and also molecules that are less energetic than the average. Suppose a demon lets only those molecules of the table hit the kettle that have greater than average energy. Since originally the average energy of the molecules in the kettle is the same as that of the table (their temperature being the same), the demon's action will certainly increase the temperature of the kettle (and the water with it). To help things even further, the demon does not allow energetic molecules from the kettle to collide with the table. Does this method of heating water in a kettle violate any laws of nature? Explain your answer.

12. Get a friend, a pack of cards, and a timer. Deal the cards. Communicate to your friend the contents of your hand and let your friend keep track of the time. Your results after a few deals should indicate that a more orderly hand needs less time to communicate. Explain why.

13. In Section 10.1 there are several statements that can be used as formal statements for the entropy law (or the second law of thermodynamics). Find as many of these statements as you can.

*14. In the text we showed that the potential energy due to an attractive force, when defined with respect to infinity, is negative. In the same vein, prove that, when defined with respect to a reference level at infinity (where the force vanishes and the potential energy is taken to be zero), the potential energy due to a repulsive force is always positive. Give an example of repulsive force.

15. Show that a particle having a negative total energy is bound.

16. Explain the concept of escape velocity. Why does Mars have a thinner atmosphere than earth's?

17. What is a black hole? How can we detect it?

18. What is meant by the heat death of the universe? What are the hang-ups that keep the universe going?

Arithmetic

1. A heat engine obtains 30% work from its heat input. What is its efficiency? What is the thermal pollution generated by the engine, expressed as a percentage of the input energy?

2. Which causes maximum thermal pollution? (a) A 40% efficient coal power plant; (b) a 32% efficient nuclear power plant; or (c) a 25% efficient geothermal power plant? Explain your reasoning.

*3. Mars has a mass of 6.5×10^{23} kg and a radius of 3.4×10^6 m; calculate an object's escape velocity from Mars.

*4. The mass of the entire universe is estimated to be 10^{53} kg. Carry out the numerical calculation to show that the Schwarzschild radius of an object of this mass is roughly 10^{10} light-years.

---

*Optional.

11

The Motion of Waves

What are waves good for? When I was a child, my favorite pastime was to watch water waves. I used to throw pebbles in a pond and watch the spreading rings of crests and troughs with fascination. My younger brother was more practical and he had no use for wave watching. "What good are the waves, anyhow?" he used to tease me. I didn't know what to answer—not then.

Later in school I was pleased to learn that sound and music and even light are wave phenomena. Certainly these are important things, not irrelevant like the water waves of my childhood seemed to be. Still later, when I learned that even matter behaves like waves in some ways, the whole universe began to have a new meaning for me.

Now after all these years I feel that I know the answer to my brother's question. I feel much like Basudeva, the boatman in Hermann Hesse's *Siddhartha,* who often said, "I learn from the river." I feel like saying, "I learn from the waves." What can the waves teach us?

> The waves can teach us
> How to listen
> And how to watch.
> Waves can teach us
> Love.

■ 11.1   What Is a Wave?

When we talk about energy a natural question that comes up is this: How is energy carried from one place to another? We are familiar with material particles doing this chore, but is there any other way? For example, what transports the energy called sound? There is energy in ocean waves, but does the water transport the energy or is energy transport associated with the wave phenomenon? This is a natural place to begin a discussion of waves: waves are a way of energy transportation.

The concept of waves crops up in other contexts as well. We talk about waves of gossip, waves of relaxation, waves of fear, and so on. What do we mean by the

FIGURE 11.1 Water waves are created on the surface of the water of a pond when a pebble is thrown into it.

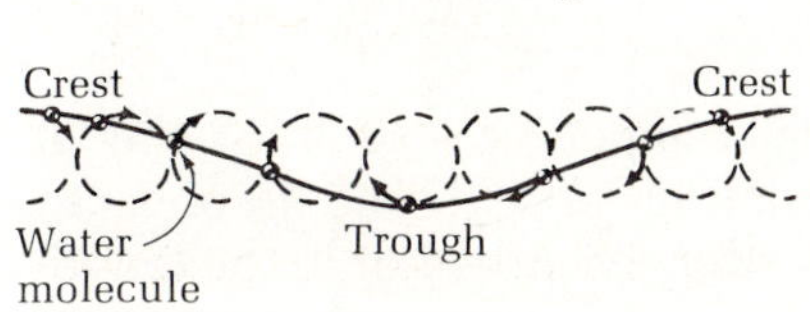

FIGURE 11.2 As the wave propagates, the particles of water display a back-and-forth and an up-and-down motion, more or less in a circle.

word "wave" in these instances? We are referring to a traveling disturbance of one kind or another. For a spreading wave of gossip, the gossip moves on, but not the people who act as the medium. Implied is also the fact that after the wave has passed through, the medium returns to its normal state.

These are the two major attributes of a physical **wave:** a wave is a traveling disturbance, and it transports energy. When a wave passes through a medium, the particles of the medium neither travel nor transport the energy through the entire distance—the wave does. For ocean waves the energy is transported by the waves, not by the water. And the water does not travel very far from its position, but the waves do.

The basic characteristics of waves

The water waves created in a pond by a pebble thrown into it are no doubt familiar to you (Fig. 11.1). Watch carefully at your next opportunity. The individual drops of water do not travel onward as the wave passes through; it is the disturbance of the water that travels. Put a cork in the path of the waves, and the cork will exhibit some sort of motion, but, on the average, it will stay where it is.

At a closer look you can determine the nature of the disturbance of the cork and hence of the water particles. The water particles move up and down and back and forth in a more or less circular fashion as the wave passes through (Fig. 11.2). As the wave progresses, particles of water farther and farther away from the center of the disturbance display the circular motion from the propagating disturbance. Clearly the wave takes a finite time to propagate from one place to another; it has a finite velocity of propagation. Also, the pattern of the wave, or its basic shape, remains the same as it travels.

For water waves the circular motion of the water particles characterizes the propagating disturbance. For other kinds of waves, other properties of the medium may be involved in the description of the traveling disturbance. For example, in the case of sound waves, the air pressure and density vary as the wave advances. In this case the changes in the air density (or pressure) can be used to describe the traveling disturbance. Quantities whose changes can be used to describe the traveling disturbance of a wave are called **disturbance coordinates.**

Let us now think of situations in which a train of waves is created by a continuously vibrating source. What is a vibrating source? It is any object that goes through a repeated pattern of motion. For example, if you dip your finger in and out of the water in a bathtub in a regular succession, you have a continuously vibrating source, and this will create a train of water waves. Try it and see.

A tuning fork is an often-used vibrating source of sound. Suppose we start a sound wave train in the air with a tuning fork. If we could take a snapshot of the air molecules in a section of the air through which the wave passes, the snapshot would show regions where density and pressure of the air are high (**condensations**), alternating with regions where the density and pressure are low (**rarefactions**), Fig. 11.3. As the tuning fork vibrates back and forth, this series of compressions and rarefactions travels through the air and sets your eardrum

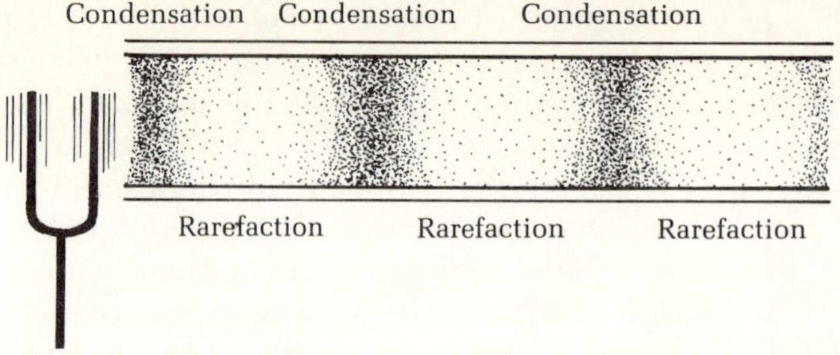

FIGURE 11.3 A sound wave from a tuning fork consists of traveling condensations and rarefactions of the air around it.

into vibration, which is interpreted eventually by your brain as sound coming from the tuning fork.

Thus sound waves consist of a fluctuating density (and pressure) pattern in the medium along the direction of its propagation. Such waves, in which the disturbance is along the direction of travel of the waves, are called **longitudinal waves.** In contrast, there are other waves where the particles of the medium move up and down as the wave travels. So the disturbance coordinate (in this case the displacements of the particles of the medium from their undisturbed position) now is perpendicular to the direction of motion of the wave. Such waves are called **transverse.** Clearly, water waves are a combination of both transverse and longitudinal patterns.

Perhaps a Slinky, a coiled spring, will help you experience such aspects of waves as their transverse and longitudinal nature (Fig. 11.4). If you fix one end of the Slinky and move the other end back and forth perpendicular to the length of the spring, the coils also will move perpendicular to the length of the spring, which is the direction of propagation of the wave. The wave in this case is transverse. If, on the other hand, you move your hand parallel to the spring to

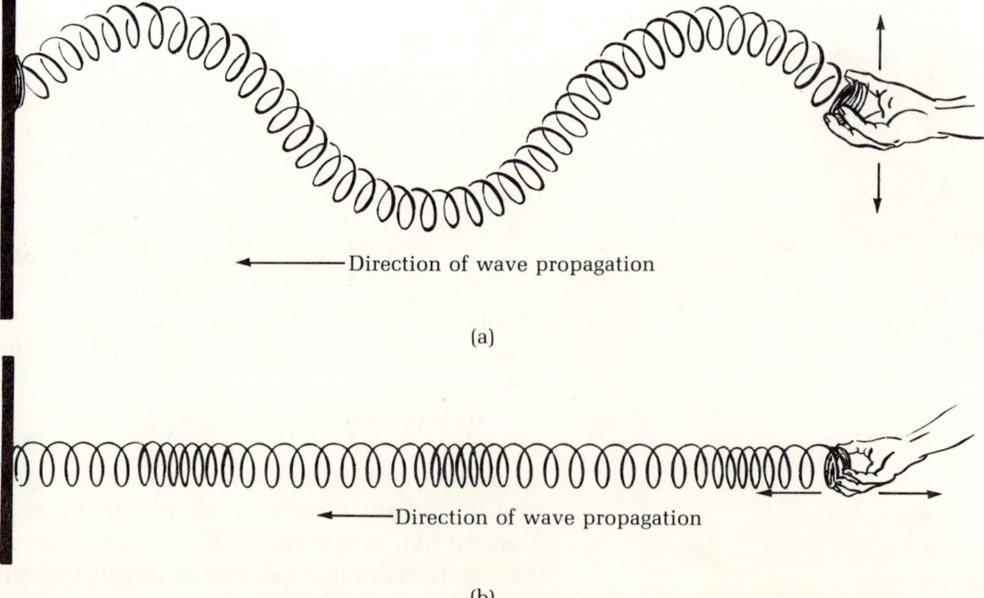

FIGURE 11.4 (a) A Slinky creates a transverse wave when it is moved up and down at one end, as shown. The coils move perpendicular to the direction of wave propagation. (b) A Slinky creates a longitudinal wave when it is moved back and forth at one end. The coils move back and forth along the direction of wave propagation.

FIGURE 11.5 Graphical representation of a wave.

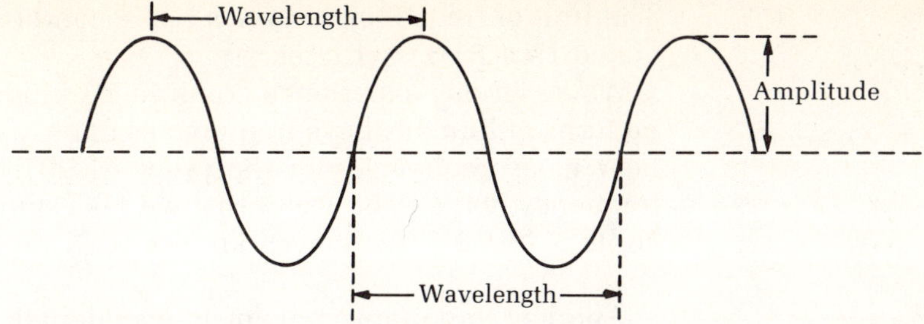

excite the waves in the Slinky, the coils will vibrate along the length of the spring, that is, along the direction of wave propagation. The wave in this case is a longitudinal wave.

One very useful way to display a wave pattern is in the form of a graph. Fig. 11.5 is a plot of the air density for the case of traveling sound waves. Don't be confused by the transverse appearance of the graph. The fact is that the graph of any wave on paper must be of this shape, because we are plotting the increases and decreases of a disturbance coordinate along the vertical axis. The disturbance coordinate (for example, air density) does indeed go up and down, but this by no means indicates that the air molecules move up and down when a sound wave passes through air in the horizontal direction.

The curve in Fig. 11.5 is one we typically refer to as "wavy." Keep in mind, though, that an object with a wave shape does not necessarily have anything to do with a traveling wave. It is also interesting to know that mathematicians call a curve of this shape a sinusoidal curve (such curves involve the sine function, which is usually discussed in the branch of mathematics known as trigonometry). Waves that look like Fig. 11.5 in graphical display are often called "sinusoidal."* All waves are not sinusoidal; in fact, most waves found in nature, and that includes most sound waves, are not.

For a train of waves produced by a continuously vibrating source, we can define some very useful concepts associated with the waves. The number of vibrations, or cycles, of the source per second is called the **frequency** of the source, which ordinarily is the same as the frequency of the waves. That is, if you counted the number of wave cycles passing a point of space per second, you will find this number to be the same as the frequency of the vibrating source.

Another related concept is the **period** of a wave, the time it takes for one wave cycle to pass a point. The period is the reciprocal of the frequency:

$$\text{period} = \frac{1}{\text{frequency}}$$

It is customary to use the symbol T for the period and the lowercase Greek letter ν (pronounced "nu") for the frequency. Then, in symbols, we get the relation

$$T = \frac{1}{\nu} \tag{11.1}$$

---

*Another name used to describe vibrations or waves that are sinusoidal is simple harmonic motion.

The unit of frequency is cycles per second (abbreviated c/sec) or hertz (abbreviated Hz). 1 Hz = 1 c/sec.

Waves in our experience come in all kinds of frequencies. The brain wave patterns of humans have frequencies of only a few hertz. Sound waves that we hear have frequencies in the range of 20 Hz to 20,000 Hz. In contrast, the frequency of the light waves that we see is much larger, in the range of 4×10^{14} Hz to 7×10^{14} Hz. Waves known as X rays have frequencies of 10^{18} Hz or so.

Another important concept is **wavelength,** the distance from crest to crest (Fig. 11.5). Wavelength is denoted by the lowercase Greek letter λ (pronounced "lambda"). Since wavelength is just a length, it is measured in meters like other lengths.

There is an important relationship between the velocity of the waves, the frequency, and the wavelength. Since it takes time, T, for the wave to travel the length λ, by definition the velocity is given as

$$v = \frac{\text{distance}}{\text{time}} = \frac{\lambda}{T}$$

But since $\nu = 1/T$, Eq. (11.1), we get

$$v = \nu\lambda \tag{11.2}$$

Thus the velocity of the waves is given by the product of the frequency and wavelength. The equation also shows that for fixed velocity the frequency and wavelength are inversely proportional to each other; if one increases, the other decreases, and vice versa.

Still another important concept in connection with waves is **amplitude.** The amplitude of a wave is defined as the maximum value assumed by the disturbance coordinate with respect to its equilibrium value (see Fig. 11.5). Amplitude is important because its square determines another attribute of the wave, its **intensity.** The intensity is directly proportional to the square of the amplitude. As the meaning of the word indicates, the intensity of a wave is the measure of the energy delivered by the waves to a point of space. Formally, intensity is defined as the power (energy per unit time) transmitted by the waves through a plane of unit area placed at the given point of space perpendicular to the incoming wave train. As an example of the amplitude-intensity relationship, consider the fact that a 4-foot-high ocean wave possesses four times as much energy as a 2-foot-high one.

The velocity of waves found in nature varies widely. Thus on one end of the spectrum, we have the open ocean waves, propagating at a speed of only about 5 m/sec. At the other end of the spectrum we have the speed of light waves traveling in vacuum, a value of 3×10^8 m/sec, which is also the ultimate speed of nature. Intermediate examples of speed are sound waves in air, which travel at about 330 m/sec (750 mi/h), and sound waves in solids, which have velocities of a few kilometers/second, depending on the type of solid. Notice that in stating the numerical value of the velocity of a wave, we are careful to state the medium of travel. We have to be, because the velocity of a wave depends on the medium.

Notice the tremendous difference in the magnitude of the speeds of light and sound. Now you can appreciate the following observation about thunder and lightning: "If you heard the thunder, the lightning did not strike you. If you saw the lightning, it missed you; and if it did strike you, you would not have known it."*

EXAMPLE 1 *The wavelengths of audible sound waves in air*

The velocity of sound waves in air is 330 m/sec,† and the range of the frequencies varies from 20 Hz to 20,000 Hz, as mentioned earlier. The range of wavelengths can be determined by using the relationship between the velocity, frequency, and the wavelength:

$$\lambda = \frac{V}{\nu}$$

If

$$\nu = 20 \text{ Hz}$$

then

$$\lambda = \frac{330}{20} \text{ m} = 16.5 \text{ m}$$

This is the upper limit of the wavelength of audible sound. The lower limit is found for $\nu = 20,000$ Hz. In this case,

$$\lambda = \frac{330}{20,000} \text{ m} = 0.0165 \text{ m} = 1.65 \text{ cm}$$

This is just about the diameter of a quarter.

EXAMPLE 2 *The wavelength of light waves that we see*

Light of each individual color that we see is characterized by a frequency and a wavelength. Red light corresponds to light waves of lowest frequency, of some 4×10^{14} Hz. The associated wavelength is given as (since the velocity of light is $c = 3 \times 10^8$ m/sec)

$$\lambda_{\text{red}} = \frac{c}{\nu_{\text{red}}} = \frac{3 \times 10^8}{4 \times 10^{14}} = 0.75 \times 10^{-6} \text{ m}$$

At the other end of the visible spectrum of light waves, we have violet light, with a frequency of about 7×10^{14} Hz. The wavelength in this case is given as

$$\lambda_{\text{violet}} = \frac{3 \times 10^8}{7 \times 10^{14}} = 0.43 \times 10^{-6} \text{ m}$$

Note one important thing. The wavelengths just calculated are to be taken as typical examples of wavelengths of red and violet light. Actually, both of these

---

*Quoted in P. Viemeister, *The Lightning Book*, (Cambridge, Mass.: MIT Press, 1961), p. 287.

†This value is correct only near 0°C. In warm air the speed of sound increases somewhat. You may have noticed that sound travels faster on a warm day.

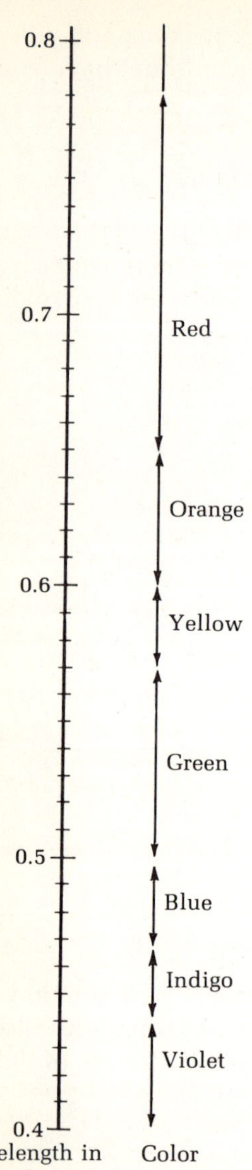

FIGURE 11.6 The spectrum of white light. The arrows indicate the ranges of wavelengths of different colors.

colors (and the other intermediate colors, too) cover a range of wavelengths and do not correspond to a fixed wavelength or frequency (Fig. 11.6). White light is a mixture of light of all these colors.

Spherical and plane waves

There is one difference between surface water waves and those of sound and light that we should emphasize. The water waves are waves on a two-dimensional surface, whereas sound and light waves propagate in three-dimensional space. Thus, whereas surface water waves form circular patterns around the source, the corresponding pattern for sound and light waves is spherical. Sound and light waves are **spherical waves.**

Consider a point source of spherical waves of sound. Then a picture that we can make of the spherical waves is this. Along any radius from the source, a wave proceeds with the appropriate velocity as if it is traveling down a tube, as in Fig. 11.3. Once you can visualize these waves along all the possible radii emerging from the source (Fig. 11.7), you can assign a **phase** to each point of space; the phase denotes the stage of a particle (of the medium) located at the point with respect to the wave. For example, points that are on the crest of a wave at the same time have the same phase; particles located on the crests are all doing the same thing. Points on the trough of the wave also have the same phase with respect to each other, but their phase is different with respect to the points on the crest. An advancing surface of identical phase for the wave train is called the **wave front.** In Fig. 11.7 we have joined all points of space that are on the condensations (the crests) of the waves shown, which gives us the circles of advancing wave fronts. If you make a three-dimensional mental image of this, you will realize that the wave fronts of the three-dimensional waves form a point source are all spherical surfaces.

Now we can go back to the water waves and identify the advancing circles of crests and troughs as the circular wave fronts of a two-dimensional water wave.

One simplification arises when we look at a circular or a spherical wave at a large distance from its source. A small portion of a very large circle looks like a straight line. Therefore, far away from its source, a circular wave can be looked upon approximately as a **straight wave** [Fig. 11.8(a)]. By the same token, a small portion of a large sphere looks like a plane, to a very good approximation. Thus sufficiently far away from the source, a spherical wave can be considered a **plane wave** when we are looking only at a small portion of the wave [Fig. 11.8(b)].

Interactions of waves

How does a wave behave when it meets a different medium? How does a wave behave when it encounters another wave? In the following sections we will examine these two basic questions.

We know from experience that waves do not change their properties when traveling through a single medium of matter. But when they encounter a different medium of matter, they exhibit very pronounced changes. These changes are as follows:

1. **Reflection.** The wave trains are at least partially turned back at the boundary

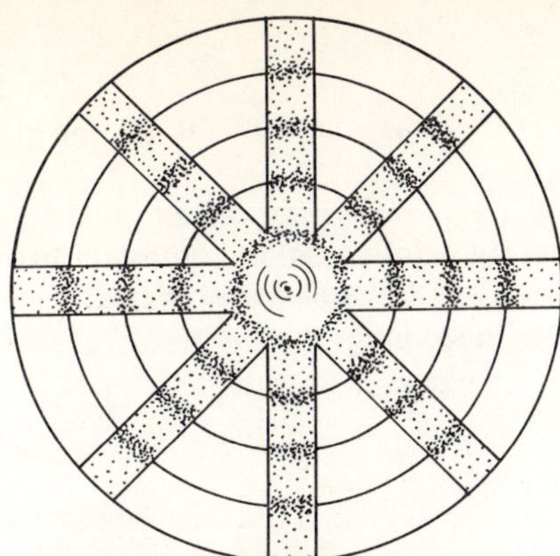

FIGURE 11.7 Propagation of a spherical wave idealized in two dimensions. The circles represent the spherical surfaces of advancing wave fronts obtained by joining all points of identical phase, in this case the points on the condensation.

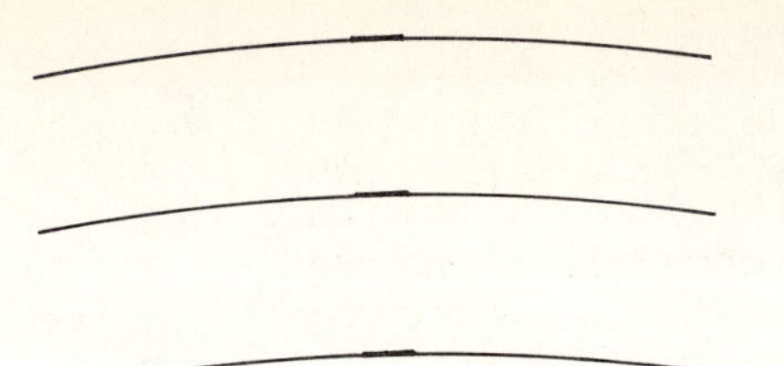

FIGURE 11.8 (a) A small part of a large circle looks straight. Thus at a large distance from the source, the circular wave fronts of a two-dimensional surface wave can be regarded as straight fronts when we look at a small section of the front.

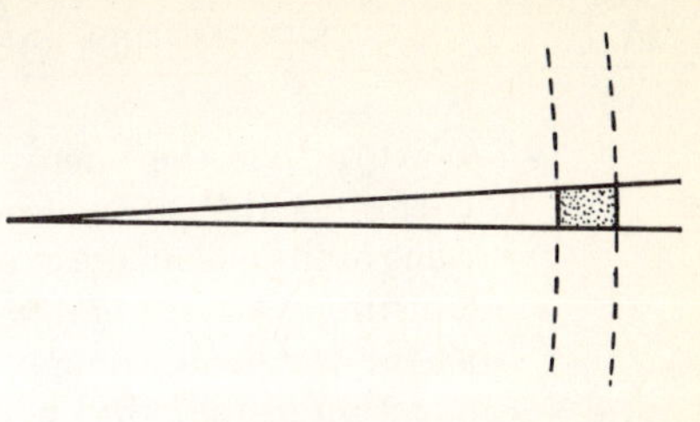

(b) At a large distance from the source, a spherical wave front can be approximated by a plane wave front when we look at a small portion of the wave front.

of a different medium; all of the wave's energy does not get through.

2. **Refraction.** The direction of travel of the wave changes as it crosses the interface of the two media.

3. **Diffraction.** The waves bend around an obstacle and to some extent appear even inside the shadow of the obstacle. A similar bending effect also occurs when a wave train passes through a small aperture.

What happens when one train of waves collides with another? Often they pass through each other without change. In that part of space in which they overlap, their effects are combined, a phenomenon known as **interference.** Thus interference is wave-wave interaction.

■ 11.2 Reflection of Waves

Reflection is a phenomenon displayed by waves when they encounter a different medium. If you have not seen the reflection of water waves before, try an experiment in the bathtub. Create a water wave by dipping your finger in and out quickly, and watch the waves reflect from the straight long side of the tub. Fig. 11.9 depicts this. The curious thing in this picture is that the reflected waves seem to originate from a point O' behind the barrier, which is as far back as the real center O is in front. O' is called the image of the source O, a language that we most commonly use in connection with light. Reflection of light is the phenomenon to which we now turn our attention.

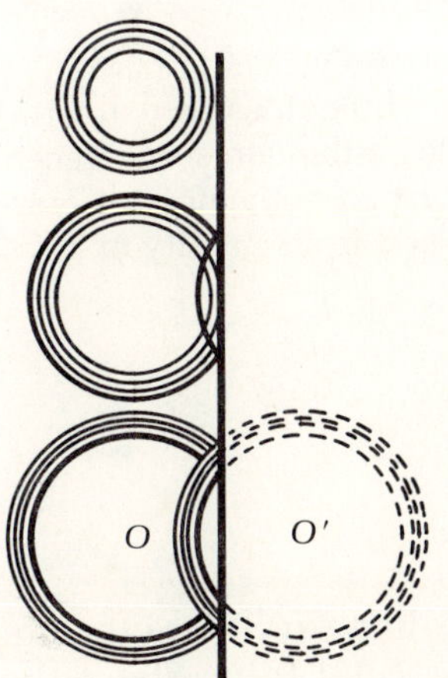

FIGURE 11.9 Reflection of water waves from the walls of a bathtub. The reflected wave seems to originate from an image source behind the wall.

Reflection of light

Suppose light waves encounter a mirror at an angle, as shown in Fig. 11.11. These waves are the incident waves. They are represented by straight lines in the figure, but they really describe plane wave fronts of light; the straight lines represent the edges of the planes. The wave fronts of the reflected light are shown also; notice how they appear to come from a place behind the mirror.

We can make an even simpler picture of the reflection of light with the following construction. We draw lines that are perpendicular to the incident and the reflected wave fronts, respectively, and that have the direction of travel

Can we utilize the energy of ocean waves? The ocean waves, since they are caused by the action of wind on the surface of the ocean, are a renewable energy resource and, interestingly, are quite reproducible. If we extract energy from the waves, the energy is replenished rapidly by the interaction of the wind with the ocean surface.

How much useful energy can we get from the ocean waves? Estimates by experts indicate a power capacity of more than a million megawatts (abbreviated MW) for the entire planet, about the same as the total amount of possible tidal power.

Let us make a rough estimate of the available wave power for a wave train of width 50 mi, for which the waves have an amplitude (height) of 5 ft and a wavelength of 200 ft. The area enclosed by the wave above the undisturbed sea level (Fig. 11.10) can be regarded very roughly as the area of a triangle, which is half the product of the height (5 ft) and the length of the base of half the wavelength (100 ft):

$$\text{area} \approx \tfrac{1}{2} \times 5 \times 100 \text{ ft}^2 = 250 \text{ ft}^2$$

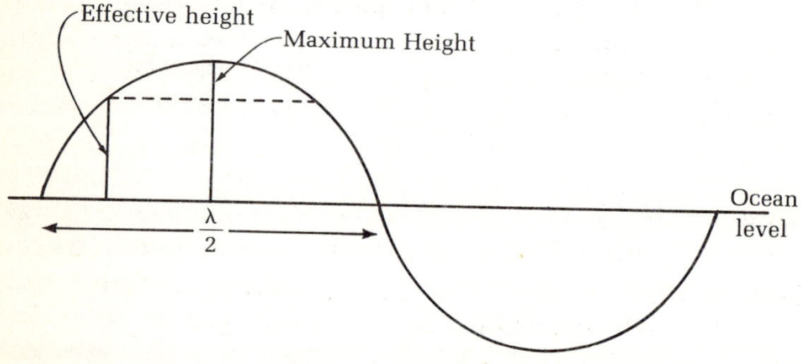

FIGURE 11.10 An ocean wave. The effective height of the wave above the ocean level is 0.7 times its maximum height.

A more accurate calculation gives 350 ft² for the area.

The volume of the water raised above sea level in a cycle is this area times the width of 50 mi (which is 50 × 5,280 ft = 26,400 ft). Thus the volume of water raised above sea level per wave cycle is

$$350 \times 26,400 \text{ ft}^3 = 9,240,000 \text{ ft}^3 = 9.24 \times 10^6 \text{ ft}^3$$

The weight of the water is its volume times its weight density, which is 62.4 lb/ft³. Thus

$$\text{weight of raised water per cycle} = 9.24 \times 10^6 \times 62.4$$
$$= 5.77 \times 10^8 \text{ lb}$$

The potential energy of the water is the weight times the effective height through which the water is raised.* The effective height in this case can be calculated to be $0.707 \times 5 = 3.54$ ft. Thus the potential energy of the water raised per cycle is given as

$$\text{weight} \times \text{effective height} = 5.77 \times 10^8 \times 3.54$$
$$= 2.04 \times 10^9 \text{ ft·lb}$$

The available power is obtained by dividing this number by the time period between cycles. For the kind of wave that we have, a rough estimate gives the time period to be 40 sec. Then the power is given as

$$\text{power} = \frac{\text{potential energy}}{\text{time period}} = \frac{2.04 \times 10^9}{40} \text{ ft·lb/sec}$$

$$= 5.1 \times 10^7 \text{ ft·lb/sec} = 5.1 \times 10^7 \times 1.36 \text{ J/sec}$$

$$= 6.94 \times 10^7 \text{ W} = 69.4 \text{ MW}$$

Clearly this is a lot of power, if only we could harness it. Unfortunately, the engineering difficulties in extracting wave power are considerable, although some recent suggestions look promising. The attractiveness of wave power lies in its renewability and in its quality of being pollution free.

---

*Since the height of the water raised in a wave is not uniform throughout, we will work with the concept of the effective height, which is given as maximum height × 0.707.

of the respective wave [Fig. 11.11(b)]. These lines are called **rays,** the incident ray and reflected ray, respectively, in this case. The special thing about the incident and reflected rays is that if we draw a line normal (perpendicular) to the mirror [the dotted line in Fig. 11.11(b)], the two rays make equal angles with the normal.

The angle that the incident ray of light makes with the normal to a boundary is called the **angle of incidence** and the angle that the reflected ray makes with

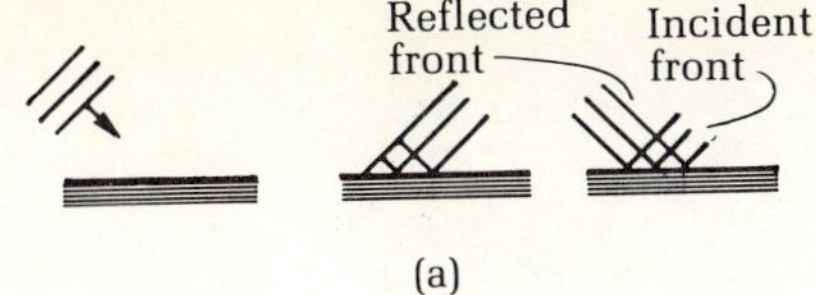

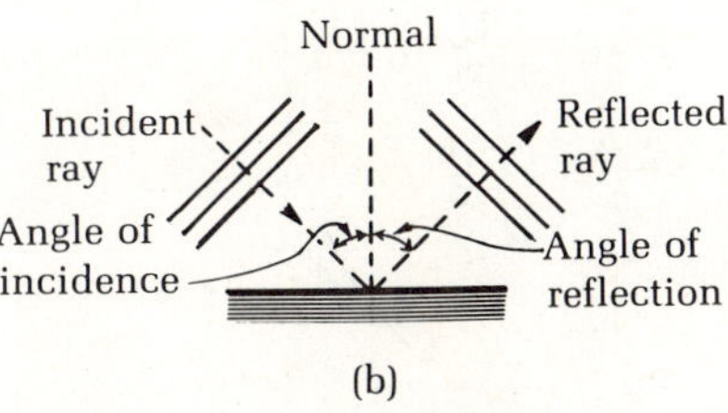

FIGURE 11.11 Reflection of light from a mirror. (a) The incident and reflected plane wave fronts. (b) The reflection of light in the ray description. Notice that the angle of incidence is equal to the angle of reflection.

the normal is called the **angle of reflection.** Thus the phenomenon of reflection of light satisfies the following law of reflection, which was known to the ancient Greeks:

$$\text{angle of incidence} = \text{angle of reflection}$$

This law is all we need to find the location of the image of an object upon reflection from a plane mirror. In Fig. 11.12 we trace some of the light rays that arrive to the eyes of an observer from a human face upon reflection from a plane mirror. Note that the rays are drawn following the law of reflection: their angle of incidence equals their angle of reflection. The rays that enter the observer's eye seem to originate from a place behind the mirror; for each point of the face there is an image point behind the mirror. The image has the same height as the original, and it appears to be the same distance behind the mirror as the object is in front.

The reflected image is not an exact look-alike of the object. If you look closely at your own mirror reflection, you can quickly discover one important difference: the left and right are reversed in the image. In the figure this is apparent from the hair part of the subject and the image. Whereas the subject parts his hair on the right, the image does it on his left.

Now we have an interesting question. What should be the height of a full-length mirror, one that enables you to see an image of all of your body? Fig. 11.13 gives the construction necessary to arrive at the right answer, which is half of your height. The answer does not depend upon your distance from the mirror.

In the beginning of this subsection, a comment was made about how at any boundary between two media, some part of the incident wave train is always turned back. There is always some reflection. This is demonstrated dramatically at night when you look out from a lighted room through your window. The glass, which seems to be transparent during the day, suddenly seems to have transformed itself into a pretty good mirror. Actually, the reflected light from inside the room is always there, but during the day we cannot discern it from the light coming from the outside. At night, against the background of the dark

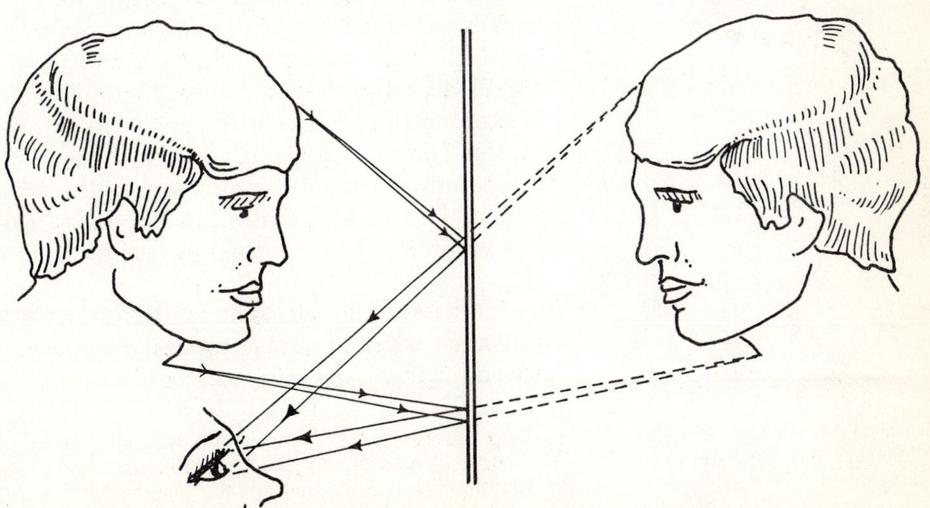

FIGURE 11.12 The image of an object in a plane mirror. The image is not an exact look alike of the object—the left and right are reversed.

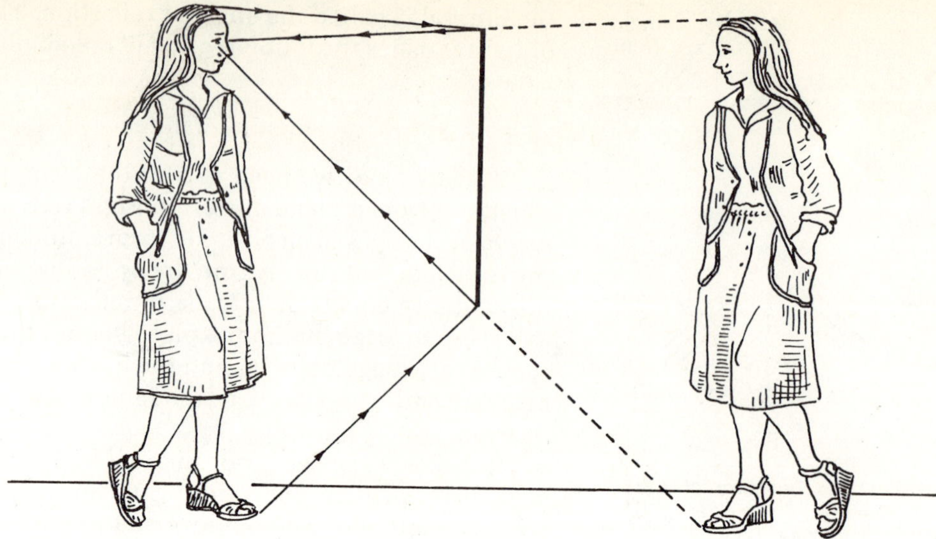

outside, these reflected rays are much more visible to our eyes, producing the mirrorlike effect.

Thus far we have talked about the reflection of light from only polished surfaces, the metal coating of a mirror or window glass. Most objects, however, have rough surfaces. When light falls on such a surface, it is reflected in many directions, as shown in Fig. 11.14. This is called **diffuse reflection.**

How do we see an object that is not luminous, as is a lamp or other source of light? We can see it only when it is illuminated, and then only by means of the light that is reflected off the object into our eyes.* Thus there is real value in the fact that the reflection off most objects is diffuse reflection. Since the reflected rays reach out in all directions, we are able to see the object from all angles.

Music in the shower

Do you enjoy singing in the bathroom shower? Many people do. Have you wondered why? The walls of the shower, often made of hard tiles, reflect the sound waves of your music back and forth many times before the sounds are finally absorbed. When you sing you hold a certain tone for a period. Before the beginning of the tone fades away, its continuation adds to it on and on for a while. The tone gains in volume because all its reflections (or echoes, as they are commonly called) contribute to the increase of volume. Such mixing of a sound with that which follows is called **reverberation.** Reverberation adds to the volume of a given tone of bathroom music, making it so pleasing to your ears.

But what is so pleasant for humming a tune in the shower can become a nuisance when it comes to designing an auditorium, where intelligibility of speech is the major concern. So the acoustical engineer tries to avoid too much

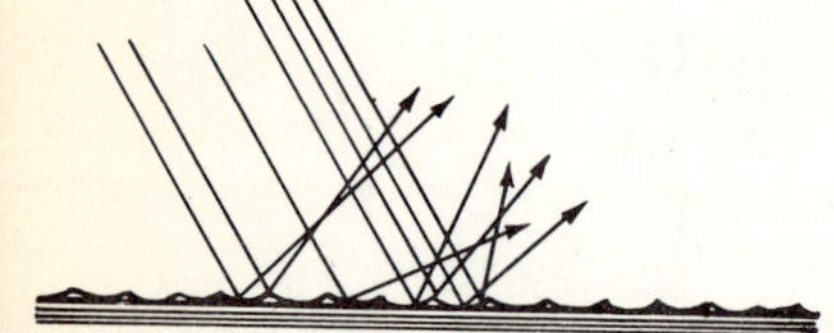

FIGURE 11.14 Diffuse reflection. The reflected rays reach out in every direction.

---

*Even cats who can "see (objects) in the dark" do so with the help of reflected light. In their case, they see lower-intensity light than humans. Rattle snakes, however, do see in the dark, with the aid of infrared waves emitted by their targets.

reverberation, often by covering the ceiling, floor, and walls with special absorbing materials.

Sonar or sound navigation and ranging

One important application of the phenomenon of echo, or the reflection of sound from a barrier, lies in the determination of the distance of a barrier. We do this by making a sound and measuring how long it takes for the echo to come back. An ocean liner can use this method to locate underwater objects, such as a submerged rock, and even to determine the depth of the ocean at a given point.

Let's look at an example. Suppose we make a sound that travels straight down underwater. The sound is then reflected at the bottom, and its echo is heard 3 sec after we made the original sound. Since the speed of sound in water is about 4,700 ft/sec, in 3 sec the sound must have traveled a distance of $3 \times 4,700 = 14,100$ ft. This, of course, is the distance of the round trip, so the bottom must be half the distance, or 7,050 ft away.

The sound waves employed in sonar ranging, as this method is called, usually have frequencies over 20,000 Hz, above the range of human hearing. Sound waves that have frequencies greater than audio frequencies are called **ultrasonic.** Curiously, although inaudible to humans, ultrasonic waves are known to be audible to dogs and to some other animals.

■ 11.3 Refraction of Waves

FIGURE 11.15 Refraction. (a) When light waves enter a medium in which they travel at a slower speed, the parts of the wave fronts inside the new medium bend, making themselves parallel, or nearly parallel, to the interface. This way they travel less distance and thus keep pace with the rest of the wave fronts, in spite of their smaller velocity. (b) The wave fronts bend away from the interface if the waves are incident on a medium in which their speed is greater.

Refraction, or change of direction of a wave train at the boundary of a medium, is due to a change in the velocity of the waves. Suppose the velocity is decreased in the new medium. The wave fronts, upon entering the new medium, are then turned in such a way that they are more nearly parallel to the boundary. This is how it happens. If the wave front arrives at the boundary slanted at an angle to the normal, as shown in Fig. 11.15(a), a part of each wave front will encounter the boundary earlier than the rest of it. On entering the different medium, this part of the wave front is forced to move at a slower pace. If it now continues in the same direction, it will fall behind. To prevent this— that is, to keep pace with the rest of the front—the part of the front inside the new medium takes on a new direction such that the distance of its travel path in

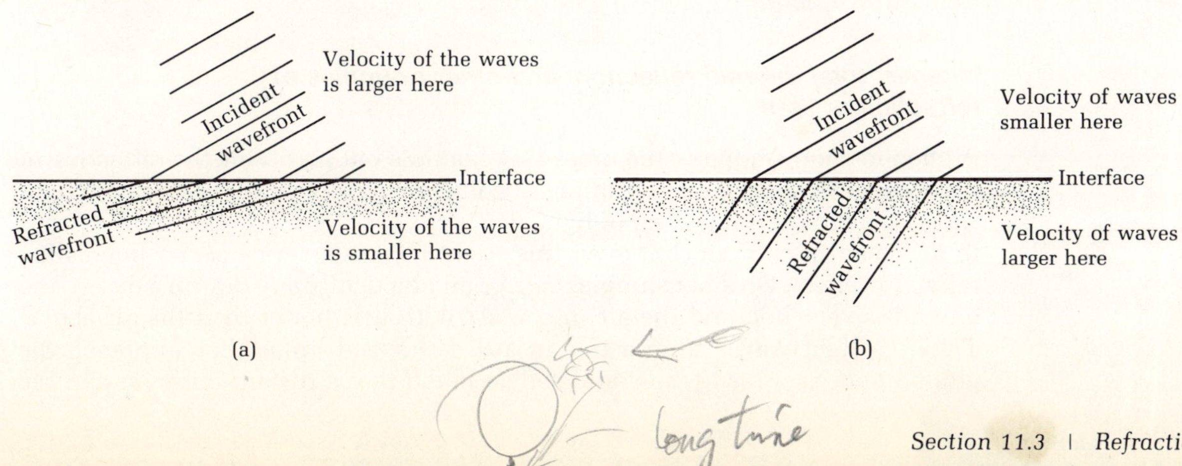

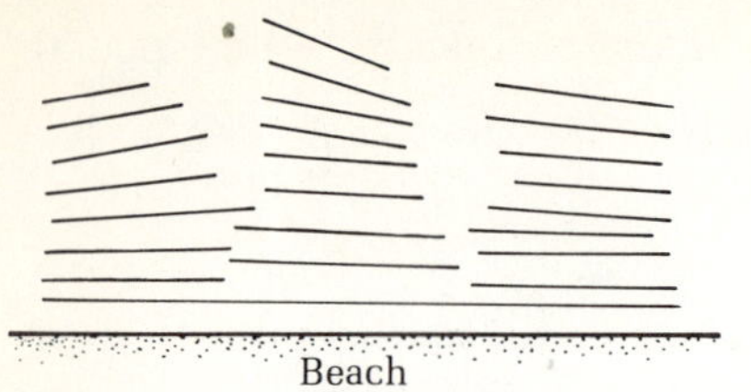

FIGURE 11.16 The ocean waves are refracted as they approach the shore, and they become parallel to the shoreline. The lines represent the straight wave fronts of ocean waves.

this medium is reduced. Thus the part in the new medium maintains its time relationship (its phase) with other points of the wave front by traveling less distance to offset its reduced speed.

The same argument applies to waves that speed up when entering a new medium. The waves will bend so that the distance of the travel path inside the new medium is increased. So the wave fronts bend away from the boundary upon entering a medium where their speed is greater [Fig. 11.15(b)].

The refraction phenomenon offers a nice explanation of something that everybody can see at a beach. If you look at faraway waves, they are found to move in all directions. Yet when they approach the beach, their crestlines are almost parallel to the shoreline. Why? The velocity of the ocean waves is continuously reduced in the shallow water near the shore. So they are continuously refracted as they approach; each refraction makes the wave fronts a little more parallel to the shoreline than before (Fig. 11.16).

An example of refraction of sound

If you are a mountain climber or a balloon enthusiast, you know that up in air you can hear people speaking on the ground, even from as high as half a mile, whereas people on the ground will swear that they can't hear you. Because of our usual expectation of a reciprocity of observations of sights and sounds, this may seem strange. Actually, there is a simple explanation, based on the refraction of sound waves. As already stressed, refraction can result only if there is a change in velocity of the wave. In this case the change in velocity is caused by the fact that air normally cools with height, and sound travels faster in warm air than in cool air. This is easy to understand. Sound waves are transmitted by the collisions of the molecules with their neighbors. If the molecules travel faster (as with higher temperature), they approach their neighbors in less time, thus transmitting the wave faster. As a result of the gradually decreasing velocity with increasing height, the sound waves originating on the ground are refracted (much as the ocean waves bend toward the shoreline) toward an overhead balloonist or mountaineer, who can hear these sounds. For sounds originating in the balloon or on the mountain, the reverse happens. Now the velocity of the waves increases as the waves approach the ground, and thus they are bent away from the ground, often missing it. This fact, augmented by other factors like high background noise, makes it very difficult to hear somebody "up there" from "down here."

Mirages, total internal reflection, and other examples of refraction of light

A phenomenon similar to the preceding example but involving the refraction of light waves is responsible for mirages, commonly seen on deserts and on highways on hot days. The speed of light also increases slightly in hot air compared to its speed in cold air, but even this slight increase can cause an interesting refraction effect. On a hot summer day, or on practically any day on a desert, the ground is very hot, and the air in contact with it is hotter than the air above. Thus the light waves from a tree in a desert speed up as they approach the ground and, as a result, are refracted so much that a distant observer can see

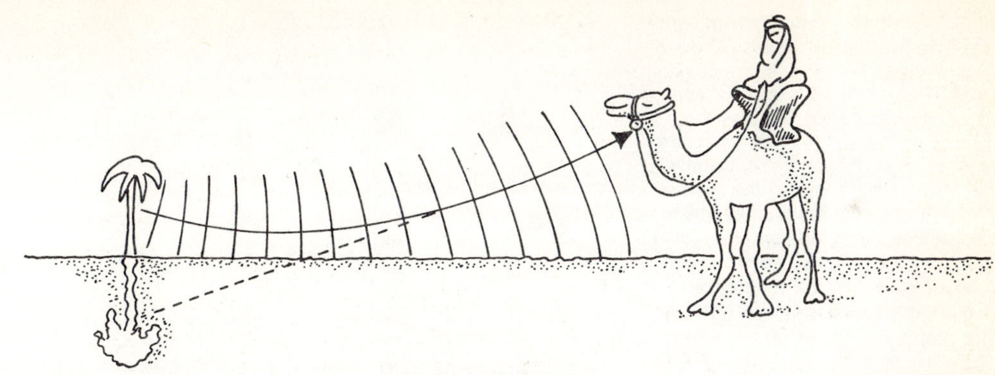

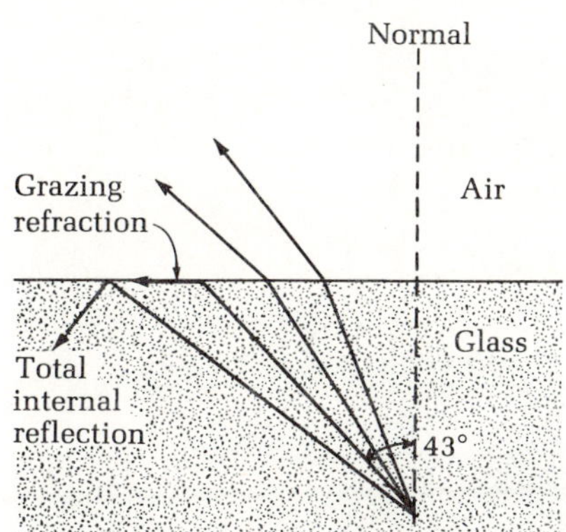

FIGURE 11.18   Total internal reflection. When light from glass is incident on a glass-air interface at an
angle greater than the critical angle
of 43°, it is totally reflected.

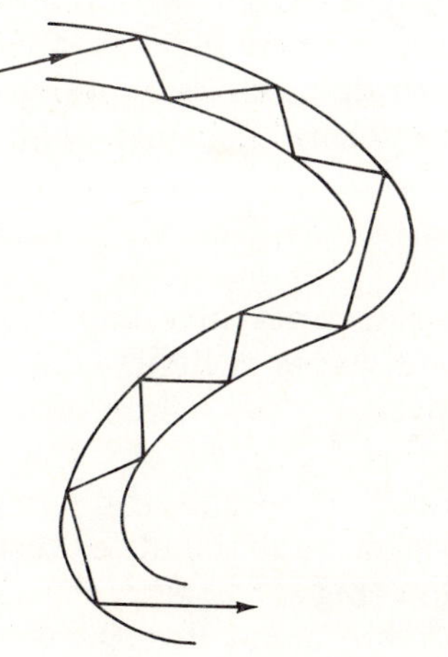

FIGURE 11.19   A light pipe. Light
reaches the other side, although the
route is circuitous, by taking advantage of total internal reflection.

them, as shown in Fig. 11.17. The observer interprets what he sees as a reflected image of the tree in water. In reality, the light he sees is only refracted light—there is no water. Similarly, when the highway is hot, we sometimes can see the image of the sky in it, which again is not a reflected image but is an image produced by refraction.

The light that reaches you in a mirage really is refracted light. All the energy of the original light from the tree is intact; if it were regular reflection, a part of the energy would have gotten lost. This phenomenon—in which refracted light is actually turned back as in reflection—is called **total internal reflection** and is always a possibility whenever light travels from a denser to a less dense medium.

To understand total internal reflection, and refraction in general, the ray description of light is particularly helpful. Fig. 11.18 shows light rays traveling from glass to air. The light waves have greater speed in air than in glass, and so the rays will be refracted away from the normal to the interface (corresponding to the fact that the wave fronts bend away from the interface). Now suppose we increase the angle of incidence, the angle that the incident ray makes with the normal. Then the refracted ray will shift further away from the normal and will become close to being parallel to the interface. Thus if we go on increasing the angle of incidence, an angle is finally reached such that the refracted ray grazes the surface of the boundary, as you can see in the figure. The angle of incidence in this situation is called the **critical angle.**

What happens when the angle of incidence is larger than the critical angle? Clearly, the light ray cannot emerge on the other side of the interface but instead is turned back; it is internally reflected. This is total internal reflection. For glass the critical angle is about 43°, depending somewhat on the type of glass. So no light can escape glass that is incident on an air-glass interface at an angle greater than 43°. Light also slows down in water, so the same argument holds except that the critical angle is 48°. For a cool-air and warm-air interface, the critical angle is much larger (close to 90°). This is why, in this case, we can see the internally reflected rays only from a distance. When we come close, the mirage disappears.

Total internal reflection has some interesting practical applications. One of these applications is the light pipe, in which light is made to follow a circuitous route of a pipe (Fig. 11.19), employing a series of total internal reflections. Light pipes are used by dentists and doctors in their attempt to see into places that are very difficult to light up in any other way.

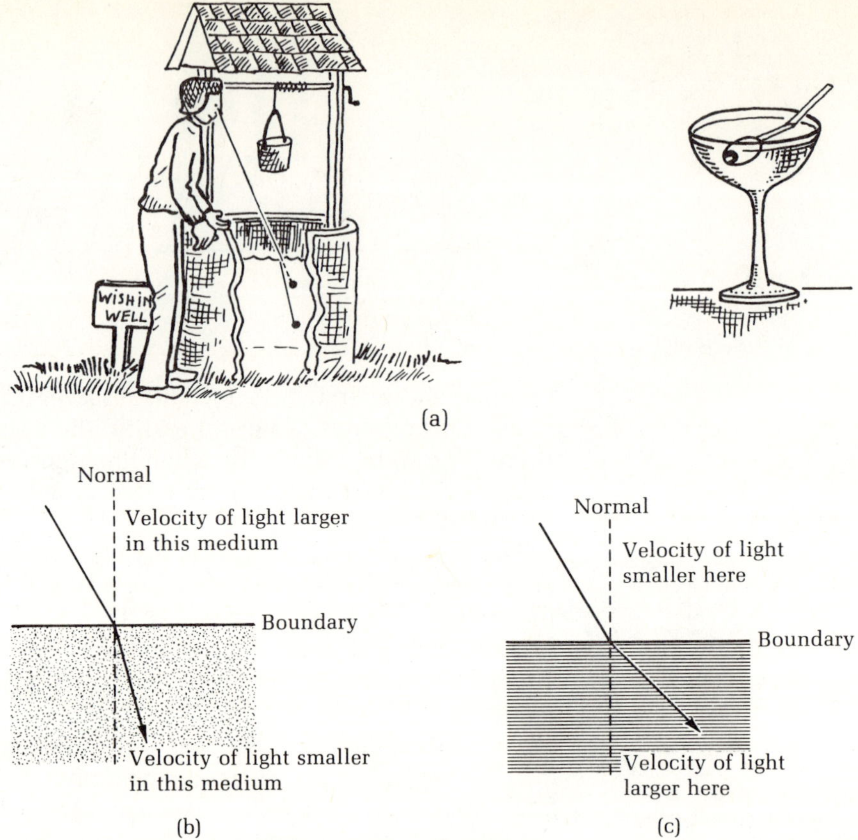

Fig. 11.20(a) shows you two other familiar examples of the refraction of light. In treating the refraction of light via the ray description, remember that whenever a light ray enters a medium that lowers its speed [Fig. 11.20(b)], the ray will bend toward the normal (as the wave fronts tend to become more parallel to the boundary of the two media. The opposite is true when light enters a medium and is speeded up; then the ray bends away from the normal [Fig. 11.20(c)].

Dispersion of light

Our perception of light comes primarily from our experience of white light from the sun. White light is a mixture of several component waves of light, each of which gives the sensation of a color in our eye. These are the colors of the rainbow: red, orange, yellow, green, blue, indigo, and violet, in order of increasing frequency. The rainbow is caused by the breakup of white light through a series of multiple reflections and refractions as it passes through the little drops of water. The breaking-up process is called **dispersion** [Fig. 11.21(a)].

Light of each of the individual colors is characterized by a wavelength and a frequency. The velocity of all light waves in a vacuum is the same, namely, 3×10^8 m/sec, and it is symbolically denoted by c. However, the velocity is different for waves of different color when they pass through a medium. This is the cause of the phenomenon of dispersion.

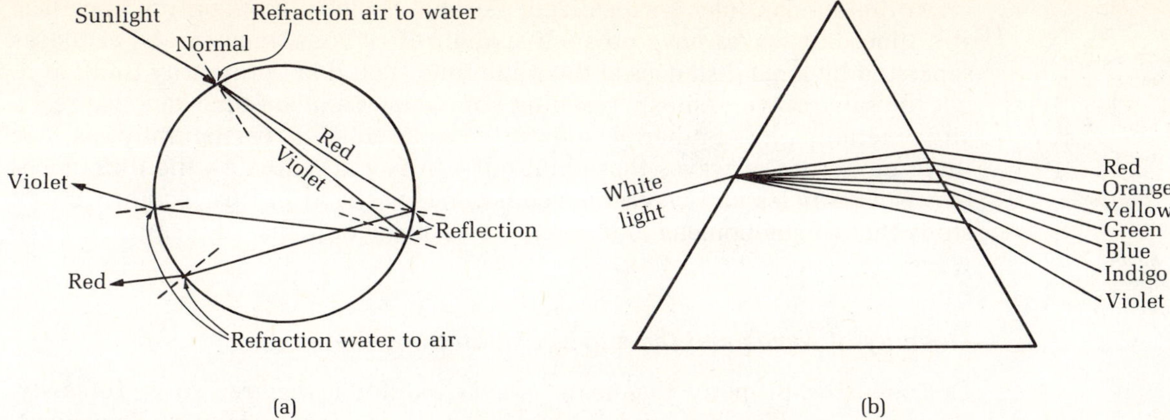

FIGURE 11.21 Dispersion of light. (a) The rainbow. Combined reflection and refraction of incident sunlight cause the light to separate into its component colors; this is dispersion. Notice that for a curved surface the direction of the normal (dotted lines) at a point of the interface is given by the line joining the point with the center of the curve. (b) Dispersion of white light in a prism.

To see this let us first explain why the speed of light changes when it passes through a medium. The velocity of light in a medium is less than c, the velocity in a vacuum. Why? A simple way to understand this is to realize that in a medium a light wave is absorbed and reemitted by particles of the medium (electrons), and therefore some of the time it is not traveling at all. Thus it loses some time and appears to have slowed down; its velocity is smaller, on the average. The next thing we observe is that the slowdown depends on the color. Some colors are slowed down more than others. These slowdown factors for a medium define the **indices of refraction**.* The indices of refraction depend on the color or wavelength of light. Violet is slowed down the most, red the least, and the other colors intermediately. Thus the index of refraction of a medium is maximum for violet light and minimum for red light.

From what we have said earlier about the way refraction occurs, it should also be clear that the larger the index of refraction is, the more the corresponding color is refracted from the incident direction. Thus violet is deflected the most and red the least. One important question is this: Why don't we see colors whenever white light passes through any medium? In air, the most common medium, the dispersion is very small, not enough to be perceptible. For a rectangular slab of glass, the dispersion is large enough, but the refracted waves of different colors emerge parallel to the incident direction and our eyes recombine them.

Thus we get perceptible dispersion effects only when the surface of the medium further accentuates the dispersion. This is what happens in a drop of rain and in a prism (Fig. 11.21).

11.4 Interference and Diffraction of Waves

Reflection and refraction are properties that are not restricted to waves alone. Indeed, a beam of particles encountering matter tends to behave in similar ways. They, too, exhibit reflection and refraction. Yet particles and waves are

---

*Exactly speaking, the index of refraction of a medium is given by the ratio

$$\frac{\text{velocity of light in a vacuum}}{\text{velocity of light in the medium}}$$

FIGURE 11.22 Linear waves. They do not change each other when they meet in a region of space.

very different. Particles are localized; particles can be present only at one place at a time. But waves have no such spatial restrictions; they can be at places separated by large distances at the same time. You don't ordinarily think of it, but the same wave of music reaching your ears at a given time can also reach and entertain a lot of other persons at the same time. So we naturally ask: Are there properties of waves that point out—in fact accentuate—the differences between particles and waves? The answer is yes, and in this section we will study the two phenomena, interference and diffraction, that can happen only to waves.

Linearity of waves and the superposition principle

One amazing property that many waves exhibit is referred to as **linearity.** Throw two pebbles in a pond and see if the waves affect each other. They don't. The waves seem to pass through each other without making any change. Although the intersecting pattern looks more complicated to the eye, it can still be seen as two expanding circular waves (Fig. 11.22). Waves that do not affect others in their path are called **linear waves.**

Water waves created by a pebble are very nearly linear. So are sound waves. Imagine how much more difficult it would be to communicate with people if when two people spoke simultaneously (which often happens in conversation), the speech patterns changed as a result of passing through each other. Light waves are also pretty close to being linear. However, not all waves in nature are linear. Ocean waves are examples of nonlinear waves.

A very powerful principle, called the **superposition principle,** holds for linear waves. According to this principle, to determine the total effect of two or more waves coexisting simultaneously at one place, the disturbance coordinates due to each individual wave are added together. The disturbance coordinates add algebraically; this is an important thing. Suppose two wave crests arrive at a place at the same time—in wave terminology we say they are **in the same phase.** They bring identical types of disturbance (both maximally positive) to the medium at that place [Fig. 11.23(a)]. The superposition principle says that the total disturbance is the algebraic sum of the two disturbances. Thus since in this case the disturbances are of the same sign, the waves reinforce each other and the effect is a bigger crest. Similarly, two troughs arriving together produce a bigger trough [Fig. 11.23(a)]. In contrast, if the trough of one wave arrives simultaneously with the crest of the other, the waves are **out of phase.** The negative disturbance of one (trough) tends to cancel the positive disturbance of the other, and the result is reduced disturbance. If the ampli-

FIGURE 11.23 Interference. (a) Waves in phase interfere constructively. (b) Waves out of phase interfere destructively and cancel each other out in that region of space.

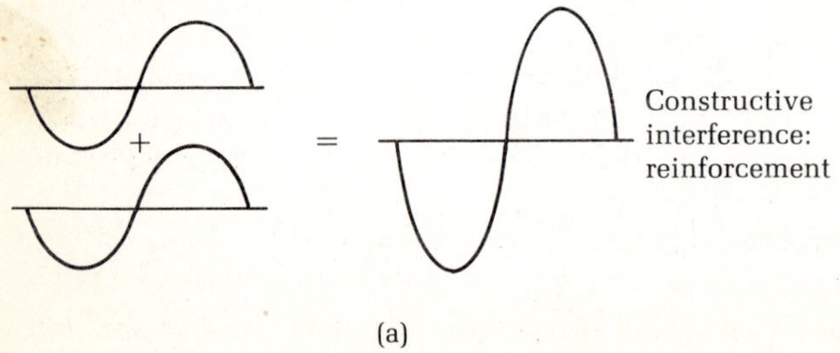

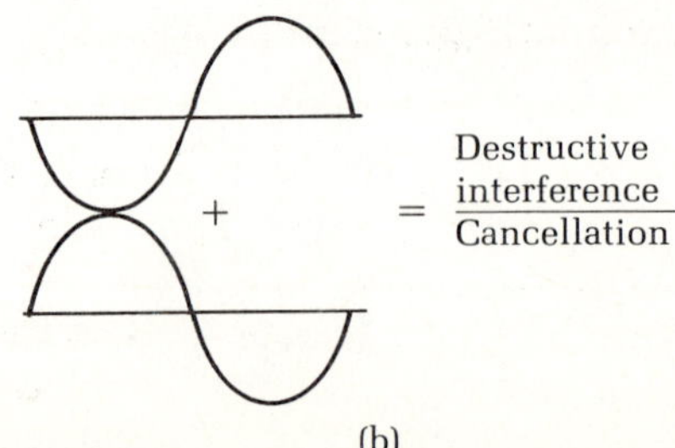

tudes are equal, as in the case shown in Fig. 11.23(b), we get zero total disturbance. The first situation is referred to as **constructive interference** and the second as **destructive interference.**

Now we begin to see some real contrasts between the behaviors of a beam of particles and a train of waves. Particles do not obey the superposition principle; they do not pass through one another without affecting each other. Most importantly, two particles—for example, two bullets—cannot meet at a place and superpose to give zero bullets. Only waves can give destructive interference.

Interference patterns

When two linear wave trains interfere under certain conditions, the resultant pattern is quite distinctive. The pattern is easy to understand by applying the superposition principle to find the total effect of two waves.

Suppose we have two wave trains that maintain a constant relative phase with each other—for example, crests in each always leave the two sources at the same time. For water waves we could guarantee such a fixed phase relationship by having two vibrators operating simultaneously inside a tank of water (called ripple tank). Now suppose in addition that the sources have identical frequencies, in which case they create waves of identical wavelengths. The two waves now will arrive in step, in phase, at some points of space if they have traveled the same distance or if the relative distance (i.e., the distance traveled by one minus the distance traveled by the other) is an integral multiple of their common wavelength (Fig. 11.24). So at these points of space the waves will interfere constructively. In contrast, if one has traveled one-half wavelength more than the other (or $\frac{3}{2}$, $\frac{5}{2}$, . . . wavelengths—in general, an odd number of half wavelengths—more than the other), then the two waves will arrive at the point of space out of phase with each other and interfere destructively. The resultant pattern does not change with time and is called a **stationary interference pattern.**

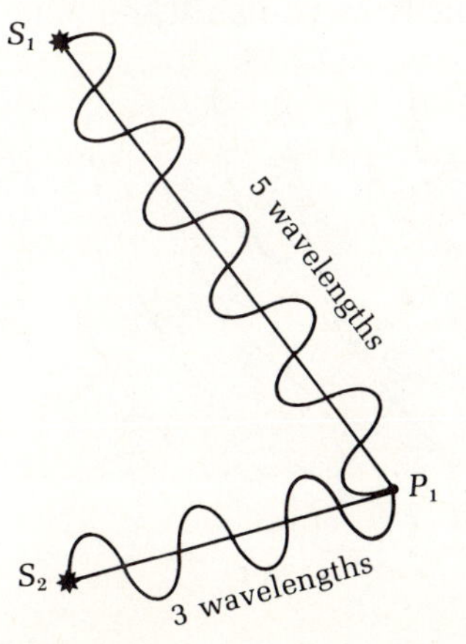

FIGURE 11.24 (a) Coherent waves from sources S_1 and S_2 traveling a relative distance, equal to an integer multiple of their wavelength, to a point P_1 arrive there in phase. (b) But if one of the waves travels an odd number of half wavelengths more than the other to reach a point P_2, they arrive out of phase and interfere destructively.

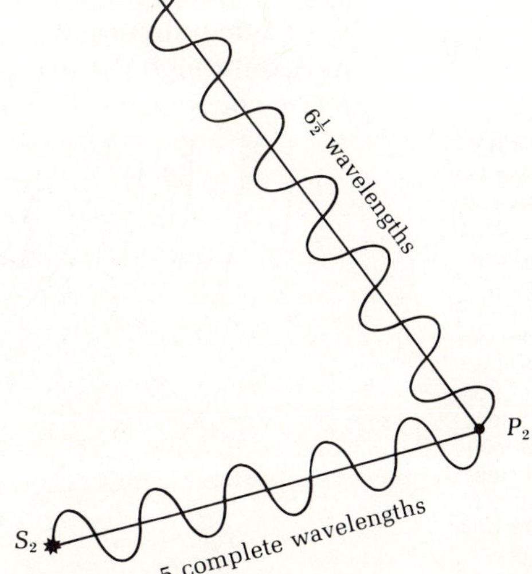

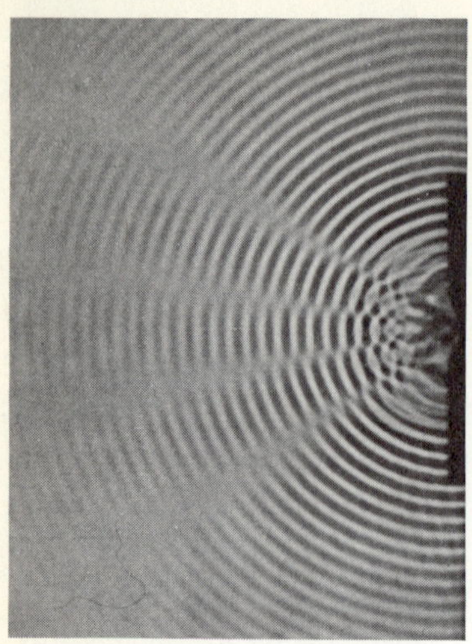

FIGURE 11.25 Interference of water waves in a ripple tank, photographed from above. For the individual ripples of each wave, the light and dark circular bands correspond to the crests and the troughs, respectively. If you tilt the page and view the pattern from a glancing direction, you will be able to see almost straight bands of intermediate shade, neither as bright as the crest band nor as dark as the trough bands of the individual waves. These are the areas where the two waves cancel each other, and they are called the nodal bands. [From *PSSC Physics* (Lexington, Mass. Heath, 1965).]

FIGURE 11.26 Analysis of the interference pattern shown in Fig. 11.25. The dark circles indicate where crests are meeting crests, and the white circles show where troughs are meeting troughs. Upon joining these circles, we get the antinodal lines A_0, A_1, and so on—the lines of reinforcement. The half circles indicate where the crest of one wave meets the trough of another, producing destructive interference. The bands joining the half circles give the nodal lines N_1, N_2, and so on. [From F.J. Rutherford, *The Project Physics Course* (New York: Holt, Rinehart and Winston, Publishers). Copyright © 1970.]

The condition for creating such an interference pattern, then, is to have two sources generating waves of equal wavelengths and constant relative phase. Such sources are called **coherent sources.** If in addition the two waves have identical amplitudes, then at the places where destructive interference occurs, the destruction is total; the crests of one nullify the troughs of the other, producing zero total disturbance. Such points are called **nodes.** Of course, the constructive interference that occurs at other points is now also specially strong, and such points are called **antinodes.**

As already mentioned, for water waves the conditions of coherence can be created by having two vibrators operating simultaneously inside a ripple tank. If the two vibrating sources of identical frequency go through their up-and-down motions together, then they will generate waves that are coherent in phase.

The pattern that we get from the interference of the two sets of water waves in such a case is shown in Fig. 11.25. For the individual ripples of each wave, the light circular bands correspond to the crestlines and the dark bands correspond to the troughs. In the interference pattern, on the other hand, there are regions that are neither as bright as the crest nor as dark as the troughs. In fact, these regions seem to form almost straight bands.

Fig. 11.26 shows how this happens. At points on the lines denoted by A_0, A_1, . . . (called antinodal lines), if you look at the phases of the individual wave patterns, you can discover that the two waves at these points are in phase. Two crests are meeting (or two troughs), giving rise to constructive interference. Thus as the waves spread, the points on these lines are going to move up (or down) more vigorously than with the passage of each individual wave. In contrast, the points on the lines N_1, N_2, . . . (called nodal lines) represent bands along which destructive interference takes place. The points on these lines are not agitated—a situation very different from the presence of only one wave.

If sound is a wave phenomenon, then we ought to be able to generate an interference pattern with sound waves, starting with a similar arrangement. Instead of trying it with two separate sound sources, what works best in this case is a single source giving off sound of a single frequency. Then we split the sound in two through the use of two loudspeakers (separated by about 1 meter). The phase coherence is then guaranteed. Now listen carefully as you shift the posi-

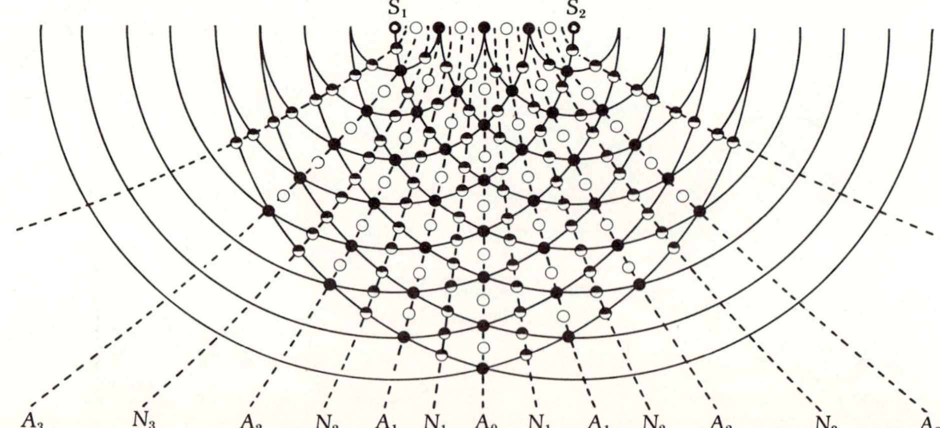

tion of your head while standing several meters in front of the two speakers. With little effort you can discover the nodal and antinodal lines. Again, the nodal lines are those along which there is no sound. The "dead spots" in an auditorium are caused by this interference phenomenon.

Beats

There is still another interference effect of sound with which you may be familiar. If two sound waves have very nearly the same frequency, they give rise to a spectacular auditory phenomenon known as **beats,** a characteristic waxing and waning of sound. Beats occur in the following way. The two sound waves initially interfere constructively, but because of the slight difference of frequency, the phase relationship changes at every point of space. Very soon there will be destructive interference at the same point where there was constructive interference before. The frequency of the beats, in fact, is given as the difference of the frequency of the two sound sources. Beats can thus be used to tune a musical instrument. When the frequencies are exactly matched, the beats will disappear.

Interference of light

If light is a wave phenomenon, then two light waves also ought to give rise to interference patterns, as occurs with water waves and sound waves. At one point in history, many people questioned the validity of the wave description of light. Then in 1801 an English physicist named Thomas Young demonstrated the interference of light, which finally convinced everyone.

Remember what we said about the conditions for interference? The two wave patterns have to have phase coherence. Since light consists of very high frequency waves, such coherence is impossible to achieve with two independent sources. Young, however, figured out a way. He split the light from one source into two by passing it through a double slit-screen. Then his system became identical with the interference of two water wave trains or that of sound coming from two loudspeakers hooked up to the same source. So the conclusions reached must be the same. The combined effect of the light waves creates alternate regions of brightness—when the waves reinforce each other (in phase)—and darkness—when the waves are out of phase and cancel. Although Young's original experiment was carried out with white light, the experiment works best when light of one color (monochromatic light) is used. The alterna-

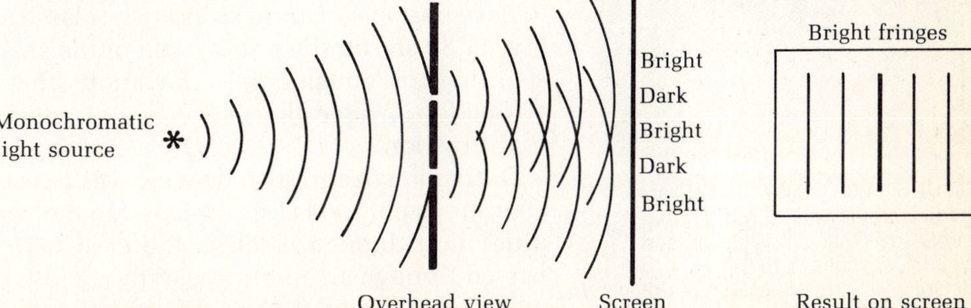

FIGURE 11.27 Young's experiment. Light from a monochromatic source is split into two by passing it through the double slit. The interference pattern of alternate bright and dark fringes is seen on the screen.

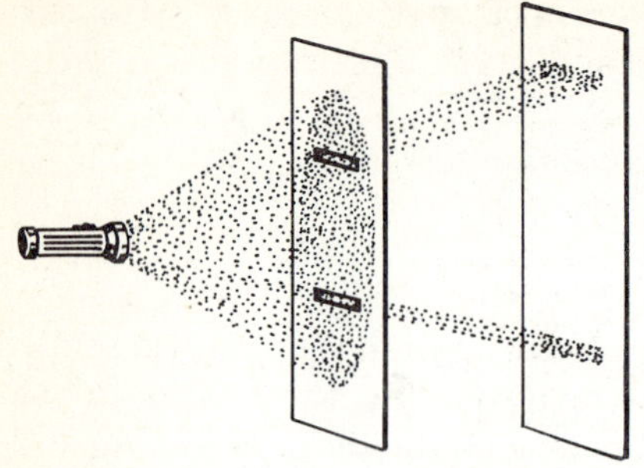

FIGURE 11.28   If light consisted of particles, this is what we would see in Young's double-slit experiment.

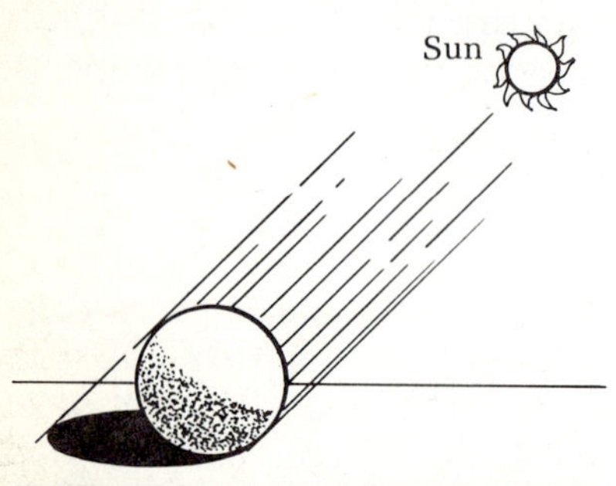

FIGURE 11.29   Are shadows cast by objects really this sharp?

FIGURE 11.30   Diffraction of light produced by a wire. (From *Fundamentals of Optics,* by Jenkins and White, Copyright © 1976 by McGraw-Hill, Inc. Used with permission of McGraw-Hill Book Company.)

tive bright and dark fringes shown in Fig. 11.27 represent a typical interference pattern for light when using a monochromatic source. Incidentally, the experimental setup of Fig. 11.27 enables us to determine the wavelength of light from a measurement of the distance between the interference fringes.

If light consisted of a beam of particles, as even Newton thought was the case, the result of Young's experiment would have been as shown in Fig. 11.28: two fringes of light on the screen right behind the slits. Instead, we find that light somehow appears at other places on the screen as well. This is easy to understand in the wave picture of light but impossible to reconcile with the particle picture. Thus Young's experiment settles the issue: light consists of waves.

Strangely enough, physics of the twentieth century has brought back the particle picture with a new twist: light is now regarded as both a particle and a wave. We will see how this is possible in Chapter 21.

Diffraction

Diffraction describes the phenomenon of the bending of waves upon encountering an obstacle. If waves were propagated in a perfectly raylike fashion, in straight lines, then the shadows cast by an obstacle would be sharp (Fig. 11.29). It is diffraction that prevents this from happening. Some of the waves bend around the corner of the obstacle and propagate into the shadow region.

Diffraction is quite common in water waves and sound waves. If you stand behind a small wall in a lake and a boat goes by making waves in the water, the waves will reach you in spite of the fact that you are standing inside the geometrical shadow of the waves cast by the wall. When your friend is calling to you from around a corner, the direct sound waves are blocked by the walls. How do you hear her, then? In both cases the waves diffract around the barrier and reach you. In fact, with water waves or sound waves, diffraction is so common that we don't ordinarily think of raylike propagation at all when referring to these waves (i.e., we don't talk about sound rays).

The situation is quite different with light. Here we do see raylike propagation. Shadows of light do look like geometrical shadows, although on a close look we can see deviation. So diffraction effects are small for light waves. The reason for this is that light waves possess much shorter wavelengths ($\lambda \sim 10^{-6}$ m) compared to a water wave or a sound wave ($\lambda \sim 1$ m). Diffraction effects are prominent only when the size of the obstacle is comparable to the wavelength of the wave. So with obstacles of everyday size, water waves or sound waves exhibit diffraction, but light waves do not. For light waves to show diffraction, we have to have obstacles whose sizes are not large compared to the wavelength of light. Fig. 11.30 shows the photograph of the shadow cast by a thin wire. As you can see, there is considerable deviation from a geometrical shadow, indicating diffraction. We see alternate bright and dark fringes, which is characteristic also of diffraction.

Diffraction occurs also when waves pass through an aperture. If the size of the aperture is not too large compared to the wavelength of the waves, there will be diffraction. Indeed, such is the case with water waves. Water waves, when passed through an aperture having a width that is several times larger than their wavelength, show a clear diffraction pattern, with nodal and antinodal lines (Fig. 11.31).

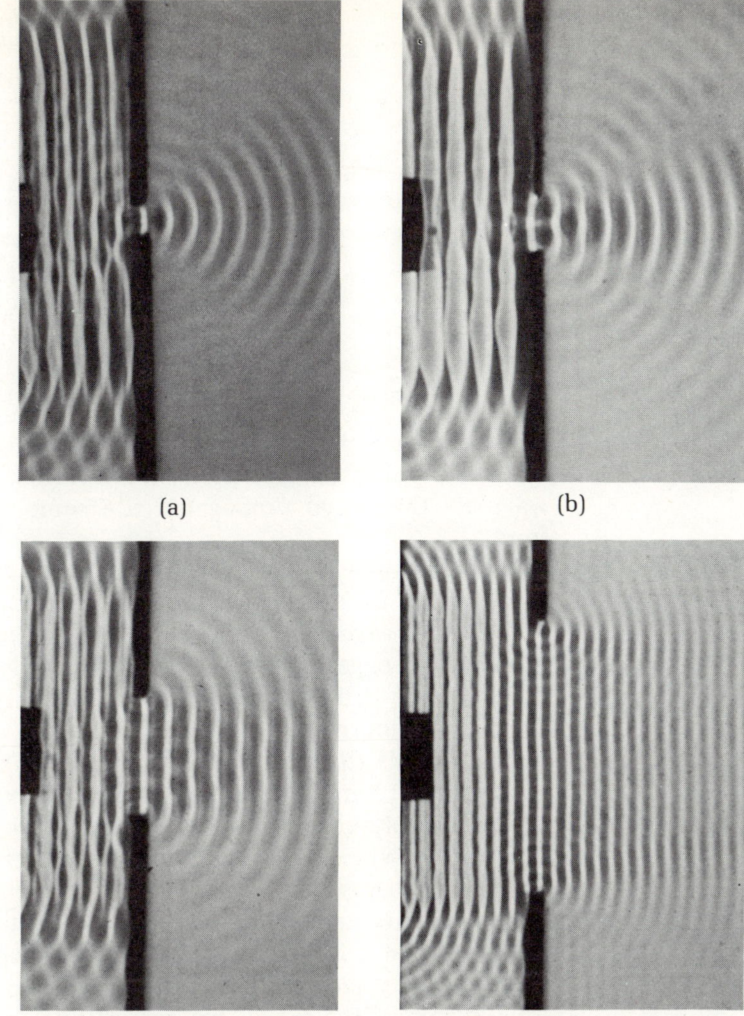

FIGURE 11.31 Diffraction of straight water waves passing through apertures of different sizes. Notice how the diffraction becomes pronounced as the aperture becomes comparable to the wavelength of the waves. (a) w (width of opening) $= \lambda$ (wavelength); (b) $w = 2\lambda$; (c) $w = 5\lambda$; (d) $w = 19\lambda$. [From *PSSC Physics* (Lexington, Mass.: Heath, 1965).]

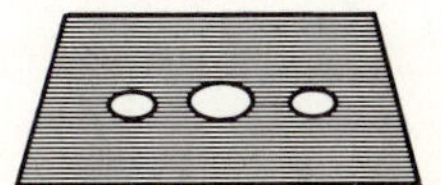

FIGURE 11.32 A single-slit diffraction pattern.

Light waves, when passed through an aperture of width less than a millimeter, also show diffraction patterns. In fact, you can do the following simple experiment to convince yourself. Make a narrow slit (as narrow as you can; definitely keep it less than a millimeter) with your fingers and look through it at a point source of light (a distant streetlight will do). The light will appear blurred. With some care you will even be able to see bright fringes on the side, something like the pattern of Fig. 11.32. Now experiment further with the pattern by varying the width of the slit between your fingers. If you make the width too large (about a millimeter), the diffraction effects will disappear. If the width is very small, the central lighted band will become very wide and drown out everything else.

To understand diffraction, it is convenient to refer to the principle formulated by Dutch physicist Christian Huygens, a contemporary of Newton. According to Huygens, points along a line of constant phase (a wave front, such as a crest or a trough of a water wave) themselves act as sources of new wavelets that fan out from these points. The subsequent position of the wave front is determined by

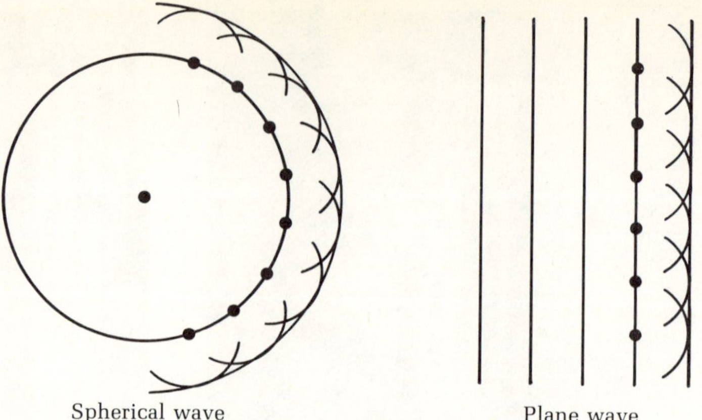

FIGURE 11.33 Illustration of Huygens' principle. The dots indicate the secondary sources of new wavelets along the wave front.

the superposition of these tiny wavelets. Fig. 11.33 shows how the principle works for spherical and plane wave fronts. In each case the envelope of the wavelets that have propagated the same distance gives us the new wave front.

Thus a wave incident on an obstacle would give rise to new sources of wavelets at its corners, which will propagate in its shadow region [Fig. 11.34(a)]. A wave front passing through an aperture would give rise to spherical wavelets propagating in all directions, as shown in Fig. 11.34(b).

But now a different problem arises. Looking at a wave from the Huygens principle point of view, it is difficult to see why diffraction should not always occur. Why is there any raylike propagation of waves at all? Fortunately the answer is not hard to find. The original shape of the wave front could be recovered if we got enough wavelets to superpose (see Fig. 11.33). Thus if the aperture is very wide, we have enough secondary sources to give adequate superposition so that we get the original shape of the wave front. On the other hand, if the aperture is narrow, the wave fronts will appear distorted.

The crucial criterion of narrowness is applied in comparison to the wavelength of the waves. If the size of the aperture (or obstacle) is of the order of the wavelength of the wave, we get diffraction. If the size is too large, we get unchanged propagation. But even in the case of wide apertures (or obstacles), some diffraction effects at the boundary regions of the shadow will be seen. All these aspects are illustrated in Fig. 11.31 for water waves passing through holes

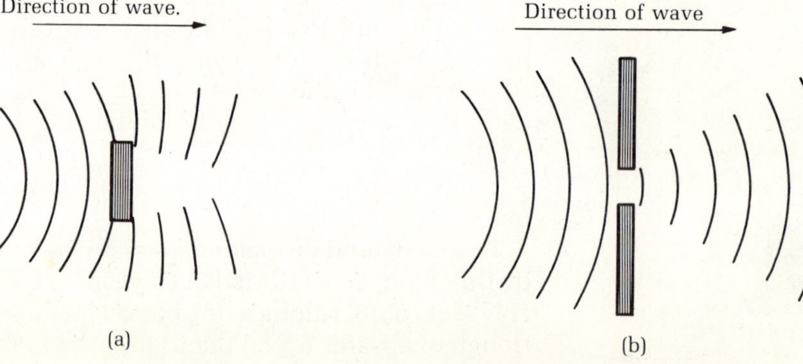

FIGURE 11.34 (a) Diffraction of waves at an obstacle. (b) Diffraction of waves at an aperture.

FIGURE 11.35 A diffraction grating and the diffraction pattern of light when passed through the grating. Notice that if we use white light, we get colored fringes.

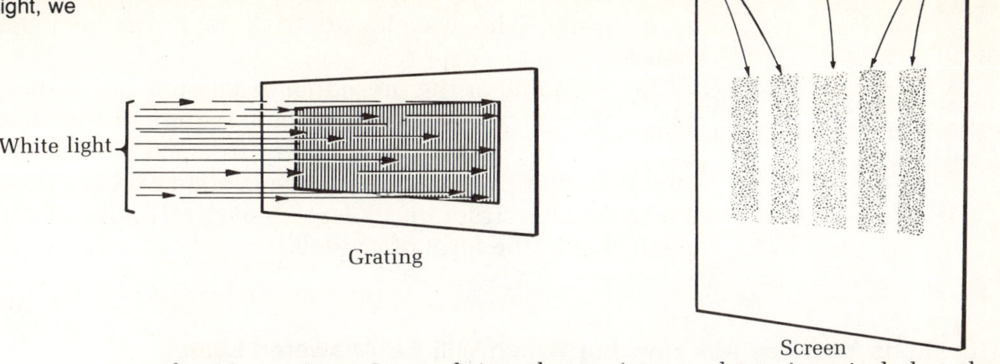

of various sizes. In studying these pictures, keep in mind that the distance between two crests (the light lines) is the wavelength of the water waves.

In Fig. 11.31 notice in particular that when the width of the aperture is of the same order as the wavelength of the waves passing through, the diffraction pattern that results looks quite similar to previous interference patterns. Why does this happen? When an aperture is of this size, it cannot be considered as a single point source of secondary wavelets; it is more like a continuous source of secondary waves. These waves can now interfere and produce a pattern similar to the interference pattern.

Thus the only difference between the diffraction and interference phenomena is that diffraction results when there is a continuous distribution of coherent sources, whereas interference results from a discrete number of coherent sources. In the latter case, we must note that if the width of the slits is wide enough, there will be diffraction also. So even for the double-slit experiment (except for very narrow slits), what we get is due to a combined effect of interference and diffraction. We can call the resultant pattern either a diffraction or an interference pattern; it really doesn't matter.

If diffraction effects are difficult to observe with waves of wavelength even as small as those of light, then how do we study waves of even smaller wavelength, like X rays? Before we answer this, we will mention one interesting thing about the diffraction pattern of light. When we use only a few apertures (but more than two), the diffraction pattern is usually blurry. But if we make a very large number of slits, separated by a distance of about a micron (10^{-6} m), the fringes of the diffraction pattern produced by light passing through become very sharp and distinct. Such a device for producing diffraction of light is called a **diffraction grating** and can be made simply by ruling many parallel grooves on a sheet of glass with a diamond point. Fig. 11.35 shows the diffraction pattern of light produced by such a grating.

Diffraction gratings work for light because we are able to make the grooves with appropriate separations, say, as small as a micron. But the wavelength of X rays is of the order of 10^{-10} m; it is impossible to make a diffraction grating with so fine a mesh size. Fortunately, nature helps us out of this predicament.

Recall our discussion about crystals. They are orderly arrangements of atoms with a regular spacing, very much like a grating—except, of course, that it is a

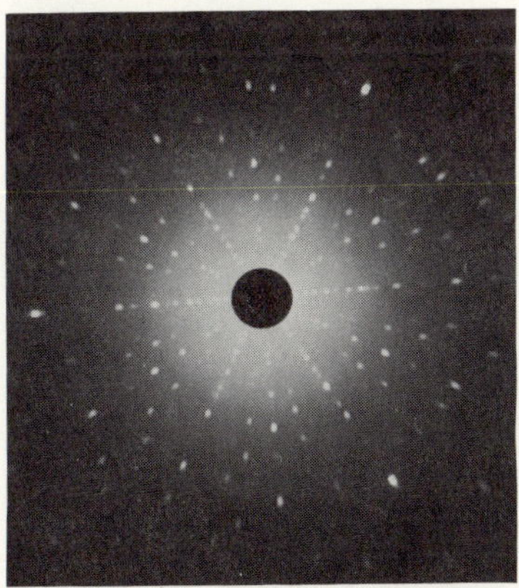

FIGURE 11.36 Diffraction pattern (of bright dots) produced by X rays when passed through a crystal. (Courtesy General Electric Research and Development Center.)

three-dimensional grating. Most importantly, the spacings between the atoms are also of the order of 10^{-10} m; so these gratings are just right for the study of diffraction of X rays. Fig. 11.36 shows a diffraction pattern of X rays produced by a crystal. The wavelength of X rays can be determined from such a pattern.

The reasoning of the preceding discussion can be reversed. The diffraction patterns must carry information about the atomic structure of the crystals themselves, and hence they can be used for such a study. Today one of the most important applications of the phenomenon of X ray diffraction is the study of the molecular structure of large biological molecules like proteins that are obtainable in the form of crystals.

■ 11.5   Questions You May Ask Now but Which Will Be Answered Later

We have carefully introduced you to light waves, omitting a few important details (we hope you have noticed). For example, we have never stated what the disturbance coordinate is for light waves. And we have not considered questions like the following: Is the disturbance coordinate parallel or perpendicular to the direction of propagation of the waves? Or, in other words, are light waves longitudinal or transverse in character? What emits the light waves? We know that something must vibrate for every wave phenomenon. What is it that vibrates in the process of emission of light? All these questions will be dealt with in Chapter 18. Because light involves electromagnetism, a discussion of these questions would be premature before you know the story of electricity and magnetism.

However, we will anticipate later development to the following extent. Some of the waves we have mentioned during the course of this chapter—infrared and X rays, for example—are examples of electromagnetic waves, as is visible light. Table 11.1 lists some electromagnetic waves you may have heard of. (The table is not inclusive.)

TABLE 11.1   Some electromagnetic waves.

Name	Approximate Frequency (Hz)
Radio waves	10^6
Microwaves	10^{10}
Infrared	10^{12}
Visible light	5×10^{14}
Ultraviolet	10^{16}
X rays	10^{19}
Gamma rays	10^{22}

The question of the disturbance coordinate also suggests the question of a medium for light waves. What is the medium for light waves, which can and do travel through a vacuum? Can light exist without a medium? Does it?

At one time many people were very perturbed about the question of the medium. In fact, it is this question that kept very intelligent people from accepting the wave nature of light for a very long time. When the wave nature was established beyond doubt, people started to search for a medium for light waves, a medium they called "ether" (the name was borrowed from the ancient Greeks; ether was one of their five "elements"). But ether turned out to be purely hypothetical; nobody ever found any evidence for it.

Today we picture light as a very special kind of wave that doesn't need a medium for propagation. This may sound paradoxical, but this is the way it is.

The situation reminds us of an episode in Lewis Carroll's *Alice in Wonderland*. Alice is talking to the Cheshire cat:

> "I wish you wouldn't keep appearing and vanishing so suddenly," replied Alice. "You make one quite giddy."
>
> "All right," said the cat; and this time it vanished quite slowly beginning with the end of the tail and ending with the grin, which remained sometime after the rest of it had gone.
>
> "Well! I have often seen a cat without a grin," thought Alice, "but a grin without a cat! It's the most curious thing I ever saw in my life!"

Our picture of light as a wave without a medium is very much like the grin of the Cheshire cat—a grin without a cat.

Matter waves

We alluded to the wave nature of matter in the introduction to this chapter. When a beam of electrons is passed through a crystal and allowed to fall on a photographic plate, the picture we get is not a blob on the plate identifying where the electrons hit it. Instead, we get a pattern as shown in Fig. 21.9 (in Chapter 21; take a look) which must be interpreted as a diffraction pattern (compare Fig. 11.36). So in this kind of experiment a beam of electrons is clearly seen to possess wavelike properties. How a particle like an electron can be a wave is a puzzling question, and an explanation is deferred until Chapter 21. In the meantime, you can look at the phenomenon in the manner my friend, science fiction writer Damon Knight, does: "Electrons are very friendly, they wave at you."

SUMMARY AND OUTLOOK

This chapter introduces you to the concept of waves, with particular emphasis on sound and light, both of which are wave phenomena. Waves are propagating disturbances that transport energy. Important basic concepts in connection with waves are frequency, wavelength, amplitude, intensity, phase, and wavefronts. Waves are classified as longitudinal and transverse, but there are waves, like water surface waves, that are a mixture of the two. Sound and light are three-dimensional waves, in contrast to surface water waves, which are two-dimensional.

The frequency and wavelengths of waves are related by the following equation, involving also the velocity of the waves:

$$v = \nu\lambda$$

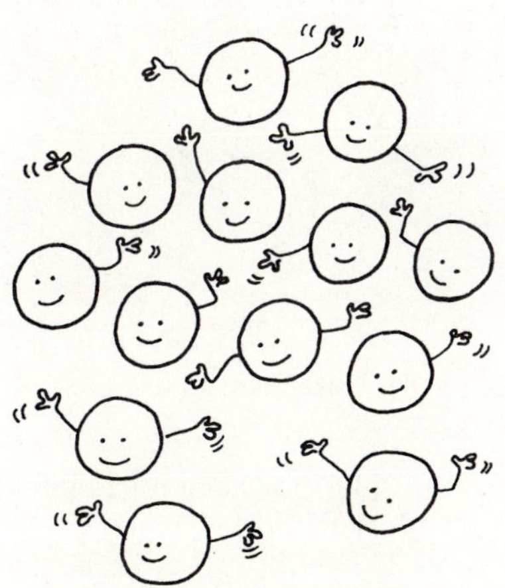

ELECTRONS ARE VERY FRIENDLY–THEY WAVE AT YOU.

When waves encounter a different medium, they undergo reflection, refraction, and diffraction. When waves encounter another wave, the result is a superposition (for linear waves), which leads to the phenomenon of interference. Typical of the interference of waves are patterns known as interference patterns. Diffraction is the bending of waves around an obstacle or an aperture. Huygens' principle greatly facilitates our understanding of diffraction.

Interference and diffraction are important because these two processes can only occur with waves, and they distinguish waves from material particles. Developments in modern physics show us that at the submicroscopic level matter is dual, both particle and wave at the same time. To grasp the meaning of this duality, you must have a clear idea of the polar characteristics of particles on the one hand and of waves on the other.

Substantial portions of this chapter are devoted to phenomena—most of them quite familiar to you—arising from the reflection and the refraction of waves. In connection with refraction, total internal reflection and dispersion are important and perhaps new aspects to you. Total internal reflection is really a refraction phenomenon. It occurs when a wave is incident on the boundary from a dense to a rarer medium at an angle greater than the critical angle. Dispersion is the phenomenon of the breaking up of white light into its different color components.

QUESTIONS

Review and reason

1. Explain the following concepts in connection with a wave train: frequency, wavelength, amplitude, and phase.
2. Distinguish between longitudinal and transverse waves. Give an example of each.
3. If the speed of sound decreased with frequency, then how would distant music sound?
4. How can you estimate the distance of a faraway thunderstorm?
5. Why do you see the lightning before you hear the thunder?
6. What sorts of things do you expect when a wave encounters another medium?
7. State the law of reflection. Illustrate the law with a diagram.
8. Suppose you look at your image in a mirror, taking a particularly close look at the size of your image. Does it seem about the same size as that of a person standing before you at the same distance from you as your image?
9. The picture in Fig. 11.37 is one of French impressionist Edouard Manet's most famous. Comment on what's different from reality in the reflections depicted in the painting. (However, don't come to the conclusion that Manet didn't know physics. The

FIGURE 11.37   Manet, "Bar at the Folies Bergére." (Courtauld Institute Galleries, London.)

reality seen by a painter does not have to resemble the physical reality in every detail.)
10. Describe your image as it would appear in a two-dimensional corner mirror (Fig. 11.38).
11. Is the reflection from the screen in a movie theater diffuse? Why or why not?

FIGURE 11.38

12. What is reverberation? Give an example. It is often easier to hear a speaker in a room full of people than in an empty room. Explain.

13. Why do echoes sound weaker than the original sound?

14. Consider a wave moving in a medium with a certain speed. Draw a diagram showing how refraction takes place when the wave passes into a different medium where its velocity is *increased*.

15. Do sounds refract? Give an example of refraction of sound.

16. We are able to see the sun for a short while before sunrise and after sunset. How is this possible?

17. In a *Prince Valiant* comic strip, Princess Karen saw pirate ships rising from the horizon. She knew that the pirate ships were not yet in her line of sight; she was seeing the images of the ships produced by a process called "looming." Can you explain the phenomenon?

18. If somebody is planning to capture an African lion in the Sahara Desert by shooting it with a tranquilizer gun, should she aim low or high at a certain spot of the lion?

19. A block of glass has a pin engraved in it. Does the pin look shorter than, longer than, or the same as it really is?

20. If a dolphin looked at you from underwater, would it see you as taller, shorter, or at your actual height? Explain.

21. Construct a fish's-eye view of the different positions of the sun from sunrise to sunset (as the sun moves from horizon to horizon). What would the fish conclude about the shape of the world from the way it sees the sun?

22. Why does a pond or a swimming pool appear shallower when full of water than when empty?

23. Why does light appear to slow down when passing through a medium like glass? Is the amount of slowing down the same for all frequencies of light? What is the index of refraction?

24. If light falls on a slab of glass perpendicular to the glass surface, is it refracted? Is it reflected?

25. What is meant by the linearity of a wave? What is meant by the superposition principle?

26. What is meant by constructive interference and destructive interference?

27. Fig. 11.39 shows a wave. Draw a wave that is completely in phase with this one. Now draw a wave that is completely out of phase with the one in the figure.

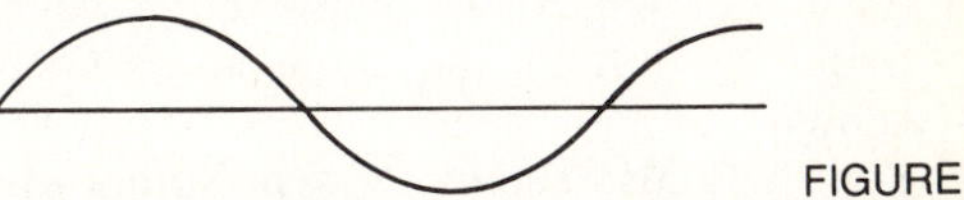

FIGURE 11.39

32. Suppose you are in a boat on a lake. However, there is a solid barrier (Fig. 11.40) in the lake between your boat and those traveling across the lake. Can you feel the splashes when a motorboat goes by, in spite of the fact that you are in the shadow of the solid barrier? Why or why not?

FIGURE 11.40

28. State the conditions necessary for two waves to produce an interference pattern. Draw a diagram showing how an interference pattern is produced.

29. Explain the phenomenon of beats. How can beats be used to tune a musical instrument?

30. Is light actually destroyed in destructive interference?

31. Describe Young's experiment. What is the significance of this experiment in the history of science?

33. Give an example of the diffraction of sound waves.

34. Explain how Huygens's principle works for waves. Draw diagrams.

35. Bring your thumb and forefinger together slowly in front of a distant streetlight. Explain what you see through one eye just before they touch.

36. Car headlights, when seen through a screen, look very different than when seen without a screen. Try it and see. Can you give an explanation?

37. Look at a distant streetlight through a dark umbrella. You will be able to see a diffraction pattern. Describe the pattern. Draw a picture.

38. Look at white light scattered from a phonograph record at an almost glancing angle. What do you see? Explain what you see.

39. What is a diffraction grating? Crystals act as a diffraction grating to incident X rays. Explain.

40. Discuss some of the differences between light waves and sound waves.

Arithmetic

1. Fig. 11.41 shows a wave of sound passing through air. Read the wavelength of the wave from the figure. The velocity of sound in air is 1100 ft/sec. What is the frequency of the sound? Sketch a sound wave with twice the amplitude (but of the same wavelength) as the one shown. Also sketch a sound wave of twice the frequency but of the same amplitude as the one shown.

2. Fig. 11.42 also shows a wave. Notice what is plotted along the abscissa; it is time instead of distance. Can you tell what the frequency of the wave is? Can you tell the wavelength? Explain both answers. What additional data do you need in order to determine the wavelength?

3. Suppose you are rocking in a rocking chair. Your head moves back and forth over a distance of 3 in about 20 times a minute. What is the frequency, period, and the amplitude of the vibration of your head?

4. The wavelength of water waves is given as 90 m and their frequency is 0.1 Hz. Calculate the velocity of the water waves.

5. A light ray is incident on a mirror, making an angle of 45° to the normal. What is the angle of reflection? Draw a diagram.

6. Suppose you make a sound by shooting a gun and the echo from a nearby hill comes back to you in 1 sec. What is the distance of the hill from you?

7. The velocity of light in a vacuum is given as 3×10^8 m/sec. The refractive index of water is given as 1.33. What is the velocity of light in water?

FIGURE 11.41

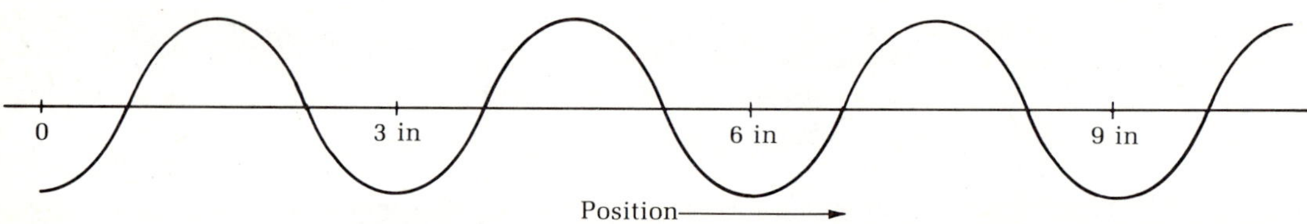

FIGURE 11.42

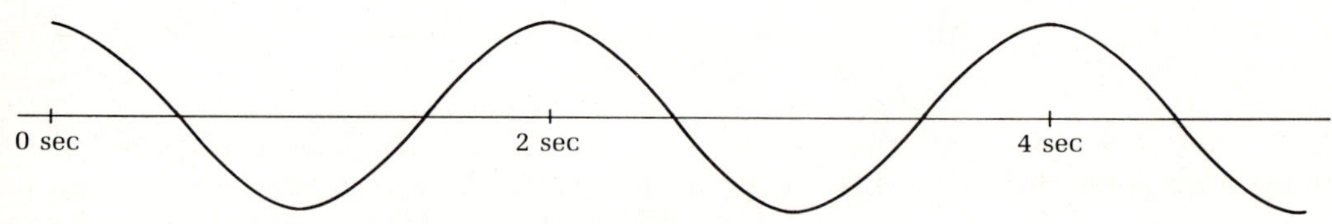

SOUND AND LIGHT

Pongileoni's bowing and the scraping of the anonymous fiddlers had shaken the air in the great hall, had set the glass of the windows looking on to it vibrating; and this in turn had shaken the air in Lord Edward's apartment on the further side. The shaking air rattled Lord Edward's membrana tympani; the interlocked malleus, incus and stirrup bones were set in motion so as to agitate the membrane of the oval window and raise an infinitesimal storm in the fluid of the labyrinth. The hairy endings of the auditory nerve shuddered like weeds in a rough sea; a vast number of obscure miracles were performed in the brain, and Lord Edward ecstatically whispered "Bach!"

ALDOUS HUXLEY

Light, my light, the world-filling light;
the eye-kissing light, heart-sweetening light.
Ah the light dances, my darling, at the
center of my life; The light strikes, my
darling, the chords of my love; the sky opens,
the wind runs wild, laughter passes
over the earth.*

RABINDRANATH TAGORE

■ 12.1 The Physics of Sound and Music

One thing should be clear. When we talk about sound, we are talking about sound waves. This is to be distinguished from another usage of the word meaning the sensation produced in the ear by a sound wave. Physiologists and psychologists interpret sound in this latter way.

*Collected Poems and Plays, by Rabinranath Tagore, from Gitanjoli, LVII. Copyright 1913 by Macmillan Publishing Company. By permission of the Trustees of the Tagore Estate and Macmillan, London and Basingstoke.

For audible sound we commonly employ the word **pitch** to signify whether a particular sound is shrill or dull. High-pitch sound is shrill, whereas low pitch has a dull sound. Pitch and frequency have almost a one-to-one correspondence. If a particular sound has a higher frequency than another, the ear will perceive the first to be higher in pitch as well.

Another characteristic of audible sound concerns the physiological concept of **loudness.** Loudness of a particular sound as it strikes the ear is connected with the intensity of the sound, but in a complicated manner. Moreover, loudness is a subjective property of sound, whereas intensity is defined completely objectively (in terms of the square of the amplitude of the sound wave). Loudness even depends to some extent on the frequency of the sound being heard. Relative loudness of sounds as heard by our ears is measured in units of decibels (abbreviated dB). In the decibel scale barely audible sound, at the threshold of hearing, is assigned zero decibel.* A one-decibel difference in loudness is pretty close to the limit of the difference of sound level that our ears can discern.

Table 12.1 shows the decibel rating of some common sounds and noises. It also gives the intensity of each sound relative to the barely audible sound at the bottom of the scale. A comparison of the decibel rating and the actual relative intensity shows the peculiarity of the decibel scale, or rather the peculiarity of human hearing. When two sounds differ by 10 on the decibel scale, their actual intensities differ by a factor of 10. A difference of 30 decibels indicates a difference of sound intensity by a factor of 10^3, or a thousand. And finally, sound with a decibel rating of 120, which can be produced by a rock band with amplifiers a few feet away from you, is actually a trillion times (10^{12}) more intense than the softest sound that we can hear. Incidentally, 120 decibels are about all we can take. Beyond this, hearing is painful. It is interesting to note that hearing on the decibel scale—that is, perceiving loudness rather than the actual intensities—enables our ears to accommodate a large range of intensities of sound.

Let us mention some other interesting aspects of the sounds involved in that basic human characteristic, speech. Speech does not correspond to a single frequency, which is hardly surprising. However, vowel sounds are closer to being of a single sound frequency than are those of consonants. It is also interesting that the average sound energy involved in giving a one-hour lecture, for example, is less than a joule. No wonder many people consider words to be cheap. They sure are, from an energy point of view.

One unusual feature of human voices is that other people hear us differently than we hear ourselves. If you don't believe this, just listen to your voice on a tape recorder. You'll be surprised how much weaker you sound. The reason for this is that when our speech sound is conducted through the air, some of the low-frequency components are lost. When we hear ourselves, on the other hand, the sound is conducted not only through the air but also through the bone of the skull, which, being a better conductor than air, does not lose any of the components. A tape recorder, of course, records what is conducted through the air and therefore records the weaker voice, which is familiar to everybody but the originator of that voice.

---

*The decibel scale is an example of "logarithmic" scale.

Source of Sound	Loudness (dB)	Intensity of the Sound Relative to the Intensity of Barely Audible Sound
Barely audible sound (threshold of hearing)	0	$1(10^0)$
Rustle of leaves	10	$10(10^1)$
Whisper	20	$100(10^2)$
Radio (playing quietly)	40	$10,000(10^4)$
Ordinary conversation	60	10^6
Busy street traffic	70	10^7
Auto interior (moving at high speed)	80	10^8
Inside of a subway	90	10^9
Noisy kitchen	100	10^{10}
Power mower	110	10^{11}
Indoor rock concert (amplified)	120	10^{12}
Threshold of pain	120	10^{12}
Jet plane (100 ft away)	140	10^{14}

Noise and music

When a dog barks, or when we slam the door or turn on a water faucet, the sound we hear is characterized by irregularity and unpleasantness. These are examples of what we call **noise.** In contrast, when we pluck a guitar string or strike the key of a piano, we create a sustained note of definite pitch which the ear enjoys, and we call it music. Thus noise is made of random irregular vibrations of a source, whereas music is made of regular repetitive sustained vibrations of a source. We cannot be any more specific than this. Obviously there is a large gray area, and the distinction sometimes gets blurred. For example, some people insist that rock music is all noise, whereas others think it musical.

Even for notes made by musical instruments, certain combinations are more pleasant to hear than others. We now know that all the well-known *harmonious* combinations of single frequency notes that are most pleasant have frequencies in the ratio of small whole numbers. One example of such a combination is a note with another an *octave* higher: the ratio of frequencies is 1:2 in this case. Another harmonious combination is the fifth, where the frequency ratio is 2:3, again a ratio of two small whole numbers. A still more important combination is the major chord, where three notes in the ratio of 4:5:6 are combined.

Musical tones are characterized by their loudness and pitch. In addition, there is a third characteristic that allows us to discern the music coming from two different musical instruments or even the music played on the same instru-

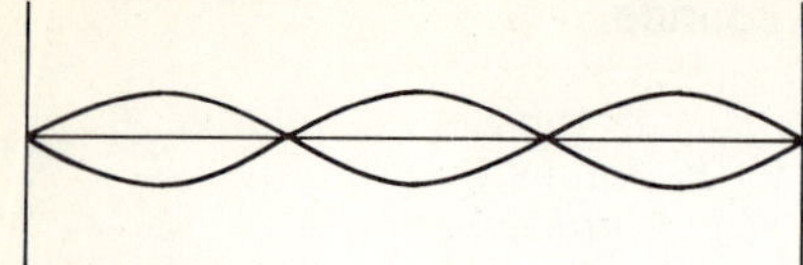

FIGURE 12.1 Only a wave that can fit an integral number of half wavelengths within the length of the string gets set up on a guitar string.

ment by two different people. This characteristic is referred to as **quality,** and the way in which it arises is most interesting.

Stationary waves

Musical instruments produce a variety of musical sounds. How is this accomplished? How are we able to distinguish between the sounds coming from two different instruments?

Let us consider a stringed instrument—for example, a guitar. The guitar string is tied down at its two ends. When the guitar string is plucked, the waves on the string travel to the ends and are reflected so that there are waves traveling in both directions on the string. These traveling waves will now encounter each other and interfere with each other. What is the result of the interference? The answer to this question provides us with the key to understanding the sounds from musical instruments.

If the waves traveling in opposite directions along the string are out of phase with one another, the resultant wave pattern should be rather small in amplitude because of destructive interference. But this is just not true! The guitar string can produce as loud a sound as is desirable. What's more, the sounds that come from guitars have a definite pitch, provided the correct length and tension of string are used.

This is the way it works. The trains of traveling waves combine in such a fashion that only vibrations of a definite frequency are sustained; all others very nearly cancel out. The vibrations that survive are the ones that can fit an integral number of half-wavelength sections (as shown in Fig. 12.1) in the given length of the string. If this happens, a simple relationship develops between the waves traveling in opposite directions.

Suppose at some instant two such waves traveling in opposite directions are in phase, as shown in Fig. 12.2, and produce a reinforced wave. A quarter of a cycle later, one wave will have moved to the right and the other to the left. Now they are out of phase and cancel each other out. But another quarter of a cycle later, the two will be in phase again and reinforce each other. The result is a wave, because the particles of the string are undergoing a periodic displacement as time passes, but the wave itself is not going anywhere. Such a wave is called a **standing** or **stationary wave.** The crucial point in the creation of standing waves is that they are waves in confinement. Whenever waves are produced in confinement, be it a string or an organ pipe or a coffee cup, stationary waves are set up.

It is easy to see now that waves that can fit one, two, three, and so on, integral number of half wavelengths between the boundaries of the string will set up stationary waves. The smallest-frequency standing wave that fits one-half wavelength within the length of the string is called the **fundamental** mode, or the first harmonic; the rest are second, third, and so on, **harmonics** (Fig. 12.3). To find the relationship between the frequency of the fundamental and that of the higher harmonics, we proceed as follows. For the fundamental mode, the length of the string, L, contains half a wavelength, $\lambda_1/2$ (subscript 1 for the wavelength λ denotes the first harmonic). Thus

$$\frac{\lambda_1}{2} = L \qquad \text{or} \qquad \lambda_1 = 2L$$

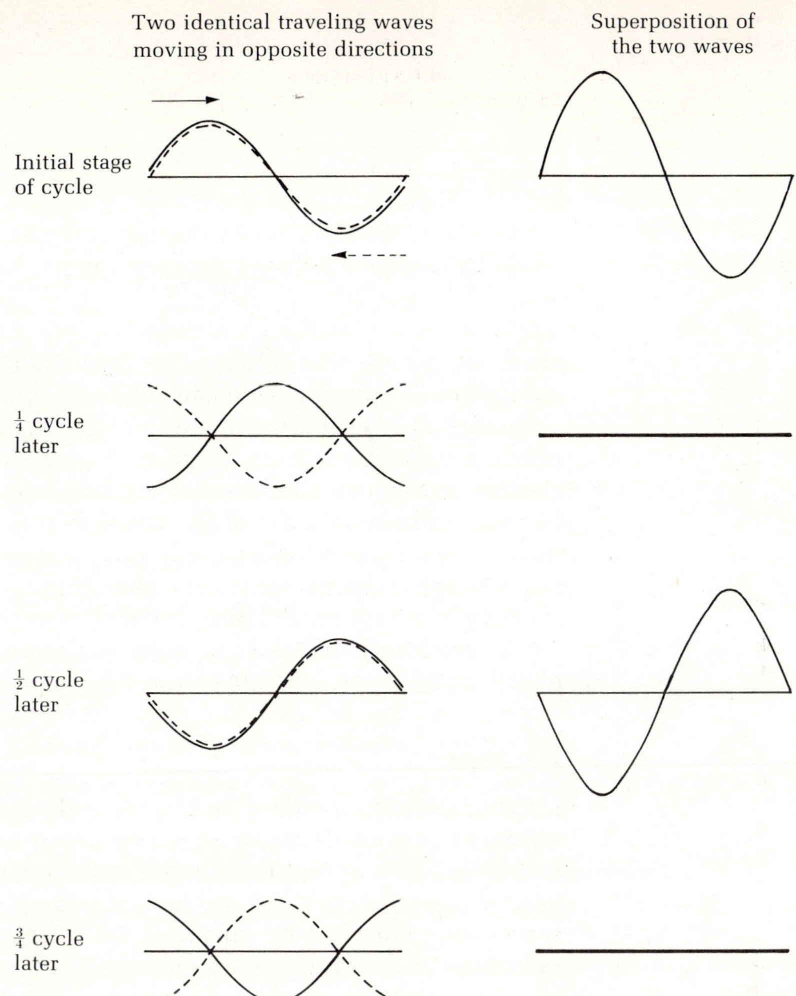

For the second harmonic, L contains two half wavelengths. Thus

$$2\left(\frac{\lambda_2}{2}\right) = L \qquad \text{or} \qquad \lambda_2 = L$$

where λ_2 is the wavelength of the second harmonic. Similarly, for the wavelength λ_3 of the third harmonic,

$$3\left(\frac{\lambda_3}{2}\right) = L \qquad \text{or} \qquad \lambda_3 = \frac{2}{3L}$$

Thus the wavelengths are related in the ratios $2L : L : \frac{2}{3}L$, or $1:\frac{1}{2}:\frac{1}{3}$, and so on. Since the frequencies are inversely proportional to the wavelengths, the frequencies are found to be related as $1:2:3$ and so on.

The quality, or timbre, of an actual musical note depends on how rich the sound is in the higher harmonics, although the pitch is determined by the fundamental mode. If you were to bow a violin string, perhaps the sound would

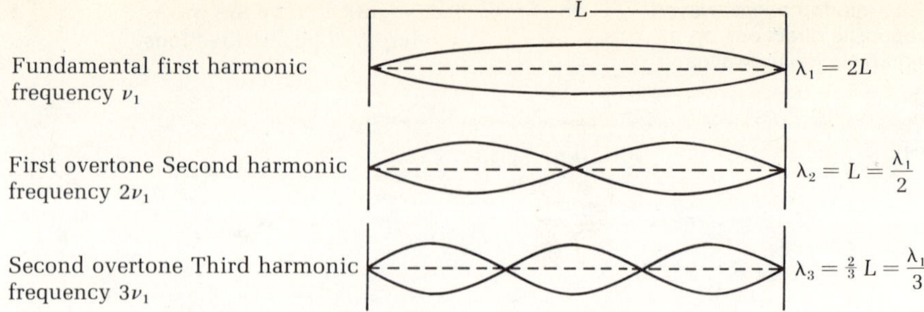

FIGURE 12.3 The first few modes of vibration of a guitar string.

not be as good as that achieved by Yehudi Menuhin. Why? Because of the rich admixture of harmonics he can produce.

Quality also is the factor that enables us to distinguish the sound of one musical instrument from another. If you strike middle C (fundamental frequency 262 Hz) on the piano and pluck it on the guitar, you can differentiate between the two because of the vast difference in quality due to the difference in the admixture of higher harmonics. However, if you listen to the same two notes after the sounds have been taken through an acoustical filter that filters off every higher harmonic of middle C (for example, a filter that cuts off all frequencies above 500 Hz will do), the notes will sound exactly alike. You won't be able to tell which note is coming from which instrument.

The laser

Light waves also interfere, so can we make stationary waves of light by confining them in a small region of space? Such devices have been made; they are called lasers. A mirror, only partly reflecting (so that a lot of light leaks out), is used at one end, the other end having a completely reflecting mirror. Stationary waves are generated by virtue of the interference of the traveling waves (as discussed in the previous subsection), but for only those waves that are exactly aligned with the axis of the device. The result is a unidirectional narrow beam (Fig. 12.4). Also, in this case, the number of wavelengths that fit inside the length of the tube is very large, of the order of hundreds of thousands.

One of the most interesting things about laser beams is that they are straight and narrow as they come out of the apparatus, and they more or less stay that way. This is very different from ordinary light beams. The intensity of the ordinary light beam falls rapidly with the distance, following an inverse square law:

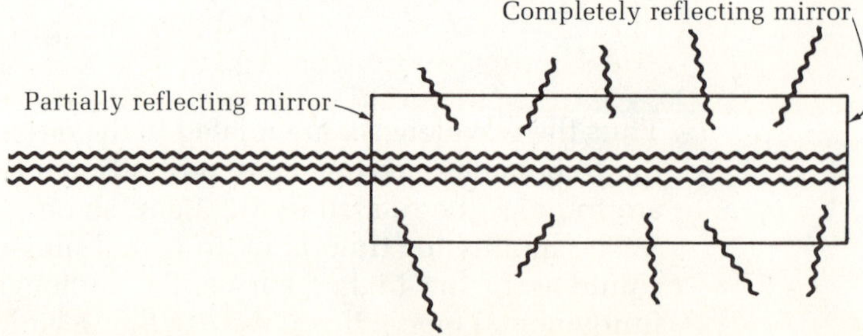

FIGURE 12.4 The laser beam is very well aligned with the axis of the laser tube because of the stationary wave principle.

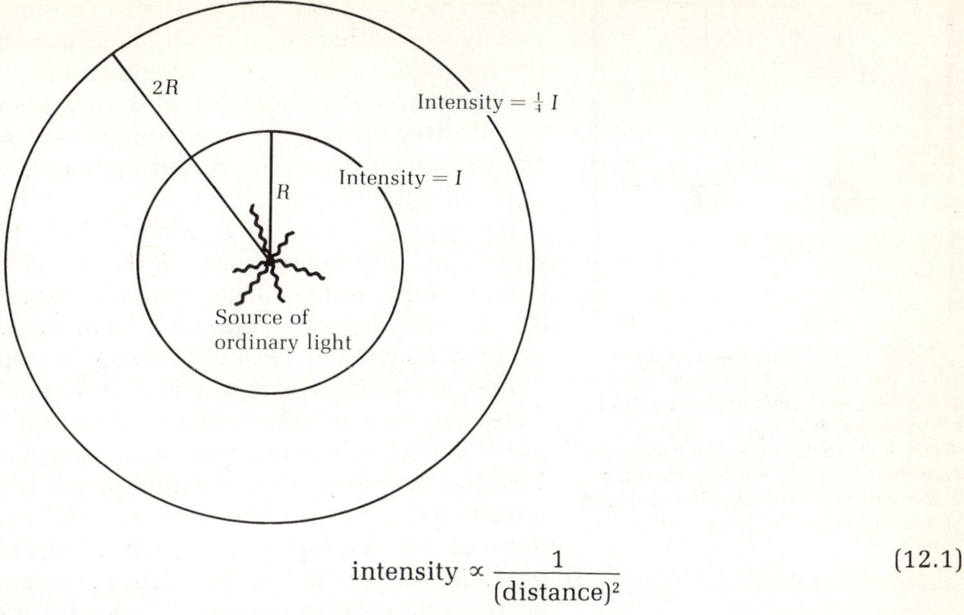

$$\text{intensity} \propto \frac{1}{(\text{distance})^2} \qquad (12.1)$$

The reason for the inverse square decrease is simply that at a distance R from the source, the power (energy per unit time) of the light beam is distributed over the surface of a sphere of area $4\pi R^2$, which is the area of a spherical surface of radius R (Fig. 12.5). Thus each unit of area at the distance R gets only $1/4\pi R^2$ of the power. Therefore, the intensity, which is power per unit area, decreases as

$$\frac{1}{R^2}$$

The intensity is inversely proportional to the square of the distance R from the center of the sphere, the light source. This is actually true for all spherical wave sources.

So one reason the laser beam retains its intensity so well with distance is that it does not spread as it travels, whereas the light beam from an ordinary source does. The range of the laser beam consequently is spectacular. A laser beam starting from earth can illuminate the surface of the moon. Experiments have been done to determine the exact position of the moon at various times by using range finding with a laser beam. We'll discuss more on lasers later.

Resonance

Have you ever pushed a child in a swing? Particularly one who insists on going higher and higher? It's not especially hard work for you. What you do is simply time your push so that the time interval (between successive pushes) is the same as the period of the natural vibration of the swing. If the frequency of your push matches that of the natural vibration of the swing, imparting energy becomes easy and the energy transfer is maximum. This is the process of **resonance.**

Resonances play an important role in the area of musical instruments, partic-

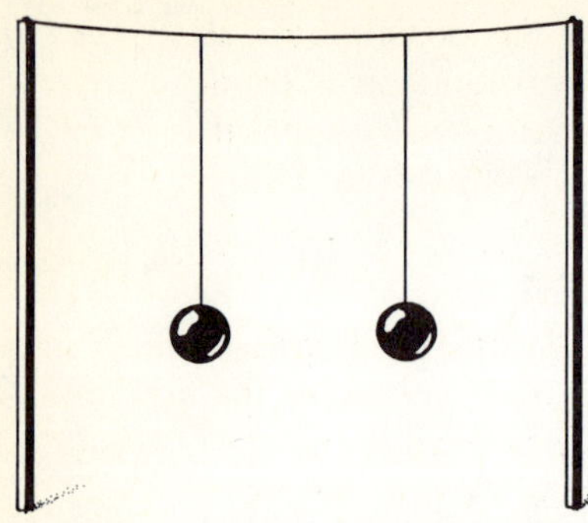

FIGURE 12.6   When the strings are equal, the frequency of natural vibration of the two pendulums are equal. They resonate and their energy sharing is maximal. If we start one into oscillation, very soon the other will take up some of the energy. The energy is then found to alternate between the two pendulums.

ularly the wind instruments. The air column of the instrument is put into vibration by some other vibrating object, such as the lips or reeds. If the vibrations of the lip and air column match in frequency, the resonance condition is satisfied and the result is a pleasing sound. New resonance conditions can be achieved by adjusting the length of the column, which changes its natural frequency. For the stringed instruments, sounding boards are used to amplify the vibrations, again through the use of the process of resonance.*

Have you seen the experiment in which the sound of a musical voice breaks a glass? This is resonance again. If the sound wave has the same frequency as the natural vibration frequency of the glass tumbler, even the tiny pushes of the air can set up resonant vibrations large enough to break the glass. Marching soldiers are usually allowed to go out of step on a bridge, in order to avoid the danger of causing resonant vibration of the bridge.

Resonance is an important aspect for other kinds of waves as well. For example, it plays a role in microwave cooking of food in suitably constructed ovens. You may know that 85% of animal tissue is water. The water molecules have a natural rotational mode of oscillation, with a frequency of roughly 10^{10} Hz, the same as that of a typical microwave. Thus when microwave radiation falls on food, the water in the food absorbs the energy via resonant absorption, thereby heating and cooking the food. The metal of the oven has no water, so it never gets hot. Of course, you have to be careful not to put your hand in the way of the microwaves, or it too will be cooked. So all microwave ovens come with the safety provision that they cannot be opened while the oven is in operation.

Now you can try an experiment. Tie two identical balls with strings of equal length and hang them from a taut horizontal rope, as shown in Fig. 12.6. Set one to oscillating. You will find that very soon the other ball will start to oscillate also. It will take up more and more of the energy of the first ball. Eventually we have a situation in which the first ball comes to rest and the second ball oscillates alone. The process is then reversed; the first ball now gains back the energy slowly, and so on. The energy goes back and forth between the two balls.

Now try the same experiment with balls tied with strings of different lengths. The energy sharing now will be very different and much less spectacular.

What is happening? The length of the string determines the frequency of a pendulum. So the two balls with identical string lengths have matching frequencies. They resonate and therefore share their energy well. On the other hand, when the lengths of the strings are different, there is a frequency mismatch, and no particular tendency for energy sharing is exhibited. The key here is resonance. If there is resonance, there is more energy sharing or interaction.

You can go one step further to philosophize and apply the resonance principle (in analogy, of course) to human interactions. Have you ever noticed that with some people you interact optimally and with others very poorly? Maybe the reason is that in one case you are resonating and in the other case you are not. Of course, human minds, being such complicated systems, should possess

---

*Actually, sounding boards often do not use a well-defined resonant frequency. Even away from resonance, a sounding board can intensify the music; this is called **forced vibration.** You may have noticed the vibration of the floor of a machine shop caused by the operation of the heavy machinery; this is another example of forced vibration.

many natural modes, and by adjustments it ought to be possible to satisfy the resonance condition between any two people, if the intention is there. To go even further, we can speculate that when one develops this capability, one could discover a feeling of oneness with all people and even all things.

■ 12.2 Optics: Images, Illusions, and Color

We have talked about images of objects produced by mirrors as well as images formed by total internal reflection. Most commonly we use lenses for making images: in the camera for photography, in the telescope for distant objects, and in the microscope for looking at very small objects. What are the properties of lenses that lend themselves to such varied uses?

The human eye itself has a lens. How does the eye perform its function of seeing? Is it true that the image formed on our retina is upside down? Is the image seen by our eyes a faithful copy of the object? Most often it is, but sometimes we see things differently, not as they are. This is the subject of optical illusions, which we will discuss in some detail. Finally, what gives rise to the color of an object? Does light of certain frequencies really have distinctive colors or is it all in our brain? These too are questions that we will deal with in this section.

Lenses and optical instruments

Lenses are a practical application of the refraction of light, as distinguished from mirrors, which use reflection of light. Lenses use refraction to produce images that in many respects are more versatile than the ones made by mirrors. This is the reason that lenses have found such diverse applications as in magnifying glasses, telescopes, microscopes, binoculars, and cameras.

Fig. 12.7 shows a cross section of a converging, or **convex,** lens. The line drawn through the centers of the two faces of the lens is called its axis. Rays parallel to the axis, after passing through the lens, converge on the other side at a point called the **focal point** of the lens. The distance of this point from the center of the lens is a very important feature of a particular lens and is called its **focal length.**

When we construct the image of an object that is produced by a converging lens, we find some interesting new features. Suppose the object is placed as shown in Fig. 12.8, beyond the focal point on one side (incidentally, the lens behaves the same way from both sides). The rays coming from each point of the object will be made to converge to a point on the other side of the lens, forming an image point. The image is the combination of all these image points. The simple recipe for finding the image of an object formed by a lens is as follows. For convenience, we use an arrow placed vertically on the lens axis as the object (Fig. 12.8). First, draw a ray from the arrowhead parallel to the axis. This ray will emerge on the other side so as to pass through the focus. Second, draw a ray from the arrowhead again through the center of the lens. This ray proceeds through the lens undeviated. The intersection of these two rays gives the image point of the arrowhead. If you drop a perpendicular from the image point of the arrowhead on the axis of the lens, you get the entire image of the arrow. Note that the central ray from the tail of the arrow also propagates undeviated since it

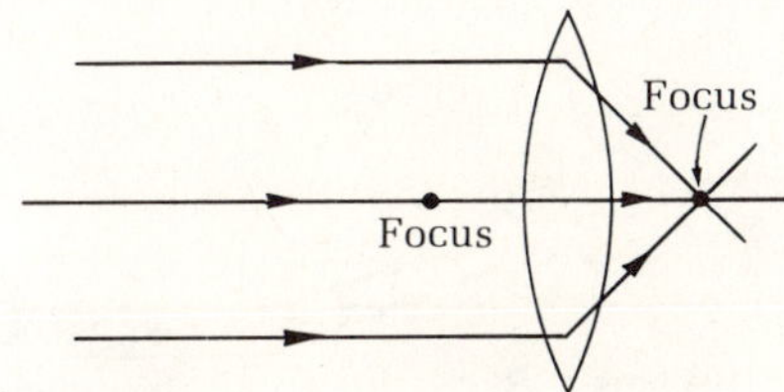

FIGURE 12.7 The cross section of a convergent lens, showing its focal points. Rays parallel to the axis converge onto a focal point.

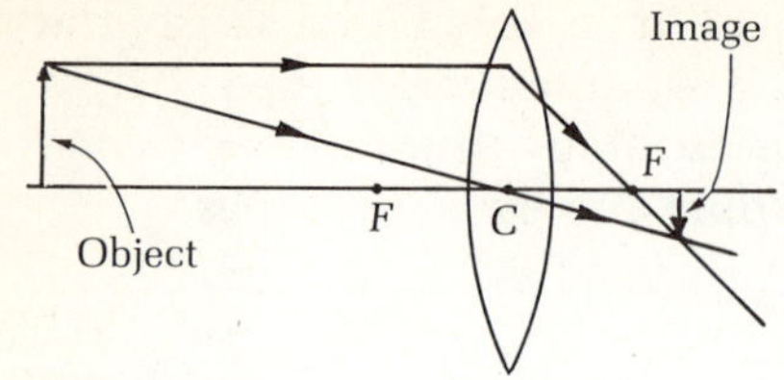

FIGURE 12.8   Ray diagram for finding the image of an object placed beyond the focal point. The image of the arrowhead is the point of intersection of the parallel ray, which passes through the focus *F* on the other side, and the ray through the center *C*, which passes undeviated. See text for further explanation.

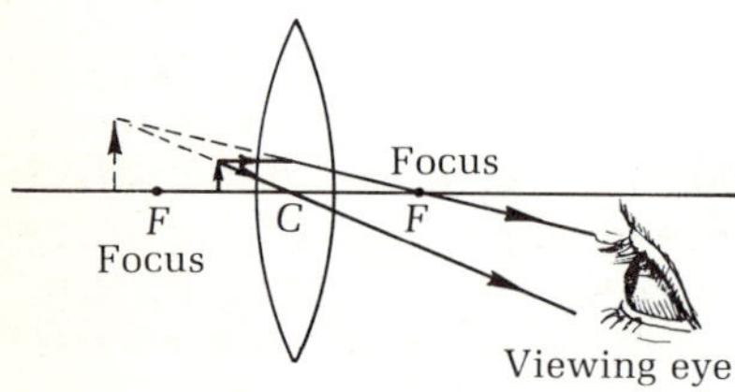

FIGURE 12.9   A converging lens as a magnifying glass. When the object is placed between the center *C* and the focal point *F*, the rays from the arrowhead, according to our previous recipe, diverge from each other on the other side of the lens. However, when extrapolated backward, these divergent rays *appear* to come from an image point on the same side as the object. The image is virtual in this case. It is also magnified and right side up.

FIGURE 12.10   The image formed by a microscope. The image formed by the objective lens acts as the object for the eyepiece. Notice the increased angle of vision offered by both the intermediate and final images, which lets us see much more detail of the specimen.

is incident on both the surfaces of the lens at right angles. Images for more complicated objects can be drawn using the same principles for each point of the object.

The image in Fig. 12.8 is a different sort of an image than that produced by a plane mirror. It is called a **real image,** one we can put on a screen. In contrast, we refer to an image that we cannot put on a screen as **virtual.** The image formed by a plane mirror is always virtual.

Note that the image formed by the lens in Fig. 12.8 is upside down. And it is true that our eye—which also uses a convex lens to make an image of the object it sees on its retina—produces an upside-down image. This was verified first perhaps by Descartes himself, who experimented with the eye of an ox. He scraped the back of the eye to make it transparent and observed the inverted image of objects on the retina. But seeing is more than the image formation. When the brain processes the information, it corrects the "fault" of the eye lens.

If the object is placed very close to the lens, within its focal length, when we construct its image we find that it is on the same side as the object. It is also a virtual image, but one quality makes it worthwhile: it is greatly magnified (Fig. 12.9). Also the image is right side up. Thus a magnifying glass, which produces a magnified image of an object, can be used as a reading glass.

The microscope combines the function of the magnifying glass (which is used as the eyepiece of a microscope) with that of another lens called the **objective,** which has a very short focal length. This lens combination produces an enlarged virtual and inverted image of a close object (Fig. 12.10). In this case the object is placed at such a distance from the objective lens that the image formed by the objective (which acts as the object for the eyepiece) is real and somewhat enlarged already. This allows a much larger angle of vision than the object itself does—and this is the key factor. The lens at the eyepiece then increases the angle of vision even further. This increased angle of vision enables us to see more details of the object.

The telescope functions almost the same way as the microscope does except, of course, that it is designed to see distant objects. This is accomplished by using an objective lens with a large focal length (Fig. 12.11). The image in Fig. 12.11 is inverted, which is fine when the telescope is used for viewing astronomical objects. For terrestrial telescopes the image is "corrected" by the use of an additional lens (Fig. 12.12).

Finally, there is also the diverging lens (Fig. 12.13), which diverges rays inci-

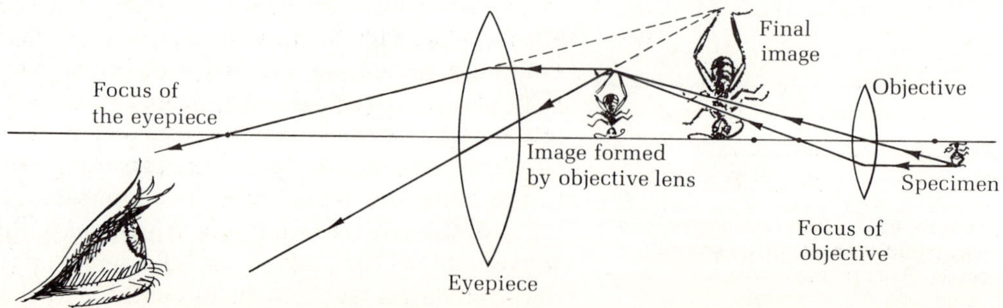

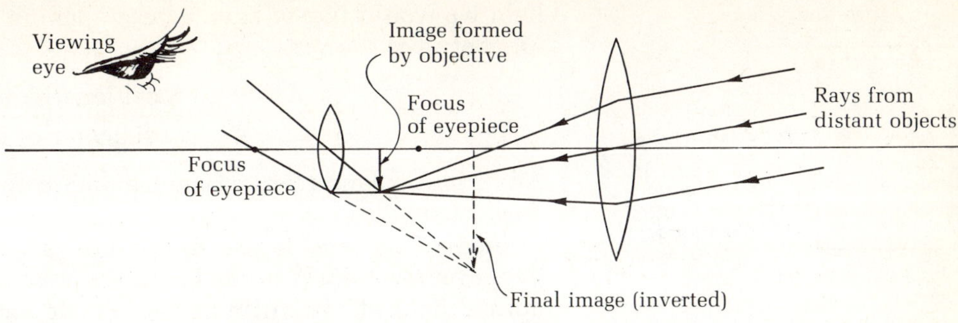

FIGURE 12.11 The image formation by an astronomical telescope. The final image formed by the lens combination is inverted.

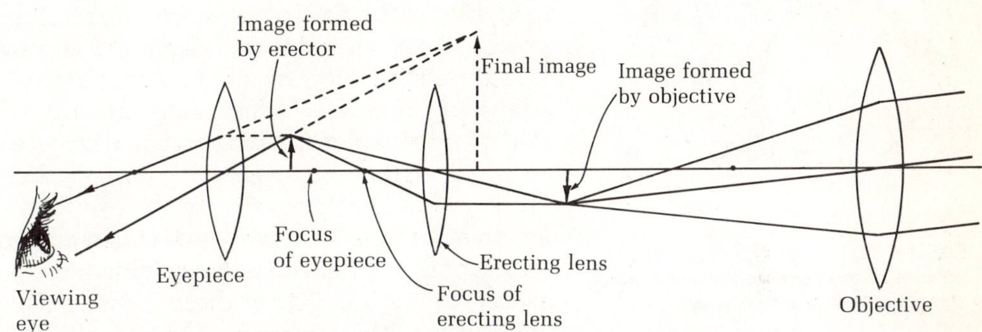

FIGURE 12.12 The image of a terrestrial telescope, or spyglass. Using a third lens, the inverted image of the astronomical telescope is straightened out.

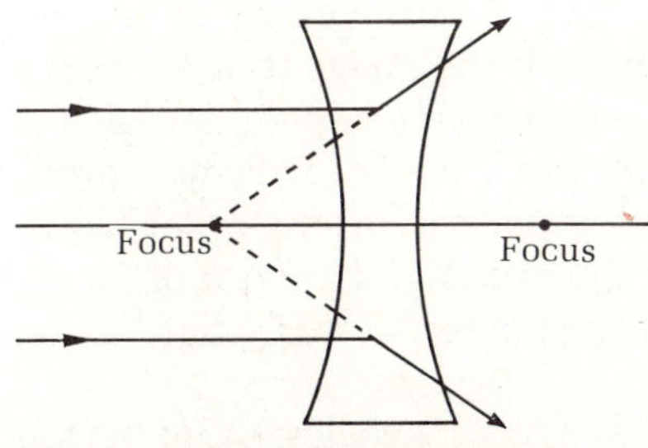

FIGURE 12.13 The divergent lens. Rays parallel to the axis incident on the lens diverge upon emerging on the other side. When extrapolated back, they appear to come from a point on the axis, which defines the focal point of the lens.

dent on it that are parallel to its axis. A diverging lens finds use as the correcting glass for nearsightedness. This will be discussed a little later.

The camera and the eye

We will focus on the camera in some detail, for two reasons. First, you probably have some experience with it. Second, the camera and the eye have some interesting similarities, which are important to point out.

In its essence the camera is a very simple device. It is a lighttight box with a shutter that can let light in only as desired. The light passes through a lens and falls on a light-sensitive film, which records it. If we are photographing an object, we want the image to be at the position where the film is in the camera. Of course, this is a slight problem, since objects at different distances from the lens give rise to images also at varying distances from the lens. So we must have a way of varying the distance of the lens from the film, a process we call focusing. Expensive cameras have it. In inexpensive cameras we have a compromise. The distance of the lens from the film is chosen so that pictures of objects within a certain range of distances can be taken.

Expensive cameras also come with several other adjustments. One is the shutter speed. Ordinarily you need to open the shutter only for a tiny fraction of a second to admit enough light into the camera—the film is that sensitive to light. But on a cloudy day when the light is dim, you need to prolong the exposure to light, and this can be done by adjusting the shutter speed.

Another useful adjustment, called the iris diaphragm, regulates the amount of light that passes through the lens onto the film [(Fig. 12.14 (a)]. The iris has an adjustable opening. In bright light we use a small opening of the iris. But in dim

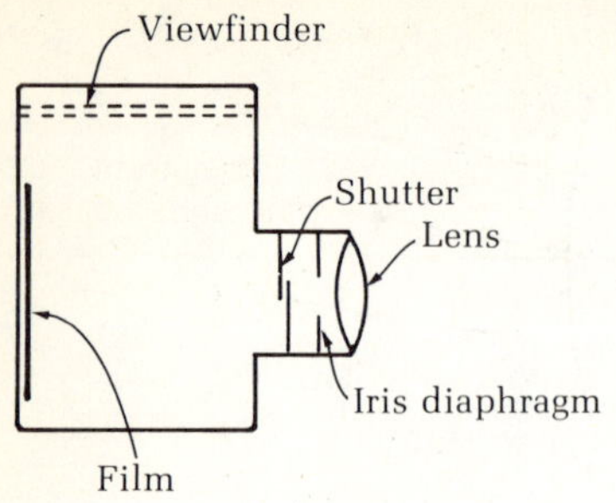

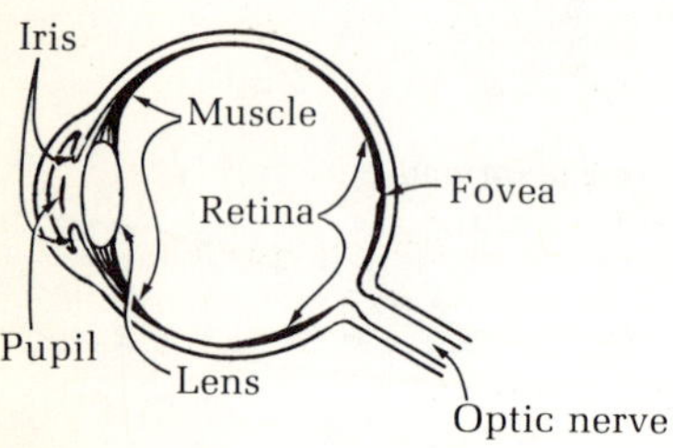

FIGURE 12.14 (a) The different components of the camera. (b) The different components of the human eye.

light we would like to have a large opening. You may have heard of the *f*-stop number; it is the ratio

$$\frac{\text{focal length of lens}}{\text{diameter of iris}}$$

An *f*/2 lens means a lens whose maximum opening is half the focal length of the lens.

Finally, the focal length of the lens is vital for determining the size of the image on the film. A normal lens has a focal length that fills the film with the normal field of view of a person. A telephoto lens, on the other hand, has a larger than normal focal length; it is designed to magnify the object with a sacrifice of the field of view. In contrast, the wide-angle lens is designed with a smaller than normal focal length and accommodates a wider field of view.

A simplified diagram of the eye is shown in Fig. 12.14(b). It has a lens, an adjustable opening—which automatically varies the amount of light let in (the iris is the name of the diaphragm whose opening is called the pupil)—and a retina to receive the image. Interestingly, here the distance of the lens from the retina is fixed by necessity, so the eye focuses its object by changing the focal length of its lens. This process is called **accommodation** and is accomplished by means of the eye muscles. Nearsightedness and farsightedness are due (to some extent) to the inability of these muscles to perform properly.

The film of the camera has light-sensitive chemicals on it. Likewise, the retina of the eye has light-sensitive receptors. But the analogy ends here. The receptors are able to do so much intricate gathering of information so fast that the process is more like a motion picture, but without the change of film. Even more amazing is the processing of the information that the receptor cells collect. It is now known that the nerve cells carry the messages in the form of electrical signals. The eye has the backing of a very complex and surprisingly compact electrical information network in performing its function.

Defects of vision and their correction

Two common defects of vision are nearsightedness (myopia) and farsightedness (hypermetropia). Whereas the normal eye produces sharp images of objects right on the retina, the eye of a shortsighted person forms images of distant objects in front of the retina, so that on the retina itself the image is out of focus (Fig. 12.15). If the object is close to the eye of a nearsighted person, the image is formed close to the retina. Thus a nearsighted person does not have any problem seeing well at close distances. For a farsighted person, on the other hand, the images of nearby objects are formed behind the retina and are out of focus. Thus the farsighted person has difficulty seeing near objects clearly but has no problem seeing distant objects.

Nearsightedness arises, for example, from having an eye lens that is too curved, producing too much convergence of the light coming from distant objects. A divergent lens corrects this defect. The combination of the divergent lens and the overconvergent eye lens restores the image to its proper position on the retina [Fig. 12.15(b)]. Farsightedness, on the other hand, arises from the eye lens lacking convergence; thus another convergent lens must be added to correct this defect of vision [Fig. 12.15(c)].

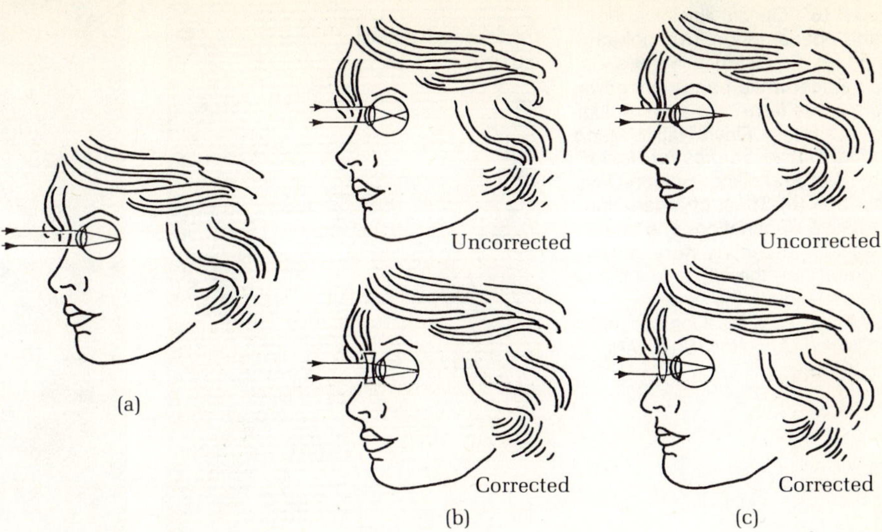

FIGURE 12.15 Defects of vision and their correction. (a) Normal vision: the image always forms on the retina. (b) The image of a distant object formed in a nearsighted eye falls short of the retina. The lens is too convergent. The defect is corrected using a divergent lens. (c) The image formed in the eye of a farsighted person falls behind the retina. There is too little convergence in the eyelens. The defect is corrected with added convergence from a convergent lens.

Optical illusions

Why are we able visually to distinguish an object from its surroundings? Only because it either reflects or—if it is a luminous object—emits an amount of light that is significantly different from its surroundings. You may have read H. G. Wells's *Invisible Man*. The man became invisible because a potion changed his body's transmission of light. If the body could transmit light—as does the surrounding air—instead of reflecting it, we could not tell that the body was there.*

Thus the contrast between the object and its surroundings gives the object a boundary. Likewise, the perceived image must have a boundary also. Now we can ask if this image is an exact copy of the object. Do the "light and shade" of the perceived image correspond exactly to that of the object? Several examples have been found by scientists which show otherwise.

Fig. 12.16 shows some of these striking examples of things the eye sees quite differently, as far as the contrast of light and dark is concerned. Fig. 12.16(a) shows the **Mach band** phenomenon, discovered by physicist philosopher Ernst Mach. Horizontal black lines are constructed with constant thickness from the middle to the left; from the middle to the right, they are gradually thickened. Look at the pattern from a distance and you will see a white vertical band right down the middle. This is the Mach band, but where did it come from? It turns out that the Mach band is a physiological effect, which originates from the fact that the light receptor nerve cells on the retina are coupled to each other, exerting inhibitory influences. When observing a contrast situation, these "lateral inhibitions," as they are called, become important and produce interesting illusions like the Mach bands. Painters, well aware of this phenomenon, are able to create in a viewer the perception of different illuminations between two parts of a picture just by having a dark contour [Fig. 12.16(b)].

---

*Of course, you realize that the "invisible man" is also blind because his eye lens, having the same refractive property as air, cannot bend light to make images.

(a)   (b)

Even more startling examples of optical illusions are found in the area of size distortion. In Fig. 12.17, which rectangle is larger? The top one looks larger, for sure, but measure them; they are exactly equal. What's happening here is an interpretation of the data by our brains, which know that the distant rail ties are as large as the near ones and which therefore automatically enlarge any object lying close to a distant tie. Of course, for real objects this would be true. The construct fools us but makes the point.

Perhaps the most familiar example of size distortion is the case of the apparent large size of the moon on the horizon compared to the size of the moon in the middle of the sky (the zenith position). In photographs the moon always has the same size, yet the eye sees the low moon as distinctly bigger, and thus this is an optical illusion.* You may have heard one popular but incorrect explanation of this, which is that the horizon moon is seen adjacent to other objects in the background (e.g., a house), which the brain knows are big, and therefore it reconstructs the image of the moon to be big also. However, this is not quite accurate, since the effect persists even when we look at the horizon moon across an ocean, in which case there isn't any background object.

The correct explanation was given by Ptolemy of second-century Greece. He suggested that any object seen across a large terrain—such as the moon when it is at the horizon—is perceived to be at a greater distance than an object which is equally distant but seen through empty space. Accordingly, the horizon moon seems further away than the zenith moon, and this produces the illusion that the horizon moon is also bigger (similar to the size illusion in Fig. 12.17).

---

*A good discussion of this is given in an article by L. Kaufman and I. Rock, "The Moon Illusion," *Scientific American*, July 1962, p. 120.

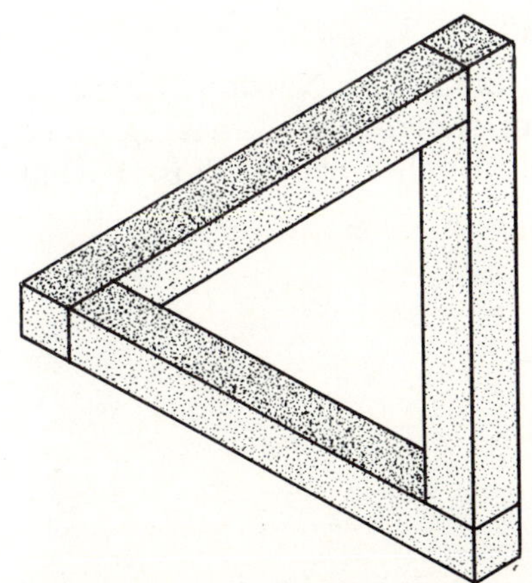

FIGURE 12.18 The impossible triangle of Penrose and Penrose. (From R. L. Gregory, "Visual Illusions." Copyright © 1968 by Scientific American, Inc. All rights reserved.)

Perception of depth

The perception of depth, the third dimension of space, is most intriguing since the eye lens forms an image on the retina that is without question a two-dimensional one. How do we get a three-dimensional image constructed out of one that is two-dimensional? Perhaps the answer is known to you: the fact that we have two eyes has something to do with the perception of depth.

That this must be so has been verified by some very interesting studies. You may have seen the "impossible triangle" devised by Lionel and Roger Penrose, mathematical physicists at the University College, London (Fig. 12.18). It is impossible perceptually to interpret this as any kind of a solid in normal three-dimensional space; indeed, it was constructed to illustrate the possibility of abnormal spaces. In one experiment a researcher constructed an actual three-dimensional solid that creates the same confusion if you see it with only one eye from exactly the right angle.* If you photograph it from that angle, you can recapture what one eye sees [Fig. 12.19(b)], whereas the real structure is as shown in Fig. 12.19(a). But we can never find any orientation from which binocular vision will miss seeing that it is an open solid.

---

*R. L. Gregory, "Visual Illusions," *Scientific American*, November 1968, p. 66.

(a)
(b)

FIGURE 12.19   (a) The actual structure, which gives the semblance of the impossible triangle when seen with one eye from a particular angle without shifting the eye. (b) This is the illusion that one eye sees. (Photograph Philip Clark.)

So binocular vision is crucial to the perception of three-dimensional objects, including their depth. With binocular vision the two retinal images from the two eyes, each individually two-dimensional, are compared texturally and correlated for the relative displacement of similar textural points.* Existence of this relative displacement, or parallax, between points of similar texture is the reason we see depth. With one eye we do this by shifting the angle of viewing.

Color

Poet James Thomson's imagination was amply tickled by Newton's work on color and light, in particular by his experiment showing the breaking up of white light into different colors when passed through a prism. This is what Thomson wrote in 1727:

> . . . First the flaming red,
> Springs vivid forth; the tawny orange next;
> And next delicious yellow; by whose side
> Fell the kind beams of all refreshing green.
> Then the pure blue, that swells autumnal skies,
> Ethereal played; and then of sadder hue,
> Emerged the deepened indigo, as when
> The heavy-skirted evening droops with frost;
> While the last gleamings of refracted light
> Died in the fainting violet away.†

---

*Bela Julesz, "Texture and Visual Perception," *Scientific American*, February 1965, p. 38.
†In "To the Memory of Sir Isaac Newton."

Newton certainly deservedly inspired this ecstasy. He clearly established several important aspects of color. He pointed out that the color of a luminous object is not created by the object itself but arises from the color of the light the object emits. Thus a red-hot piece of iron *is* red because it emits the wavelength of light that we perceive as red.

How about nonluminous objects? Here also Newton gave a very simple explanation for their colors:

> That the colors of all natural bodies have no other origin than this, that they . . . reflect one sort of light in greater plenty than another.

Thus grass is green because, of all the colors of white light, it reflects only the green wavelengths, absorbing most of the rest. Red paint is red because it reflects red; the rest of white light is absorbed by it. If red paint is seen in blue light, there is no red light for the paint to reflect; it absorbs the blue light, so we don't see any light at all and the paint appears to be black. Black color is the result of the absorption of the light of all colors by a body, while an object that reflects all colors appears as white.

All this essentially is correct but it is not the complete story. For example, we can suggest cases where it is clear that color is in the eyes of the beholder rather than in the light itself. A color-blind person, for example, does not see a red light as red.

If you look at a disk of orange light surrounded by an intense white background, it appears to be brown. So color depends also on the intensity of illumination in the region adjacent to the perceived object. The change in the appearance of the color of the moon from daytime (when it is seen in the intense background of the blue sky) to the evening is perhaps the most familiar example of this. The moon doesn't change; our perception of it does.

So rather than being a property of the light itself, the color we see is the sensation the light generates in our eye. And the sensation *can* vary, sometimes drastically. For example, all objects look colorless in very dim light.

If you project beams of light of red, green, and blue on a white screen so that they partially overlap, you can see the various colors of the spectrum in the different parts of the overlapping regions (Fig. 12.20). So all these colors can be produced just by adding red, blue, and green, which is the reason these three

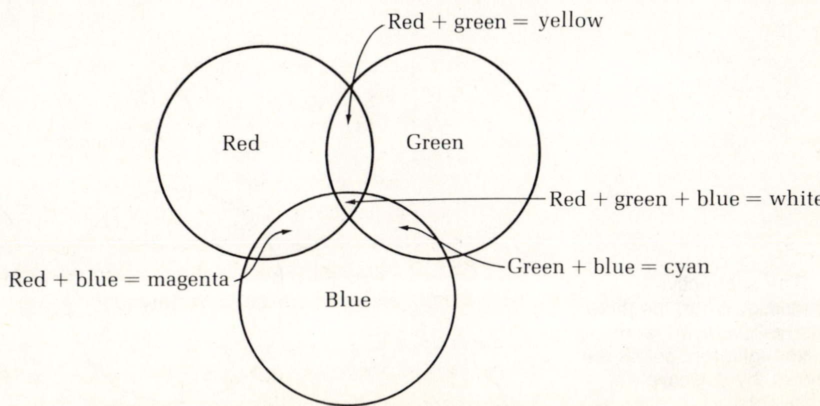

FIGURE 12.20 The additive method of color mixing. When red, green, and blue light beams are projected simultaneously on a white screen, their overlapping sections produce different colors, as indicated. Other hues and shades can also be produced by mixing these additive primaries in different intensities.

colors are called the **primary colors** (sometimes the **additive primaries**). This experiment makes clear that, although a single wavelength appears to our eye as a definite color in normal conditions, the reverse is not true at all. We can produce light of different colors as an additive mixture of others.

Interestingly, notice that a particular blend of red, green, and blue lights also produces the same sensation as white (Fig. 12.20). Actually, we now know that light of only two colors is needed to produce the sensations of white, if they are mixed in the right combination. Two colors that give white are called **complementary** colors. For example, red and cyan (a hue between blue and green, or turquoise), blue and yellow, green and magenta are complementary colors; their combination always produces the sensation of white in the eye.

Before we go on, let's talk about the subtractive method of color mixing, which is what painters do when they mix their paints. Above we said that yellow and blue make white when added, but anybody who has mixed paints knows that yellow and blue make green. So mixing paint colors is different from mixing lights of different colors. It is actually easy to understand why. A yellow paint gets its color because it reflects yellow and absorbs all else, as we said before. It actually also reflects a little bit of the wavelengths adjacent to yellow, namely, orange and green. Similarly, blue paint reflects mainly blue and a little of green and indigo. But now when you mix them up and look at the mixture in white light, as we ordinarily do, the mixture can subtract additional colors from the white light than can the individual paints. For example, now blue will mostly be absorbed by the yellow paint, as will yellow by the blue paint. The color that will be predominantly reflected by the mixture is green, and so the mixture appears green. Thus mixing paints is a subtractive method of color mixing. Cyan, magenta, and yellow are called the three **subtractive primaries;** combining pigments of any two can produce all the additive primary colors (Fig. 12.21).

All these things can be summarized by saying that when we are talking about color, we really are talking about color vision; there is no color without vision. How does the eye see color? First of all, the receptor cells on the retina that respond to the light are divided into two classes, the rod-shaped ones called **rods** and the cone-shaped ones called **cones** (Fig. 12.22). The cones are the cells responsible for seeing color; they reside in the central portion of the retina, the

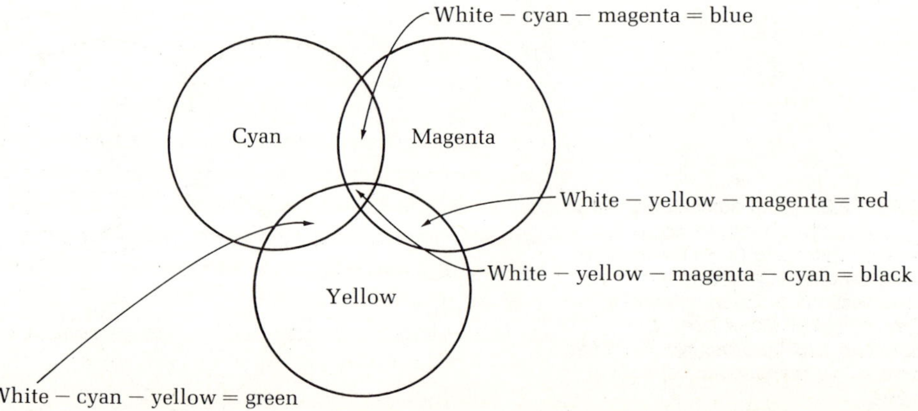

FIGURE 12.21 The subtractive method of color mixing. When the three primary pigments (yellow, cyan, and magenta) are mixed, different colors are produced, as shown, by selective absorption from white light.

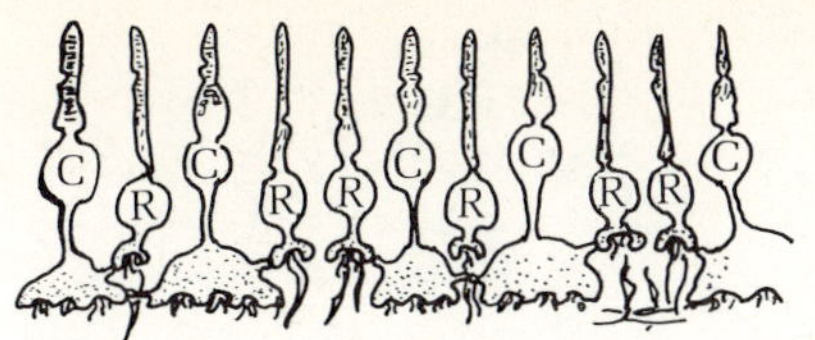

FIGURE 12.22 Rods and cones. C denotes cones and R rods.

fovea. The rods are concentrated at the periphery of the retina, which is one reason we do not see colors in peripheral vision. In dim light only the rods are active, so we do not see colors in dim light. And color-blind people have some sort of lack in their cone cells, producing a varied degree of color blindness.

How do the bright-color-sensitive cones manage to put together the information regarding color? A theory proposed by physicists T. Young and H. Helmholtz suggests that the cone cells have pigments, one for each of the three primary colors, red, green, and blue. When we shine light on the eye, it results in different amounts of absorption of the three different primary colors. This information is now interpreted in the brain and in part in the eye itself to decide which color the eye will see.

The idea that the eye itself does some of the deciphering of the input information is interesting. One of the readily verifiable phenomena it leads to is the enlargement of the pupil (which lets more light in) when the eye sees something interesting. For example, when a woman looks at a picture of a beautiful male, her pupils enlarge (and vice versa for the males).

■ 12.3 Sound and Light Waves from Moving Sources

When a wave train is emitted by a moving source, two rather interesting effects occur. One of them is an apparent change in the frequency of the wave as perceived by an observer. After all, the source is chasing down the waves it emits, so you would expect some changes. The apparent change of frequency is called the Doppler effect.

Another spectacular effect occurs when the source of the waves moves with a speed exceeding that of the waves it emits. This results in the phenomenon of shock waves; sonic booms are perhaps the most well-known example of this. We will discuss this second effect first.

Shock waves and sonic booms

Let us specifically talk about sound waves, although the effect occurs with other waves as well. Because of the advent of the supersonic transport (SST), which travels at speeds greater than the speed of sound (750 mi/h), there has been much discussion of the **sonic boom** recently. How do sonic booms result?

The supersonic transport flies faster than the waves it produces. The crests of a sequence of waves such a source emits overlap in the manner shown in Fig. 12.23 and form a single crest along the lines shown in the figure. This total effect is a wave of very large amplitude and is called a **shock wave.** Since sound waves are three-dimensional, what we get from a supersonic transport is an entire cone of shock waves. This cone is called a sonic boom. What causes the louder sound of the sonic boom? It is the effect of receiving the crests all together rather than a single wave crest at a time, as would occur with a regular sound wave. The energy transported by a sonic boom is so large that it can cause damage to fragile structures in its way.

The velocity of the SST is usually rated in "Mach number," named for Ernst Mach, the scientist philosopher we have mentioned before. The Mach number of an object is the ratio of its speed to the speed of sound. So an aircraft with a

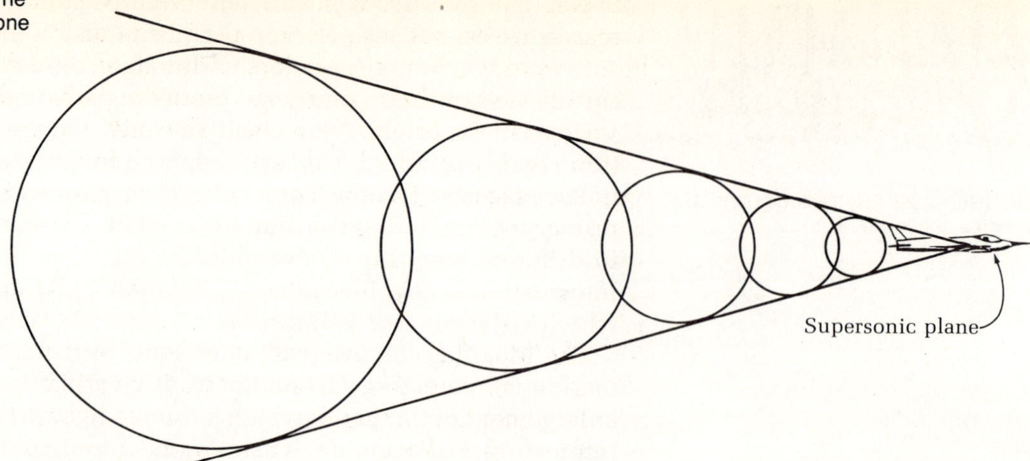

FIGURE 12.23 The sonic boom. The successive wave crests pile up on one another, producing the big boom.

speed of 1500 mi/h is a Mach 2 object; it has a speed twice the speed of sound.

Now we will clear up one common misconception. Sonic booms are not produced only as an aircraft attains a supersonic speed. It is true that an aircraft needs extra energy to overcome the sound barrier while overtaking the speed of sound, but this has nothing to do with sonic booms. The sonic booms are emitted continuously by any SST in the form of the spreading cone of shock waves that it drags along with it. As this cone reaches the ground, an observer encountering the cone will hear the sonic boom. In Fig. 12.24, observer No. 1 has already heard the sonic boom; the cone has passed her. Observer No. 2 is just encountering the cone, and observer No. 3 will shortly receive it.

Doppler effect

Austrian scientist Christian Doppler in the year 1842 discovered a very important effect concerning the perception of light when there is a relative motion

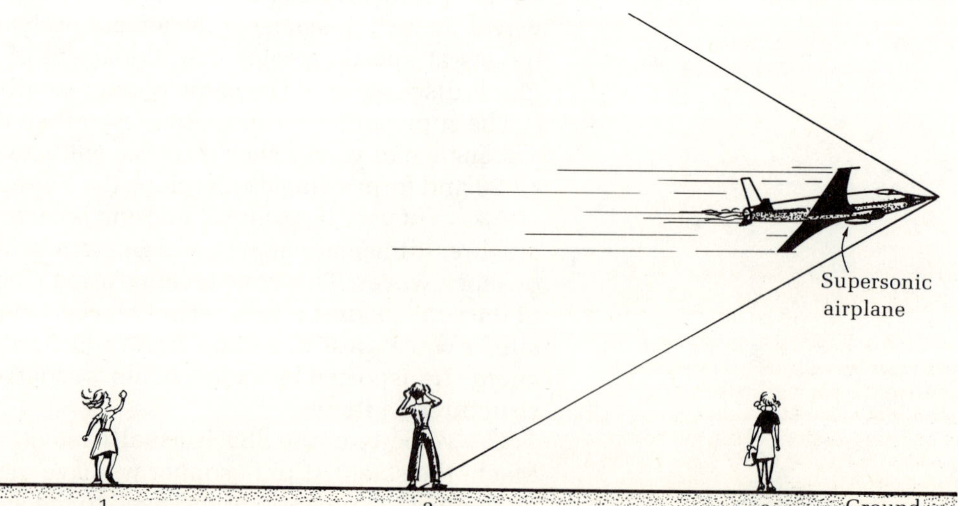

FIGURE 12.24 Observer No. 1 has already heard the sonic boom, observer No. 2 is hearing it, and observer No. 3 will hear it shortly.

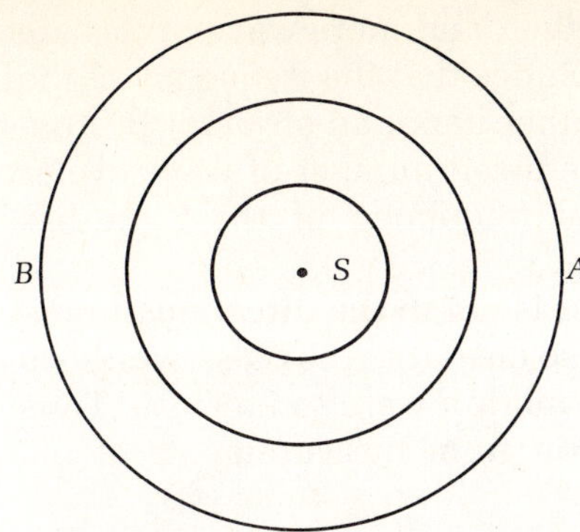

FIGURE 12.25 Stationary source. Waves reach each observer at *A* and *B* at equal intervals.

between the light source and the observer. If the relative motion is such that the source and the observer are receding from each other, then the frequency of light as measured by the observer is found to have decreased. The wavelength of the light is correspondingly increased; the light is shifted towards the red end of the spectrum, and this is called a **red shift.** On the other hand, if the relative motion is one in which the source and observer are approaching each other, then the observed frequency of light is found to be increased. Thus the wavelength is diminished, and we get a **blue shift,** a shift toward the blue end of the spectrum.

Thus receding relative motion gives a red shift, and an approaching relative motion produces a blue shift. This is called the **Doppler effect.**

Notice that only relative motion matters. We can have red shift when the light source is moving away from a stationary observer or when an observer is moving away from a stationary source or when both the source and observer are moving (with respect to a third stationary object) away from each other. This is because only relative motion can be detected anyway. The same kind of situations hold for the blue shift.

To understand the Doppler effect, first suppose that a stationary source emits a light beam. We can think of the light waves spreading from the source just as the ripples created by a splashing stone in a pond spread away from the center of the splash. We show this in Fig. 12.25. We find that the crests of successive waves come toward the observer, one behind the other, separated exactly by the distance of one wavelength. The distance between two successive wave crests is given by the wavelength of the wave pattern.

Now suppose that the source is moving. In this case the successive waves are emitted by the source from different positions, 1, 2, 3, and so on, as indicated in Fig. 12.26. To an observer A, toward whom the light source may be approach-

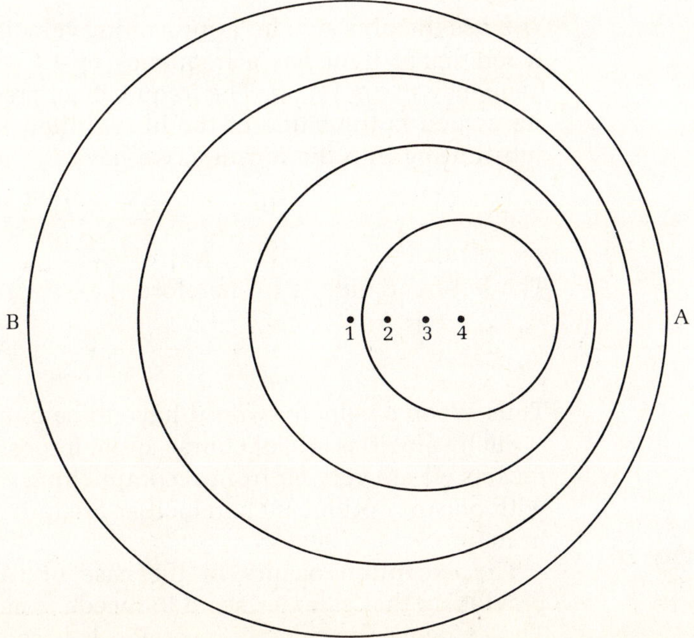

FIGURE 12.26 Doppler effect. Waves are emitted by the moving source at positions 1, 2, 3, 4, and so on. Observer A, whom the source is approaching, receives more wave cycles per second than when the source is stationary; the frequency is found to increase. In contrast observer B, from whom the source is receding, receives fewer wave cycles per second. The frequency of the waves has decreased for this observer.

ing, the waves appear to be more crowded together, and more wave cycles are received per second. Observer A therefore concludes that the frequency of the light beam has increased: a blue shift. On the other hand, an observer B, from whom the light source is receding, receives a reduced number of wave cycles per second and therefore concludes that the frequency of the light has decreased: a red shift.

The figure also makes clear that if the observer is not in the direction of relative motion, the effect is less pronounced. It is also clear that the effect would be more pronounced if the velocity of the relative motion were to increase. Thus we can measure the velocity of relative motion from measurements of the Doppler shift of light of a known wavelength.

The Doppler effect is a property shared by other waves too. You probably are familiar with the sound of passing cars changing from a higher to a lower pitch, which is due to the Doppler shift.

Coming back to the Doppler shift of light, you may enjoy the following story. American physicist R. W. Wood was once caught merrily going through a red light. At court Wood argued that since he was approaching the red light, his relative motion of approach produced a blue shift of the light, which thus appeared to be green. And so in his perception there was no violation. The judge thought for a little while and decided to accept Wood's plea; but he fined Wood anyway, for speeding.

Let's figure out what speed Wood would have to use in order to have red light shifted to green. An approximate formula for calculating the Doppler shift of light, when the velocity is v, is given as (ν is the original frequency and ν' is the shifted frequency)

$$\frac{\nu'}{\nu} = \left(1 \pm \frac{v}{c}\right) \qquad (12.2)$$

You use the plus sign for approaching velocity and the minus sign for receding. A red signal light has a frequency of 4.4×10^{14} Hz. This is ν, the original frequency in Eq. (12.2). The frequency of green light can be taken as 5.5×10^{14} Hz, which is the value of the blue-shifted light, ν', according to Wood. Now substituting into the formula, we have

$$\frac{\nu'}{\nu} = \frac{5.5 \times 10^{14}}{4.4 \times 10^{14}} = 1 + \frac{v}{c}$$

The left-hand side is $\frac{5}{4}$. Therefore, $1 + (v/c) = \frac{5}{4}$ and so

$$\frac{v}{c} = \frac{5}{4} - 1 = \frac{1}{4}$$

Thus Wood's velocity would have to be one-fourth the speed of light.

Such velocities are, of course, quite impossible to achieve with cars. Even for rockets we are very far from accomplishing velocities of this magnitude. In fact, with one exception, such velocities are quite uncommon for large objects even in nature.

The exception occurs in the case of runaway galaxies. We have noted previously that galaxies seem to recede from us. How do we know about this? The evidence for galactic recession is based on the observed Doppler shift of light that comes to us from these galaxies. We look at light of a known characteristic color and find that when this light comes to us from a runaway galaxy, it is shifted toward the red.

The evidence is inescapable. The faraway galaxies are moving away from us with velocities as large as a few tenths of the speed of light. The farther away the galaxy is, the larger is its recession speed.

The biggest surprise in the field is provided by objects known as quasars, which have been found by their red shift to be receding from us with a velocity of as much as 80% of the speed of light.

SUMMARY

The first section of this chapter deals with sound waves. Sounds are characterized by loudness and pitch. Musical sounds are characterized by an additional attribute called quality. Quality of sound involves stationary waves—standing patterns of waves in a guitar string, for example, as opposed to traveling sound waves in air. Stationary waves are set up any time there are waves in confinement. The frequencies of different waves in such confinement occur in multiples of a fundamental frequency related to the dimensions of the confining system. The standing wave of lowest frequency is called the fundamental mode, and the waves of progressively greater frequency are called second harmonic, third harmonic, and so on. The quality of a musical sound is due to the richness of the admixture of the higher harmonics in a given sound.

Stationary waves also occur in light and are used in the laser to produce a unidirectional beam, which does not spread appreciably, even far away from the source, as does ordinary light. The intensity of an ordinary light beam follows the inverse square law because of spreading:

$$\text{intensity} \propto \frac{1}{R^2}$$

In contrast, the intensity of laser light does not diminish rapidly with distance. This is one of the many useful features of laser light.

Also important in connection with waves is the phenomenon of resonance, which is a matching of frequency of two coupled systems of vibrations or waves. The intensity of vibration is greatly increased as a result of the frequency match.

The second section of this chapter deals with light, particularly with aspects connected with the ray description of light. The important subjects discussed are optical instruments, the eye, and the perception of images and color by the human eye. (Some further related aspects of color are given in Section 14.5.)

The final section of the chapter deals with waves from moving sources. When the velocity of the source exceeds that of the waves it produces, we get shock waves. The shock waves of sound are responsible for the sonic boom, a phenomenon that has received much publicity with the advent of the supersonic transport. A less spectacular event—but, as it turns out, an event that is even more important to physics—in connection with moving sources is the change in the frequency of the waves emitted in the perception of a stationary observer. This is known as the Doppler effect. One famous manifestation of the Doppler effect of light is the red-shift phenomenon.

Review and reason

1. Explain the difference between loudness, pitch, and quality in connection with musical sound.
2. What is the difference between traveling waves and stationary waves?
3. Explain how stationary waves are formed when waves are confined in a given space.
4. What is meant by resonance? Tie strings to two balls of the same size and shape and hang them from a taut clothesline. Now set one ball vibrating. Write down your observations. How do you explain the observations? Try the experiment again, but this time tie the balls with strings of different lengths. Explain your observations.
5. The double bass has a longer strip compared to the violin. Its sounding board also has larger dimensions. Can you speculate why?
6. What problem can arise if a large band marches in unison across a bridge?
7. Explain how an image is formed by a convergent lens.
8. Fish in a spherical bowl appear larger than they really are. Explain.
9. Why are slides turned upside down before putting them in a slide projector for viewing?
*10. The performance of a lens is said to be limited by diffraction effects. Explain.
11. Can we photograph a virtual image? Why or why not? Can we photograph a real image? Why or why not?
12. Explain the workings of the following two optical instruments with the help of diagrams: (a) the telescope; (b) the microscope.
13. In a close-up photograph, the background looks out of focus. Explain why.
14. How does the eye work? Draw the parallels between the eye and the camera.
15. Discuss some of the defects of vision and how they are corrected with lenses.
16. Why do you see better underwater with your goggles on?
17. You can see the Mach band phenomenon in your shadow cast by the sun if you look carefully. You will see an inner and an outer shadow. Look at the shadow boundaries, and describe what you see and where you see it.
18. In a B.C. cartoon Peter asks Thor (or is it the other way around?): "Why does the sun get so big just before it goes down?" The answer: "Look at the daylight it has to suck up." Can you give a better explanation of why the sun looks bigger when it is at the horizon?
19. Why do stars look white to the eye although many of them are really colored?
20. Why can't we see any color in peripheral vision? In dim light?
21. Distinguish between the additive and the subtractive methods of color mixing.
22. Ordinarily smoke looks slightly bluish. But if you look at smoke in direct sunlight, it looks slightly red. Explain why.
23. Discuss some of the functions of the rod and cone cells of the retina.
24. How is a sonic boom produced? What is meant by the Mach number of a supersonic transport? Can there be sonic booms when the Mach number exceeds the value of 1? Why or why not?
25. Boats produce a bow wave (Fig. 12.27) when traveling at high speed. Discuss the similarity of this phenomenon with the sonic boom.

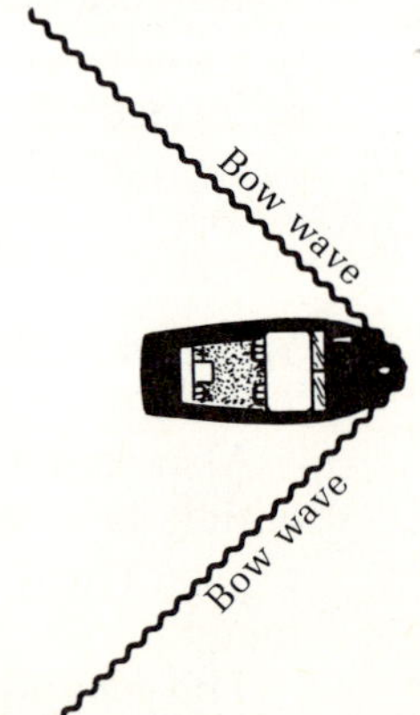

FIGURE 12.27   A bow wave.

26. What is Doppler effect? How do you explain it? How do we know that distant galaxies are receding from us at enormous speeds?
27. The whistle of a train moving away from an observer seems to have a lower pitch than when the train is stationary. Explain.
28. Christian Doppler, in experimenting with the effect that is named after him, put a few musicians on board

---

*Optional.

a moving train and had them play a certain note. As observers on the ground, he chose some of those very gifted musicians who are said to have "perfect pitch," musicians who can identify the frequency of every note they hear. Suppose the train was approaching the observers in the course of one of these observations. How would a ground observer perceive the note played on the approaching train—higher, lower, or the same in frequency—compared to when the note is played on the ground?

29. Suppose you are approaching a red giant star with a speed of one-fourth that of light. Will you perceive the star as red? If not, can you guess what color you will see?

Arithmetic

1. The smallest frequency of light that we see is about 4×10^{14} Hz and the highest is something like 8×10^{14} Hz. In terms of octaves, how many octaves of light waves do we see?

2. The sounds coming from two separate sources differ in loudness by 30 decibels. What is the ratio of their intensities?

3. The fundamental frequency of a certain guitar string is 250 Hz. Calculate the frequency of the second and third harmonics of the string.

4. A clarinet and a guitar are played using exactly the same note of frequency of 440 Hz. The sounds of the instruments reach your ears through an acoustical filter that filters out every sound above 500 Hz. Will you be able to tell which sound belongs to which instrument?

5. Compare the intensity of a streetlight when seen from 25 ft away as opposed to 50 ft away.

*6. Light of frequency 8×10^{14} Hz coming from a certain galaxy is found to be red-shifted to the value of 4×10^{14} Hz. Is the galaxy approaching us or is it receding from us? From the Doppler shift formula given in the text, calculate the speed of the galaxy.

---

*Optional.

13 Thermal Physics and Technology

You can't win,
And you can't break even,
And you can't get out of the game.*

■ 13.1 Some Basic Facts about Heat

Although all objects are made of atoms, their macroscopic behaviors are quite diverse. A solid object does not respond to heat in the same quantitative way as gaseous matter does. What causes the difference? When heat is applied to a solid object, sooner or later it will transform into a liquid; further application of heat will transform the liquid into a gas. This change of the state of matter is a fundamental effect of heat. What causes it? Where does the heat energy of fire—or of chemical reactions in general—come from?

And then there is water, with its very special thermal properties so essential to the maintenance of life. What makes water so special?

These are the kinds of questions we will begin with.

Specific heat

American physicist Count Rumford (born Benjamin Thomson), who is credited with the discovery of many of the fundamental aspects of heat, once made the following observation:

> When dining I had often observed that some particular dishes retain their heat much longer than others; and that apple pies, and apples and almonds mixed (a

---

*From "The Dealer (Down and Losing)." Words and music by Bob Ruzicka. © Copyright 1972 Lions Gate Music Ltd., New York, N.Y. Devon Music, Inc., controls all publication right for the U.S.A. Used by permission.

dish in great repute in England), remained hot a surprising length of time. Much struck with this extraordinary quality, which apples appear to possess, it frequently occurred to my recollection; and I never burnt my mouth with them, or saw others to meet with the same misfortune, without endeavoring, but in vain, to find out some way of accounting, in a satisfactory manner, for this surprising phenomenon.*

We can empathize with the Count. Most of us remember burning our tongues at least once in our impatience to enjoy some gastronomical delight. But have you wondered why such things as onion soup or the cheese on a pizza take a long time to cool (and of course to heat, too)? Most important of all, pure water has this property, too.

Today we refer to objects that have this capacity for retaining their heat as objects of high **specific heat** (or specific heat capacity). Apples and onion soup are examples of materials of high specific heat. In contrast, objects that are easy to heat as well as easy to cool—such as iron and aluminum, of which most kitchen utensils are made—have low specific heat.

Quantitatively, we define the specific heat of a substance as the amount of heat (in Calories) needed to raise the temperature of 1 kg of the substance through a temperature of 1°C. Thus the unit of specific heat is Calorie/kilogram °C, also often expressed as calorie/gram °C. Note that the two ways of writing the unit are exactly equivalent.† In this unit the specific heat of water is 1.0, whereas the specific heat of iron is only 0.1. This means that in order to heat 1 kg of iron by 1°C, you need one-tenth of a Calorie, but ten times as much energy, or 1 Cal, is required to heat the same amount of water through 1°C.

Table 13.1 gives the specific heat of some important materials.‡

TABLE 3.1   Specific heat of a few substances.

Material	Specific Heat (Cal/kg °C)
Aluminum	0.22
Copper	0.09
Hydrogen	3.4
Iron	0.11
Helium	0.75
Oxygen	0.16
Ice	0.5
Steam	0.5
Water	1
Window glass	0.2

It is an experimental fact that when gases are heated, they tend to expand. Thus the numbers in Table 13.1 for the gases refer to the specific heat determined under the constraint of constant volume—that is, the gas is confined in a

---

*Count Rumford, *Collected Works*, vol. I (Cambridge, Mass: Harvard University Press, 1968), p. 122.

†Recall that 1 Cal = 1 kcal = 1000 cal, and 1 kg = 1000 g.

‡Actually, specific heats do vary some with temperature, but we will ignore this fact in the present discussion.

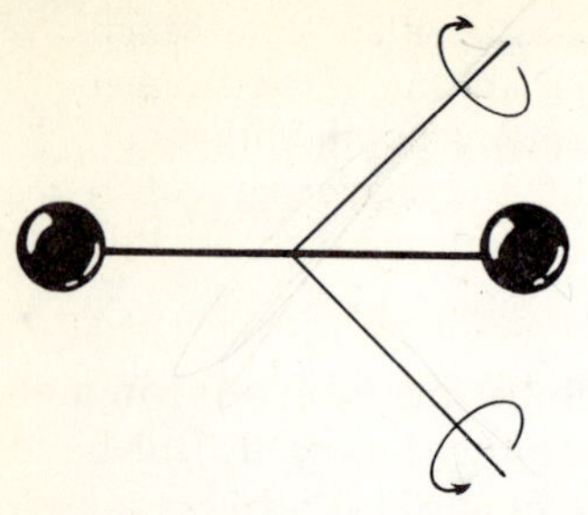

FIGURE 13.1   A diatomic molecule can rotate about the two axes, as shown.

container while it is heated. In contrast, the specific heat of solids and liquids is most conveniently determined under conditions of constant pressure. It is important to note, however, that the specific heat of a gas changes substantially if the pressure is kept constant instead of the volume. For liquids and solids it does not make much difference what is kept fixed, volume or pressure.

The values in the table make it clear that the specific heat capacity varies widely from one substance to another. Yet when we look at specific heat from an atomic point of view, we quickly discover some underlying simple facts.

We can think of a gas as consisting basically of free atoms or molecules, free meaning that the interaction among the atoms (or molecules) is negligible. Thus in helium, which is a monatomic gas, the atoms are so free to move that when you apply heat, they just jiggle faster and faster in their random translational motion. In oxygen the situation is slightly more complicated. Oxygen is a diatomic gas. Its molecules are free to move about, but the two atoms forming each molecule must remain confined within the molecule. Yet the atoms are not entirely without freedom. For example, they can rotate without changing the configuration of the molecule (Fig. 13.1); this rotation is not unlike that of a dumbbell. Thus when you apply heat to oxygen gas, not only do the molecules jiggle faster in their translational motion, but also some energy goes into their rotation. As you can see, the rotational motion has an element of order built into it. So for diatomic molecules not all the applied energy goes into the random heat motion of translation. The result is that the diatomic gases have a larger specific heat *per molecule*.

This last item is important when comparing the specific heat of two substances. From a fundamental point of view, we must compare the specific heat per molecule. Or, since a mole of every substance has the same number of molecules, we can compare the specific heat per mole, called the **molar specific heat.**

The molecular weight of helium is 4; thus a kilogram-mole of helium has a mass of 4 kg. Since the specific heat of helium is 0.75, it takes 0.75 × 4 or 3 Calories to heat 1 kg-mole of helium through 1°C. So the molar specific heat of helium is 3 Cal/kg-mole °C. Since the molecular weight of oxygen is 32, a similar calculation (0.16 × 32) gives the molar specific heat of oxygen to be 5.1 Cal/kg-mole °C. As you see, oxygen does have a larger molar specific heat, just as we expected.

Let us now look at the case of a simple atomic solid like iron. As you know, in a solid the atoms are not quite free to move; they are fixed to specific locations of a crystal structure. When the atoms in a solid are agitated, they are capable of moving away from their centers, but only to a limited extent. There are restoring elastic forces acting on them, which tend to bring them back. The result is a back-and-forth vibration, very much like the vibration of masses attached to springs (Fig. 13.2).

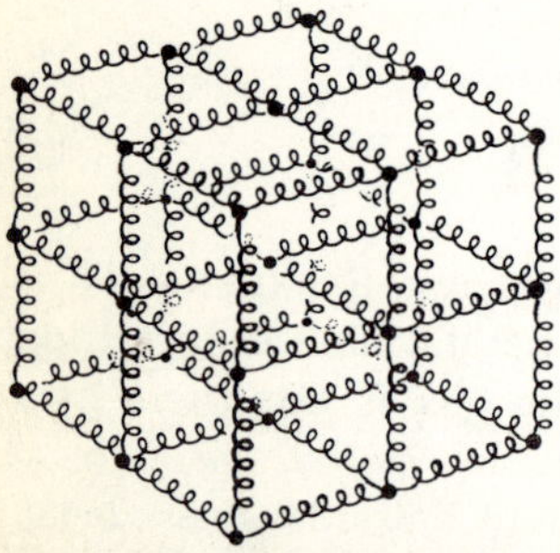

It is instructive to look at the composition of the total energy of such an oscillator. When the spring is extended maximally, all the energy is potential. Then it starts its return trip. Just as it passes through its equilibrium position, its energy is all kinetic (the elongation of the spring is zero at this point—no force, no potential energy). At intermediate positions, it has both kinetic and potential energy (Fig. 13.3). The similarity of this motion with that of the simple pendulum should be evident (Section 9.3). Both of these are examples of the so-called

FIGURE 13.2   The model of a solid. The atoms in a crystal can be looked upon as masses attached to springs.

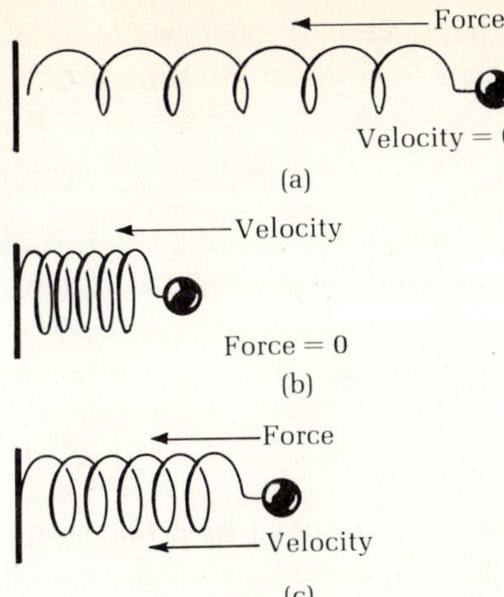

FIGURE 13.3 In the motion of a mass attached to a spring, the energy changes back and forth between kinetic and potential. (a) Spring maximally extended: the energy is all potential. (b) As the mass passes through the equilibrium position, the energy becomes all kinetic. (c) At an intermediate position the energy is partly kinetic and partly potential.

harmonic oscillator.

The important thing to notice about the energy of a harmonic oscillator is that most of the time it is partly kinetic and partly potential. In fact, when we do detailed mathematics, we find that, on the average, half the energy is kinetic and the other half is potential.

Suppose we picture solid iron as consisting of atomic oscillators, as in Fig. 13.2. We conclude that when we apply heat to such a system, only half the heat goes into kinetic energy, which increases the temperature of the solid, but the other half goes into increasing the potential energy of its atomic oscillators. Therefore, mole per mole, solid iron should take twice as much heat compared to a monatomic gas like helium to raise its temperature by the same amount. In other words, we should expect that the molar specific heat of iron should be twice that of helium. The atomic weight of iron is 56 and its specific heat is 0.11 Cal/kg °C. Thus its molar specific heat is given as $56 \times 0.11 = 6.2$ Cal/kg-mole °C, which is indeed roughly twice that of helium gas.

Of course, simple considerations like this apply to only the simplest of cases that we have looked at. But perhaps if we make a general observation at this stage you will have no trouble accepting it. In complex solids and liquids, only a part of the applied heat energy goes into the kinetic energy of random motion. The rest of the energy is taken up by the orderly energy of rotational motion of even potential energy, which does nothing to the temperature of the body. Thus such materials, in comparison to a monatomic gas, need a lot more heat in order for their temperature to be raised by a specific amount. They have a large specific heat. However, a detailed calculation of the specific heat requires a detailed knowledge of the molecular structure and interactions.

Thus the simple question that Count Rumford asked about the specific heat of apple pie is not such a simple question after all. We hope our discussion has helped you to understand why such substances have a large specific heat. But what we really want you to see is the necessity for getting a clear picture of the molecular structure of materials. Macroscopic material properties like the specific heat are actually manifestations of the underlying behavior of the atoms and molecules.

Here is another point. The preceding discussion makes it clear that in the consideration of the total "internal energy" of a body in matters of thermodynamics, we must account for not only the kinetic energy of random heat motion but also all other forms of energy, like rotational and potential, of the constituent atoms or molecules.

EXAMPLE 1

Tea making is a rather refined art. For one thing, you have to heat the water to 100°C, its boiling point, but you must not let the water boil, which spoils everything. So our question is, how much heat do we need to make half a kilogram of tea water?

The amount of heat, Q, in such problems can be calculated from a simple formula based on the definition of specific heat. If C is the specific heat, which is the heat needed per kilogram per °C, for m kg, we will need an amount of heat mC. If the temperature difference between the final and initial temperatures is $\Delta\theta$ (pronounced "delta theta"; Δ (delta) is a capital Greek letter and θ (theta) is a lowercase Greek letter), then we need $mC\,\Delta\theta$ of heat. Thus to raise the temperature of m kg of the substance whose specific heat is C by an amount $\Delta\theta$, the amount of heat needed is given as

$$Q = mC\,\Delta\theta \tag{13.1}$$

In the situation here, $m = \frac{1}{2}$ kg, $C = 1$ Cal/kg °C, and if we assume the initial room temperature of water to be 20°C, then $\Delta\theta = 100 - 20 = 80$°C. Therefore,

$$Q = \tfrac{1}{2}\ \text{kg} \times 1\ \frac{\text{Cal}}{\text{kg °C}} \times 80\ \text{°C} = 40\ \text{Cal}$$

EXAMPLE 2

As a more practical problem, we can estimate the amount of heat necessary for an average hot bath in a tub. The volume of water used in a tub can be roughly estimated by noting that a bathtub is about 1.3 m long, 0.5 m wide, and the water probably would stand about 0.2 m deep in an average bath. So the volume of hot water is $1.3 \times 0.5 \times 0.2 = 0.13$ m³. The mass is the volume times the density of the water, which is 10^3 kg/m³. So the mass is 0.13×10^3, or 130 kg. Suppose you like a temperature of about 50°C (122°F) for the water in your bath. The quantity of heat needed to heat the water, starting from 5°C (41°F) is [using Eq. (13.1)]

$$Q = 130 \times 1 \times 45 = 6{,}850\ \text{Cal}$$

This is a lot of energy, more than twice the average daily energy of food intake (which is about 3000 Cal for Americans).

Actually, water heaters do not operate at 100% efficiency, so the resource energy spent for a bath is even higher. If the efficiency of the water heater is only 60%, the amount of energy spent for a bath is actually $6{,}850/0.6 = 11{,}417$ Cal. If you use electricity to heat the water, and since electricity is obtained from an energy resource at an efficiency of 35% on the average, the energy cost is a soaring $6{,}850/0.35 = 19{,}571$ Cal for a single bath.

Latent heat

On a hot summer day it is very pleasant to have a whirring fan directed at you; the air feels cool. What makes it feel cool? Is the air really cool? No; the fan is just circulating air at the temperature of the room. It's the evaporation of the perspiration on your body that causes the cooling; the fan just facilitates the evaporation process.

An even more dramatic demonstration of the cooling effect of evaporation is to pour some alcohol on your palm; the alcohol evaporates very quickly and the cooling effect is striking.

Thus evaporation causes cooling. This is the most common demonstration of the **latent heat** phenomenon. It takes energy to convert a liquid to its gaseous form.* When water evaporates from your body—becomes vapor—it takes heat energy from your body to make this happen, and that's why you feel cool.

---

*Quantitiatively, the latent heat of vaporization of a substance is defined as the amount of heat (in Calories) needed to evaporate 1 kg of liquid into the vapor form.

This latent heat of vaporization—or, simply, the heat of vaporization—does not just disappear. It stays inside the steam in the form of potential energy (chemical potential energy). Indeed, if the steam condenses back, we get the latent heat back too. Heat is given up when steam condenses.

Have you wondered where the tremendous energy comes from that drives a hurricane originating in the Atlantic Ocean? Originally the sun's heat evaporates water from the ocean, which condenses as it rises, releasing back all that latent heat, the chemical potential energy. This is the energy that drives the hurricane. Incidentally, the whirl of the hurricane is an example of the cyclonic phenomenon discussed in Chapter 7.

Coming back to evaporation, let's consider a few more examples. If you visit some countries, you may be surprised to find that people drink a lot of hot tea or coffee, even in the hot summer. Shouldn't tea be uncomfortable in the summer—all that heat pouring into your body? The surprising thing is that it actually has a net cooling effect. This is how it works. The latent heat of vaporization of water is about 540 Cal for each kilogram evaporated.* Suppose the tea you drink has a temperature of 60°C on the average. When you drink it, it will surely add heat to your body in the process of cooling itself to your body temperature of 37°C. The amount of heat added to your body is exactly equal to the heat lost by the tea in cooling to 37°C, starting from 60°C. For each kilogram this heat turns out to be 23 Cal (figure it out). This is much less than the 540 Cal that the body has a chance to lose by evaporation of that kilogram of water in the form of perspiration. Even if you lose some of the water in one of the other natural ways of excretion, there may be enough left in the evaporation mode to make tea drinking worthwhile in the summer.

At the other extreme, the heat loss by evaporation can occasionally be a hazard. Hikers are warned always to carry a raincoat to avoid getting soaked by rain or snow. Once soaked, hikers lose heat so quickly from their bodies through evaporation that they have the extreme danger of contracting hypothermia, a condition in which the body can maintain only a subnormal temperature. Uncontrolled, the body acquires the temperature of its environment (much like a cold-blooded animal).

If you step out of a shower while still wet, you feel chilly. It is uncomfortable to step out to dry yourself. This is evaporation again. However, you may have discovered that if you dry yourself while standing in the tub or shower stall, it is not so bad. What's the difference? The tub area is full of moisture from your shower, which has a lot of latent heat to donate to you as it condenses on your body, almost as much as you lose by evaporation. This is the important thing: the molecules can go both ways, provided the energetics are right. In order to escape, the molecule in a liquid has to have enough energy to overcome the force that holds it to the liquid. It also has to be on the surface, with its velocity pointed outward. On the other side, if a molecule in the vapor comes close enough to the surface to be under the spell of the attractive molecular force, and it does not have enough kinetic energy to escape, it can be captured with a release of energy.

Evaporation is enhanced as the temperature is increased, because there are

*This is the correct value for evaporation at 100°C. At lower temperatures the latent heat of vaporization of water is somewhat larger, depending on the temperature.

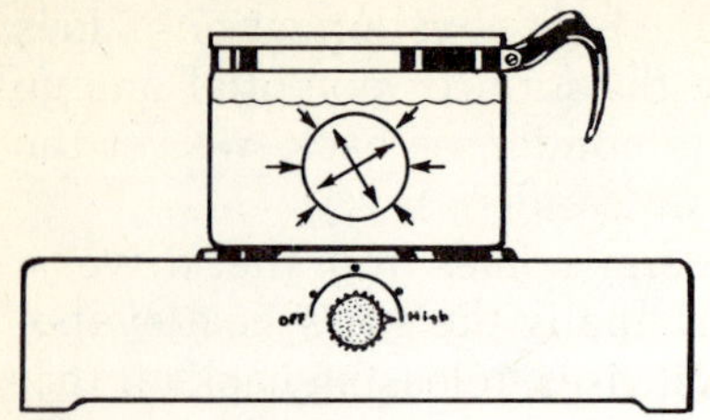

FIGURE 13.4   Formation of a bubble in a pot of water on a hot burner. The bubble is shown much enlarged. The motion of the water molecules creates a vapor pressure, which overcomes the external pressure of the atmosphere and the water pushing inward on the bubble.

more fast molecules on the surface. When the temperature of water reaches 100°C, the water starts boiling, which is very vigorous evaporation. Bubbles of gas form even inside the water when it starts to boil, and the bubbles manage to come up to the surface and escape. This is what distinguishes boiling from evaporation. (Fig. 13.4). From the molecular view, the speed of the molecules is now great enough to exert as much pressure from within on a vapor bubble as exists on it externally from the atmosphere and the layer of water above the bubble.

Thus pressure is very important in the phenomenon of boiling. If the atmospheric pressure is lower than usual, the vapor pressure within the bubble does not have to be as high for the bubble to rise to the surface and escape. So boiling can occur at a lower temperature if the atmospheric pressure is low. Anybody who has been to Colorado can vouch for this.* Rice takes longer to cook there because the water boils at a lower temperature due to the lower atmospheric pressure at the high altitude.

Just the opposite happens when the pressure on the liquid is higher than the atmospheric pressure. Then the liquid will not boil until its temperature is raised above the normal boiling point—how much above depending on how high the pressure is. This is the underlying principle of a pressure cooker. Due to the high pressure inside, the water boils at a considerably higher temperature than 100°C, and, of course, cooking is much faster at the higher temperature.

We can study boiling to observe the underlying nature of latent heat. Suppose we put a pot of water on a heater and apply heat. Initially, most of the heat will go into raising the temperature of the water (Fig. 13.5). But as soon as the water temperature reaches 100°C, the boiling point of water under normal atmospheric pressure, an interesting thing happens. The temperature of the water stays at that 100°C; it does not increase further until all the water has boiled away. After this the steam can be further heated if desired, as shown in the figure, and temperature will increase again.

During the process of boiling, then, all the applied heat goes into breaking up the bonds that keep the molecules together as a liquid. The energy needed for this is the latent heat; so latent heat is really the energy required to overcome the binding of the molecules of the substance to form a liquid. For each kilogram of water, the amount of energy necessary to free the molecules in conversion to

FIGURE 13.5   The energetics of heating and the change of state from liquid to vapor for 1 kg of water. The temperature remains constant at 100°C until all the water becomes vapor.

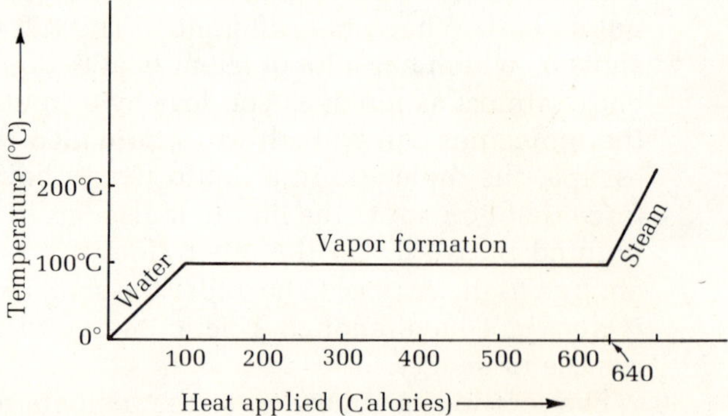

---

*In Denver, the "mile-high city," water boils at 95°C.

steam at 100°C is 540 Cal. This energy, of course, does not disappear; it stays on as potential energy in the steam. When the steam is condensed, we recover the energy.

Latent heat in melting and freezing

It should now be clear that whenever there is a change of state of a substance—from a state of greater to lesser binding energy—work must be done or heat must be applied to accomplish the change. In the reverse process of conversion to a state of stronger binding from one of less binding, energy is given off as potential energy is converted into heat. Thus when a solid melts, we need to supply an amount of energy or heat to the solid. Conversely, when the liquid is frozen into the solid form, the latent heat is released. For water this latent heat of freezing or melting is 80 Cal for every kilogram.

Let's consider a few examples connected with the latent heat of the water-to-ice transition and vice versa. You may have heard of the olden-day practice of putting a large pot of water near vegetables to prevent them from being killed during a frost. How does the water help? When the nighttime temperature is lowered enough for the water to freeze, kilogram per kilogram the freezing gives up an amount of 80 Calories of heat to the environment, which helps prevent the frosting of the vegetables.

As in a liquid-to-vapor transition, pressure also has some effect, albeit smaller, on the melting temperature. Under increased pressure, ice melts at a temperature lower than 0°C, the usual melting point. The lowering isn't much, only 0.0072°C for each atmosphere of pressure added (14.7 psi), but even this much lowering is an advantage in the skating rink. Under the intense pressure of the skates, the ice below melts, so the skater really skates in a thin film of water. Of course, as soon as the skates are removed, the freezing point is back to zero degrees and the water freezes again. This principle also operates in the making of a snowman. By applying pressure we melt some ice to manipulate the shape. As soon as we remove the pressure, the water refreezes, preserving the shape (until the weather warms up, of course).

How does antifreeze keep water in the automobile radiator from freezing (as long as the temperature doesn't fall too far below zero)? The molecules of antifreeze interfere with the water molecules and prevent them from forming the hexagonal crystals of solid ice. The net result of this is an extreme lowering of the freezing point. Interestingly, the addition of antifreeze also raises the boiling point of the solution, and thus antifreeze is also effective in the summer in keeping the water from boiling away. However, you should note that the water in the radiator acts as a coolant for the engine by virtue of its high specific heat. So it is not advisable to add more than 50% antifreeze to the radiator content.

Similarly, salt drastically reduces the freezing point of water, which is the reason it is used so much on winter roads in spite of its adverse effects on the metallic parts of automobiles.

Heat of chemical reactions

Suppose we continue to heat the steam beyond 100°C, the temperature at which the bonds holding the water molecules as a liquid are broken. Can we supply enough heat to break the bonds that hold the two hydrogen atoms to the oxygen atom? Indeed, as the temperature goes higher and higher, some of the energy is taken up by the atoms inside the individual molecules, causing them to vibrate. At a temperature of a few thousand degrees, the water molecules break down into their constituent atoms.

Perhaps you see a parallel between the latent heat phenomenon—and now this breaking down of the molecules—and the escape velocity from the gravity pull of the earth discussed earlier. By heating we provide individual molecules and atoms enough escape velocity to achieve freedom from the binding forces. In a solid the molecules are in a state of greater binding than in a liquid. By providing the material the heat of fusion, we supply the individual molecules enough energy to break away from the solid state. The same is true of the liquid-to-gas transition, all the way to the breaking up of water into oxygen and hydrogen.

From what we have said before (Chapter 10), a bound state is a state of nega-tive energy for the constituent particles. Positive kinetic energy has to be supplied to the atom until its total energy becomes zero, or positive, for it to break away. In the reverse situation, the heat of fusion must be released when liquid water solidifies. Why? Because the molecules are entering a state of increased negative energy of a solid-state crystal structure, and an equivalent amount of positive energy must evolve to balance the energy ledger—conser-vation of energy.

Looking at things this way is most advantageous in understanding the heat that evolves in chemical reactions. When hydrogen is put in a flame in the presence of oxygen, it burns in an explosive brilliance. It is giving up an equiva-lent amount of positive energy as it enters the bound state of negative energy that water represents for the oxygen and the hydrogen atoms. Likewise, the carbon in coal or wood combines with oxygen to form the bound state of the molecule carbon dioxide (CO_2 in chemical notation) and again energy comes out of the burning process (Fig. 13.6).

The energy produced in a chemical reaction is in the form of the kinetic energy of the molecule that is formed. These energetic molecules can now bounce around, sharing energy with every other molecule in the vicinity, and thus everything gets heated, including the remaining fuel and the oxygen mole-cules in the surrounding air. Now it is a fact that usually we have to supply some kind of an ignition energy to make a chemical reaction happen—for exam-ple, the fire applied to the kindling of the wood. This is to enable the carbon and the oxygen atoms to go over the so-called reaction barrier. They must have a minimum amount of energy before they can come together. But once the reac-

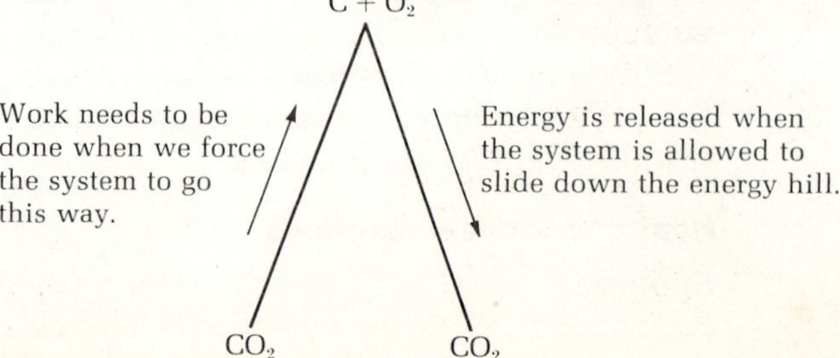

FIGURE 13.6   The energy hill. It costs energy to go uphill, but the energy re-mains stored in the system. When pushed downhill, the stored energy is released.

TABLE 13.2 Energy of some fuels.

Fuel	Energy Produced by Burning
Wood	5,600 Cal/kg
Coal	4,600 Cal/kg = 4.6×10^6 Cal/ton
Oil	30,000 Cal/gallon = 1.26×10^6 Cal/barrel*
Natural gas	252 Cal/ft^3

*1 barrel = 42 gallons.

tion occurs, the heat of the reaction itself provides enough energy for the carbon and oxygen atoms to overcome the barrier, and so the reaction can go on without any further rekindling. It is a chain reaction—the reaction itself keeps the whole thing going.

Energy-producing chemical reactions are called **exothermic** reactions, to distinguish them from **endothermic** reactions, which need a constant supply of energy to take place. An important example of an endothermic reaction in nature is the process of photosynthesis in plants, where the net result of a series of reactions is the conversion, with the help of sunlight energy, of carbon dioxide and water into carbohydrate molecules and oxygen.

How can we tell which reaction is exothermic and which is endothermic? We can look at the reactants and products and determine which correspond to a state of stronger binding energy. Roughly, the rule of thumb is that if we go from a less bound system to a more bound system, energy will have to evolve. In the opposite case of a transition from a more bound to a less bound system, energy has to be supplied.

Table 13.2 gives the energy production characteristics of our important chemical fuels: wood, coal, oil, and natural gas.

EXAMPLE 3 *Annual energy expense for your car's fuel*

An average American drives his car about 12,000 miles a year, perhaps getting 16 miles/gallon (abbreviated mi/gal). Thus his annual expense for the fuel of the car comes to

$$\frac{12,000}{16} = 750 \text{ gallons/year}$$

Since 1 gallon of gasoline represents 30,000 Calories of energy, the energy expense for all that driving comes to

$$750 \times 30,000 = 22,500,000 \text{ Calories} = 2.25 \times 10^7 \text{ Calories}$$

Since there is a car for about every 2.25 Americans, we can forget about the factor 2.25 and take 10^7 Calories as a fair estimate of the per capita annual automobile fuel energy expense in the United States. This is just about the same as our per capita household annual heating bill and about three times as much as our energy bill for food.*

---

*Taking into account all the energy that goes into preparing and producing the food, not just the energy we get out of it.

Our personal driving contributes something like 10% to the total energy bill. It also contributes substantially toward the energy crisis that you hear about today.

The special properties of water

"Why in winter is the day short and the night long, and in the summer the way round? The winter day is short because like all other visible and invisible things it contracts due to cold; meanwhile the night expands—it is warmed up when lights and lamps are lit." This magnificently "reasonable" but false explanation was offered by a character created by the Russian author Anton Chekov. Actually, the fellow was even wrong in the assertion that all substances contract due to cold. One of the many special properties of water is that between the temperatures of 4°C and 0°C, it actually expands in cooling.

Most substances behave the other way—they expand on heating and contract on cooling. The column of mercury rises in a heated thermometer; steel rails expand when heated, which is the reason it is customary to leave occasional gaps in railroad tracks, and so forth. Yet water has this strange and, it turns out, very important property.

As you know, if a given mass of a substance is expanded to fill a larger volume, then its density is correspondingly lowered. Thus the density of water decreases as we cool it from 4°C downward. (Actually, the density also decreases if we heat water from 4°C to a higher temperature; water has a maximum density at 4°C, Fig. 13.7.) If water were a regular substance, then in the winter, when the water of lakes cools, it would very likely freeze from the bottom up. But since the 4°C water is the most dense, it goes to the bottom, while the water at the top is actually colder. All the freezing, then, occurs from the surface and never reaches the lowest levels of water. Thus aquatic animals can survive even the most severe winters.

The reason for this interesting behavior of the density of water involves the same property of ice that makes it less dense than water, namely, the hexagonal crystal structure, with large amounts of empty space between the molecules.

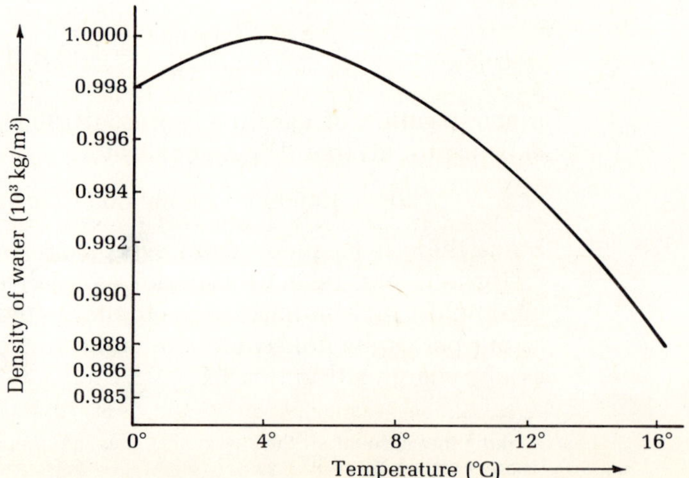

FIGURE 13.7   The density versus temperature graph for water. Water attains maximum density at 4°C.

Even after ice melts, this structure lingers on for awhile, making water increase in density with temperature. The collapse of the crystal structure is almost complete at about 4°C, after which water behaves in the usual way.*

The other very important property for living objects is the high heat of vaporization of water that we mentioned earlier. This enables us to get rid of large amounts of body heat as necessary by perspiring.

Considerations of scaling

The metabolic rate of every living being is limited by the rate of how quickly it can get rid of heat. Many living creatures lose heat by perspiration. Now heat loss by perspiration, which is an evaporation process, occurs throughout the surface of the animal. Therefore, it scales as the surface area, as L^2, the square of the linear dimension L. Thus the basic metabolic rate must also scale as L^2.

This is surprising, since this means that the maximum power delivered by animals also scales as L^2. Power is proportional to the metabolic rate, the rate at which energy is produced in the body. The surprise is due to the fact that in Chapter 9 we showed that the work done by the muscles scales as L^3.

The puzzlement is resolved when we realize that power is muscle work divided by time:

$$\text{power} = \frac{\text{work}}{\text{time}}$$

Thus power can still scale as L^2, even though muscle work scales as L^3, if time scales as L. Now time, of course, is the ratio of distance and velocity [see Eq. (1.1)]. So the only way time can scale as distance, as L, is if velocity does not scale at all—if it is independent of the size of the animal.

So animals of similar shape should be able to run at similar speed, which is true. In fact, the scaling law seems to hold even when the shapes are not exactly the same. For example, a rabbit can run at approximately the same speed as a horse.

■ 13.2 The Gas Laws and the Molecular Interpretation of Temperature

Since the molecules in a gas are more or less free, with rather small interaction between them, we can expect the behavior of gases to be the simplest to understand. Indeed, this expectation is borne out very well. Historically, the laws governing the behavior of gases were discovered experimentally at just about the time that the idea of atoms and molecules was taking shape. Some of the first triumphs of the molecular theory† were achieved with the explanation of the gas laws. In addition, there was a big dividend.

It was found that, for the quantitative success of the theory, it is essential to interpret temperature as a measure of the average kinetic energy of the molecules. The theory also led to a natural scale of temperature and to the concept of "absolute zero."

---

*For additional reading, see an article by Bruce Chalmers, "How Water Freezes," *Scientific American*, February 1959.

†Often referred to as the kinetic molecular theory since the molecules are in continual motion.

Boyle's law

Boyle's law, discovered by Robert Boyle, was the first of the gas laws. The law can be stated simply as follows:

If the temperature of a gas is not allowed to change, then the volume of the gas is inversely proportional to the pressure applied to it (which is the same as the pressure exerted by the gas on the walls of the container).

In symbols, we have

$$V \propto \frac{1}{p}$$

where we designate volume by V and pressure by p. The law says that if you put some gas in a cylinder and compress it with a piston, keeping the temperature constant in some way (Fig. 13.8), then if the pressure is doubled, the volume decreases to one-half the original volume. Another way of expressing this inverse proportionality of the volume and pressure is to notice that, corresponding to any change in pressure, the volume changes in such a way that the product of pressure and volume always remains the same, provided the temperature remains unchanged:

$$p_1V_1 = p_2V_2 \qquad \text{if temperature is constant}$$

And, finally, a third way of expressing the same thing is to say that the product pV is a constant:

$$pV = \text{constant, if temperature remains constant}$$

If you blow up a balloon at sea level in the United States and carry it with you on an expedition to the Himalayas, where the pressure is 15 inches of mercury (half of what it is at sea level), the volume of the balloon will have doubled (provided the temperature remains the same and no gas escapes from inside). Nobody has checked the validity of Boyle's law in quite this way, but we are convinced that this would happen.

In the kinetic molecular theory, the pressure of a gas is due to the thermal collisions of the gas molecules with the container wall. For example, a head-on collision and rebounding of a gas molecule imparts twice its original momentum to the wall in the process. The change in momentum imposes an impulsive force on the wall; the pressure on the wall comes from this. If the volume is reduced and the temperature remains constant, there should be more collisions. This is because the temperature determines the average kinetic energy and, therefore, the average velocity of the molecules. If the temperature remains

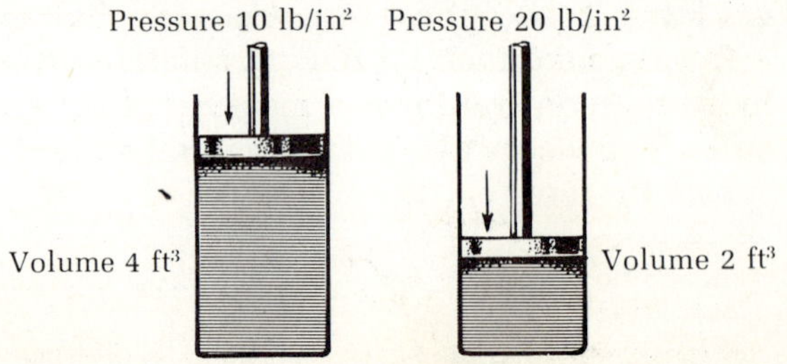

FIGURE 13.8 Boyle's law. If the pressure doubles, the volume is reduced to half.

fixed, so does the mean velocity of the molecules. But with decreased volume, the molecules have less distance to travel between collisions. Hence an increased number of collisions occur, producing the increase in pressure.

Charles' law

Boyle's law provides us with a nice application of the kinetic molecular theory. It is easily derived from our assertion that temperature is the determining factor for the average velocity of gas molecules, thus indirectly proving the assertion itself. But there is another law that is even more instructive about the nature of temperature–Charles' law, formulated by French physicist Jacques Charles:

> *If the pressure is kept constant, the volume of a gas increases in direct proportion to the increase of temperature.*

For example, if you increase the temperature by 10°C and the increase in volume is 2 cubic meters, then if you increase the temperature by 20°C, the increase in volume will be 4 cubic meters, in direct proportion to the increase of temperature.

The molecular explanation again depends crucially on the assertion that the mean velocity of the gas molecules is determined by the temperature. If the temperature is increased, so is the mean velocity. The gas molecules are now able to travel a greater distance in a given time. This will increase the number of collisions with the walls of the container and hence increase the pressure. This is the law of Joseph Gay Lussac, a contemporary of Jacques Charles:

> *If volume is kept constant, pressure increases in direct proportion to the increase of temperature.*

Alternatively, we must allow the gas molecules to travel larger distances by increasing the volume if we want the same number of collisions with the wall (which assures us that pressure remains constant). This proves Charles' law.

Charles' law, however, implies something far more exciting than a verification of the relationship of molecular velocity and the temperature of a gas. Quantitatively, what Charles found is that the volume decreases by 1/273 of its original volume at 0°C per degree Celsius of decrease of temperature. So if we started with 273 m³ of gas at 0°C, the volume would decrease to 272 m³ at −1°C, to 271 m³ at −2°C, and so forth. Suppose we extrapolate this as far as we can. At −273°C, the volume would presumably reach zero. If the temperature could be lowered any further, the volume would be negative, which is impossible. Thus it seems that −273°C gives us a limit of temperature, the lowest temperature possible. This lowest temperature is called the **absolute zero.**

The argument above depends on the validity of Charles' law for gases all the way down to −273°C. However, in reality all gases become liquid before this temperature is attained. Nevertheless, it is very compelling to believe that there is something special associated with this temperature, especially after looking at Fig. 13.9. As you can see, the volume-temperature curves for several gases are plotted and extrapolated to zero volume. Note that all the curves extrapolate back to the same temperature of −273°C.

Today we believe that −273°C is indeed the lowest definable temperature, or absolute zero. Furthermore, it is believed as a scientific law, the **third law of**

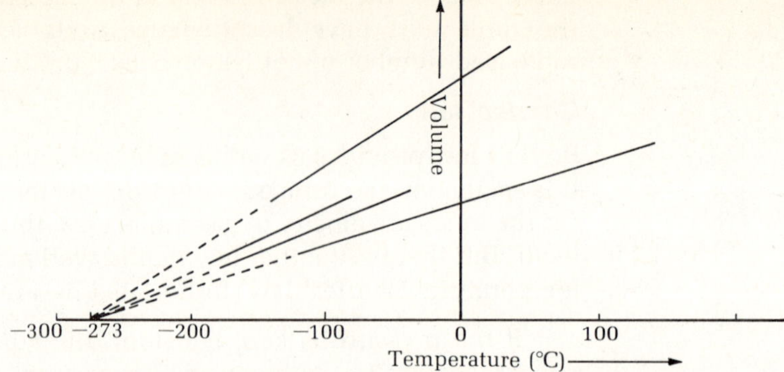

FIGURE 13.9 The volume versus temperature curves of gases (idealized). When extrapolated, all gases seem to attain zero volume at -273°C.

thermodynamics, *that we can never attain the absolute zero.* Experimentally, scientists have been able to reach within a millionth of a degree of absolute zero (which is actually placed at $-273.16°C$, but we will go on using the round number -273).

We assert that absolute zero is the temperature where molecular motion is at a minimum. So it is also the temperature where maximum order is achieved by all physical things and where their entropy tends toward zero. Thus the concept of absolute zero enables us at least to define a zero of entropy, although we must be aware at the same time that we can never attain it. You can't get out of the game.

William Thomson, an English physicist who became Lord Kelvin, clarified much of the haze surrounding the concept of absolute zero. One of Kelvin's fundamental works was to establish a natural scale of temperature with absolute zero as its zero point. This **absolute scale** of temperature is also called the **Kelvin scale.** The temperature on the Kelvin scale (°K) differs from the Celsius scale only by virtue of the zero-point difference of 273°. Thus given a temperature in degrees Celsius, the corresponding temperature on the Kelvin scale is obtained by adding 273°.

$$\text{temperature in degrees K} = \text{temperature in degrees C} + 273$$

Thus 27°C is 300°K, $-10°C$ is 263°K, and so forth.

Now we come to a very important point. If we measure temperature in this

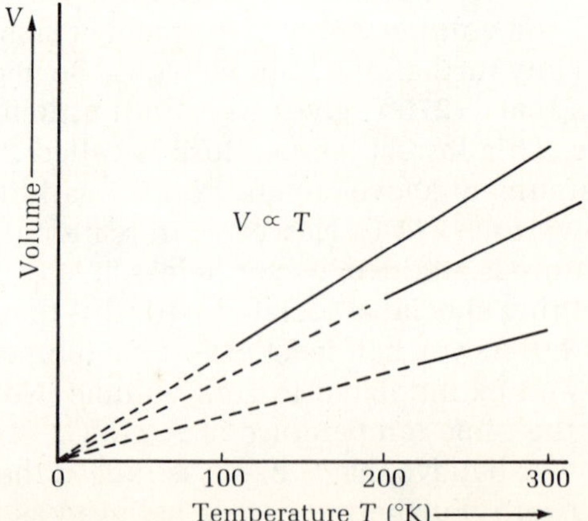

FIGURE 13.10 The Charles' law curves (idealized) plotted as volume versus absolute temperature graphs. Now they become straight lines passing through the origin.

natural Kelvin scale of temperature, all the Charles' law curves become lines through the origin (Fig. 13.10). Mathematically, this means that the quantity on the y-axis, volume, is proportional to the quantity along the x-axis, temperature:

$$\text{volume} \propto \text{temperature}$$

or
$$V \propto T$$

where we use capital T to denote temperature on the Kelvin scale. Notice that this is a much simpler statement than the previous one about increases of volume and the corresponding increases of temperature.

Perfect gas law

We now have the following two laws:

$V \propto 1/p$, if temperature is kept constant.
$V \propto T$, if pressure is kept constant.

A third proportionality is intuitively easy to guess. The volume of a gas must be proportional to the number of gas molecules present if both the pressure and temperature are held constant; that is,

$$V \propto N, \text{ if } p \text{ and } T \text{ are constant.}$$

(Also, needless to say, the number of molecules N remains constant in the first two rules we presented.) Under a rule of mathematics, it follows that the volume must be proportional to the product of all three, N, T, and $1/p$, or

$$V \propto \frac{NT}{p} \qquad \text{when nothing is held constant.}$$

Introducing a constant of proportionality k_B (k_B is called **Boltzmann's constant;** Ludwig Boltzmann was an Austrian physicist who made fundamental contributions to theories relating to the molecular nature of thermodynamics), we can write:

$$V = \frac{Nk_B T}{p}$$

or

$$pV = Nk_B T \tag{13.2}$$

This relationship is known as the **perfect gas law,** because it is truly valid only for an ideal or perfect gas. However, it is quite useful for dealing with real gases as well. Of the four variables of a gas—p, V, N, and T—if three are known, the fourth can be determined by using this formula. Note that the product pV has the unit of energy:

$$\frac{\text{newton}}{\text{meter}^2} \times \text{meter}^3 = \text{newton·meter} = \text{joule}$$

Since N is just a number, it follows that $k_B T$ also has the same unit as energy. Boltzmann's constant k_B thus relates the units of energy and of temperature. Its numerical value is given as

$$k_B = 1.38 \times 10^{-23} \text{ J/}°\text{K}$$

It is also very useful to know that the kinetic energy of a microscopic particle (atom, molecule, or a subatomic particle) existing in equilibrium with an environment of temperature T is given by $\frac{3}{2} k_B T$.

EXAMPLE 4

Have you ever wondered how many air molecules there are in your room? You can figure it out from Eq. (13.2):

$$N = \frac{pV}{k_B T}$$

Suppose the pressure is 1 atm (thus $p = 10^5$ N/m^2) and the temperature is 27°C (thus $T = 300$°K). If the room is 4 m $\times$ 4 m $\times$ 3 m, then the volume is

$$V = 4 \times 4 \times 3 \text{ m}^3 = 48 \text{ m}^3$$

Substituting, we get

$$N = \frac{10^5 \times 48}{1.38 \times 10^{-23} \times 300} = \frac{48}{1.38 \times 3} \times \frac{10^5}{10^{-23} \times 100}$$

$$= 11.6 \times \frac{10^5}{10^{-21}} = 1.16 \times 10^{27}$$

So there are approximately 10^{27} air molecules in your room.

EXAMPLE 5 *Speed of oxygen molecules at room temperature.*

In the classic science fiction novel *Dune*, author Frank Herbert imagines a kind of energy shield that protects the wearer from a fast-moving projectile such as a dagger or even a bullet but lets "slow-moving objects like oxygen" in so that he can breathe. We can be sympathetic to his requirement for such a shield for the sake of a good story, but it still must be pointed out that a shield that lets oxygen molecules in will also let a bullet in.

Let's calculate the speed of an oxygen molecule at ordinary room temperature. The formula for the speed v can be figured out from the fact that the average kinetic energy $\frac{1}{2}mv^2$ of a molecule of mass m at an absolute temperature of T is given as $\frac{3}{2}k_B T$:

$$\tfrac{1}{2}mv^2 = \tfrac{3}{2} k_B T$$

$$v^2 = \frac{3k_B T}{m}$$

$$v = \sqrt{\frac{3k_B T}{m}}$$

In this formula v must be regarded as an average speed. We now can calculate a number for it; $T = 300$°K, $k_B = 1.38 \times 10^{-23}$ J/°K, and the mass of the oxygen molecules of the air can be taken to be 5.3×10^{-26} kg. Therefore,

$$v = \sqrt{\frac{3 \times 1.38 \times 10^{-23} \times 300}{5.3 \times 10^{-26}}} \text{ m/sec} = \sqrt{2.34 \times 10^5}$$

$$= \sqrt{23.4} \times \sqrt{10^4} = 4.84 \times 10^2 = 484 \text{ m/sec}$$

Thus the oxygen molecule is not a slow-moving object at all; its speed is quite comparable to the speed of a bullet.

■ 13.3 Heat Engines and the Efficiency Question

Heat engines play a crucial role in modern technology. Most electric power plants use the intermediary of a heat engine. The energy of the fuel is first converted into heat, which then is converted into mechanical energy through a heat engine called the turbine. The final step is the conversion of the mechanical energy of the turbine into electricity (Fig. 13.11). The automobile engine is also a heat engine. Here the fuel energy is converted into heat inside the engine itself (that's why the automobile engine is called an internal combustion engine).

The entropy law greatly restricts the efficiency of heat engines. The man who first recognized this was Sadi Carnot, a French engineer, who around 1824 derived an expression for the maximum efficiency that can be achieved from a heat engine. Carnot's expression is very useful in any discussion of the question of the efficiency of the heat engine.

The efficiency of an engine can be defined as

$$\text{efficiency} = \frac{\text{work output}}{\text{energy input}} \tag{13.3}$$

Suppose the heat input to the engine is Q_1 and the work output is W. From the law of conservation of energy, $W = Q_1 - Q_2$, where Q_2 is the energy dumped into the heat sink, the environment (Fig. 13.12). Thus the efficiency can be written as

$$\frac{W}{Q_1} = \frac{Q_1 - Q_2}{Q_1} = 1 - \frac{Q_2}{Q_1}$$

This expression is valid for all heat engines. For an ideal engine—one that gives maximum possible efficiency under the restrictions of the entropy law—Carnot was able to prove the following. If the engine operates between a heat source of Kelvin temperature T_1 and a heat sink of Kelvin temperature T_2, then the expression for the efficiency reduces to

$$\text{Carnot efficiency} = 1 - \frac{T_2}{T_1} \tag{13.4}$$

Note that the Carnot efficiency is zero if $T_2 = T_1$, as it must be since no work can be obtained from an engine that operates in a system in which all the parts have the same temperature. Additionally, notice that the efficiency attains the value of 1 (100%) only if $T_2 = 0$ or if $T_1 = \infty$ (the last condition is due to the fact that

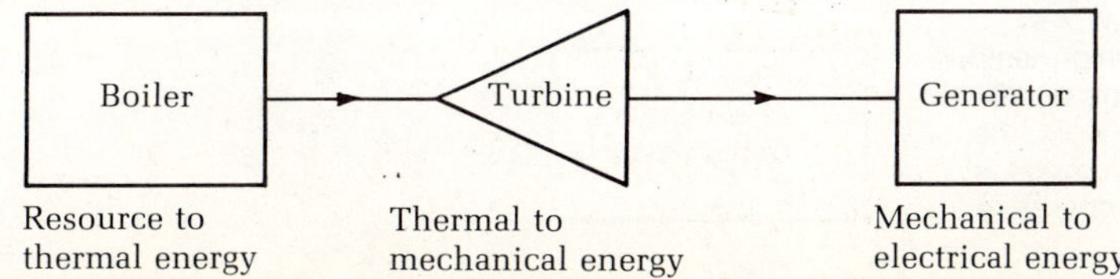

FIGURE 13.11 A schematic diagram of the components of an electric power plant.

The concepts of heaven and hell have dominated many of the theological beliefs of humanity. Hell is portrayed as the ultimate of disorder, a place where maximum entropy prevails always. On the other hand, heaven is the place where order prevails.

This picture of hell as the place where chaos reigns suggests a very uncomfortable place to live, since no energy is available there to bring in even the slightest comfort. Incidentally, if you are sold on poet John Milton's picture of Satan rebuilding a "kingdom" of hell where he will reign, reconsider.* Without utilizable energy he couldn't build anything. So hell is sort of the end of it all, according to physics as well as to religion.

In contrast, if heaven is accepted as the state of infinite order, or zero entropy, thermodynamics tells us that it is unattainable. It's only an eternal quest for us, forever unseen and unknown.

Physics likewise resolves an age-old mystery that has baffled many Eastern philosophers. I am talking about the attainment of Nirvana—perfect order again, but this time in the Buddhist sense. Nirvana is often portrayed as the end of one's self. Philosophers have wondered if it is desirable to give up the self, because once Nirvana is attained, one won't even know what one gave up. Physics, in its claim that the "absolute zero" is unreachable, may easily be extrapolated to imply that Nirvana is equally beyond our reach.

This really leaves us at a fine place as far as I am concerned, since reaching Nirvana or heaven or perfect order is also an end of the game. I am content knowing that I can stay in "the game," forever seeking and finding, never having found the final "it."

---

*In *Paradise Lost*.

■

any number divided by an infinitely large number gives zero). But both of these values of temperature are impossible to attain. Thus we can never have 100% efficiency even for an ideal heat engine. You can't break even.

To give you some idea of the maximum theoretical efficiency we can expect from a steam turbine in a coal-powered electrical power plant, let's estimate a number. The steam used is superheated to a temperature of about 500°C (773°K $= T_1$). The waste heat is dumped into the environment with a temperature of about 15°C (288°K $= T_2$). Thus the maximum efficiency obtainable from such an

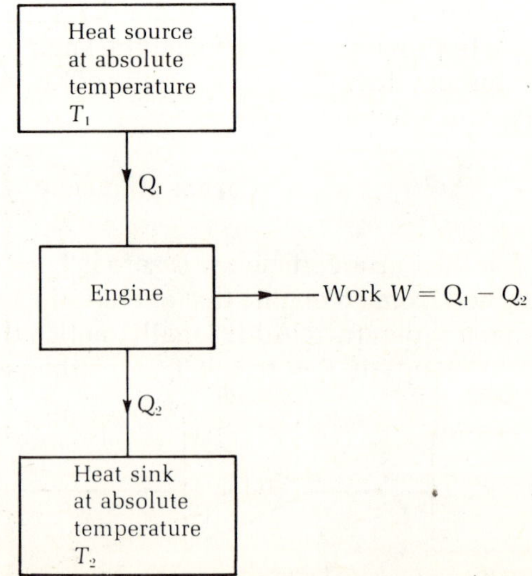

FIGURE 13.12 The schematic diagram of a heat engine. Heat is extracted from a source at a high temperature T_1; the engine converts part of the heat energy to mechanical work; the rest of the heat energy is dumped into the heat sink at a lower temperature T_2.

arrangement is given from the Carnot formula as

$$1 - \frac{288}{773} = 1 - .38 = 0.62$$

roughly 60%. In practice, the efficiencies attained for a steam turbine are only about 45%.

How can we improve the efficiency of heat engines? A glance at the Carnot formula suggests a couple of possibilities. We can try to use as low a value of T_2 as possible. Obviously we do not want to expend energy to attain a low value of T_2—for example, using a refrigerator for this purpose is self-defeating. But perhaps we should have all our power plants located in places where the year-round temperature is the lowest in the country, perhaps Alaska. One catch (among others) with this kind of thinking is that it costs energy to transfer electricity over long distances, much more than we gain from the increase of efficiency.

We can also ask this: Why not increase T_1 to higher values than the 773°K used in the estimate above? This is a technical question and the experts tell us that not much improvement is foreseen. In fact, in nuclear power plants, T_1 is even lower than the value used above, and so the efficiencies attained are somewhat below that of coal power plants. Even lower efficiencies are obtained from geothermal power plants, where natural underground steam is used with a value of T_1 of only about 500°K.

Perhaps at this point we should give you some idea of how a steam turbine works without going into too much technical detail. Fig. 13.13 shows a multistage turbine, which consists of a set of fixed blades (plain in the figure) that direct the entering steam (from the left) onto a set of blades (striped in the figure) that are attached to a shaft and rotate with it. Since the steam expands as it does work on the blades, the turbine chamber also expands, to the right. When the steam has done all the work it can, it is directed into the condenser to dump its remaining (waste) heat to a lake or river. The condensed steam is fed back into the boiler to start the cycle again.

At this point you may suggest a couple of cost-saving propositions. First, why go through the trouble of condensing the spent steam? Why not let it escape to the atmosphere and use fresh water to restart the cycle? After all, water is cheap.

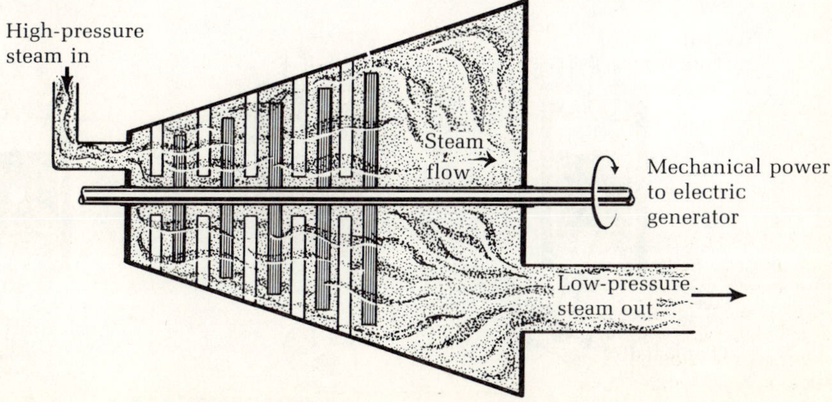

FIGURE 13.13   A multistage turbine.
See text for description.

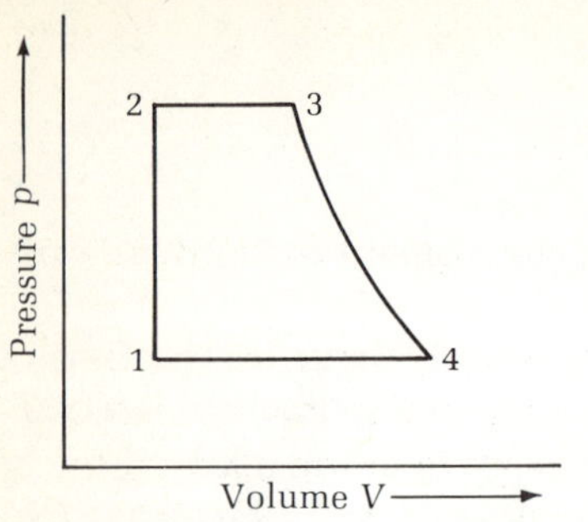

FIGURE 13.14 A *p-V* diagram of the conventional steam cycle. See text for explanation.

FIGURE 13.15 The four-stroke internal combustion engine. (a) The gas-fuel mixture rushes into the cylinder. (b) The upward motion of the piston compresses the gas; the spark plug fires at the end of this stroke, raising the gas to a very high temperature. (c) The gas expands, pushing the piston; this is the power stroke. (d) The burnt gases are pushed out the exhaust pipe. After this the cycle begins again.

Or is it? Actually, the water in a steam turbine must be purified through expensive processes to prevent corrosion and mineral desposits in the system. So it is actually more economical to recycle the water. Also, if large amounts of steam were to be released in the vicinity of all heat power plants, the weather patterns of many localities would be affected.

You may still ask: Why not recycle the steam without condensing it? Before answering this question, let's study the working of a conventional steam cycle through the use of a diagram, where the pressure of the fluid is plotted against its volume at all stages of the cycle. Such a diagram is called the *p-V* diagram (Fig. 13.14).

The first stroke of the cycle corresponds to the vertical line 1-2 in the diagram. Water is being heated in the boiler, the volume is constant, and the pressure builds up rapidly. At the second stage 2-3, steam is being formed; presssure remains constant at this stage, but the volume increases somewhat. The third stage is the working stroke. The line 3-4 shows the steam expanding rapidly as the pressure drops to the starting value. The turbine moves, and external work is done at this stage. The fourth stage is the return stroke; the spent steam is condensed and the water is pumped back to its starting point.

Now to answer the original question: why not reuse the steam without the condensation? Look at the *p-V* diagram. In order to use the steam for external work, we must return to point 3 on the cycle. The only way to do this, aside from condensation, would be to compress the steam back to point 3. But this is a losing proposition because then we are doing as much work on the steam as the steam did for us.

Having to dump some heat into the environment in order to get work from a heat engine is a requirement of the second law of thermodynamics. There is no escape from this law.

The automobile engine

The automobile engine operates with the following four strokes (Fig. 13.15): (a) intake, in which the intake valve opens, admitting the fuel-air mixture; (b) compression, in which the fuel-air mixture is compressed by the piston, now driven by a flywheel; at the end of compression the mixture of fuel and air is

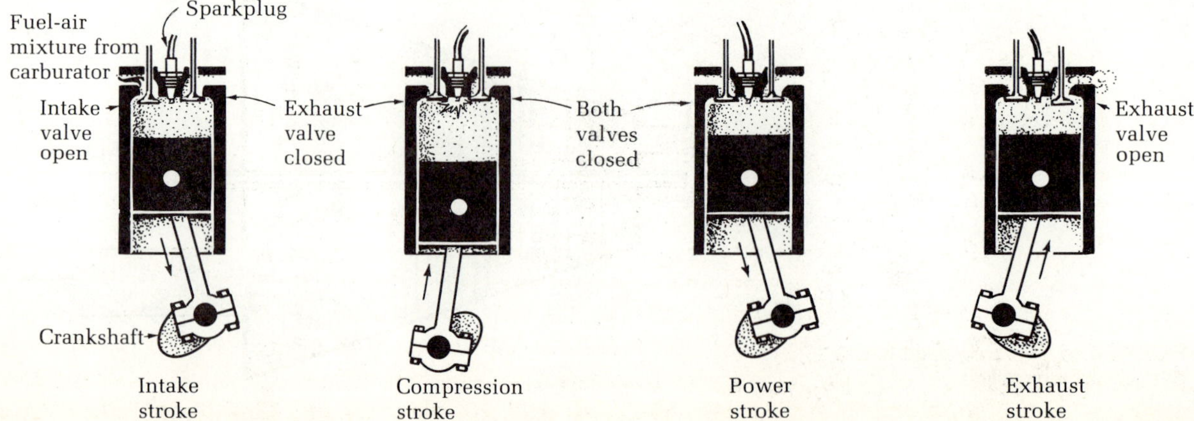

ignited by means of sparks from the spark plugs; (c) the working stroke, in which the hot gas expands and drives the piston so that the car gets a thrust; (d) return, in which the piston returns, driving the burnt gas out through the exhaust valve, which is now open.

Let's calculate the Carnot efficiency of such an engine. The combustion temperature of the fuel-air mixture is very high in the internal combustion engine, something like 2000°C, or 2273°K. Unfortunately, the exhaust temperature is also high, about 500°C, or 773°K. Thus the Carnot efficiency is given as

$$1 - \frac{773}{2273} = 1 - .34 = 0.66 = 66\%$$

In actual automobile engines, the fuel is not burnt completely. Moreover, there is a lot of leakage. Thus the actual efficiency never reaches a value even close to the Carnot limit. A typical car engine efficiency is only about 20%.

What is the possibility of improvement? The experts tell us there is not much, at least not in engine efficiency. Fortunately, the efficiency of the car as a transportation device must be defined as the number of miles per gallon achievable by the car, and this we can improve by using compacts and subcompacts, which have lower mass. Another idea for improved efficiency is to use a flywheel, which can store some of the energy (in the form of kinetic energy of its rotation) otherwise lost as heat when the car makes a stop.

There is still another way to look at the matter of automobile efficiency. Instead of miles per gallon, we can talk about an efficiency in terms of passenger-miles per gallon. We can immediately see that there is a real advantage to using car pools and other mass transport for most of our transportation needs. For example, an average compact that carries 2 passengers perhaps gets 16 miles/gallon, which is 32 passenger-miles/gallon. This efficiency is doubled with carpooling, thereby increasing the average number of passengers by a factor of two. And for an inner-city bus, which may get only 4 miles/gallon but carries 32 passengers, the passenger-miles are 128 per gallon, and the efficiency has doubled again.

The heat pump

It was not long ago that utility companies advertised the benefits of clean all-electric homes in which electricity would be used for space heating. At first thought, electrical space heating may seem desirable, but we really lose in terms of efficient use of the resource energy. Oil furnaces for household heating operate on the average with a 60% efficiency, whereas electricity is produced from the same oil with an efficiency of only 40%. Aside from efficiency, the electricity is not even clean (except for hydroelectricity). As you know, conversion of a fuel into electricity at an overall 40% efficiency also converts 60% of the fuel into thermal pollution. And production of electricity from oil involves almost as much air pollution as direct burning of oil for space heating.

However, there is a way to use electricity for space heating that gives us more for our money: the use of a heat pump. What is a heat pump?

Do you remember our discussion of the refrigerator? A refrigerator is a heat engine in reverse: heat is taken from inside a sink—a reservoir at low temperature—and delivered to a reservoir at higher temperature. An amount of work

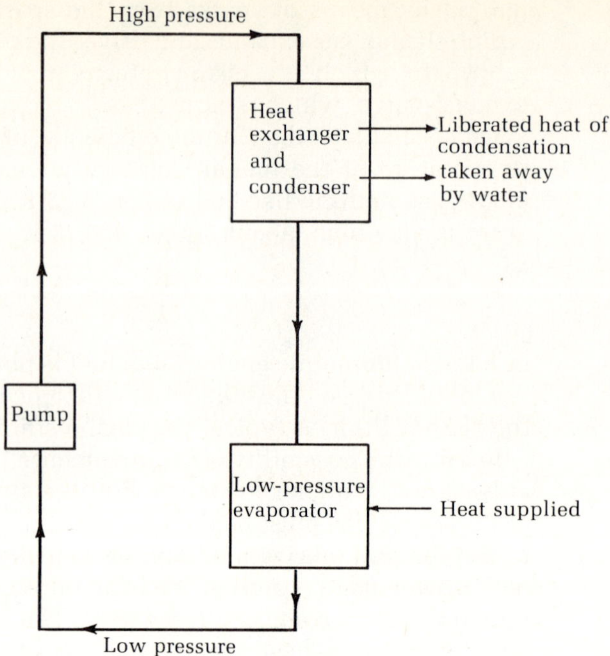

FIGURE 13.16 A schematic diagram of the operation of a heat pump. See text for description.

W is done on the machine to make this possible. The heat delivered to the high-temperature reservoir Q_1 is the sum of the heat taken from the cold reservoir Q_2 and the work W:

$$Q_1 = Q_2 + W$$

Suppose we use the outdoors around a house as the low-temperature reservoir (in the winter) and the inside of the house as the high-temperature one. A machine operating on the same principle as the refrigerator can now take heat from the outside and deliver it to the inside. Since this is like pumping heat from a cold place to a hot one, it is customary to call such a machine a heat pump.

Thus the heat pump is really a refrigerator, but it is more versatile. In the summer we can use it to pump heat from the cold inside to the hot outside, as an air conditioner.

How does a heat pump work exactly? The basic idea is to evaporate a liquid, which requires heat energy. The vapor carries the energy in the form of chemical potential energy. When the vapor condenses back (inside the house to be heated), the energy is released. But isn't the outside environment too cold to evaporate the liquid? First of all, there is always some heat energy. Second, we could put the vessel containing the liquid to be evaporated underground for better operation, but this is not always necessary. However, there is one essential factor: all liquids evaporate at a lower temperature if the pressure is lowered. So the outside evaporator always contains a liquid like ammonia at a low pressure.

After evaporation, the vapor is compressed to a higher pressure and passed through the coils of a condenser, which are cooled by circulating water. The vapor condenses, giving the released energy to the water, which can be used for heating by means of regular household radiators, for example. The condensed liquid is, of course, recycled. It is passed through a valve to expand and evaporate at low pressure to start the cycle over again (Fig. 13.16). As you can see, electrical energy is needed to run the compressor.

What kind of restrictions, if any, does the entropy law put on the performance of a heat pump? Since the heat pump is designed to supply heat, which is disordered energy, the entropy law does not impose very severe limits on the performance of the machine. We can define a coefficient of performance as

$$\text{coefficient of performance} = \frac{\text{heat delivered } Q_2}{\text{work done } W}$$

Then for an ideal heat pump, considerations like Carnot's show that the coefficient of performance reduces to the following expression:

$$\text{ideal coefficient of performance} = \frac{T_2}{T_1 - T_2}$$

Suppose that on a winter day the outside temperature is a cold 0°C (273°K), and you would like to have the inside at a comfortable 21°C (294°K). Then the coefficient of performance of an ideal heat pump is

$$\frac{273}{294 - 273} = \frac{273}{21} = 13$$

This tells us that we can never design a heat pump with a higher coefficient of performance than 13 when operating under these conditions. In practice, heat pumps give only a coefficient of performance of 3.

But even the value of 3 is pretty good. This means that for every unit of electrical energy it takes to run the heat pump, we get 3 times as much heat delivered. So even if the electricity comes at an efficiency of 40% (0.4), buoyed by the heat pump, we get $0.4 \times 3 = 1.2$, or 120% heat from it, more energy in terms of resource energy than we put in. The extra energy, of course, comes from the environment.

Can we get work out of the heat energy of the random motion of molecules in the oceans of air and of water?

We may have given you the impression that it is impossible to tap the energy of random motion of such vast reservoirs as our atmosphere or our huge oceans of water. These are systems in which all parts are, by and large, at the same temperature. So according to the second law of thermodynamics, it is not possible to extract useful work from such a system. However, it is also true that neither the atmosphere nor the oceans ever achieve a complete equilibrium of one temperature, due mostly to the action of solar radiation. The laws of thermodynamics apply to systems in equilibrium; thus there is a loophole. How large a loophole?

Actually, we can construct a heat engine that operates taking advantage of the nonequilibrium nature of the atmosphere. One such heat engine is the famous toy, "the dunking duck" (Fig. 13.17). As the figure shows, in the lower bulb of the duck there is some liquid, which has to be a highly volatile one. The idea is

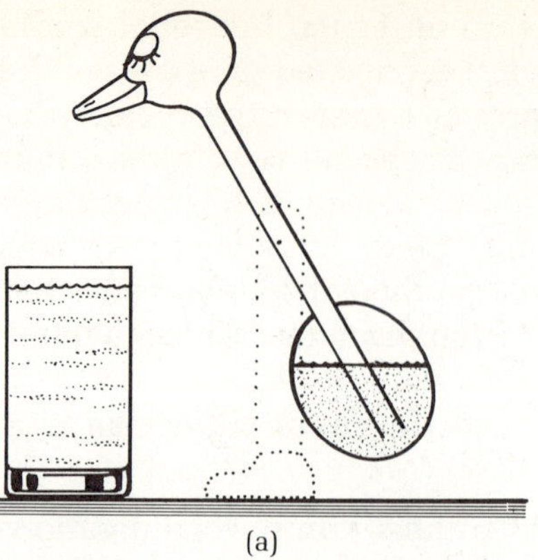
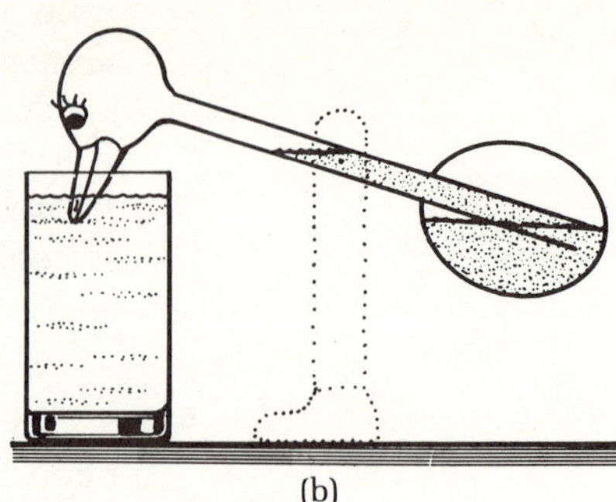

FIGURE 13.17 (a) A dunking duck in the upright position. (b) When sufficient liquid is sucked up the tube, the bird dunks into the water of the beaker.

to generate a lot of vapor pressure even at room temperature. Often ethyl ether is used. The upper tube is sealed to the lower bulb and extends into the volume of liquid as shown. Notice that in this position, where the bird is erect, the upper region of the bulb is isolated from the upper region of the system (the tube).

To start off, we put a little water on the felt head. The evaporation of the water cools the head and the upper part of the tube, and thus the pressure of the vapor there is reduced. But the vapor pressure in the space above the liquid in the bulb remains the same as before, which is now higher than that in the tube. This higher pressure will push the liquid up inside the tube. When sufficient liquid goes up the tube, the center of gravity shifts and the tube tilts, causing the bird to dunk into the water [Fig. 13.17(b)].

As the bird dunks, the lower end of the tube rises to the surface of the liquid in the bulb so that a series of vapor bubbles can now pass up the tube to equalize the pressure. The arrangement is such that the tube never becomes quite horizontal, so now the liquid can return to the bulb and the bird becomes erect, to start the cycle over.

We can run an engine using the movements of the duck (Fig. 13.18).* The energy comes from the atmosphere. Since the atmosphere is indeed cooled a little locally, energy is conserved. A more baffling question is this: Is there a violation of the second law of thermodynamics?

As stated in the beginning of the subsection, there is no violation because the system is not in equilibrium. There is obviously a one-way motion in the system: the water only evaporates one way—an equal number of water molecules do not condense back on the head of the bird. The vapor as it forms outside the head is removed by local wind (indeed, a fan will increase the performance of the engine, but then you will be using more energy than you can get out of the system). Therefore, a complete equilibrium is never established between the water vapor and the water on the bird's head.

So the operation of an engine depends crucially on the local fluctuation of equilibrium conditions in the atmosphere. Since these fluctuations are never very large, the energy we can get out of such a system operating in air is rather small, which is the reason nobody has tried a system like this on a large scale.

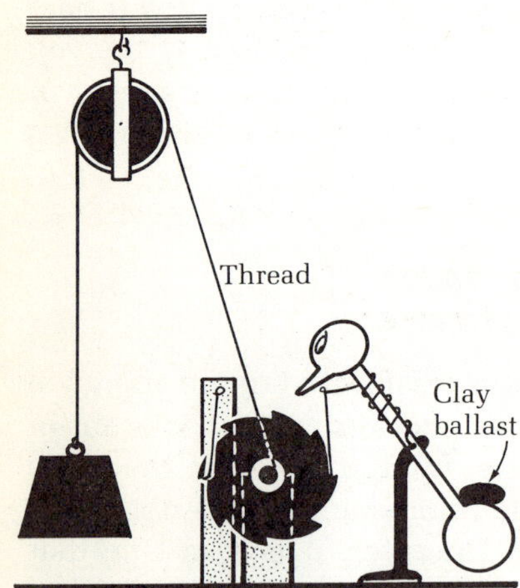

FIGURE 13.18 A heat engine using the dunking duck idea. [Reprinted with permission from Kemp Bennett Kolb, "Reciprocating Engine," *Physics Teacher*, vol. 4, No. 3, (1966), p. 121.] Copyright 1966 by The American Association of Physics Teachers.

---

*Kemp B. Kolb, "Reciprocating Engine," *Physics Teacher*, vol. 4 (1966), p. 121.

How about the ocean water? Does nonequilibrium rule there as well? The answer is, yes, and on a large scale—large enough, in fact, to make an engine worthy of active consideration. The temperature of ocean water at many places varies considerably with depth, as much as 20 degrees (Celsius) between the surface and a lower depth of just 1 km. So the ocean isn't really a system of identical temperature everywhere in the sense of the second law of thermodynamics, and we could run an engine between the surface and the lower level, taking advantage of the existing difference of temperature. Since the difference is really caused by uneven heating of the sun, obviously the donor of energy in this case is the sun (as in the atmospheric engine discussed above). We will not go into the details here of the proposals under study for use of solar sea power.*

SUMMARY

This is the basic chapter concerning the subject of heat and, as such, many different concepts are introduced without a major theme. The first section discusses such basic thermal properties of objects as specific heat and latent heat. The latter arises in connection with the change of state of matter from solid to liquid, or liquid to gas. The heat of chemical reactions is another important topic covered.

The dominant theme of the second section is the molecular understanding of the concept of temperature via the gas laws. Among the gas laws discussed are (1) Boyle's law, $p \propto V$; (2) Charles' law, $V \propto T$, where T is the absolute temperature of a body (Charles' law leads to the concept of absolute zero); (3) the perfect gas law, $pV = Nk_BT$, Eq. (13.2).

The third section deals with heat engines and their efficiencies. The most important consideration in connection with the efficiency question is Carnot's expression, which gives the maximum theoretical efficiency of a heat engine as

$$1 - \frac{T_2}{T_1}$$

Also discussed in Section 13.3 is the functioning of actual heat engines and that of the heat pump.

---

*If you are interested in doing some further reading on the subject, see C. Zener, "Solar Sea Power," *Physics Today*, vol. 26 (1973), p. 48.

QUESTIONS

Review and reason

1. Define the specific heat of a substance. What is meant by molar specific heat? Why do diatomic gases have a greater molar specific heat than monatomic gases? Why do monatomic solids have a greater molar specific heat than monatomic gases?

2. Ocean water is warmer than surrounding land areas at night. Why?

3. If you have a potbelly stove, you can utilize its heat even better by putting a heavy steel block on the stove while the stove burns fuel. Explain how the steel block helps.

4. An empty room seems to cool faster than a room full of furniture. Why?

5. The temperature at ground level during the day does

not reach a maximum at exactly 12 noon but at nearer to 2 o'clock in the afternoon. Why is that?

6. You can cool your hands by blowing hard on them. Explain why.

7. Give a molecular explanation of the cooling produced by evaporation.

8. When you take a shower with the windows and doors closed and no fan blowing, the bathroom mirror gets foggy. But with proper ventilation, the mirror stays clear. Explain the difference.

9. It is possible to severely damage a frying pan while cooking food, but only after all the water has evaporated away. Why is this?

10. Households often are heated through hot-air and hot-water systems. However, for very large living areas—for example, a large apartment building—it is preferable to use a steam-heat system to convey heat from the furnace to the living areas. Explain.

11. In Death Valley, California, would it be easier or more difficult to cook rice, or does it make any difference?

12. On cold nights it is quite common for orange growers to flood their irrigation canals. This is supposed to keep their orange trees from freezing. Explain how this works.

13. Why is it difficult to ice-skate when it is very cold?

14. Define the latent heats of vaporization and of melting of a substance. Give a molecular explanation (in terms of binding energy) of why the latent heats of vaporization of the liquid form of substances are usually greater than the latent heats of melting from their solid state.

15. Distinguish between endothermic and exothermic reactions. Explain, in terms of binding energy, why certain chemical reactions are exothermic and others endothermic.

16. If you pour boiling water on a thick-walled glass, it probably will break, but a thin glass may survive the hot water. Explain.

17. Why do lakes freeze in the winter from the top down instead of from the bottom up?

18. State Boyle's law and give an explanation for the law based on kinetic molecular theory.

19. A child has an inflated balloon on an airplane. For some reason the aircraft depressurizes its chamber. What happens to the volume of the balloon?

20. There is a story of a wedding party given in an underwater tunnel. The party was somewhat flat because the champagne did not pop. The guests, most of them at least, drank the champagne anyway. Later, upon coming to the surface, a few of them complained about popping in their stomachs. Why didn't the champagne pop in the tunnel? And why did the guests suffer from popping stomachs when they surfaced?

21. State Charles' law. How do you explain his law from the molecular point of view?

22. After driving a few hours, the pressure of the tires on a car changes. Give a kinetic molecular explanation of this.

23. What is absolute zero? How is the absolute scale of temperature defined?

24. Write down the mathematical relationship known as the perfect gas law. Explain the notation.

25. You have a small container of hydrogen gas and another container of oxygen gas of identical volume. The two gases in the two containers are also at identical temperature and pressure. Can you tell how many times more molecules there are in the oxygen container than in the hydrogen container? Explain your answer.

26. Write down Carnot's equation for the maximum theoretical efficiency of a heat engine and explain the notation.

27. Do you expect the efficiency of a power plant to be greater in the winter or in the summer? Explain your answer.

28. Would it be advantageous to refrigerate the water used for cooling in a steam power plant? It certainly reduces T_2 in Eq. (13.4), but would the overall efficiency improve? Explain.

29. Describe the cycles of an internal combustion engine.

30. What is a heat pump? Explain how it works.

31. Explain the operation of the dunking duck. Does a heat engine based on a dunking duck violate the second law of thermodynamics? Why or why not?

32. If we run a heat engine between the warmer top layer and a colder layer of water, 1 km below the surface of the ocean, the temperature of the two layers will tend to equalize. Why? What can we do to ensure a continuous operation of the engine at a steady efficiency?

Arithmetic

1. Calculate the amount of heat needed to heat a 100-kg iron block through a temperature of 50°C.

2. Calculate the amount of heat needed to convert 1 kg of ice at 0°C to 1 kg of steam at 100°C.

3. A child inflates a balloon on board an airplane. Suddenly the pilot announces that due to unforeseen circumstances the cabin pressure is being lowered. As the cabin pressure begins to decrease, the volume

of the balloon starts to increase. Why? Finally the volume of the balloon stabilizes after it reaches a value twice its original volume. Can you figure out from this by what factor the pressure was changed?

4. What is the maximum possible efficiency of a gasoline engine in which fuel is burnt at a temperature of 1200°C and the exhaust temperature is 200°C?

*5. It is likely that a 1000-MW electric power plant, which uses coal as its fuel, runs at an efficiency of 33%. This, of course, means that it releases 2000 MW, or 2×10^9 J/sec, of heat energy into a local river. Assuming that the rate of the flow of water in the river is given as 10^5 kg/sec, calculate the rise of temperature of the river from all that heat.

6. A power plant burns oil and generates electricity delivered to the users at an efficiency of 35%. Some possibilities for heating a home near the power plant are to (a) use the electrical energy directly for heating, with an efficiency of 100%; (b) use oil for heating by means of an oil furnace, which runs at an efficiency of 70%; (c) use a heat pump that delivers three times as much heat to the house as the electrical energy required to run it. Which of the three possibilities consumes the least amount of oil? Which of the possibilities consumes the most oil? Explain.

7. Suppose the top layer of the ocean at a certain place has a temperature of 27°C and the temperature at a depth of 1 km is given as 7°C. Calculate the maximum theoretical efficiency of a heat engine which uses the warm layer as the heat source and the colder layer of water as the sink.

---

*Optional.

14 Thermal Physics and the Environment

O wind, rend open the heat,
cut apart the heat,
rend it to tatters.

. . .

Cut the heat—
plough through it,
turning it on either side
of your path.*

H. D.

14.1 The Transmission of Heat: Conduction, Convection, and Radiation

Two objects in thermal contact tend to equalize their temperature. How do they do it? That's easy to answer—by the transmission of heat from one body to another. There are three modes of such heat transfer: **conduction, convection,** and **radiation.**

Put a metal rod in an open fire while holding on to the other end of it. You won't be able to hold on for long—the heat will reach your end very quickly. The metal of the rod has conducted the heat from the fire to your hand.

When somebody says, "heat rises," what does that mean? What rises is the warm air—or more precisely, the convection currents of warm air. Hold your hands over an open fire. You can feel the hot-air molecules bringing your hands warmth. This convection is different from conduction. Convection occurs through actual transfer of energetic molecules from one place to another. In conduction, on the other hand, heat is taken from place to place by successive

*"Heat," from H. D. (Hilda Doolittle), *Selected Poems.* Copyright © 1957 by Norman Holmes Pearson. Reprinted by permission of New Directions Publishing Corporation, Agents.

FIGURE 14.1 The three means of heat transfer between bodies. (a) Conduction: heat is conducted by the metal rod, but no mass is transferred from one place to another. (b) Convection: the convection currents of warm air transfer heat by the actual transfer of air molecules from one place to another. (c) Radiation: heat transfer via infrared and other electromagnetic radiation; no material medium is needed.

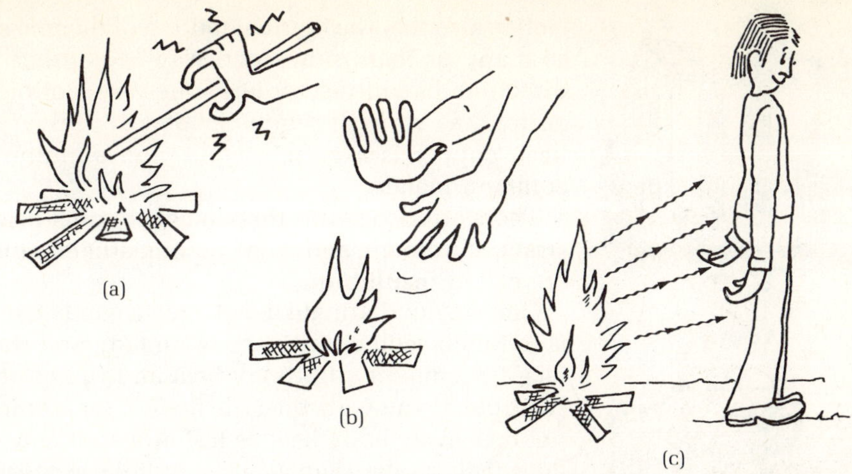

collisions; vigorously jiggling molecules collide with their neighbors and share energy. The game is joined by other neighbors until every molecule of the body (or bodies) in contact with the hot object is sharing energy.

Solids are conductors of heat, but since the molecules in solids move about fixed positions, solids don't allow any convection. In contrast, liquids and gases are poor conductors; but since their molecules are free to move, they can transfer heat by convection.

Both conduction and convection need a material connection of some kind in order to transmit heat from one body to another (Fig. 14.1). Then how does the heat from the sun reach us? The space between us and the sun is practically empty; the solar heat does not reach us either by conduction or by convection. It reaches us through radiation; the light that we receive (along with some infrared and ultraviolet radiation) brings us the heat. If you stand in front of an open fire, the heat you receive is reaching you mainly through (infrared) radiation. (Air is a poor conductor, and the hot convection currents of air rise; they don't go sideways.) As a process of heat transfer, radiation consists of three distinct steps. First, the hot body emits radiation, converting heat energy into radiant energy. The radiation then travels. Finally, when the radiation encounters another object, this object may absorb all or part of the radiant energy, which then becomes heat again. Also, notice that through radiation two objects can be in thermal contact without any material contact.

Conduction

Metals are the best **conductors** that there are. The reason was suggested by German physicist Paul Drude, who pointed out that some of the atomic electrons in a metal are perfectly free to move about within the metal. In other words, these free electrons are not bound to one atom in particular but to a piece of metal as a whole. The mobility of the free electrons is crucial to the thermal conductivity of a metal. Whereas the atoms in the metal are restricted to vibration about fixed positions, the free electrons do not have any such restrictions. It is the random collisions of these electrons with one another and with the atoms that makes a metal so effective in transporting heat from one place to another.

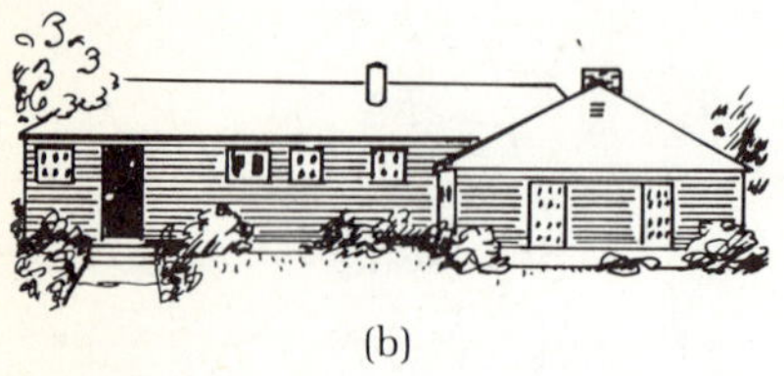

FIGURE 14.2 (a) A heat-conserving house is noted for its compactness. (b) This house loses a lot of heat because of its large surface area.

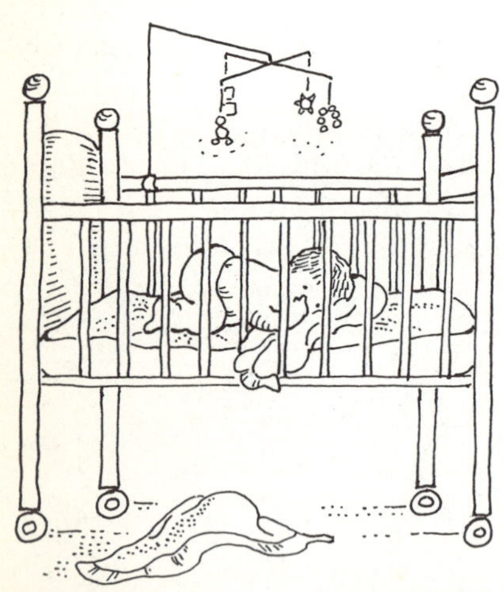

FIGURE 14.3 A baby instinctively will minimize its surface area in order to keep warm. This reduces its conduction heat loss.

Notice also that the thermal motion of the free electrons is completely random—as many of them move one way as another. There is no net motion in one direction. In contrast, orderly one-way motion of electrons gives rise to electric current, as we will see in Chapter 16. Silver is the best metal for conducting heat, copper is a close second, and aluminum and iron are next among the common metals.

There are also solids that do not conduct heat well at all, solids such as wood, straw, styrofoam, and wool, among others. Substances that are poor conductors are called **insulators.**

One way to distinguish between a good heat conductor and an inefficient one is to touch each at ordinary room temperature. A good conductor will quickly take up some of your body heat and therefore will feel cool to the touch. For example, if you walk on a tile floor, it feels cold. But an insulating material does not take away body heat as fast and so it feels comfortable. Even so, a wooden floor feels cooler than a carpeted floor because of its slightly better conductivity.

The conduction of heat depends on the surface area of contact. A house with a large surface area of exposure to the outside loses more heat (most of it by conduction through the walls and the ceiling to the outside atmosphere) than a more compact house (Fig. 14.2). Even babies seem to know this fact about compactness. A baby sleeping under cold conditions commonly will roll itself up in a ball, exposing a minimum surface area (Fig. 14.3). Heat conduction depends also on the thickness of the conducting material (the thicker the material, the lower the rate of conduction) and the temperature difference between the two bodies in contact (a greater temperature difference results in more conduction).

The automobile radiator is an example in which we make good use of all these factors affecting the conduction of heat. As you know, a car engine accumulates a very large amount of heat, which needs removal as fast as we can manage it, otherwise the engine just might melt down. A stream of water is circulated within the engine (through specially designed pipes) to take up the heat. The water is pumped through the radiator, where it loses its heat. It's then recirculated back to the engine to begin the cycle again.

Inside the radiator, the water flows through a series of vertical tubes. These tubes are flattened and supported by a large number of horizontal metal sheets; both factors contribute to increase the effective area of heat loss by conduction. The tubes and their supports together give the radiator its unique honeycomb appearance.

All parts of the radiator are made of copper, one of the best possible conductors of heat. The thickness of all the metal walls is kept at a minimum to increase conductivity. Finally, there is the matter of temperature difference. Initially the temperature of the water inside the radiator tubings is about 180°F, and the outside temperature where the heat is dumped is ordinary air temperature. But as the radiator conducts heat to the outside, this situation changes drastically. The air in contact with the outside wall of the radiator gets hot, inhibiting the conduction. A fan continuously has to blow away this air for the radiator to function properly.

As you can see, the name radiator is really a misnomer; the device is designed successfully to get rid of most of the heat by conduction. But then, what's in a

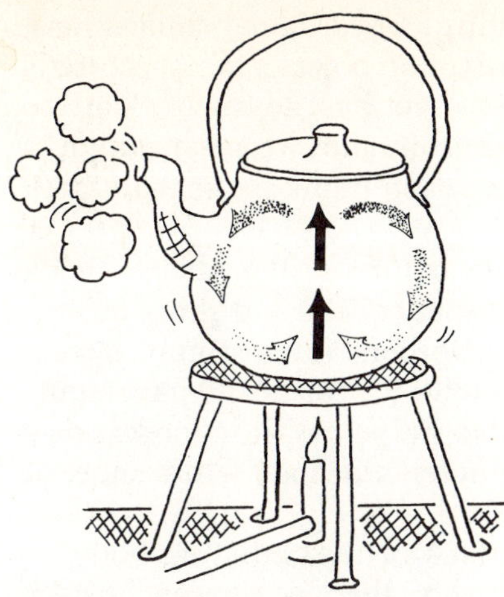

FIGURE 14.4 Convection currents in water heated from underneath. The dark arrows indicate hot water rising; the light arrows are those of descending cold water. Gray arrows indicate water of intermediate temperature.

name when the job is done well; didn't Shakespeare say something to that effect?

Designers of an energy-conserving house have just the opposite objective in mind: they want to cut down on heat loss by conduction. So they want thick walls made of insulating material, with extra insulation in the attic. They want the house to be compact, with as little surface area as possible exposed to the outside. Of this exposed area they will minimize the portion that faces directly into the prevailing wind. Finally, the occupants of such a house will try to live with as low an inside temperature as they can comfortably tolerate.

Convection

Convection is a fairly simple process. When an object is heated, it expands in volume. For a given mass of a fluid, this means a reduction of density. Since mass = volume × density, then if volume increases, density must decrease to give the same mass. Now recall the buoyant force; less dense objects are buoyed up by fluids that have a larger density. So warm air rises, as does warm water.

When we heat water in a kettle, the bottom of the kettle is in direct contact with the stove, so the water at the bottom gets hot first (Fig. 14.4). Its density reduces, buoying it upward, and its place is taken by colder water moving down from above. The upward current of warm water and the downward current of cold water make up the **convection cell.** These currents mix the water, distributing the heat uniformly, and soon the tea water is ready.

Electrical baseboard heating of homes takes advantage of convection currents of air to produce the desired temperature (Fig. 14.5). The air above the heater gets hot and rises. Cooler air from above descends, setting up the convection cells. The same principle works in setting up local winds and even global wind patterns, as we will see later.

Radiation

All objects radiate, but the wavelength of radiation depends on the temperature of the object. An ordinary hot object emits **infrared radiation,** invisible electromagnetic radiation of wavelength even longer than that of red light. As an object gets hotter and hotter, it will emit radiation of shorter and shorter wavelength. Turn on an electric burner on your stove: it first radiates in the infrared wavelength; you can feel the heat but cannot see it. As the temperature of the burner goes higher, you can see it become red; it is now emitting red in addition to infrared. With an incandescent solid like charcoal, you can see the charcoal become progressively red, orange, yellow and even white (if forced air is used) when it is emitting all the visible wavelengths. The sun's surface temperature is so high (6000°C) that it not only emits infrared and visible light, but also the even shorter wavelength of ultraviolet. All these radiations from hot objects are electromagnetic waves, so a complete treatment of these radiations will be delayed until a later chapter.

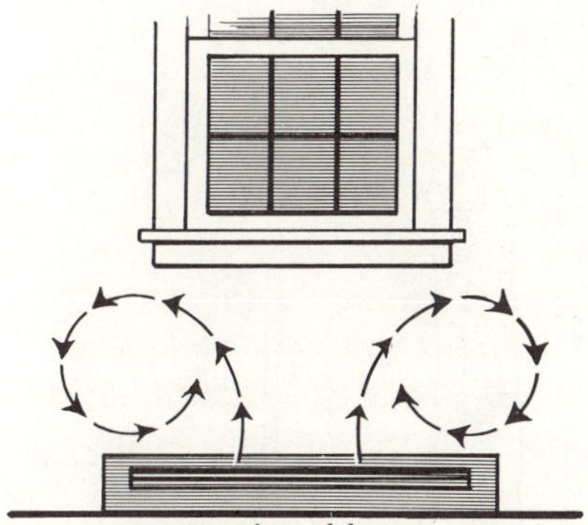

One question is often asked: If an object is always radiating, why doesn't its temperature continuously decrease? The answer is that if an object is at the same temperature as its environment, then it is also receiving heat radiation

FIGURE 14.5 Convection currents of air near a baseboard heater.

from the environment equivalent to what it is giving away. There is no net loss; therefore its temperature remains unchanged. Only if an object's temperature is higher than that of its surroundings will it suffer a net heat loss by radiation. When we say a hot object is radiating heat to its environment, we are speaking of this net heat loss by radiation. And the rate of radiation increases sharply with the temperature of the radiating body.

The amount of radiation also depends on the nature of the object. Black objects are the best absorbers of heat radiation, whereas white and shiny objects reflect the radiation to some extent and are poor absorbers. Interestingly, a good absorber of radiation is also a good emitter. Cities with icy roads and pavements often use sand to provide better traction; this also helps melt the ice because sand, being darker, absorbs the solar radiation much better than white snow or ice.

Radiation is emitted and absorbed at the surface of the radiating body. A rough surface promotes radiation more effectively than a smooth surface because of the greater surface area. So polishing can decrease the radiation loss from an object.

All three modes of heat transfer are important. Confusion about these different modes of heat transfer (or forgetting one while remembering another) can lead to some misunderstandings. Let's look at two examples. First, since air is as much of a nonconductor as cotton or wool (of which shirts are made to keep us warm), you could argue that roaming around naked should not change your heat loss. You would be correct in thinking that, for a naked person, little heat loss occurs by conduction—no more than it would for a clothed person. However, you would be forgetting about the convective process. The air adjacent to one's warm body gets hot and rises while cool air replaces it, making room for more heat loss from the body, and so on. A shirt holds the warm air in its place, and this reduces convection loss. Two shirts work better than one shirt of the same total thickness for the same reason.

A second example involves the conventional household fireplace, which

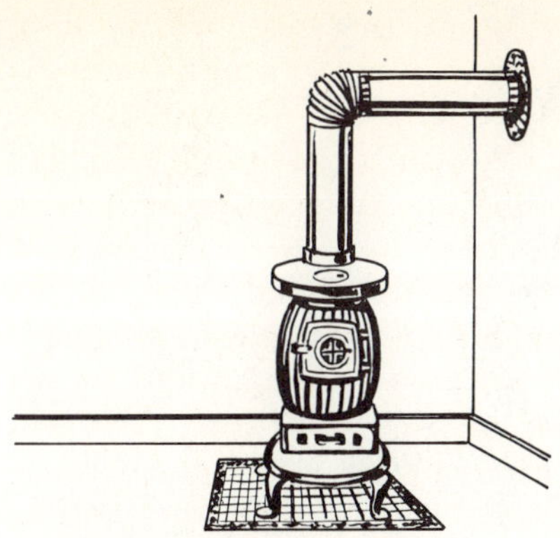

FIGURE 14.6   A potbelly stove.

many people think provides a source of heat when they build a fire. It probably does not, unless it is specially constructed. Here's why. The heat you receive from the fireplace is via radiation, and it's there, directly in front of the fire, that you can feel it. But of course there is also convection: the warm air rises through the chimney and cold air comes in, usually from the outside, to replace it and also to replenish the oxygen consumed by the fire. You can also "feel" this convection or cold-air draft in the room. So it's doubtful that there is a net heating of the room by having a fire in the fireplace.

The old-fashioned potbelly stove is a far better heat source than a fireplace, because it allows radiation into the room from all sides (Fig. 14.6). There is currently a resurgence of interest in metal stoves that, with a suitable design, even make use of some of the convection heat. The convection current goes up through metal pipes exposed to the room. The metal pipes are heated by the convection current and by conduction, and they radiate heat back to the room.

The Thermos bottle is a system in which all three modes of heat transfer are minimized in an attempt to create a highly heat-insulated environment (Fig. 14.7). It is a double-walled glass bottle with a partial vacuum between the walls. Of all things that inhibit conduction and convection, vacuum is the best. Since it contains no material at all, a perfect vacuum provides no convection or conduction. Some heat conduction still occurs because of the connection between the glass walls and the stopper. But the stopper is made of cork, which, like glass, is an insulating material, and the area of contact is small. To minimize radiation loss, the inner surfaces of the glass are silvered and polished so they reflect back, rather than absorb, most of the radiation from the contents of the bottle. Likewise, when a cold substance is placed in the bottle, since the walls are bad emitters, the contents will not be heated by excessive radiation from the Thermos wall.

Wind and the chill factor

One interesting thing about heat transfer is that our feeling of hot or cold comes from whether and how quickly we gain or lose heat. Thus the wind is very important to our comfort. If there is a wind, the number of air molecules that can in cold weather take away heat from our body or that in hot weather can add heat to our body increases, and this directly adds to our feeling of being cold or hot. (There is also increased evaporation from the body in the presence of a wind.) Often we use the concept of a "chill factor" to incorporate the effect of wind on cold weather. The midwestern city of Chicago is quite cold in the winter, and its winds add a large chill factor that make it seem much colder than the temperature suggests. Look at the attire of people (e.g., the Eskimos) in very cold weather or (e.g., the Arabs) in very hot weather. Their clothes are always designed to protect against the wind.

Speaking of clothing, fur is an excellent insulator against cold, as you may know. However, its effectiveness wears off when the wind is high unless the fur is used as the inside lining of the clothing. So for warmth people really should wear their fur coats inside out.

Here is one more interesting thing. You may have heard the following very

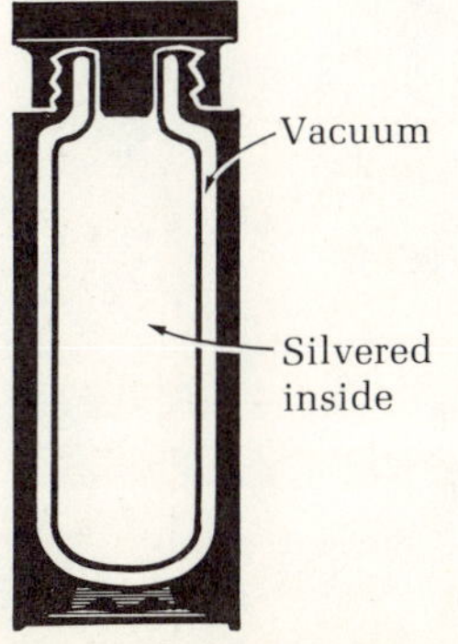

FIGURE 14.7   A Thermos bottle.

startling news. The upper atmosphere, more than 100 miles up, is very hot. This is true if the speed of the air molecules is the indicator of temperature, as we decided in the last chapter. By that definition the temperature of the atmosphere at such heights comes out to be some 25,000°C.* But it is also true that if you were to expose yourself to this high temperature in the upper atmosphere, you wouldn't burn. The reason is that the atmosphere at such heights is very, very thin, so that many fewer air molecules would impinge on you in a second than at ground level. In the absence of many collisions, there is very little heat exchange and you would not be able to feel the heat of the air at all.

---

*The molecules at the outer edge of the atmosphere get extra energy from the unfiltered (by the lower levels of the atmosphere) ultraviolet component of solar radiation.

Physics in the kitchen

What kind of pots and pans should you use for cooking, or does it make any difference? Many a chef will claim that cast iron pots are the best, but they are heavy to carry. Is there any physical basis for the chef's claim?

The purpose of the cooking pan is to take heat from the burner and distribute it evenly to the food inside it. The material must have high heat conductivity for this purpose. Of the three common materials used—aluminum, stainless steel, and cast iron—aluminum is the best heat conductor, cast iron is next best, and stainless steel is a poor third. Isn't aluminum preferable then to cast iron?

Maybe, and maybe not. Aluminum is light, and that counts against it, at least to some extent. More important, however, is one property of cast iron that compensates for its slightly inferior (to aluminum) conductivity, particularly when frying is the objective. Frying, of course, is done in oil; and cast iron is somewhat porous to oil. This porosity increases the effective surface area for the conduction of heat. Incidentally, you never want to wash a cast iron pan in *detergent* because it dissolves the all-important oil.

Knowledge about conductivity is useful in baking potatoes or even roasting a big piece of beef. Often potatoes are pierced by a pronged metal holder in order to conduct heat more efficiently (while baking) to the inside of the potatoes. Actually, the same can be done with the beef. Put a few metal skewers through it (as is done with shish kebab); this will cook the inside faster and so reduce the cooking time. Incidentally, the skewers used for this or for shish kebab should have as much surface area as possible to improve the conduction.

Thus knowledge of conduction is handy around the kitchen. Let's see how useful convection is.

So far we have only discussed heating things. Now let's look at the subject of cooling. We can start by asking this: Is cooling done most efficiently in the same way as heating? Do we put the ice under what we want to chill? If you are tempted to say yes, it's probably because you think cold flows from one object to another, as does heat, which is not true. Heat flows from the object you want to cool to the ice, not the other way around.

Remember the properties of convection: cool fluid descends while warm fluid rises. If the ice is beneath the object you are cooling, say a carton of milk, the bottom part of the milk will cool first and then that's it. The cooler milk just sits there at the bottom of the carton because it's the coldest part of the fluid and therefore the densest. On the other hand, if you put the ice on top of the carton, the top of the milk gets cold first, descends, and warm milk takes its place from underneath, setting up convection currents and thermal mixing within the milk, and the whole carton of milk will cool. What further helps the cooling process here is that the air surrounding the ice also gets cold and sinks, being replaced by warmer air from below, which is then cooled, and so forth. Very soon this creates a cool envelope for the milk.

Now you can see why the freezer compartment is most often on the top of the refrigerator rather than at the bottom. This way the layer of air in the refrigerator section adjacent to the freezer gets cold, descends, and is replaced by warmer air from underneath, setting up the convective mixing, so that air in all parts of the refrigerator attains a uniform low temperature. If it were the other way, with the freezer at the bottom, only the air in the bottom of the refrigerator section would cool, but there would be no convective mixing.

In addition to conduction and convection, radiation also plays a role in the kitchen. One example of this is aluminum foil. Have you noticed that one side of it is shiny and the other side dull? Now if you wrap baked potatoes in it to keep them warm, do you put the shiny side in or out? Does it make any difference? The shiny side reflects radiation and thus is a bad absorber, so it preserves the heat of the potatoes better when it is on the "in" side.

Now that you have looked at some of the physics in your kitchen, you may want to relax. So you settle down to enjoy a freshly brewed cup of coffee in a leisurely fashion. Oops, suddenly you remember that you have a class in a few minutes. But your coffee is too hot to drink before going to class. You could, of course, take it with you, but it would be too cold by the time you made it to class. What do you do?

Relax. The physics of this section can come to your rescue. In fact, you have a choice; you can cool it just enough to drink before you leave, or you can keep it fairly warm for the time it takes you to run to the class. Here's what you do if you want to cool it quickly. Dip a metal (preferably silver) spoon into the cup; it will conduct away some of the heat in a hurry. Leave the coffee black, because black objects radiate better. Stir the coffee frequently to promote thermal mixing; natural convection is too slow in this case.

And if perchance you prefer to take the coffee with you, then don't do any of the above. Instead, add some powdered cream or nondairy substitute to the top but don't stir; the white surface will reduce radiation loss.

■ 14.2 Degree-Days

What are **degree-days?** Instead of starting with a definition right away, let's start with some background. How much heat do you use in heating your house? It depends crucially on the heat loss by conduction to the outside; the furnace, or other heating arrangement you have, merely replenishes that loss. You can guess that the heat loss will be proportionately larger if the difference between the indoor and outdoor temperature is high. Similarly, the heat loss should also be proportional to the time interval under consideration. Thus we can write a simple equation for the heat that flows out of your house as follows:

$$Q = k \, \Delta\theta \, \Delta t \tag{14.1}$$

Here k is a constant of proportionality that is a property of the house (insulation, number of windows, etc., are critical factors determining k) called the **thermal conductance.** $\Delta\theta$ is the difference in temperature between the indoors and the outdoors ($\theta_{indoor} - \theta_{outdoor}$). Δt is the time interval.

How do we figure out $\Delta\theta$? Suppose you set your thermostat at the usual 68°F, and the average temperature during a day is 38°; then $\Delta\theta = 68 - 38 = 30$°F for a value of Δt of 1 day. Thus we see that the product of $\Delta\theta$ and Δt measured in this way, in degree-days, is a useful quantity. In our example here we get 30 degree-days for this product. If we know the value of k, we can calculate the heat loss (to be replenished by your heating system) during that day.

It is customary to use the British engineering system of units in cases like this. Thus Q is to be measured in a unit called the British thermal unit (abbreviated

Btu): this is the amount of heat necessary to heat a mass of water that weighs 1 lb through a temperature of 1°F (1 Btu = 1055 J $\approx \frac{1}{4}$ Cal). The unit for k now can be seen to be the ratio of Btu and degree-day, or Btu/degree-day.

For an average house with good insulation, k is typically 20,000 Btu/degree-day. So during a day, if the number of degree-days is 30, as above, the heat loss is

$$Q = 20,000 \times 30 = 600,000 \text{ Btu}$$

If we know the average daily temperature during an entire winter (the heating season), we can figure out the heat bill for the entire period. Sum up the degree-days for each day to figure out the total number of degree-days during the season. This is how oil companies decide when customers will run out of heating oil and thus when to replenish their supply.

The larger the number of degree-days during the heating season in your area, the more you gain through savings of energy by minimizing the heat loss—for example, by putting extra insulation in the attic.

As an example, let's consider storm windows. A typical storm window reduces the value of thermal conductance k by some 200 Btu/degree-day. New York City has about 5000 degree-days in a year, so the savings in energy for one storm window is

$$200 \times 5000 \text{ Btu} = 1 \text{ million Btu}$$

But if the efficiency of your oil furnace is only 50%, the actual saving of energy is 2 million Btu (and for electrical heating, the efficiency is more like 33%, so the saving is 3 million Btu).

From Table 12.2, one gallon of oil amounts to

$$30,000 \text{ Cal} \approx 30,000 \times 4 \text{ Btu} = 120,000 \text{ Btu}$$

Thus the energy saving in gallons of oil is

$$\frac{2 \times 10^6}{120,000} = 16.7 \text{ gal}$$

At 40¢ per gallon, this has a dollar value of $6.68, so the storm window perhaps can pay for itself in 5 to 6 years.

Table 14.1 gives you the typical annual number of degree-days in some American cities.

TABLE 14.1   Number of degree-days/per year in some U.S. cities.

City	Annual number of degree-days (average)
Atlanta, Ga.	3000
Boise, Idaho	5800
Boston, Mass.	5650
Cleveland, Ohio	6350
Indianapolis, Ind.	5700
New York, N.Y.	5000
Salt Lake City, Utah	6000
Washington, D.C.	4200

Suppose you are sitting on a winter evening by a closed window, which very effectively shuts out the outside air. Suddenly you feel a draft. Does this mean there is a leak in your window? Not necessarily; very likely the draft is due to a convection current. The air close to the window is getting cold and must find its way down. Warm air from below replaces it, to be cooled in turn. This sets up the drafts of a convection current. Wherever there is uneven heating of a fluid, there is a convection current.

In this section we will look at some of the important convection currents of nature. Generally speaking, convection plays the crucial role in the shaping of both local and global wind patterns. We will discuss some of the details. More recently it has been discovered that convection currents in the interior of the earth play a crucial role in the shaping of geological processes. This will be our other concern.

Convection and local winds

A very common form of local wind is due to the uneven heating of land areas near an ocean or other large body of water. During the day the land is heated more relative to the ocean water, because water has a large specific heat. Thus the air over the land rises, to be replaced by cool air from the sea. This is called a sea breeze (in talking about winds, we always name them according to the direction they come *from*). At night the situation is just the opposite. The land cools more quickly than the water of the ocean. The warmer air on the ocean now must rise, to be replaced by air rushing in from the land. This is a land breeze. Fig. 14.8 shows the convection cells for the sea and land breezes.

You should recognize what the wind does. Originating from uneven heating, it tends to distribute the heat more evenly. The wind is the atmosphere's heat engine; it takes the heat from the hotter places and dumps it in the colder regions.

Next time you are on an ocean coast, light up a fire and watch the smoke. During the day the smoke will go inland. And in the evening as you sit watching the fire, the smoke will move seaward. You can look at it and sing:

> You don't need a weatherman
> To know which way the wind blows.*

The global wind pattern

You know that the sun bestows more heat on the equatorial regions of the earth than on the subtropical and polar regions. This uneven heating at the global level sets up the global wind patterns.

If the earth were at a standstill, with no rotation, then the structure of the convection cells would be very simple. They would be similar to the convection cells of the sea breeze, only the cold air would move from the poles to the

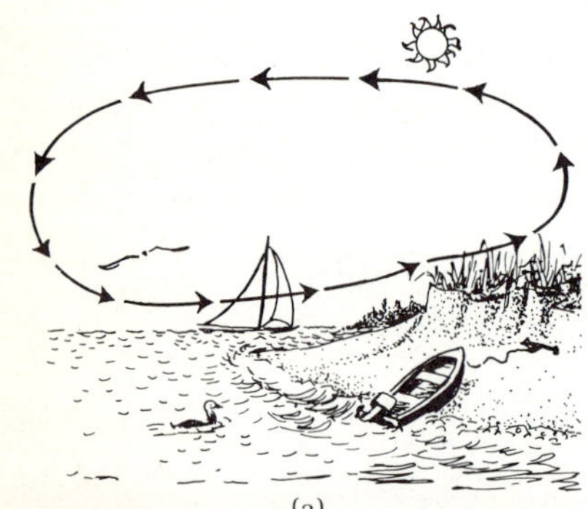

(a)

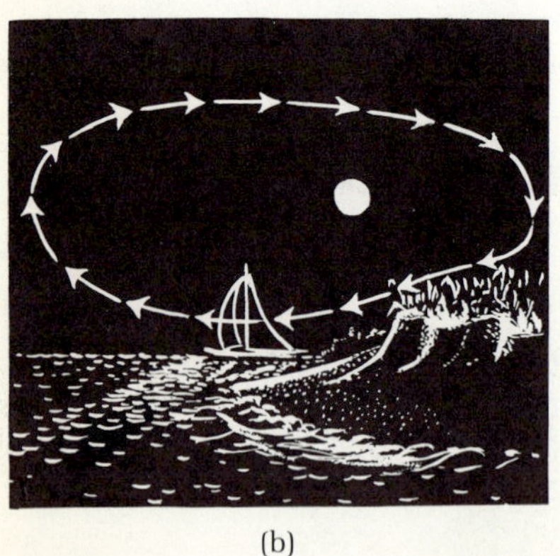

(b)

FIGURE 14.8 (a) Sea breeze. (b) Land breeze.

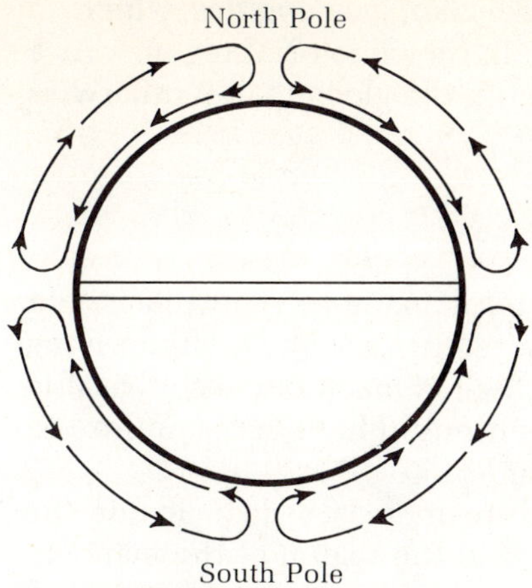

FIGURE 14.9 Global convection currents if the earth were stationary. There is basically one convection cell in each hemisphere.

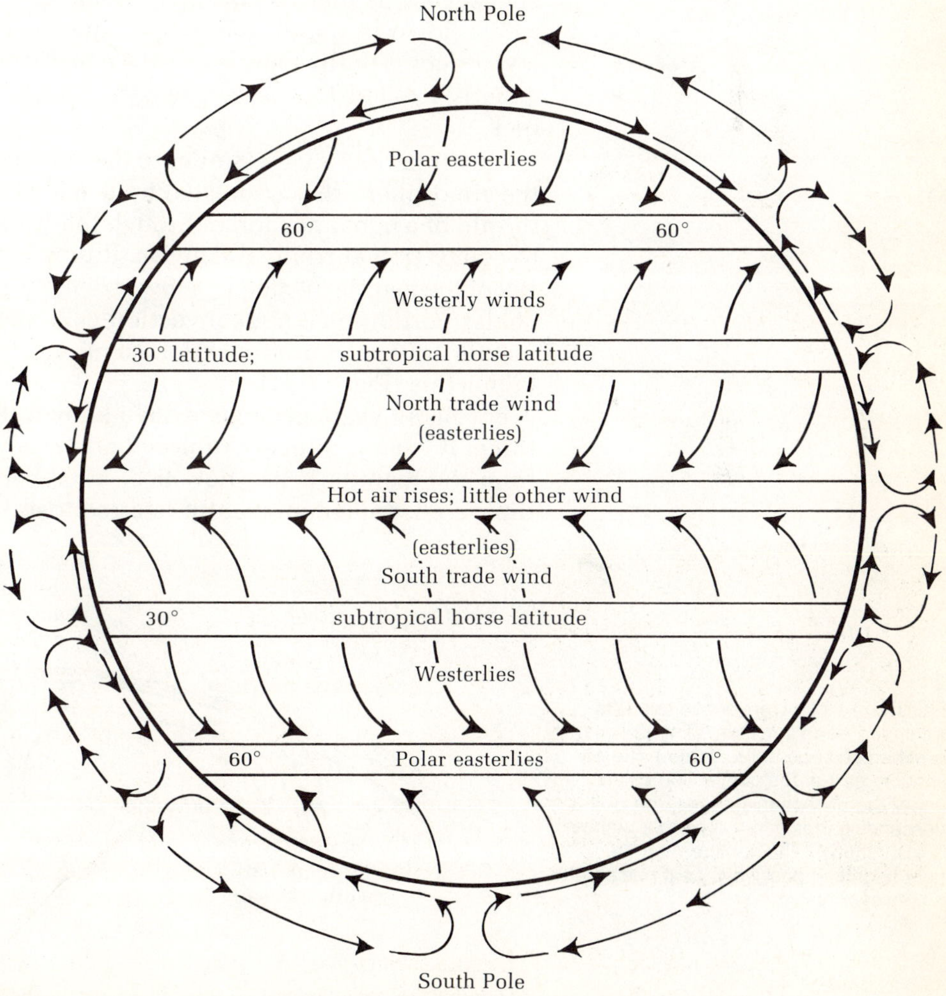

FIGURE 14.10 The effect of the Coriolis force on the convection cell in each hemisphere, shown in the Fig. 14.9, is to break it up into three cells: one cell between the equator and a region around 30° latitude; one cell between 30° latitude and 60° latitude; one cell between 60° latitude and the pole.

equator at ground level, and the warm air would move from the equator to the poles at the high level of the atmosphere (Fig. 14.9).

But as we all know, the earth rotates. The chief effect of this as far as the convection cells are concerned is the rightward deflection of winds due to the Coriolis force in the northern hemisphere (in the southern hemisphere the deflection is to the left). As a result of the Coriolis force, the simple cell in each hemisphere breaks up into three convection cells (Fig. 14.10). Furthermore, within each convection cell the Coriolis force still operates and turns the winds toward their right (in the northern hemisphere). So winds that originate at 35° latitude, just about at the Florida coast, do not go straight north but instead turn and become southwesterly (moving from southwest to northeast). In general, in the continental United States we get westerlies, a west-to-east movement of the wind. Some of the other globally important wind patterns are shown in Fig. 14.10. The actual wind situation is more complicated because of the distribution of land and sea areas, with their different heating characteristics, and because of mountains and valleys—in short, all local factors.

Let us emphasize again that global winds act like giant heat engines, which do a fantastic job in distributing the excess heat from the equatorial regions. As a result, the climate is more nearly temperate over all the globe than it otherwise would be.

Convection within the mantle

The earth's lithosphere (crust and upper mantle) is made of rigid plates, not unlike the pieces of a shattered glass slab. The plates float in the underlying layer of the mantle called the asthenosphere, which is made of denser basaltic rocks and yet behaves llke a fluid under the conditions of high temperature and pressure prevailing at those depths of up to 500 km.

Perhaps you know that the earth's temperature increases deep inside the earth. There is considerable radioactive material in the region of the mantle.* As these radioactive materials give off radiation, a lot of heat is released. This heat is responsible for setting up a rather steep gradient of temperature in these layers, even as high as 20°C per kilometer. This means that for every kilometer we go down in depth, the temperature increases by 20°C. So the rocks in the lower depths of the mantle are at a considerably higher temperature and consequently expand to a lesser density, when compared to the upper and colder rock.

This situation is very similar to the one in the atmosphere. The lower layers of the atmosphere also are heated up to a lowered density and then travel up with the aid of a buoyancy force through the colder and denser air above, giving rise to convection currents. Likewise, it now is believed that the underlying material of the mantle, heated to a lower density, is buoyed up through the upper and colder portion, making convection cells that extend vertically throughout the mantle (several hundred kilometers) and horizontally perhaps thousands of kilometers (Fig. 14.11).

For many years geologists have known that the seafloor is not a permanent thing; it is continuously replaced and renewed by molten material coming up from the mantle. It has now been established that the seafloor expands with time, a phenomenon usually referred to as seafloor spreading. There are ridges

FIGURE 14.11 Convection currents within the earth's mantle. (The drawing is schematic and not to scale.) The ascending part of the convection cell is under the midoceanic ridge. One of the descending legs into the earth's mantle is also shown. The continent on the right is drifting passively, as if placed on a conveyor belt.

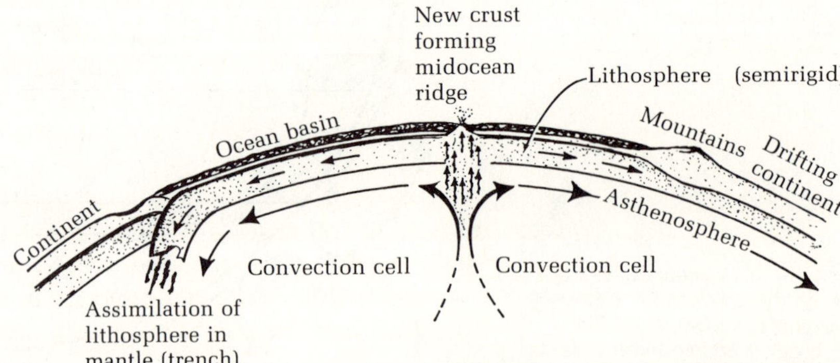

---

*Material that spontaneously releases energy in the form of energetic particles and radiation is called radioactive. See Chapter 23.

in the middle of the ocean which mark the loci of the rising columns of convection currents. The hot mantle material makes its way up vertically along these ridges and then spreads horizontally. The spreading of the mantle material causes the seafloor to move away from the ridges in both directions. At other places the cold rock above descends back into the mantle; these are the sites of the deep oceanic trenches. These trenches are the cold legs of the convection cells.

Actually, seafloor spreading is just one of several kinds of motion of crustal plates that is made possible by the energy of the convection currents. A closely related crustal movement is the continental drift, in which plates carrying entire continents move away from each other. A most startling example of this, put together by geologists, is the case of Africa and South America. Apparently some 200 million years ago, these two continents were joined together. Where was the Atlantic Ocean? As hard as it is to believe, the Atlantic Ocean did not exist at that time. Since then, though, the convection currents in the underlying mantle, working like giant conveyor belts, have carried the two continents apart, in the process creating the Atlantic Ocean.

The most important implication of this theory is that the boundary between any two plates is geologically the most active. The theory, now well established, is referred to as the theory of plate tectonics.

Geothermal energy

Every great scientific achievement usually has given us something tangible and important. Plate tectonics theory is no exception. One of its many applications is the possibility of the location of concentrated sources of geothermal energy.

What is geothermal energy? As the word itself indicates, it is the thermal energy trapped inside the earth. The total magnitude of the heat in the interior of the earth is staggering, but unfortunately much of it is buried deep inside and is quite inaccessible. The thermal conductivity of rocks inside the earth is so small that is takes a very long time for the heat inside to migrate to the surface by thermal conduction. So at one point people thought there was not much chance of our tapping this vast energy resource.

But plate tectonics changed all that. We know that conduction is not the only process that can bring out the interior heat. And plate tectonics says that convection currents are the dominant process of heat transfer within the upper body of the earth. And these currents bring the heat up (at specific places) at a much faster rate, making geothermal energy a viable resource.

Actually, it is common knowledge that some of the earth's interior heat finds its way to the surface through natural steam fields and hot springs (also geysers). So there are places on or near the surface where geothermal energy already exists in a concentrated form. Furthermore, it has been known for some time that underground water often finds its way to such superheated rocks slightly underneath the earth's surface and becomes superheated itself. If we know the location of these accumulations of superheated water, we can dig wells to the water level, extract it, and bring it to the surface (Fig. 14.12). At the reduced pressure of the surface, part of the water will convert to steam, which we can use to run a turbine to generate electricity.

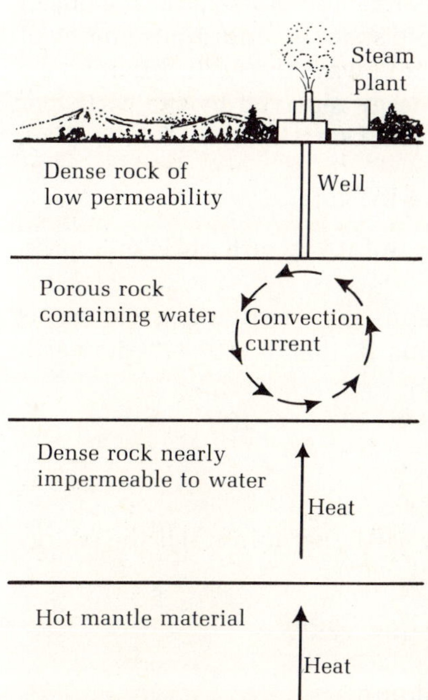

FIGURE 14.12 Geothermal well. Water in the region of porous rock is heated convectively from the underlying bed of hot rock. By drilling a well to the porous rock, we can extract the geothermal energy contained in the water.

Plate tectonics theory can help us in finding the locations of these underground water accumulations. The broad aspects of the theory tell us that the plate boundaries are geologically most active and suggest that these are the places to look for sites of geothermal wells. Further fine tuning of the theory can lead to better prospecting.

It is quite likely that the theory will tell us the location of the underground hot rocks, which may or may not have a natural water accumulation. Research is now in progress to devise means of extracting the energy of these hot spots near the crust. One idea, for example, is to blast a hole all the way through to the hot rock and pour water in it. When the water becomes heated, it can be brought back to be used.

Two things should be noticed before we get carried away with these prospects. Geothermal energy is not likely to be an exclusive solution to our energy shortage today. There isn't that much of it available near the surface. Second, the efficiency of the geothermal plant is not high (only 25%) due to the relatively low temperature of the steam obtained, so there is a lot of thermal pollution. Furthermore, the water is contaminated with corroding minerals. So it is more expensive to use this water for power than it seems to be at first glance.

■ 14.4 Solar Radiation

Life on earth depends on solar radiation as its primary source of energy. So the sun is of special importance to us, although it is like any other radiating object whose surface temperature is 6000°C. It sends into space an enormous energy of 9.1×10^{22} Cal every second. From 92 million mi away, the earth intercepts a small portion of this energy. The amount of energy received by the earth per unit area per unit time is called the **solar constant.** Denoting it by S, it is given as

$$S = 20 \text{ Cal/min/m}^2 = 1.4 \text{ kW/m}^2$$

As we will see, this is a very important number and it is worthwhile memorizing it.

The earth would be seen as a disk from the sun—just a circular area, not as half of a sphere. The area of the circle of radius R_e (6.4×10^6 m), the earth radius, is given as

$$\pi R^2_e = 3.14 \ (6.4 \times 10^6)^2 = 3.14 \times 40.96 \times 10^{12}$$
$$= 129 \times 10^{12}$$
$$= 1.29 \times 10^{14} \text{ m}^2$$

The total energy from the sun that flows into the earth (per minute) is the energy incident per unit area (per minute) times area:

$$S \times \pi R_e^2 = 20 \times 1.29 \times 10^{14} \ \frac{\text{Cal·m}^2}{\text{min·m}^2}$$
$$= 2.6 \times 10^{15} \text{ Cal/min}$$

Since 1 year $= 5.3 \times 10^5$ min, we get as the energy flow to the earth in one year

$$2.6 \times 10^{15} \times 5.3 \times 10^5 = 13.7 \times 10^{20}$$
$$= 1.37 \times 10^{21} \text{ Cal/year} = 5.75 \times 10^{24} \text{ J/year}$$

What happens to all this energy? Earth's atmosphere acts as a translucent window to the incident radiation, turning back some, letting some through, and absorbing the rest. The fraction that is turned back or reflected is called the **albedo;** the albedo of the earth as a whole is around 35%. The atmosphere absorbs another 15% or so. One important thing about the atmospheric absorption is that most of the highly energetic components are taken away by a layer of ozone, which acts as a protective sheath. The absorption of the ultraviolet radiation from the sun by earth's ozone layer is a matter of major importance to life on earth. The rest of the atmospheric absorption is due to the presence of water vapor and carbon dioxide.

The water vapor and carbon dioxide, particularly the latter, play another interesting role in the maintenance of earth's average temperature. They are transparent to visible radiation but opaque to infrared. Now the earth, upon being heated by the sunlight during the day, must radiate back some of the energy to space at night. Actually, of course, the radiation from the earth goes on all the time; a hot object always radiates. But during the day earth receives more than it gives away; it has a net influx of radiation. At night it suffers a net loss of radiation.

Now when the earth reradiates, the radiation is all in the form of the infrared. The atmospheric carbon dioxide absorbs some of this radiation and does not let it pass on to outer space. So it makes a sort of thermal blanket above the earth. Without the carbon dioxide, the earth's night temperature would be a lot colder. This effect is called the **greenhouse effect** since greenhouses operate on the same principle (Fig. 14.13). The glass or plastic cover of a greenhouse also has the property of being transparent to visible light but opaque to the infrared. The incoming solar radiation is mostly visible and therefore gets into the greenhouse, which gets hot as a result. But when it reradiates back, all the radiation from inside the greenhouse is in the infrared, which glass can block pretty effectively. This is why the inside of the greenhouse is so hot.* An everyday example of the greenhouse effect is the heating of the inside of a parked car when it sits in the sun all day with all its windows closed.

Cloud and water vapor in the sky also can produce extra heating on the ground through the greenhouse effect. Thus the green vegetation of the earth really lives in a sort of global greenhouse, a greenhouse whose cover is made of carbon dioxide and water vapor. The green plants not only get their food from the carbon dioxide by means of photosynthesis, but they also are warmed by it.

Let us come back to the accounting of the incoming solar energy. The albedo of the earth and the atmospheric absorption take away roughly 50% of the incoming flux; the rest reaches the ground, where it is absorbed. This absorbed energy is the driving agency for many phenomena on earth. For example, during the day the ground warms up rapidly and conveys some of the heat to the air through convection, a process that reverses during the night. Different parts of the earth receive somewhat different amounts of solar radiation, causing an uneven heating of the surface. However, the wind circulation over the globe tends to even up this distribution to a large extent.

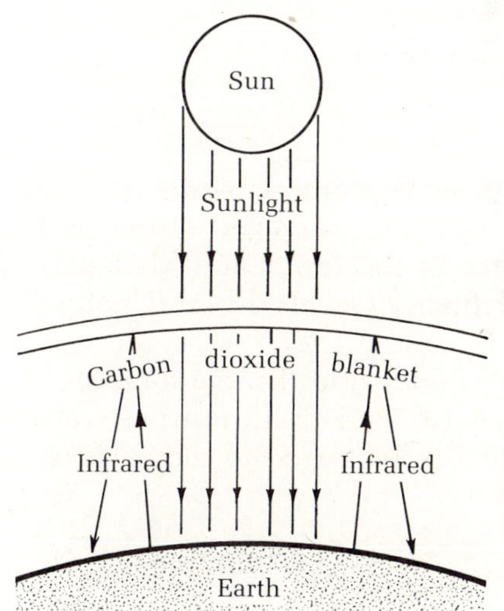

FIGURE 14.13   The greenhouse effect of the carbon dioxide blanket over the earth. It lets in all the sunlight but reflects back part of the infrared radiation of the earth.

*The glass also cuts down convection.

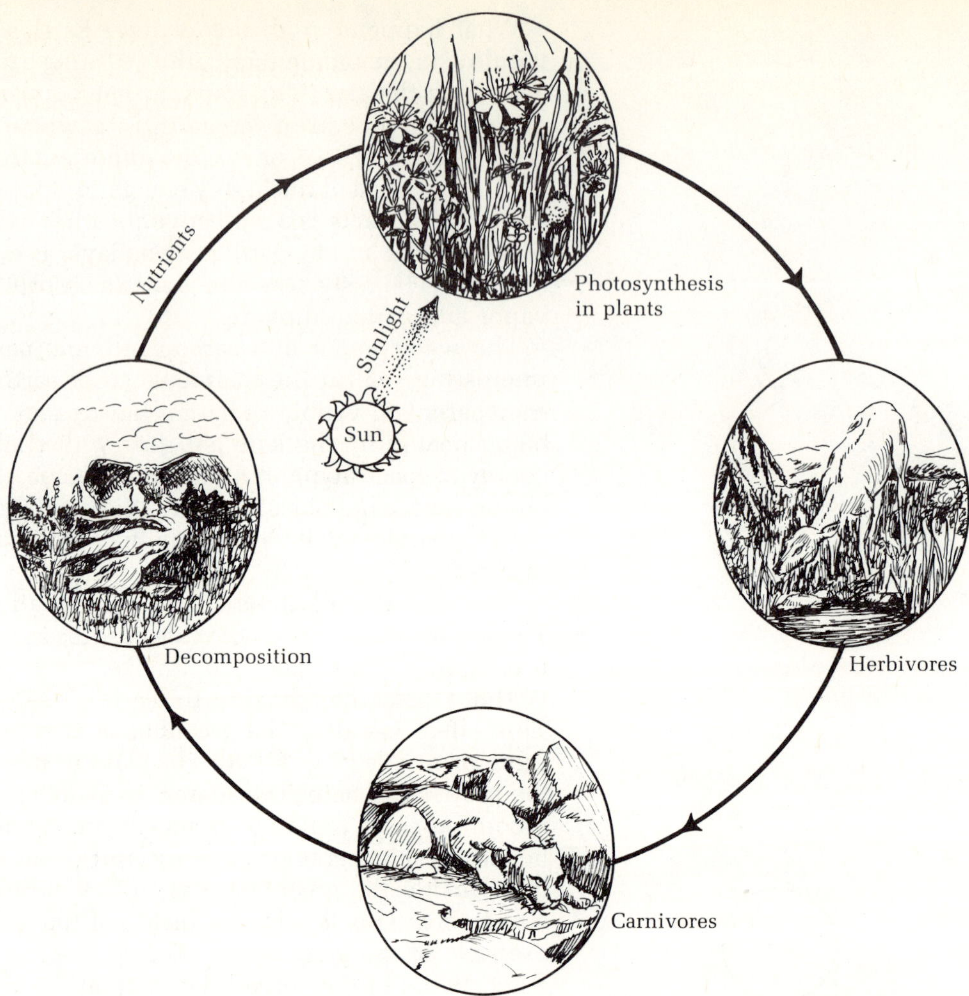

FIGURE 14.14   The web of life.

Part of the energy at the surface goes into the green vegetation to form the base of the web of life (Fig. 14.14). A small portion of this biomass gets buried and, after being worked over for millions of years by bacterial and geological processes, becomes converted into the fossil fuels of coal, oil, and natural gas.

A substantial part of the absorbed energy also goes into the evaporation of water. It is the solar energy that drives the water cycle. Part of the returned water forms rivers. So the water power of river dams can be traced to the solar energy.

Space heating with solar energy

> And pluck till time and times are done
> . . .
> The golden apples of the sun.*

*Lines from "The Song of Wandering Aengus," reprinted with permission of Macmillan Publishing Co., Inc. from *Collected Poems* of William Butler Yeats. Copyright 1906 by Macmillan Publishing Co., Inc., renewed 1934 by William Butler Yeats. Courtesy Macmillan Publishing Company and A. P. Watt Ltd., London.

About 20% of all the energy used in America is for space heating. Can we use solar energy for this chore?

We often hear that solar energy is so dilute that there just isn't enough of it falling on the roof of an average house in the colder regions, where the heating needs are most severe. Let's examine how far this is true.

Let us mention at the outset some of the problems associated with the use of solar energy for household heating. First, heating is needed more in those places where sunlight intensity in the winter months (when it is most needed) is not high. The question is is it enough? Second, in order for solar heating to be a practical alternative, we must be able to store enough energy to use on cloudy days.

Let's deal with the first question. In the peak of the winter, a typical place in the Midwest with hard winters would get sunlight energy at a much reduced rate of 100 W/m². This can be improved by a factor of two by suitably tilting the roof, which collects the sunlight, and making it face south. Therefore, we can work with a 200-W/m² figure. Suppose we have a collector area of 60 m² (about 645 ft², which is just about one side of the roof of an average-sized house). The total solar power collected by the house is

$$200 \times 60 \text{ W} = 12,000 \text{ W}$$

What power is needed to heat an average house? We previously mentioned a value for the thermal conductance for the average house as 20,000 Btu/degree-day. In the midwestern winter, the average outside temperature is something like 20°F. So the number of Btus needed for one typical winter day is given as (assuming a thermostat setting of 70°F)

$$20,000 \times (70 - 20) = 20,000 \times 50 = 1,000,000 \text{ Btu}$$

Now we need to convert units. Since 1 Btu $\approx$ 1,000 J

$$10^6 \text{ Btu/day} \approx \frac{10^6 \times 10^3}{10^5} \text{ J/sec}$$

where we take 1 day to be approximately 10^5 sec for an order-of-magnitude calculation. This comes out to be 10,000 W. As you can see, the sun's power of 12,000 W estimated above is quite adequate.

Now to the question of storage. If we store the heat in water tanks, how much water must we store to have a reserve of heating energy for just one day? We can get the water as hot as 170°F without too much trouble. The water will provide the heat by cooling from 170°F to the house temperature of 70°F. For each pound of water, the heat delivered by cooling 1°F is 1 Btu, so for a cooling of 170°F to 70°F, or 100°F, 1 lb of water can provide an amount of heat of 100 Btu. Thus for a million Btu, we need 1,000,000/100 = 10,000 lb, or 5 tons. What's the volume of this much water? The weight density of water is 62.4 lb/ft³; thus the volume is

$$\frac{10,000}{62.4} = 160 \text{ ft}^3$$

which can be stored in a cubical tank with one side equal to roughly 5.5 ft, not an unreasonable value at all.

Thus it seems that solar energy passes both crucial tests. However, there are still problems. The most severe one is that if we want to store the energy we

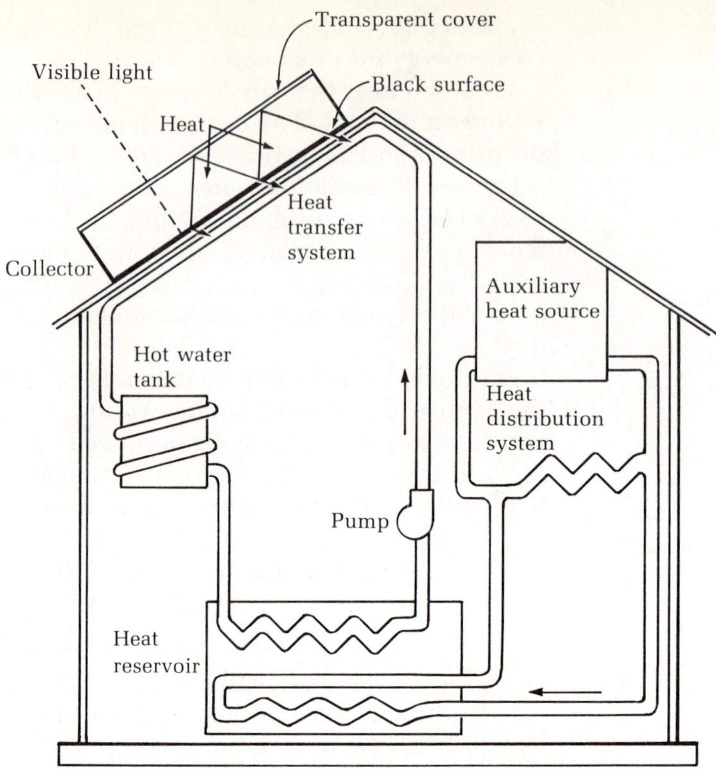

might need for more than a day or so, the storage tank has to be impractically big. Thus for unusually severe weather we need some auxilliary heating system.

Fig. 14.15 shows a solar-heated house, just a minimum system. It can be improved in a variety of ways, notably by using reflectors, but the details are too technical to go into here.

■ 14.5 Physics and Weather Folklore

Physical laws are an objective summary of human experiences in nature. In contrast, weather folklore is subjective. Yet, it too represents human attempts to describe and predict nature. Perhaps you have heard the Zuni Indian saying: "With your eyes in the sky you may walk lively in a path of beauty." Weather folklore was conceived and written by people with their eyes in the sky, and its teachings certainly are beautiful. Most interestingly, even when subjected to an objective analysis of physics, they often make a lot of sense. Moreover, weather prediction remains an art to some extent and is not an exact science in spite of physics. In many cases, physics has given us prediction rules that fare no better than folklore; in a sense, the new "rules" can be looked upon as "scientific folklore."

With this preamble in mind, let's get on with our subject. The ultimate objective of weather prediction is the forecast of "rain or shine." So most folklore

Peter Higginbothum is a good friend of mine. He is an ecologist. (Behind his back some people call him an ecology freak; I think Peter knows this and takes it with good humor.) Lately he has been trying to convince me to become a vegetarian.

"You are the one who should convince me," he insists. "You come from a vegetarian country."

"All the more reason for me to enjoy all the beefsteak I missed as a kid," I protest as usual. But today Peter will not be put off.

"Do you know why Hindus are vegetarians?"

"Religious prejudice, what else? Cows are sacred, you know."

"But why did killing them become taboo? You have told me there are allusions to beef eating in the Vedas. Why did they give it up?"

Peter has a point, but I don't give in so easily. "What's your point?" Taking the offense is often the best defense.

Peter replies solemnly, "They had an energy crisis."

"Don't tell me you have new historical evidence for this!" I banter, amused. Energy is my business; I am the physicist.

"Who needs evidence? All you need is common sense and arithmetic." He shows me a matchbook with some numbers on it.

I look at his calculation curiously—arithmetic by an ecologist! It is not their usual tool, you know. But going through it, I can't help but be impressed. Although I question some of his figures, his basic logic is sound.

"Well, suppose some scientist in the Indus Valley figured all that out on the back of a matchbook? Wouldn't that lead to a total ban on beef?"

"They didn't have matchbooks then," I stall, feebly.

"So they used the back of a banana leaf. My argument still stands."

We argue back and forth until he has to leave. I don't give in, but it is difficult to hold on to my steak.

I don't think I am ready to give up meat yet. After all, one can make a good case for so many things—for example, for adopting the Japanese custom of taking family baths, which can save a lot of energy. Also for giving up cars and depending entirely on bikes and mass transit (one more thing I haven't gotten around to yet).

I hope you are curious enough to take a look at Peter's calculations yourself. Here they are, with some added elaborations to make them more comprehensible.

The calculations start with the solar constant, 20 Cal/ min/m², which is the energy received by the earth from the sun per unit area per unit time at the top of the atmosphere. At the bottom, the figure is halved to 10. Now 1 mi² = 2.6 × 10⁶ m². So the energy per square mile per minute comes out as

$$10 \times 2.6 \times 10^6 = 2.6 \times 10^7 \text{ Cal}$$

If we assume that solar energy is utilized only through photosynthesis of plants, which occurs with an approximately 2% efficiency, the number for the available solar energy per square mile comes out as

$$2.6 \times 10^7 \times 0.02 = 5.2 \times 10^5 \text{ Cal/min}$$

Assuming a 12-hour day and 6 months of growing season— rather reasonable assumptions—then the solar energy stored by plants in an entire year comes out as

$$5.2 \times 10^5 \times (6 \times 30) \times (12 \times 60) \text{ Cal/mi}^2$$

The factors 12 × 60 and 6 × 30 come from the 12 hours/day and the 6 months/year. When you do all the multiplying, you get

$$6.7 \times 10^{10} \text{ Cal/year/mi}^2$$

For an agricultural society, the energy need is 12,000 Cal/day per capita (see Table 9.1), which is

$$12{,}000 \times 365 = 1.2 \times 3.65 \times 10^4 \times 10^2$$
$$= 4.38 \times 10^6 \text{ Cal/year/person}$$

Thus if such a society depends on solar energy through photosynthesis for all of its energy, the number of people that could survive on each square mile is given as (assuming a 10% utilization of the stored energy on the average)

$$\frac{6.7 \times 10^{10}}{4.38 \times 10^6} \times 0.1 = 1.5 \times 10^3 = 1500$$

This is a soberingly small number. A very strong case can be made for vegetarianism in that particular primitive society. Vegetarianism would reduce the per capita energy need substantially (perhaps even by 100%), so more people could be supported on their usable land.

Suppose we in America ran out of all energy resources, had only solar energy to live on, and had no advanced technology to improve some of the numbers above. Even assuming present-day consumption levels of 230,0001 Cal/day/capita, which is 230/12 times the rate for the primitive society, we get the survival rate reduced to

$$\frac{1500}{230/12} = \frac{1500 \times 12}{230} = 75 \text{ people/mi}^2$$

This is just about our present population density. A very sobering fact.

touches upon this subject. For a starter, consider the following:

> When the morn is dry
> The rain is nigh
> When the morn is wet
> No rain you get.

There is quite a bit of physics in this, although nothing so complicated that you could not figure it out just using common sense. Wet morning refers to a heavy dew. Now, why does dew form? Air always has moisture in it; the term "humidity" is used to denote how much moisture a given volume of air holds. If we find that 1 m³ of a given sample of air has 0.2 g of water vapor, we say that the **absolute humidity** of the sample is 0.2 g/m³. The moisture-retaining capacity of air increases with an increase of temperature, and at every temperature a certain amount of water vapor is all that the air can hold. When this limit is reached, we say that the air is **saturated** (with water vapor).

Perhaps you are more familiar with another way of expressing humidity—the **relative humidity.** It tells us the amount of water vapor present in the sample as a percentage of the amount that would saturate the air at the given temperature. So when we say that the relative humidity is 80%, what we mean is that the air contains 80% of the amount of moisture it would take to saturate it. If relative humidity is high, our perspiration does not evaporate, we feel uncomfortable, and we call the weather "muggy." People with long hair in (not natural) curls are even more sensitive to humidity—it uncurls their hair.

Since warm air holds more moisture than cold air, as the colder night replaces the day, the cooling of the air forces it to give up some of the water vapor, which condenses in the form of dew. On cloudless nights there are no clouds to hinder the cooling of the ground below through the greenhouse effect, and the dew collects at a faster pace. Thus if there is a lot of dew—if "the morn is wet" with it—there is a clear sky and we don't expect any rain. Unfortunately, the verse is not perfect in its predictive power; the absence of dew does not necessarily mean rain.

When warm moisture-laden air rises via convection to the heights and cools, the water vapor condenses and coalesces, forming clouds. (Clouds can form also at ground level, in which case they are called fog.) There are many types of clouds, some of them so dark that anybody can predict rain if he or she sees them. There is, however, a type of high cloud called cirrostratus—a thin veil of ice crystals, really—that sometimes covers the whole sky or a significant portion of it (Fig. 14.16). When you look at the moon or the stars through these clouds, the optical effects are quite spectacular. Weather folklore talks about these:

> The moon in halos hid her head.

Or in a slightly different form,

> A ring around the moon is a sure sign of rain.

Both of these sayings are talking about the halo of the moon, an optical effect seen when we look at the moon through the cirrostratus clouds. A third form is more poetic:

> When the stars begin to huddle
> The earth will soon become a puddle.

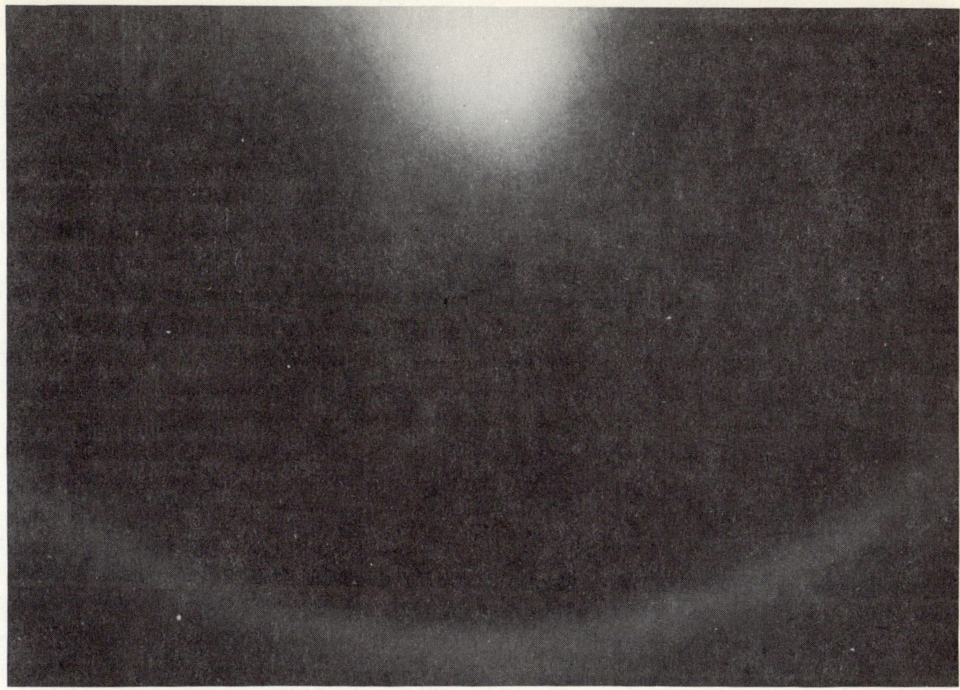

Indeed, viewed through the cirrostratus clouds, the stars look closer together. So when the stars huddle, we know clouds are around. And the author of the folklore knew this as a forebearer of rain. But why is this a sure sign of rain?

This is the way it works. The high clouds themselves do not produce rain, but they tell us rain clouds are coming (according to weather jargon, we would say that a warm front is approaching). The cirrostratus clouds are the upper edge of a whole mass of warm moist air moving into and over a mass of stationary cold air (Fig. 14.17). The warm air gently rises over the shoulder of the cold air mass. In the process water vapor condenses and gives us low rain clouds and rain.

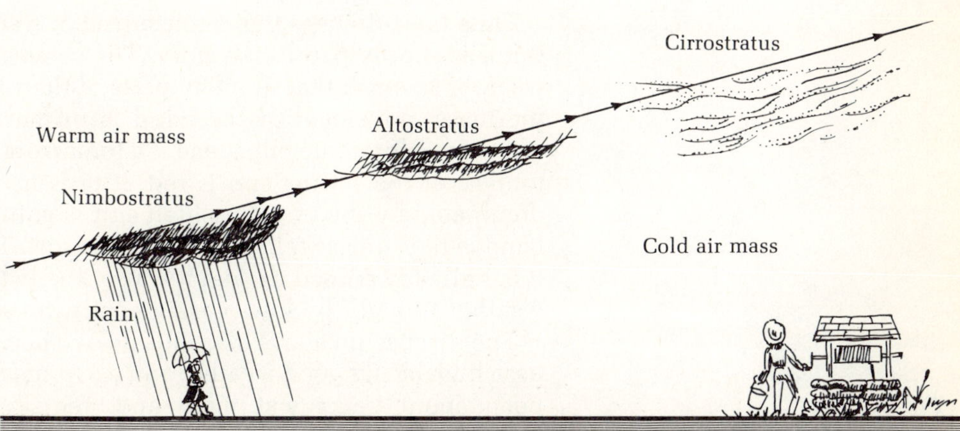

FIGURE 14.17   Cirrostratus clouds are the bearers of advance notice of rain.

The last bit of folk knowledge we will discuss goes as follows:

Red sky at night, sailors delight
Red sky in morning, sailors take warning.

Again there is an astonishing amount of physics in this little two-line poem.

First, we'll discuss the matter of the color of the sky. Normally during the day when we look at the sky, we look away from the sun and we see sunlight scattered from the air molecules and other impurities of the air. Physics tells us that long-wavelength light is not scattered much by air and other such molecules compared with those of short wavelength. So the scattered light consists mainly of short-wavelength components of sunlight, blue and beyond. Our eyes are more sensitive to the blue than to the violet, so all we see is blue.

The story is different when we look directly at the sun. Now we receive the direct rays; the low wavelengths have scattered away to some extent, and so the sun may look slightly yellowish at midday. The scattering effect increases as the day lingers on and the direct sunlight has to travel longer and longer paths of the atmosphere in order to reach us. So as the sun sets, its color progressively becomes yellow, orange, and then red. It is fully red when all the shorter-wavelength light has been scattered away by the intervening air molecules and dust particles and so forth. Thus a lot of dust particles in the atmosphere near the ground enhances the red color of the setting (or rising) sun.

Now the presence of a lot of dust particles in the air is facilitated by the presence of a cell of high pressure. In Chapter 7 we talked briefly about this and mentioned that air rushes to the surrounding areas from such a high-pressure zone and that the outrushing air is bent clockwise by the Coriolis force. There is also a downward motion of air from the cell. And the descending air currents, when they reach the higher pressure of the lower levels, are compressed and warmed up. This is just the reverse of the situation that produces condensation, so clearly high pressure means good weather. For a low-pressure cell, on the other hand, there is counterclockwise air movement toward the cell and also an upward rush. As moisture-filled air rushes upward, it cools upon expanding in the lower-pressure regions up high, and the moisture can condense into the clouds, bringing rain.

Thus logic dictates that a red sunset or a sunrise brings the news of a "high." But this is only part of the story. The verse also recognizes one primary rule of weather science: that in many parts of the world (including the United States), the mass movement of the wind is primarily from the west to the east; the westerlies dominate our scene. So tomorrow's weather is to be found in the sky tonight: if the setting sun is red, it tells us that tomorrow's air is going to be dominated by the traveling high and is going to be nice and dry. On the other hand, a red sunrise tells nothing of the sort. The air in the east is yesterday's air; it has already passed us, and it is very likely to be followed by moist air and foul weather brought by low-pressure cells. So sailors take warning.

The discussion above also tells us two new "rules" weather science has given us: a high brings good weather and a low means foul weather. (But somehow the poem about "red sky at night" and "red sky in the morn" seems more appealing.)

Utilization of energy resources for the production of useful work has this unavoidable problem: pollution of the environment. There are three major types of environmental pollution today that are of grave concern: air pollution due to burning fossil fuels (oil, coal, and natural gas); thermal pollution, which is present anytime an energy resource is used; and radiation pollution, due mainly to the use of nuclear energy. In this section we will examine some of the physics associated with the first two, air and thermal pollution. Radiation pollution will be discussed in Chapter 24.

Besides these three pollutants there is also water pollution and the pollution due to solid wastes. There is not much physics in the subject of water pollution. But one thing can be said about solid wastes: recycling is a very sound physical idea. Previously we have talked about the entropy law, how order gets obliterated by disorder. This is a law of nature. However, there is a catch. Nature does not tell us any time limit. So it is perfectly possible to arrest the progress of entropy. Remember that the purpose of recycling wastes is precisely that.

The cartoon shows a young man kicking garbage in his attempt to get rid of it. Clearly he takes the motto "Dilution is the solution to pollution" too seriously. The space in the room is already saturated with garbage and he is not getting anywhere. But, aren't we in today's society behaving the same way? Our earth is also a closed and very finite environment. So putting tons and tons of garbage out will sooner or later saturate this environment, too.

Air pollution

First we need a definition. We use the term "air pollution" to signify the addition of unwanted airborne matter (such as smoke or sulfur dioxide gas) to the atmosphere by specifically human activities. On the west coast of America, the air pollution is produced mainly by automobiles in the form of unburned gaso-

DILUTION IS THE SOLUTION TO POLLUTION

line and nitrogen oxides, among other things. On the east coast, the main source is the burning of coal, which produces the toxic gas sulfur dioxide. Sulfur is one of the impurities of coal. When coal is burned, the sulfur combines with oxygen to produce the harmful sulfur dioxide.

The production of nitrogen oxides in an automobile engine is an interesting phenomenon. Ordinarily, nitrogen is rather inert and does not mix with other elements to form compounds. However, this fact changes at the very high temperatures of the internal combustion engine of the automobile. We saw previously that the use of very high temperatures in the car engine helps us to get a better efficiency, but now we see the other side of the coin. Nitrogen burns with oxygen, making nitrogen oxides at these temperatures.

Sunny California! There was a time that this meant something desirable to most Americans. It is as likely now to be a statement of irony. Sunlight helps the nitrogen oxides combine with the unburned hydrocarbons and atmospheric ozone to make vicious pollutants which go by the name of oxidants. The visible ingredient of this pollution is the notorious photochemical smog found in some sunny urban areas like Los Angeles. This smog is different from the original version (sometimes called "London smog") which is a mixture of smoke and fog (smoke + fog = smog).

Temperature inversion

We previously have discussed the horizontal movement of air masses. In the dispersion of air pollution, the vertical motion of the air plays a more crucial role. Essentially it is a question of volume; if there is a lot of vertical movement and mixing, more space is available for the dilution of the air pollution. In contrast, with little or no movement in the vertical direction, the polluted air gets stuck in the low levels of the atmosphere, causing conditions of smog.

The factor that most controls vertical air movements is the variation of temperature with altitude. The rate at which temperature changes with the increase of height is called the **temperature gradient,** or the **lapse rate.** Normally, the first few miles of the troposphere (the lowest stratum of the atmosphere) display a continuous cooling of the air with altitude. So successive ascending layers of air would be like this:

$$\text{warm} \rightarrow \text{cold} \rightarrow \text{colder} \rightarrow \text{still colder}$$

and so forth. Displayed graphically, this temperature profile looks like the one shown in Fig. 14.18. The temperature decreases continuously (at a constant rate in the figure) with height. If this condition prevails, then the warm pockets of polluted air can rise uninhibitedly, since they are less dense than the surrounding air and thus there is a buoyancy force pushing it up.

If for some reason there happens to be a layer of warm air trapped between the layers of cold air—that is, if the successive layers look like this:

$$\text{warm} \rightarrow \text{cold} \rightarrow \text{warm} \rightarrow \text{cold}$$

then the warm polluted air cannot rise. It will get trapped at the base of the trapped warm layer. No vertical mixing will occur, and stagnancy will prevail. The temperature profile of this kind of layering is shown in Fig. 14.19. The temperature decreases with height initially, but a short distance later it starts

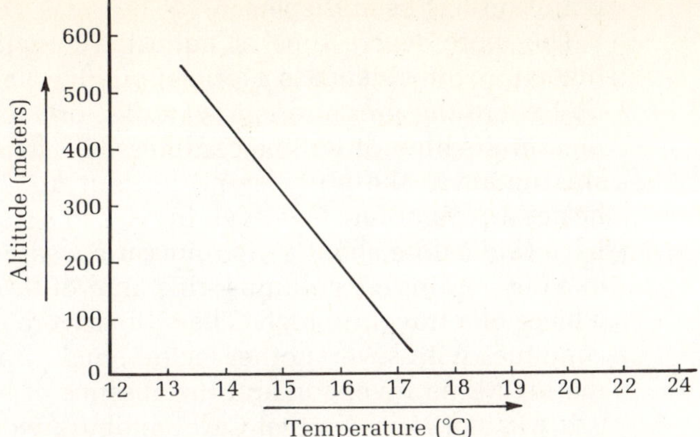

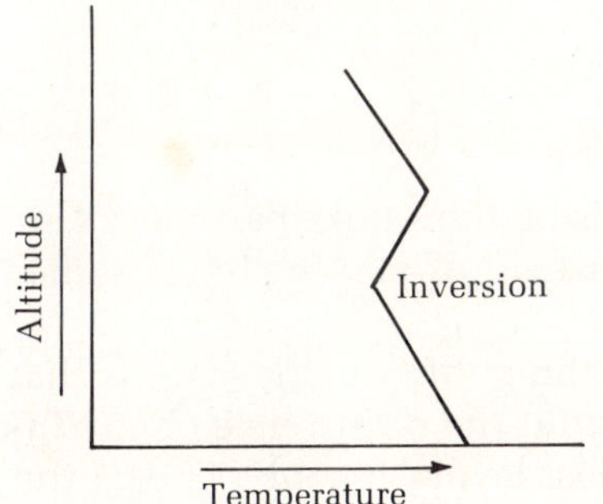

FIGURE 14.19   Altitude versus temperature graph showing temperature inversion.

increasing with height. This situation is known as **temperature inversion.**

What causes temperature inversion? There are two major types of inversion that cause bad pollution situations. The most common is the ground-based normal nighttime inversion. The ground gets hot during the day and reradiates the energy back at night. If there is no cloud cover, this nighttime cooling of the earth can go unhindered. The ground level of the air cools first, which then cools the next layer, and so forth. But as you can see, the process takes a finite amount of time. So the lower layers could cool before some of the upper layers, causing temperature inversion. With the sun coming up in the morning, the process reverses itself. The ground layer gets warmed up first, this warms the next layer, and so forth, and very soon we are back to the normal temperature profile again (Fig. 14.20).

Clearly much of the adverse effect of this kind of inversion could be avoided if no pollutants were released into the atmosphere during the night. Another factor that contributes to this pollution is the early morning traffic in urban areas, which releases a fresh supply of pollutants before the trapped nighttime

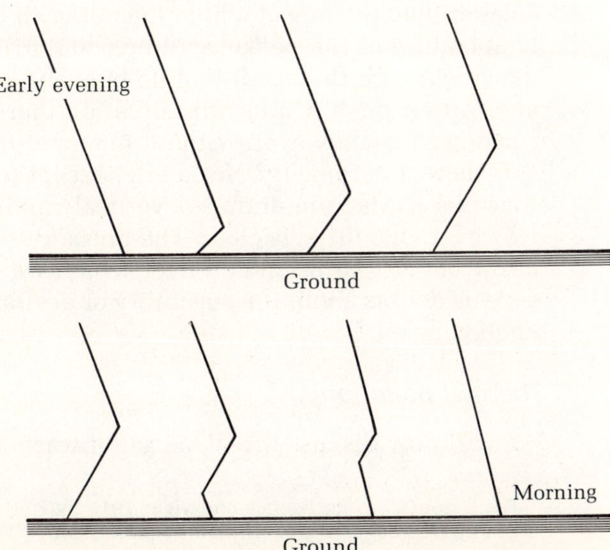

FIGURE 14.20   The temperature profile at various stages of a normal nighttime temperature inversion, starting in the evening (top left) and ending in the morning (bottom right).

pollution has been dispersed.

The more severe type of pollution situation is caused by the **subsidence** inversion. Subsidence is a term applied to a sinking mass of air. If air from high up (where the pressure is low) sinks, or subsides, to a lower level where the pressure is high, it will be compressed. The surrounding air does work on the sinking air in the process and heats it up. This will form a trapped layer of heated-up air at low levels, giving rise to temperature inversion. From what we have said before about a high-pressure weather system—namely, the existence of a descending air column—this kind of thing can routinely happen with the advent of a traveling high. The more severe conditions result when this effect combines with several other factors, including primarily the lingering effect of the nighttime inversion and the absence of horizontal air movement.

It was not long ago that Carl Sandburg wrote the following beautiful lines:

> The fog comes
> on little cat feet.
> It sits looking over harbor and city
> on silent haunches
> and then moves on.*

Alas, we must add an epitaph to this. The fog sits on the smoke particles of air pollution trapped by a temperature inversion. Instead of moving on, it sometimes makes deadly smog.

Although temperature inversion has gotten all the publicity, the truth is that vertical air movement can be inhibited even without the occurrence of an "inverted layer." When warm air goes up to the highest levels, it cools. The reason it does this is that it encounters lower and lower pressure, and a given volume must expand under such a condition. As the air mass expands, it does work on the surrounding air, loses energy, and cools. If it cools at a rate smaller than the lapse rate of the atmospheric air, then the pocket will continue to rise. Eventually it reaches a height where the temperature of the surrounding air is the same as its own. If, on the other hand, the lapse rate of the atmospheric air is smaller than the rate at which the rising air pocket cools, then soon enough the temperature of the pocket will be equal to its air environment, equalizing the densities, with the result that no buoyant force will be available for providing any further push. Under this situation there is little or no vertical movement, causing stagnancy even without temperature inversion.

Is there any remedy? Not really, except for reducing the air pollution in the first place. Machine-induced vertical mixing is too costly even to consider. However, one thing is clear. The presence of such atmospheric conditions that aggravate air pollution and over which we have very little control raises very serious doubts about the feasibility or desirability of coal as our major source of energy.

Thermal pollution

We will now discuss the dispersal of waste heat into the environment and the

---

*"Fog," from *Chicago Poems*, copyright 1916 by Holt, Rinehart and Winston, Inc.; Copyright © 1944 by Carl Sandburg. Reprinted by permission of Harcourt Brace Jovanovich, Inc.

FIGURE 14.21 Cooling tower.

effect it has on the ecosystems. Most of the waste heat generated by factories and power plants is dumped into the water systems of rivers or lakes. This raises the average temperature of the water. The life of the aquatic environment, especially fish, does not respond well to this temperature increase. First, warmer water contains less dissolved oxygen, which the fish must breathe. Second, at the higher temperature the fish's metabolic activities go up, so that it actually needs more oxygen, not less.

One alternative to this situation is to use air for the dispersal of at least some of the heat. Unfortunately, the specific heat capacity of air is too small to make complete air cooling feasible. So the hot water from the power plant is taken through a cooling tower (Fig. 14.21), giving off some of its heat to the air before it is dumped into the water system.

Since thermal pollution is inevitable, it would be nice if we could find some use for the waste heat. Unfortunately, the concentration of heat is not large enough to use for a profit, nor is it small enough for the safety of the environment.

The earth's heat balance

The fact that the earth maintains a constant average temperature has an important implication. The heat flow into the earth is equal to the heat flow out of it; the earth is in a condition of **heat balance.** It reradiates into space practically the same amount of heat it receives from the sun.

The heat balance is a rather delicate matter complicated further by the presence of large amounts of water, a portion of which exist as the polar ice caps. Suppose the earth had more heat than it could disperse through radiation into space. The extra heat would raise the earth's average temperature until the temperature reached a level such that the extra amount of heat could then be radiated away. The amount of radiation from a hot body increases rapidly with its temperature. So with a rise in its average temperature, the earth would be able to radiate away the excess heat, and the heat balance would be reestablished. However, there would be a price to pay. With the temperature higher

MALCOLM'S LOGIC

than before, some of the polar ice cap would melt and flood the oceans.

The opposite case, in which the earth receives less heat than it radiates away at its present average temperature, would be equally devastating. In this case the earth's average temperature must decrease until heat balance is restored. Suppose the temperature were lowered just by a couple of degrees as compared to the present. The snow that accumulates during the winter would not completely melt away in the summer. The snow lines would slowly advance toward the temperate zones, and eventually an ice age would have arrived.

So tampering with the heat balance is a complicated matter, and we should stay away from it. Unfortunately, we add to the problem through both air and thermal pollution.

Let us first discuss the case of air pollution. One of the pollutants produced by the combustion of fossil fuels is carbon dioxide gas. Ordinarily this gas is not regarded as a pollutant since it is a natural ingredient of the atmosphere. But an excess of carbon dioxide released into the atmosphere has serious repercussions on the environment, due to the greenhouse effect induced by carbon dioxide. As discussed earlier, the carbon dioxide naturally present in the atmosphere acts as a thermal blanket over the atmosphere. Putting more carbon dioxide into the atmosphere increases the thickness of this blanket. One possible result is a net increase in the average temperature of the earth.

Some calculations have shown that a raise in the carbon dioxide concentration of the atmosphere by a factor of two would produce a temperature rise of 2°C, which is enough to melt some of the ice caps. Many scientists today believe that, ultimately speaking, carbon dioxide pollution is perhaps the most dangerous air pollution there is.

The conventional modes of air pollution put a lot of small particles (e.g., smoke) in the atmosphere. Referred to as particulate pollution, this has the effect of increasing the albedo or reflectivity of the earth. The earth will receive less energy from the sun as a result, which means its temperature will be decreased. Thus particulate pollution tends to work in the opposite direction to that of carbon dioxide pollution. This doesn't mean that we are safe in randomly creating both. In practice, these matters are extremely delicate, and an exact offsetting of the opposite effects would be unlikely to achieve.

Now let's discuss the thermal pollution we inject into the environment through the burning of energy fuels. Almost every unit of energy we use ultimately ends up as heat. So every year we put about 2×10^{20} joules of extra heat into the environment (at the present rate of energy consumption). This is miniscule compared to the 6×10^{24} joules that the earth receives (and disposes of) every year from the sun.

Even so, the following consideration should be kept in mind. If the energy consumption rate were to double itself every 22 years or so, as it does now, it would take only several hundred years for the thermal pollution to reach amounts comparable to the energy we receive from the sun. This will have the effect of heating up the earth, causing the disaster we mentioned earlier. In this sense, then, thermal pollution is the ultimate limit of growth.

SUMMARY

The themes of this chapter are the processes of heat transfer and, in general, the ideas of thermal physics as they relate to our environment. The first section deals with the basic modes of the transmission of heat: conduction, convection, and radiation. The subsequent sections are elaborations of one or more aspects of these basic processes.

Perhaps you hadn't heard about degree-days before you read Section 14.2. But now that you have read about it, you can see the importance of such a consideration in coping with a big chunk of our energy bill: household heating. In the last chapter you read about heat engines; in this chapter you read about nature's own atmospheric heat engine. The global wind patterns created by the process of convection really do the same job as a heat engine.

Section 14.4 dealing with solar radiation talks about the flow of solar energy in the earth's biosphere and the prospect of using solar energy for such things as household heating. Section 14.5 discusses some aspects of our weather: humidity, clouds, and rain. As a side issue, some aspects of optics are also discussed— for example, the reason why the sky is blue. The final section concerns the subject of air and thermal pollution of our environment.

QUESTIONS

Review and reason

1. Distinguish between conduction, convection, and radiation—the three processes of heat transfer between objects.
2. In a cold basement the temperature of a throw rug is as low as the temperature elsewhere in the basement. Why then do people usually prefer to stand on the rug rather than on the bare floor, particularly if they are barefoot?
3. An automobile radiator is not really a radiator. What is it, then? Explain the working of an automobile radiator.
4. Discuss some of the features of the design of better thermal insulation for a house.
5. See if you can fill in the blanks in the following sentence: Oil slicks are dangerous for aquatic birds; feathers keep a bird insulated from cold, but soaking the feathers with oil removes this _______ of the feathers. As a result the bird loses ____ quickly and may contract hypothermia, which is __________.
6. Is it really preferable to wear dark-colored clothes in the winter, or is this just a fable?
7. If a house is air-conditioned, an exhaust fan in the attic will help. Why?
8. Farmers usually don't worry about their crops dying from a sudden severe cold if there is a blanket of snow on the crops. Explain.
9. Explain why smaller animals suffer from cold more than larger animals.
10. People in the tropical and equatorial regions usually prefer white clothing in the summer. Explain.

11. What can be ignited quicker by means of sunlight focused through a converging lens, a black or a white piece of paper? Explain your answer.
12. On warm nights it is preferable to sleep as stretched out as possible on the bed. Why?
13. A Chinese wok (Fig. 14.22) is a very useful cooking utensil for high-heat cooking. Does its shape have anything to do with this? Explain.

FIGURE 14.22   A wok.

14. Why are Thermos bottles good both for keeping hot liquids hot and cold liquids cold? Explain.
15. If windows of double-paned glass, with a layer of trapped air between the panes, are used in buildings, the cost of heating and cooling can be significantly lowered. Explain.
16. Light-colored roofs are better in areas that require air conditioning in the summer. Why?
17. You may have seen the science fiction movie *2001, A Space Odyssey*. In this movie the hero is shown

stepping into outer space without the protection of a space suit, and he survives. Ignoring questions like oxygen or air pressure, just consider the following question: Would he feel cold? Why or why not?

18. Distinguish between a land breeze and a sea breeze. How are they caused and at which times?

19. Describe the global wind pattern. Draw a diagram. Why do we have westerlies in America?

20. Discuss briefly the convection currents in the earth's mantle, and name a few geological processes related to these currents.

21. Write a short essay on the prospects of geothermal energy in America.

22. Snow on grass does not melt as fast as snow on concrete or even snow on bare soil. Explain.

23. What is meant by the solar constant?

24. Discuss some of the aspects of the flow of solar energy in the earth's environment.

25. What is the greenhouse effect? Why is it that carbon dioxide released in the atmosphere may prove to be dangerous eventually for the global environment?

26. If you sit before a fireplace and the heat is a little too much for you, you can reduce the amount of radiant energy you receive by putting a glass screen in the fireplace opening. Why does the screen prevent some of the heat radiation from reaching you? Why is the radiation only partially blocked?

27. Why is the color of the sky blue? Why are the sunsets and sunrises red?

28. When you fly on a plane into a city, you often see the city streetlights as green. But when you are on the ground, you find them to be white, as usual. What is the reason for the green color?

29. If you lived on a planet of a red-colored star, and the planet's atmosphere were exactly the same as earth's, what color sky would you see?

30. If the atmosphere were very thick, direct sunlight would look red even during the day. Is this true? Explain.

31. In severe fog conditions does it help you to see the car in front of you if its back lights are red? Why or why not?

32. Christmas lights look all red from a distance. But when you come close, often you discover that there are other colors, too. Why don't you see the other colors at a distance?

33. Here is a quotation from Chapter 16 of Matthew: "When it is evening, ye say, it will be fair weather; for the sky is red. And in the morning, it will be foul weather today; for the sky is red and lowering." Is the Bible correct in its weather prediction? Explain the weather wisdom expressed in these lines in terms of physics.

34. What is meant by temperature inversion? How is temperature inversion caused by subsidence? Describe the mechanism by which normal nighttime inversion is produced.

35. Here is a hypothetical way to go up from the earth to the moon:

> Smoke having a natural tendency to rise
> Blow in a globe enough to raise me.*

How far can smoke rise? What happens if there is a temperature inversion?

†36. What causes fog? The fog in London decreased substantially after open burning of coal was stopped. Explain.

37. Come frost time, why do orchard owners put smudge pots in their orchards at night? Is it just for the heat, or are some other things involved? Discuss some of the factors you can think of.

38. In many science fiction stories, the earth of the future is assumed to have become completely urbanized. Would the average temperature of such an earth be higher than, lower than, or the same as the present temperature of the earth? Explain your answer.

39. In his classic science fiction trilogy, *Foundation*, American author Isaac Asimov imagines a small planet named Terminus which supplies the technological necessities for all nearby planetary systems. What is the one severe and inescapable environmental problem that Terminus could have? Be specific.

40. Discuss the reasoning behind this statement: Thermal pollution is the ultimate limit of growth.

---

*From *Cyrano de Bergerac*—Brian Hooker translation by Edmund Rostand. Copyright 1923 by Holt, Rinehart and Winston. Copyright 1951 by Doris C. Hooker. Reprinted by permission of Holt, Rinehart and Winston Publishers.
†Optional.

1. Perhaps it is winter now in your town. Keep track of the daily average temperature for a week. From the knowledge of the thermostat setting in your house (or dormitory), calculate the number of degree-days in the week. Assuming that your residence has a conductance of $k = 20,000$ Btu/degree-day, calculate the amount of heat needed to heat it for the entire week.

2. Do some arithmetic to convince a friend in Indianapolis that she gets enough solar energy on her roof to provide her with just about all the energy needed for household heating in the winter. Write down the steps of your arithmetic and discuss the results.

15

Static Electricity and the Electric Field

The victory over the concept of absolute space . . . became possible only because the concept of material object was gradually replaced as the fundamental concept of physics by that of the field. Under the influence of the ideas of Faraday and Maxwell the notion developed that the whole of physical reality could perhaps be represented as a field.

ALBERT EINSTEIN

■ 15.1 The Subject of Electricity

When somebody mentions electricity, what comes to your mind? Perhaps the electric current that flows through the circuits of your household and powers the lights and various appliances? Or perhaps you are romantic and the word "electricity" recalls the sights and sounds you enjoy in a thunderstorm. Possibly you have been curious enough about electricity to experiment with it: to charge a glass rod by rubbing it with silk and then to observe how the glass rod makes small pieces of paper dance around. Human civilization has been aware of this last demonstration of **static electricity** for more than two thousand years.

All the phenomena described above arise from the motion of the atomic electrons we have previously discussed. Bulk matter is composed of two kinds of electrical matter; we distinguish between the two kinds by ascribing a negative charge to one and a positive charge to the other. The charge of the electron is negative. The nucleus of an atom contains a particle, the previously mentioned proton, which carries an equal amount of positive charge.* Actually, which one we call negative charge and which one positive is arbitrary, entirely a matter of convention.

---

*There are also other subatomic particles carrying positive or negative charges as well as neutral ones. See Chapter 23.

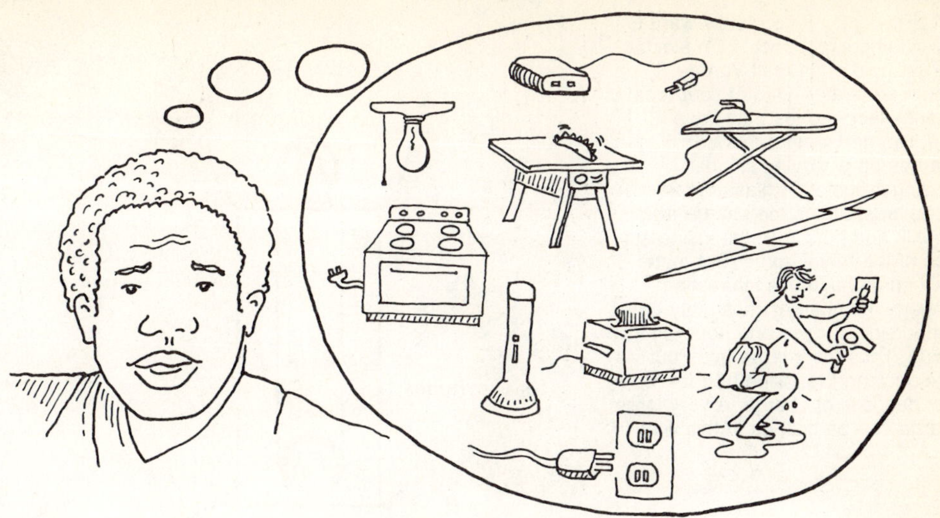

The two kinds of charges are distinguished by the manner in which they interact. Like charges—for example, two electrons—are found to repel each other. Unlike charges—for example, an electron and a proton—attract each other electrically. In fact, it is this attraction between the nuclear protons and the orbital electrons that keeps the atom together (Fig. 15.1).

Ordinarily, an atom contains an equal number of electrons and protons and thus identical amounts of positive and negative charges. The atom as a whole is thus neutral, and so are the molecules and bulk matter they compose.

Electrical charging

When we rub our hair with a comb made of hard rubber, the rubber comb manages to rub off a few electrons from our hair. In the process it acquires an excess of negative charges, whereas the hair suffers a deficit of the negative charges, leaving an excess of positive charges. This explains the mutual charging of two objects by rubbing. In case you are wondering why the protons don't move from one body to another in rubbing, one answer lies in their inertia. The protons are almost two thousand times heavier than the electrons and cannot be forced to move from one object to another by such maneuvers as rubbing.

Thus when the comb picks up electrons from our hair, the hair becomes positively charged. Strands of hair, now having like charge, repel each other, and the hairs now literally try to stand up on their ends. (See Fig. 15.2.)

Once an object is charged, its charge can be transferred to another body through simple physical contact. There are substances called **conductors,** such as a metal, in which charges can flow fairly freely. On a conductor the charges are distributed uniformly all over the body through their mutual repulsion. But nonconductors, or **insulators,** have only localized charges, which do not flow from one part of the insulator to another. So if you touch your charged comb with a metal rod that has a wooden handle, some of the charge from the comb will flow into the metal rod but not into the insulating wooden handle.

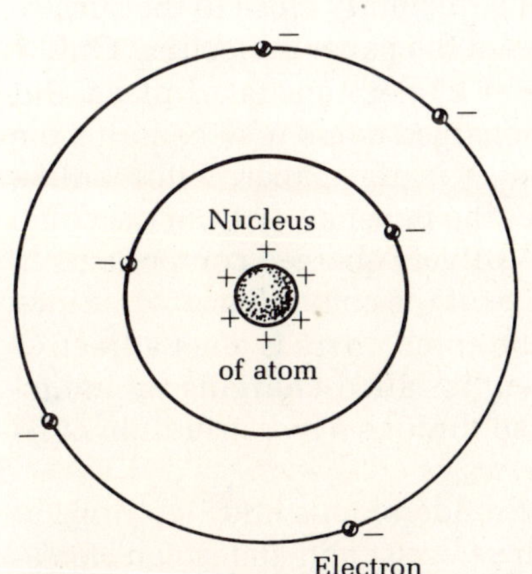

FIGURE 15.1 The atom consists of a positively charged nucleus and negatively charged electrons orbiting around the nucleus. The negative charge of all the electrons combined exactly cancels the positive charge of the nucleus. Thus the atom as a whole is electrically neutral.

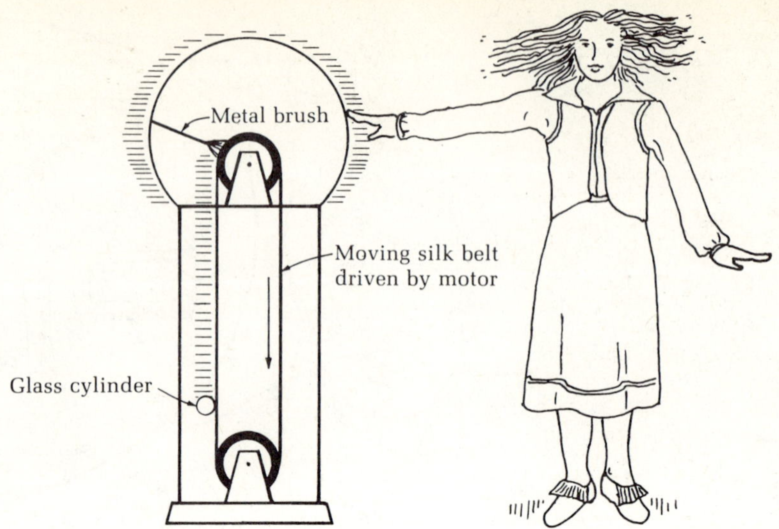

FIGURE 15.2   If you really want a hair-raising experience from electricity, get hold of a small Van de Graaff generator. One of your local science laboratories may have it. The Van de Graaff generator charges up a moving silk belt by rubbing it against a glass cylinder. A metal brush takes the charge from the silk and puts it on the inside surface of the metal dome. But metal enclosures have this interesting property: all charges must stay on the outside surface and none on the inside. Thus the charge goes outside, leaving the inside free for further deposits of charge. A very large charge can be built up in this way.

A convenient device for measuring charges is the **electroscope.** It consists of two thin leaves of gold foil at the end of a metal rod (Fig. 15.3). The rod is mounted in a box to protect the leaves from air current and is itself insulated from the box with rubber. Now if a negatively charged comb were to touch the metal knob, some of its electrons would flow into the metal all the way to the leaves. Since like charges repel, the leaves must separate. If the comb is charged with a greater amount of charge, the leaves will separate even further.

There is still another way of charging a body. You may have wondered why a charged comb attracts small pieces of paper, which are supposed to be neutral. The negative charge of the comb, when brought sufficiently close to the surface of the paper, forces a reorientation of the charges of the paper molecules. On the whole the paper remains neutral, and no transfer of electrons takes place. But the surface of the paper close to the negatively charged comb now contains the positively charged components of the molecules in the paper, since unlike charges attract. By the same token, the surface of the paper away from the comb will become negatively charged, because the negatively charged components of the molecules are repelled by the negative charge of the comb (Fig. 15.4). Since the positive charges of the paper are closer to the comb, there is a net attractive force toward the comb. This method of charging is called **charging by induction,** and the paper is said to have been **polarized** through a redistribution of its molecular charges.

We now can understand the mechanism of thunderclouds and lightning, at least to some extent. It was the great American scientist and statesman Benjamin Franklin who discovered the basic theory. The bottom surface of the thunderclouds facing the ground is heavily charged with excess electrons. This negative charge polarizes the ground immediately under it, which becomes positively charged as a result (Fig. 15.5). If the attractive force between the electrons of the cloud and the positive charge of the ground beneath is great enough, the electrons will take a jump through the air onto the ground, producing the lightning sparks.

FIGURE 15.3   A gold leaf electroscope. If a charged body touches the metal knob, the leaves separate.

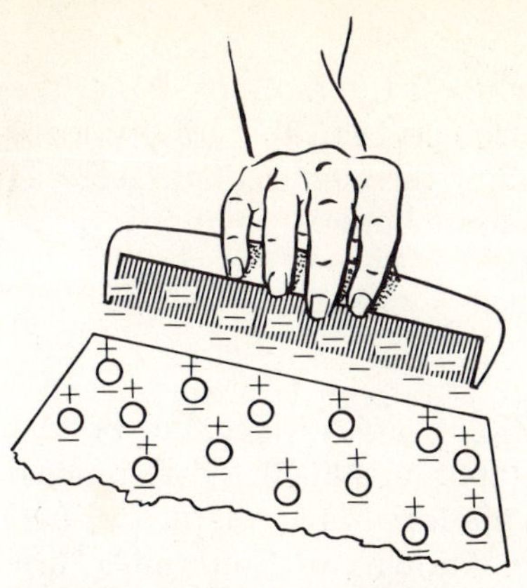

FIGURE 15.4   In the presence of a negatively charged comb, the charges inside the piece of paper are reoriented. Negatively charged electrons tend to be repelled from the negative charge of the comb, leaving the comb side of the paper slightly positive. Since the positively charged side of the paper is closer to the comb, there is a net attraction between the paper and the comb.

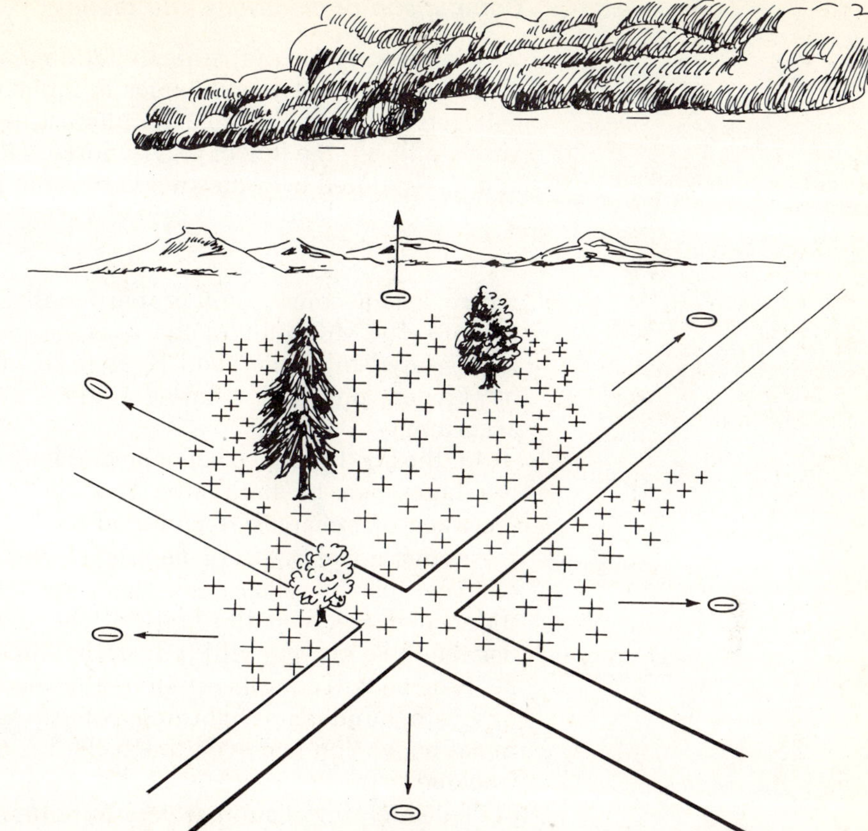

FIGURE 15.5   The negative charge of the thundercloud facing the ground polarizes the ground underneath. If the attraction between the positively charged ground and the negative charge of the thundercloud is large enough, a lightning spark will result.

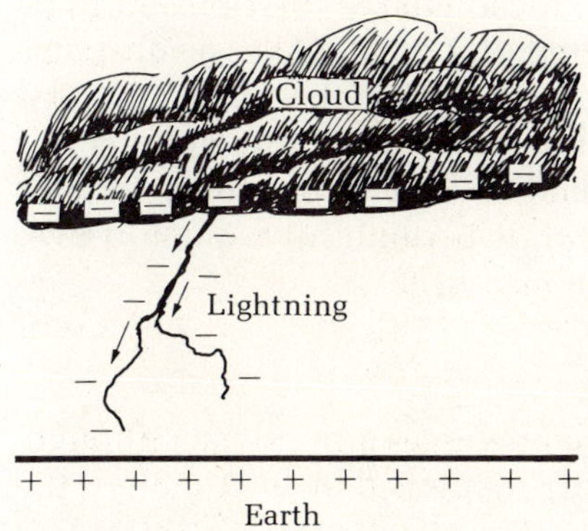

FIGURE 15.6   The "step leader" of lightning, establishing a path toward the ground.

The preceding picture of lightning is an oversimplified one. We now know quite a few other fascinating pieces of the lightning story. For example, the lightning discharges start with what is called a "step leader," which starts as a bright spot at the cloud and then moves downward, establishing a path, in steps (Fig. 15.6). Once the path is established there usually are several lightning strokes along the same path. Also, when the step leader gets close to the ground, say within a hundred meters or so, there is evidence that a discharge rises from the ground to meet it. This is why a grounded, sharp-pointed object like a lightning rod can protect a building; discharge from the rod will reach up to the leader. The lightning rod thus facilitates the lightning discharge by providing an easier route.

Charging by friction and by induction are examples of "electrostatic" phenomena. In contrast, an electric current consists of a flow of electrons through a conducting wire like copper. Such a flow needs a source of energy, an availability of "free" electrons, and a creation of a certain condition so that the electrons can flow, as we will see later.

Comparison of electricity and gravity

It is interesting to compare gravitational and electrical interactions. First, the electrical force between charges is immensely stronger than the gravity force between them [see Eq. (15.2 a)]. Interestingly, though, the electrical force law is very similar to the law of gravity force. Thus the force between two charges q_1 and q_2 separated by a distance d is given as

$$F_{el} = \frac{kq_1q_2}{d^2} \tag{15.1}$$

where k is a constant of proportionality. In MKS units,* the value of k is 9×10^9. The MKS unit for charge is the coulomb (abbreviated C); the distance d is measured in meters, and F_{el} is in newtons. This force law was discovered by French physicist Charles Coulomb and is known as **Coulomb's law** accordingly.

But the big difference between gravitational and electrical interactions is that the latter can be both attractive (between unlike charges) and repulsive (between like charges). Gravity, of course, is always attractive.

Another novel aspect of electrical forces is that if there is matter between two charges, the forces between them are reduced. This is due to the effect of induced charges on the body of the interfering matter. Thus water placed between two charges will reduce the force between them by almost a factor of 80. A metal will completely shield the electric force. As a result, in the presence of severe thunder and lightning, you can stay safely in the confines of your car's metal body. You can even touch the inside ceiling of the car and not be electrocuted.

This shielding is another very important difference between the electric and gravity forces; only electrical forces can be shielded, not gravitational. The difference is due in part to the existence of two kinds of charges in connection with electric forces, positive and negative, versus only one kind of mass in connection with gravity.

Perfect shielding depends crucially on the presence of substances like metal, which we call conductors. In conductors some of the electrons are quite free, and their mobility makes it possible for the induced charge distribution to be precisely such that no electric force is transmitted. But what makes a substance a conductor as opposed to an insulator? It turns out that the answer to this question comes from "quantum effects," the dynamics of the interactions at a very small scale (see Chapter 22 for further details). Since the gravity force is known to be very small at the "quantum" level, it is doubtful we could ever make a gravity shield, even if we had negative masses.†

Electrical interactions

It is the electrical interactions that dominate the processes around us, although, because of the tendency toward charge neutrality displayed by bulk matter, the

---

*Since $k = F_e \times d^2/q_1\,q_2$ a unit of k can be constructed by multiplying the unit of force with that of the square of the distance and dividing by the unit of the square of the charge. In other words, newton meter²/coulomb².

†The question of negative mass is discussed in Chapter 25.

validity of this statement is far from obvious. To see the truth of the statement, we have to look at the manifestation of these interactions at the microscopic level. Submicroscopic objects like molecules and atoms are electrically neutral, but the electrical character of their constituents nevertheless is important. When an atom is brought into close proximity with another atom, the negatively charged electrons of one are, on the average, a little bit closer to the positively charged nucleus rather than to the electrons of the other. Since the force varies inversely as the square of the distance, this means that the attractive forces are slightly bigger than the repulsive ones, and thus there is a net attractive force between the two atoms. Therefore, the attractive atomic force that binds atoms to make molecules is really due to electrical forces between the atomic constituents, arising from slight charge imbalances. The same can be said about the forces between molecules and the whole hierarchy of forces at the macroscopic scale, like friction and stresses and so forth.

Take a solid like sugar, a complex compound made from carbon, oxygen, and hydrogen atoms. The binding force between the atoms is again due to the electrical forces between the atomic constituents. When we put sugar in water, these electrical forces are weakened in strength as a result of the shielding effect of water. And so we find the sugar dissolves away because the force has become too weak to bind the atoms. Water is a very good solvent because it acts this way.

There are two more important things about charges that we can discuss. Charges have a minimum operating currency. The smallest charge ever detected is that of the electron; all other charges exist only as multiples of this electronic charge. Numerically, the charge of the electron is $e = 1.6 \times 10^{-19}$ coulomb. Nobody quite knows what gives rise to this discrete, or "quantum," nature of the charges.* Another important fact is that the charges obey a conservation law: the total number of charges (positive plus negative added algebraically) always remains a constant. This conservation of charges plays an important role in interactions between elementary particles.

EXAMPLE 1 *Calculation of the Coulomb force between two charges*

Suppose we have two glass balls 2 cm apart, each carrying a charge of 3 microcoulombs [1 microcoulomb (abbreviated μC) = 10^{-6} coulomb]. What is the magnitude of the electric force between them?

From Eq. (15.1),

$$F_{el} = \frac{9 \times 10^9 \times 3 \times 10^{-6} \times 3 \times 10^{-6}}{(2 \times 10^{-2})^2}$$

$$= \frac{81 \times 10^{-3}}{4 \times 10^{-4}} = \frac{20.25}{10^{-1}} = 202.5 \text{ N}$$

EXAMPLE 2 *Comparison of electrical and gravitational forces between two electrons*

It is instructive to verify that the electrical force between two electrons is

---

*There is now some indication that charges of $\frac{2}{3}e$ or $\frac{1}{3}e$ may exist.

indeed very, very much larger than the gravitational force between them. Let us take the distance between the two electrons to be 10^{-10} m, which is a typical distance on the atomic scale. Of course, for the comparison, any distance can be chosen; the distance doesn't matter since both forces vary the same way with distance. From Eq. (15.1), since the charge of the electron is 1.6×10^{-19} coulombs,

$$F_{el} = \frac{(9 \times 10^9) \times (1.6 \times 10^{-19}) \times (1.6 \times 10^{-19})}{(10^{-10})^2}$$

$$= \frac{2.30 \times 10^{-28}}{10^{-20}} = 2.30 \times 10^{-8} \text{ N} \sim 10^{-8} \text{ N}$$

In comparison, the mass of the electron is 9.1×10^{-31} kg and the gravity force constant, $G = 6.67 \times 10^{-11}$ MKS units. Thus the gravitational force is given as, [Eq. (6.4)]

$$F_{grav} = \frac{(6.67 \times 10^{-11}) \times (9.1 \times 10^{-31}) \times (9.1 \times 10^{-31})}{(10^{-10})^2}$$

$$= \frac{5.52 \times 10^{-71}}{10^{-20}} = 5.52 \times 10^{-51} \text{ N} \sim 10^{-50} \text{ N}$$

Thus the ratio of the orders of magnitude of the gravitational force to the electrical force is found to be

$$\frac{\text{gravitational force}}{\text{electrical force}} \sim 10^{-42} \tag{15.2a}$$

a very small number indeed.

Here is an interesting thing to note. This ratio is a pure number and does not depend on which system of units is used to calculate it. We would like to believe that there is some hidden meaning to such a ratio and that one day, when we have a complete physical theory, such a theory will explain the ratio.

English physicist Paul Dirac has suggested that the ratio of Eq. (15.2a) does not remain constant at all but varies with time. One way to obtain such a small ratio is to divide the smallest elementary time interval with the largest time that occurs in nature. Now the smallest observed time is the lifetime of certain elementary particles, some 10^{-24} sec, and the largest time is the age of the universe. The big bang theory estimates the latter to be 20 billion years, which is about 10^{18} sec. The ratio of these two times is then

$$\frac{10^{-24}}{10^{18}} = 10^{-42} \tag{15.2b}$$

That is of the same order of magnitude as the ratio of the strength of gravity and electromagnetic forces.

Suppose we assert that the two ratios are the same. But the second ratio, Eq. (15.2b), is changing, becoming smaller as the universe becomes older. So the gravitational force or the electrical force or both must change with time. There is good reason to believe that the electrical force does not change. So Dirac suggests that the gravity force (i.e., the gravitational force constant G) decreases with time. G may have been larger than its present value by a factor of two or so at the time of the formation of the solar system, according to Dirac.

All this is fancy numerology, very difficult to prove or disprove, but recently some measurements have revealed that the value of G may possibly have changed during historical time. We will have to wait for further tests of Dirac's idea through many more measurements.

The situation brings to mind a very beautiful poem by Norma Farber in which the poet equates permanence with the constant of gravity:

> Constant of gravity that's how I look
> On love which lets my lively seasons
> Stand
> On law so certain that it cannot be broken.*

However, if Dirac is right, Norma may have to rewrite her poem.

■ 15.2   The Electric Field

Perhaps the most amazing aspect of a charged comb making little pieces of paper dance with no strings attached is the fact that the action (the force) reaches out. The space surrounding the charged body must be special: it contains a potential force. Any other charge located in this space feels this force, the electric force. This special state of tension in the space around a charge we ascribe to the **electric field** created by the charge.

We can describe the electrical interaction of two charges as follows. A charge creates an electric field in the surrounding space. A second charge in this field feels the action of the field in the form of a force. The field of a charge q_1 at a distance d is given as

$$E = \frac{kq_1}{d^2} \tag{15.3}$$

Here k is the same constant as the one in Eq. (15.1). If we place a charge q_2 in a field E, the charge experiences a force given by

$$F = q_2 E \tag{15.4}$$

The electric field is thus an electric force per unit charge, $E = F/q_2$. Its MKS unit is newton/coulomb. Like the force, the field is also a vector.

In effect, we can say also that there is a force between the two charges given by F. Clearly, $F = q_2 E = q_2 k q_1/d^2 = k q_1 q_2/d^2$, exactly the same as F_{el}, the coulomb force of Eq. (15.1). The field description of the interaction of two charges is thus entirely equivalent to the force description.

So nothing has changed. Or has it? From the point of view of the proponents of absolute space, a lot has changed. Because if the concept of field represents reality, then empty space loses its meaning. Space adjacent to electrical matter, even in vacuum, is not empty but is pervaded by the electric field. The existence of empty space is an important element in the picture of absolute space; therefore, its elimination must be regarded as significant.

---

*"The Gravitational Constant," by Norma Farber, in *The Christian Science Monitor*, June 19, 1976. Reprinted by permission from The Christian Science Monitor. © 1976 The Christian Science Publishing Society. All rights reserved.

Notwithstanding the objection of the advocates of absolute space, the field idea presents a solution to a couple of otherwise "nagging" problems. One of these is the difficulty one encounters conceptually in trying to imagine how a force acts on a distant object. How does the "action at a distance" work? As you know, gravity is another such force which reaches onto other bodies through seemingly empty space. Now read what the discoverer of the gravity law says about this issue: the following is an excerpt from Newton's letter to theologian Richard Bentley.

> It is inconceivable that inanimate brute force should without the mediation of something else, which is not material, operate upon and affect other matter without mutual contact; as it must do, if gravitation . . . be essential and inherent to it. And this is one reason why I desired you would not ascribe innate gravity to me. That gravity should be innate, inherent, and essential to matter, so that one body may act upon another at a distance through a vacuum, without the mediation of anything else, by and through which their action and force may be conveyed from one to another, is to me so great an absurdity, that I believe no man who has in philosophical matters a competent faculty of thinking, can ever fall into it.*

Newton's consternation is obvious.

One other problem stems from our conviction that even a force should take a nonzero time to propagate through space; the speed of propagation is limited by the maximum of c, the speed of light. Then the force that acts on a particle at a given moment of time is not determined by the position of the source at that time but its position somewhat before then. The time lag or retardation is introduced due to the nonzero travel time of the interaction. Now we are forced to think of interaction of a particle with another one at a point of space where it was found some time ago, maybe yesterday. This is very unsatisfying if the action-at-a-distance principle is insisted upon. Where was the force between yesterday and now? Should we not expect a continuity in the action of forces between objects?

The concept of field provides us with a way out of such problems. The field description introduces an element of continuity to the subject of forces. The time lag for the interaction to reach from one point to another is no longer difficult to reconcile. A traveling particle creates a field. The field propagates with a finite speed and acts on other particles as it arrives at the positions of the particles.

Aside from electric and gravitational phenomena, in which the field concepts are useful, another familiar area that employs the field concept is magnetism. Perhaps experimenting with a magnt will most quickly give you some intuitive feeling for what a field is. Watch a compass needle. See how it reacts without any visible material connection to the magnetic field of the earth reaching out to it. Surely the space around earth to which a magnetic needle responds is special.

It is likely that even after reading this section (and perhaps all of this chapter) you will be a little uncertain as to exactly what these fields are, whether they be

____
*This letter was written by Newton on 25 February 1693. It is reprinted in *Newton's Philosophy of Nature*, ed. H. S. Thayer [New York: Macmillan (Hafner Press), 1953], p. 54.

fields of gravity or electricity or magnetism. Richard Feynman once said in a lecture, after posing a question of how to make a picture of the electromagnetic fields:

> I am sorry I can't do that for you. I don't know how. I have no picture of this electromagnetic field that is in any sense accurate.

But then he gave some reassurance.

> If you have any difficulty in making such a picture, you should not be worried that your difficulty is unusual.*

So here you come across an idea of physics that transcends your experience. If you have a philosophical bent of mind, there is a Zen koan that bears some similarity to the concept of a field. The koan is:

> "What is the sound of one hand clapping?"

Field lines

One way we can find the value of the electric field at a point is by actual measurement. For example, we can bring a small positive charge (call it the test charge) to the point and measure the electrical force on the charge. The measurement can be done using the electroscope (Fig. 15.3). Once the force is measured, the value of the field strength is given by the ratio of the force and the magnitude of the test charge. The direction of the field at that point is the direction of the force. There is one implicit assumption here: the field due to the test charge itself is assumed not to have any disturbing effect on the field we are measuring.

Now that the electric field has been determined, both magnitude and direction, we can represent it by an arrow of a certain length: the length of the arrow indicates its magnitude. If the given position of space is specified by the coordinates x, y, z, then this arrow gives us a picture of the electric field vector $\mathbf{E}$ at (x, y, z). We can measure the field at other points in the neighborhood, too, and very soon we will have an entire map of the electric field in an entire region of space. An example of this is shown in Fig. 15.7 in two dimensions; extend it to three dimensions in your imagination and you have a picture of a **vector field.**

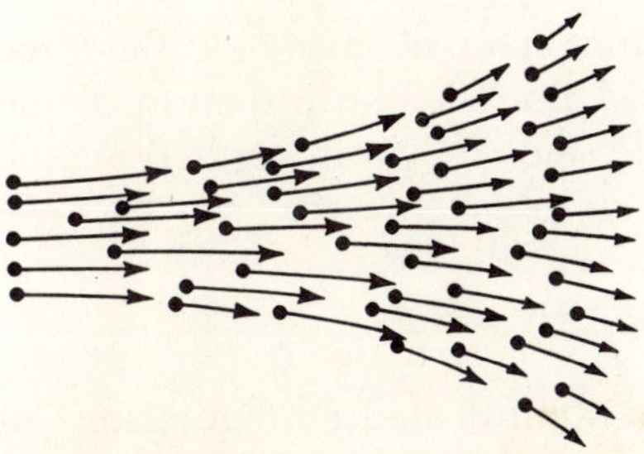

Maybe another example will help clarify what we are doing. Suppose we measured the ground temperature all over the United States and wrote down those numbers beside every location. What we would have then is a representation of the temperature field of America at a given time (Fig. 15.8). The difference between the temperature field and the electric field is that the latter is more complicated on two accounts. First, when we have to keep score for $\mathbf{E}$, which is a vector, we have to take into account both its magnitude and direction. For the temperature, which is a scalar quantity, we just need the magnitude; the temperature field is a **scalar field.** Second, since we collected only the ground temperature, we have a two-dimensional map, whereas in the case of

FIGURE 15.7   A vector field.

---

*Feynman Lectures in Physics, Vol. II, R. P. Feynman, R. B. Leighton, and M. Sands (Reading, Mass.: Addison Wesley, 1964), p. 20-g.

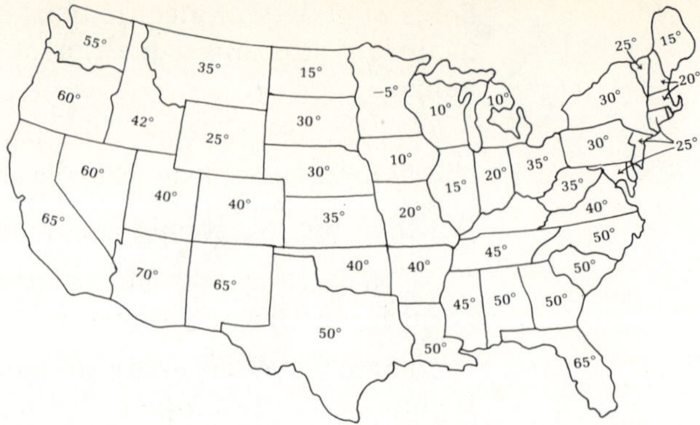

the electric field, we have a three-dimensional map, which is a little more diffi-cult to visualize.

Actually most people agree that this procedure—keeping track of arrows of different sizes in an entire region of space—is too complicated. So following an idea of English physicist Michael Faraday, we join up the arrows as shown in Fig. 15.9. The curves that we get in this way are called the **field lines** (or **lines of force**). The direction of the electric field at any point is now given by the tangent to the field line at that point. However, in achieving this we are forced to give up the information on the magnitude of **E.** To indicate the field strength, what we do now is to draw the field lines more densely where the field is strong and draw them somewhat apart where the field is weak. We adopt the convention that the number of field lines per unit area intersecting a plane perpendicular to the field lines is proportional to the strength of the field.

Consider the field lines of a single positively charged particle for which the field is known to be given by Eq. (15.3). We construct the field lines in such a way that our pictorial description of the field is consistent with Eq. (15.3). The field lines are shown in Fig. 15.10. The arrows are directed away from the center. This is a convention (due to the use of positive test charges); we always draw the arrows directed away from a positive charge and going toward a nega-tive charge. Since the surface of a sphere is proportional to the square of its radius, the number of lines per unit area in a distribution like this in three dimensions does fall off inversely as the square of the radius or the distance from the charge at the center. Thus the proportionality of the field strengths given by Eq. (15.3) and the number of lines per unit area are maintained in our picture.

Note also that according to our convention, in a given electric field the force on a positive charge is directed along the field lines in the direction of the arrows. But the force on a negative charge in the same field, although along the field lines, is directed opposite to the arrows.

Superposition of fields

We can figure out the field lines due to more complicated configurations of more than one charge just by using Eq. (15.3), which defines the field of a single charge. Electric fields obey a superposition principle: the total field at a point

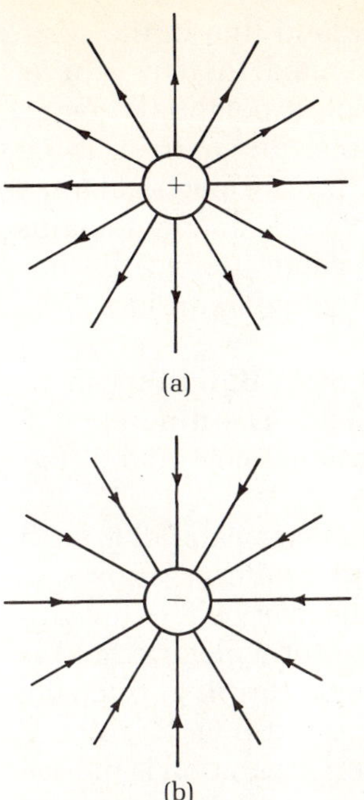

FIGURE 15.10   (a) The field lines of a single positive charge. (b) The field lines of a single negative charge. Notice the difference in the direction of the arrows.

due to several charges is the vector sum of the individual fields. Fields coexist in a region without affecting one another. So to get the field lines due to a more complicated configuration of charges, we use the superposition principle. Add up the electric field vectors at several points of space. Then draw the lines of the total field in such a way that the direction of the total field is tangential to the field lines at each of the points and the density of the field lines has a proportionate correspondence with the strength of the resultant field.

Fig. 15.11 shows the field of the **dipole** configuration of charges (a positive and negative charge separated by some distance) constructed with the aid of the superposition principle. Direct experiments confirm that the field lines in this case are indeed of this shape. The electric dipole configuration is quite common in nature. Even in an extended system for which the total charge is zero, a situation sometimes develops where there is an excess of positive charges on one end and an equal excess of negative charges on the other; this is, of course, a dipole configuration. A molecule, for example, sometimes becomes polarized in this fashion, as we discussed earlier.

Electric potential

As was mentioned in Chapter 9, there is always potential energy associated with conservative forces. Electric forces give rise to electric potential energy. The sign of the potential energy is negative for attractive electric forces and positive for repulsive ones.

In connection with fields, it is convenient to define a quantity called the **electric potential** as the potential energy per unit charge. We can then associate with every point in a field a certain value for the electric potential. The understanding is that if a charge is placed at the point, it will acquire a potential energy given by the product of its charge and the electric potential at the point.

As in the case of potential energy, it is often more convenient to talk about the potential difference instead of the actual value of the potential. We can think of surfaces of equal potential in the given field; such surfaces are called **equipotential surfaces.** The difference of potential between two such surfaces is called the **potential difference.**

The lines of force give us a "picturesque" way to visualize these equipotential

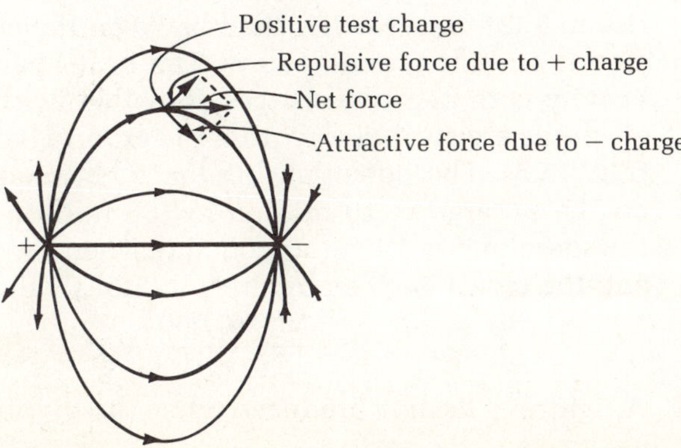

FIGURE 15.11   The field lines of an electric dipole.

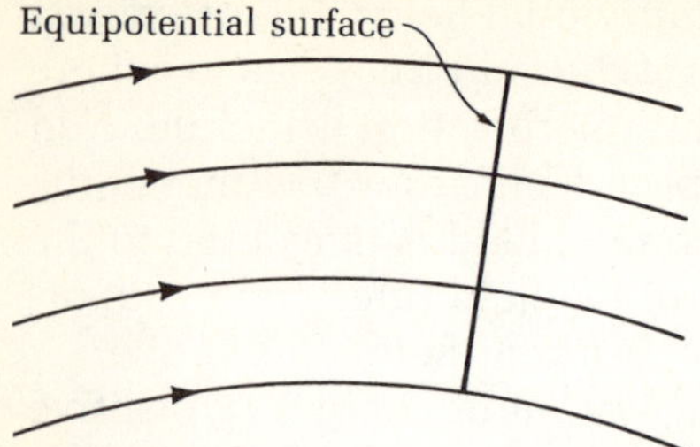

FIGURE 15.12 The equipotential surface of an electric field; it is perpendicular to the field lines.

surfaces. Suppose a charged particle is moving along a field line in the same direction as the force that acts on it. The particle is accelerated, thereby gaining kinetic energy—at the expense, of course, of the particle's potential energy, which thus undergoes a change from one point to another. In contrast, if the particle moved in a direction perpendicular to the field lines, no electrical force acts on it; its kinetic energy does not change and the potential energy remains constant. Thus a charged particle has constant potential energy along a surface that is perpendicular to the field lines: this is the equipotential surface (Fig. 15.12).

Think of gravity. The different levels of the Empire State Building can be thought of as equipotential surfaces in earth's gravity field. The difference of potential energy per unit mass would give us the gravitational potential difference in that case.

For a uniform electric field, equipotential surfaces are planes parallel to each other, like the different floors of a building. Here the word "uniform" is used to mean specifically an electric field that has the same value everywhere (magnitude and direction) in a given region of space. Such a uniform electric field is produced, approximately, between two parallel plates carrying opposite charges, an arrangement referred to as a parallel plate capacitor (Fig. 15.13). If V is the potential difference between the two plates and their separation is d, then the magnitude of the electric field in the region between the plates is given as

$$E = \frac{V}{d} \tag{15.5}$$

The field is practically zero everywhere else. Clearly the electric field is the change in potential per unit distance: it is a potential gradient.

The unit of electric potential is the volt (abbreviated V). As an example, a potential difference of 110 volts is maintained between the two ends of a household electric terminal.

Using Eq. (15.5), we can define another unit of the electric field as volt/meter. Of course this unit is exactly equal to the previously defined unit of newton/coulomb.

EXAMPLE 3 *Electric fields for thunderclouds*

One interesting example of a parallel plate arrangement of charges that gives rise to a uniform electric field between them comes from nature: the thunderclouds and the ground below. The lower parts of a thundercloud are charged heavily with negative charges. Repelled by these electrons of the cloud, many electrons are swept away from the ground below, leaving it positively charged (Fig. 15.5). The potential difference between the earth and the thundercloud can be as large as 10 million to 100 million volts. Suppose the height of the thundercloud is 1 km (or 1,000 m). Then the electric field between the ground and the cloud is given by the enormous value

$$\frac{10,000,000}{1,000} \text{ V/m} = 10,000 \text{ V/m}$$

A lightning flash is produced when the thundercloud discharges, returning the electrons to the earth.

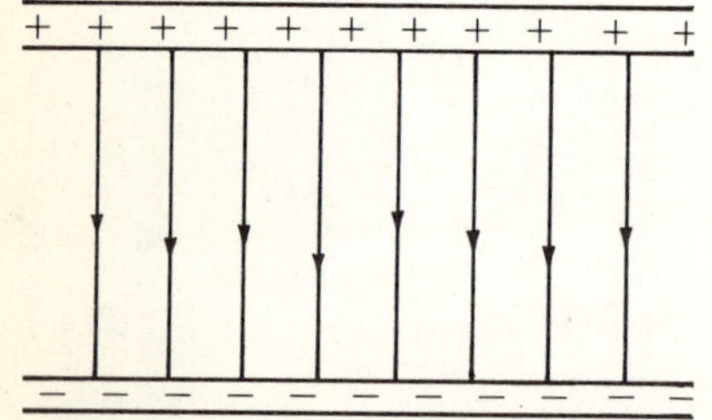

FIGURE 15.13 The electric field lines of a parallel plate capacitor with equal and opposite charges on its plates.

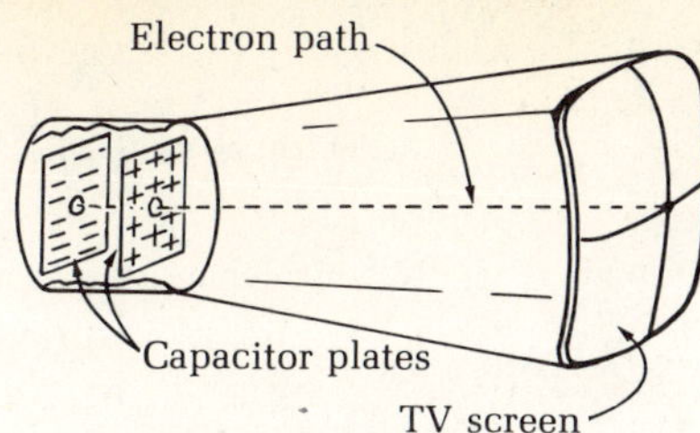

FIGURE 15.14 The acceleration of the electrons by an electric field inside the TV picture tube.

Electric fields in a TV set

The TV screen, when the TV is on, is swept continuously by a beam of electrons. The picture information is contained in this beam. In the TV picture tube, electrons have to be accelerated and deflected both in the vertical and horizontal directions so as to hit in any desired spot on the fluorescent screen. All these feats can be performed by electric fields between suitable capacitor plates. How the TV actually works is described in a later chapter, but we can tell you here how the electric fields do the jobs described above.

First we'll consider the linear acceleration. The electrons "fall" from a negatively charged plate to a positively charged one (Fig. 15.14) through a potential difference. During the travel they are accelerated by the electric field and gain some kinetic energy (at the expense of electric potential energy, of course) as they emerge through a hole in the positive plate. There is no further force acting on them, so they do not undergo any further acceleration as they proceed forward toward the screen.*

How is the turning of the electron beam accomplished? For turning horizontally (left to right of the screen), we need a force in that direction, and this can be provided by an electric field in a parallel plate arrangement, as shown in Figs. 15.15(a) and (b). Similarly, vertical deflection can be obtained by the parallel plate arrangement shown in Figs. 15.15(c) and (d). In both cases the motion of the electron while between the parallel plates is similar to that of a falling body, with a velocity in one direction and an acceleration in the perpendicular direction.

The oil drop experiment: measurement of the electronic charge

If we spray oil in the gap between the plates of a parallel plate capacitor, oil droplets are formed, which pick up some of the electrons that exist in the atmosphere. If there is a uniform electric field E in the gap, as shown in Fig.

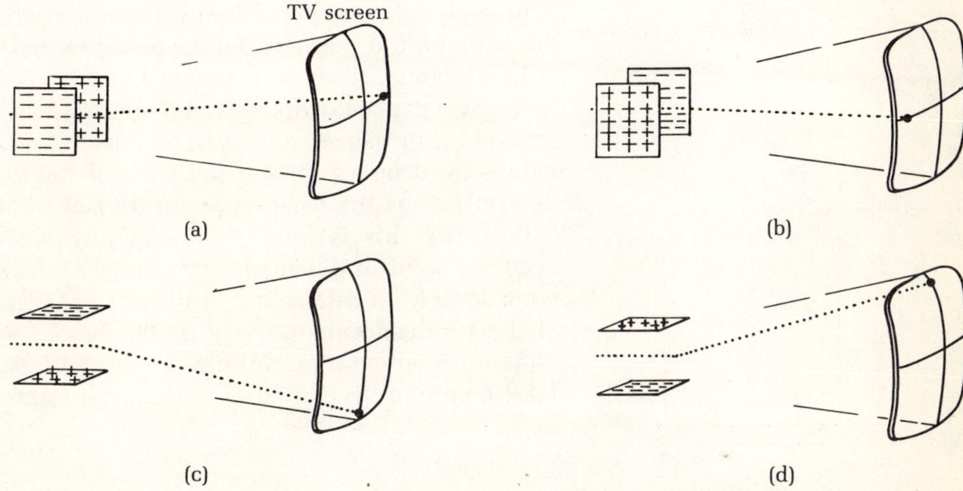

FIGURE 15.15 (a) The parallel plate arrangement for deflecting the electrons toward their left. Notice that the electrical force on the electron is toward the positively charged plate when it passes through the plate. (b) The arrangement for deflecting electrons toward their right. (c) and (d) The arrangements for vertical deflection of the electrons.

*The force of gravity on the electron is neglected; it is much too small in this case to have any appreciable effect.

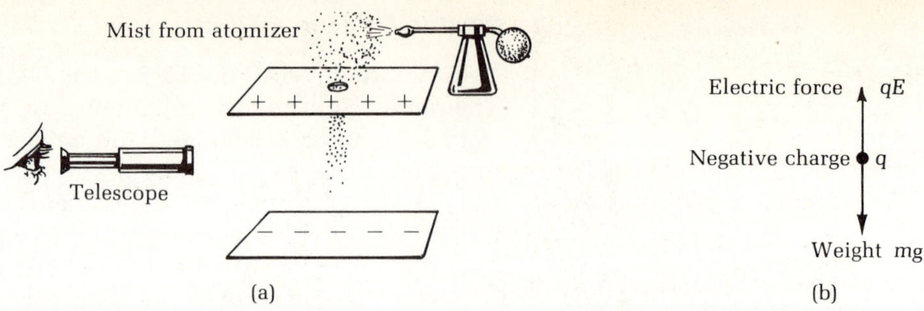

FIGURE 15.16 (a) The oil drop experiment. (b) A negatively charged droplet can be held motionless if the electric force on the charge cancels the weight force on the droplet.

15.16, it will exert an upward force qE on a negatively charged droplet of charge q. The charged droplet is also acted on by the downward force of gravity, mg, where m is the mass of the droplet. If the electric field is adjusted so that the electric force exactly cancels the gravity force, a charged droplet can be suspended motionless. The condition for suspension, then, is

$$qE = mg \qquad (15.6)$$

Suppose the droplet's net charge arises from picking up n free electrons (n is an integer like 3 or 5 or 29). Then $q = ne$, e being the electronic charge (the charge of an electron). Now we can write Eq. (15.6) as

$$neE = mg$$

or

$$ne = \frac{mg}{E} \qquad (15.7)$$

The mass m can be measured for a particular oil droplet by measuring its downward velocity in the absence of the electric field. Since the droplet is small, its fall through air under gravity is so greatly affected by air resistance that it acquires its terminal velocity very quickly and basically falls its entire course at this velocity. From a measurement of the terminal velocity, which depends on the mass of the droplet, the mass can be determined. Thus in Eq. (15.7), the quantities on the right-hand side can be experimentally determined. Now if we can get droplets small enough, n will be a small integer. A very large number of measurements will give us values of ne with various n; each of the values must be an integer multiple of the electronic charge e. Then e can be determined as the least common divisor of all these numbers we get for ne.

Basically this is how American physicist Robert Millikan determined an accurate value of the electronic charge e, although the details of his experiment were somewhat different. Significantly, Millikan's experiments also prove that charges indeed exist only in multiples of the electronic charge; e is the basic "quantum" of charge. None of his numerous measurements ever revealed any least common divisor smaller than the numerical value of e.

SUMMARY

The first section of this chapter deals with static electricity and includes a discussion of the elementary notions of charge, charging by friction and the like. The difference between a conductor and an insulator is pointed out. Charging by induction is explained, and the subject of electrical shielding is briefly mentioned. The law of force between charges is described by Coulomb's law:

$$F_{el} = \frac{kq_1q_2}{d^2}$$

The similarity of electrical and gravitational interactions of matter are discussed.

The second section of the chapter introduces the very important concept of field. Although it is defined simply enough [the electric field is (potential) electric force per unit charge], in actuality the field concept is a little difficult to grasp; it transcends our immediate experience. Physicists, following Michael Faraday, have learned a very useful way to make a picture of the field. This is the picture in terms of the lines of force, or field lines.

The electric field of a charge q at a distance d from the charge is given as

$$E = \frac{kq}{d^2}$$

If a charge q is placed in a field E, it experiences a force of

$$F = qE$$

Another important concept is that of electric potential, which is the electric potential energy per unit charge. The unit of electric potential is the volt. The last topic of the chapter, the oil drop experiment—which measures the charge of the electron—is intended to give you some sense of the reality of these particles that we call electrons.

QUESTIONS

Review and reason

1. Explain the concept of charge.
2. You can often get a shock while walking across a carpet (if you touch something) but not while standing. Why is this? If the air is humid, we seldom get such shocks. Why is this?
3. After a party one of the guests walks across the carpet (everybody else is standing still) and opens the door. Just before he touches the doorknob, a spark jumps across from the doorknob to his hands. When all this is over: (a) Is the person still charged? (b) Is the carpet still charged? Explain your answers.
4. Bathrooms are often found to have a negatively charged atmosphere. Can you explain what causes this? Would you expect the air near a waterfall to contain negative charges? Why or why not?
5. A balloon, when charged negatively, seems to cling to many of the objects around it. Does this mean that all these objects are positively charged since they seem to attract the negatively charged balloon? Explain your answer.
6. What is the difference between a conductor and an insulator?
7. Describe a gold leaf electroscope. Explain what happens when its metal knob is touched by a positively charged rod.
8. How does a lightning rod work?
9. Write down the mathematical expression for Coulomb's law of electric force between two charges. Explain the notation.
10. When lightning strikes an airplane (which often happens), is there any danger to the passengers inside? Perhaps one should not touch the ceiling while in an airplane during times of a thunderstorm. Discuss this idea.
11. Why is water a good solvent?
12. What are the similarities and differences between gravitational and electric interactions?
13. Explain the difference between a scalar field and a vector field. Draw diagrams.
14. Given the electric lines of forces, how can you tell the

direction of an electric field at a given point? How can you determine the relative magnitude of the field at two different points?

15. Draw the lines of a field of (a) a positive electric charge, (b) a negative electric charge, (c) a dipole, and (d) a parallel plate capacitor whose plates are charged with equal and opposite charges.

16. How can we measure the electric field due to a given distribution of charges at a given point of space?

17. Can the field lines of an electric field ever intersect? Why or why not?

18. Suppose the sun's gravity underwent a sudden unforeseen change. In the "action-at-a-distance" theory, we would know about it instantly. Is it the same in the field theory? Explain.

19. Explain the concept of electric potential. What is meant by an equipotential surface? Indicate the equipotential surfaces on a line of force diagram of the field of a single positive charge.

20. Explain the mechanism by which a large potential difference is set up between a thundercloud and the ground.

21. Describe how suitable electric fields can be used to deflect a beam of electrons, both in vertical and horizontal directions.

22. Describe how it is possible to determine the charge of the electron from the oil drop experiment. What is the value of the electronic charge? What do we mean by saying that this is the value of the "quantum" of charge?

Arithmetic

1. Suppose you charge up two combs with an identical charge of 2 microcoulombs (μC) each and place them 10 m apart. What is the electric force between the two combs? Is the force attractive or repulsive?

2. In Fig. 15.17 charges of $+1~\mu$C and $-2~\mu$C are placed at points A and B as shown, which are separated by a distance of 1 m. Determine the sign and the magnitude of the charge that you would have to place at point C in the figure so that the net electric force on the charge at B is zero.

3. An electron of charge 1.6×10^{-19} C is placed in an electric field of 10,000 N/C. What is the value of the electric force on the electron?

4. There is a potential difference of 100 V between two plates of a parallel plate capacitor placed 1 cm apart. What is the value of the electric field in the region between the plates? What is the value of the electric field elsewhere?

5. Show that if an electron is allowed to accelerate in a vacuum through a potential difference of 1 V, it will acquire an amount of kinetic energy of 1.6×10^{-19} J. Incidentally, this much energy is called an electron-volt (abbreviated eV) of energy:

$$1~\text{eV} = 1.6 \times 10^{-19}~\text{J}$$

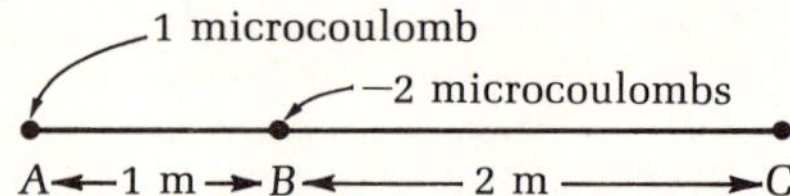

FIGURE 15.17

16 Electric Currents and Magnetism

The discovery of electric currents has had far-reaching consequences on human civilization. This remarkable feat is also one of the first recorded cases of a scientific discovery shared (after a fashion) by a man-woman team, in this case, Professor and Mrs. Luigi Galvani of Italy. Eager to please his wife, Lucia, who had expressed a desire for a dinner of frog legs, Professor Galvani bought some frogs. Since Lucia was ill, he decided to prepare the meal himself. After skinning a frog, he put the legs and the knife temporarily on his laboratory table near an electric machine (a generator of static electricity).

It is said that Mrs. Galvani was hungry and came downstairs to her husband's laboratory to find out how things were progressing. Why were the frog legs moving by themselves? Being very perceptive she noticed the electric machine and decided that the electricity from the machine must somehow be making a current flow through the frog legs, causing them to twitch. She told her husband so. Her husband at once verified that she was right. Indeed, every time he touched the knife to the frog leg while the machine was on, the leg moved. With the machine off, nothing happened.

However, this event was a red herring. The big discovery was yet to come. Continuing with the dinner preparations, Galvani hung the legs from a kitchen iron rail with some copper hooks. Guess what? As soon as the legs touched the railing, again there was violent movement. And this time there was no electric machine around. The conclusion was that somehow there was a sustained flow of charge (a current) through the frog legs.

■ 16.1 Electric Currents

When a thundercloud discharges, there is a temporary flow of charge and a momentary current. However, the current does not last—it is transient. Think of another situation. Suppose we have a parallel plate capacitor with an electric field (and a potential difference) between the two plates (Fig. 15.13). If we connect the two plates with a conducting wire, electrons will flow through the wire, momentarily producing a transient current, which will last only until the charges on both plates are neutralized.

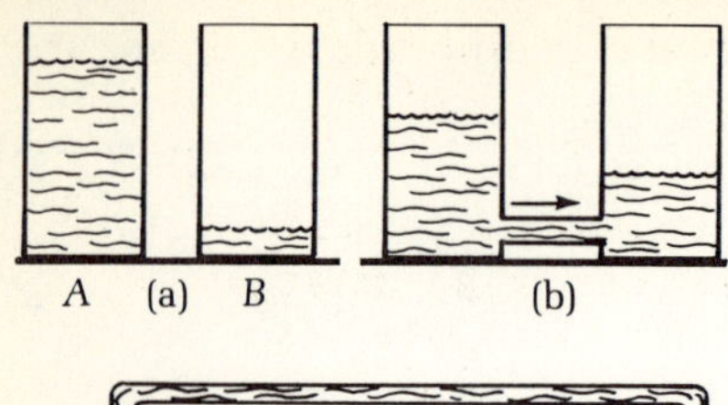

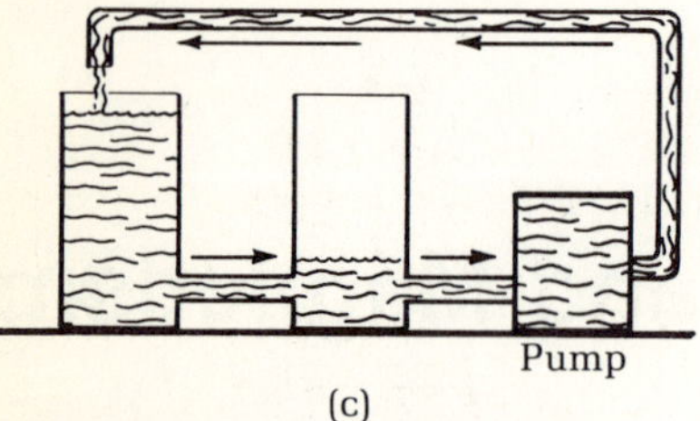

FIGURE 16.1   One way to get a continuous flow of water between two reservoirs. (a) The first step is to achieve a difference in the level of the water. (b) The second step is to connect the two reservoirs with a pipe. (c) The third important step is to maintain the different levels of water.

Thus if there is a separation of charge and therefore an electric field (or potential difference) established which can "push" the charges, and there is a conductor through which the charge can flow, we can generate a transient current. (Note that in the case of the thundercloud, the atmosphere acts momentarily as the conductor.) The next question is this: How can we generate a steady current, a steady flow of charges?

Perhaps an analogy will focus the problem. Suppose we want to create and maintain a steady flow of water. First, we need some water. Second, we need a driving force. For example, the difference of fluid pressure between two reservoirs with different water levels, as shown in Fig. 16.1(a), can provide the driving force. As you can see, the pressure at point A in the "high" reservoir is larger than the pressure at point B of the "low" reservoir. If we connect a pipe between them [Fig. 16.1(b)], water flows from the high to the low reservoir. But as the water flows, the level of water in the two reservoirs tends to equalize. And the flow of water will come to a halt when the pressure at the two ends of the pipe is equal, just as the flow of charge between the two plates of Fig. 15.13 stops when the charges on both plates are equalized.

Can you think of a way to get a continuous flow of water? One way is to pump water back continuously from the low to the high reservoir and thus maintain the difference in water levels [Fig. 16.1(c)]. Of course, this costs energy, but we didn't say we could do this kind of thing without paying for it.

So the trick in having a steady current is to pump the charges that make up the flow (electrons usually) back to the higher potential, over and over again. We do work on the electrons in the process. Later, as the electrons flow back to the low point of the potential, we extract work from the electrons.

Thus an ordinary **electrical circuit** carrying a steady current is a convenient way to transmit energy.* It has a device that acts as the source of energy; this device uses some form of resource energy to do work on electrons to lift them to the higher potential. The electrons flow through a conducting wire, and somewhere in the circuit there are devices or energy sinks that extract work out of the electrons (Fig. 16.2). Examples of the energy source in an electrical circuit are the battery and the electric generator. Examples of energy sinks are many: a light bulb, an electric motor (as in a household blender), an electric heating element, to mention a few common ones. Actually, since current is a flow of charge

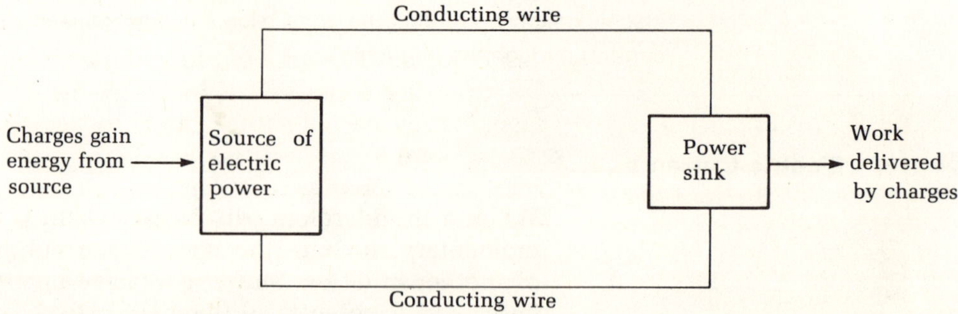

FIGURE 16.2   A schematic diagram of the components of an electrical circuit.

---

*What's a circuit? In connection with electricity, it denotes a continuous path (or array of paths) along which an electric current can flow.

involving an element of time, it is more customary and appropriate to talk about a power (energy/time) source and a power sink.

A quantitative definition of an **electric current** is given by the number of charges that flow across a point of the circuit in 1 second. If 1 coulomb of charge flows past a point each second, we say that there is 1 ampere (abbreviated A) of current. Thus 1 ampere = 1 coulomb/second.

The battery: a power source

How does a power source work? The electric generator uses mechanical energy (obtained from falling water, for example) to work on the electrons; the exact mechanism will be explained in the next chapter. Here, let us go through some details of the battery, which is the other major source of electric power.

The principle of the battery was invented by Italian scientist Alessandro Volta around 1800. It uses the chemical potential energy of a substance to "pump" electrons across a potential difference. The chemical reactions inside the battery create and maintain the separation of charges, giving rise to the potential difference. The battery has two dissimilar metals dipped in a chemical solution of some kind (Fig. 16.3). As a result of the chemical reactions, one of these metal strips becomes positively charged; this one is called the **positive electrode** or the **anode** (the + terminal). Simultaneously, the other metal strip becomes negatively charged and is called the **negative electrode** or the **cathode** (the − terminal). When an external wire is connected between the two terminals, electrons flow from the negative electrode toward the positive one through the wire, giving a current.

With the start of a current flow, electrons move into the positive terminal to neutralize the positive charge accumulated there, but the continuous chemical reactions in the battery produce additional positive charges at the positive terminal and simultaneously an equal number of electrons at the negative terminal to go into the circuit. In effect, the battery "pumps up" the electrons coming in at the positive anode to the place of higher potential of the negative cathode. It does this through the intermediary of chemical reactions inside the battery and at the expense of the chemical energy of these reactions.

In the D cell battery (Fig. 16.4) used in a flashlight, with which you no doubt are familiar, there is a central rod made of carbon, which acts as the positive terminal. The case of the battery is made of zinc and acts as the negative terminal. The case holds a paste containing a chemical compound named ammonium chloride. The ammonium chloride molecules in the paste partially break up, releasing some free chlorine atoms. However, these chlorine atoms are special: they are not neutral but carry an extra electron each. An atom carrying one or more extra electrons is called a negatively charged **ion.**

Two chlorine ions react with one zinc atom of the case forming a compound and also releasing the two excess electrons to the case. So, every time a zinc atom of the case is removed through the chemical reaction, the net result is an accumulation of two electrons on the case. Thus the case becomes negatively charged and acts as the negative terminal.

This is only half of the explanation of how the charge separation occurs. When the ammonium chloride breaks up, giving up the chlorine ions, the rest of the molecule has a deficit of two electrons and is thus positively charged (it is

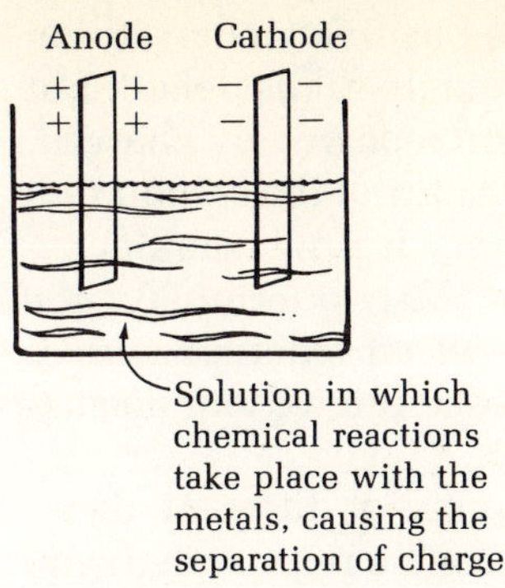

FIGURE 16.3   An electric battery.

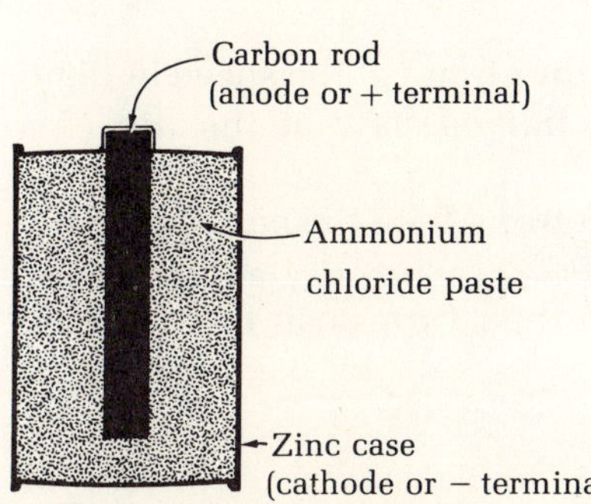

FIGURE 16.4   A flashlight battery.

called a positively charged ion). At the carbon rod, this positively charged ion now reacts with a carbon atom, taking two electrons from it to fill its deficit. But now the carbon rod has lost two electrons and has become positively charged. The positive charge accumulates every time a reaction occurs at the carbon rod, and hence the rod acts as the positive terminal. The story is now complete.

As a result of this separation of the charges between the two terminals, the battery has created a potential difference of 1.5 volts. If an external wire is connected, electrons can flow through it. In the next subsection we will discuss the details of the current flow.

All batteries work much the same way except for the use of different chemicals and metals for the reactions. Also, some batteries can be regenerated; you can pass an electric current through it and the chemical reactions will be reversed to return the battery to its original condition. Your automobile battery is one of this kind. The D battery is not rechargeable and runs out when its supply of chemical energy is gone.

Since potential difference is measured in volts, we often use the word **voltage** for it. The phrase "electromotive force" (abbreviated EMF) is also used sometimes to denote potential difference. But note that potential difference is not a force; it is the *source* of a force.

The current

Now let us concentrate on current flow. First, we'll consider the direction of the current. We have been saying that the flow of electrons in the external circuit is along the wires in the direction from the negative terminal toward the positive terminal.* However, you may be accustomed to thinking that the current flows from the positive to the negative terminal, which is the direction of current agreed upon by convention. Is the convention wrong? Not really; the convention regards the current as a flow of positive charges from the positive to the negative terminal, and mathematically this is entirely equivalent to the actual situation, where negative charges flow the opposite way.

As to the exact nature of the electric current, many beginning students have a variety of pictures or models of their own. It is a confusing subject, and some of these models reflect this confusion. (Perhaps one or all of them have occurred to you.) Here is a sampling.

Model No. 1. The negative terminal of the power source is the sole supplier of electrons making up the current.

Model No. 2. The electrons must flow through the circuit at the speed of light. This explains why the electric light turns on immediately at the flip of a switch.

Model No. 3. It happens like this. First, the electrons from the negative terminal bump into the "free" electrons of the connecting wires of the circuit and thus give them a push. Then those electrons in their turn push the electrons ahead and so forth, and, lo, we have a current.

Is one of these the way it works?

---

*"External" circuit means all parts of the circuit *excluding the inside of the power source*. The flow of charge inside the battery is more complicated, as indicated previously

No, the negative terminal is not the sole source of electrons for the current. The free electrons in the conductors elsewhere in the circuit also participate in the flow.

There is no need to think that the electrons move with the speed of light through the circuit. They don't, and they don't have to. There are electrons in the filament of the light bulb itself. When the circuit is complete, these electrons flow too, instantly supplying energy to the light bulb, which glows as a result. Actually, what travels with the speed of light is the electric field. As soon as the switch is thrown, the electric field gets established everywhere in the circuit and pushes the electrons, which are already there.

So the bumping or thermal collisions among the electrons have no direct role in the progress of the current in a circuit. However, they do exist, and they play a very important adverse role. The thermal collisions bring into effect resistive forces to the motion of the electrons. Just as a sky diver does not accelerate indefinitely when he falls under the influence of gravity, because of the action of air resistance, the electrical resistance prevents indefinite acceleration of the electrons under the action of the electric field. Very quickly the electrons acquire a terminal velocity and move with this terminal speed the rest of the way. So what we call an electric current is due to a net forward drifting at this terminal speed, superimposed on the random thermal motion. The terminal speed is something of the order of 10^{-4} m/sec.

Let us dwell some more on this last subject. The continual thermal motion of the free electrons of the conductors is always there, even in the absence of the electric field. However, there is no current due to the thermal motion because as many electrons go one way as the other; there is no net flow. The electric field creates the net flow and hence the current.

The velocity of the thermal motion of the electrons can be calculated from a formula given in Chapter 13:

$$\text{velocity} = \sqrt{\frac{3k_B T}{m_e}}$$

where m_e is the mass of the electron, k_B is Boltzmann's constant, and T is the absolute temperature of the wire. $m_e = 9.11 \times 10^{-31}$ kg and $k_B = 1.38 \times 10^{-23}$ J/°K; we can take T to be 300°K. So the velocity is found to be

$$\sqrt{\frac{3 \times 1.38 \times 10^{-23} \times 300}{9.11 \times 10^{-31}}} = \sqrt{1.36 \times 10^{10}}$$

$$= \sqrt{1.36} \times \sqrt{10^{10}} = 1.17 \times 10^{10/2}$$

$$= 1.17 \times 10^5 \text{ m/sec}$$

This is huge compared to the orderly drift velocity of the electrons. Yet this little orderly motion on top of such huge disorder is all it takes to produce the miracle of the electric current.

To summarize, this is our picture of the electric current. As soon as there is a complete circuit, an electric field is established at light speed throughout the circuit. The field accelerates electrons everywhere in the circuit only to a specific terminal speed, due to the existence of forces that resist the motion of the electrons. As a result, the electrons drift forward (along the conductor toward

the positive terminal) everywhere in the external circuit. The drift motion is a net forward motion superimposed on the random thermal motion of the electrons.

You may be curious about whether the electrons from one end of the circuit ever make it to the positive terminal. In a circuit in which the current always flows in one direction (such a current is called a **direct current** or **dc**), as we have described here, they can, given sufficient time. We do not know for sure (because of the random uncertainty of the thermal motion) whether a particular electron reaches the other end or how long it takes to do so. But, on the average, we can predict that electrons from one end do reach the other end in the time it takes them to travel the distance of the external circuit with a speed of some 10^{-4} m/sec. (In the analogous case of water flow, if you put fluoride in a town's water reservoir, it takes some time, but the fluoridated water does reach the households of the township after a while.) However, remember this is true only for a dc circuit. (In household electrical circuits, which use an **alternating current** (**ac**), the electrons move alternately one way and then the other, changing directions 120 times each second.) Therefore, they never go very far from where they are initially. We will have more to say about ac circuits in the next chapter.

So when we say that there is a 1-ampere current in a conductor connecting the two terminals of a power source, what do we mean? By definition, a 1 ampere current involves the flow of 1 coulomb of charge each second past every point of the conductor. Since the charge of one electron is 1.6×10^{-19} coulomb, 1 coulomb contains

$$\frac{1}{1.6 \times 10^{-19}} = 6.25 \times 10^{18} \text{ electrons}$$

So 6.25×10^{18} electrons each second flow by each point of a 1-ampere, current-carrying conductor. Again, this number does not mean that this many electrons flow from the negative terminal to the positive terminal each second. It means that each second this number of electrons enters one end of the wire and the same number leaves the other end. Also, exactly the same number flows per second through every cross section of the wire.

What determines the value of the current through a conductor? The water flow analogy can help us find the answer. For two reservoirs of water connected through a pipe, the rate of flow is directly proportional to the difference of the water levels: the relative height of the water in the high reservoir determines the flow. The height of water is analogous to the potential difference in the electrical case. Thus in the electrical case we can expect the current to be proportional to the difference of electric potential between the two terminals.

German physicist George Ohm experimentally found this to be the case:

> *The current through a conductor is directly proportional to the potential difference (voltage) between its ends.*

$$\text{current } I \propto \text{voltage } V$$

Introducing a constant of proportionality $1/R$, we get the equation

$$I = \frac{V}{R} \tag{16.1}$$

The constant R is called the **resistance** of the conductor. The relationship of Eq. (16.1),

$$\text{current} = \frac{\text{voltage}}{\text{resistance}}$$

is referred to as **Ohm's law.**

Notice that the larger the value of the resistance R, the smaller is the current for a given voltage, and vice versa. The resistance of a conductor originates from the resistive forces offered to the flow of electrons by its material, that is, the resistive forces that prevent the unlimited acceleration of the electrons under the influence of the applied electric field. The resistance depends on the material; copper has less resistance than aluminum and iron but more than silver. Predictably, nonconductors have much larger electrical resistance than conductors.

The value of the resistance of a conducting wire is also in direct proportion to its length (the longer the wire, the more its resistance). Furthermore, the resistance is inversely proportional to the cross-sectional area of the wire. Thus a thick wire has less resistance than a thin one. Interestingly, these properties of electrical resistance are quite analogous to the behavior of the frictional resistance offered by the pipe in the water flow example. The longer the pipe is, the greater the friction is. And a fat pipe offers less resistance than a thin one, leading to an increased flow.

The unit of resistance is the ohm (abbreviated as Ω, the capital Greek letter "omega"), which is equal to 1 volt/ampere. As an example, suppose a current of 2 amperes flows through a conductor when a potential difference of 110 volts is applied to its ends. What is the resistance of the conductor?

From Eq. (16.1),

$$R = \frac{V}{I} = \frac{110}{2}\,\Omega = 55\,\Omega$$

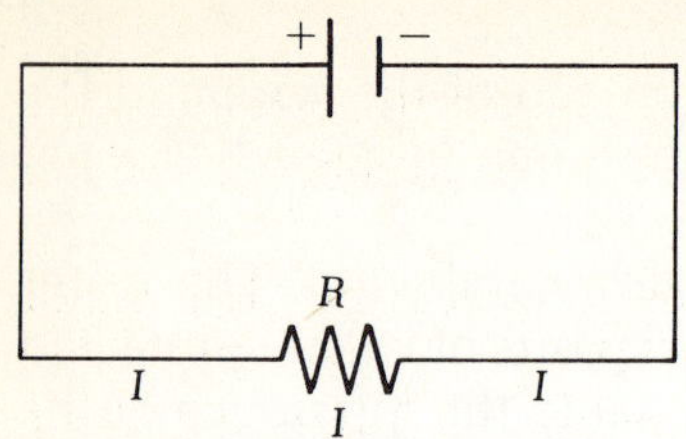

FIGURE 16.5   A simple circuit. ⊣⊢ denotes the battery, with the long bar always denoting the positive terminal and the short bar the negative terminal. ⌁ denotes a resistance.

Power sinks

Now we want to talk about the other important ingredient of an electrical circuit: the power sink. This is where the charges deliver their energy. Of course, the total amount they can deliver is limited by the amount of work done on the charges by the power source. Actually, the electrons lose some part of their energy in every segment of the circuit. By suitable design factors we keep the share of the electric power consumed by the connecting wires as small as possible and maximize the power delivered at the machines we want to run. The power dissipation arises from the electrical resistance offered by the conductors. Electrical resistance is quite analogous to mechanical friction; the work done in maintaining a current against the resistance is dissipated—as in mechanical friction—as heat. In a circuit we use very low resistance material in the connecting wires to keep this power dissipation in that segment at a minimum. On the other hand, we maximize the dissipation in the power sinks in devices where we can use the heating effect such as in light bulbs and toasters.

If a power sink has a resistance R, and a current I flows through it (Fig. 16.5), from Ohm's law the potential difference across the device is

$$V = RI \qquad (16.1a)$$

Since the voltage has decreased by this amount across the device, we often refer to this product RI as the **voltage drop.** Notice that in a circuit diagram such as Fig. 16.5, a resistance is denoted by the symbol ⌁ and the battery by $+ \dashv\vdash -$.

The power dissipation can be calculated from the voltage drop and the current according to the simple formula

$$\text{power} = \text{voltage drop} \times \text{current} \qquad (16.2)$$

or

$$P = VI$$

In this formula, if the voltage is measured in volts and the current in amperes, the power is given in the standard MKS unit of watt.

Electrical devices used in the household are designed by necessity to operate on a particular voltage drop of either 110 volts or, in a few cases, 220 volts (the clothes dryer, for example). All household electrical equipment comes with a definite power requirement. It is designed in such a manner that, when connected to the appropriate household line, there is just enough current through it to satisfy the power specification.

EXAMPLE 1

What is the current that flows through a 100-W light bulb? What is the resistance of the filament in the bulb?

All household lighting fixtures are connected to the 110-V line. From Eq. (16.2),

$$\text{current} = \frac{\text{power}}{\text{voltage}}$$

From its definition, power is the total energy expended divided by the time.

$$\text{power} = \frac{\text{potential energy expended } W}{\text{time } t}$$

The potential energy expended, W, is the voltage drop V times the charge q that has flown in the time t through the sink. This follows from the definition of the electric potential, which is potential energy per unit charge.

Thus potential energy expended $W = V \times q$, and the

expression for power becomes

$$\text{power} = \frac{V \times q}{t}$$

But q/t is the charge flowing in unit time, which is the current I, by definition. So finally we get

$$\text{power} = V \times I$$
$$= \text{voltage} \times \text{current}$$

we get

$$\text{current} = \frac{100}{110}\,\text{A} = 0.91\,\text{A}$$

Knowing the current, we can calculate the resistance by using Eq. (16.1):

$$\text{resistance} = \frac{\text{voltage}}{\text{current}} = \frac{110}{0.91} = 120.88\,\Omega$$

EXAMPLE 2

What is the current through an electric motor with a $\frac{1}{4}$-hp rating?

Since 1 hp = 746 W, the power of the motor in watts [we must have it in watts, in order to use Eq. (16.2)] is

$$\frac{746}{4} = 186.5\,\text{W}$$

Now from Eq. (16.2),

$$\text{current} = \frac{\text{power}}{\text{voltage}} = \frac{186.5}{110} = 1.7\,\text{A}$$

Series and parallel connections

Sometimes we want several devices to extract energy from the same source. We can hook up the devices in two ways. One way is to connect them as in Fig. 16.6, where the same current flows in each one of the devices. This is called a **series connection.** One disadvantage of such a connection is that, if for some reason one of the devices gets turned off (if a wire burns out or something), charges stop flowing, the entire circuit becomes open, and no current flows through any part. Perhaps you can remember a time when one of your Christmas tree lights burned out, causing all the bulbs to go off. This happened because your Christmas tree lights were connected in the series fashion.

But we can connect the power sinks another way, in a **parallel connection** (Fig. 16.7). Here each of the power sinks provides a separate pathway for the

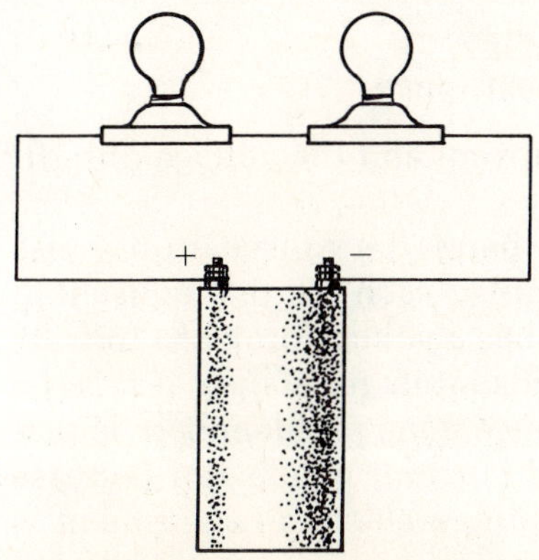

FIGURE 16.6   A series circuit.

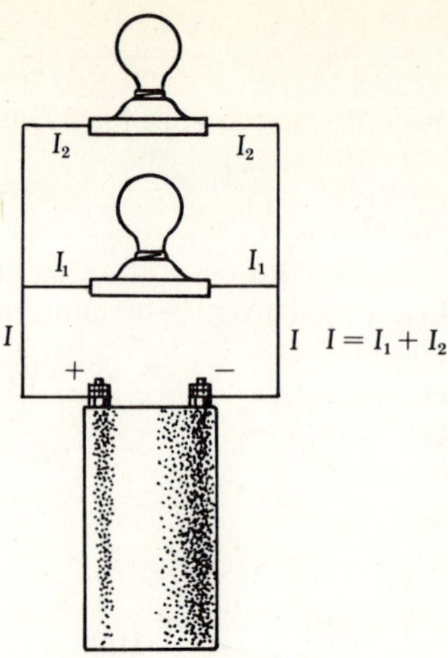

FIGURE 16.7   A parallel circuit.

current; the current literally splits up into branches. Each branch extracts all the electrical energy of the charges that flow through it. The advantage is obvious. A break in one path does not affect the flow of current in the other paths. The electric circuits in our households are connected in this way.

In a series circuit the current anywhere in the circuit is the same. Thus if I is the current, then in Fig. 16.8 the three devices of resistance, R_1, R_2, and R_3, cause voltage drops of R_1I, R_2I, and R_3I, respectively. Since the total voltage drop must be the potential difference V maintained by the power source, we get

$$R_1I + R_2I + R_3I = V$$

or

$$(R_1 + R_2 + R_3)\, I = V$$

Thus the current I is given as

$$I = \frac{V}{R_1 + R_2 + R_3} = \frac{\text{voltage}}{\text{total resistance}} \qquad (16.3)$$

Including more sinks in a series decreases the current and the voltage drop RI across each sink.

In the case of parallel connection, on the other hand, the voltage drop across each electrical device is the same. For example, each of the household appliances connected to the 110-volt power line has a voltage drop of 110 volts across it. The current that each one draws depends on its resistance: if R is the resistance, V/R is the current through the appliance. One problem here is that putting in more power sinks increases the total current, which can increase beyond the capacity of the household lines if enough electrical equipment is turned on simultaneously. To cope with this problem, a fuse or a circuit breaker

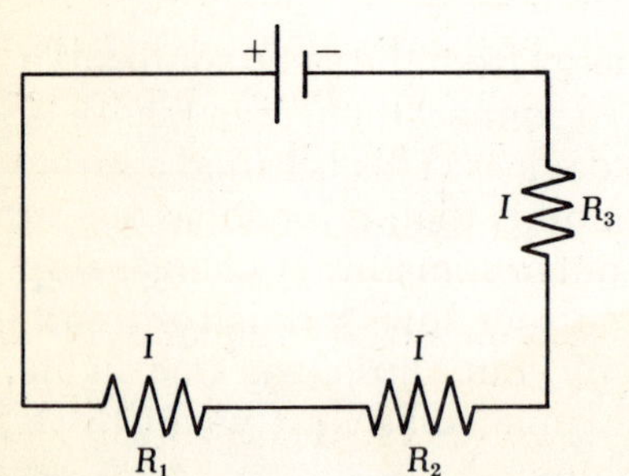

FIGURE 16.8   A series circuit containing three resistances.

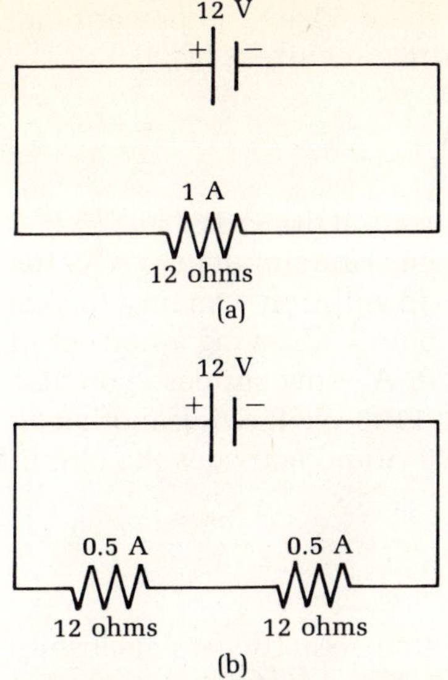

FIGURE 16.9 (a) A lamp of resistance 12 Ω connected to a 12-V battery draws a 1-A current. (b) But when two lamps are connected in series to the same battery, the current through each is halved, and the power drawn by each as compared to (a) is quartered.

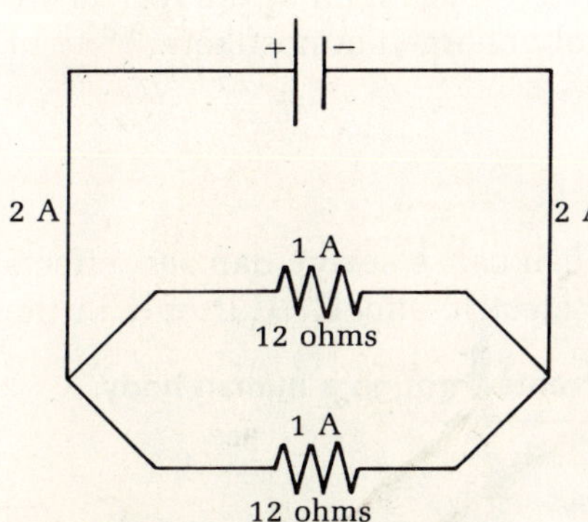

FIGURE 16.10 When two lamps are connected in parallel to a battery, they draw the same current as they would if connected individually to the battery, so their power remains the same. However, the current in the main circuit increases.

is usually connected in series to each appliance; also, a fuse is usually connected to the main circuit that brings the transmission line into the house. The fuse or circuit breaker puts a limit on the current through each circuit. If the limit is exceeded, the circuit is turned off. The main circuit breaker disconnects the household line every time there is an overload, that is, every time the total current exceeds the limit the household line can handle.

EXAMPLE 3   *Lamps in a series*

Suppose we connect a lamp of 12-Ω resistance to a 12-V automobile battery, as in Fig. 16.9(a). The current through the lamp is

$$\frac{\text{voltage}}{\text{resistance}} = \frac{12}{12} = 1 \text{ A}$$

Therefore, the lamp glows with a power of voltage drop $\times$ current = 12 $\times$ 1 = 12 W.

Now suppose we connect an identical lamp in series with the first [Fig. 16.9(b)]. The total resistance is now 12 + 12 = 24 Ω, and so according to Eq. (16.3), the current is

$$\frac{\text{voltage}}{\text{total resistance}} = \frac{12}{24} = 0.5 \text{ A}$$

The voltage drop across each of the lamps is the current times the resistance, which is 0.5 $\times$ 12 = 6 V. The two lamps must share the total voltage drop, and thus each gets half. The power with which each lamp glows now is only

$$\text{voltage drop} \times \text{current} = 6 \times 0.5 = 3 \text{ W}$$

So the lamps are dimmer when connected in series.

EXAMPLE 4   *Lamps in parallel*

Let's do the previous example again but this time with the two lamps connected in parallel to the battery (Fig. 16.10). Since for parallel connections the voltage drop across each lamp remains the same, as does the current, each of the lamps still draws a current of 1 A (as if they were connected to the battery singly). So their power (current $\times$ voltage) also remains unchanged. However, not everything is unchanged. Consider the total current in the main line: it is the sum of the currents through the two parallel branches. And that gives a current of 1 + 1 = 2 A. Thus for parallel connections, the current drawn in the main circuit increases every time we add a new parallel component.

EXAMPLE 5   *Overloading your household line*

Figure 16.11 shows how several appliances are connected in parallel to the household line. Suppose the maximum tolerance of the line is 20 A. If the total current reaches this value, the circuit breaker will be activated and will turn off the current. Since each of the appliances is connected in parallel, there is a voltage drop of 110 V across each. The current flowing through each can then be

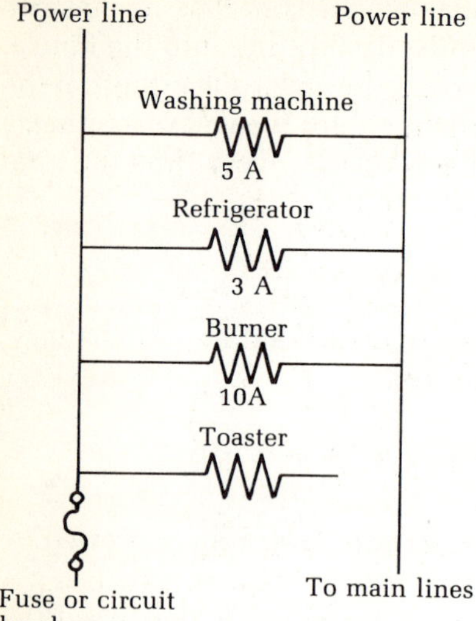

FIGURE 16.11 The electrical appliances of a household connected in parallel to the power lines. In the example given in the text the fuse would blow if the toaster were connected.

calculated if we know the power requirement of the device. The power of an automatic washing machine is 550 W, so it will draw a current of

$$\frac{\text{power}}{\text{voltage}} = \frac{550}{110} = 5 \text{ A}$$

The refrigerator runs at a power of 330 W, so the current it draws is $330/110 = 3$ A. Thus if the washing machine and the refrigerator run simultaneously, the total current in the house is $5 + 3 = 8$ A. Suppose, in addition, you turn on one of the stove's burners. That's another 1100 W in power, drawing a current of $1100/110 = 10$ A. The total of all three comes to 18 A. Now suppose you also decide to turn on the toaster, with a power rating of 1100 W, drawing another 10 A. The total current now overshoots the limit of 20 A and activates the circuit breaker.

Sensing electricity

Nature gives us the example of the electric fish, which actually uses electricity as one of its major sensing devices. The electric fish maintains a potential difference of a few volts between its head and its tail. The potential difference gives rise to an electric field, with lines of force as shown in Fig. 16.12. Charges can flow along these lines and give a current. Now these fields and currents in the vicinity of the fish will be modified if they are entered by objects different from water in electrical properties. The fish has sensory cells in its skin, which are able to detect these changes in the current. This is how the electric fish uses electricity as sensors.

We humans do not possess any useful sensory capacity via electricity. There is no built-in potential difference between the parts of our body. However, there is some evidence that people feel specially happy if the surrounding space is filled with negative charge. Since bathroom showers produce an environment heavy with negative charge, it has been suggested that much of the reason we feel happy in a bathroom is due to its negatively charged atmosphere. Why or how this happens is not known.

What causes an electric shock?

An electric current passing through our body can cause severe damage, effects that we summarize with the blanket phrase "electric shock." Currents in the

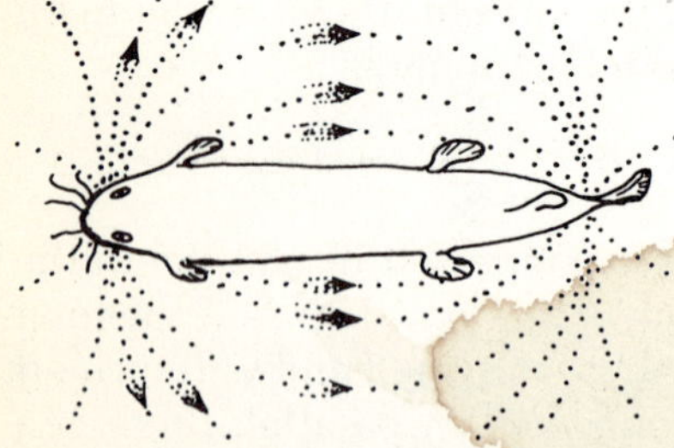

FIGURE 16.12 The field lines of an electric fish.

TABLE 16.1 **Physiological effects of electric currents through a human body.**

Amount of Current (amperes)	Physiological Effect
0.002	Threshold of sensation
0.01	Mild sensation
0.02	Difficult to let go
0.05	Severe shock
0.08–0.1	Extreme breathing difficulty
0.1–0.2	Death
>0.2	Stoppage of breathing, severe burns

FIGURE 16.13 Nothing happens to the bird because there is no appreciable current through its body. Current takes the low-resistance path of the wire rather than the high-resistance path of the bird's body.

range of 0.1 to 0.2 A are lethal, but currents that are not necessarily lethal, either stronger or weaker, also affect the body in various adverse ways (Table 16.1).

For a given voltage, the amount of current that will flow through the body depends on the body's resistance. The dry skin of a human body is a pretty good insulator, offering as much as 100,000 Ω of resistance to the flow of current. Thus if you touch a faulty household line and make a "short circuit"—so the electrons flow between the earth and the line through your body—the shock is quite negligible if your body is dry. The problem of a shock arises because the resistance of the body is vastly lowered in the presence of water, such as wet shoes and wet floors, sometimes to a value as low as 1000 Ω.* A simple Ohm's law calculation shows that the current in this case from a household voltage of 110 V is

$$\frac{110}{1000} = 0.11 \text{ A}$$

right in the lethal range. This is why many cases of electrocution occur when a person standing in the bathtub touches a faulty switch.

Here is one interesting item. You may have seen birds sitting on a live wire; why don't they get electrocuted (Fig. 16.13)? Actually, you could hang from a live wire; nothing would happen to you. The reason is this: in order for current to flow, there has be a potential difference. If you touch a live wire while standing on the earth, it is the potential difference between the live wire and the earth that drives the current through your body.

Another way to look at the situation is as follows: when more than one path is available for the current to follow, more current flows through the path of least resistance. In the case of the bird sitting on a live wire, the current chooses the path of least resistance, the wire. The resistance of the body of the bird is just so much greater than the copper wire that practically no current flows through it.

Electrical gadgets become much safer to use when the manufacturers provide an extra path for the current in case of an accidental short circuit. Most power tools come with this added safety in the form of a round post in the plug of the instrument (Fig. 16.14). If you look at your electrical outlets, many have a hole to accommodate this extra round post. The round post is called the system ground, and its purpose is to connect the metal case of the instrument directly to the earth in case of a short circuit (if the "hot" wire inside the gadget breaks loose and makes contact with the metal case). The current, following the path of least resistance, will choose to flow through the copper post rather than the person who is holding the gadget.

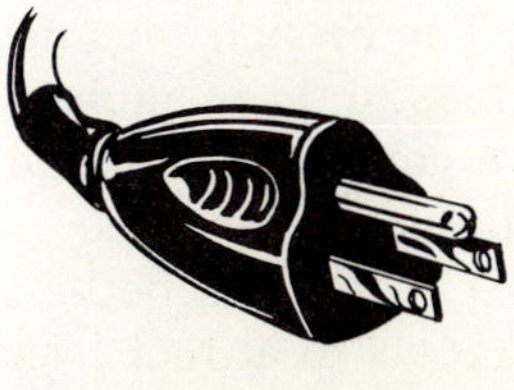

Currents and magnetic fields

Is there a field due to the current-carrying wire in the space around the wire? No, there is no electric field, since the wire as a whole is neutral (don't forget the positively charged atomic nuclei sitting at their places along with the "core

FIGURE 16.14 The round post is the system ground.

---

*Even more important is that electrical contact is more easily established when there is water on your body.

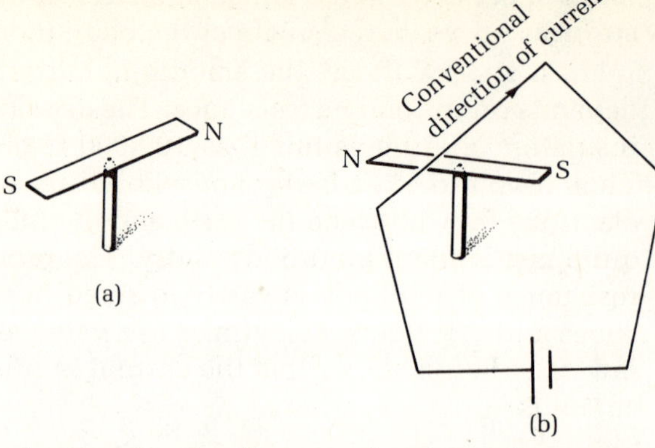

electrons''; the free conduction electrons do all the drifting). It turns out, however, that there is a field of a different kind. A Danish physicist named Hans Christian Oersted discovered this in 1819 while giving a public lecture. Oersted found that a magnetic compass needle, when placed near a wire carrying a current, swung to a new direction indicating the existence of a field that affects magnets (Fig. 16.15). Such a field naturally is called a **magnetic field.** So an electric current generates a magnetic field.

Oersted's experiment opened up new vistas of scientific research in the fields of electricity and magnetism. Let us mention just one application here. You may have been curious about how we measure such a thing as an electric current; surely we are not going to count electrons going by a point in the wire carrying a current (we could not do that anyhow). No, indeed; such things as current are almost always measured today by measuring the magnetic effect they produce. The devices that measure the electric current and the voltage are called the ammeter and the voltmeter, respectively.

■ 16.2 Magnetism

Magnetism involves electric charges in motion. This is true of all magnetic phenomena. Even the magnetism of the so-called permanent magnets, with which you may be familiar, is due to charges in motion. The presence of many current loops of atomic electrons inside the body of the magnet does the trick. This is perhaps the first surprising thing about magnetism: it is a secondary attribute of electricity.

Historically, the magnet and the study and use of some of its ''mysterious'' properties can be traced back as far as the ancient Chinese, who used to call it ''chu shi,'' meaning ''lovestone.'' Interestingly enough, the idea of associating magnets with love occurred to others besides the Chinese. The French seem to have had similar thoughts. The most important use of magnets in those ancient days was as direction finders, as compass needles. So we can wonder if the ancients thought of love also as a direction finder (path finder) in life.

In more recent times William Gilbert, an English scientist of the Elizabethan period, was very interested in magnets. The ability of magnets to attract iron through space without any material connection was given particular notice. And we find ''magnetic'' explanations suggested by many scientists for other

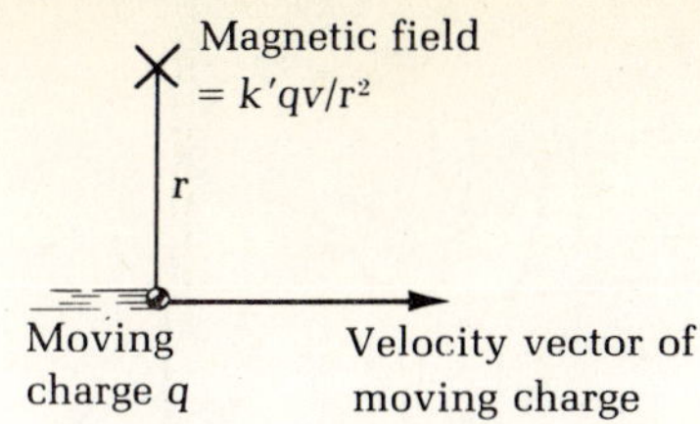

FIGURE 16.16 Magnetic field of a moving charge at a distance r directly across from it is $k'qv/r^2$.

phenomena that have this characteristic of "action at a distance." Both Descartes and Kepler looked on gravity as a "magnetic phenomenon." Italian biologist Luigi Galvani (see the beginning of this chapter) speculated that the electric current is due to "animal magnetism." Interestingly, nobody has succeeded in understanding gravity and magnetic phenomena in the same theoretical framework. So as far as we know, there is no connection between these two. On the other hand, magnetism and electricity are very intimately connected, although not in the way pictured by Galvani.

Magnetic field of a moving charge

As the Oersted experiment first suggested, a moving charge gives rise to a magnetic field. However, the field that is produced has some rather unusual characteristics, quite different from the electric field (or the gravitational field). For example, the magnitude of the field, unlike gravitational and electric fields, depends on the velocity of the charge. Also, the gravitational and electric fields at a point in space are always directed along a line or curve joining the point with the source. But the direction of the magnetic field at a point is not along the line joining the point with the moving charge.

Let us write down an expression for the magnetic field of a moving charge. If the magnitude of the charge is q and its velocity v, then the field at a point directly sideways from the moving charge and at a distance r from it (Fig. 16.16) is given as

$$\text{magnetic field} = \frac{\text{constant of proportionality} \times (\text{amount of charge}) \times (\text{velocity of charge})}{(\text{distance})^2}$$

Or, using symbols,

$$B = \frac{k'qv}{r^2} \tag{16.4}$$

We use the symbol B for the magnetic field. Like the electric field, the magnetic field is also a vector field. The constant k' is 10^{-7} in MKS units. The unit for the strength of the magnetic field B is called the tesla (abbreviated T). For some idea of what 1 tesla is, note that the magnitude of the field near a small bar magnet is about 10^{-2} tesla.

The direction of the magnetic field vector **B** is best visualized by drawing the lines of the field, as shown in Fig. 16.17. Here we have assumed the charge to be

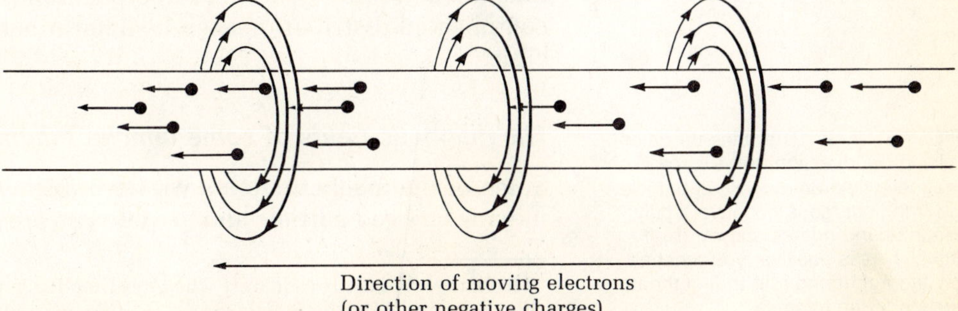

FIGURE 16.17 The magnetic field lines due to a current of moving electrons. The direction of arrows of the magnetic field line can be determined from the left-hand rule (see text).

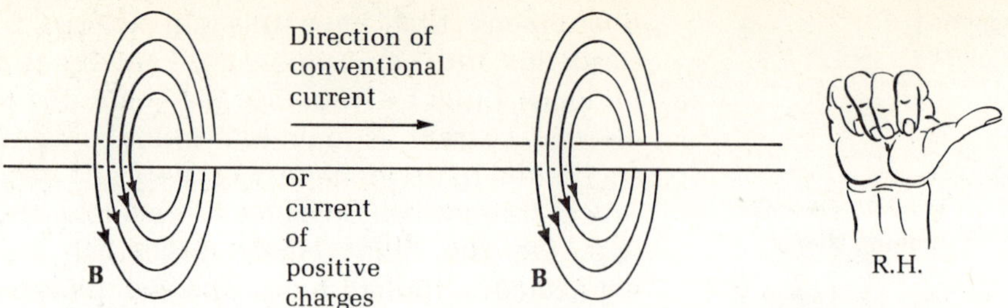

FIGURE 16.18 If you use the conventional direction of current (from positive to negative terminal), then use the right-hand rule to figure out the direction of the arrows of the magnetic field lines.

negative (for example, an electron). The sense of the arrows can be memorized by using the **left-hand rule.** Imagine that you hold the electrons (or any other negative charges) in the palm of your left hand. Stretch your thumb (as in hitchhiking) to point in the direction of the velocity of the moving electrons. Then your fingers will curl in the direction of the magnetic field. For positive charges the arrow would be in the opposite direction.

An electric current is a flow of electrons. So we can figure out the direction of the magnetic field due to a current by using the same rule given above. However, if you use the regular convention for the direction of the current (i.e., from the positive to the negative terminal), you must use a **right-hand rule** (Fig. 16.18) for finding the direction of the field. Convince yourself that both methods give the same direction for the field. Such fields as described by the field lines in Fig. 16.18 are called **circular fields.**

Just as a moving charge creates a magnetic field, it is also acted on by a force when placed in an already existing magnetic field. If q is the magnitude of the moving charge, v is its velocity, and B is the magnetic field, then the force is given as*

$$\text{magnetic force} \sim qvB \qquad (16.5)$$

The direction of the force is perpendicular to the plane containing both of the vectors **v** and **B**; it can be determined by using the left-hand rule if the charge q is negative. Imagine the velocity vector of the charge to be along (toward the fingers) the palm of your left hand, and let your fingers curl to denote the direction of the magnetic field. The direction of your outstretched thumb will then indicate the direction of the force [Fig. 16.19(a)]. The magnetic force is zero if the velocity vector **v** of the charged particle has the same direction as the field.

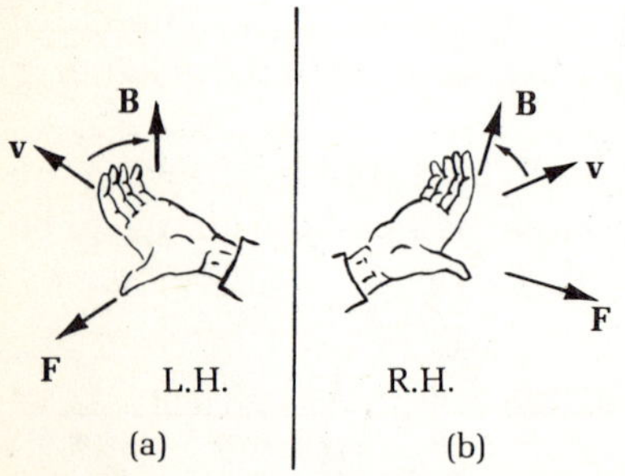

The magnetic field shares one very important property of the electric field. Like the latter, it too obeys a superposition principle. So the fields due to more complicated distributions can be determined by using the superposition principle.

FIGURE 16.19 The rules for determining the direction of the force **F** of magnetic field on a charged particle. (a) If the charge is negative, apply the left-hand rule as shown. (b) If the charge is positive, you must apply the right-hand rule to find the direction of the force.

The magnetic fields of some familiar magnets

Some of the magnetic fields we have discussed so far, namely those due to a moving charged particle and to a current, are not the kind of thing you are likely

---

*The formula is exactly right only when the direction of the velocity vector **v** is perpendicular to the direction of the magnetic field **B**.

to be familiar with. In fact, very likely this is the first time you have heard that a moving charge or a current has a magnetic field. You are probably much more familiar with small permanent magnets—bar or horseshoe—and perhaps electromagnets. However, in our picture these latter are actually more complicated cases of magnetism; the former are the fundamental types of magnets.

But now we must have some understanding of the nature of permanent magnets. First, just for the record, let us recap some of the properties of a permanent magnet, which you probably already know.

A bar magnet has two poles. If a bar magnet is pivoted to rotate freely, then its north pole will seek out the North Pole of the earth and the south pole will likewise point towards the south.* A magnet does this due to the fact that the earth itself is a magnet, having a magnetic field in the space around it of magnitude 5×10^{-5} tesla.

The terminology of the paragraph above suggests that the bar magnet is a magnetic dipole (which means "two poles"). With two such magnetic dipoles, you can verify another remarkable thing. The north pole of one when brought close to the north pole of the other will repel it, whereas unlike poles will exhibit an attraction. Apparently we have rules of magnetic poles—namely, like poles repel and unlike poles attract—that are similar to the rules of the force between charges.

Furthermore, the force between two such poles can be written in the form (as determined experimentally)

$$\text{magnitude of force between two poles} = \frac{k' P_1 P_2}{d^2}$$

where P_1 and P_2 are called the **strengths** of the two poles and d is the distance between them. k' is the same constant that appeared in Eq. (16.4). Does this remind you of Coulomb's law for the force between two charges? Fig. 16.20 shows the lines of field of a bar magnet. Do you notice the similarity of these lines with those of an electric dipole?

All these facts might make you think that a bar magnet is made up of an excess of one kind of magnetic pole at one end and another kind at the other. But you would be incorrect if you thought that. Suppose you try to separate the two poles and you cut the original bar into two halves. Surprise! You now will have two dipoles instead of two isolated monopoles (a single isolated pole is called a monopole). It is a fact that no subdivision ever leads to an isolated monopole. It appears that magnetic monopoles do not exist.

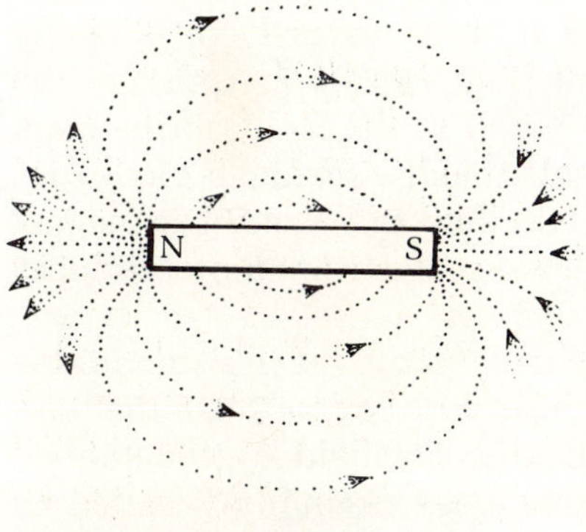

If magnetic monopoles existed, then magnetism would truly be on the same footing as electricity. Much stronger magnetic effects could be generated than are now possible. We would perhaps have magnetic currents of monopoles that would give electric fields, just as an electric current of charges gives a magnetic field. And so forth.

It is possible that the magnetic monopole is a rarity only in our part of the physical universe, and that there are other parts of the universe where magneic

FIGURE 16.20 The magnetic field lines of a bar magnet. Conventionally the arrows are drawn directed away from the north pole N and toward the south pole S.

---

*So in actuality what we call the magnetic North Pole of the earth is really a magnetic South Pole. Also, the earth's magnetic poles do not exactly match geographic poles.

monopoles are the common feature while electric charges are collector's items. If this is so, should we not be able to get at least a glimpse of the magnetic monopole in our laboratory? It has been suggested that a pair of opposite magnetic monopoles might crop up every once in a while in high-energy collisions in the laboratory, or at least in cosmic ray events. In 1974 an experimental group at Berkeley made news all over the world with the announcement of the observation of one magnetic monopole. Unfortunately, closer examination of the data revealed many loopholes in the validity of the claim.

In spite of such disappointments, physicists keep searching. One reason for their persistence is that a very clever physicist named Paul Dirac has shown that the concept of the magnetic monopole helps us to achieve an understanding of the quantization of the charge—that is, that charges exist only in multiples of e, the electronic charge. Another reason is that a perfect symmetry between magnetism and electricity would be esthetically more satisfying.

Actually, there is a third very compelling reason for some. Many physicists have begun to believe in a principle proposed by an English author named T. H. White in a novel entitled *The Once and Future King*.* The story is about the adventures of King Arthur, accomplished with the help of magician Merlyn. In one incident Arthur had been in bed with an injury for some three days. Finally the boredom got to him, and so young Arthur asked Merlyn for help. There was a glass case containing an ant colony in his room. Arthur wanted Merlyn to make him into an ant so that he could explore the fascinations of the ant colony. Merlyn consented begrudgingly and, lo and behold, Arthur found himself as an ant outside an ant castle. Above each tunnel leading into the castle there was this notice written boldly: "Everything Not Forbidden Is Compulsory."

King Arthur in the story did not like his compulsory ant routines and eventually was rescued by Merlyn before getting too involved. Many physicists, however, are convinced that the principle plays an essential role in the shaping of physical reality. Thus they continue to look for objects like the magnetic monopole which, although a product of human conceptualization, is not forbidden to exist by any natural law.

Nevertheless, whether monopoles exist or not, the magnetism of magnets as we know it is not due to monopoles but arises from circulating electric currents inside the body of the magnet. To understand this, let us first consider the magnetic field due to a ring of electric current [Fig. 16.21(a)]. Applying the right-hand rule as before, you can reconcile yourself to the fact that the magnetic field lines around the loop are as shown. Note that the current is clockwise when seen from above. Now look at the magnetic field of a small bar magnet [Fig. 16.21(b)]. Do you see the similarity? *At a distance* a current loop produces the same kind of magnetic field as a short bar magnet.

Now suppose we take a whole stack of these current loops, as in a solenoid (a cylindrical coil of wire) (Fig. 16.22). The field now will be more like that of a long bar magnet placed along the axis of the solenoid. The field is strengthened by placing an iron bar along the axis, and the resultant assembly is called an **electromagnet.**

---

*(New York: Berkeley, 1966), p. 122.

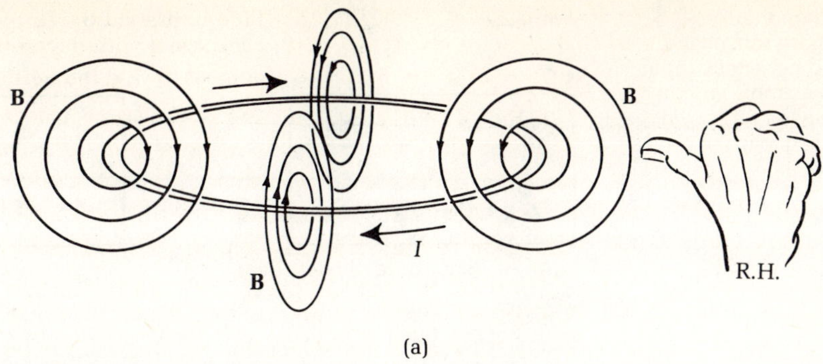

(a)

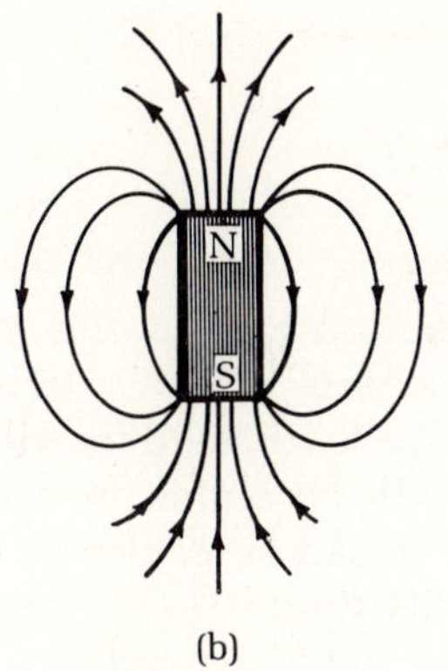

(b)

The physicist who discovered all this was André Ampère, after whom the unit of current was named. Ampère suggested that bar magnets owe their magnetism to internal stacks of current loops. Where do these loops come from?

Using the atomic picture we now know that these currents originate from the atomic electrons spinning around their axes. Now visualize zillions of these current loops. In ordinary matter the current loops will produce fields in random directions, and the net effect will very nearly cancel out. But in a few magnetic substances, like iron, there is a special kind of ordering tendency that operates, at least if the temperature is not too high (Fig. 16.23). This ordering tendency is responsible for the cooperative phenomenon called **ferromagnetism,** which is an alignment of all the current loops in the same direction with the production of a large field.

Charged particles in a magnetic field

The magnetic field exerts a force on a moving charge that is perpendicular to the velocity vector of the moving charge (as well as to the field). This is the same

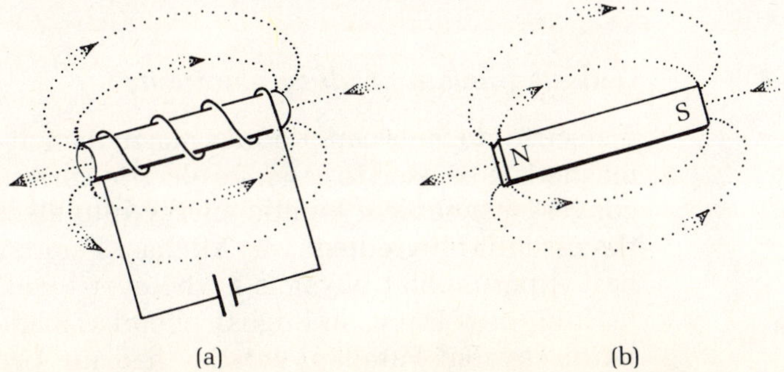

(a)

(b)

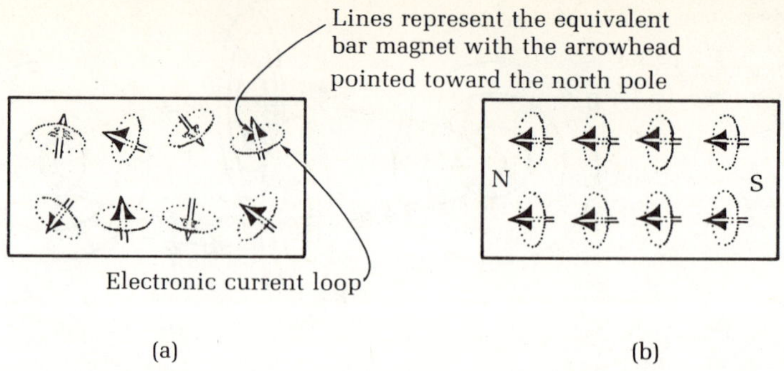

FIGURE 16.23 When the electronic current loops are randomly oriented, there is no net effect. (b) In a bar magnet the electronic current loops are largely aligned as shown, producing a large magnetic field.

situation as that of the centripetal force in circular motion; indeed, the magnetic force can provide the centripetal force needed for a charged particle to move in a circular orbit. If a charged particle is incident on a magnetic field in a direction perpendicular to the field lines (i.e., the field), it will be under the spell of a force and be deflected. Under suitable conditions it can be captured into an orbit, just as a comet is captured by the sun's gravity field.

Such particles trapped in orbits in a magnetic field showed up in one previously unsuspected place—the upper atmosphere of the earth. The charged particles in this case perhaps belong to those we call cosmic rays, rays consisting of various elementary particles (one major component is the proton) that bombard us from outer space. Possibly some of the particles also may be coming from the sun. At any rate, these charged particles, upon encountering the magnetic field of the earth (which is a dipole field like that shown in Fig. 16.20), experience its magnetic force. If the conditions are right, they get trapped in orbits. All these orbits together make belts around the earth (Fig. 16.24). These belts are often called **Van Allen belts,** after the name of their discoverer, American physicist James Van Allen.

Thus the earth's magnetic field acts as a protective shield against bombardment by the charged particles, at least to some extent. Of course, particles that are incident toward the poles, since they descend along the field lines (**v** parallel to **B**), are not deflected; the magnetic force is zero for them. They come right down to the earth through the atmosphere, sometimes causing brilliant displays of light and color. This is the phenomenon of the **aurora**—called the aurora borealis in the northern hemisphere and the aurora australis in the southern hemisphere.

Can magnetism produce electricity?

If electric currents can produce magnetism, it is natural to ask this question: Can magnetism be used to generate electric currents? The man who found the way to convert mechanical kinetic energy into electrical energy by using magnets as the essential ingredient was Michael Faraday, regarded by many as one of the best experimental physicists who ever lived.

Humphrey Davy, an English chemist, made this one contribution to physics: he discovered Faraday. Faraday had no formal education in science; he was

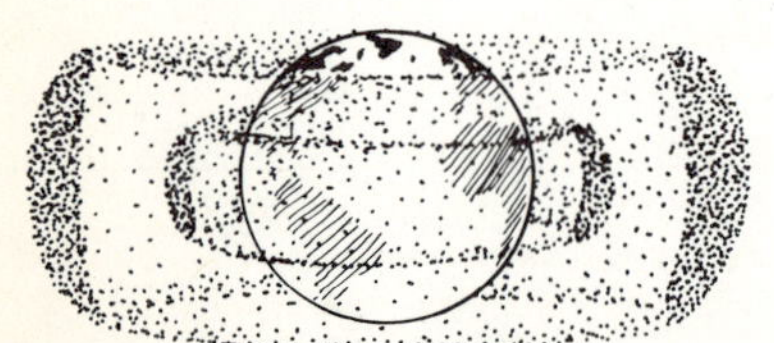

FIGURE 16.24 The Van Allen belts.

working as an apprentice to a bookbinder when he attended a series of popular lectures by Davy. After hearing the lectures, Michael knew what he wanted to do with his life. He sent Davy an application for an assistantship to work with him, along with some notes he had taken of Davy's lectures. The notes must have been impressive, because Faraday got the job.

Initially he was doing chemistry, but then he was attracted by a brand new field on the horizon, electricity and magnetism, which he pursued with an unusually fresh point of view. His lack of formal education turned out to be a boon in this case; he had no preconceived beliefs to overcome. He did not know mathematics, so the Newtonian ideas of action at a distance were too difficult for him to apply to the study of electricity and magnetism, as was the fashionable thing to do in those days. So he began to look at the electromagnetic phenomena with the aid of constructs like the lines of field that we have introduced previously.

Soon Faraday's attention was attracted to the problem of generating a current by using a magnetic field. It took him ten years of experimentation before he hit upon the idea that if he created a changing magnetic field in the vicinity of a loop of metallic wire, a voltage could be induced in the latter, giving a current. The changing magnetic field could be produced simply by moving a magnet. Very soon it was discovered that only the *relative* motion matters. You can also move the loop of wire; the wire still experiences a changing magnetic field, and that's what it takes to induce a current. This is the phenomenon of **electromagnetic induction** (Fig. 16.25). Further details of this subject are given in the next chapter.

Thus the great discovery of Faraday that revolutionized our society was really one more step toward showing that there is much in common between magnetism and electricity. A changing magnetic field produces an electric field, which drives the current that is induced in a circuit. Nowadays Faraday's method is almost universally used for the large-scale generation of electricity. The energy of the resource is used to move the loop, the motion of the loop in the field of a magnet subjects it to a changing magnetic field, and a current flows in the loop and any circuit connected to it.

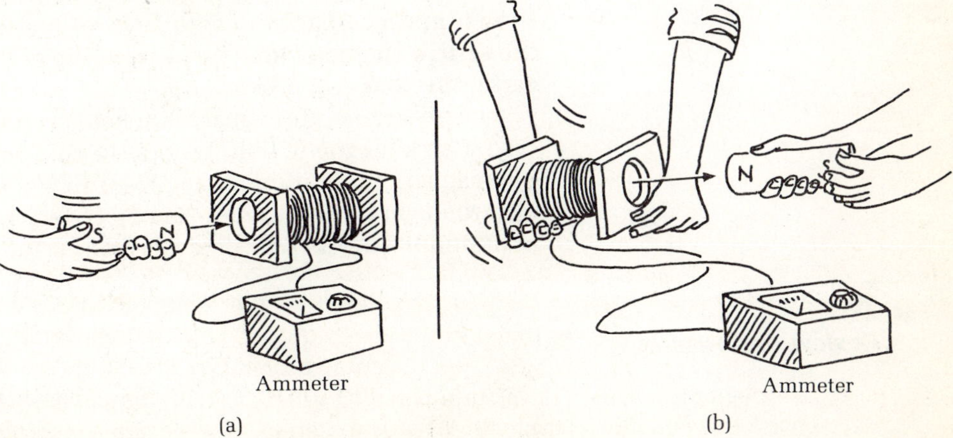

FIGURE 16.25 (a) If you move a magnet toward a wire loop, a current will be induced in the wire, as indicated by the ammeter. (b) Since only relative motion counts, a current is also induced when you move the coil of wire toward the magnet.

SUMMARY

Whenever there is a flow of charges, there is a current. But a steady potential difference must be maintained in order to get a steady current. This is what the generator at your local power company does. On a small scale, the battery performs the same function.

What exactly is the picture for the flow of an electric current? Many people have a confused notion about this, and Section 16.1 mentions some of these notions as well as giving an accurate description of current flow. In a complete circuit there is also equipment that extracts the electrical energy. How much power such equipment extracts depends on the voltage and also on how much current it draws:

$$\text{power} = \text{voltage} \times \text{current}$$

And how much current the equipment draws depends on a property called resistance. Ohm's law relates the voltage, current, and the resistance of a conductor:

$$I = \frac{V}{R}$$

Other important notions of electrical circuitry are also discussed, such as series and parallel circuits.

The second section of this chapter deals with the subject of magnetism. We stress the fact that magnetism is due to moving electric charges. Thus it is a secondary phenomenon of electricity. Even a bar magnet gets it magnetic properties from circulating currents of electrons in its body.

Magnetism is also best studied in terms of the field concept, the magnetic field. The magnetic field of a moving charge is given by the expression

$$B = \frac{k'qv}{r^2}$$

The magnetic field exerts a force on the moving charge, a force proportional to the velocity of the charge:

$$F \sim qvB$$

Our earth has a magnetic field whose field lines look pretty much like those shown in Fig. 16.20, the field lines of a magnetic dipole. One consequence of the earth's magnetic field is the trapping of charged particles in belts around the earth, the Van Allen belts.

A current generates a magnetic field around it, a fact that Oersted first discovered. Can a magnetic field generate a current? Faraday discovered that a changing magnetic field induces a current in a conducting wire: this is the subject of electromagnetic induction. The electrical generators of today use electromagnetic induction, as will be described in the next chapter.

QUESTIONS

Review and reason

1. Describe how a battery develops a potential difference between its two terminals. What is the direction of flow of the electrons when the two terminals are connected? What is the "conventional" direc-

tion of the current? How do you explain having two directions for the "same" thing?

2. What is meant by a power sink? Give an example.

3. Describe the flow of electrons that gives a current in a completed electrical circuit. Why does the current stop when the circuit is "broken"?

4. In an electrical circuit, the electric potential and hence the potential energy of the electrons decreases from point to point. Why, then, does the kinetic energy of the electrons remain a constant? Is this in violation of the law of conservation of energy? Explain.

5. State Ohm's law. Discuss the origin of the resistance of a conductor. Suppose you are given a long wire and a short wire of identical thickness made of the same conducting material. Which has the greater resistance?

6. Of two wires of identical material and length, one thin and the other thick, which has the greater electrical resistance?

7. Distinguish, both in writing and through diagrams, between a series connection and a parallel connection of two pieces of electrical equipment.

8. In a household a dishwasher, a TV, and a couple of lights are on. If the dishwasher is turned off, will the total current in the house increase or decrease? Explain.

9. Describe the function of a regular light switch in its on and off positions, respectively. (It is not given in the text, but you can figure it out.)

10. This is what American author James Thurber writes about his grandmother on his mother's side (in *My Life and Hard Times*)

> . . . (she) lived the later years of her life in the horrible suspicion that electricity was dripping invisibly all over the house. It leaked, she contended, out of empty sockets if the wall switch had been left on. She would go around screwing in bulbs, and if they lighted up she would hastily and fearfully turn off the wall switch and go back to her *Pearson's* or *Everybody's*, happy in the satisfaction that she had stopped not only a costly but a dangerous leakage. Nothing could ever clear this up for her.

Perhaps you would like to make an attempt to clear this up. Discuss what's wrong with her contention.

11. What causes an electrical shock, the current or the voltage?

12. Is voltage enough to give you a shock? One aspect of atmospheric electricity is that there is a potential difference of about 200 volts between your feet and your nose. Why don't you get a shock?

13. Two women touched the same faulty electric switch. The first woman was unharmed and walked away, but the second one got a severe shock. How is this possible?

*14. Suppose we take Malcolm's thinking (see the cartoon on p. 350) one step further. Since motion is relative, if an observer is on a conveyor belt, the electrons in a stationary wire will appear to be moving relative to the observer. Since moving electrons constitute a current, the observer should see a current flowing through the wire. However, no observer has seen a current in just this way. Can you guess why? (Hint: Do not forget that a conductor is, overall, electrically neutral.)

15. A current-carrying conductor gives a magnetic field outside its body but no electric field. Explain why.

16. Draw the lines of magnetic field due to a moving negative charge.

17. A charged particle is placed at rest in a magnetic field. Will it start moving? Why or why not?

18. You are told that inside a certain room there is either an electric or a magnetic field. How can you make sure which one it is?

19. Draw the field lines of a magnetic monopole, assuming it exists.

20. You are given two steel bars and told that one of these is magnetized but the other is not. With no other machinery can you figure out which is which?

*21. "The space or neighborhood about a magnet must have a concrete structure," declared Faraday. To prove this he sprinkled an ample amount of iron filings on a sheet of paper that was laid over a bar magnet. The filings arranged themselves in the manner shown in Fig. 16.26. What did this prove? What do the curved lines signify?

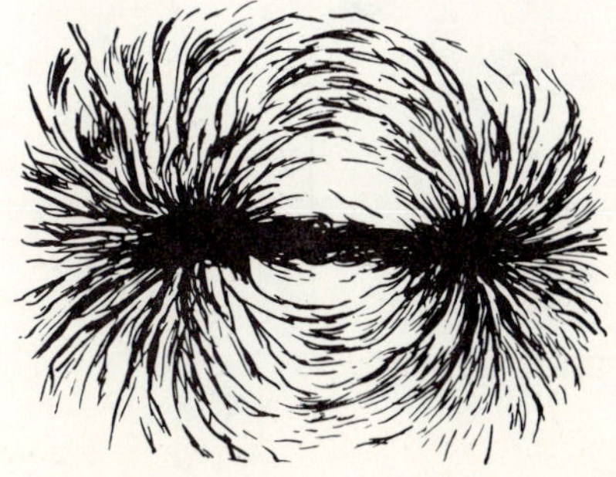

FIGURE 16.26

---

*Optional.

22. Bringing a magnet near a TV will affect the picture. Explain. (Don't try it, though; the picture can be damaged permanently!)
23. How is the magnetism of a bar magnet explained in terms of moving charges?
24. Make a sketch of the field lines of the earth's magnetic field. What is the explanation of the Van Allen belts around the earth?
25. What are the similarities and differences between electric and magnetic fields?
26. Discuss the phenomenon of electromagnetic induction. What is the basic principle for the operation of an electrical generator?

Arithmetic

1. How many electrons flow by every second across a cross section of a wire that carries a current of 2 A?
2. Estimate the electrical resistance of a 15-W electric shaver.
3. Two light bulbs, each of resistance 110 Ω are connected in series to a household power line of 110 V. What is the magnitude of the current passing through each of the bulbs. What is the voltage drop across each bulb?
4. Estimate the total current in a household with the following equipment turned on: 3 light bulbs with a power rating of 100 W each; a radio with a power rating of 100 W; and a toaster with a power rating of 500 W.
*5. The six resistances shown in the circuit diagram (Fig. 16.27) are each equal to 5 Ω. The potential difference at the wall outlet is 110 V. (a) What is the potential difference between A and B? (b) What is the voltage between F and E? (c) What is the current at D? (d) What is the current at X and Y? Explain how you got each of your answers.

6. The electrical resistance of a certain person, after taking a bath, was found to be 880 Ω. If he or she touched a live wire in the household line, what would be the value of the current flowing through his or her body. Is such a current harmful to the body? If so, in what way?
7. The south poles of two identical bar magnets are placed 0.2 m apart and the force is found to be 10^{-1} N. Calculate the pole strength of each magnet. Is the force attractive or repulsive? (Note: The unit of magnetic pole strength is the ampere meter.)

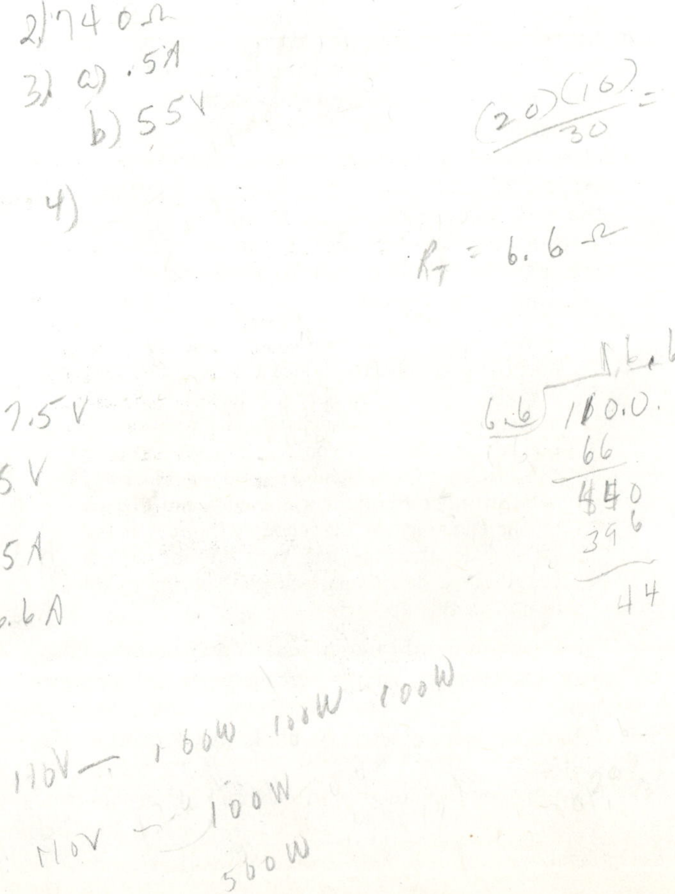

*Optional.

Electric Energy

Stars face stars in the sky,
Under the beam of every house hangs an electric lamp.
The stars are dim against the lamplight,
The electric lights shine in everybody's heart.

The above is a portion of a nursery rhyme from the People's Republic of China. It may sound somewhat naive, particularly from your perspective, but the Chinese poet seems to get across this one important message: Electricity is technology's greatest gift to humanity.

■ 17.1 The Large-Scale Generation of Electricity

When Benjamin Franklin began his historic experiments with lightning and electricity, one of his major motivations was to make electricity work for people. With Faraday's discovery, Franklin's dream became reality. It was also fitting that an American physicist, Joseph Henry, simultaneously made the same discovery as Faraday—that if a coil of a conductor encounters a changing magnetic field, a current is induced in it. Today Faraday's and Henry's discovery is the basis for the large-scale generation of electricity.

One major reason for the success of this method for getting a current is economics. Compared to the battery, electricity made by large-scale generation is extremely cheap. Let's look at a couple of figures. The 12-volt battery of your automobile gives you electric energy at a cost of about $1.20 per kilowatthour. For comparison, look at your monthly electric bill from your local power company; you get electric energy at a rate of a few cents per kilowatthour. This is the reason that even the automobile has a generator of electricity that uses Faraday's technique; we cannot afford to use the battery except for the first few essential moments of starting the car.

The machine that makes electricity by using electromagnetic induction is called the **generator.** The generator makes a current that changes its direction

369

back and forth in a regular rhythm. We call this type of current **alternating current,** or **ac.** Alternating current has one big advantage over the direct unidirectional current (dc): it can be transported over a long distance with little loss of power. Thus the power generated in one locality does not need to be used locally—it can be transmitted to other places according to demand. This makes possible the giant electric power plants of today.

These are the topics of this section. How does the modern electric generator work? And how do we transmit electricity from the power plants to the power sinks?

The details of the working of an electric generator

The basic principle of the **electric generator** is amazingly simple. We rotate a coil of wire in the field of a magnet; a voltage is thus induced in the coil due to the changing magnetic field it experiences. If we increase the number of turns of the coil, the induced voltage will be larger. Iron turns out to be a big booster of electromagnetic induction, so we wrap the coil around an iron cylinder. This assembly of the coils plus iron is customarily called the **armature.** Since a large magnetic field increases the effect even further, we use an electromagnet. We keep the armature rotating by connecting it to the moving shaft of a turbine. We now have a modern generator of electricity.

What kind of a voltage and current does this machine produce? We can draw a picture of the field lines of the magnet that the coil is in (Fig. 17.1). As the armature rotates, it cuts across these field lines, which has the same effect as varying the strength of the field itself. This rotation, too, puts the armature loop under the influence of a changing magnetic field. But now here is the interesting thing. The armature does not cut across the same number of field lines in all its positions as it describes a complete circle. In fact, sometimes it does not cut through any [Fig. 17.2(a)]. Thus it is clear that the induced voltage varies with time, depending on the position of the armature at the given time.

We can be more specific. Suppose we start with the armature in the position of Fig. 17.2(a); the armature is moving parallel to the field lines, intersecting none. The induced voltage is zero. Time moves on, and so does the position of the armature loop. It is now moving perpendicular to the field lines [Fig. 17.2(b)], momentarily cutting across a maximum number of them. The induced voltage is now maximum. With a further shift of position, the armature moves parallel to the field lines again [Fig. 17.2(c)], giving zero induced voltage. After this the voltage becomes nonzero again, but now the coils are cutting the field lines from a reverse direction. The direction of the induced voltage now reverses; the voltage has become negative. Further on, in the position of Fig. 17.2(d), the voltage is maximum on the negative end. Finally, the loop completes one full circle in the position of Fig. 17.2(e). The voltage is zero and the cycle can begin again. What we have now is a graphical picture of the induced voltage versus time. It is called an **alternating voltage** and the current it generates is alternating current. Since the current through a conductor is the ratio $I = V/R$, and since resistance does not change with time, it is clear that the current must vary with time in exact correspondence with the induced voltage (Fig. 17.3).

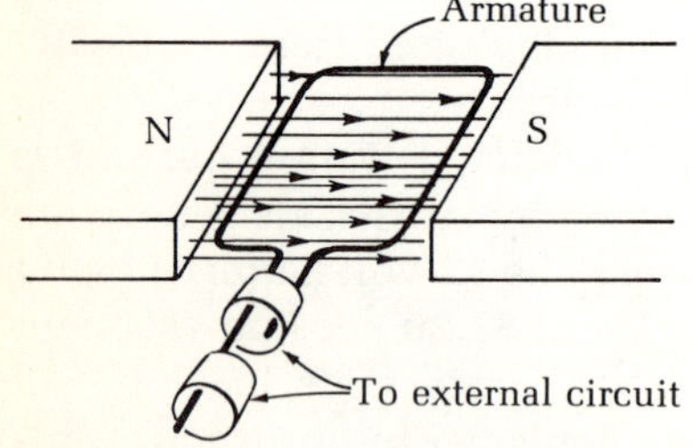

FIGURE 17.1   A generator basically consists of a rotating coil of wire (the armature) in a magnetic field. N and S denote the north and south magnetic pole faces. The lines from N to S are the magnetic field lines.

FIGURE 17.2 Development of an alternating voltage in a generator. See text for explanation.

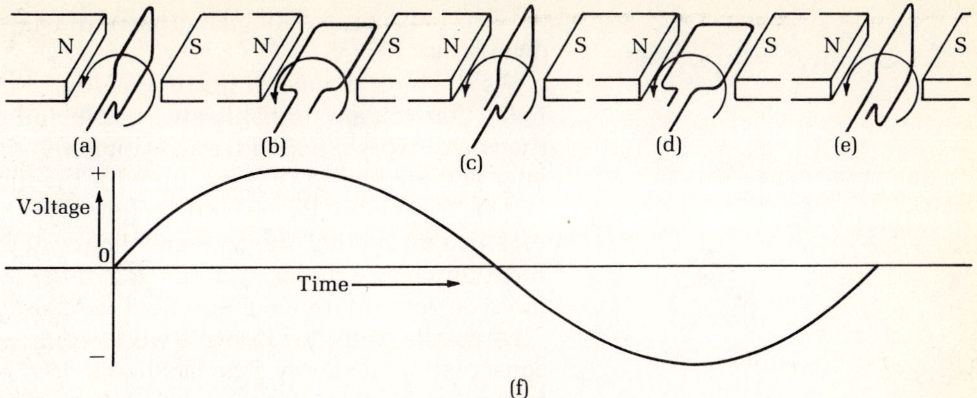

The number of cycles per second is the frequency of ac voltage and current. The household alternating current in America oscillates at 60 cycles per second. During this period it goes to zero exactly 120 times without our ever noticing it. It all happens that quickly.

Fig. 17.1 also shows how the external circuit (or circuits) is connected to the armature. The two terminals of the armature loop connect to separate metal rings. The wires from the external circuits establish contact with the rings by means of carbon brushes (not shown).

Let us briefly discuss the **electric motor,** which in some sense is a reverse generator. In this device a current is passed through a conducting wire loop (which can also be called an armature) that is placed in a magnetic field. The passing of the current through the loop makes it a magnet, as Oersted first discovered. This magnet is repelled by the stationary magnet, causing the loop

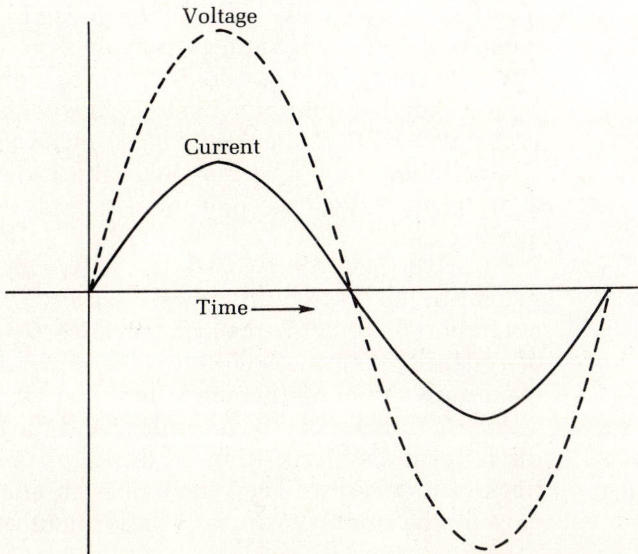

FIGURE 17.3 The current follows the same general pattern as the voltage.

to turn. This is the basic principle of the motor (we will not go into any further details here).

We mention the motor for two reasons. First, it is used in so many machines today that you are probably aware of it, and perhaps you are also curious about the way it works. Second, once you know the function of the motor, sooner or later you are bound to come up with the idea of a perpetual motion machine. Why not combine the motor principle with that of the generator? A rotating wire loop in a magnetic field generates a current through the loop, which now turns, functioning as a motor. The turning of the loop in the magnetic field generates more current, more magnetic field, and so forth. Does it go on forever?

Of course, the idea doesn't work—that would be a violation of the law of conservation of energy. The fact that it does not work in this particular case was pointed out by Russian physicist Heinrich Lenz. Lenz showed experimentally that the induced current in the loop, when the loop is functioning as a generator, has a direction that opposes the loop's initial motion instead of adding to it. Thus we have to keep on adding external energy to the system; otherwise the system will stop.

Induced effects always act in such a fashion as to oppose the cause that produces them.

This is **Lenz's law.**

Power in an alternating current circuit

The power dissipated across the conductor in any circuit is given by the product of the current and the voltage drop across it. The product of positive current and positive voltage is, of course, positive. But even when the voltage is negative, since the current is also negative, the product is still positive. So ac power is always positive.

In Fig. 17.4 we make a plot of the power in an ac circuit against time. Initially, when both the voltage and current are zero, so is the power. A quarter of a cycle later, the voltage and current increase to their maximum values, and so does the power. Another quarter cycle later, the voltage and the current are zero, and so is the power. Still another quarter cycle later, the current and voltage have attained their maximum negative values, and their product, the power, is again at its peak. After this, one more quarter cycle takes us to the beginning of another cycle.

Since all these changes for the power occur at a high speed—60 times per second in the household circuits, for example—the details of the variation are not important but only the average power over time. Now if you take the average of a quantity that varies with time, as shown in Fig. 17.4, what you get is half the maximum value. Mathematically it just works out this way.

As you now know, the ac voltage and the current actually fluctuate—change direction and so forth—but all this happens so quickly we don't need to keep track of it. What we keep track of is an **effective value** of the voltage and the current; the effective values of these quantities are such that their product gives

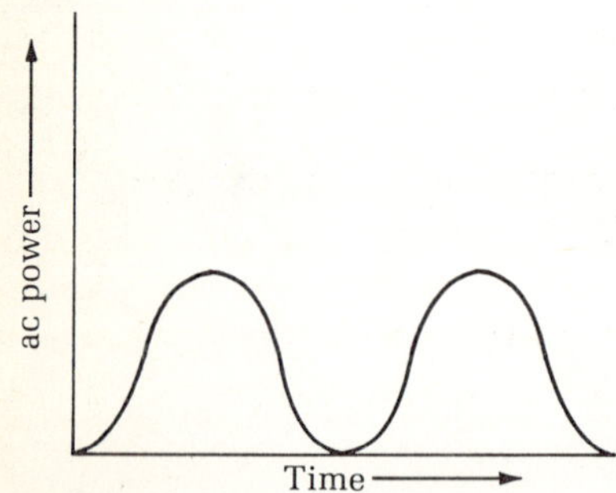

FIGURE 17.4 The ac power is always positive.

the average value of the power.*

The 110 volts that we quote for the voltage of the household power line is really the effective value of the ac voltage. Its actual value fluctuates between -155.59 volts and $+155.59$ volts, passing through zero twice in every $\frac{1}{60}$ of a second. $(110/0.707 = 155.59.)$

In the last chapter, in the household circuit examples, when we calculated the current we were really calculating the effective values. The mathematics of ac and dc circuits works out exactly the same way when we use the effective values of the ac voltages and currents. And unless otherwise stated, this is always to be understood.

Now we will discuss a slightly different expression for the power dissipation in a conductor, to make a point. If there is a voltage drop V across the conductor and it draws a current I, then the power is

$$P = I \times V$$

But if R is the resistance of the conductor, then the voltage drop V is the product RI. Substituting $V = RI$ in the equation above, we get

$$P = I \times V = I \times RI = RI^2 \tag{17.1}$$

So the power dissipation is proportional to the square of the current.

Looking at the power dissipation in this way, we immediately realize the following: When we want to minimize the power dissipation, it pays to have a low current. Now in the long-distance transmission of electricity, we would like to keep the power loss in the transmission line at a minimum. Thus there is a real advantage in transmitting the electricity in the form of as low a current as possible. We cannot do this with direct current, but with alternating current there is a device named the transformer that performs this specific function. Let's take a look at this device.

The transformer

The **transformer,** like the generator, works on the principle of electromagnetic induction (Fig. 17.5). But in the transformer, instead of creating a changing magnetic field by means of moving the magnet or the conducting loop, we change the magnetic field directly. If we use an alternating current to activate an electromagnet, the current varies with time, as does the magnetic field it produces. If there is a coil in this changing magnetic field, a voltage will be induced in this coil.

This is the basic principle of the transformer. A coil of wire is connected to an

---

*Since average power is half the maximum power, the effective value of both the voltage and the current is $1/\sqrt{2}$ (or 0.707) times their maximum (absolute) value.

$$\text{average power} = \text{effective voltage} \times \text{effective current}$$
$$= \frac{1}{\sqrt{2}} \times (\text{maximum voltage}) \times \frac{1}{\sqrt{2}}(\text{maximum current})$$
$$= \frac{1}{2} \text{ maximum voltage} \times \text{maximum current}$$
$$= \frac{1}{2} \text{ maximum power}$$

FIGURE 17.5   A transformer.

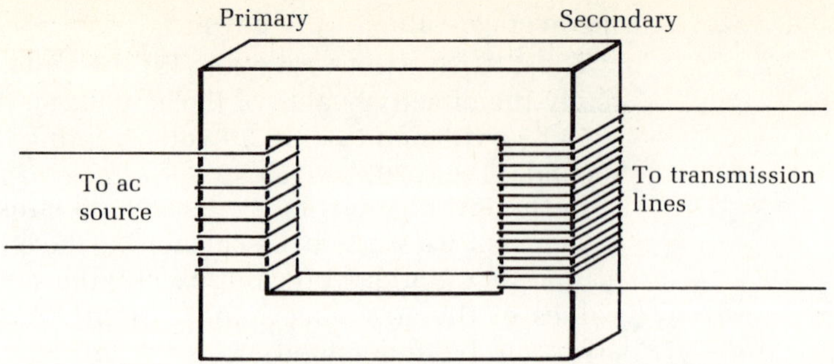

alternating voltage (this is called the **primary coil**) and another coil (the **secondary coil**) is placed nearby but not touching the first. Actually, since iron is such a help not only in increasing the strength of an electromagnet (the primary coil, in this case) but also in boosting the induced effects, both coils are wrapped around a rectangular iron core. When a current is passed through the primary coil, we get an induced current in the secondary coil.

The advantage of using electromagnetic induction is that each turn of the secondary coil acts as an individual generator and the turns in the coil act cumulatively, as in a set of batteries in series (such as in a flashlight). Their voltages add to produce a large total effect. The net result is that we can generate as large a secondary voltage as we wish, subject only to technical limitations; all we have to do is add more loops to the secondary coil.

The voltage generated in each loop of the secondary is the ratio of the primary voltage V_p and the number of turns in the primary N_p:

$$\frac{V_p}{N_p}$$

Thus if there are N_s turns in the secondary, the total secondary voltage, V_s, is the voltage above for one turn multiplied by N_s:

$$V_s = N_s \left(\frac{V_p}{N_p} \right)$$

This relationship is also written as

$$\frac{V_s}{V_p} = \frac{N_s}{N_p} \tag{17.2}$$

The ratio of the secondary voltage to the primary voltage is equal to the ratio of the number of turns in each, respectively. If the secondary has a hundred times more turns than the primary, the induced secondary voltage will be a hundred times the primary voltage.

Eq. (17.2) also tells us the following fact: If N_p is less than N_s, then the induced voltage created in the secondary is proportionately less than the primary voltage.

So the device can be used both to step up voltage (to get an increased voltage at the secondary) or to step down voltage (to get a reduced voltage at the secondary). The transformer accordingly is called a step-up transformer or a step-down transformer.

In all this discussion we have not talked about the current. The situation with the current can be figured out by using the law of conservation of energy, which holds for all energy transfer. Power transferred to the secondary must equal the primary power. Since power = current × voltage, we get the relationships

$$\text{secondary current} \times \text{secondary voltage} = \text{primary current} \times \text{primary voltage} \tag{17.3}$$

We can rewrite Eq. (17.3) in the form

$$\frac{\text{secondary current}}{\text{primary current}} = \frac{\text{primary voltage}}{\text{secondary voltage}}$$

Using the symbols I_s and I_p for the secondary and primary currents, respectively, we get

$$\frac{I_s}{I_p} = \frac{V_p}{V_s} \tag{17.4}$$

Thus the currents behave inversely as the voltages. When the voltage is stepped up, the current is stepped down, and vice versa.

Now you can see the advantage of a transformer in transmitting ac power. With a step-up transformer, we can arrange to transmit the power with a very large voltage (up to 300,000 V) and a very small current. This will keep the dissipation loss at the power lines at a minimum. When the time comes to deliver the power, we can use a step-down transformer to reduce the voltage (for example, to 110 V for delivery into a house). How the electric power comes to your home from the power plant is schematically shown in Fig. 17.6.

EXAMPLE

An electric power plant produces 200,000 W of power at 20,000 V. The current that flows from the output of this plant is

$$\text{current} = \frac{\text{power}}{\text{voltage}} = \frac{200,000}{20,000} = 10 \text{ A}$$

Using a step-up transformer, the voltage is stepped up to 300,000 V for transmission. What is the current in the transmission line?

From Eq. (17.4), since $V_p = 20,000$ V, $I_p = 10$ A, and $V_s = 300,000$ V, we get for I_s, the secondary current that goes into the transmission line,

$$I_s = \frac{V_p}{V_s} \times I_p = \frac{20,000}{300,000} \times 10$$

$$= \frac{2}{3} \text{ A} = 0.67 \text{ A}$$

We should consider some things about the transmission lines. First, in spite of the low currents used, there is some heat loss, amounting to 1 to 2 percent of the energy. So transmitting electricity over a long distance is wasteful. Second, the metal usually used, copper, is not an unlimited resource. Third, the towering transmission lines overhead certainly are not a welcome addition to the environment, either for safety or esthetics.

FIGURE 17.6 How the electric power comes to your home from the power plant.

Underground transmission lines are better, environmentally speaking. However, this is not particularly good for the dissipation of the heat that is generated in the lines, since ground is not as good as air in dissipating heat. One way to overcome this difficulty is to use a special property of some metals at the very low temperatures of a few degrees Kelvin. For example, at 18°K a niobium-tin alloy loses all its electrical resistance and becomes what is called a superconductor. Such an alloy can carry a current without any heat dissipation at all and is therefore ideally suited for underground transmission. However, don't forget that energy is needed to cool the metal to such a temperature, so there is no gain in terms of the energy spent in transmission. More research will be necessary before superconducting transmission lines come into use.

Pumped storage

The giant power plants of today have one more problem. Usually a power plant supplies the electrical needs of an entire metropolitan area. As is well known, in a city there are certain peak hours when the demand for electricity is maximum. And there are other times when there is little demand, as in the late night hours.

It generally is not considered economical to shut down some of the power generators during times of low demand. Instead, the power company would rather store the electrical energy in some way during the slack hours; this energy could then be used later to generate electricity when the demand is high.

Usually water is pumped up into a high reservoir, using the excess electric energy of the nighttime slack hours. Come the peak hours of the following day, this water can generate electricity in the same way as any other hydroelectric

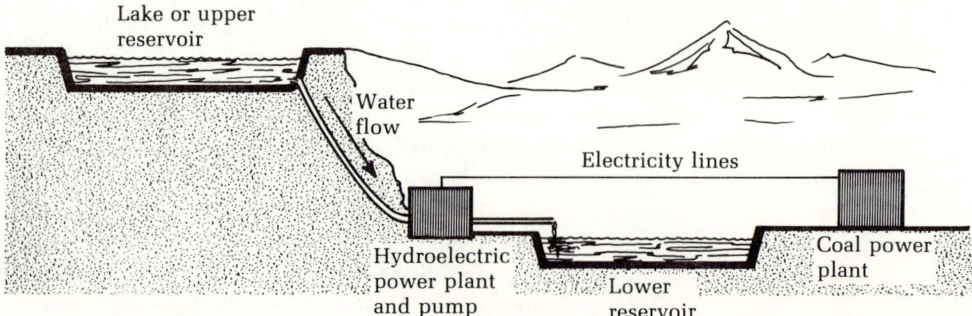

FIGURE 17.7 Pumped storage system.

plant (see Section 17.3). Pumped storage is now being used in many places in the United States (Fig. 17.7). Because of the auxiliary pumped storage plant, the main plant itself does not have to supply all the excess electricity needed at the peak hours. This is highly economical and makes up for the efficiency loss of energy involved in storing and reusing it.

Another possibility is to dissociate water electrically and store the hydrogen that results. Hydrogen is an excellent fuel, and research is underway to design batteries (called fuel cells) that generate electricity from hydrogen directly. We'll talk more about this later.

■ 17.2　How We Use Electricity: The Growth in its Consumption

When American inventor Thomas Edison, around the turn of the century, got into the game of electricity production, his motive was very simple. He wanted to sell his light bulbs, which he had just invented. For a while the light bulb was just about the only use of electricity at the residential level. But, of course, this has changed dramatically since then.

Today we depend on electricity to assist with most of the chores in and around the house. To light, to cook, to heat or cool (depending on the season), to clean, to be entertained, and so forth. As somebody has said, to own a garbage disposal unit for your kitchen sink has become a symbol of accomplishment. We can wonder how this dependence came about. But perhaps the answer is as simple as the Shakespearian doctrine: "How use doth breed a habit in man."

Let's consider some details of the household consumption of electricity. The electric power required to run each of your electrical appliances is something worthwhile to learn, particularly if you are interested in conserving energy, to save it for better use. Table 17.1 gives the electric power rating of most household machines and appliances.

Actually, knowing about the electrical power requirement of a piece of electrical equipment is not enough. Recall that power is the energy spent per second; ultimately what counts most is the total energy spent. If P is the power rating of an electrical device and it runs for a time interval t, then the total energy spent is

$$\text{energy} = \text{power} \times \text{time}$$
$$E = P \times t \tag{17.5}$$

So the time that you use a device is very important. By multiplying the average annual time of use of an electric device with its power requirement, we can calculate what its annual energy requirement is. This also is given in Table 17.1. The total energy spent is given in the convenient (in this case) unit of kilowatthours (kWh). To refresh your memory, if a machine of 1 kW power runs for an hour, the energy spent is 1 kWh. Also, 1 kWh = 3.6×10^6 J.

When you study the table, you can discover one important thing right away: all gadgets that purposely produce heat (or cold) consume the most power, and often the most total energy, because they are used for extensive periods of time. Much less power is used by those machines that have motors in them. The washing machine and the vacuum cleaner are among the best utility machines for their cost of energy.

Appliance	Power (W)	Annual Energy (kWh)
Air Conditioner (window)	1,600	1,400
Clock	2	17
Clothes Dryer	4,900	1,000
Dishwasher	1,200	360
Fan (circulating)	90	45
Food Blender	400	20
Freezer (frostless, 15 ft^3)	450	1,800
Frying Pan	1,200	200
Heat Pump (electric heating system)	12,000	16,000
Radio	70	85
Radio-Phonograph	100	110
Range	12,000	1,150
Refrigerator-Freezer (frostless, 15ft^3)	600	1,800
Shaver	14	18
Television (color)	330	660
Toaster	1,150	50
Toothbrush	7	5
Vacuum Cleaner	650	50
Washing Machine (automatic)	500	100
Water Heater (standard)	2,500	4,300

Let's do a calculation to see how to obtain the last column in the table. A color TV runs at a power of 330 W. An average time of TV viewing of 5.5 h a day amounts to 5.5 × 365 = 2007.5 h of running in an entire year. Multiplying the two, we get

$$E = P \times t = 0.33 \text{ kW} \times 2007.5 \text{ h} = 662.48 \text{ kWh}$$

If you are interested in conserving energy, you can put this kind of calculation to good use. When you buy a new electrical appliance, take an interest in knowing the power rating and its projected annual energy consumption. If you don't find the total energy from the seller, try to estimate it yourself. Estimate the number of hours per day you intend to use it. Multiply this by 365; this gives you the annual hours of usage. Then multiply this last number with the power requirement of the device. Compare the value of the energy consumption of this device with that of others that you may be considering.

Saving electrical energy is important for a variety of reasons. First, it is expensive energy—you really pay for the convenience. For example, take cooking. If you use natural gas, you get practically 100% return for the energy. In contrast, you get electricity only at a 40% efficiency at best.

Second, although electricity is still advertised as "clean energy," the truth is that it is far from clean. The dirt is hidden under a rug, so to speak. If the power plant that supplies electricity uses coal, it is producing air pollution and thermal pollution. If your electricity comes from a nuclear plant, it is simultaneously producing radioactive wastes.

Third, and perhaps the most important reason, is that the growth in our expenditure of electrical energy is a very rapid one. If we continue to grow this fast, it is doubtful that we can sustain the growth. Exactly how fast is this growth? We will look at the details below; it is a very revealing story.

But before we go into that, perhaps you can take some consolation from the fact that the household is not the largest consumer of electricity. Industry uses the biggest chunk of all the electricity in the United States, about 45%. The next biggest chunk is in households, where another 30% is accounted for. The share of the commercial sector is some 20%, and the rest is used for various purposes, customarily labeled "other uses," such as street lighting.

Interestingly, of the total industrial usage, two industries are responsible for almost half the total: the chemical and the primary metal industries. Also of interest is the small percentage (less than 5%) of the total industrial usage devoted to electrical mass transit systems.

One more thing is noteworthy about the electrical energy use of industry. Several studies have found that if steel, aluminum, copper, and paper were manufactured from recycled scrap, there could be significant savings. One study showed that, in 1970, if only one-half of the total production of steel, aluminum, and paper came from scrap rather than raw materials, there would be a savings of electricity amounting to 3% of the entire amount of electricity consumed during that year. And, of course, all municipal trash is quite abundant in steel, aluminum, and paper.

The growth in the consumption of electricity and energy in general

The price for technological progress is the increasing energy bill it brings. How fast has the rate of energy consumption been increasing? In the answer to this question lies much of our present energy predicament.

We can imagine many types of growth rate. For example, the growth rate may remain constant in time, the amount of growth in a given amount of time always being the same. This is called linear growth. A baby growing an ounce in weight each day is an example of linear growth (Fig. 17.8).

If growth were always linear, our life would be quite simple. Unfortunately, the rate of growth often depends on the size of the growing object—how much quantity is present at the moment. Take the case of a growing population; its

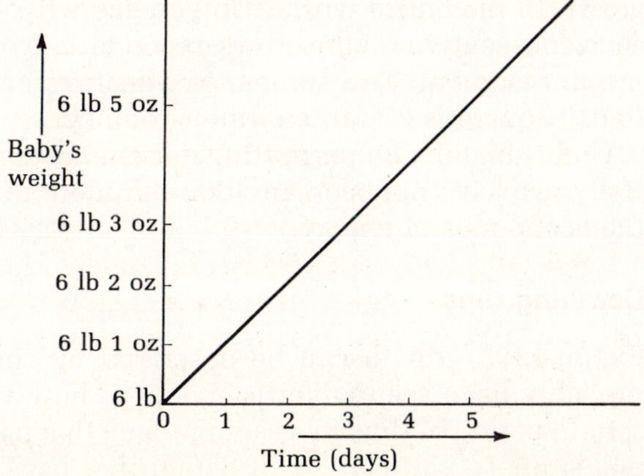

FIGURE 17.8   Increase of a baby's weight with time: an example of linear growth.

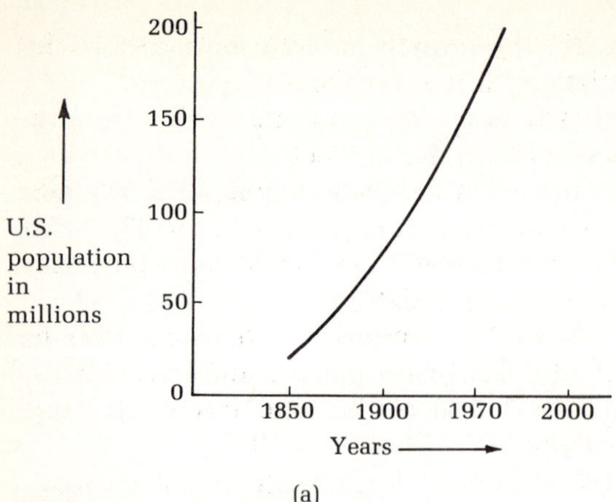

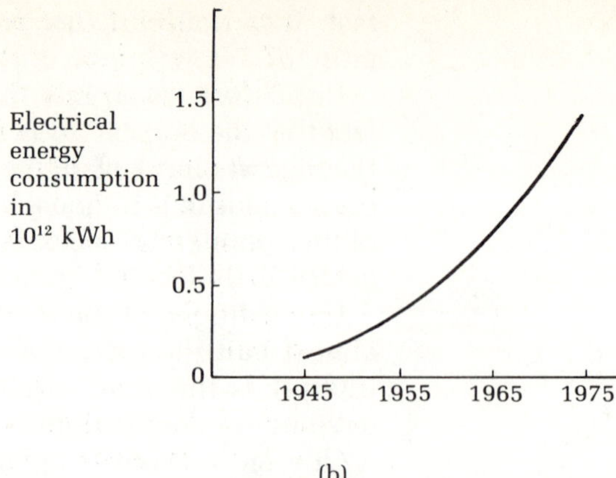

FIGURE 17.9 Two examples of exponential growth curves. (a) The growth of the U.S. population. (b) U.S. electrical energy consumption since 1945.

growth rate at any particular time depends on the size of the population at that time. If the growth rate is a constant percentage of the (changing) size of the growing quantity, we call it an exponential growth rate. Exponential growth is a fairly common occurrence in nature. Fig. 17.9 shows two exponential growth curves, one for the population and one for the growth in the consumption of electricity in America. Notice how the curve becomes steeper as it gets higher. If you look carefully you will find that the steepness of the curve increases in proportion to the height. Such is the nature of an exponential curve.

The economy in America and the energy consumption (including electrical energy) have grown exponentially. If you don't appreciate how explosive exponential growth can be, perhaps the following story will be helpful.

A King of India wanted to give a reward to his prime minister, the grand Vizier, for introducing him to the game of chess. Now this Vizier was a mathematician. As you know, there are 64 squares on a chess board. So the Vizier asked for the following very innocent-sounding gift. He wanted a grain of wheat on the first square; for each subsequent square, his wish was that the number of grains of wheat should be double that of the previous square. Thus for the second square he would get 2 grains of wheat, for the third, 4 grains, and so forth. The king did not hesitate even the blink of an eye to grant the Vizier's wish, thinking that he had a bargain. Did he?

Actually the king had promised the prime minister more wheat than can be grown in the entire world. Do you see why? The number of wheat grains for each consecutive square grows exponentially in proportion to the number at the present moment. The number of wheat grains that the Vizier gets for the sixty-fourth square is 2^{63}, an enormous number.

Unfortunately, by permitting our energy consumption rate to grow exponentially, we have not been any more prudent than this king in our dealings with the economics of energy.

Doubling time

Exponential growth can be described by considering "doubling time." You probably have seen advertisements of how you can double your money in a certain period of time in a savings bank that pays compound interest. For example, suppose you start with $1000 in a bank that pays a rate of 6% interest,

compounded yearly. At the end of the first year, you will have earned interest
of

$$6\% \times \$1000 = 0.06 \times 1000 = \$60.00$$

So your present capital is $1060. Note that we can get this number directly by
multiplying 1000 by 1.06 [i.e., 1 + (percentage rate/100)]. At the end of the
second year, you will have

$$1060 \times 1.06 = 1124$$

If we keep calculating this way, we have the figures shown in Table 17.2.

TABLE 17.2 The growth of $1000 at 6% interest rate, compounded annually.

Time (years)	Amount ($)	Time (years)	Amount ($)
0	1000	7	1504
1	1060	8	1594
2	1124	9	1689
3	1191	10	1791
4	1262	11	1898
5	1338	12	2012
6	1419		

The money thus has doubled in 12 years and will continue doubling every 12
years. This is an example of exponential growth, since the rate of growth at any
time is a fixed percentage of the instantaneous value of the capital. Thus exponential growth is characterized by a **doubling time.** You can easily convince
yourself that in the case of linear growth there exists no such thing as a constant
doubling time.

For relatively small values of the annual percentage rate of growth, there is a
simple formula that enables us to calculate the approximate period of the doubling time for exponential growth:

$$\text{doubling time in years} = \frac{70}{\text{annual percentage rate}}$$

In our example of 6% annual compound interest, the formula gives a doubling
time of $\frac{70}{6} = 11.67$ years, which is very close to the correct value.

Table 17.3 shows the average doubling time of some of the indicators of our
technological and economic growth.

We have created today's crisis by increasing our energy consumption at an
exponential rate for almost the past one hundred years. Occasionally the doub-

TABLE 17.3 Growth rates and doubling times.

Item	Percentage Rate of Growth	Doubling Time (years)
Population	1.6	44
Total energy consumption	3.2	22
Industrial production	4.0	17.5
Motor fuel consumption	4.5	16
Electric power consumption	7.6	10

ling times have changed, but except for the short period of the depression in the thirties, the exponential nature of the growth has prevailed.

And as you can see, electricity is the fastest growth item in Table 17.3. Today our total electric power need is a towering 250 billion W; a little more than 1 kW for each person. Twenty-four percent of our total consumption of energy goes to electricity. It is also interesting to note that the increase in the consumption of energy or electricity is not primarily due to the increase of population.

What are the resources from which all this electricity comes? Do we have enough resources to cope with the exponential growth? What are the environmental effects? What holds for the future? These are the questions we will consider in the next two sections.

■ 17.3 The Sources of Electric Power

In previous chapters we have mentioned various energy resources—the fossil fuels, geothermal energy, energy of the waves, and so forth. When traced to their ultimate origin, we find that basically there are only four fundamental sources of energy for us. First, there is solar energy. This certainly is the most important energy source on the earth. The fossil fuels, the wind and the waves, the energy of the biomasses like plants, food, and animal muscle energy, the energy of the running water of the rivers—all these can be traced back to the sun. The second biggest source of energy is nuclear fuel, the sources of highly concentrated nuclear potential energy. Third is geothermal energy, a vast amount of energy (only a little of which is available to us) trapped inside the earth. Fourth is the energy of the earth's rotational motion, which also is largely inaccessible to us except for the little portion that is dissipated in the form of tidal energy.

There is a more practical way to look at the energy resources, that is, in terms of how we use them today. Table 17.4 gives the energy resources used in the United States today for the generation of electricity. Coal is the major source, but natural gas, hydro, and nuclear resources make significant contributions to electrical production.

TABLE 17.4 Resources for U.S. electrical energy (1972 figures).

Resource	Percentage of Total
Coal	46
Natural Gas	23
Hydro	16
Oil	11
Nuclear	4

Then there are some prospects that are promising but have not materialized yet. The prime example is nuclear fusion power, nuclear energy released when very light nuclei are fused to make bigger ones, in the manner of processes occurring in the stars. Another important example is solar energy and the direct use of it for generating electricity. Geothermal, tidal, wind, and wave energy have not made any major impact on the electrical scene.

The thermal power plants that use fossil fuels

The fossil fuels make the following contributions to the production of electricity:

Coal (46%)

Natural gas (23%)

Oil (11%)

We have already indicated most of the components of a thermal plant using a fossil fuel like coal. There is a boiler where the coal is burned, producing high-pressure, superheated steam. The typical efficiency of this phase of the operation is 90% (0.90). The energy of the steam is used to turn the shaft of a steam turbine. Power plants using natural gas and low-grade oil use a somewhat different turbine called the gas turbine, where the gases resulting from the burning itself do the work of turning the turbine shaft. In either case, the turbine is a heat engine and is severely limited in efficiency. The best efficiencies, obtained from steam turbines, are around 45% (0.45). The turbine shaft is connected to the generator to keep its armature rotating. The generator has a high efficiency: 99% (0.99). Combining all these efficiency factors we can calculate the total efficiency of a thermal power plant:

$$0.90 \times 0.45 \times 0.99 = 0.40 = 40\%$$

The more recent coal power plants are truly mammoth in their power generating capacity. A 1000-MW power plant is quite common. To give you some idea of how big an operation that is, let's calculate how much coal is used per day in a 1000-MW power plant. From Table 13.2, the energy value of coal is

$$4.6 \times 10^6 \text{ Cal/ton} = 4.6 \times 10^6 \times 4.2 \times 10^3 \text{ J/ton}$$
$$= 19.32 \times 10^9 \text{ J/ton}$$

For 1000 MW, or 10^9 W, the amount of coal needed is

$$\frac{10^9 \text{ J/sec}}{19.32 \times 10^9 \text{ J/ton}} = \frac{1}{19.32} \text{ ton/sec} = 0.05 \text{ ton/sec}$$

This amounts to $0.05 \times 86,400 = 4,320$ tons of coal per day (since 1 day = 86,400 sec).

The shipment of such large amounts of coal is one of the factors influencing the high operating costs of coal power plants. The shipment of oil and natural gas is simpler—pipelines can be used, for example, but they too are expensive.

One problem with the large amounts involved and the necessary expenses for shipping is that the power may have to be manufactured locally. This brings in new environmental concerns.

Do we have enough of a coal reserve? It is clear that the day of the use of natural gas and petroleum for the generation of electricity are numbered. The lifetime of the entire oil supply of the world is estimated to be something like 50 years even by the most enthusiastic optimist. Most of the world's natural gas will be gone even sooner. Thus it is generally agreed that prudence demands that these two valuable resources be saved for those users for whom they are most convenient or indispensable. Perhaps coal can pick up some of the slack.

Do we have enough of a coal reserve? The United States is the second largest producer of coal in the world, with huge reserves still left underground. The estimate for our recoverable coal reserve is around 1600 billion tons. Our annual expenditure of coal toward electrical and other uses presently is about 600 million tons. So at the present rate, our coal can last us for

$$\frac{\text{total reserve}}{\text{yearly depletion}} = \frac{1600 \times 10^9}{600 \times 10^6} = 2.67 \times 10^3 \text{ years} = 2670 \text{ years}$$

Unfortunately, there are several things wrong with this kind of an estimate of the lifetime of coal. The foremost one is the question of consumption rate: it is almost a certainty that the rate is not going to stay put at the present level. If the rate goes up exponentially, as present trends indicate, the lifetime comes down to around 300 years. On top of this we have to allow for the fact that the estimate of recoverable reserves could prove to be an overestimate.

It should also be stressed that coal is not an immediate solution of the energy crisis, either, because the industry cannot convert to coal as fast as is necessary to move us away from a dependence on oil and natural gas.

The worst problem comes from the environmental impact of the widespread use of coal. It is one of the worst producers of air pollution, as well as of thermal and solid waste pollution. Furthermore, mining the coal (and strip-mining is more economical as well as practical from the industry's point of view) destroys the environment.

There is a lot of talk about desulfurization of coal to reduce air pollution and about recovery of the land after strip-mining. But the expense of these things can add to the cost of electrical energy, and should. Electricity is simply not as cheap as we have been led to believe.

Water power

Solar energy, by evaporating the ocean water, starts the water cycle. The water vapor rises with the convection currents to the high level of the atmosphere, condenses, and gives rain and snow. The rain and melting snow combine into the running water of rivers. Finally, the rivers return the water to the ocean, and the cycle is complete.

Human civilization has known for a long time that the energy of the water can be used for its benefit instead of letting it go to waste (it is mostly converted into heat as the water enters the ocean). Long before the harnessing of water energy for electricity, water wheels had been used to obtain the water's energy, although on a much smaller scale. The advent of hydroelectric power plants has enabled us to use the energy of running water not only on a large scale but also at a greatly increased efficiency.

The technology of harnessing water power consists of the following steps: (1) the water is stored in a dam at a high level; (2) the water is released as needed and allowed to fall on a hydraulic turbine (this is a turbine designed to rotate from water falling on its blades, just a generalization of the old water wheel); (3) the shaft of the turbine is coupled to the electric generator, where its mechanical energy is converted into electricity.

How much power can we get from a typical plant? It depends on the water flow rate and the height of the water, which initially depend on the size of the reservoir and the conditions of the water flow in the river. The power of the water is the weight of the water flowing per second times this height. The weight of the water can be calculated by multiplying the volume of water falling per second with its weight density.

Let's consider a typical case. Suppose the water flow rate is 2000 ft³/sec and the height is 300 ft. The power of the water is given as

$$\begin{aligned}
\text{power} &= \text{flow rate} \times \text{height} \times \text{weight density} \\
&= 2000 \text{ ft}^3/\text{sec} \times 300 \text{ ft} \times 62.4 \text{ lb/ft}^3 \\
&= 2000 \times 300 \times 62.4 \text{ (ft}^3/\text{sec)} \times \text{ft} \times \text{(lb/ft}^3) \\
&= 3.74 \times 10^7 \text{ ft·lb/sec} = (3.74 \times 1.36) \times 10^7 \text{ J/sec} \\
&= 5.01 \times 10^7 \text{ W} = 50.1 \text{ MW}
\end{aligned}$$

(Note: 1 ft·lb $= 1.36$ J.) This is adequate electricity for a population of 50,000 people.

The total hydropower available in the United States, based on stream flow records across the country, is estimated to be 161,000 megawatts. More than half this available hydropower is in use.

The world situation is as follows. There is a total of 2857 billion watts of available hydropower, of which only 8% has been developed, mostly in the industrial countries. Thus there exists a large amount of water power in reserve in the Third World countries of Africa, Asia, and South America.

The efficiency of hydroelectric power plants is very high, because heat energy is not used as an intermediary. Hydraulic turbines run at efficiencies of over 90% (the slight loss coming from unavoidable friction) and electric generators have an efficiency of 99%.

However, not everything is so bright. Wherever there is a river dam, there is also irrevocable damage to the natural ecosystem (the living organisms in a community and their environment together are called an ecosystem; the subject of the interaction of living organisms with their environment is called ecology). This damage can be so unpredictable that many specialists today think that the establishment of large hydroelectric plants in underdeveloped countries is not the solution to their energy problems. An example of such environmental disaster is the case of the Aswan Dam in Egypt.

The other thing to remember is that river dams are not a permanent source of power. The reservoirs get a continuous supply of sediments along with the water. Although it takes perhaps a couple of hundred years, eventually the sediments fill up the reservoir and end production.

Nuclear power

Nuclear power, though still in its infancy, may never grow up because of the controversy surrounding it. The energy of nuclear power comes from the splitting of the atomic nuclei of the element uranium, a process usually referred to as nuclear fission. We will discuss the entire nuclear controversy in a later chapter.

There is no doubt that the demand for electric energy will continue to exist and, most probably, grow at today's rate for some time to come. In the next few decades, as oil and natural gas become scarce, other energy resources must pick up the slack. It is more convenient with some of these resources to convert them first into electricity before channeling them toward the desired use. For example, nuclear energy, both fission and fusion, must be converted into electricity before it can be used.

Considering this, it is clear that improving the efficiency of the generation of electricity should be a high-priority matter. In this section we will look at some new ideas in this direction.

Of course, our energy problem would be much easier to handle if we learned to use solar energy directly to generate electricity. There is some interesting research in this area, which we will also look into.

Future electricity at improved efficiency

The magnetohydrodynamic (MHD) generators. In the turbine, the energy of steam must be converted into the mechanical energy of the turbine shaft before we can generate electricity. Could we eliminate the turbine and convert the energy of steam directly into electricity? The MHD machines attempt to accomplish this.

Instead of getting hot steam from the combustion of the resource, these machines use certain gases that are readily ionized when heated to very high temperatures. At the high temperature some of the gas molecules will split into atoms, and some of these atoms will each lose an electron. The remaining portion of each of these atoms becomes a positive ion in the process. The electrons are picked up by some other neutral atoms or molecules, producing negative ions. Thus a very hot gas consists of some negative as well as positive ions. The thermal ionization is facilitated by "seeding" the gas with a metal like potassium.

Now the ionized gas, with its ions moving at high speed, passes through a magnetic field, whose lines of forces are perpendicular to the direction of motion of the ions (Fig. 17.10). The magnetic field bends the charged particles

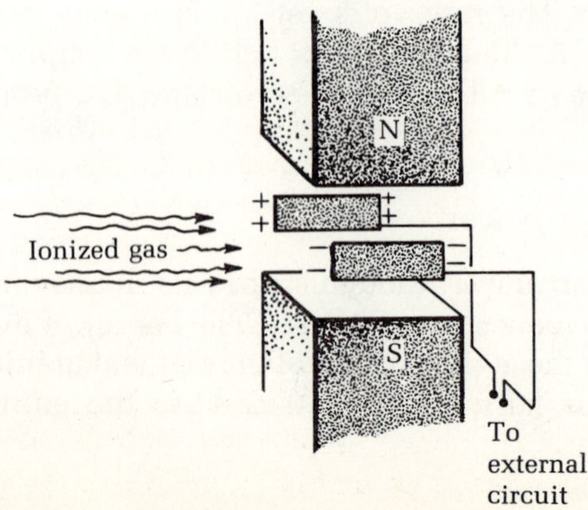

FIGURE 17.10 The MHD generator. See text for explanation.

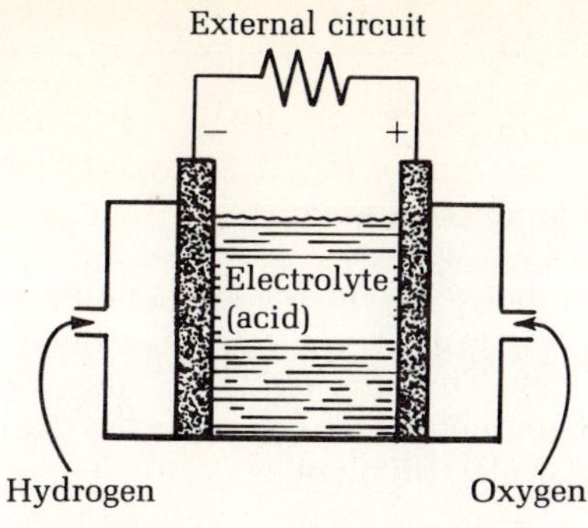

FIGURE 17.11 The fuel cell. See text for explanation.

moving perpendicular to itself; also, it bends positively charged particles one way and negative charges the other way. In effect, then, a magnetic field separates the charges, which can then be collected onto two electrodes. The electrode receiving the positive ions becomes positively charged and the one receiving negative ions will be negatively charged. We will get a current when we connect the two electrodes; electrons will flow from the negative to the positive electrodes through the external circuit as long as we maintain the separation of charges between the plates.

Another way of looking at how the charge separation occurs is to recognize that the ionized gas behaves pretty much like a conductor moving in a magnetic field (the seeding helps to make the gas even more like a conductor). So now there will be an induced electric field, which moves the positive charges one way and the negative charges the other. Thus there is a charge separation, a positively charged electrode and a negatively charged one.

Greatly improved efficiencies can be obtained with the MHD because of the use of the very high temperatures that we can get with the gases. The estimated efficiency is 60% for the entire MHD cycle, as compared with 40% in conventional steam plants. The Russians have already constructed a pilot plant for MHD generation, and the United States is not far behind.

Fuel cells. We previously discussed electric batteries, which convert the energy of chemical reactions directly into electricity. **Fuel cells** operate on the same idea except that the chemical reaction used is the combination of hydrogen with oxygen. The arrangement is shown in Fig. 17.11. Hydrogen and oxygen come in from opposite ends of the cell. At the negative terminal the hydrogen becomes ionized, releasing two electrons for every molecule so ionized. The H^+ ions now move on toward the positive electrode. The presence of the electrolytic acid facilitates the motion of the hydrogen ions. The electrons also move toward the positive electrode, but through the external circuit. At the positive electrode the electrons ionize the incoming oxygen molecules, converting them to negative oxygen ions. The oxygen ions now neutralize and react with the hydrogen ions, forming water.

Notice that in this kind of a battery, the electrodes do not take part in the chemical reaction. In fact, the operation described is the exact reverse of the electrolysis of water (the breaking up of water into oxygen and hydrogen when an electric current is passed through it). Since the electricity is produced without the intermediary of heat, we can easily expect efficiencies of 60% or higher.

Of course, widespread use of the fuel cell depends on the availability of hydrogen as a fuel. Hydrogen is not a natural fuel and is currently quite expensive to manufacture. However, this may change in the future when some of the then prevalent energy resources may perhaps be most conveniently converted and stored in the form of hydrogen. Hydrogen is quite cheap to transfer from place to place. There may be some safety problems, since hydrogen is an explosive, but none that could not be solved in advanced technologies.

The fuel cell is a dc machine, but since there is no need to transfer the electricity through transmission lines (large fuel cell plants do not seem feasible anyhow), there is no need for ac. In fact, this may be the biggest advantage of the hydrogen economy. With it we could have household fuel cells and eliminate the dependence on utility companies.

The value of the solar constant is 1.4 kW/m², half of which, 0.7 kW/m², is available at ground level. Thus for 1000 MW of incident solar power, we need the area of

$$\frac{1000 \times 10^6 \text{ W}}{0.7 \times 10^4 \text{ W/m}^2} = 1.43 \times 10^6 \text{ m}^2 = 1.43 \text{ km}^2$$

since 1 km² = 10⁶ m².

But, of course, we can harness the incident solar power with, at best, an efficiency of only 30%. Thus the land area needed is increased to

$$\frac{1.43}{0.3} \text{ km}^2 = 4.8 \text{ km}^2$$

We must also remember that we get only about 8 hours of good sunlight in a day, but the usage of electricity extends to at least 14 hours a day. Thus the land area must be increased by a factor of another $\frac{14}{8}$. This gives us the total area for a 1000 MW solar electric plant as

$$4.8 \times \tfrac{14}{8} = 8.4 \text{ km}^2$$

Electricity from the sun

Can solar energy be utilized on a large scale? One often heard answer to this question is, no, it's too dilute. The specialist then brings out a pocket calculator and pushes some buttons. Finally, with a sardonic smile, she shows you a number. "See, I told you. It takes 8 km² for a 1000-MW plant. It's preposterous to use so much land for generating electricity."

No, don't doubt her calculations, they are correct. But you can question her assertion. Is it really preposterous to use this much land area? Well, we do it for growing food. We even do it for highways, which essentially are for the consumption of energy. Why, then, so much reluctance to reserve adequate space for a solar "farm"? Food energy is manufactured by plants with an efficiency rating of only 2%. The solar plants with present technology can perhaps be made as efficient as 30%.

One suggested scheme is to collect the solar energy in a specially constructed glass pipe (called a collector) in which a suitable liquid flows (liquid sodium was proposed at one stage). The glass pipe consists of layers of thin films, each of which has the property of the greenhouse effect built into it. These films are transparent to incoming solar radiation, but almost opaque to the radiation they themselves emit when hot. Thus a very hot environment can build up within the pipe. An even better effect can be obtained by using focusing mirrors overhead to direct the sunlight onto the pipes. If the liquid inside the pipe gets as hot as 500°C, then we can get as high as a 30% efficiency.*

Aside from the large land area needed, we can point out other flaws in this scheme to use solar energy. The collectors are expensive. And how long will they last? Also, sunlight is available in its full glory only in places like Arizona. So this is perhaps a solution to the electricity problem only for such desert areas. Electricity is expensive and environmentally hazardous to transfer over very large distances.

Another attractive plan for getting solar electricity is by means of the solar cell, a device about which we will have a lot to say later.

All these things must be checked out, and research is the only way to check them out. Such research is continuing. Can solar energy be used on a large

---

*For further details read an article by A. B. Meinel and M. P. Meinel, "Physics Looks at Solar Energy," *Physics Today*, February 1972, p. 44.

So far we have talked about large-scale production and savings. Now we will consider what I call the "little ways" of producing electricity.

I am reminded of a story in *Ramayana*, the great Indian epic. Rama, the hero of the story, is building a big bridge requiring great effort from many strong men. All of a sudden, a squirrel comes and rolls itself on the sand, then carries the sand over to the bridge and deposits it. It does this repeatedly. Everybody around begins to laugh at the squirrel's effort except Rama. He says something very valuable: lots of little efforts together can solve even the biggest of problems.

We have mentioned some sources of energy that, although abundantly available, cannot be harnessed and utilized on a large scale: wave energy, tidal energy, geothermal energy. Add to this wind energy and energy from garbage burning, and you have a pretty good list.

Some of these sources have been tapped. There is a successful tidal power plant in France. There is a fair-sized geothermal plant in the geyser area of California near San Francisco. Wind mills have been used in remote areas of Sweden for many years for communication purposes. Even garbage and trash have been used in a town in Ohio, among other places, to generate electric power.

The picture that emerges is that these small-scale energy productions are viable but not economically attractive enough to interest the profit-motive sector. However, this may change as conventional means of power become more costly.

But there is also the possibility that these are viable means right now if we do our figuring right. The large-scale power plants have produced irreparable damage to our environment. The transmission of electricity from the large-scale plants costs us in both efficiency and environment. It is time that we put environmental effects into the cost-benefit analysis for power plants.

There is also the overall psychological effect of doing everything on a big scale, which economist E. F. Schumacher has talked about.* Schumacher makes a cogent point that people working in big businesses develop a consumer attitude toward life and cease to give expression to the creative urge, which may be what life is all about. This consumer attitude is then fed back to the society in the demand for more consumer items, and therefore more energy and more consumption of electricity. On a finite planet with finite resources and a finite space for disposal of pollution, a spiraling consumer attitude may be disastrous.

Who says that "bigger and better" is the only solution and the "frontier expansion spirit" will forever dominate the American scene? When I was a teenager, somebody told me a joke about this, which caught my fancy. The joke goes like this:

The United Nations undertook as a research project an exhaustive study of the elephant. Each nation was asked to make a contribution on one important aspect, of their choice, pertaining to this animal. England chose to report on Her Majesty's Navy and the elephant. The French promised to find out everything about the elephant's love life. India decided to look into the role of the elephant in nonviolence. The Russians picked the subject of the elephant and dialectic materialism. And, finally, the Americans proposed to devise methods of making "bigger and better" elephants.

Today as I look at the different parts of the joke, I'm impressed with how much the world has changed since then. Once-mighty Britain hardly has a navy to be proud of. The French are not too keen on the subject of love; other nationalities such as the English, Scandinavians, and the Americans are making many of the new moves in this area. India has long since abandoned the path of nonviolence. Even Russia is no longer dominated by the idea of dialectic materialism.

How about it, America? Are we ready to make some changes?

---

*E. F. Schumacher, *Small Is Beautiful* (New York: Harper & Row, 1975).

scale? We don't know. But we are going to try and find out.

In the fourteenth and fifteenth centuries, people in England were faced with an energy crisis coming from a dwindling supply of wood (the principal fuel then). They went ahead to find various ways to use coal and, in the process, gave rise to the industrial revolution. Perhaps there is a lesson in that. Perhaps we will learn to use the abundant solar energy for all or most of our energy needs. And then today's energy crisis would be considered worthwhile by posterity in that it pointed us in this direction.

SUMMARY

The theme of this chapter is electrical energy, technology, and energy and society, in general. We begin by discussing aspects of the large-scale generation of electricity—the generator, the transformer, the pumped storage plants, and the like.

This nation has a tremendous appetite for electrical energy. How is electrical energy used in this country? And at what rate is the production of electricity growing? These questions are discussed in Section 17.2. The next section goes into the matter of available energy resources. And we sum up the chapter with an outlook for the future.

If you are interested in conservation, learning to calculate your spending of electrical energy is a very good idea. Perhaps you read the comic strip *Gasoline Alley*. One of the characters in the strip, Skeezix, got into the "fun" of finding out the cost of electrical spending in his family, like how much it costs to toast a piece of bread. But his wife did not take this kindly. So be careful. Anyhow, the formula that can help with such calculations is given as

$$\text{energy} = \text{power} \times \text{time}$$

QUESTIONS

Review and reason

1. A wire loop is placed between the pole faces of a large magnet. If the loop is rotated, a voltage is induced across the loop. Why? At what position of the loop is the voltage maximum and at what position is the voltage zero? Draw pictures.
2. How is it possible that the ac voltage and ac current are both positive and negative, but the ac power supplied is always positive?
3. Compare the function of an electric motor with that of an electric generator.
4. State Lenz's law.
5. What is the reason that direct current is not supplied to households?
6. What is a transformer? What does it transform? Explain the difference between a step-up and step-down transformer.
7. Make a cogent argument in favor of conservation of our energy resources.
8. Briefly discuss one reasonable and *different* way of conserving energy in each of the following economic sectors:
   (a) transportation
   (b) residential
   (c) commercial
   (d) industrial
9. List a few things that you personally could do to conserve energy.
10. Power companies sometimes lower the voltage of their electric supply ("brown out") at times of high demand. What do they save by doing this? Is "brown out" a proper name for it? Explain.
11. What is the difference between linear and exponential growth? Give an example of each.
12. Read the following riddle and see if you can answer it:

    Suppose you own a pond on which a water lily is growing. The lily plant doubles in size each day. If the lily were allowed to grow unchecked, it would completely cover the pond in 30 days, choking off the other forms of life in the water. For a long time the lily plant seems small, so you decide not to worry about cutting it back until it covers half the pond. On what day will that be?*

13. Which is in shorter supply in the United States, coal or oil? Do some outside reading and make some suggestions as to what we should do to solve the problem of the short supply of the important fossil fuels.

---

*"Riddle of the Lilies," quoted in an article by R. W. Comstock, in *Energy Crisis*, V. J. Yannacone (New York: West Publishing, 1974), p. 343.

14. Name two energy resources that, in the process of conversion to electricity, add practically no heat to the environment.

15. Why is a hydroelectric power plant more efficient than a thermal plant using coal? Why do we not turn to hydroelectricity for solving the energy crisis?

16. What would happen to the energy that is converted into electricity in a hydroelectric power plant if the plant were never built?

17. Explain how the magnetohydrodynamic generator works.

18. Explain how the electric fuel cell works. Is the efficiency of a fuel cell limited by Carnot's maximum efficiency formula?

19. Hydroelectric energy is used to produce hydrogen by electrical dissociation (called electrolysis) of water. The hydrogen is taken by a pipeline to a building where a fuel cell uses it to produce electricity for the building. The electricity is used to operate all the lights in the building. Trace the energy from the water behind the dam to its ultimate form. Mention each transformation, giving the name of the form of energy at each stage of the energy conversion pathway.

20. Describe a plan for the large scale generation of electricity from solar energy.

21. Can the harnessing of solar energy in large-scale electric power plants cause thermal pollution? Discuss.

Arithmetic

1. You have a pocket calculator that operates at 10 V. You can plug the calculator into the household power lines, but you have to use a transformer to bring down the voltage first. What is the ratio of the number of turns in the secondary and that in the primary in the transformer that you have to use?

2. Consider the following sacrifices for the sake of conservation of energy. (a) You have a dishwasher with a power rating of 1000 W. You use the dishwasher every day, with a running time of an hour. Instead, suppose you use the dishwasher every two days. How much energy do you save annually? (b) You have a color TV set that runs at a power of 300 W. Instead of watching an average of 6 hours a day, suppose you decide to watch only 3 hours a day on the average. Calculate the annual energy saved by this maneuver.

3. The number of automobiles in the United States is growing according to the following table:

Year	Number of cars (millions)
1940	28
1950	40
1960	62
1970	89

Draw a graph of these data. Is the growth exponential?

4. A farmer has a fish pond in which some water lilies begin to grow. If the lilies cover the entire pond, the fish will die. The farmer notices that the coverage doubles every day. One day he sees that the pond is one-eighth covered. He decides that the problem is not urgent and puts off any action for a week. How long did he really have in which to act? Show how you get your answer.

5. We wish to compare the total efficiencies of several possible systems of powering an automobile. Assume that the weight of each system is the same, and that the same power reaches the wheels. An electrical power plant that burns gasoline is available, and it delivers electrical energy to the customer with an overall efficiency of 35%. Work out the total efficiency for each of the following systems: (a) An auto with a gasoline engine of 30% efficiency, but half of the engine's power output is lost in transmitting the power to the wheels. (b) An electric auto with a battery that is recharged with electricity from the electric power plant. The battery delivers back 80% of the electrical energy used to charge it, and all its power output reaches the wheels. (c) Electrical power is used to produce hydrogen by electrolysis of water, with an efficiency of 90%. The hydrogen is used to operate a fuel cell, which has an efficiency of 80%. All the power output of the fuel cell reaches the wheels.

Let There Be Electromagnetic Waves

> . . . the great success of field theory is that it enables two completely different branches of physics, optics and electromagnetism, to be described in terms of the same theory. Initially the field description of phenomena was introduced as a means to understanding the mechanical description of nature. However, as it became clear that it was a description of the field rather than its sources that was essential for a complete understanding of the behavior of the sources, it was realized that a new concept was arising for which there was no place in the old mechanistic philosophy.
>
> R. J. BLIN-STOYLE

■ 18.1 Maxwell's Synthesis: Electromagnetism and Light

English physicist R. E. Peierls once remarked that if you wake up a physicist in the middle of the night and whisper to him "Maxwell," he would immediately respond "electromagnetic theory."* Indeed, electromagnetic theory is the most famous achievement of this great physicist of the nineteenth century; James Clerk Maxwell. If we turn Peierls's quip around and say to a physicist "electromagnetic theory," it is most likely that his response will be "Maxwell," so profound is Maxwell's contribution to the field of electromagnetism.

Let's summarize the different bits and pieces of electric and magnetic knowledge that were available around 1860 for everybody to look at. The experiments of Oersted and Ampère had established that an electric current flowing in a conductor produces a magnetic field around itself. The experiments of Faraday gave us the complementary fact that if there is relative motion between a conductor and a magnet (i.e., if a conductor experiences a changing magnetic field), a current flows in the conductor. Both of the experiments were analyzed and summarized in the form of laws. Given a current one knew how to find the magnetic field, and given a changing magnetic field, one knew how to find the

---

*In *Clerk Maxwell and Modern Science*, ed. C. Domb (London: Athleone Press, 1963), p. 26.

induced current. But one word frequently used in describing these experiments in modern terms was far from accepted in pre-Maxwell days. This word is "field." The reluctance of the physicists of the time to look at the concept of field as something real and novel is what held up the progress of the subject of electricity and magnetism for a while.

Faraday, just before Maxwell's time, had already constructed the picture of fields in terms of field lines. But he was ridiculed for his simplistic picture. Here is a quotation from the contemporary literature.

> I declare (said the astronomer Royal of England of the time), I hardly imagine anyone who knows the agreement between observation and calculation based on action at a distance to hesitate an instant between this simple and precise action on the one hand and anything so varying and vague, as lines of force on the other.*

Part of the reason that Faraday's picturesque way of looking at electromagnetic phenomena was unacceptable particularly to the mathematically minded physicists of the time, was that Faraday did not make any attempt whatsoever to communicate his findings by using even the slightest bit of mathematics. Even algebraic symbols are not to be found in much of his scientific work. Thus Faraday irritated the "regular" scientific scholars; it is not clear if Faraday was aware of this particular cause of his unpopularity among his more mathematical peers. Of course, even if he was cognizant of the cause, there was not much he could do since he did not possess much expertise in mathematics.

Thus all future progress had to wait in suspense for a mathematical genius who at the same time liked to make physical models of ideas. Clerk Maxwell was precisely this kind of a genius. He took over where Faraday left off. He started by establishing the mathematical proof that Faraday's "field" theory gave exactly the same results as the action at a distance theory for the case of nonchanging or "static" fields. And thus he convinced others, and most importantly himself, of the essential validity of the field idea, at least as an alternative approach. And now he was ready to move on further than anybody had gone before, even Faraday.

If the field idea was really one of nature's important manifestations, then, Maxwell felt, there must be some phenomena that only fields would be able to explain. Intuitively Maxwell started looking into the subject of changing fields.

Maxwell started from the known facts: a current in a conductor produces a magnetic field, and a changing magnetic field produces a current in a conductor. Thus electricity and magnetism seem to be intimately connected. But is this true only in the presence of conductors and currents? Or does the connection of electricity and magnetism survive even away from current- (and charge-) carrying conductors?

How about insulators? An insulator like wood or paper has its positive and negative charges in perfect balance normally. But this balance is slightly upset when an insulator is placed in an electric field. The electrical charges inside the insulator are now acted on by forces. The positive charges and negative charges

---

*Quoted by J. Randall, in *Clerk Maxwell and Modern Science*, p. 8.

feel an equal and opposite force due to the field. So in response to the field, the positive charges would move one way and the negative charges would move the other way. Of course, none of the charges are free to move very far. They are brought to rest quickly by the restoring forces operating within an insulator. Yet, momentarily, the charges do move, and this by definition constitutes a current. If the electric field doesn't change, the charges don't move. However, whenever the electric field changes (it can be either a switching on or a switching off or just a change in the magnitude of the field), the charges respond by moving into a new equilibrium position. So clearly there is a current inside an insulator in a changing electric field. Maxwell called this a **displacement current** to distinguish it from **conduction currents** inside conductors that are caused by the motion of "free" electrons.

If there is a current, there has to be a magnetic field. So the displacement current in an insulator must also be surrounded by a magnetic field. And the origin of the whole thing is the changing electric field. So an idea began to take shape in Maxwell's consciousness: a changing electric field produces a magnetic field; one doesn't have to have a conduction current.

But then does one even need an insulator and the displacement current to produce a magnetic field? In a flash of sudden enlightenment the answer came to Maxwell. And he dared to put forward the seemingly absurd idea: a changing electric field always generates a magnetic field, even in empty space. Mind it, *even in empty space.*

Maxwell also knew that one way—in fact the only way—to interpret Faradays observation of induced current in a changing magnetic field is to say that a changing magnetic field generates an electric field, which in turn causes the electrons to move in the conductor. Again he realized there was no necessity for the phenomenon to be restricted to space with conducting matter in it. A changing magnetic field will always give an electric field, he said, even in empty space.

Now came the final piece of the synthesis. Maxwell could think of the following situation. Suppose initially for some reason a charge shakes—vibrates, that is—and produces a changing electric field. The changing electric field generates a magnetic field, according to Maxwell, even though it is in empty space. But behold! This magnetic field is being created starting from zero, so it is really a changing magnetic field. Thus it also in its turn must produce an electric field, forcing the original electric field to change further with this new addition. This new change in the electric field produces a change in the magnetic field by creating a new piece of magnetic field, and so forth. Clearly the process propagates in space. A changing electric field at a point of space produces a magnetic field in its vicinity, which in turn gives rise to an electric field a little further on. Thus the original disturbance of the oscillation of the charge propagates in space, and we have a traveling disturbance of the electromagnetic field. This is an **electromagnetic wave.** But do such waves exist? Yes, Maxwell said, *light is such an electromagnetic wave.*

One needs proof for making such unprecedented, earth-shattering claims. The proof was provided by mathematics: Maxwell's equations.

While making the field picture more cogent to his contemporaries, Maxwell was led to use a mathematical description of the fields. He transcribed all the known laws (like Coulomb's law for static electricity, Ampère's law of the magnetic field due to a conduction current, and Faraday's law for the current due to a changing magnetic field) into mathematical equations containing only the fields **E** and **B**, the currents, and the charges. In writing these equations he included his own ideas about the extra contribution to the magnetic field due to a changing electric field. And he also wrote Faraday's law in the light of his more fundamental interpretation, namely, that a changing magnetic field gives an electric field. These equations are Maxwell's equations, mentioned above. The solution of these equations determines the electric and magnetic fields **E** and **B** completely. Mathematicians call such equations partial differential equations and the method for finding a solution for equations like these was already known.

To find the nature of his electromagnetic waves, Maxwell applied his equations to the case of empty space, put all the charges and currents to zero, and found the solution. Not suprisingly to him, the solution came out as a wave, his electromagnetic wave. The solution also told him something more about the waves, just what he hoped for. The velocity of the electromagnetic wave, according to his theory, was given by the square root of the ratio of the electric and magnetic field constants, k and k',

$$\sqrt{\frac{k}{k'}}$$

Even before Maxwell developed his equations, both k and k' had been determined experimentally. The values were not much different from the modern accepted values (the experimentors of those days were very good). So let us use the values quoted earlier in Chapters 15 and 16 for k and k'. In MKS units, k is 9×10^9 and k' is 10^{-7}. Therefore, the velocity of electromagnetic waves is given as

$$\sqrt{\frac{k}{k'}} = \sqrt{\frac{9 \times 10^9}{10^{-7}}} \text{ m/sec} = \sqrt{9 \times 10^{16}} \text{ m/sec} = 3 \times 10^8 \text{ m/sec}$$

The velocity of light was also determined very accurately before Maxwell's equations were derived. Maxwell saw immediately that the electromagnetic waves travel with the velocity of light. This could not be a coincidence. This was all the proof Maxwell needed to proclaim that light is an electromagnetic wave.

Great ideas in physics are remarkable for the amount of synthesis they bring about. Newton's gestalt joined the two sciences of astronomy and physics. And now Maxwell's gestalt made it possible to look at electromagnetism and optics in the same light. Newton's theory of universal gravitation combined with his laws of motion enabled us to solve the motion of all objects under gravity. Maxwell's theory supplemented by the knowledge of the forces on a charged particle in an **E** or a **B** field [Eqs. (15.4) and (16.5)] enables us to understand the motion of all charges. So Maxwell's work must be regarded as important as Newton's.

Let's discuss one last point. Maxwell's gestalt is also Faraday's. Faraday also perceived light as an electromagnetic phenomenon, completely independently of Maxwell, as documented in a public lecture he gave once. However, Faraday did not have the mathematics to back him up, so it was left to Maxwell to immortalize the ideas that can be attributed to both himself and Faraday.

FIGURE 18.1 An electromagnetic wave. The **E** and **B** fields are the disturbance coordinates. Note that **E** and **B** are perpendicular to each other and to the direction of propagation.

Light as an electromagnetic wave

In Chapter 11 we examined the evidence that light is a wave phenomenon. Now we find that it is a very special kind of wave: it is the traveling disturbance of a field rather than that of a medium.

We can be more specific. We can seek an answer to the question: What is the disturbance coordinate for light and other electromagnetic waves? It must be the changing **E** and **B** fields themselves. Thus we can picture a light or any other electromagnetic wave as shown in Fig. 18.1. The electric field vector **E** and the magnetic field vector **B** are both perpendicular to the direction of propagation of the wave and are perpendicular to each other. The complete electromagnetic wave can be looked on as a combination of an electric and a magnetic wave. Since the disturbance coordinates are perpendicular to the direction of travel of the wave, the waves are transverse in character.

Light as a transverse wave: polarization

One way to study waves at home (or anywhere else) is to get hold of a Slinky. A Slinky is a toy that is particularly susceptible to creation of wave patterns, both longitudinal and transverse, as shown in Figure 11.4. Now imagine putting bricks, as shown in Fig. 18.2, in the direction perpendicular to the coils of the Slinky. The longitudinal wave pattern will not be disturbed, of course, but the transverse pattern will certainly stop.

This is the way we can determine that light waves are transverse. Nature provides the kind of "bricks" that are necessary for the experiment. These bricks are called crystals. A crystal, though transparent in appearance, can block out every direction of displacement except for one for which the crystal has a "window."

To see the details, let's consider a light wave. The electric field vector **E** can be aligned in any way in the plane perpendicular to the direction of propagation (Fig. 18.3). After transmission through a crystal suitably placed, the **E** vector of the transmitted waves is aligned along a definite direction in the plane perpendicular to the direction of the propagation. The **B** vector is, of course, perpendicular to **E**. Such a beam of light is called a **plane-polarized beam.** If we now put a second crytsal in the path of the beam, with the window perpendicular to the direction of **E**, there will be no light transmitted. Such a complete blackout of light by passing through apparently transparent crystals is possible only

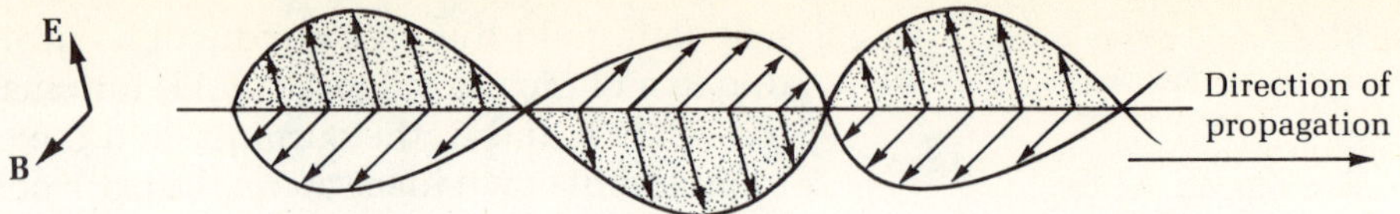

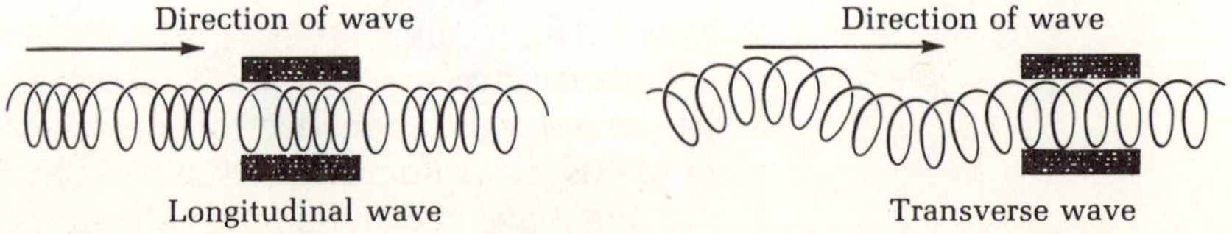

FIGURE 18.2 Bricks placed as shown will stop the transverse wave in a Slinky.

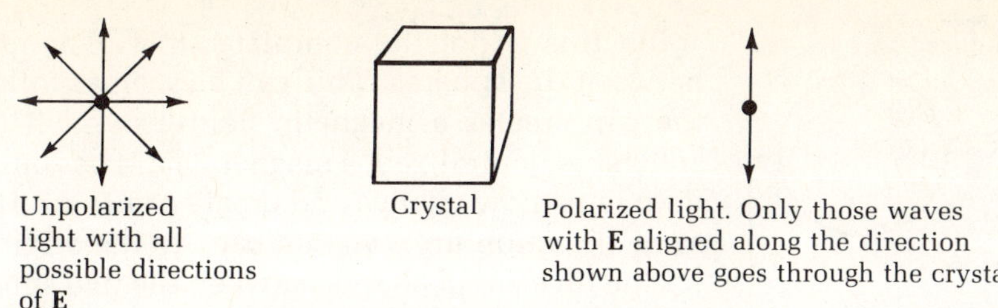

FIGURE 18.4   The transmitting window of Polaroid sunglasses is vertical. Therefore, the horizontally polarized reflected light cannot go through.

because light waves are transverse in nature. Polaroid, discovered by Land, uses the polarization principle in various commerical uses. In one of the gadgets, Polaroid sunglasses, the fact that reflected sunlight from a horizontal surface is polarized mainly in the horizontal plane is made use of (Fig. 18.4). The transmission direction of the window of Polaroid sunglasses is vertical, which does not allow transmission of any of the horizontally polarized reflected light, or "glare."

■ 18.2 Electromagnetic Waves

No message since the Biblical one, "Let there be light," has had more impact on human communication than that of Maxwell's theory, which can be summarized by the statement, "Let there be electromagnetic waves." It took scientists a while to capitalize on Maxwell's predictions, but in 1887 German physicist Heinrich Hertz made the breakthrough. Hertz produced and detected electromagnetic waves in the radio frequency region and started the modern era of electromagnetic wave communication.

Hertz connected two polished metal spheres, separated by a small air gap, to a specially designed transformer, also called an induction coil (Fig. 18.5). Since the transformer builds up a very large voltage in its secondary, a very large voltage between the spheres results. When the voltage difference is sufficiently large, sparks fly from one to the other, carrying charge. Momentarily the air between the two conductors has become ionized and a conductor of a current; air acts this way under a high voltage. The flow of the current tends to neutralize the charge separation, decreasing the electric field in the gap.

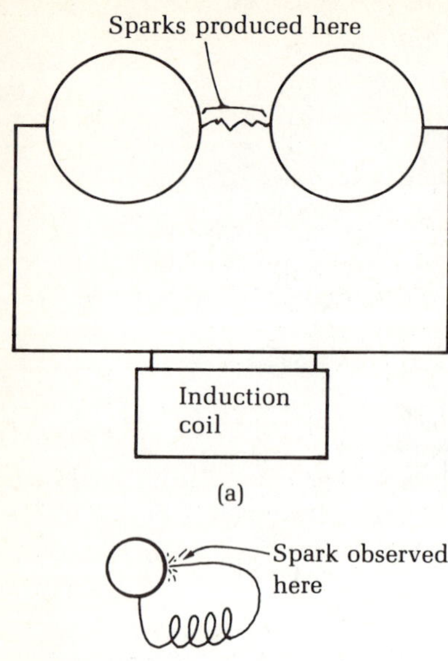

FIGURE 18.5   Hertz's transmitter and detector. (a) Transmitter. (b) Detector. Note the spark gap in the detector wire.

FIGURE 18.6   (a) The two spheres are charged with opposite charges by the induction coil. (b) The air in the gap has become conducting and an electric current has started, tending to neutralize the spheres. (c) The flow of charge, which is a current, gives rise to a magnetic field. (d) The changing magnetic field leads to a continuation of the current in the same direction and (e) leads to the reverse of the original situation, with charges reversed between the spheres. Sparks now flow again in the reverse direction. Charges thus flow back and forth until equilibrium is established.

But this is not the complete story. The charges also move back and forth between the spheres. This can be seen as follows. The current flowing through the gap creates a magnetic field around it (Fig. 18.6). As the charge in the spheres is neutralized, a magnetic field has built up [Fig. 18.6 (c)]. This magnetic field must now change and create a current. Hence the current continues in the same direction until we are back to the original situation [Fig. 18.6 (e)] except for the reversal of charge between the two spheres. At this time all the energy of the magnetic field has gone into creating the separation of charges, as shown in Fig. 18.6 (e). The cycle now repeats itself with the current in the opposite direction, and the charges oscillate back and forth between the two spheres until the air gap ceases to be conducting and equilibrium is restored. The potential difference then builds up once more between the spheres by the action of the induction coil and is followed by another oscillatory spark as described above; and so forth it continues.

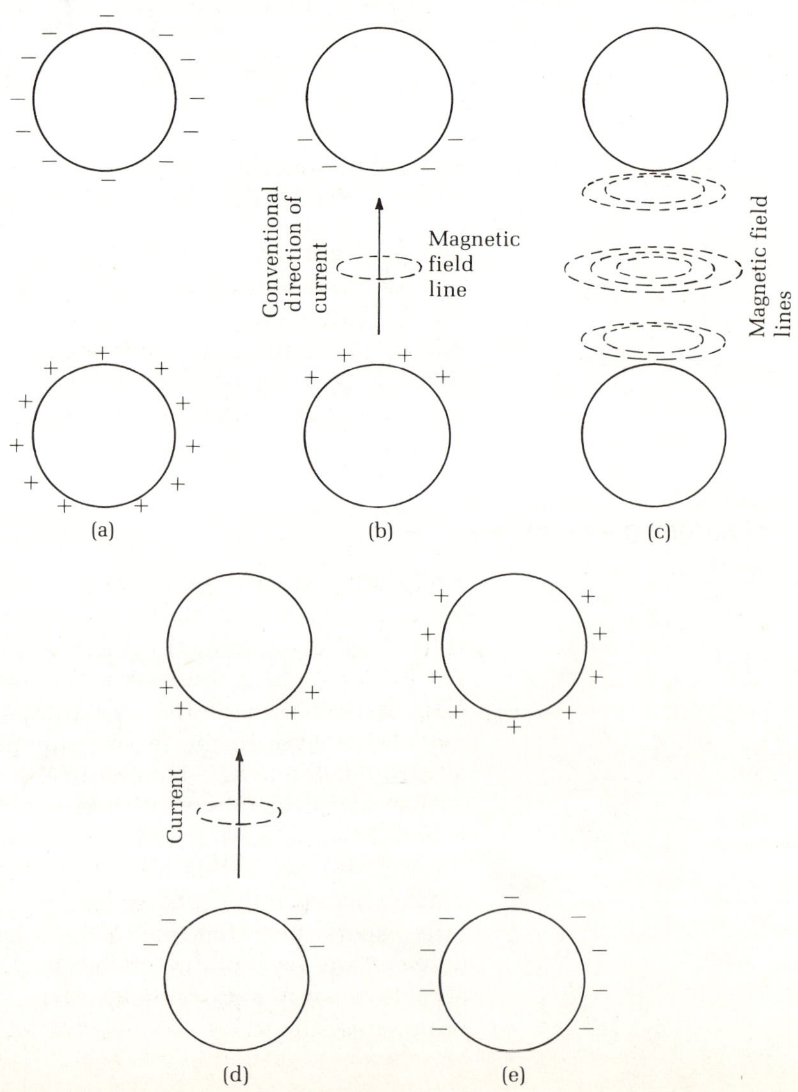

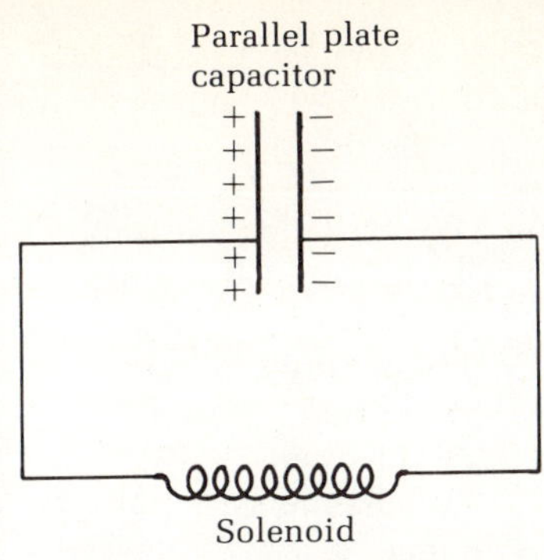

FIGURE 18.7   An oscillator circuit.

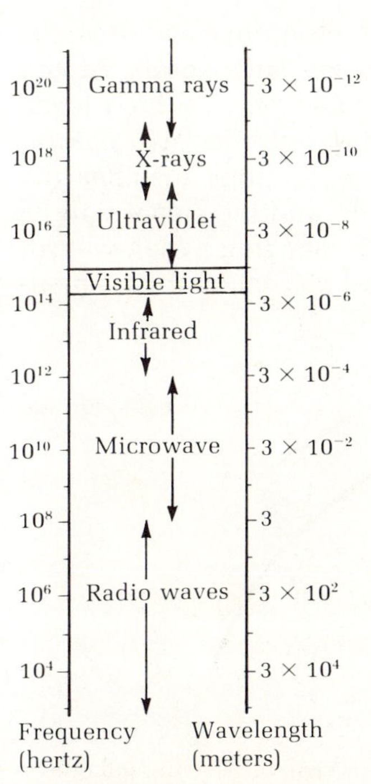

FIGURE 18.8   The spectrum of electromagnetic waves.

Hertz perceived that these oscillatory sparks are pretty much like the back-and-forth swing of a tuning fork. Just as the latter sends off sound waves into air, the oscillating electrons should send off propagating electromagnetic waves in the surrounding space. The changing electric and magnetic fields carried by the wave should be detectable at a distance.

Hertz had a little circuit [Fig. 18.5(b)], a bent wire with a short air gap between its end, set up at a distance from his transformer and the spheres. Indeed, when a spark jumped across the spheres, there was a spark also across the two ends of the wire. The wire thus acted as a receiver of the electromagnetic waves.

A modern **oscillator** circuit, which produces electromagnetic waves in the radio, television, and microwave frequency regions (10^6 to 10^{10} hertz), operates pretty much the same way as Hertz's oscillator (Fig. 18.7). The two spheres are replaced by a capacitor, but its function is the same. In order to have a better magnetic field, we connect the plates of the capacitor with a coil of wire, or solenoid. In addition, a vacuum tube (not shown) is used, whose function is to give steady oscillations. As you know, a tuning fork will stop vibrating after awhile when its energy is dissipated—unless, of course, you keep on hitting it. The vacuum tube carries out this function.

The spectrum of electromagnetic waves is shown in Fig. 18.8. Radiation higher in frequency than the microwaves is called infrared (actually the two overlap somewhat) and is produced by the vibration of charges inside molecules. This is a lucky coincidence, since it is very difficult to make oscillator circuits that generate waves as short as the infrared. Visible light and ultraviolet radiation are due to the oscillation of electrons inside the atoms. When energetic electrons are suddenly stopped by a metal, the result is their deceleration, which also leads to radiation in the form of highly penetrating X rays. Finally, the highest frequency electromagnetic radiation is called the gamma ray. Gamma rays are generated by the jiggling of charges inside atomic nuclei, and they are produced also when high-energy particles (as in cosmic rays) suddenly encounter a decelerating blow from a target.

Electromagnetic radiation is due to the acceleration of charges. According to Maxwell's theory, whenever such acceleration occurs (oscillation and deceleration are just special cases of accelerated motion), there is radiation of electromagnetic waves.

The spectrum of radio waves

Electromagnetic waves in the frequency range of 10^4 to 10^{10} hertz are used in one form or another in wireless communication. Table 18.1 summarizes these. The longest wavelengths, because of diffraction, more or less follow the ground as they travel, and thus they can be received over distances as great as a thousand miles. But the higher-frequency waves tend to travel in straight lines, as diffraction effects decrease, and the only reception is along the so-called lines of sight (if your receiver is on the line of sight of the transmitter, you will receive the signal; otherwise you will not).

Fortunately, there is one favorable condition that occurs for the part of the electromagnetic spectrum used by AM radio. These frequencies are reflected by a layer of the upper atmosphere called the ionosphere and therefore can be received at the longest distance (Fig. 18.9). That's why you receive so many out-

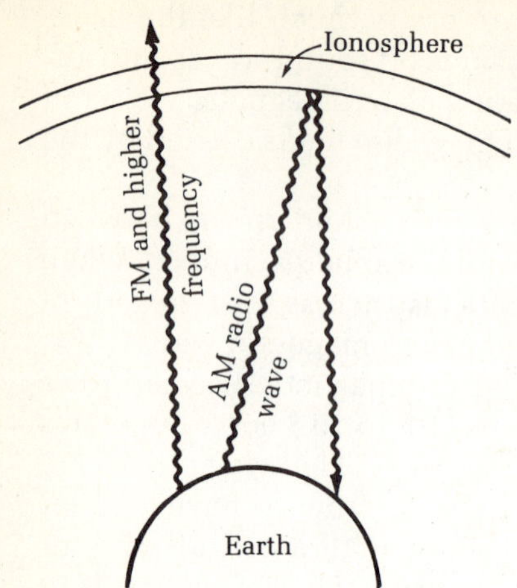

FIGURE 18.9 The reflection of AM radio waves at the ionospheric layer.

TABLE 18.1 Table of radio frequencies.

Name	Frequency	Wavelength (meters)	Use
Very low frequency	10–30 kHz (kilohertz)	30,000–10,000	Long-distance telegraphy
Low frequency	30–300 kHz	10,000–1,000	Navigation, CB radio
Medium frequency	300–3000 kHz	1,000–100	AM radio
High frequency	3–30 MHz (megahertz)	100–10	Broadcasting
Very high frequency	30–300 MHz	10–1	FM radio, TV (channels 1–13)
Ultrahigh frequency	300–3,000 MHz	1–0.1	TV (channels 14–82)
Superhigh frequency	3,000–30,000 MHz	0.1–0.01	Radar

of-town stations on your radio in the evening. All these waves come to you after being reflected by the ionosphere.*

TV and FM radio use waves of higher frequencies than AM radio, and the ionospheric reflection is no longer available to these waves; they go straight through the ionosphere. So their transmission is strictly "line of sight." This is the reason we have TV networks of stations to relay the messages (Fig. 18.10). Alternatively, we can use the reflection from a communication satellite to transmit a television message. Another alternative is to use the cable.

Radar (*radio detection and ranging*), which, among other applications, is used to control air traffic, uses the microwave end of the spectrum.

The radio

The electromagnetic waves generated by an oscillator circuit drive a transmitting antenna, which radiates the electromagnetic waves into space. At the receiving end, these waves are intercepted by another antenna, which feeds them into a receiving set. The receiving set has a circuit with its own natural oscillations. If the frequency of one of its natural oscillations matches the frequency of the incoming signals, there is a resonance, and the signal will be received strongly. Usually the receiver has a tuner: by changing a dial, we can vary its natural frequency and thus "tune" it to receive any incoming frequencies in the right range.

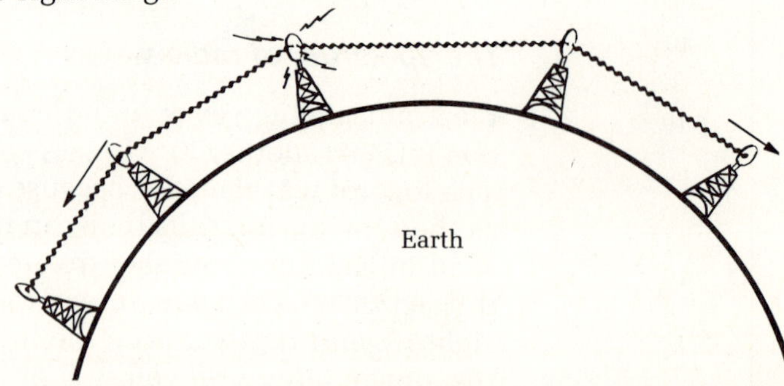

FIGURE 18.10 A network of TV relay stations.

*The ionospheric layer that reflects AM waves is more stable and dense in the evening and reflects better.

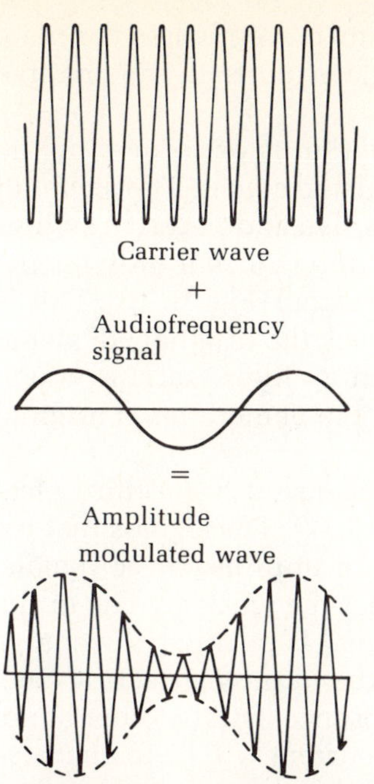

FIGURE 18.11 Modulation. High-frequency carrier wave (top) is modulated by the audio frequency wave to produce the modulated signal (bottom). In AM shown here, the amplitude is modulated.

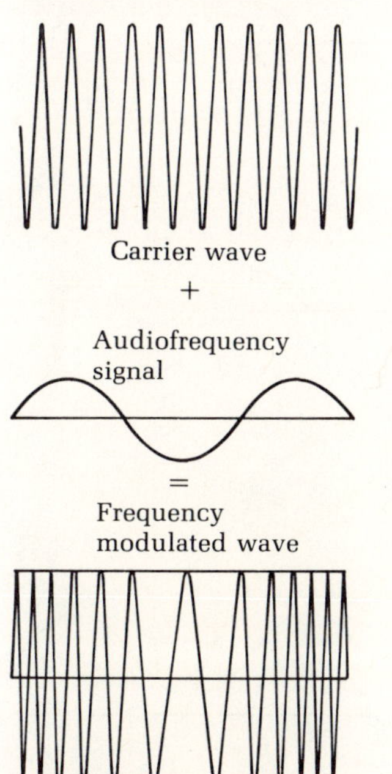

FIGURE 18.12 Frequency modulation (FM).

The waves created by the oscillator circuits are just repetitions of the same thing over and over again; there is no message in it. For a message there must be variations. The simplest kind of message we can think of is made by making and breaking the circuit for short and somewhat longer signals. If we call the short ones dots and the long ones dashes, we have a Morse code. In Morse code every letter of the alphabet is coded in a combination of dots and dashes, and with just these two signals, entire messages can be sent.

Morse code works fine for written messages, but how do we send things like voices or music? The answer is **modulation** (the "M" of AM and FM). The sound waves are first converted into electrical signals by a device called the microphone. These signals are then used to modify the radio signal created by the radio frequency (RF) oscillator before they are sent out into space. It is like giving a princess a magic carpet to ride on. Superimposed on the RF signal, the audio frequency (AF) signal can make it to the receiving end. At the receiving end, special devices take it off the shoulder of the carrier waves. Now this message, still in the electrical waveform, is fed into the "loudspeaker," which converts it back into the sound that it originally was.

Now we can understand the difference between AM and FM. In AM, which stands for **amplitude modulation,** the amplitude of the RF signal is modulated, the amplitude varies with time (or space) as shown in Fig. 18.11, but the frequency of the RF wave remains the same. In FM, or **frequency modulation,** it is the frequency of the RF wave that is modulated (Fig. 18.12).

So these are the things that must happen in order for good music to reach your ears from the radio. Of course, we have not told you in detail about the devices that carry out specific chores. With a few exceptions, these devices are too intricate to go into here. In connection with the physics of communication devices, certain equipment is too important to slight, however: electronic equipment.

Electronic devices

Electronic devices discussed here* are based on this one property of some metals: when heated, these metals emit electrons. This process is known as **thermionic emission** (Fig. 18.13). If we take a glass tube from which air has been evacuated, put a wire thread (the filament) in it, and heat the wire, the wire will emit electrons. Or it can heat another electrode, called the cathode, and drive electrons off that. (Such a tube is called a **vacuum tube.**) This produces a supply of excess electrons on the surface of the electrode. If there is a second electrode (called the plate or the anode) in the tube which is positively charged, then the electrons can flow through the vacuum to the anode, making a current. This arrangement is called a **diode** (Fig. 18.14).

What does the diode do? The usefulness of the diode comes from the fact that the anode attracts the electrons from the cathode only when it is positively charged. If the anode plate is negatively charged, no electrons can be attracted (on the contrary, the electrons will be repelled), and no current will flow. The plate of the diode, if connected to an ac voltage, will be alternately positively

---

*Solid-state devices are discussed in Chapter 22.

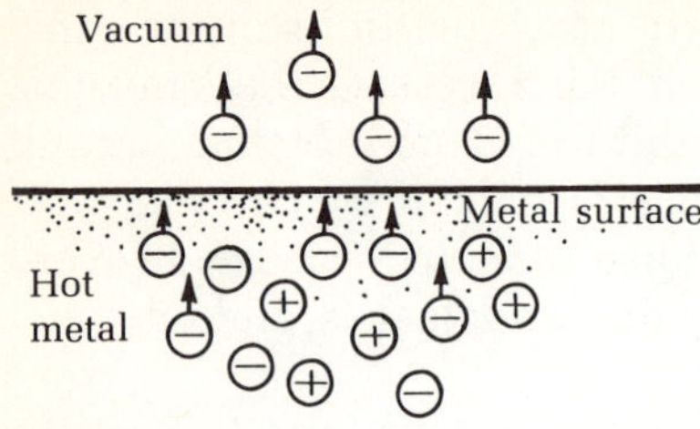

FIGURE 18.13   Thermionic emission.

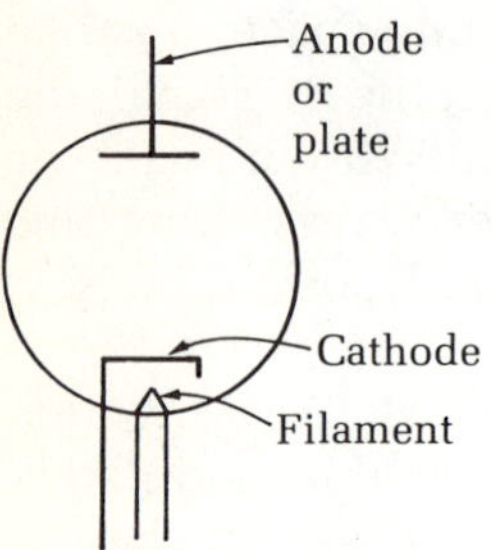

FIGURE 18.14   A schematic diagram of the diode.

and negatively charged. But the diode does not pass any current when the plate is negatively charged. So the diode just keeps half the ac current, the positive half, and thus converts it to dc (Fig. 18.15).

One of the most important functions in radio reception is performed by the diode. The diode can demodulate an incoming modulated signal. The signal of Fig. 18.11 enters the receiving antenna with both positive and negative swings of voltage. The diode in the receiving circuit will admit a current only for the positive swings, cutting off the negative portion of the signal (Fig. 18.16). This is the half wave signal. Of course, we have not gotten back the original AF signal yet, but in one final step we pass the signal through a "filter" circuit which removes all those rapid ups and downs and gives us the demodulated original message.

Another type of electron tube, called the **triode** because it has a third electrode, also performs a very important function (Fig. 18.17). The signals that we deal with in communication are often weak. The triode amplifies these signals, makes them strong. This is how. Suppose we connect the weak signals to the middle electrode, or grid, of the triode. When the signal is positive, the charge on the grid is positive. This promotes the current in the tube by attracting electrons even more copiously than would occur in its absence. On the other hand, if the charge on the grid is negative, the effect on the current will be to decrease it. The fluctuations of the current thus induced are proportional to the signal but much magnified. In effect, we now have a current that follows the signal except that it is stronger; it has been amplified.

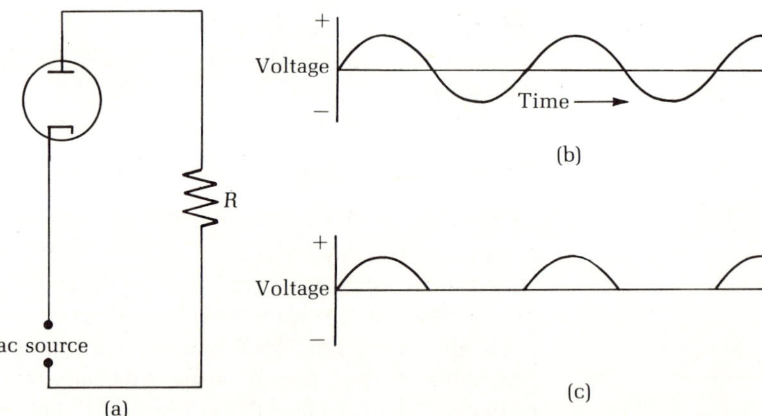

FIGURE 18.15   (a) The rectifier circuit. (b) The input ac voltage. (c) The output voltage. It is rectified, that is, all positive.

FIGURE 18.16   Demodulation of a modulated signal. See text.

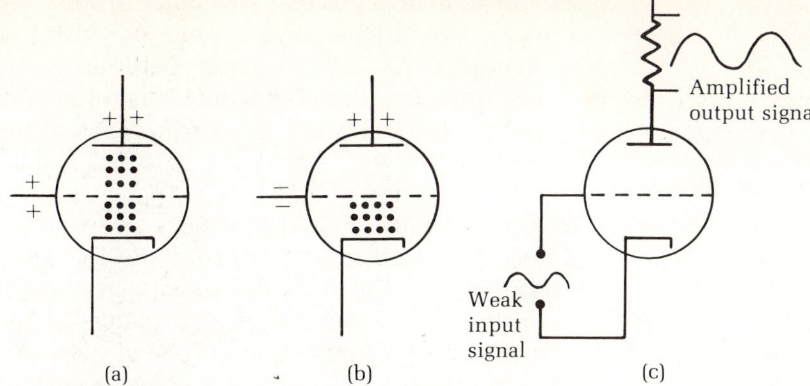

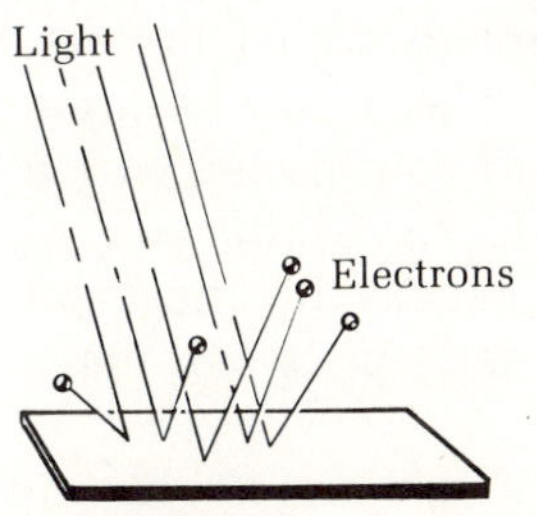

FIGURE 18.18 Photoelectricity. Light incident on a metal can cause ejection of electrons.

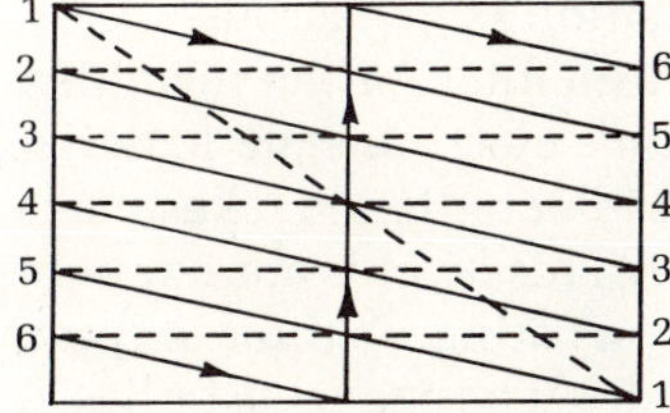

FIGURE 18.19 The sweeping of the signal plate by an electron beam.

The television

The production and reception of the audio portion of a TV message is exactly the same as in FM radio. We will discuss the production and reception of the video portion.

When we see an object, what we really see is a pattern of light; the bright spots and the dark and shaded areas together make up the pattern of light. The television camera changes this pattern of light into a pulsating current of electrons; these electrical impulses modulate carrier electromagnetic waves of about 10^8 hertz through the use of frequency modulation. These modulated waves are then sent through the air to be received on our TV sets. The function of the receiving set is to demodulate the signal and to change the electrical signals back into the original pattern of light that they represent.

In the simplest of television cameras, the image of the subject is focused on a signal plate, which has a mosaic of photosensitive dots on its surface. Unlike regular camera film, in which the photosensitive material reacts chemically under the light, here the material of the mosaic is a metallic compound that gives off electrons when light falls on it. The phenomenon of emission of electrons from a body when light shines on it is called **photoelectricity** (Fig. 18.18). The number of electrons emitted depends on the strength of the light: the stronger the light, the greater is the number of electrons emitted.

When light falls on this mosaic, it knocks off a varying number of electrons from each droplet, depending on the intensity of the light that falls on a particular droplet. So different parts of the mosaic will be charged positively by different amounts. This is very similar to how a photographic film works. Different parts of the film are affected differently by the incoming light, except in this case, instead of a regular image, we have an electrical image. Thus we have painted a picture of positive charges on the mosaic.

The entire electrical image on the plate cannot be transmitted all at once. It has to be transmitted bit by bit after being broken up into hundreds of thousands of picture elements. An electron beam, forming a scanning spot on the signal plate, sweeps over it rapidly line by line, a complete sweep taking only $\frac{1}{30}$ of a second, (Fig. 18.19). Each of the droplets on the mosaic thus discharges thirty times a second, giving an electric current whose strength corresponds exactly to the illumination of that particular droplet at that instant. The left-to-right and up-and-down sweeping is done by means of electric fields, as described in Chapter 15. Actually, magnetic fields also can bend the path of electrons and are most often used in television transmitters and receivers.

At the receiving end, the screen of the cathode ray tube, or the picture tube, is also scanned by a beam of electrons. The screen is fluorescent; when hit by electrons, every spot on its glows in direct proportion to the strength of the electron beam. Thus if this beam is designed to work synchronously with the electrical impulses coming from the TV camera (after demodulation), the original picture painted on the signal plate will be reconstructed.

Although the picture is also reconstructed bit by bit, our eyes see it as a continuous picture because of the property of **persistence of vision.** (Our visual images persist for about $\frac{1}{10}$ of a second). Motion pictures "fool" us the same way.

The screen of the color TV receiver is more complicated. It consists of an orderly array of tiny spots, each containing three phosphors of red, green, and blue (it imitates the color perception of the cone cells of the human eye). When hit by the electrons, there is a selective absorption, depending on the color of the video signal, which reproduces the original color of the subject.

More than likely you don't think of such things as electromagnetic waves when you watch your favorite TV program. But imagine the following scenario. You move into a new house, turn on the TV, and find that the picture has ghosts (a shadow beside each image on the screen). Where is the shadow coming from, you wonder with irritation after an unsuccessful attempt to adjust the set. Maybe it is time now to think of electromagnetic waves. There may be a big building or some large structure between the TV station and your home, which is giving you an added reflected wave signal for every direct signal. Because of its longer path, the reflected signal arrives a little late and causes the ghosts on your set. Realizing this will not solve your problem, but perhaps it will make you feel less inept.

Ultraviolet radiation from the sun and the earth's ozone layer

Ultraviolet radiation, the component of sunlight that gives us sunburn, can initiate biologically harmful photochemical reactions. An intense dosage can produce serious burns, can lead eventually to skin cancer, and can cause such genetic effects as mutation. In short, ultraviolet radiation is dangerous.

Fortunately, the earth's upper atmosphere, acting as a filter for the incoming sunlight, removes most of this harmful part of the sun's electromagnetic radiation. Most of the atmospheric absorption of the ultraviolet radiation takes place in a layer of the atmosphere known as the ozone layer. This layer is abundant in ozone, a heavy molecule (symbol O_3) of oxygen consisting of three oxygen atoms (instead of two as in the normal oxygen molecule). The ozone layer lies at a height of 20 to 30 km above the earth's surface.

Both the formation and dissociation of ozone continuously take place in the ozone layer, and both occur through chemical reactions initiated by the energy of ultraviolet light. Symbolically, we can write the first of these important chemical reactions as

$$O_2 + \text{ultraviolet radiation} \rightarrow O + O$$

indicating that the oxygen molecule breaks up into its constituent atoms under the influence of the energy of the ultraviolet light. Oxygen in the atomic state is highly reactive and immediately combines with an oxygen molecule to form an ozone molecule:

$$O + O_2 \rightarrow O_3 + \text{infrared radiation}$$

In this chemical reaction, since a more bound system (O_3) is being formed from less bound objects, energy is released. However, the emitted radiation is in the form of a harmless infrared.

Once the ozone is formed, it absorbs ultraviolet radiation, breaking up into an oxygen molecule and an oxygen atom in the process:

$$O_3 + \text{ultraviolet energy} \rightarrow O_2 + O$$

Thus ozone is continuously formed and destroyed by the sun's ultraviolet radiation. As a result of these reactions in the earth's atmosphere, the concentration of ozone in the ozone layer has reached a certain equilibrium value.

Any chemical pollution that reacts with ozone will affect this equilibrium of the ozone layer and thus reduce or destroy its effectiveness in removing the ultraviolet rays from sunlight. Recently much controversy has focused on two such pollutants. The prime offenders in this controversy are the fluorocarbons used as propellants in aerosol cans. The fluorocarbons, especially Freon, are (chemically) perfectly harmless at the lower levels of the atmosphere. But when they migrate to the ozone layer, they readily dissociate with the help of ultraviolet energy, giving up a chlorine atom. It is this chlorine atom that chemically reacts with and destroys the ozone.

The controversy centers around whether the effect is big enough to cause us concern. Regardless of the answer to this question, why should we take any unnecessary risk with something as vital as the ozone layer?

The other pollutant that can deplete the ozone layer is nitric oxide (formula: NO), a compound of nitrogen and oxygen that results from the high-temperature combustion in the automobile and other internal combustion engines. The automobile's production of nitrogen oxide does not make it to the ozonosphere and, therefore, does not concern us. However, supersonic transports (SST) fly close to the ozone layer. Nitric oxide pollution from the SST can be a serious threat to the ozone layer in the future if it becomes the common carrier in air transport.

X rays

When a new kind of radiation was discovered in 1895 by German physicist Wilhelm Röntgen, the nature of this radiation was completely unknown, therefore it was called X rays. Soon it was established that X rays are high-frequency electromagnetic radiation. They are highly penetrating and can pass through low-density biological tissue, such as flesh, but not so easily through hardened tissue like bone or a tumor. This is what makes them useful in diagnostics; we can take a photograph of otherwise inaccessible portions of the body with the help of X rays.

On the other hand, X rays are even more dangerous than ultraviolet rays in causing harmful biological reactions. Because of this, reckless use of X rays in medicine has been increasingly questioned recently.

Gamma rays

Gamma rays are the most penetrating portion of the entire electromagnetic spectrum. They occur as a component of the radiation of many radioactive substances, both natural and manufactured. They are the result of nuclear transformations, or decay, a subject that will be discussed in full later.

■ 18.3   The Reality of the Electromagnetic Field

Are electromagnetic fields real? Certainly if their disturbances are propagated in space, they have to be regarded as something real. There are some additional convincing facts pointing to the reality of the fields.

Consider two charged particles q_1 and q_2 moving perpendicular to each other. Suppose, in the course of their motion, they get in a position as shown in Fig. 18.20, one in front of the other. Let's consider the forces on each of the two particles when they are in this position. First, there is the electrical force with which each pushes the other. But since they are moving charges, they can also exert magnetic forces: each moving charge creates a magnetic field, which the other particle experiences. But the given configuration is a very special one. Since the magnetic field created by a point charge is a circular one, there is no field in the line of its motion. So the particle q_2 has no magnetic force coming from the magnetic field of q_1, since the field of q_1 is zero at the position of q_2 the way we have arranged things. On the other hand, there is a magnetic force on q_1, coming from the field of q_2. The forces on each particle are also shown in the figure. The electrical forces are equal and opposite, as they must be in order to satisfy Newton's third law, but there is a magnetic force on one and none on the other. Is Newton's law being violated for the magnetic forces?

We previously stated that momentum conservation for a system is guaranteed in the absence of external forces, because the internal forces are all action-reaction pairs and therefore cancel out for the entire system. So if the internal forces do not cancel, as seems to be the case here, the momentum conservation does not hold: the system would acquire new momentum even in the absence of external forces.

Are we to give up Newton's third law and momentum conservation? No, that's not really necessary. The way to resolve our puzzle is to realize that the electromagnetic field is being ignored in our discussion so far. The field takes up an equal and opposite momentum, so that the momentum of the particles, together with that of the field, is conserved. Thus the field is as real as the particles.

Previously we have seen how the fields have an associated potential, a way of looking at things that gives reality to the concept of potential energy itself. Whenever a field is present, there is energy distributed in the entire space of the field. Thus we are dealing with something that carries energy and momentum

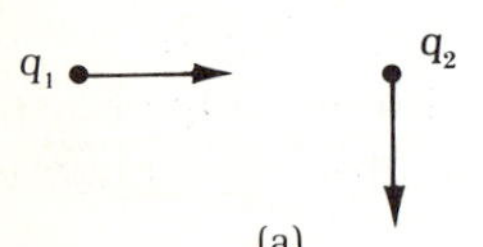

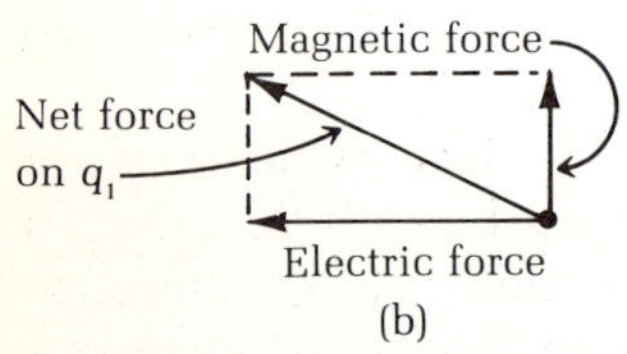

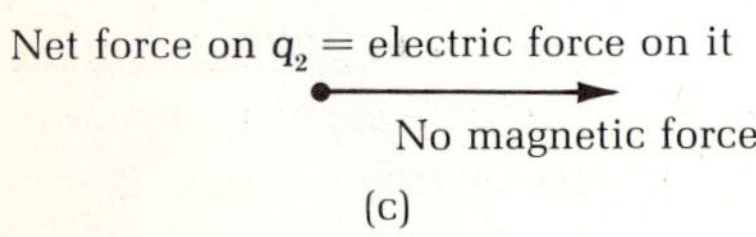

FIGURE 18.20   (a) What are the forces exerted by a moving charge q_1 on another moving charge in the position q_2? Does q_2 exert the same net force on q_1? (b) The net force on q_1: the sum of an electric and a magnetic force. (c) The net force on q_2: only an electric force acts on q_2. But shouldn't forces on interacting partners be equal? The riddle is solved when we consider the electromagnetic field.

and (we can show) also angular momentum, in other words something as real as material objects.

Maxwell and the idea of the ether

Now we'll talk about a final question. If the field is real, what is it made of? In a purely materialistic frame of reference, it is difficult to conceive of something as real as a field without making a mechanical model for it. Thus it is not surprising that the nineteenth-century physicists, including Clerk Maxwell, spent so much time and effort in building a mechanical model for the electromagnetic field.

Light waves are transverse in nature. So the material of the field, which was called "ether" (the name was borrowed from the ancient Greeks, who had an element named ether; the two ethers have not much to do with each other, though), had to have a very special character. For example, it could not be like air, since air cannot carry a transverse wave. One model represented it like a jelly. And this jelly had to pervade all space if it was to be the carrier of light waves that apparently travel over all space.

Is there any evidence for this ether? In the next chapter we will see that the ether has been discarded due to experimental results. In the process of that elimination of the ether theory, Einstein was led to the revolutionary theory of relativity.

Maxwell's toying with mechanical models for the field is an interesting process to note. Originally he made a rather complicated mechanical construct for the fields. He gave that up when he published his equations, realizing that a mechanical construct is unnecessary. But then a few years later, he made up the theory of ether. Clearly he was experiencing an internal struggle. He recognized that his equations were supreme; they alone are necessary to determine the fields. Yet he was constantly persuaded by the prevalent beliefs that said: everything real must be material.

SUMMARY AND OUTLOOK

In the last four chapters we have discussed the subjects of electricity and magnetism. You probably noticed that our coverage of the subject centered around two major objectives. The first objective was to familiarize you with the technological aspects of electricity—electric current, electric circuits, generators, transformers, radio, TV, and the like. The second objective was to familiarize you with the ideas of electric and magnetic fields, the ideas that have played a central role in the development of physics as a whole. This chapter, presenting Maxwell's equations, is the culmination of the field development of electricity and magnetism.

The novel insight that Maxwell brought to the picture of fields is their reality, that they exist independently of matter. The field exists even when there is no matter or circuits to test it. Maxwell's equations clearly demonstrated that when a charge oscillates, the oscillations generate changing electric and magnetic fields, which are then propagated in space in the form of an electromagnetic wave. Light is such a wave. And following Maxwell's ideas, we have produced in the laboratory electromagnetic waves that have revolutionized our commulication systems.

In Chapter 11 we introduced the wave nature of light. Now we find that light is a very different kind of wave; it is the wave of the electromagnetic field. The changing electric and magnetic field vectors are the disturbance coordinates of the waves. But it took a long time for physicists to accept this picture of the electromagnetic wave as a wave without a medium. The search for a medium for light waves, and the theorizing about such a medium (which was called ether), went on until Einstein in 1905 published a few papers that revolutionized physics and, permanently eliminated the idea of the ether as a medium for light.

Maxwell's equations are the highlight of nineteenth-century physics and one of the milestones of our physical understanding of the universe. It was almost the final step of the classical physicist's world view, a view that was now complete with its electromagnetic and gravitational fields, its equations to determine the fields, and Newton's laws of motion to determine the effect of the forces that the fields exerted on matter. There was only one nagging thing that worried the classical physicists: a mechanical explanation of the electromagnetic field was still lacking. Such a mechanical explanation was never to come, because it doesn't exist.

The final eight chapters of the book will deal with what we call modern physics. And we start modern physics with the work of Einstein, which demolished all hopes of the classical physicist for finding a mechanical construct, an ether, for the electromagnetic field.

QUESTIONS

Review and Reason

1. What is the difference between a displacement current and a conduction current?
2. Describe Maxwell's work, which combined electromagnetism and optics within one theoretical framework.
3. Are light waves transverse or longitudinal? How can you tell experimentally?
4. What is the difference between unpolarized and polarized light?.
5. In Fig. 18.1, what is the direction of polarization of the light wave?
6. If you are lying on your side on the grass (Fig. 18.21), Polaroid sunglasses do not protect you from the glare due to the reflection of sunlight from the ground. Explain.

7. How can you find out if a pair of sunglasses is made of polarizing material or just tinted glass?
8. Suppose you hold two Polariod sunglasses at right angles to each other and look through them. Will the world appear less dark or darker or the same? Explain.
9. Can we polarize sound waves? Why or why not?
10. Long before Heinrich Hertz came along, American physicist Joseph Henry used to say that sparks from a static electricity machine send something out in space. Do you agree? Why or why not?
11. Describe Hertz's experiment on the production and detection of electromagnetic waves.
12. An atomic bomb test in a Pacific island was televised

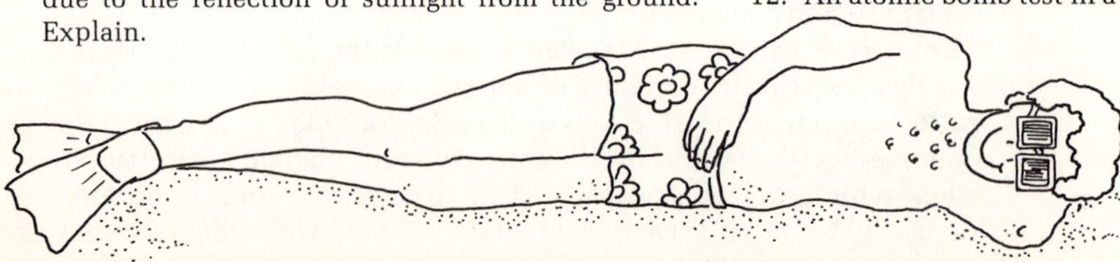

FIGURE 18.21

nationally in the United States. People on the mainland heard the explosion on TV even before the observers near the island did. Is this possible? Explain.

13. Arrange these in order of increasing wavelength:

    red light, microwaves, X rays, blue light, ultraviolet

14. Why does the Federal Communications Commission (FCC) require many AM radio stations to cut down their power during the evening?

15. Do you expect AM radio transmission to be any different on the moon than it is on the earth? Explain your answer.

16. Even in a locality where you don't receive TV waves very well, radio reception is usually quite good. Can you explain this?

17. What is meant by modulation and demodulation? Draw diagrams. What is the difference between amplitude modulation and frequency modulation?

18. An ac power source is connected to a diode. What kind of voltage do you get at the output? Draw a diagram.

19. Explain how a triode vacuum tube can amplify a signal.

20. Briefly describe the steps of a complete radio broadcast, starting with the sounds that are broadcast and ending with the sounds that reach your ears while listening to the broadcast.

21. What are thermionic and photoelectric effects? What is the difference between the two effects?

22. How is the photoelectric effect used to make an electrical image of an object?

23. What is meant by "persistence of vision"?

24. What is the principal function of the ozone layer of the atmosphere? Discuss briefly the dangers a few pollutants present to the earth's ozone layer.

25. Do you expect to be bombarded by more, less, or the same amount of ultraviolet radiation on the moon as on the earth? Explain your answer.

26. Why do X-ray pictures show only our skeletons and not our flesh?

27. What can you say for the reality of the electromagnetic field?

19

Einstein and Relativity

Old order changeth, yielding place to new.

ALFRED TENNYSON

■ 19.1 Ether or no Ether, That Is the Question

Waves are quite common in nature; who hasn't appreciated the romance of ocean waves? But could they exist without the water? Ridiculous? Or consider sound waves for a second; they, too, need a medium in which to travel. Indeed, we can block off sound by creating a vacuum in its path.

When we talk about light waves, however, we make no reference to a medium. We say that light waves are traveling disturbances of the electromagnetic field. We imply that they don't need a medium.

Are light waves truly so different? Are they really like the Cheshire cat, a smile without a body? As we mentioned in the last chapter, many people were bothered by this aspect of light waves at one time. They insisted that there should be a medium even for light waves; they called this medium ether. But experimentation usually provides the final solution. The most famous experiment designed to prove or disprove the existence of the ether was that of two Americans, a physicist named Michelson and a chemist named Morley.

Their experiment was simple in concept but quite difficult to perform, particularly when you consider the status of experimental technology during the 1880s. Perhaps it is significant that the two scientists worked in America, where the study of physics was rather young and unusually bold.

The idea behind their experiment is that if there is an ether medium, then earth's motion through it must cause a wind—like the wind you create while driving an automobile in the medium of air. Such a wind has about the same velocity as the car, so we can guess that the velocity of the ether wind created by the earth's motion should be about the same as that of the earth in its orbital motion around the sun, which is about 30 km/sec. If light waves are waves in the ether medium, then an ether wind will have a profound influence on the time of travel of light waves through a given distance. A careful study of any influence should enable us to detect the ether wind, if it exists.

What kind of influence should we be looking for? Compare the travel time of a swimmer going downstream and back a certain distance versus the time he takes going cross-stream the same distance (Fig. 19.1). Which time is longer? The upstream-downstream time is. (If you doubt the answer, take a swim yourself.) The reason is that the swimmer loses more time swimming upstream against the current than he can make up by getting a better time during the downstream swim. It is true that when he swims cross-stream he has to swim at a slight angle in order to compensate for the velocity of the stream and thus travel a slightly greater distance than in the up-down swim. Even so, the cross-stream time is smaller.

Thus if there is an ether wind, light waves will take a longer time to travel downwind and back than cross-wind. To prove that there is an ether wind, then, we need only detect a difference in the two travel times. A measurement of the difference in travel times will even give us the actual speed of the ether wind.

Sound simple? In practice, it isn't, not exactly. There is one big difference between the example of the swimmer in the stream and the passage of light through an ether wind. In the former case the speed of the swimmer and the speed of the stream are comparable, and so the difference of travel times is quite appreciable. But this is not true for the latter case. The velocity of light is 300,000 km/sec, which is some 10,000 times greater than the expected velocity of the ether wind of 30 km/sec. The difference of travel time is thus extremely small.

How can we measure such a minute difference in travel time between two light rays, one of which travels cross-wind and the other one upwind and downwind along the path of the ether wind? This is where Michelson's ingenuity came in. He split a light ray in two with the aid of a half-silvered mirror, sent one ray in one direction and back, and the other ray in the perpendicular direction and back again to the half-silvered mirror, where they then recombined (Fig. 19.2). Both rays traveled as nearly equal distances as was possible to

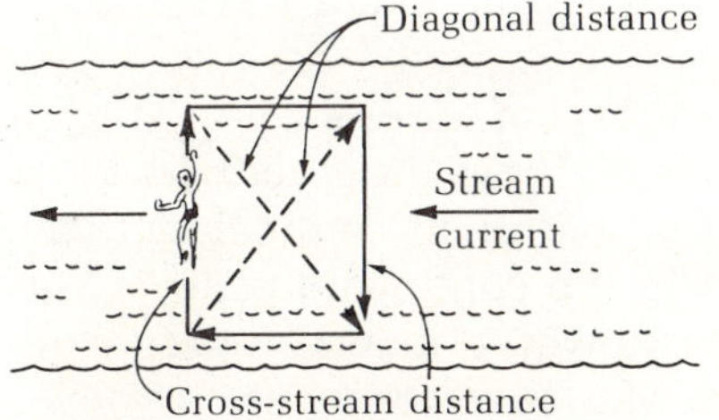

FIGURE 19.1   For a swimmer the upstream-downstream time is greater than the time it takes to swim cross-stream (as shown) over equal distances. See text.

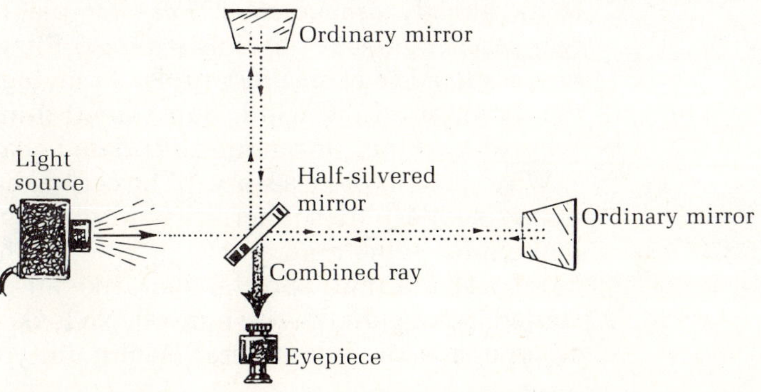

FIGURE 19.2   The light from the source is split and recombined by the mirror arrangement shown. The recombination produces an interference pattern at the eyepiece. If there were an ether wind, the interference pattern would shift. No such shift was observed, even when the apparatus was rotated.

attain experimentally. What happened?

Suppose we looked at the recombined light through an eyepiece. What would we see? From what we have discussed about light waves in Chapter 11, we would expect the two waves to interfere and form a pattern of alternate light and dark fringes. This pattern should shift if we rotated the apparatus, since rotation changes the direction of travel of the two rays with respect to the direction of the ether wind, which in its turn changes the travel time of the two waves. Then the relative phase of the two waves at the time of recombination should also change, altering the interference pattern.

This last point is crucial. Regardless of the direction of the ether wind, which we couldn't know with certainty, from one alignment of the apparatus to another rotated one, there should be a shift of the interference pattern because of a relative change in the travel time of the two waves in an ether wind. Thus we don't need to know the direction of the wind in order to find the evidence of it.

To their surprise Michelson and Morley did not observe any shift; the interference pattern never changed. Thus there is no ether wind. Could there have been an experimental error? Their apparatus was so accurate that even a velocity of 1 km/sec for the ether wind would have been detectable. No, the conclusion was inescapable: there was no experimental evidence of an ether wind.

What could explain the nonexistence of the ether wind? The proponents of the ether got busy explaining the null result of the Michelson-Morley experiment. This is the common human practice: when you start with a belief and put it to tests, a negative result seldom dissuades you from the basic belief. Chances are you end up with a modified belief, modified to accommodate the results of the tests.

This is exactly what happened in the continued saga of the ether. One suggestion was that there is no ether wind because the earth drags the ether with it. As an analogy consider a jet plane, which drags the air within its confinement. The motion of the plane does not cause any wind in the confined air. So, it was suggested, is the case with the ether. Other tests, however, managed to rule out this particular rationalization of the Michelson-Morley experiment.

The ether drag hypothesis had to attribute an unusual property to the ether, namely, that it stays glued to the earth as the earth moves, so that there is no relative motion between them. Other models assigned even stranger properties to the already strange ether. For example, in one of the models developed by two very intelligent physicists named Fitzgerald and Lorentz, the ether wind was assumed to exert pressure on a moving object so that the object shrunk in the direction of the wind. Subsequent thinking led to even further modifications of the ether. But none considered abandoning the ether altogether.

Why? Let's explore a story.* The central figure in the story is Mulla Nasrudin, a very subtle philosopher, as you will see. One day a man found Nasrudin looking for something on the ground. "What have you lost?" he asked. "My key," was the reply. So the man, like all good helpers, got on his knees and started looking for the key himself. No luck. After a while the man said, "Mulla, are you sure the key is here? Where did you drop it?" "In my house," Mulla

---

*Quoted in R. Ornstein, *The Psychology of Consciousness* (San Francisco: Freeman, 1972), p. 176.

FIGURE 19.3 Albert Einstein (1879–1955). What is he making faces at? (Albert Einstein by Photoworld.)

replied. "Then why are you looking here? You stupid fool!" the man snapped angrily. "Because there is more light here," answered Mulla complacently.

It is easier to look under the light. There was a lot of light associated with the ether idea, a lot of vested time and interest. And a potential challenger always felt welcome to look under this light, young scientists like Fitzgerald. They did not find the key, but they got a lot of pats on the back for their clever ideas of looking. And in the meantime everybody forgot about where the key was lost.

Where was the key lost? Remember that initially we were worrying about light; then gradually the worry shifted to ether. But light was the key. The man who finally figured this out was Einstein.

The speed of light stays the same

Einstein was born in 1879, about the time the Michelson-Morley experiment was performed. So he did not grow up with a lot of preconceived notions about ether and the like. More importantly, he started doing his research in physics away from all vested interests, in the isolation of a Swiss patent office. If Einstein had found a job at a university and worked under the umbrella of a believer in ether, who knows what would have happened?

When Einstein was a teenager he used to wonder about light, in a variation of the old catch-up game. Only in his case the object of the catch was a beam of light. Young Einstein would wonder, suppose he was on a rocket ship traveling with the speed of light. Would he not then catch up with light and be in a position to take a close look? What would he see? An electromagnetic field at rest oscillating in space? Somehow Einstein did not think so; his intuition told him otherwise. The reason is that even a rocket traveling at the enormous speed of c is an inertial reference frame, just like an earth-fixed one. Einstein knew that physical laws remain the same from one inertial reference frame to another, and this included Maxwell's equations of electromagnetism. But Maxwell's equations dictated that light travel with a velocity c in vacuum on earth, a fact that had been verified experimentally. The same equations must hold for any other inertial reference frame. So even if you look at light from a rocket traveling with velocity c, the velocity of light will still measure c.

The same reasoning applies to a light beam shot from a moving rocket. The light beam will not travel any faster. The velocity of light is always the same, irrespective of the velocity of your frame of reference, never more and never less.

Somehow light does not obey the usual kind of addition rules for velocity in everyday experiences. Like magic, $c - c$ isn't zero and $c + c$ isn't $2c$. In both cases we get c, according to Einstein. Very strange!

The situation is reminiscent of a story that, to make a point, will be told in a slightly modified version. The story is about an executive who needs a secretary. He goes to the personnel director for help, and they put an ad in the newspaper; three people come in for an interview. The first one is a blonde—pretty, of course. The personnel director does the questioning. "What's one and one?" he asks. "Why, sir, one and one is two, of course," comes the reply.

Next comes the second applicant, a practical-looking brunette. The same question is asked. "Well, it depends," she says thoughtfully. "One plus one is

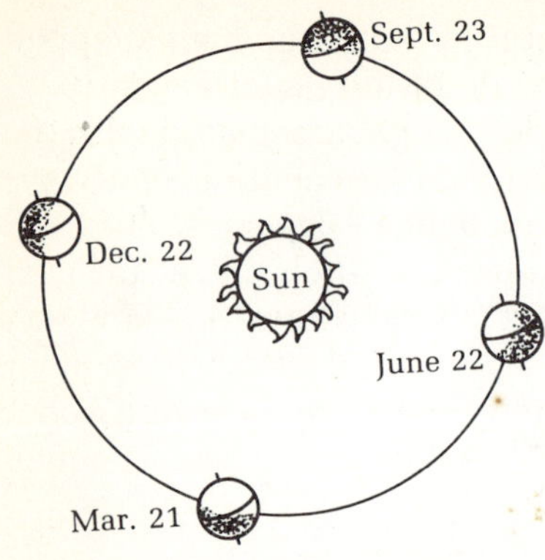

FIGURE 19.4 At different times of the year, earth represents different "inertial" reference frames in space.

two, one times one is one, and one to the power of one is also one. And you can also make eleven out of one and one in the decimal system." The interviewers are impressed.

The third applicant is Einstein. In response to the same question, Einstein answers, "Depends on what you are adding. I happen to think that for the velocity of light one and one just makes one." His dumbfounded interviewers exchange looks of utter disbelief.

Guess who gets the job? The blonde, of course. And not for her looks, but for her ability to give programmed answers.

Perhaps your first reaction to Einstein's reply would also be one of disbelief. It is very hard to give up believing in something you "know" in favor of something outside the realm of your experience!

The Michelson-Morley work provided the experimental support for Einstein's intuitive guess. If we forget the ether, what the Michelson-Morley experiment tells us is that the speed of light relative to the earth is always the same, no matter in what direction light travels compared with the direction of motion of earth in space. Specifically, consider the following fact. Michelson and Morley verified that their null result persists even when the experiment is repeated at different times of the year. But during the whole year the earth continuously changes its direction of travel in space as seen from the rest of the universe (Fig. 19.4). Thus an observer on earth is at different times in a different inertial frame. (If it occurs to you that earth is not an inertial frame at all, it should be pointed out that during the time it takes to perform the Michelson-Morley experiment, the direction of motion of the earth and hence its velocity does not change appreciably. So during such a short period, we can certainly regard our earth-fixed lab as an inertial frame.) An obvious generalization is this: the speed of light remains the same with respect to all inertial reference frames; so said Einstein.

It turns out that the rules of communication between two different inertial reference frames regarding observation made from each frame have to be reformulated carefully. The change in the addition rule for velocity is only one particular example. Many more surprises will come.

Before we go into some of these surprises, here is a particularly appropriate excerpt from Lewis Carroll's *Alice Through the Looking Glass*.

> Alice laughed: "There is no use trying," she said. "One cannot believe in impossible things."
>
> "I dare say you haven't had much practice," said the queen. "When I was younger, I always did it for half an hour a day. Why sometimes I have believed in as many as six impossible things before breakfast."

You may have to do just that.

■ 19.2 The Relativity of Time

We are now ready to plunge into the wonderful and adventurous world of relativity. Following Einstein, we start with the following premises: (1) *physical laws are the same from all inertial frames of reference*: this is the *principle of relativity*; (2) *speed of light remains the same irrespective of the velocities of the reference frames of observation*. We then examine the consequences of these premises on our experiences with time.

After understanding the constancy of the speed of light, Einstein found he could now attempt answers to some of his childhood questions about the nature of time. His answers led to the theory of relativity. Before we go into the relativity of time, however, I want to dwell on this intriguing question: What makes an Einstein?

We have previously dealt in some detail with the works of such giants as Newton and Maxwell but we have never asked this particular question. Why is Einstein so special?

Einstein is special because the questions about time (and later space) that he sought to answer are unusual in this one unprecedented respect: there was no pressing need to ask them. There was no glaring discrepancy in our understanding of time; there were no experimental data to be explained. Nobody else, even philosophically, was pondering time; no significant questions had been asked about time for more than two hundered years, since Newton in his *Principia* proclaimed the absolute nature of time (with perhaps one exception: Mach, who certainly had an encouraging influence on Einstein). In the case of Newton and Maxwell, we were curious about how, in their perception of their own gestalts, they found answers that consolidated the work of their contemporaries. In the case of Einstein, we want to know "How come" he thought of such questions.

Let me be systematic and discuss a few methods that are available to us when we search for knowledge. As it turns out, we need to discuss here only three methods. At the bottom of the ladder is the method employed by most people (including many scientists), which goes through the following two basic steps:

test

⇅

belief

The upward arrow indicates the progression "belief→ test" and the downward arrow indicates the regression back to belief based on the tests. The results of the tests usually make no difference; if you start with a belief, you will end up with one.

Let me give you a real life example of how this works. Suppose you have a belief that you don't understand physics. Then you come to the university and get an opportunity to put this belief to a test: you take a physics course. In my experience you will do your best to prove you were right in the first place (that you have no talent in physics) and usually you will succeed, too. Once in a while, however, you may come across a course that you really like, in spite of its being physics, and you are successful. But what is your reaction toward your suc-

cess? The course was too easy. You did well not because you have an ability to understand physics but because the course was too watered down. Thus you will maintain your belief that you have no talent in physics.

We saw in the ether story that many scientists are equally vulnerable to this pitfall. However, a few geniuses like Newton and Maxwell rose above this pitfall and followed a different path.

Newton and Maxwell were guided by observations, and by experimental laws established by their predecessors, while they were making their most important discoveries. For example, Newton had the experimental laws of Galileo and Kepler to ponder, while Maxwell's thoughts must have been dominated by the experimental results of Ampère and Faraday. Both men bestowed a fresh point of view on the topic and were able to put the evidence all together in sudden outbursts of sheer brilliance. Once they realized the truth, they communicated it to the world with the uncanny assurance that comes only with total confidence in what they knew. And subsequently, the certainty of their knowledge grew. Newton's refusal to make up mechanical models for understanding the force of gravity ("Hypothesis non fingo," he used to say. "I feign no hypothesis.") testifies to his complete certainty that his theory of gravitation was correct, as is.

Let's schematically indicate the steps that lead Newton to his discovery:

gestalt, or realization

↑

observation

Notice that in this case there is only an upward arrow. Once realization is achieved, the enlightened know that they know, and never does the certainty of that knowledge decay. This is what makes this technique radically different from the previous one. If you start with a belief, knowledge can grow only for a short time; eventually the seeker ends up with belief again, indicating a decay of certainty.

But if there are no observations from which to build your gestalt, then what? Fortunately there is a third method that has been used numerous times in our search for truth. This is how this method works:

intuition, or natural knowing

↑

don't know

In this method you begin with the assertion: I don't know. Your mind is one of no inherent beliefs. If you start with this "don't know" mind, the strangest thing happens: knowledge enters the consciousness from nowhere, a process very familiar to people as intuition. I don't think I can give you any logical explanation or model about why this happens. The important thing again is that once knowledge is obtained this way, then you know. You have experienced the truth. Your knowledge is forever—it can only grow.

To illustrate, let me quote the following description of an encounter of Ilse Rosenthal-Schneider with Einstein. The incident took place right after Einstein's theory of curved space had been confirmed experimentally.

> Once when I was with Einstein in order to read with him a work that contained many objections against his theory . . . he suddenly interrupted the discussion of the book, reached for a telegram that was lying on the windowsill, and handed it to me with the words, "here, this will perhaps interest you." It was Eddington's cable with the results of measurement of the eclipse expedition (1919). When I was giving expression to my joy that the results coincided with his calculations, he said quite unmoved, "But I knew that the theory was correct"; and when I asked, what if there had been no confirmation of his prediction, he countered: "Then I would have been sorry for the dear Lord—the theory is correct."*

Such is the conviction of natural knowing. You can stand up to God and say, "My theory is correct."

I now can tell you more about why I think Einstein is so special. It is not at all because his discoveries were any greater than Newton's or Maxwell's. They weren't, as Einstein himself acknowledged many times. He is special because he could maintain a "don't know" state on space and time, concepts that almost everybody encounters very early in childhood, quickly picking up beliefs about them. How did Einstein miss picking up the commonplace beliefs about space and time? He himself gives some hint:

> I sometimes ask myself, how did it come that I was the one to develop the theory of relativity? The reason I think is that a normal adult never stops to think about problems of space and time. These are things which he has thought of as a child. But my intellectual development was retarded, as a result of which I began to wonder about space and time only when I had already grown up.†

*Quoted in G. Holten, *Thematic Origins of Scientific Thought* (Cambridge, Mass.: Harvard University Press, 1973).
†Quoted in Ronald Clark, *Einstein: The Life and Times* (New York: Avon Books, 1971), p. 27.

The second hint comes from the almost universal agreement among people who knew him in his childhood that he was a slow learner. Einstein's father once asked the school principal what profession he would recommend for Einstein. The principal's reply was, "It doesn't matter. He'll never make a success of anything."

Talking about slow learners, I have a story to tell that is rather revealing in this connection. This is the story about a crown prince named Yudhisthira, one of the main characters in the Indian epic "Mahabharata."

Yudhisthira, his brothers, and his cousins were all being tested by their Guru (teacher), who wanted to find out how much each had learned. "So far the results are encouraging," the Guru was saying to himself, as he approached the crown prince. He looked at the prince, and his confident face assured him that here was a boy who must have learned a lot.

When asked, Yudhisthira said, "I have learned the alphabet and also the first sentence."

"Is that all?" the Guru was disappointed.

"The second sentence, too, maybe," the boy said.

The calm self-assurance with which the prince expressed himself angered the Guru. "Not only is he a slow learner, he does not even feel guilty about his inadequacy in learning," the Guru thought solemnly. Obviously he must be straightened out before it is too late. A firm believer in the old wisdom "spare the rod and spoil the child," he gave the prince a sound thrashing. But the boy was taking his punishment very calmly and looked as confident and pleasant as ever.

The teacher dropped his cane. Something was peculiar. The boy was neither angry nor resentful, in spite of the humiliation he had suffered. The Guru wanted to understand what was going on. Accidentally, he happened to look at the open page of the primer in front of the prince. The first sentence: "Never be angry." Then he looked at the second one. It read, "Always tell the truth."

The teacher looked at the first sentence and looked at the peaceful face of his student. Suddenly, he understood. The boy had really learned this first sentence. He had experienced the meaning of the sentence and learned to practice it. Now he felt very proud to be the Guru of this young man who obviously knew the true meaning of learning. India will be in good hands when this prince becomes king, he thought.

He himself could not help feeling self-pity, though. Here was a six-year-old who was already on the right track about learning, and he, the Guru, at the very prime of his career, honestly could not claim that he had really learned a single piece of knowledge. Sadly he expressed his feelings to the prince as he congratulated him.

"Teacher, teacher, please," the boy said, "please do not congratulate me yet. I must confess that I have not learned the first sentence thoroughly enough. You see, when you started beating me, at first I did feel anger."

It was clear that the boy was telling the truth. Obviously he didn't want to take credit under even the slightest of false pretenses. And something else—the prince had also successfully learned the second sentence, his way.

What's your own experience? Most people would look at a sentence like "Always tell the truth," compare it with what they may have heard from parents or other respected authorities, quickly make it a part of their belief systems and move on to something else (perhaps for the rest of their lives they would suffer guilt every time they told a lie!). It is only a few who learn experientially. But it is slow, learning that way.

Einstein was a slow learner because he too, like Yudhisthira, insisted on learning by experiencing. Unfortunately, much of the education that one gets in schools is too structured to suit this way of learning. Thus Einstein excelled only in physics, mathematics, and music, all three experiential subjects, and usually did miserably in the more structured fields of study that require more memorization than experiencing. Later on, he once said, "I want to know how God created this world. I am not interested in this or that phenomenon, in the spectrum of this or that element. I want to know His thoughts, the rest are details."

A final question. How does one learn experientially things that have not yet been discovered? Answer: Imagination. Einstein took recourse in a method called "gedanken experiment," imagined or "thought" experiment, to make his discoveries. This method is as experiential as real experimenting. In the next section you are going to get a few samples. Also, now you know why I quoted *Alice in Wonderland*: to push your imagination button.

One word of caution. Many people somewhat mistakenly look at the relativity of time as just a philosophical generalization. They will, for example, make statements to the effect that "things depend on your frame of reference" and be satisfied at that; that's what relativity means to them. I hope you recognize that these kinds of statements are obvious truisms without much content. Einstein's work goes way beyond this. Einstein's relativity of time has astounding experimental consequences that have been verified. In the realm of elementary particle physics, people observe the tangible consequences of the relativity of time as fairly routine aspects of the physical world. This is very different from just saying "time depends on your frame of reference."

The first question that we will deal with in our new framework is this: Does simultaneity of events depend on the frame of reference?

Another consequence of the relativity of time that we will study in detail is a very famous one, and it is usually stated in the form of this oft-quoted phrase: "moving clocks run slower."

Simultaneity

Imagine a spaceship moving with a uniform velocity in a straight-line path. It is then an inertial reference frame, albeit moving with respect to an earth-fixed inertial reference frame. We want to examine two events that are simultaneous but occur at two different places on the spaceship. Will these events appear to be simultaneous from the earthbound frame?

First we have to make sure that the two events that take place at two different places on the spaceship are really simultaneous. After a little thought we can come up with the following setup. Suppose you stand in the middle of the spaceship and arrange for two of your friends to stand at the two ends at equal distances from you. Then you flash a light signal, say with the help of a flashlight. The light travels the same distance toward both of your friends. The velocity of light is the same in both directions irrespective of the motion of the spaceship (remember the Michelson-Morley experiment). Therefore, light should take the same amount of time to travel to either end and the two friends should receive the signal simultaneously. The events of their reception of the light flash are simultaneous events.

The question now is this: Are the two events that we have just described simultaneous in the perception of the ground observer? Remember that the ground observer sees your ship to be moving (which you cannot see, not without looking outside) and this must affect his perception. In his view the spaceship is moving forward. The front end is moving away from the light beam as it tries to catch up with it. Thus the signal will travel a little further, and therefore a little longer, before it reaches the man at the front end compared to what would happen if the ship were stationary. On the other hand, in the perception of the ground observer, since the rear end man is moving toward the light signal, light has to travel a shorter distance to reach him. Thus the light signal must appear at the rear end first and then at the front end. The two events are not simultaneous, not according to the ground observer.

This is an example of a gedanken experiment and its result is simple to appreciate. Why, then, did it take people so long to comprehend this? In fact, everybody before Einstein had always implicitly accepted the simultaneity of two such events from all frames of reference. The key to that misunderstanding lies in their belief that time is absolute and in their lack of knowledge of the constancy of the speed of light. People before Einstein thought that light traveling toward the front end would have a velocity of $c + v$, the sum of light speed and the forward speed of the ship. On the other hand, the light beam going backward would travel with a decreased velocity of $c - v$. Thus they thought that the longer distance that light has to travel to reach the front end is made up by its having a greater velocity. The fact that light speed is the same irrespective of the fact that you are shooting it from a spaceship and irrespective of the direction compels us to a greatly different conclusion.

By the same token, two events that are simultaneous but occur at two different places on earth will not experimentally appear to take place at the same time from an inertial frame that is moving at a uniform velocity with respect to the earth. The conclusion from this is that the measurement of a time interval depends on the reference frame. Time is relative.

It is interesting to analyze a similar but rather silly experiment that involves interchanging time and space. That is, we now consider two events that occur at the same place but at two different times on the spaceship. Do they appear to occur at the same place in the perception of the ground observer? For example, suppose you are sitting in a chair on the spaceship and smile. A little later you frown (for the sake of our experiment, if not for any other reason). You are still at the same place as far as the spaceship reference frame is concerned but certainly not according to the earth observer. He sees you to have moved away some distance in the mean time. Thus the measurement of a spatial interval, or distance, is also relative.

Take notice that this second experiment is sort of "silly," because everybody

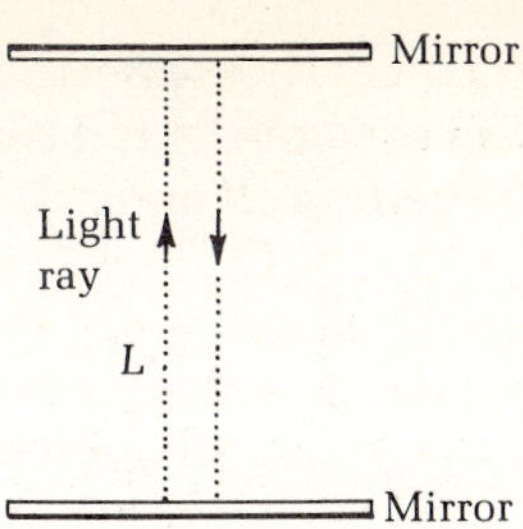

FIGURE 19.5   A light clock.

knows and can visualize the result. The first experiment, on the other hand, is more difficult to reconcile with our senses. Nevertheless, both conclusions are correct: space and time are both relative. Their perception depends on the reference frame; they are intertwined with each other.

Time dilation

The gedanken experiment regarding simultaneity makes it clear that the time between events is measured differently by an observer in a moving reference frame compared to an observer in a stationary frame. It is now natural to ask this: How does the measurement of the rate of flow of time differ between the two observers? The answer is this: moving clocks run slower.

To demonstrate this we have to scrutinize the performance of actual clocks. It is to our benefit to employ very simple clocks. We will use the so-called light radar clock for our purpose. You know that the radar that is used to trace aircrafts sends a microwave signal to and fro between the radar station and the aircraft. The light radar clock operates on the same principle. It consists of a rod with mirrors at each end so that a light signal can be kept going back and forth between the two mirrors (Fig. 19.5). We can imagine that a flashbulb is attached to the lower mirror (in the figure); this creates the signal. In addition, the lower mirror also has to have some sort of a ticking mechanism attached to itself (for example, a photoelectric device), which gives a tick every time the light signal comes back down after being reflected by the upper mirror.

Suppose such a clock is placed in a moving spaceship and the clock's performance watched from an earth-fixed reference frame. We want to know how the result of this observation differs from that made from the spaceship itself. To the observer on the spaceship, the clock is stationary. For the ground observer it is a moving clock. To avoid any controversy, we must insist that the ground observer uses an exactly similar light clock for timekeeping and also that the two clocks are synchronized in the beginning of the gedanken experiment.

One more caution is needed. The rod of the light clock in the spaceship must be placed perpendicular to the direction of motion of the ship. This way we avoid any complication arising from the shrinkage of the length of the rod, which would occur were it placed in the same direction as the ship's motion. Such a shrinkage is itself a consequence of the theory of relativity, as we will see shortly.

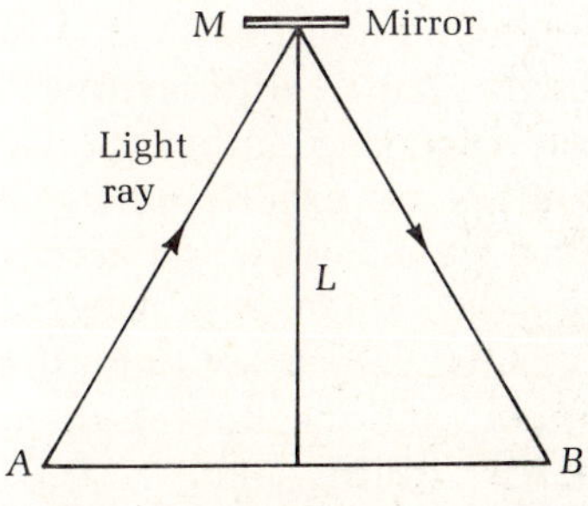

The person in the spaceship, who sees his light clock as stationary, perceives the signal traveling straight up and down between the two mirrors (Fig. 19.5). If the length of the rod is L, light travels a distance 2L between ticks, taking an amount of time given as

$$t = \frac{2L}{c} \tag{19.1}$$

This, then, is the time between two ticks of the clock as measured by an observer in the spaceship.

The ground observer does not see the light signal going straight up and down because of the motion of the spaceship carrying the clock. To him the signal appears to have taken the zigzag path AMB of Fig. 19.6, where AB is the distance the ship has traveled during the up-and-down motion of the light. Let us denote

FIGURE 19.6   This is how a ground observer sees the motion of the light beam in the light clock of the moving spaceship. The path (according to the ground observer) is *AMB* and is not straight up and down.

by t' the time taken by light to complete the zigzag course. Then t' is the period of the moving clock, the time between two consecutive ticks, as measured by the ground observer. Since light travels with velocity c, the time t' must be given as

$$\text{time} = \frac{\text{distance traveled}}{\text{velocity of light}}$$

or

$$t' = \frac{\text{length of } AMB}{c} \qquad (19.2)$$

One thing is already clear from just looking at Fig. 19.6. The distance AMB is greater than $2L$, the up-and-down distance. The t' must be longer than t, a conclusion very easily reached upon comparing Eqs. (19.1) and (19.2). This is the phenomenon of **time dilation.** The time between ticks of a moving clock as measured by an experimenter on earth is longer than what is observed by somebody in the spaceship, to whom the clock is stationary. Thus moving clocks slow down.

With the help of a little mathematics, we can find the extent of the slowing down of the moving clock—that is, the relationship between t and t'. We are going to skip the mathematics and only give the final result:

$$t' = t\left(\frac{1}{\sqrt{1-(v^2/c^2)}}\right) \qquad (19.3)$$

where v is the velocity of the spaceship relative to earth. The factor

$$\frac{1}{\sqrt{1-(v^2/c^2)}}$$

is often referred to as γ, the lowercase Greek letter "gamma." Thus

$$t' = \gamma t \quad \text{where} \quad \gamma = \frac{1}{\sqrt{1-(v^2/c^2)}} \qquad (19.4)$$

It is clear from the expression for γ that for any nonzero velocity, $v > 0$, gamma is greater than 1. Thus for any $v > 0$,

$$t' > t$$

It is interesting to study the value of γ for different values of the velocity of the moving spaceship. For this purpose we plot γ against v/c in Fig. 19.7. The figure shows that for small values of v/c, up to 0.1, the value of γ hardly differs from 1. This means that the relativistic effect of time dilation becomes significant only when $v/c > 0.1$. Since one-tenth of the velocity of light is not exactly what we encounter in everyday life, it is understandable why we do not come across time dilation in everyday life activities.

As the velocity of the spaceship exceeds one-tenth of the velocity of light, the effect of time dilation becomes more and more prominent, and the moving clock runs slower and slower. Table 19.1 gives you some idea of this.

We have proven the slowing down of clocks only for one kind of clock. How do we know that the result holds true for all other kinds of clocks as well (which it does)? We can prove it like this. Assume that we have two much more complicated clocks, which we synchronize with each other and with the first

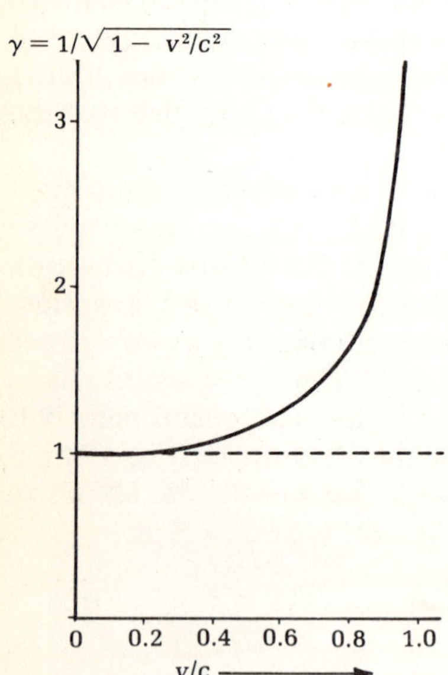

FIGURE 19.7   The quantity gamma, $1/\sqrt{1-(v^2/c^2)}$, is plotted against v/c. Gamma is appreciably greater than 1 for values of v/c greater than 0.1.

v/c	*On a clock moving with velocity* **v**, *1 hour is dilated to**	
0	1 hour	(no time dilation)
0.1	1 hour and	18 seconds
0.3	1 hour and	2 minutes and 53 seconds
0.8	1 hour and	40 minutes
0.99	7 hours and	5 minutes

*Interpret the table as follows: when 1 hour has passed on a clock moving at 0.8c, 1 hour and 40 minutes will have passed on a stationary clock.

two. Let the spaceship person now take one of these aboard along with a light clock. We know that the light clock slows down. But then the other clock must, too, because if it doesn't, by looking at the mismatch of the two clocks, an inside observer of the spaceship can tell that his ship is moving. This is impossible; it is a violation of the basic principle of relativity. A person cannot tell if he is moving if the motion is at constant velocity in a straight line. Therefore, all types of clocks must slow down in a moving spaceship—pulse rates, metabolic rates, perception times, aging, every clock.

And all this occurs without the passengers of the spaceship having the slightest sensation of what is going on. For them life goes on as before. It will be only when they return to the earth that they will discover the disparity of their clocks compared to earth's. In effect, they will have completed a one-way time travel to the future.

Does this sound like science fiction? So did the idea of manufactured earth satellites when Newton's gravity theory made the prediction. However, the technology for accelerating to near-light velocity so that the voyagers get the advantage of time dilation is still very far away, way beyond our reach right now, but few think that it is impossible.

Contraction of length (Lorentz contraction)

If moving clocks slow down, it is easy to see that the length of moving objects appear to shrink along the direction of motion. Let's give a simple outline of the logic. We'll go back to the observer in the spaceship, who is now holding a rod along the direction of motion of the spaceship. The rod is stationary with respect to the space observer, who gives the true length of the rod to be L. How does an observer on the ground measure the length of the rod? He knows the velocity of the rod, which is v, is the same as the spaceship. Then if he measures the time it takes the rod to go past a reference point in his system, say the tower of the Empire State Building, he could get a measure of the length of the rod. However, we have to be careful here. The ground observer's clock is moving with respect to the reference frame of the rod (the spaceship) and therefore he is making the measurement with a slow clock. So the time that he will measure will be too short. And thus the ground observer will measure the length to have shrunk.

Thus moving rods appear to be contracted along the direction of motion (Fig.

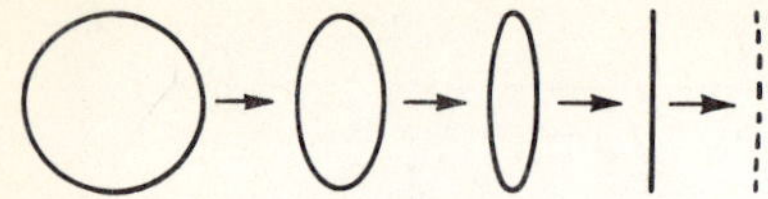

FIGURE 19.8 Lorentz contraction of a moving sphere. As the speed gets closer and closer to the speed of light, the sphere looks more and more like a flat disk. Note that only the length in the direction of motion is contracted.

19.8). This contraction is named after Lorentz since he was the first to propose the contraction idea (although in the context of the wrong ether theory).

Since all objects are embedded in space, we are in effect dealing with an underlying contraction of space itself when it is in motion relative to an observer.

What is the shrinkage factor? Because of the close connection of the Lorentz contraction and time dilation, you can readily envisage that the shrinkage factor is γ. Thus the formula for length contraction is

$$L' = \frac{L}{\gamma} \tag{19.5}$$

where L is the original length (measured in the reference frame in which the rod is stationary) and L' is the contracted length.

The following is a limerick by an unknown author who obviously was amused by the Lorentz contraction phenomenon:

> There was a young fellow named Fisk
> Whose fencing was exceedingly brisk
> So fast was his action
> The Lorentz contraction
> Reduced his rapier to a disk.

Suppose you could travel at the speed of light

This is another of Einstein's teenage dreams. How would time tick for somebody who is rushing through the universe at the speed of light?

Eq. (19.4) gives us the answer. If we substitute $v = c$ in this equation, we get

$$\gamma = \frac{1}{\sqrt{1-(v^2/c^2)}} = \frac{1}{\sqrt{1-1}} = \frac{1}{0} = \infty$$

As γ goes to infinity, t' also becomes infinity. The time between ticks of a moving clock becomes infinitely long at light speed. This is true timestop! Since motion is relative, to the traveler at the speed of light, everything else in the universe appears to travel with light speed, and all their moving clocks appear to him to be at a standstill. Everything seems to be part of the Madhatter's tea party, the clock always showing 6 o'clock.

Furthermore, Eq. (19.5) for the Lorentz contraction tells us that to a traveler moving at the speed of light, all spatial distances in the direction of his motion contract away to zero. This enables him to travel anywhere he likes, instantly in his time.

Unfortunately, relativity theory quite firmly also tells us that it is impossible for any material object to accelerate to the speed of light, as we will see later. Speed of light is a limit that material particles can only approach but never reach.

■ 19.3 The Twin Paradox: Which Twin Is Older?

No matter how real Einstein's experience was to him, how can we know that it coincides with physical reality? Such a concern is expressed best in a poem by

Wallace Stevens:

> They said: you have a blue guitar
> You don't play things as they are.
> The man said: things as they are
> Are changed on my blue guitar.*

How do we know that Einstein's "music" is real, not the music of a blue guitar, a figment of human-created reality? The **twin paradox** reflects the concern of scientists regarding the basic validity of the concept of time dilation.

Time dilation tells us that moving clocks slow down. The principle of relativity asserts that motion is relative. How can we tell which clock is moving? To each, the other would seem to be moving, so each observer would claim it is the other's clock that is slowed down. Who is right?

Suppose we have twins, a brother and a sister, one of whom, let's say the sister, sets off in a spaceship toward the nearest star outside our solar system, a star named Alpha Centauri. Alpha Centauri is about 4 light-years away from us; we assume she travels with a speed 80% that of light, so she makes the round trip in 10 earth years. Her brother and everybody else on earth have aged 10 years in the meantime. How much will she have aged?

According to her brother, the space twin will have aged much less since she has been traveling close to the speed of light and thus had enjoyed the advantage of the dilation of time. "Not so," says his twin. "In my perception, you and the earth were moving away from me. Therefore you ought to be the one to benefit from the slowing down of moving clocks, brother dear." Who is right? When the space twin returns to earth after her round trip, what will she find? Will she find (a) herself older than her brother, or (b) herself younger, or (c) that both are the same age?

This is the twin paradox, also called the clock paradox. Which twin is older? Or maybe time dilation is just in the mind, an idea from a blue guitar. Perhaps the reality is that they both age at the same rate?

Since such a dispute cannot be resolved by actual experimentation (at least not for human subjects) with present technology, we must resolve the paradox by logic. Or else the entire theory of relativity is brought into question.

Einstein himself recognized the importance of resolving the paradox. He came up with the following logic. The paradox exists only if the space twin can truly claim that she did not know—or rather could not tell by physical means—that she is the one who did the traveling. Einstein said that in any real case of space travel, the space traveler can tell and still not violate the principle of relativity. The point is that when she starts from the earth, turns around Alpha Centauri, and stops back on earth, she is either accelerating or decelerating (Fig. 19.9). In either case she will feel the action of pseudoforces. Thus the twin who feels the action of pseudoforces during parts of the period under question is the traveling twin. Thus she is subject to time dilation and will have aged less. So according to Einstein, a complete symmetry of the relative motion of the twins is a mirage. If the twins were ever to meet again, one of them would

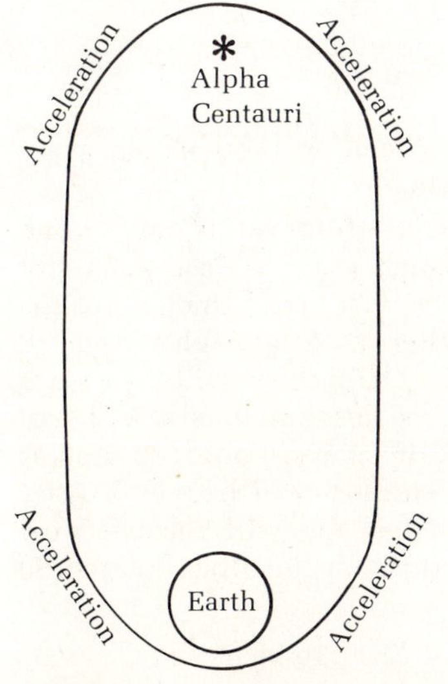

FIGURE 19.9 A round trip to Alpha Centauri has to involve periods of acceleration, as shown.

---

*From "The Man with the Blue Guitar." *The Collected Poems* (New York: Knopf, 1964). Courtesy Alfred A. Knopf, Inc.

The twin paradox is a special relativistic effect.* But in its resolution in the last section we have taken into account the effects of acceleration. Does this mean that the theory of special relativity is an incomplete theory? One feels uneasy if the twin paradox cannot be resolved by the use of special relativity alone.

We will now show that a resolution of the paradox indeed exists within the confines of special relativity.† We will use modern atomic clocks for measuring time in this instance. In such a clock the quantum transition of an atomic electron, which results in the emission of light of a definite frequency, is used as the standard of frequency. Since the frequency and period are reciprocally related to each other by

$$\text{period} = \frac{1}{\text{frequency}}$$

this also gives a standard for the period of a clock. But now a very interesting thing happens with these clocks. In the presence of relative motion between the atomic clock and an observer the frequency of the emitted light changes because of the Doppler effect. So we now have a way of pursuing the behavior of moving clocks rather closely.

Suppose the twins equip themselves with two such identical atomic clocks, which are synchronized before the space twin departs on her trip. Also suppose that each clock is geared to strike hours and that at the strike of an hour a light flash goes out, which the other twin can receive and log against the hour strokes of his or her own atomic clock.

So that acceleration does not enter into our consideration, we can instruct the space twin to turn off her clock during periods of acceleration, which she can determine from the presence of pseudoforces. This will, of course, introduce an error in total time registered by the space twin, but the error is a systematic one and will always be the same whether the trip is between the earth and Alpha Centauri or to the center of our galaxy. Any such systematic error can be eliminated. We can thus disregard the effect of acceleration.

So let's imagine again a trip to our nearest neighboring star, Alpha Centauri, which is 4 light-years away. Let's assume that the trip is made with a velocity of $v = 0.8c$.

Thus according to earth time, the journey to Alpha Centauri and back will take a time of

$$\frac{\text{distance}}{\text{velocity}} = \frac{8}{0.8} \text{ years} = 10 \text{ years}$$

We can calculate γ to be

$$\gamma = \frac{1}{\sqrt{1-(v^2/c^2)}} = \frac{1}{\sqrt{1-(0.8)^2}} = \frac{1}{\sqrt{0.36}} = \frac{1}{0.6}$$

Thus the path of 8 light-years will have contracted away to only $8/\gamma$ [which is $8 \div (1/0.6) = 8 \times 0.6$] or 4.8 light-years. Moving at the speed of $0.8c$, the space twin will complete the journey in

$$\frac{4.8}{0.8} \text{ years} = 6 \text{ years}$$

according to her clock. Her log will verify this.

Now let's examine how each twin will preceive the performance of the other's clock. Initially the twins are moving away from each other, and the Doppler effect will produce a shift toward lower frequency (i.e., longer period), and so each will observe the other's flashes to have slowed down. The Doppler factor is given by the following expression in relativity theory:

$$\frac{v'}{v} = \frac{\sqrt{1 \pm (v/c)}}{\sqrt{1 \mp (v/c)}}$$

(with $\pm$, use the upper sign for approach and the lower sign for recession). Substituting $v/c = 0.8$, we see that the factor is $\frac{1}{3}$ in our case of recession:

$$\frac{v'}{v} = \sqrt{\frac{1 - 0.8}{1 + 0.8}} = \sqrt{\frac{0.2}{1.8}} = \sqrt{\frac{1}{9}} = \frac{1}{3}$$

Thus each twin will receive the other's signal once every 3 hours of his or her own time.

This state of affairs will not last forever, of course, and the slow flashes will change over to fast ones (for approach, the Doppler factor is 3; verify this), since the space twin will turn back after reaching Alpha Centauri. The crucial point is that the changeover will take place differently for the two. For the space twin, she will start receiving the fast signals (3 times every hour) as soon as she starts back toward the earth. It will happen exactly midway in her journey. Thus she will receive slow flashes for 3 years and fast ones for the other 3 years. So she will count

$$(3 \times \tfrac{1}{3}) + (3 \times 3) = 10 \text{ years}$$

worth of flashes from her earth brother, who, she therefore concludes, will have aged by 10 years.

Her earth twin, on the other hand, will observe the

changeover quite differently. The flashes that the space twin sends him after turning toward the earth will reach him $5 + 4 = 9$ years after she left earth, according to his clock. The 5 years is the time it takes his space sister to reach the star and the 4 years is due to the time light takes to reach him from the star. So he will receive fast flashes only after 9 years have gone by and thus for only 1 year.

Thus he will count that

$$(9 \times \tfrac{1}{3}) + (1 \times 3) = 6 \text{ years}$$

have gone past for his space twin.

Thus you see they both appraise each other's aging quite correctly and in accordance with the time dilation theory. And the paradox is resolved.

be subject to acceleration and would know herself to be the traveling twin. And what if they never meet? Well in that case, who is going to raise the paradox?

Actually, Einstein demonstrated more than this. Using the theory of general relativity, which accounts for accelerated motion, he was able to show that both twins can come to complete accord as to each other's age, when they take all effects into account. And after doing this they find that Einstein's first hunch is right: it is the space twin who ages less.

What do experiments have to say about time dilation?

Another way to convince ourselves of time dilation is through experimentation. Fortunately there is now plenty of experimental evidence in support of the dilation of time that moving clocks are predicted to show. One example of time dilation has been known for some time. Charged subatomic particles named muons are known to be produced in the upper levels of the atmosphere through the action of cosmic rays. The muon particles have a very short lifetime, on the average of the order of a microsecond. So even if they are known to move very fast, with speeds close to that of light, most of them would have decayed away before reaching ground level were it not for the time dilation effect. Because of their fast motion, their lifetimes are extended by factors of ten, or something of that sort, and thus they have enough time to reach ground level before undergoing decay. The observation of large numbers of muons at sea level—muons that were created at the top of the atmosphere—thus confirms the validity of the dilation of time.

Recently the same phenomenon has been verified with muons produced in the laboratory. Again, fast-moving muons are found to live longer than stationary ones and exactly according to theoretical prediction.

The most interesting experimental verification of the theory has been achieved recently when actual "atomic" clocks were flown in jets and found to behave in accordance with the theory of relativity.* The time dilation effect that one gets with the velocities of modern-day jet travel is indeed very small, about 10^{-7} seconds. The detection of effects of this order must be regarded as a marvel of modern experimental technology. We have indeed come a long way since the days of Galileo's water clocks.

---

*J. C. Hafele and R. E. Keating, "Around the World Time Clocks: Observed Relativistic Time Gains" *Science*, 14 July 1972, p. 168.

You have now learned about one of the most important properties of time—namely, that time is relative. Until Einstein corrected everybody, all humanity seemed to be quite content with the picture of an "absolute time" following Newton who had written: ". . . absolute true and mathematical time of itself and from its own nature flows equably without relation to anything external." Einstein's work radically altered this view of time. It recognized that the perception of time depends crucially on one's frame of reference.

Let me caution you against some popular pictures of the phenomenon of time dilation. In trying to give you an idea of what is meant by the statement, "moving clocks slow down," popular books sometimes give examples like this: "When you are bored, time passes very slowly for you, whereas when you are happy, time passes rather quickly." Are these examples any help in understanding the dilation of time for moving clocks? Not quite. These are examples of subjective experience. Curiously enough, our memory retains the experiences above with an interesting twist. The slow passage of time when you are bored is retained in the memory as a very short instance, with hardly any detail. On the other hand, the seemingly fast and happy time is retained in the memory with a large amount of detail, with the impression of the passage of a large segment of time.

Compare this with time dilation as encountered by our traveling twin in the twin paradox example. The traveling twin has managed to age little compared to the earth twin by virtue of the slowing down of her clocks. But ask her about her perception. She has no idea that her clocks were slower. She lives an absolutely normal 6 years of life during the travel, with normal memories of it later. Only when she comes back and compares does she discover that she aged much less than her brother. Most importantly, time dilation is a completely objective phenomenon, true for every object in the traveling frame.

Nevertheless, time dilation is a very strange phenomenon and therefore difficult to reconcile with our normal experience. It is because of this difficulty that you will often hear people say that Einstein's relativity is a very difficult theory and that very few people understand it. There is an amusing anecdote on this matter. Eddington, a famous English physicist, was once asked by a newspaperman, "Professor Eddington, is it true that only two people besides Einstein understand his theory?" To this Eddington replied with his usual sense of humor, "Who is the other one?"

Let me assure you that you can understand the relativity of time if you are patient and if you use some imaginative effort. I can guarantee you that the pleasure of this understanding is quite spiritual. When I went through the process, I was swept into profound emotions, which I expressed in the following poem:

Suddenly I understand relativity.
My eyes shine;
For a moment, I think
I am Einstein.

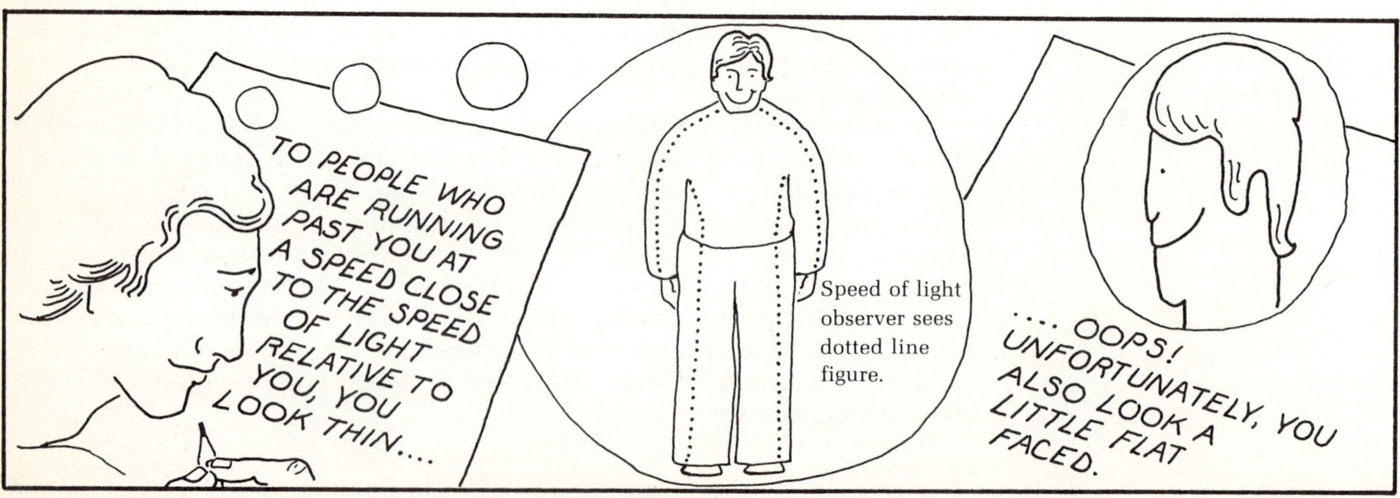

Speed of light observer sees dotted line figure.

SUMMARY

The theme of this chapter is relativity. The major result of the chapter is that the measured value of a certain time interval depends on the frame of reference of the observer. This new premise for the concept of time leads to the ideas of time dilation and the Lorentz contraction, both of which have been verified experimentally. Both time dilation (moving clocks run slower) and Lorenz contraction (the contraction of the length of a moving rod) occur with the following factor γ:

$$\gamma = \frac{1}{\sqrt{1 - (v^2/c^2)}}$$

The twin paradox is a prime example of how puzzling time dilation phenomena can be. When the paradox is resolved, it becomes clear that, however contradictory to our everyday experiences these new concepts may be, they are nevertheless correct.

QUESTIONS

Review and reason

1. What did Michelson expect to observe on the basis of the ether wind idea?
2. How does the Michelson-Morley experiment show that the speed of light remains the same, independent of the velocity of the reference frame?
3. If there is no ether, as the Michelson-Morley experiment seems to indicate, then what carrries the waves of electromagnetism?
4. Here is another one of Einstein's childhood fantasies. A runner running close to the speed of light holds a mirror in front of himself. Will he be able to see himself in the mirror?
5. How does the simultaneity ("gedanken") experiment show that the measurement of the time interval between two events depends on the frame of reference?
6. Explain time dilation in your own words.
7. A subatomic particle named the pi meson has a lifetime of only about 2×10^{-8} sec. A shower of pi mesons created by cosmic rays is found to live 10^{-7} sec, on the average. What can you conclude about the velocity of the pi mesons in the shower? A qualitative answer will do here.
8. What is the Lorentz contraction?
9. Robert Heinlein wrote a science fiction story called *Time for the Stars* in which a twin brother leaves earth in a high-speed spaceship, then comes back to earth and marries his grandniece. The "real" age of the newlyweds was about the same, of course. Do you agree? Why or why not?
10. Relativity is famous for arousing the philosopher in many of us. Write a few sentences conveying what comes to mind for you.
11. Suppose there were a universe where the speed of light is infinity. Would time still be relative in such a universe? Why or why not?
12. Suppose the speed of light were rather small in a certain hypothetical universe, say 100 mi/h. Imagine some of the amazing things that would be part of everyday life for the people of this universe, and write a short essay based on your imagination. You may consult American physicist George Gamow's *Mr. Tompkins* (it's in paperback) for ideas.
*13. In the text it is stated that light originating from a moving spaceship travels with the velocity c with respect to a "ground" observer. Light does not travel with the velocity $c + v$ (where v is the velocity of the spaceship), as would be expected from Newtonian rules of velocity addition. Thus the addition rule is changed. For a projectile with velocity u thrown from the ship moving with velocity v, the velocity that the "ground" observer measures is given by the formula

$$\frac{u + v}{1 + (uv/c^2)}$$

From this formula, find the velocity that the ground

---

*Optional.

observer measures for a bullet that leaves the muzzle of a gun at a velocity of $c/2$ from a spaceship that is traveling at a speed of $c/2$.

14. Ship A approaches ship B, each moving with velocity $0.75c$ with respect to the ground. Thus in the reckoning of the ground observer, the relative velocity of ship A with respect to ship B is $0.75c + 0.75c = 1.5c$. This is greater than c. Does this violate the theory of relativity, which says that no material particle can exceed the speed of light? Why or why not? (Hint: Is reckoning the same thing as measurement? Can the ground observer measure the velocity of one spaceship with respect to another?)

15. Two identical spaceships approach each other with a velocity close to the speed of light with respect to each other. State and explain if the measurement of the following quantities will produce different results in the different spaceships:
    (a) the velocity of the other ship relative to its own
    (b) the velocity of a cannonball fired in the other ship
    (c) the velocity of light flashed from the other ship
    (d) the rate of time clocks of the other ship
    (e) the length of the other ship

Arithmetic

1. Neutrons have a lifetime of about 13 minutes when existing as free particles. (Although neutrons can be quite stable when inside the atomic nucleus, free neutrons are unstable and undergo a decay.) If a neutron moves at a speed of $0.9c$ (corresponding to $\gamma = 2.3$), what is the value of the lifetime of a neutron as observed in an earth-fixed laboratory?

2. The distance from us to the Vega star system is about 27 light-years. To a traveler at $0.9c$ ($\gamma = 2.3$), this distance will shrink to _______________ light-years.

Fill in the blank.

3. Calculate the numerical value of the relativistic factor γ for the following two cases: (a) $v = 0.8c$ and (b) $v = 0.6c$.

4. A rocket ship from earth makes a round trip to the Vega system (27 light-years away) and back at a speed of $0.8c$. How much time will have elapsed on earth in the meantime? How much will the passengers of the spaceship age during the round trip?

20 The Relativistic World View

Parsifal: I pace hardly at all, nevertheless I feel I have come far already.

Gurnemanz: You see, my son, that here time transforms into space.

RICHARD WAGNER IN "PARSIFAL"
(Tr. R. M. SINCLAIR)

■ 20.1 Space-Time

When Hermann Minkowsky first met him, Einstein was a student in one of Minkowsky's courses. Minkowsky was not too happy with the performance of his student, who behaved like a "lazy dog" who never "bothered about mathematics at all." Yet when Minkowsky saw Einstein's paper on relativity, which was published in 1905, there was no doubt in his mind that this work should be looked at from a mathematical point of view. You know the fable about Mohammed and the mountain: If Mohammed does not go to the mountain, the mountain will go to him. So, Minkowsky decided, if Einstein "doesn't bother about mathematics," maybe mathematics will bother about Einstein, or at least about his work. The phrase "space-time" came out of the great mathematician's study of Einstein's relativity.

This is what Minkowsky declared in 1908 to a group of scientists in a talk entitled "Space and Time":

> Gentlemen! The ideas on space and time that I wish to develop before you grew from the soil of experimental physics. Therein lies their strength. Their tendency is radical. From now on, space by itself and time by itself must sink into the shadows, while only the union of the two preserves independence.*

For three years Minkowsky had quietly pursued a marriage between two seemingly diverse concepts of space and time. Einstein's paper suggested a

---

*Reprinted in A. Einstein, H. A. Lorentz, H. Weil, and H. Minkowsky, *The Principle of Relativity* (New York: Dover, 1923), p. 75.

courtship between space and time. Minkowsky mathematically united them for a final consummation of their relationship. And it is very romantic, this wedding of space and time.

The four-dimensional world of space-time

We have talked about the three-dimensional nature of space and about reference frames consisting of three mutually perpendicular rods tied together at an origin. The coordinates for the location of an object in space depend on the specified reference frame.

Physical events involve both spatial and temporal location. An event refers to an occurrence at a given moment of time and a given point in space. Before Einstein, people assumed that time is measured the same for all reference frames, moving or not. Einstein's work changed all that. Now we know that different frames of reference with relative motion between them judge time differently. Thus it is imperative that each frame carry its own clock and regard the time read by its clock as a fourth coordinate; time is as essential for a description of an event as are the three spatial coordinates.

From Descartes to Minkowsky, from three to four coordinates—our concept of the universe takes one more leap forward. From the safe cocoon of the familiar experiential world of three-dimensional space, we are forced to plunge into a new dimension of reality. What are the rules of our four-dimensional world? Can we comprehend them? How do we communicate with other voyagers in four dimensions?

Let's deal with the communication problem right away. Think of the following situation. Suppose you are located in a particular reference frame in four-dimensional space-time, and you measure the coordinates of an event as x, y, z, and t. A friend of yours is traveling in a spaceship along your x-axis with velocity v and measures a different set of coordinates x', y', z', and t' for the same event. What are the rules for converting one set of coordinates to the other? The advantage of knowing these rules is obvious. It will make your communication with him more reliable because each of you will know what the other is talking about when he is using the coordinates x', y', z', and t' to describe his observations and you are using a different set, x, y, z, and t.

Here is a trivial example. Suppose you live in San Francisco and have a friend who lives in Los Angeles on your x-axis (we can always orient the axes that way). One day she visits San Diego, which is also on your x-axis but further south, and tells you (on the telephone) that San Diego is 120 miles away. Now you too have been to San Diego and know from your own experience that it is 535 miles away. You two are assigning different x-coordinates to the location of the same city. Of course, in this case you can resolve the discrepancy very quickly. You realize that the origin (LA) of your friend's coordinate system is shifted south from your origin (SF) by a distance of 415 miles. So to compare with your results, she must subtract the shift (415 miles) from your measured distance to get her measured distance. Thus the rule for converting one set of coordinates to the other in this case is

$$x' = x - a, \qquad y' = y, \qquad z' = z, \qquad t' = t$$

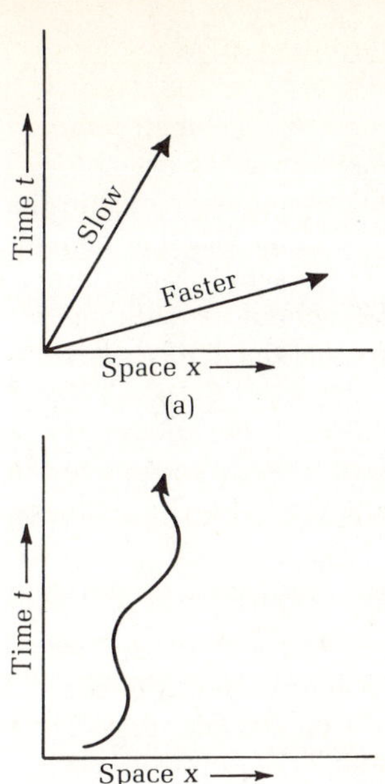

FIGURE 20.1   The world lines of particles. (a) The world lines of free particles are straight lines. (b) The world lines of particles in accelerated motion are curved.

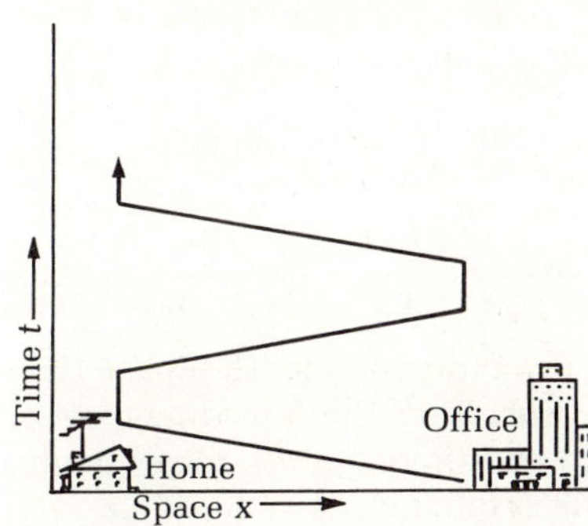

FIGURE 20.2   The world line of an average American.

where a is the shift of the origin, your friend's coordinates are called x', y', z', t', and yours are x, y, z, t.

So this is the general idea. In the more complicated case of relative motion, we find that the x' of the spaceship is connected to both your x and t; likewise his t' is also a combination of your x and t. The actual connecting equations for the two sets of coordinates are (assuming their origins coincide at $t' = t = 0$)

$$x' = \gamma(x - vt)$$

$$t' = \gamma\left(t - \frac{vx}{c^2}\right)$$

$$y' = y, \qquad z' = z \tag{20.1}$$

Here v is the velocity of the spaceship relative to you. Notice that our old friend γ which is $[1/\sqrt{1 - v^2/c^2}]$, has made an appearance again.

This set of equations is called the **Lorentz transformation.** We are not going to use them here. But perhaps you will find it comforting to know that the mode of communication in the four-dimensional world is at your disposal.

Here is one more thing to note. The principle of relativity tells us that physical laws remain the same in all inertial reference frames. But different frames use different four-dimensional coordinates related by the Lorentz transformation. Thus all physical laws must remain unchanged under the Lorentz transformation. This is the symmetry principle of the four-dimensional world of relativity.

The path of a particle in space-time is called its **world line.** To get some idea of this, we can plot the x-axis horizontally and the time axis vertically (and forget the other two space axes momentarily). In such a plot [Fig. 20.1(a)] the world line of a free particle is a straight line: the familiar natural motion. If there is acceleration, we get a curved world line [Fig. 20.1(b)]. Fig. 20.2 shows the approximate world line of an average American who goes back and forth every day between work and home (where the TV is).

20.2   The Dynamical Consequences of Relativity: $E = mc^2$

The discussion so far may give you the erroneous idea that relativity is just a theory of time and space. Actually, this theory has also given us a revolutionary new understanding of matter in motion. The most famous aspect of this is the mass energy relation $E = mc^2$, which adds a new perspective to the concept of the mass of an object.

When we apply relativity to the Newtonian framework, we find that the first and third laws of Newton are not affected by this new point of view of space-time. How about the second law? Recall that at the end of the discussion of

Now let's go back to the lesson that we learned from the gedanken experiment regarding simultaneity. There we saw that the time interval between two events is not measured as the same from two different frames when there is a relative motion between them. Two events observed to occur simultaneously in the spaceship were considered to be separated by a certain time interval by an earth-based observer. The same was true for the spatial interval between events; this too is judged differently from different frames of reference.

Now this can pose a problem. If neither observer agrees to the other's perception of the spatial or temporal interval between two events, how are they going to reach a consensus about the description of the events? Minkowsky discovered that although spatial and temporal intervals separately lose their importance, a certain combination of them remains invariant, that is, is the same for all observers. This combination is called the space-time interval or the four-dimensional interval between two events.

Before we introduce the invariant interval of Minkowsky, you need to be familiar with some elementary concepts regarding spatial and temporal intervals. The case of temporal intervals is easy. If one event has a time coordinate t_2 and the other one t_1, the time interval between the two events is $t_2 - t_1$. Likewise, for one-dimensional space, we could say that the spatial interval is simply $x_2 - x_1$. But how do we find the spatial interval for three-dimensional space?

Let's look at the case of two dimensions first. In two dimensions—for example, on a plane surface—suppose there are two points P_1 and P_2 with coordinates (x_1, y_1) and (x_2, y_2), respectively (Fig. 20.3). The spatial interval between P_1 and P_2 is the length of the line P_1P_2. How do we find this length in terms of the coordinates?

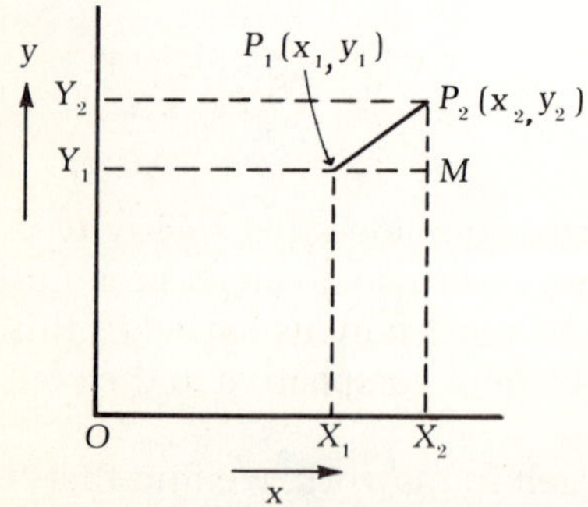

FIGURE 20.3   The two-dimensional distance between two points P_1 and P_2 on a plane.

For this purpose we use a theorem of geometry known as Pythagoras's theorem. Pythagoras was a very colorful personality of ancient times, a contemporary of Buddha and Confucius. He founded a religious sect that considered it wicked to eat beans (Guess why). Most importantly for us, he also founded a school of mathematics that studied right-angle triangles, just the kind we have in P_1MP_2.

According to Pythagoras's theorem, the square of the length of the hypotenuse of a right triangle is equal to the sum of the squares of the other two sides. In our case, then,

$$(P_1P_2)^2 = (P_1M)^2 + (P_2M)^2 \qquad (20.2)$$

Now from Fig. 20.3, it is clear that $P_1M = X_1X_2$ and $P_2M = Y_1Y_2$ and furthermore, from the definition of the coordinates, $OX_1 = x_1$, $OX_2 = x_2$, $OY_1 = y_1$, and $OY_2 = y_2$. Thus $X_1X_2 = OX_2 - OX_1 = x_2 - x_1$ and $Y_1Y_2 = OY_2 - OY_1 = y_2 - y_1$.

Going back to Eq. (20.2), we get

$$\begin{aligned}(P_1P_2)^2 &= (P_1M)^2 + (P_2M)^2 \\ &= (X_1X_2)^2 + (Y_1Y_2)^2 \\ &= (x_2 - x_1)^2 + (y_2 - y_1)^2\end{aligned}$$

Thus the spatial interval between two points in two dimensions is given as

$$\sqrt{(x_2 - x_1)^2 + (y_2 - y_1)^2}$$

It is easy to generalize this to three dimensions: the three-dimensional distance between two points is given by

$$r = \sqrt{(x_2 - x_1)^2 + (y_2 - y_1)^2 + (z_2 - z_1)^2} \quad (20.3)$$

We now can give you the expression for the space-time interval between two events with the four-dimensional coordinates (x_1, y_1, z_1, t_1) and (x_2, y_2, z_2, t_2), respectively, customarily noted by the symbol s:

$$s = \sqrt{\begin{aligned}&(x_2 - x_1)^2 + (y_2 - y_1)^2 \\ &+ (z_2 - z_1)^2 - c^2(t_2 - t_1)^2\end{aligned}} \quad (20.4)$$

Now we must notice several things. First, it is not t but the product of the speed of light c and t that appears in the expression for s. This is not hard to understand. Time is measured in seconds and spacial distances in meters, so we must do something so that everything under the radical ($\sqrt{\ }$) is measured in the same units (we cannot add things that are measured in different units). Actually, in everyday life we do sometimes use units of time to speak of distance. People in Boston will say that Washington, D.C., is 8 hours away instead of saying it is about 400 miles away. What they are doing, by approximation, is dividing the 400-plus miles by the average speed of a car on the highway, 50 miles/hour. In Eq. (20.4) we use the same principle, but instead of dividing the distances by a velocity, we multiply time by a velocity, which converts the time unit to a unit of space. Since c is a fundamental constant of nature, forever unchanging and independent of the frames of reference, we naturally use c as the velocity multiplier to convert the time unit to a unit of distance.

There is a second interesting fact in Eq. (20.4) that is also very profound. Take particular notice of the minus sign of the c^2t^2 term. It introduces a basic asymmetry in the expression: the time and space coordinates do not occur with exact symmetry in the expression of s, the invariant four-dimensional interval between two events. What is the implication of this?

Let's go back one step and look at Eq. (20.3) for the three-dimensional distance between two objects. Here all the coordinates appear symmetrically, and this implies that all of them are completely interchangeable. For earth people, it is not all that easy to accept that the up-down direction is equivalent to the two horizontal directions. But then Newton discovered gravity and showed that the up-down asymmetry is removed when we switch off gravity. Thus it is really Newton who should be credited with the discovery of three-dimensional spatial symmetry of our world. All directions of the three-dimensional space are equivalent.

In contrast, the world of four dimensions of Einstein and Minkowsky does not have perfect symmetry; all directions are not equivalent. There are some directions in the four-dimensional world that are different from others.

In effect, what this means is that space and time are not completely interchangeable, (i.e., a spatial distance cannot completely be converted into a time interval and vice versa). Even Einstein's magic cannot do away with clocks in preference to meter sticks (Fig. 20.4).

If time and space were completely interchangeable, one could travel in time by a variation of space travel technology, something like that imagined by English novelist H. G. Wells in his classic science fiction novel, *The Time Machine*. The time traveler of the story argues that time is nothing but a fourth dimension of space, and just as civilization has learned to defy gravity and travel freely in the up-down direction, one day we should be able to travel freely in the time direction also—both future and past.

Unfortunately, this is quite impossible in the four-dimensional world of relativity, and all because of the negative sign of the time term in the equation for s. Minkowsky has given us a very interesting way of looking into this. Mathematicians have a quantity called i, which is defined as the square root of -1:

$$i = \sqrt{-1}$$

The square root of any negative number can be written as the product of i and a real number, for example, $\sqrt{-16} = \sqrt{16} \times \sqrt{-1} = 4i$. Now if you thought about numbers, and in particular square roots of numbers, then surely you will say that the square root of a negative number is meaningless. Actually, mathematicians sort of agree with

(a)   (b)

FIGURE 20.4 If you start with a time interval (a), you can end up with a part time and a part space interval (b), but never just a space interval. Even Einstein's magic can do no more than what is shown.

you, so they call such numbers **imaginary.** Yet imaginary numbers have tremendous usefulness. Things don't have to be meaningful to be useful.

Minkowsky used the concept of i to make the nature of the time dimension a little bit clearer. Since $i^2 = -1$, we can write $-c^2t^2$ as $i^2c^2t^2$, which is the same as $(ict)^2$. Put $\tau_1 = ict_1$, and $\tau_2 = ict_2$ (the symbol τ is the Greek lowercase letter "tau"). Then we can write the expression for s as

$$s = \sqrt{\begin{array}{c}(x_2 - x_1)^2 + (y_2 - y_1)^2 \\ + (z_2 - z_1)^2 + (\tau_2 - \tau_1)^2\end{array}} \qquad (20.5)$$

and now there is no minus sign under the square root. There is now, seemingly, a complete symmetry between the coordinates x, y, z, and τ. But at what price? Our fourth coordinate is imaginary; τ is really ict.

This is our dilemma. The fourth dimension is not a real dimension like the space dimensions but has an inherently different character. And since an imaginary number can never be converted into a real number, we can never convert a time interval entirely into a spatial one.

Within this limitation, relativity gives us some leeway. From Eq. (20.4), we see that if

$$(x_2 - x_1)^2 + (y_2 - y_1)^2 + (z_2 - z_1)^2 < c^2 (t_2 - t_1)^2$$

s is imaginary. Such imaginary intervals are called time-like intervals and we can always find a reference frame from which such an interval is seen as a purely time interval. On the other hand, if

$$(x_2 - x_1)^2 + (y_2 - y_1)^2 + (z_2 - z_1)^2 > c^2 (t_2 - t_1)^2$$

then s is real and as such can be viewed as a purely spatial interval from some reference frame. We have seen both of these aspects in the simultaneity experiment.

So reality has a four-dimensional character. What appears in our consciousness is a three-dimensional projection of the four-dimensional reality. Three-dimensional shadows appear differently to different observers.

Lorentz transformation, we made a comment that all physical laws must be formulated in such a way that they display the four-dimensional symmetry. This means that their content must remain unchanged under the Lorentz transformation: that's the mathematical requirement of four-dimensional symmetry. When we attempt to shape Newton's second law from this point of view, we discover a very interesting thing. Newton's second law survives in only one of the forms; $F = ma$ is not true anymore. The other form of Newton's second law,

$$\text{force} = \text{rate of change of momentum}$$

remains valid in relativity theory with one twist. In the definition of momentum,

$$\text{momentum} = \text{mass} \times \text{velocity}$$

mass now must be allowed to vary with velocity according to a relation dictated by relativity. Eventually, these considerations lead to the equivalence of mass and energy. It is interesting that the form of Newton's second law that survived is the form chosen by Newton. Did he have an inkling that someday someone would find the mass of an object to vary with speed?

The variation of mass with velocity

Can mass really increase with velocity? It does and it must according to the theory of relativity and, most importantly, according to experiment. Remember the Newtonian definition of mass as the quantity of matter contained in an object? Now we find that such a definition of mass is not adequate. When an object moves, the quantity of matter in it certainly does not change, but mass does. Thus mass and quantity of matter are not the same thing.

We have outgrown the usefulness of Newton's definition of mass. But do not despair. A much more beautiful understanding of the concept of mass will result from the relativistic way of looking at it.

Actually, mass determines the resistance of an object to acceleration by a force. Imagine that a force is acting on an object, thus giving it a constant acceleration. The velocity of the object will increase continuously. But for how long? Is it without limits? Relativity has a new absolute—the speed of light. Nothing can be accelerated beyond the light speed. The question, then, is this: What's going to keep the object from accelerating right through the speed of light if there is a force to provide the acceleration?

Relativity gives the answer to this question by pointing out that mass increases with velocity. The resistance of an object to acceleration is not a fixed property of the object, like the quantity of matter in it, but is also dependent on the velocity of the object. For low velocity there is no appreciable increase of mass. Like time or length, the changes in the relativistic perception of mass become important only when the velocity of light is approached.

What is the relativistic factor for the increase of mass with speed? You probably will guess it—it is γ. So we have this formula for the relativistic mass of an object:

$$m = m_0\gamma$$

$$= m_0\left(\frac{1}{\sqrt{1 - v^2/c^2}}\right) \tag{20.6}$$

What is m_0? It is the nonrelativistic mass, something we can still identify with the quantity of matter of the body. More precisely, it is the mass of the object when the object is at rest. It is called the **rest mass** for this reason. It is the mass measured from a rest frame, a reference frame that travels with the body so that the object is at rest with respect to this frame.

With the help of the relativistic expression for the mass of an object, we can understand what happens if we try to accelerate an object to and beyond the speed of light. As the velocity becomes appreciably close to c, say greater than $0.1c$, γ starts to take on larger and larger values compared to 1 (see Fig. 19.6), and the mass starts becoming greater and greater compared to the rest mass. So now the same force will not produce an acceleration equivalent to its original acceleration; since the amount of acceleration depends on the mass, the acceleration will decrease. Another way of describing this state of affairs is to say that only a part of the force is usable now toward changing the velocity. The other portion is spent toward increasing the mass. And as the velocity becomes closer and closer to c, this latter portion becomes dominant; more and more of the force is expended to produce a mass increase.

Can we accelerate an object to the velocity of light? In Eq. (20.6), if we substitute $v = c$, γ becomes infinity and so does the mass. It takes an infinite force to accelerate an object of infinite mass. Since we do not have such a force at our disposal, we give up. So the velocity of light is truly insurmountable for all material objects of finite rest mass.

Eq. (20.6) has been verified experimentally. When we try to accelerate high-speed elementary particles, we find that their mass has the exact behavior predicted by this equation.

What about light itself? Why does light travel with the speed of light? If this is a riddle for you, don't worry. The answer is simply that light does not have any rest mass. Light cannot be observed from a rest frame; it never comes to rest or even slows down. This fact is what got the whole relativistic thinking started in the first place. Light is pure energy with zero rest mass.

Is there any other object that has this property of zero rest mass? There are particles called neutrinos, neutral particles that seem to have no rest mass, in which case they too are destined to live always in motion at the speed of light. Thus neutrinos also seem to be pure energy.

Equivalence of mass and energy

Now we come to a puzzling question: Where does the increase of mass come from? Let's ask a slightly different question: What does a moving object have that an object at rest does not possess? The answer to this latter question is simple—a moving object has kinetic energy. So we are led to conclude that the increase of mass of a moving object really comes from the kinetic energy. Kinetic energy and mass are entirely equivalent.

One question leads to another. Are other forms of energy also equivalent to mass? Einstein and relativity say yes. Thus light energy has mass by virtue of its energy.

We can use this fact to get an expression for the connection of mass and energy. Since momentum is mass times velocity, the momentum of light of mass m is mc, since c is the velocity. If you are surprised that light has momentum, recall that the electromagnetic field has momentum. Therefore light, which is the propagating disturbance of the field, should also possess momentum. Because of its momentum, light exerts pressure on an object it illuminates. Sometimes such radiation pressure gives rise to spectacular effects. At least part of the reason that a comet has a long tail, which is always directed away from the direction of the incident sunlight, is attributed to the pressure of light. Now for light, experiment tells us that energy and momentum are related by a simple formula:

$$\text{energy} = \text{momentum} \times c$$

Substituting mc for the momentum, we see that the energy of light can be written as

$$E = mc^2 \tag{20.7}$$

The energy is the product of the mass (actual relativistic mass, not rest mass) and the square of the velocity of light. Since the velocity of light is a fundamental constant of nature, we can easily interpret its presence in the expression of equivalence of mass and energy as a scale factor, a factor for the conversion of two different currencies for the same thing.

What is true for kinetic energy and light is also true for all other forms of energy. All energy is equivalent to mass, and $E = mc^2$ is valid for all of them.

Thus a heated iron rod has more mass than a cold one. An object at the top of a hill (with larger potential energy) has more mass than one at the bottom. And so forth.

Why does this fact escape our experience in everyday life? It is the smallness of the exchange rate. Energy is a very depreciated currency. For every unit of energy, we get the corresponding unit of mass but divided by the enormous factor of c^2, which makes the equivalent mass very small. Thus an object on the top of the hill has an increase of mass, but only one part in 10^{13}.

Perhaps a numerical verification will help us here. Suppose we send a 100-kg person to the top of a 1000-m hill. The increase of his potential energy can be calculated from Eq. (9.3). It is mgh. Since $g \approx 10$ m/sec^2, we get for the potential energy gain

$$100 \times 10 \times 1000 \text{ J} = 10^6 \text{ J}$$

The equivalent mass is

$$\frac{E}{c^2} = \frac{10^6}{(3 \times 10^8)^2} \text{ kg} = \frac{10^6}{9 \times 10^{16}} \text{ kg} \approx 10^{-11} \text{ kg}$$

And this is 10^{13}th of his mass. Thus the increase of mass is one part in 10^{13}. No wonder it escapes our senses and even the most accurate measuring apparatus for masses.

Now let's look at the other side of the coin. An object at rest having a rest mass m_0 has an energy given by

$$E_0 = m_0 c^2 \tag{20.8}$$

just by virtue of its rest mass. We can call E_0 an object's rest energy. But now the scale factor c^2 works to our advantage. Even an object of small rest mass has an enormous amount of energy.

Let's get an estimate for this energy. Suppose we consider a 1-kg mass of uranium. That 1-kg mass has an energy of

$$1 \times c^2 = (3 \times 10^8)^2 \text{ J} = 9 \times 10^{16} \text{ J} \approx 10^{17} \text{ J}$$

In uranium fission, which is used in nuclear power plants, roughly one part in a thousand of the mass is converted into energy.* Even so, we get a thousandth of 10^{17}, or 10^{14}, joules of energy just from 1 kg of fissionable uranium.

Perhaps you are getting grandiose ideas of satisfying all our energy needs in various forms from the conversion of mass energy into these other forms. Unfortunately, this won't work. Nature has made sure of the basic stability of matter by means of overriding principles. The relevant overriding principle that prevents our converting 100% of the uranium into other forms of energy is called the conservation of **baryon number.** Baryons are a family of particles of which the proton is the least massive member; all other baryons are heavier than the proton. The conservation of the number of baryons in the entire universe tells us that we can never destroy protons without creating other

---

*Nuclear fission is the name of the process in which a heavy atomic nucleus, like that of uranium, breaks up into two or more lighter nuclei, giving off an enormous amount of energy.

baryons. The number of baryons we start with must match the number we end up with. This puts a tremendous constraint on the convertibility of the mass currency into energy currency.

But now you may wonder whether we really get energy at all from mass. Actually, in all processes of energy harnessing, whether or not we convert mass into energy is more of a question of point of view. If we start with some protons and make a bound system out of them (a process called **fusion,** which is responsible for the energy generation in stars), energy is liberated. Where did the energy come from? Did any of the protons become converted into pure energy? No, because the number of baryons always remains the same. The protons combine into a heavier particle (helium nucleus) containing exactly the same number of baryons that we started with. What happens here is that the protons form a bound system, a system of negative energy (see Section 10.3). Negative energy is negative mass, so the mass of the bound system we end up with has less mass than the original number of protons that we started with; the bound system has a "mass defect." Thus we can say that part of the initial mass has been converted into energy; that is why we have a mass defect. But we can also say that we get energy from a process like this because we are making a negative energy system, and so an equivalent amount of positive energy must be given out as a result of conservation of energy.

One thing is clear. There used to be a principle called the principle of conservation of mass, which the chemists particularly coveted for a long time. From our discussion it is obvious that this principle is only approximately correct: even in chemical reactions a small but finite amount of mass changes into energy.

Antimatter

Any discussion like this could not be regarded as complete without a mention of antimatter. (In case you don't know, antimatter is the fuel used by the starship *Enterprise* of the TV series "Star Trek.") Antimatter is really a consequence of the relativity theory (in combination with quantum mechanics), and its existence has been verified experimentally. **Antiparticles** of the electrons, called positrons, have been observed and so have the antiparticles of protons and neutrons, the components of the atomic nuclei. So there is no doubt that these antiparticles could form antiatoms and antimatter composed of these antiatoms.

The antiparticle of a charged particle has the opposite charge; thus the positron (which is the name given to the antielectron) has a positive charge. The antiproton has a negative charge. One interesting question is how to identify antiparticles of neutral particles like the neutron or the neutrino. Although the neutron is neutral, it has magnetism and the antineutron can be distinguished by the reversal of the magnetic property.

The case of neutrino is more subtle, since neutrinos have no electromagnetic properties at all. But even so researchers have found a way to distinguish between a neutrino and its antiparticle, the antineutrino. The neutrino spins around its own axis as it moves; for a regular neutrino, the spin is in the opposite direction to the motion. The neutrino spins in the sense of a left-handed screw. On the other hand, the antineutrino spins like a right-handed screw; its spin is in the same direction as the motion.

Not all elementary particles have been found to have antiparticles. Examples of this are elementary particles called pions, a species of subnuclear particle.

Now we come to the most important thing. It is an experimental fact that if an antiparticle comes close to a particle, they annihilate each other. The baryon number conservation does not prevent a baryon from being destroyed by its antibaryon. This is explained by recognizing that the antibaryons have negative baryon numbers. So if we start with an equal number of baryons and antibaryons, the initial number of baryons is really zero. And so we are allowed to end up with pure energy, which also has zero number of baryons.

So why can't we use antimatter as our energy source in a matter-antimatter converter, "Star Trek" style? The catch is that antimatter is a rarity in our part of the universe. If we manufacture antimatter—which we can do by starting from radiation (a process called pair production, in which a particle-antiparticle pair is produced)—it costs energy, more than we get back. So antimatter won't work as a solution to our energy problem.

Still, the "Star Trek" idea is basically sound. Antimatter is the best possible storage of energy known to humans, since practically 100% of the mass energy can be converted into pure energy. For rockets one problem is the huge mass of the fuel the rocket has to carry. For example, in chemical rockets of today, only one part in 10^9 is converted into useful energy, and so we have to carry a billion times the mass we actually use. We can gain on this enormous factor of a billion by using antimatter as our rocket fuel. We may be very far away from such rocket technology, but it is mind boggling even to think that such a rocket is not impossible theoretically. Also, since radiation is the result of matter-antimatter annihilation, the spent fuel goes out the back of the rocket at the speed of light. Thus the rocket itself can be accelerated most economically, depending, of course, on how efficiently we can direct the outgoing light beam. In such a rocket, getting the velocities close to the speed of light may be possible. Then we can take advantage of time dilation and can plan a trip to outer space.

■ 20.3 Beyond the Speed of Light?

We can never achieve a velocity equal to or beyond the speed of light, so relativity tells us. A theory that got rid of absolute time ironically gave us a new absolute—speed of light. Is the speed of light really an absolute?

For matter that we are familiar with, the argument is compelling; there is no way we can overcome the light barrier, since such an effort necessitates an infinite force. But three scientists—O. M. Bilaniuk, V. K. Deshpande, and E. C. G. Sudarshan—among others, have pondered the problem in a different way.

"Can there be particles that always move with a speed greater than the speed of light?" they asked. (One way to fight an absolute is to bring out another absolute.) The strange thing is that the answer relativity gives us is that there is nothing in relativity itself that prevents the existence of such faster-than-light particles—particles that cannot slow down to and below the light speed. The name **tachyon** is given to such possible particles.

How is it that such particles can exist without violating relativity? The point that Sudarshan and colleagues made is that none of the *observable* properties of such objects contradict relativity. For example, consider the expression [Eq. (20.6)] for the relativistic mass of a particle:

$$m = \frac{m_0}{\sqrt{1 - (v^2/c^2)}}$$

For $v > c$, the denominator becomes the square root of a negative number, which we have previously designated (see optional material in Section 20.1) as an imaginary number. At first glance it may seem that all is lost; mass becomes imaginary and therefore meaningless. But there is a way to keep m from being imaginary. Suppose the rest mass m_0 is also imaginary. Then both the numerator and the denominator in Eq. (20.6) have the factor i, which neatly cancels out: the ratio of two imaginary numbers is a real number. Since m_0 is not observable, it can be imaginary without violating any physical principle.

Tachyons, if they exist, turn out to have some observable unusual properties. For example, if you accelerate a tachyon, it loses energy. Finally, when the velocity of the tachyon reaches infinity, its total energy becomes zero.

The most interesting aspects of tachyon physics concern the question of tachyons moving backward in time. Before dealing with backward-going tachyons let's consider several things about faster-than-signal travel. Imagine that you are talking with a friend who is walking toward you at a speed faster than the speed of sound. You will hear her in reverse order. Since she overtakes the sound she emits, the later sounds will reach you first.

Now imagine that your friend is coming toward you with a speed faster than light. She emits signals, light signals, during her travel, and you see her when this light reaches you. But since again she is overtaking the light coming from her, you see her backwards; the light emitted by her later will reach you first. Thus her entire travel will be revealed to you backwards. This is the basis of the following limerick:

> There was a young lady named Bright
> Who traveled faster than light.
> She started one day
> In the relativistic way
> And came back yester night.

But this is time travel of a different sort, the type of time travel that involves problems with causality, the cause-effect relationship. For example, if Ms. Bright returned last night, she would have met her own self. But this gives a contradiction, since the traveling Ms. Bright did not start with a recollection of meeting herself the night before. The contradiction is a causal contradiction and is usually very popular with science fiction writers who love to find ingenious devices for avoiding the contradiction.

Now we are ready to talk about tachyons again. Tachyonic paradoxes are created because a moving observer can see a tachyon moving backward in time in addition to forward-going tachyons. For example, suppose two friends make up the following maze of events. Friend A sends a tachyon signal to his friend B, who will travel away in a rocket ship at a certain predetermined time unless A

hears from B before A sends his message. B's part in the game is to send A a tachyon message as soon as he has received A's. The problem arises as follows: suppose B sends A his return signal. But A could see it as going backward and therefore receive it before his preset time for sending the original signal. So then he does not send a signal. Now then, why did B send one back?

The resolution of the paradox involves the detection devices for tachyons. It can be shown theoretically that detection of backward tachyons is hampered by the fact that the detector itself spontaneously emits a normal forward-going tachyon. So A will not be able to distinguish B's backward-going tachyon from one created by his detection machine. This answer seems to be begging the question, but the alternative is a breakaway from causality.

Finally, we sound one pessimistic note. Nobody has yet observed a tachyon despite thorough searching. So we do not have any evidence whatsoever for particles that travel at a speed on the other side of the light barrier.*

SUMMARY

The relativistic way of looking at time leads to the unified concept of space-time, a four-dimensional universe in which space constitutes three of the dimensions and time the fourth dimension. The rules of communication between observers in the four-dimensional world are called the Lorentz transformation. The trajectory of a particle when it is plotted in space-time (with time being the ordinate) is called the world line of the particle.

Two important dynamical consequences evolve from the space-time way of looking at the motion of objects. The first result is that the mass of an object no longer can be regarded as remaining constant during motion; the mass actually increases with velocity, by the relativistic factor $\gamma = 1/\sqrt{1 - (v^2/c^2)}$. An even more fundamental consequence of relativity is that mass and energy are found to be entirely equivalent. This equivalence is expressed by the mass-energy relation

$$E = mc^2$$

*You can read more on tachyons from the following source: Gerald Feynberg, "Particles that Go Faster than Light," *Scientific American*, February 1970, p. 69.

QUESTIONS

Review and reason

1. What is an event? How many coordinates do we need to specify an event?
2. What are the Lorentz transformations? What do they transform?
3. What is meant by a particle's world line? Draw one.
*4. What is an imaginary number? Give an example.
5. What is meant by three-dimensional symmetry of space?

*6. Is the four-dimensional symmetry of nature a perfect one? Why or why not?
*7. Can time be totally converted into space or space into time? Explain.
8. How does the mass of an object vary with velocity? Write down an expression.
9. Why is it that material particles can only approach the velocity of light but can never reach or surpass it?

*Optional.

10. In the following situations, figure out if the mass of the system increases, decreases, or does not change. (a) A hot metal rod cools down. (b) A bucket of water is taken to the top of a hill. (c) A chemical reaction is allowed to take place in a perfectly sealed container. (d) A car slows down and comes to a stop. Explain your answers.

11. Why can we not convert mass-energy of ordinary matter into other forms of energy, like heat or electricity, at will?

12. What is antimatter? What happens if matter and antimatter collide? What is the process of pair production?

13. What do we mean by "zero mass" particle? Name a couple.

14. What are tachyons? What do you find the strangest about them?

Arithmetic

1. An electron in a TV picture tube travels at a speed of $0.2c$. If the electron's rest mass is 9.1×10^{-31} kg, calculate its relativistic mass. (A velocity of $0.2c$ corresponds to $\gamma = 1.02$.)

2. A subatomic particle traveling at a speed of $0.9c$, for which the value of γ is given to be 2.3, is found to have a mass of 4×10^{-27} kg. What is the value of the rest mass of the particle?

3. Calculate the mass energy of a kilogram-mole of hydrogen gas.

4. If one kilogram-mole of hydrogen gas is annihilated completely by one kilogram-mole of antihydrogen gas, how much energy is released?

*5. The energy released from the explosion of a ton of TNT is 5×10^9 J. Atomic bombs are often rated in terms of their energy equivalent in so many tons of TNT. What is the energy release in terms of joules from a 2-kiloton atomic bomb? Calculate the mass equivalent of this much energy. Assuming that in an atomic explosion only one part in a thousand of the mass of the active ingredient is converted into energy, and furthermore only 15% of the material of the bomb constitutes the active ingredient, calculate the mass of the 2-kiloton atomic bomb.

---

*Optional

21 The Quantum Mechanical Way of Nature

This is the way the physicist rides
A quantum, a quantum, a quantum.*
FREDERICK WINSOR AND MARIAN PARRY

■ 21.1 The Quantum Nature of Light

Some ideas never die but are reincarnated in different forms. One such idea, the particle nature of light, was tried by the great Newton himself. The experimental situations that Newton attempted to explain with his corpuscles of light were equally well—or sometimes even better—explained by the wave picture. In addition, some light phenomena like interference and diffraction cannot be explained at all in the particle picture. Thus the particle picture went into apparent oblivion because it was unable to explain what was then known experimentally about light.

Toward the end of the nineteenth century and the first few decades of the twentieth century, new experimental facts were uncovered that could not be explained by the wave picture. But strangely enough, these facts found a natural explanation in terms of a new particle picture. In this picture light exists as little bundles of energy called quanta. A **quantum** of light is called a photon, a word first used by Einstein. The German physicist Max Planck first thought of quanta and thus made a decisive departure from the classical picture of nature. Around 1913 Danish physicist Niels Bohr developed a model-picture of the atom that made consistent use of the quantum idea. After another decade of frantic search, finally a young disciple of Bohr, named Werner Heisenberg, put together the

---

*From "The Space Child's Mother Goose," by Frederick Winsor and Marian Parry. Copyright © 1956, 1957, 1958 by Frederick Winsor and Marian Parry. Reprinted by permission of Simon and Schuster, a Division of Gulf & Western Corporation.

first consistent set of mathematical equations, which were to replace the Newtonian framework in the microscopic domain of atoms and the like.

The most perplexing feature of the new picture that emerged was a duality in the behavior of light and matter. Light came to be regarded as photons, but the old wave picture did not fade away—because the wave picture is also correct, as validated by experiments. So the same entity, light, now came to be regarded as both a particle and a wave. This is the picture of light that we will develop in this section. The dual nature of matter will be the subject of the next section.

The energy quanta of Max Planck

Once in Soviet Russia a professor wrote down the equation

$$E = h\nu \tag{21.1}$$

and asked one of his students: "What is ν?" "Planck's constant," was the answer. The puzzled professor asked, "But then what is h?" "The length of the plank," the student answered unabashedly.*

The student was slightly mistaken, as you no doubt suspect without even the slightest idea of what the equation is all about. The symbol ν almost always denotes frequency; it is not hard to figure out that E denotes energy. So the only unknown is h. But then if you know the name of Planck, surely you would know of h; h is a fundamental constant of nature known as **Planck's constant.** The equation $E = h\nu$, introduced by Planck, was the first (but not the last) appearance of this constant h in our picture of the microworld. The entire quantum revolution can be traced to the potential power of this one mathematical equation.

What's so phenomenal about this equation? Before Planck's discovery, energy had always been regarded as a continuous variable. Energy was assumed to be emitted or absorbed in as small a dosage as you can imagine, all the way down to zero. But this equation says otherwise. For each frequency of the radiation emitted, the size of the energy bundle is h times the frequency: no more, no less. Energy can be exchanged from the system only in multiples of these bundles, or **quanta.** Energy is quantized. It is like having a minimum currency: we cannot exchange coins in any denomination less than a cent; all exchanges must occur in integer multiples of cents.

The constant h has the same units as the product of energy and time, or what physicists refer to as action. So h is often referred to as the quantum of action. Its value has been established from very careful experimentation to be

$$h = 6.62 \times 10^{-34} \text{ J·sec}$$

As you can see, h is a very small number. This is the reason that the quantum nature of things does not play much of a role in the behavior of the macroworld, that is, in the motion of large objects.

Let us now briefly sketch how Planck's discovery came about. First, we will ask you to recall what you may know of the radiation emitted by an incandescent substance. The color of the emitted light depends on the temperature of the

---

*From *Physicists Continue to Laugh* (Moskow: Mir Publishing, 1968), L. T. Kapitanoff, Tr.

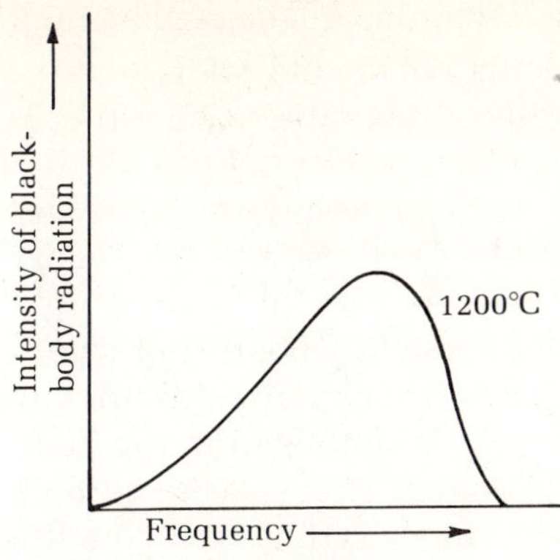

FIGURE 21.1 The spectrum of radiation emitted by an incandescent solid. The intensity rises to a maximum as the frequency increases and then tapers off at even higher frequencies.

substance. At low temperature the color will be dull red; then with increases in temperature the color changes to orange, yellow, white (signifying a mixture of the colors) and finally blue. For a given apparent color, light of other wavelengths are also present. If we make a graph of the intensity or brightness of the radiation emitted versus the frequency, the graph looks like Fig. 21.1. Such a graph is often referred to as the frequency distribution curve of **blackbody radiation.**

Why this particular name? Blackbody radiation is the light that comes out of a perfectly black enclosure, one that absorbs all the electromagnetic radiation that falls on it. When heated to a given temperature, the radiation of the blackbody theoretically can be shown to be dependent only on the temperature and on no other property of the body. It is thus an ideal source. The radiation spectrum of an incandescent solid is a close approximation to that of a blackbody.

Notice that the peak of the distribution curve shifts to higher frequencies as the temperature increases (Fig. 21.2), in accord with what we see for the color. The peak frequency of maximum emission is indeed a signature of the temperature of the emitting body and is in fact used to determine the temperature of stars. As you may know, the stars come in different colors depending on their temperature: the coolest ones are dull red, and then we find orange stars, yellow (our sun), white, and blue, in order of increasing hotness.

The model that people in Planck's time developed for blackbody radiation is based on thermodynamics, the science of interchange of heat and other forms of energy. Now Maxwell's theory tells us that radiation must be due to the jiggling motion of a few charges among the atoms of the body. Ultimately, the charges can be identified with the electrons, but there are so many interactions that we can forget about the individual electrons and assume that the radiation is due to a few charged oscillators, the detailed character of which is not important. Now these oscillators move up and down and radiate and ordinarily lose energy and cool down. But if we keep them in a closed box with perfectly reflecting walls, the oscillators can keep on moving. They lose energy by radiation, no doubt, but some of that light comes back and returns the energy to the oscillator. Thus we can achieve a state of thermal equilibrium (no net one-way flow of energy)

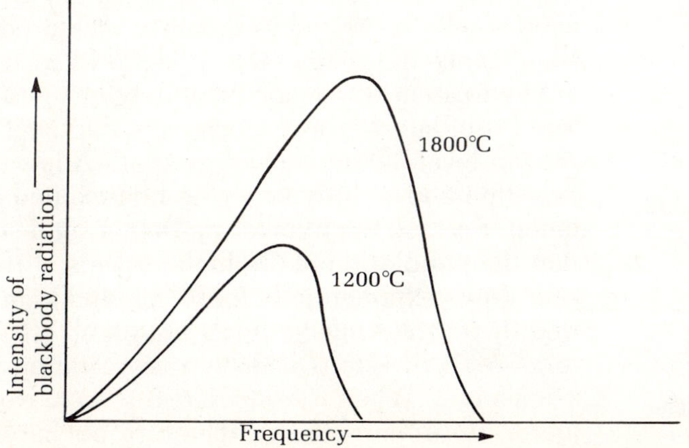

FIGURE 21.2 At a higher temperature, the intensity is much greater than at a lower temperature. Also the spectrum (and the maximum) shifts toward higher frequency at a higher temperature.

between the oscillators and the confined radiation. The average energy of such an oscillator is the same irrespective of the wavelength of the radiation it emits; the average energy is just proportional to the absolute temperature, according to thermodynamics.

Unfortunately, the number of oscillators of a given frequency increases rapidly with the frequency. And this means trouble. One result of the theory above is that most of the energy inside the black box will be in the form of high-frequency radiation—ultraviolet and so forth. Of course this is just not true if we review the experimental data of Fig. 21.1. According to experiments, for any given temperature only a very finite amount of energy is contained in the high-frequency portion of the spectrum. If the thermodynamic theory were correct, we would get a sunburn from the ultraviolet radiation everytime we had a fire going in the fireplace.

What is more catastrophic is that in even a small part of the space inside a black enclosure, the total energy would be infinite—if the arguments above were right. This would happen because of the high-frequency oscillators that contribute to the total energy. So if the thermodynamic theory were correct, energy would just drain out of matter and go into radiation, and everything would become absolutely cold. We all know this doesn't happen. So what's wrong with this theory?

Obviously the theory needs some way to avoid the high-frequency catastrophe, some way to suppress the contribution to the total energy coming from the high-frequency components of the radiation spectrum. Since the number of oscillators increases rapidly with frequency, if the average energy of oscillators somehow decreased rapidly with frequency also, then the product of the two could be finite. This is what Planck was able to show with his $E = h\nu$. That, and the **quantum jump.**

Planck's sparkle of new insight into the problem was the realization that the charged oscillators of a blackbody can only absorb energy in quantum jumps, $h\nu$ at a time. The oscillator exists in states of quantized energy resembling a staircase: the steps of the staircase are called **energy levels** (Fig. 21.3). This analogy is really a very good one. If you imagine yourself on a staircase, you can appreciate that your potential energy in earth's gravity field can change only by discrete quantum jumps; the difference of course is that in everyday life we can build the staircase with steps of varying length. Nature, on the other hand, has a fixed length for its quantum jump, a length determined by h. This is in fact the significance of the quantum of action h.

So we can now imagine the following picture. For each frequency ν there are a lot of oscillators whose steps have the length $h\nu$. Many of them will be at the ground level, taking no energy at all. A few others will absorb one quantum of radiation and will move to the first excited state, but the number of these are much less than the number in their ground state (Fig. 21.3). And this is the way that the population of the higher levels of the staircase occurs; the population goes down exponentially for higher and higher steps of the ladder. Now we can find the average energy by summing up all the energy of these various oscillators on all the steps and then dividing by the total number of oscillators of frequency ν. When we calculate this way, we find that the average energy is not just a multiple of the temperature but now has an additional factor, which

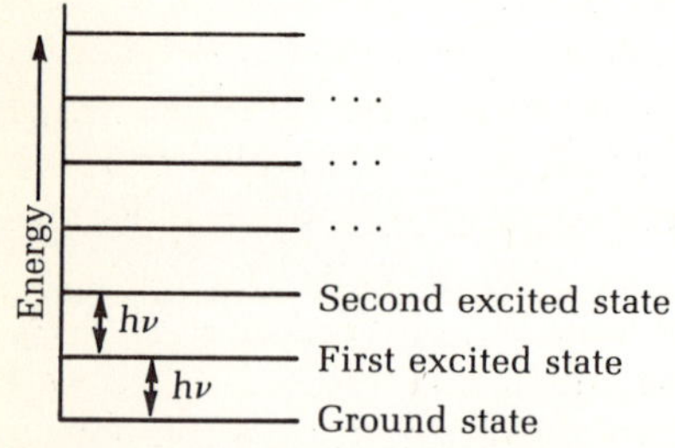

FIGURE 21.3 The quantized energy levels of an atomic oscillator. Energy can be transferred to and from such an oscillator only in quantum jumps of $h\nu$.

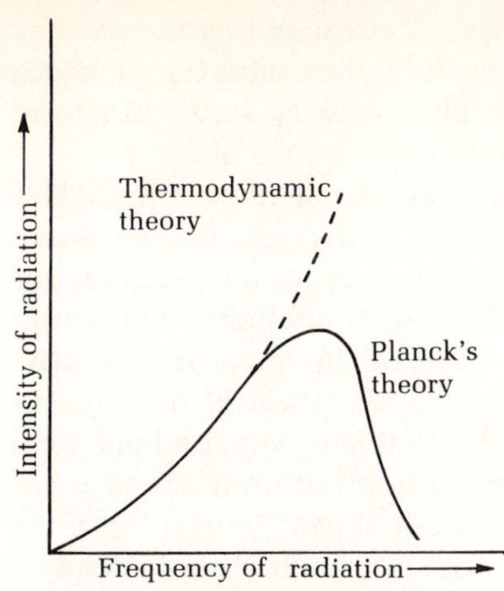

FIGURE 21.4 Planck's theory (the solid curve) agrees with the experimental curves of Figs. 21.1 and 21.2, but the thermodynamic theory fails dismally at high frequency.

basically acts as the high-frequency cutoff that is needed to get agreement with experiment.

The radiation distribution law that Planck discovered following essentially this kind of reasoning (the details were somewhat different), is the famous Planck's law. The spectacular agreement of his law with the experimental data left no doubt in anybody's mind that the law was right (Fig. 21.4). The question that remained was this: Was this the only derivation of the law that was possible? In other words, is quantum jump an essential ingredient law of nature?

Understandably, Planck's work did not have all that much immediate impact among his contemporaries. It was too revolutionary, too much against the grain of the contemporary belief system based on continuity. The idea that mass may exist in quanta of various sizes called the elementary particles was not in vogue yet, nor was the fact of quantization of charge known for sure. We don't know if such knowledge would have made a difference, but it might. Ironically, even Planck himself did not like the quantum jumps, so he spent a substantial portion of his later life trying to disprove his own work. Breaking with a belief you grow up with is not easy.

Einstein's photons

The development of the quantum ideas themselves occurred in discrete jumps. The next significant progress came in 1905, five years after Planck had proposed the energy quanta, from a man named Einstein in a patent office in Switzerland.

If one shines monochromatic light on a metal, electrons come out of it but in a very odd way. The energy of the outcoming electron has nothing to do with the intensity of the wave train of radiation incident on the metal. Now this is not the way waves behave. You can imagine ocean waves imparting energy to a piece of wood on a beach. The more energy the waves have, the more energy they will impart to the wood; it is the total energy of the wave that counts. In contrast, if you fire bullets at a piece of wood and a bullet hits the wood, the energy the bullet imparts to the wood depends entirely on the energy of the bullet. If you have a large number of bullets, you increase your chance of hitting the wood, but the energy change of the wood upon being hit still depends on just that individual energy of a bullet. Thus in the case of interaction of a beam of light with a metal, the imparting of energy to the metallic electrons takes place as with bullets, not ocean waves. Indeed, having a more intense light beam just enhances the chance of the light "bullet" to hit an electron. More electrons are found to come out, but all with the same energy so long as you keep the frequency of the light fixed.

Einstein called these light "bullets" **photons,** and the phenomenon of ejection of electrons from a metal is the **photoelectric effect.** Einstein was quite well aware of Planck's energy quanta while he developed this quantum explanation of the effect. He could now identify the individual energy quanta of light of radiation itself as photons; and for each photon the energy is connected with the frequency through the relation of Eq. (21.1):

$$E = h\nu$$

Higher-frequency light has more energy. This aspect of Einstein's theory was also experimentally true. For higher frequency of the light beam, we get more energetic electrons in the photoelectric effect. Thus it was really Einstein, guided by Planck's work, who established the quantum nature of light.

So light behaves as particles in collisions with electrons, as in the photoelectric effect. This has to be reconciled with light's already established wave nature. Physicists resisted the idea of such a reconciliation for a long time, but evidence mounted. The next great evidence came from Niels Bohr: light emission in atoms could be understood only on the basis of the quantum picture.

Before going into Bohr's work, let's consider an application of the photon idea. There is something about vision in dim light that can be explained very simply by using the photon picture, but it is very difficult to understand if we insist that light is continuous energy. When we look at objects in dim light, we don't see sharp outlines; their shapes seem to merge into each other instead of being distinct. If the light energy coming to the receptors of our eyes were continuous, there would always be some light, albeit of low intensity, coming from an object and falling on all the receptors. Granted, the contrast between light and dark would not be very great in dim light, but this would not affect the sharpness of the outline.

Instead, what happens is that the receptors of our eyes work on the principle of photoelectricity—they respond to the graininess of light, to the individual photons. Since dim light gives fewer photons, this means that too few receptors will be stimulated at one time to define an outline or shape of the object; the picture is too fragmentary. If the receptors stored the information, of course, there would be enough fragmentary pictures for the brain to put together to make the whole. Unfortunately, the receptor information is retained for only a tiny interval of time, so the result is that in dim light not enough receptors will fire at any one time to give a sharp picture of the shape of an object. Next time you say goodbye to a friend in twilight, watch how, as she moves away, her silhouette becomes more and more obscure, and then you can think of photons.

The Bohr atom

Somebody once said about the Greek language that Greek flies in the writings of Homer. The quantum idea started flying with the work of Danish physicist Niels Bohr in the year 1913.

Scientists have the advantage in their creative work—they get to look at nature from the vantage point of the "shoulders of giants." Besides Planck and Einstein, Niels Bohr had access to the shoulder of another giant, Lord Rutherford of Nelson, who dominated British physics in the first few decades of the twentieth century. Bohr was a postdoctoral research associate who came from Denmark to work with Rutherford.

Rutherford had a model of the atom, established in part from experimental observations, that intrigued Bohr no end. In this model the electrons went around the central massive nucleus (almost all the mass of the atom resided in the nucleus; this was proven experimentally by Rutherford) to form the atom. But now the problem starts.

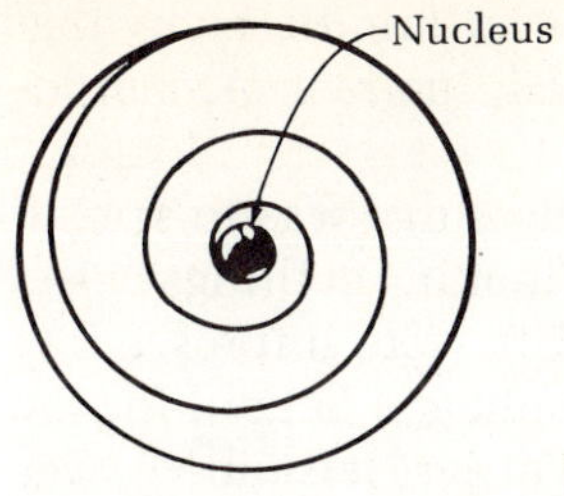

FIGURE 21.5 In Rutherford's original model of electrons going around the central nucleus, the electrons are destined to lose energy by radiation and spiral inward, eventually to fall into the nucleus.

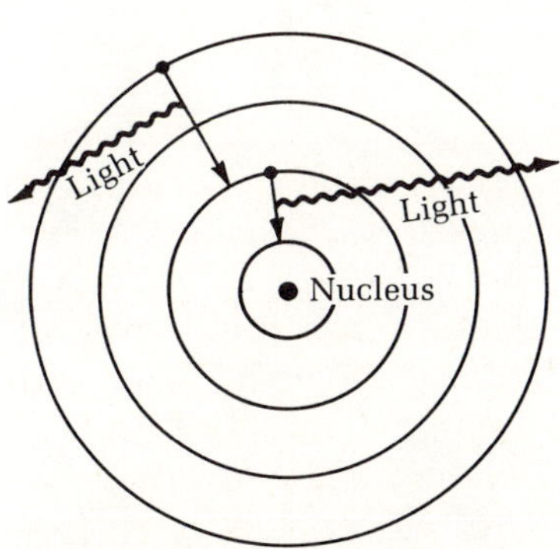

FIGURE 21.6 The electronic orbits of the Bohr atom. In this model the electrons do not radiate while they are in one of these "stationary orbits." Radiation occurs only when an electron jumps from a "higher orbit" to a "lower" one, as shown.

Since the electons are drawn to the nucleus with an attractive electrical force following the inverse square law, it is very tempting to make a tiny solar system model, where the electrons indeed revolve around the nucleus like planets do around the sun. If the electrons moved with just the right range of velocities in their orbits, the electrical Coulomb force has the right magnitude to make them go around. So this works out. The problem is this: an orbiting electron is an accelerated one. According to Maxwell's theory, an accelerated charge must radiate. Even that is all right—atoms do radiate light. But now as the electrons radiate, they must lose energy and spiral inward toward the nucleus. Eventually this leads to a collapse of the atom. So Rutherford's atom is not stable (Fig. 21.5).

Niels Bohr made it stable by injecting the quantum idea. Just as in the quantum oscillator, the electrons of an atom must also exist only in quantized energy levels, or **stationary states,** as Niels Bohr called them. An electron does not radiate when in one of these stationary states, but only when it takes a jump from a higher to a lower energy level. The difference of energy between the levels involved in the transition comes out as light. If the electron is in the lowest (ground) energy level, its ground state, it is stable because it has no other lower energy level to go to. Thus in this way the atom can be stable.

But how does one describe the motion of an electron when in a stationary state, or while it is making a transition from one level to another? The stationary state is not exactly like a planetary orbit, contrary to what you may have been led to believe. And Niels Bohr was quite aware of this. He carefully called the electronic states in the atom "stationary"; an object in an "orbit" is not stationary, it changes with time. In contrast, for a stationary state nothing changes with time, and Bohr was clear on that. Likewise, when an electron takes a jump, could Bohr describe that jump in space-time? No. So the model does not lend itself to classical pictures.

Thus if you ask questions like "What does the electron do when in one of these stationary states?" you are wasting your time. We don't know.* Maybe it improvises; maybe it is granted a certain amount of freedom of improvisation like somebody performing jazz music. Actually, we don't even much care to describe the electron while it is in one of these stationary states, because the electron doesn't do anything that is observable while in these states. So we give ourselves permission to be a little ignorant (which a classical physicist abhors), because, as we will see later, the ignorance is imposed on us by nature.

In order to get people excited about his intuitive picture, Bohr had to do some other things that weren't very kosher. He took the simplest atom, the hydrogen atom, and, in the absence of anything better, applied the planetary model to calculate its energy levels, the **Bohr orbits** (Fig. 21.6). He combined classical and quantum rules in ingenious but somewhat mysterious ways, and his method worked for the hydrogen atom, but applied less and less well for more

---

*The situation is best described by the following quotation from American physicist Robert Oppenheimer: "If we ask, for instance, whether the position of the electron remains the same, we must say 'no'; if we ask whether the position of the electron changes with time, we must say 'no'; if we ask whether the electron is at rest, we must say 'no'; if we ask whether it is in motion, we must say 'no.'" [Quoted from *Science and the Common Understanding* (New York: Simon & Schuster, 1954), p. 69.]

complicated atoms. But Niels Bohr was not upset by this failure. He knew that this part of his picture of the atom was a stopgap method; the real mechanics was yet to be discovered.

One of Niels Bohr's great contributions to human knowledge was to spread this word: that there was a new mechanics, a new way of looking at things based on the quantum principle. Many talented young scientists were impressed by his mission and gathered around him, forming the Copenhagen School for the study of the atom. And one of them, German physicist Werner Heisenberg, was the first to discover the equations of the new mechanics that were to replace Newton's mechanics in the microdomain of nature.

Heisenberg discovered his equations in an intuitive "gestalt" way that was typical of him. However, the significance of the new mechanics becomes clearer in a different formulation based on the idea of matter waves. We turn our attention to this in the next section.

■ 21.2 Matter Waves: Duality of Matter

A prince of the de Broglie family in France had a very clever idea for his doctoral thesis in physics. The prince, named Louis, was familiar with the stationary states of the Bohr atom, and they reminded him of the stationary waves of musical instruments. One of the most intriguing aspects of the stationary waves on a guitar string is the discreet nature of the allowed frequencies that can be set up. For a given length and tension of the wire, the music of the guitar string displays the same kind of discontinuity that is characteristic of the electronic energy levels of an atom. But if there is a connection, the electron has to be a wave, and the atomic electrons must be confined waves.

The stationary states of the electron in the Bohr atom could now be pictured in the following way. The electron is no longer a localized particle moving in an orbit. Instead, its charge and mass are spread out, forming a stationary wave that

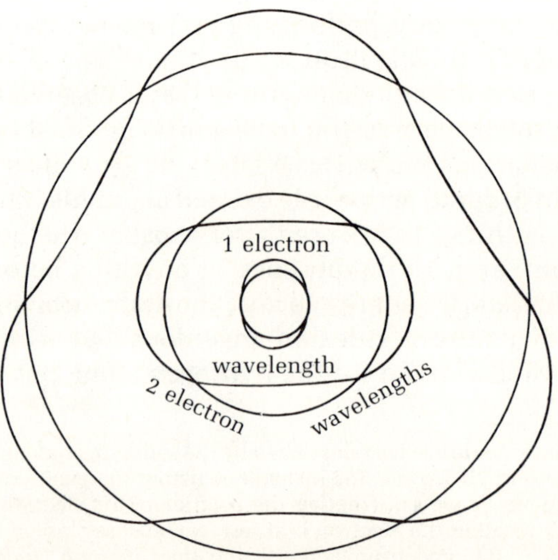

FIGURE 21.7 Stationary waves of electrons in the confinement of an atom. The electron wave fits an integral number of wavelengths in each of the successive Bohr orbits.

surrounds the nucleus of the atom. The lowest stationary state is one in which one electron wavelength fits the circumference of the Bohr orbit. Excited atomic states of the electron correspond to fitting two or more (but always an integral number of) wavelengths within the circumference of the Bohr orbit (Fig. 21.7).

The quantum picture of the atom that now emerges is remarkable. It explains in simple terms three most important properties of the atoms, namely, stability, their identity with one another, and their ability to regenerate themselves. Their stability derives from the fact that it takes a fairly large amount of energy to lift an atom from its lowest stationary state (the **ground state**) to another. This was already pointed out by Niels Bohr. With de Broglie's electron wave idea, we can understand the identity of atoms as the result of the identity of wave patterns in confinement: the stationary pattern is determined by the way the electrons are confined, not by the properties of their environment. Thus an atom of hydrogen on earth must be identical to one found on the sun. Just like a guitar's music, the music of the atom is the same, wherever you find it.

Also, once the atom is recognized as a vibrating system, it becomes clear why it has the remarkable property of returning to its previous state after it has undergone a period of distortion and excitement. The stationary patterns depend only on the conditions of the confinement in which the electron executes its motion; the pattern has no recollection of the past history.

A most important aspect of the de Broglie way of looking at the electron, which everybody knows is also a particle, is that it imparts to the electron the same duality that light has. The electron is a wave as well as a particle. And not only light and electrons but all matter is dual, if we make an obvious generalization. This was recognized by de Broglie as a consequence of Einstein's work: that mass of matter and energy are the same. It seemed right to him to attribute physical properties to matter and light in a symmetric fashion.

De Broglie also postulated that the wavelength of the wave associated with a particle is related to the momentum of a particle. If p is the momentum of a particle, the wavelength of its wave is given as

$$\lambda = \frac{h}{p} \tag{21.2}$$

Consider the electron in the ground state of the hydrogen atom. The momentum of the electron can be calculated from $p = mv$. The mass of the electron is 9.1×10^{-31} kg; the velocity of the electron in the lowest orbit was previously estimated by Bohr:

$$v = 2.3 \times 10^6 \text{ m/sec}$$

Thus the **de Broglie wavelength** of the electron is given as

$$\lambda = \frac{h}{p} = \frac{h}{mv} = \frac{6.62 \times 10^{-34}}{9.1 \times 10^{-31} \times 2.3 \times 10^6}$$
$$= 3.3 \times 10^{-10} \text{ m}$$

Bohr's estimate also gave the radius of the lowest Bohr orbit: 5.3×10^{-11} m. The circumference of the orbit is $2\pi \times \text{radius} = 2 \times 3.14 \times 5.3 \times 10^{-11} = 3.3 \times 10^{-10}$ m, exactly equal to the de Broglie wavelength calculated above. This shows that the electron wave snugly fits the Bohr orbit, as we stated earlier. The electrons indeed set up stationary waves inside the atoms.

Actually de Broglie's argument was more elaborate and convincing than this. But even so he had a hard time getting his PhD thesis approved. The story is that the thesis was finally sent to Einstein for appraisal. Now Einstein, who originated the duality idea for light, had no difficulty in seeing that de Broglie was right. Matter is also dual. Furthermore, he himself had put a big hole in the integrity of the concept of matter by showing that mass and energy are the same thing. Now de Broglie's work carried the same theme a little further by showing that matter had wave properties too. So he wrote back about the thesis: "It may look crazy, but it really is sound." After that de Broglie was given a degree.

De Broglie's formula for the wavelength of matter waves also makes it clear that for macroscopic matter the wave properties are suppressed, because the wavelength turns out to be extremely small. For example, for a bullet whose mass is 2 g, or 2×10^{-3} kg, and whose velocity can be taken to be 300 m/sec, then

$$\lambda = \frac{h}{mv} = \frac{6.6 \times 10^{-34}}{2 \times 10^{-3} \times 300} \approx 10^{-33} \text{ m}$$

This is really, really, small and nature does not have any holes or obstacles of this size. So we are never going to encounter interference or diffraction with the waves associated with macroscopic matter.

Schrödinger equation

If there are waves, there should be a wave equation, a mathematical equation that determines the waves completely for any given situation—so said one physicist to an Austrian-German physicist named Erwin Schrödinger. The physicist who made the comment promptly forgot about it. But the comment made an impression on Schrödinger, who went ahead and derived the wave equation for the de Broglie waves. He applied the equation to the case of the motion of the electron in the hydrogen atom and found that it worked. All the quantum properties of the electronic motion that Niels Bohr had postulated now came out as a solution to the **Schrödinger equation** (the name given to his wave equation).

Soon Schrödinger was able to prove that his theory was identical to the one developed earlier by Heisenberg: his and Heisenberg's are two representations or pictures of the same thing. There is only one mechanics, which is now called **quantum mechanics**. The calculations of this mechanics are more easily done using the Schrödinger picture, so we almost always solve a Schrödinger equation to find the motion of microscopic particles. And the agreement with experiment is always excellent, at least when the particles are nonrelativistic, (i.e., move with velocities less than 0.1c). Thus in the microdomain, the new quantum mechanics has completely replaced the Newtonian framework for calculating motion of objects.

Classical versus quantum physics

Many a time students will ask: Now that I have discovered the new physics, what happens to the old physics? To their mind it is an either-or question; if the new physics is correct, then the old physics must be wrong. It does not quite work that way.

There is a beautiful drawing by cartoonist W. E. Hill, named "My wife and my mother-in-law." Look at the picture (Fig. 21.8); what do you see? Initially, probably just the wife (or the mother-in-law). It may actually take you a while to discover the other picture among the same lines of the drawing. Slowly, if you keep at it, the other picture will emerge, the mother-in-law if you started with the picture of the wife. The wife's chin will transform into the nose of the mother-in-law, her neckline into the chin of the older woman, and so on. What's going on? The lines are the same. But suddenly a new way of perceiving the picture has become possible for you. And very soon you find you can even go back and forth between the old picture and the new. You can still look at just one of the two images at a time, but your consciousness has enlarged so that you are aware of the duality.

This is the relation of quantum mechanics to the perception of classical physics. It is not replacing something wrong by something right. It is gaining a wider perspective, thus enlarging the validity of the perception into a new domain. You can look at nature separately as particles and waves, like the face of the wife in the drawing. Or you can discover duality, both waves and particle all in one thing, the wife and mother-in-law, all part of the same picture that you are trying to understand.

Experimental observation of matter waves

We will end this section with a discussion of the experimental evidence for matter waves. The evidence consists in observing diffraction or interference patterns due to matter waves. Since an electron wave has a wavelength of atomic dimensions, to see diffraction we must find slits as small as atoms. Fortunately, nature provides us with such slits in the form of crystals. Crystals are regular three-dimensional arrangements of atoms; the distances between the atoms are just the right size. In fact, this was determined by looking at diffraction patterns that X rays produced when passing through a crystal. American physicist Clinton Davisson was the first to discover that a beam of electrons passing through a crystal gives a pattern similar to that of X rays—conclusive proof of the wave nature of an electron (Fig. 21.9).

Since the days of Davisson, literally thousands of different experiments have shown the wave nature of matter in the small scale.

FIGURE 21.8 "My wife and my mother-in-law" (by W. E. Hill, originally published in *Puck* in 1915).

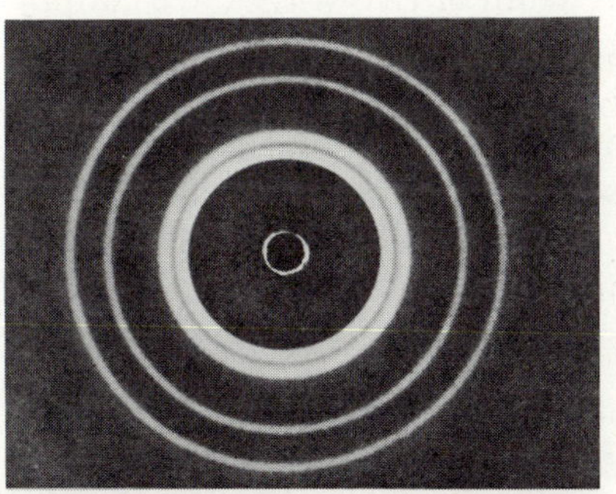

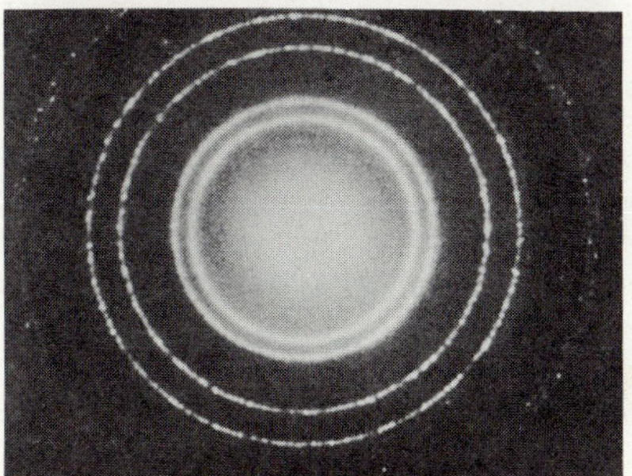

FIGURE 21.9 Diffraction pattern produced by electrons when passed through an aluminum foil *(below)* is qualitatively the same as that produced by X rays on aluminum *(above)* (Courtesy J. T. Shipman.)

What kind of waves are matter waves? When is an electron a wave and when is it a particle? Particles are localized concentrated objects with clearly defined tracks, whereas waves spread through large distances even in a short time. These are very different properties. Can the same object have such diverse properties?

Clearly we are in a tight situation. We are trying to describe what is in a new domain of nature with images developed for another domain. Logic in the familiar form is of very limited use here. So we will use imagination, a gedanken experiment to get an understanding of this strange affair. Some of the most startling aspects of the dual picture will emerge from our analysis.

The gedanken experiment that we choose is the Young's double-slit experiment, which we have already used for demonstrating the interference of light (see Section 11.4). This time we use the setup (Fig. 21.10) with an electron beam; since both light and electrons have dual properties, the results turn out the same for both.

The electron waves passing through the two slits will interfere at the screen. If crest and crest arrive together, we will get reinforcement; crest and trough will act destructively and give us dark spots on the fluorescent screen. The interference pattern is one of the alternate bright and dark fringes, exactly similar to the case of interference of light.

But now suppose we think of making the electron source very weak, so weak in fact that at any one time only one electron arrives at the slits. Will we still see the interference pattern? Since the dual property extends to individual electrons, our answer must be yes. But now we can ask a paradoxical question: How can we split one single electron? Without a split beam or two waves, we cannot imagine interference. On the other hand, it is equally impossible to imagine that an electron, a particle, can pass through two slits at the same time.

Does it? We can attempt to find out by looking. We can imagine a strong light beam that illuminates the electrons so we can see at any given time which hole an electron is passing through. Or is it through both holes, or what? So we turn the light on, and as we see an electron pass through a slit, we also look to see where the flash appears on the fluorescent screen. Every time an electron goes

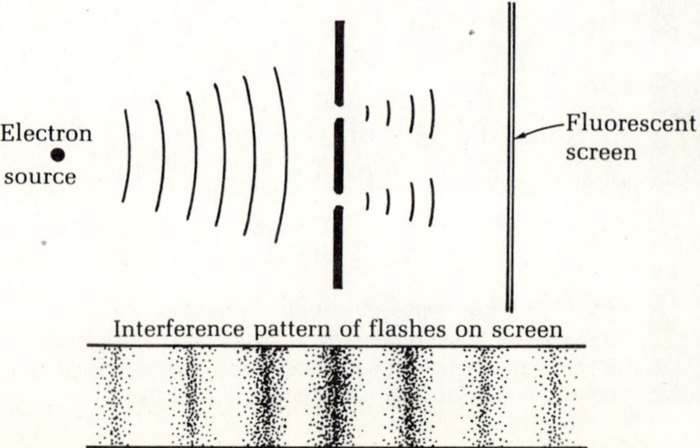

FIGURE 21.10 Result of a double-slit experiment with an electron source. We get an interference pattern on the screen.

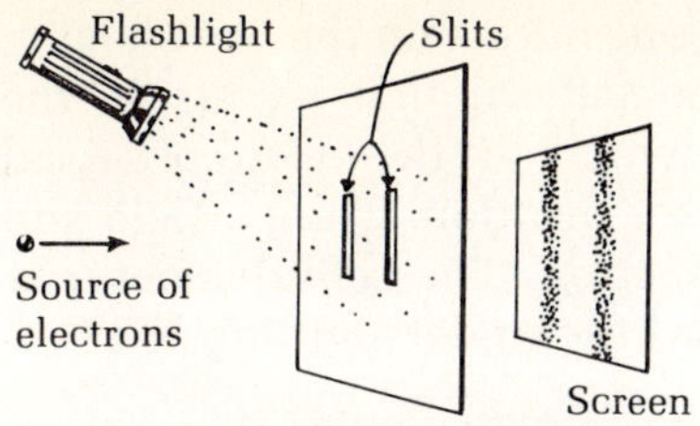

FIGURE 21.11 If we look with a flashlight and keep track of which slit an electron is passing through, we get the pattern on the screen shown here. The interference pattern has disappeared. When we look, the electrons behave as particles.

through a slit, we find that the flash appears behind the slit it just passed through. So after a time the electrons will make a pattern like Fig. 21.11; the interference has disappeared.

So now we have answered our first question. The electron passes through only one hole at a time; it doesn't split. It must be this way because the electron is a particle and a particle cannot split. When we observe, our observations bear this out. But when we don't observe, what happens?

You can easily let yourself get upset by thinking that an electron always passes through one of the slits (which we can determine through looking), and yet when we are not looking, the presence of the other slit affects the outcome of the experiment so that we get an interference pattern. However, you can't quarrel with what is. The experimentally established dual nature of the electron tells us that if we have a setup for detecting wave interference for the electron, we will observe such a pattern as long as we make no attempt to observe which slit the electron is passing through. We can interpret this. It means that the relation of the electron wave to the finding of the particle at a specific spot on the screen is a probabilistic one. Where the wave is strong, we can expect that the electron is likely to be there. Wherever the wave is weak, the electron is unlikely to be found. The density of spots on the screen corresponds to the probability of finding the electron. This probability interpretation is due to German physicist Max Born. The electron waves are **probability waves.**

On the other hand, the particle nature of the electron assures us that whenever we have a setup geared to see the electron as a particle, that's what we are going to see. When we observe each electron passing through a particular slit, we get the pattern on the screen that is characteristic of the particle behavior; the interference pattern is destroyed.

We can understand the destruction of the interference pattern in the following way. Our observation of the electron as it passes through a slit affects it so that its phase coherence with the other probability wave is disturbed. Without the phase coherence, there is no interference pattern.

So the electrons are very delicate. Actually, not only electrons but all microscopic objects are delicate, so much so that even putting a light beam on them disturbs their states. Specifically, the collision of the photon of light changes the momentum of the electron and thus its wavelength (through de Broglie's relation) by an unknown amount, which is enough to destroy the coherence of the two waves as they arrive at the screen. So instead of the interference, we see the pattern due to electrons passing through individual slits independent of the other.

This result—that if we localize the electron, we necessarily introduce an uncertainty in the momentum of the electron— is called the **uncertainty principle,** which was discovered by Heisenberg. We cannot observe an electron without affecting it. The uncertainty principle thus is a restriction on the measurement process: it restricts our ability to measure the momentum and position of a particle simultaneously.

Another way to understand the dual aspects of our gedanken experiment is to realize that the position and the momentum measurements are really complementary, mutually exclusive processes. We can concentrate on the position, in which case we lose track somewhat of the momentum and cause an admixture

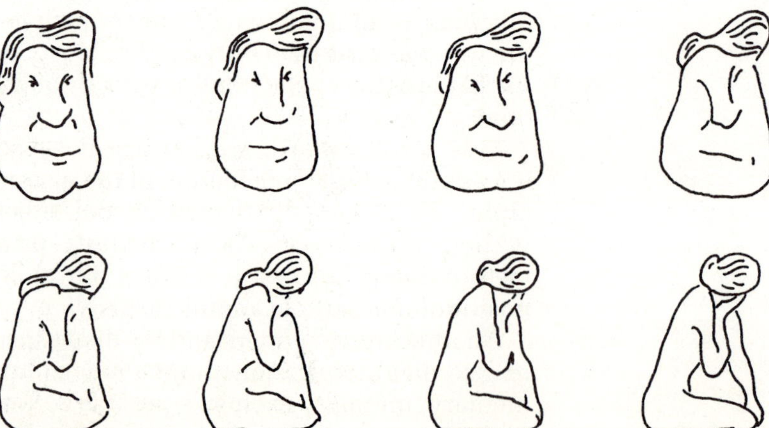

of wavelengths in the electron wave. Or we can concentrate on the wavelength, which we can determine from the interference pattern that results in the double-slit experiment, but now we cannot tell which slit the electron passed through; we have somewhat lost track of its exact position. The particle and waves are thus complementary aspects of the electron. The **complementarity principle** is the contribution of Niels Bohr toward the understanding of quantum mechanics.

To see more clearly how complementarity operates, suppose in the experiment above we make the light that we see the electrons with somewhat dimmer. When we do the experiment with dimmer and dimmer light, we find that some of the interference pattern starts to appear again, becoming more and more prominent (Fig. 21.12). When the light is turned off completely, the full interference pattern comes back.

What is happening now is this. As the light gets dim, the number of photons (don't forget that light, too, has a complementary particle nature) scattering off the electrons decreases, so some of the electrons entirely escape being "seen" by the light. The reappearing interference pattern is due to these "unseen," and therefore undisturbed, electrons. Those electrons that are "seen" still appear behind either slit No. 1 or slit No. 2, just where we expect it. So the result of this series of experiments can be described by saying that as the light becomes dimmer and dimmer, more and more electrons escape undisturbed and unobserved, so what we see on the screen becomes more and more like a wave interference pattern. In the limit of strong light, the electron is a particle; only its particle nature is seen. In the limit of no light, only its wave nature is seen. In the case of various intermediate light situations, both aspects show up to varied degrees.

We will take advantage of a series of pictures to illustrate what is going on (Fig. 21.13). In the picture on the top left, you can see the man only; this corresponds to our strong-light situation—you see only the particle nature of the electron. Then as we progress to the other pictures in the top row, we begin to see the girl—just as, when we start to make the light dimmer, some electrons defy observation and we start to see the wave nature of the electron. Then finally in the last picture (the right in the bottom row), only the girl can be seen;

FIGURE 21.13 The man-girl sequence. (Gerald H. Fischer, "Measuring Ambiguity," *American Journal
of Psychology 80,* 541–557 (1967).
Published by the University of Illinois
Press. Copyright 1967 by Karen
Dallenbach.)

this is when the light has been effectively turned off and it's all wave now for the electrons.

Complementarity

Still all this may seem strange to you. Is it really necessary to describe reality in terms of polar opposites? Just contemplating that contrapuntal characteristics like "waveness" and "particleness" can be attributed to the same object is very difficult for our minds, let alone adopting such a viewpoint. The complementarity principle was invented to mollify this consternation. Complementarity assures us that when we see one of the characteristics, we do not see the other. So in each individual observation we see, the reality is expressed as one of the two complementary pictures, each of which is accessible to our senses. The reality, of course, transcends the two complementary pictures.

Note the parallel between this teaching of modern physics and that of the Eastern tradition as described in Herman Hesse's *Siddhartha:*

> Everything that is thought and expressed in words is one-sided, only half truth; it all lacks totality, completeness, unity. When the illustrious Buddha taught about the world, he had to divide it into Sansara and Nirvana, into illusion and truth, into suffering and salvation. One cannot do otherwise, there is no other method for those who teach. But the world itself, being in and around us, is never one-sided. Never is a man or deed wholly Sansara or Nirvana; never is a man wholly a saint or a sinner.

The Eastern thinker appreciates the complementarity of opposite viewpoints for the description of the "one," and learns to transcend beyond duality. The quantum physicist learns the same thing. The same principle operates, although the realms of experience are different.

Probability and uncertainty

Thus in a description of the microworld, we cannot talk about both where the electron is and what the momentum of the electron is at the same time with arbitrary accuracy. All we can do is talk about the probability of finding the electron at a given point of space or the probability that the momentum has a certain value. Quantum mechanics is the mathematical framework for determining these probabilities.

The probabilities are given according to the rules of quantum mechanics. Usually the results of the calculation can be given in the form of a probability distribution curve. The probability is found to be maximum at some point (x_0 in Fig. 21.14), but it has finite value in a region around that point and approaches zero outside this region. The width of the probability curve can be called the uncertainty in the position of the particle, Δx (pronounced "delta x"). Similarly, we can plot a probability distribution for the momentum of the particle (Fig. 21.15), which gives an uncertainty Δp (pronounced "delta p") in the momentum of the particle. Heisenberg showed, and nature seems to demand, that the product of the two uncertainties Δx and Δp be greater than or equal to

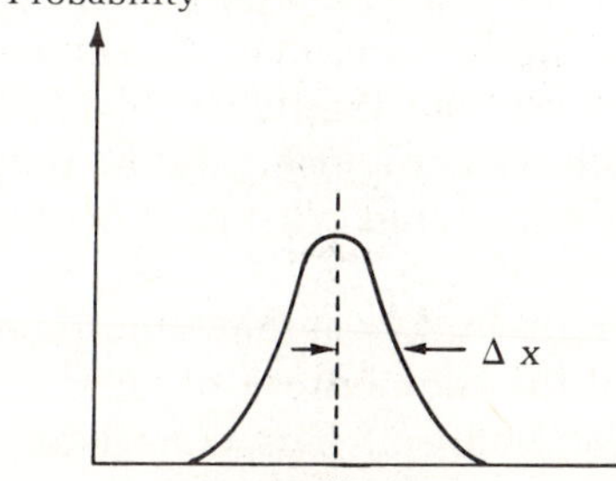

FIGURE 21.14 Probability and uncertainty. The probability that the electron occupies the position x is plotted against x. The uncertainty in the position of the electron is Δx, as shown.

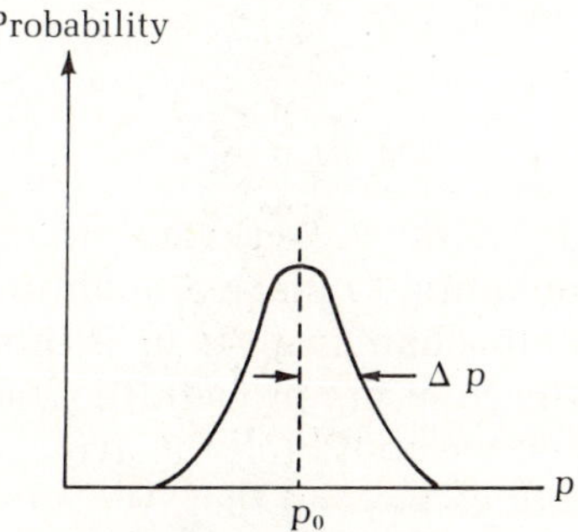

FIGURE 21.15 The probability that an electron has momentum p is plotted against p. The momentum uncertainty is given as Δp, as shown.

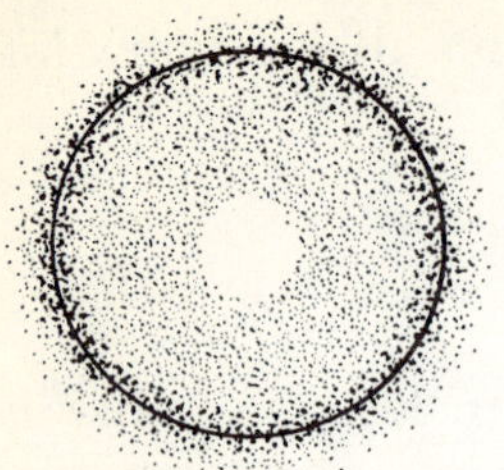

FIGURE 21.16 The "electron cloud," or probability density of the electron, in the lowest stationary state (the ground state) of the hydrogen atom. The circle denotes an orbit in the average sense.

Planck's constant h. This is the mathematical statement of the uncertainty principle:

$$\Delta x \, \Delta p \gtrsim h \qquad (21.3)$$

where the symbol $\gtrsim$ denotes greater than or approximately equal to.

With the help of the uncertainty principle, we can understand why the atoms come in a certain size and why we cannot make them any smaller. There is a science fiction story, *Fantastic Voyage,* in which it is contemplated that all sizes can be miniaturized to arbitrarily small sizes, a sort of "Lilliput" effect in modern form.* The uncertainty principle tells us that such feats are quite impossible. The atomic size represents the uncertainty in the position of the atomic electrons. If we try to localize the electrons further in a smaller region of space, thus reducing the position uncertainty, the momentum will increase so much that the electrons will escape from the atom.

The uncertainty principle also makes it clear that there is a definite quantum fuzziness concerning where the electron is inside the atom. We should really talk about an electron cloud around the atomic nucleus to get a better picture of what is, as in Fig. 21.16, it being understood that the heavier-shaded regions have a greater probability of finding the electron there than the lighter-shaded regions. Notice that there is no orbit really, although we can loosely define one (the circle in the figure), with its radius equal to the average distance of the electron from the nucleus.

There is one more important feature of the uncertainty relation that should be pointed out. Suppose you made a measurement of the momentum of an electron. There is no longer any uncertainty in its momentum; it is completely known. What can we say about the position of the particle in such a situation? The answer is nothing, absolutely nothing! From Eq. (21.3), if $\Delta p = 0$,

$$\Delta x \approx \frac{h}{\Delta p} = \frac{h}{0} = \infty$$

The position is completely uncertain. You might say, "But surely the electron must be somewhere. It's just that we don't know where." No, it is more than that; we cannot even define the position of the electron.† Thus quantum logic is very different from the logic you may be familiar with.

Indeterminacy and causality

We have referred to an electron in an excited atomic energy level several times. Suppose we ask the question: When is the electron going to make a quantum jump to a lower energy level? The answer quantum mechanics gives us is this: we do not know. All we can figure out from quantum rules are probabilities; so if we had a very large number of excited atoms, then we could tell the average time of decay for the entire ensemble. But we cannot go beyond that. We have

---

*Isaac Asimov, *Fantastic Voyage* (New York: Bantam, 1966).

†Perhaps we can say, quoting from the Upanishad,

    It is within all this
    And it is outside of all this.

no idea about the precise moment of time a particular electron is going to take the jump.

There is a phenomenon named radioactivity in which atoms of a certain species undergo a transformation, or decay. They become atoms of a different species, emitting energetic subatomic particles in the process. We can detect the radiation with an instrument known as the Geiger counter. Thus the number of ticks of the counter in a given amount of time will tell us the rate of decay of a given radioactive sample. Can we predict such rates of decay of radioactive atoms from quantum mechanics?

For a large assembly of radioactive atoms, quantum mechanics is all we need to calculate the decay rate. That is, from quantum mechanics we can calculate the probability of the decay. So long as we are interested in a large number of atoms, the actual experimental situation, the probability also gives us the rate of decay, on the average. But if you consider the decay of a single radioactive atom, precisely when the atom will decay cannot be predicted, only the chance that it will decay in such and such time.

Quantum mechanics with its probabilistic laws is thus basically acausal; no precise cause can be determined for the jump of a particular electron from one stationary state to another. We don't know the specific time of a particular jump, and therefore we cannot assign a specific cause for the jump. The relationship of cause and effect is fundamental in our minds, so this may give some consternation. However, we can take some consolation from the fact that the cause-effect relationship is still preserved in the statistical sense.

One of the most important features of the uncertainty principle, and quantum mechanics in general, is that the observer has started to play an active role in determining physical outcome. It is the disturbance created during the act of observation that gives us the uncertainty principle. This, at least to some extent, destroys the objectivity of the classical physicist; the complete separation of matter and mind is given up. This is perhaps one of the most forward-looking aspects of quantum mechanics.

Psychologists have known ever since the time of French philosopher Auguste Comte (around 1900) that a strict deterministic account of human experience is not feasible. Whenever we try to keep track of processes going on in our consciousness, the process of observation itself changes the sense and direction of the mental process. This is not qualitatively different from the indeterminacy of Heisenberg.

Schrödinger's cat

Many people still retain some healthy skepticism about quantum mechanics as being the ultimate physical theory. Certainly quantum mechanics can tell us correctly the result of an experiment, but it is sometimes ambivalent about the state of the system between experiments, in the sense that it is hard to interpret what it says. Schrödinger, who was a critic of the statistical probabilistic interpretation of quantum mechanics all his life, put it beautifully in the form of a paradox. This paradox is now called "Schrödinger's cat." Suppose we put a cat in a cage with a radioactive atom and a Geiger counter, which will tick if the atom undergoes a decay. The triggering of the Geiger counter breaks a bottle of

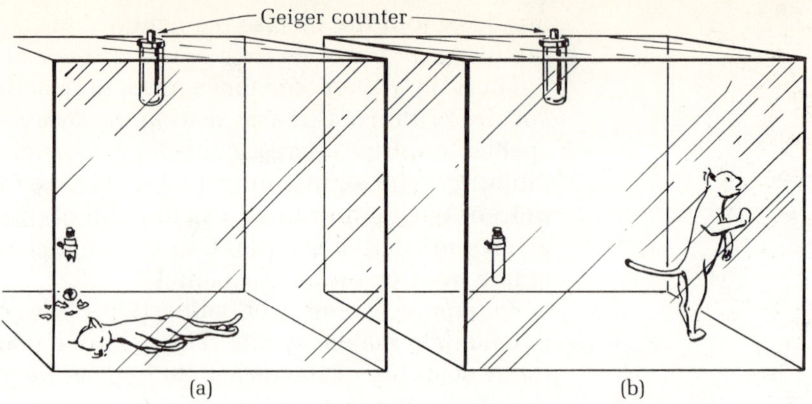

FIGURE 21.17   Schrödinger's cat. After an hour will the cat be dead, as in frame (a), or will it still be alive, as in frame (b)? Or will it end up half alive and half dead, which a literal application of the quantum mechanical probability rules would suggest as the answer? This is the cat paradox.

poison, which kills the cat. The radioactive atom is chosen so that chances are fifty–fifty that it will decay within the hour. How would quantum mechanics describe the state of the cat after this hour? Of course, if we made an observation, the cat would either be alive or dead. But if we didn't? If we talk in terms of probability, the probability that the cat is dead is 50%. On the other hand, the probability that the cat is alive is also 50% (Fig. 21.17). So is the cat a combination of these two states, half dead and half alive, after the hour? Does it make sense to talk in these terms?

One way of getting out of this kind of predicament is to say that quantum mechanics makes predictions about experiments involving a large number of objects. If there were 10^{10} cats, each in individual cages, quantum mechanics could tell us that half of them would be dead in the hour and the other half alive, and it would be right. Maybe for a single cat the theory does not apply. But isn't this just admitting that quantum mechanics is not a complete theory?

The cat paradox also brings to light another aspect of the quantum way of looking at the physical world with which Einstein could never reconcile himself. What actually happens to the cat in one hour? Will it be dead, or will it be alive? Who or what determines the outcome? It is as if somebody is throwing dice to decide what to do. Einstein used to say, "God does not play dice."

The debate about such things still goes on. There is some added spice to this issue if you bring in the question of conscious beings interfering with a physical experiment. And the answers so far have not been unanimously accepted by physicists.

If you cannot solve a problem, you can make a joke about it. Here is an artist's concept of the quantum mechanical method of catching an African lion.* Just set up your cage and wait, he says, because, "at any given moment there is a positive probability that there is a lion in the cage. Sit down and wait."

There is also a positive probability that we will eventually understand such things to everybody's satisfaction. Until then, sit down and think.

---

*H. Petard, in *American Mathematical Monthly*, vol. 45 (1938), p. 446.

MALCOLM'S LOGIC

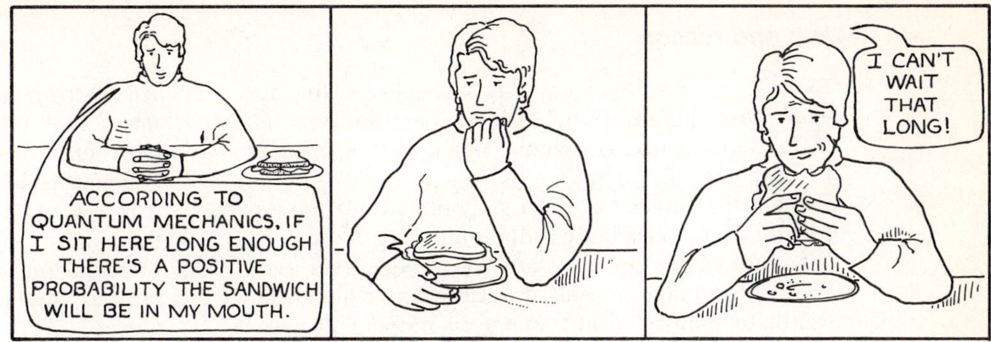

SUMMARY

This is the introductory chapter to the quantum mechanical view of nature, a view that is essential to the understanding of the behavior of submicroscopic particles. It began with Max Planck who, with his formula

$$E = h\nu$$

succeeded in developing a theory of blackbody radiation that agreed with experiment. Then Einstein developed the photon idea in order to explain the photoelectric effect. Niels Bohr injected the idea of quantum states of the electron and built a successful model of the atom on that basis.

The final piece of the puzzle was the work of three scientists. The first, de Broglie, introduced the idea of matter waves obeying the relation

$$\lambda = \frac{h}{p}$$

The other two, Heisenberg and Schrödinger, completed the quantum mechanical picture by developing the equations obeyed by the new mechanics, the equations that must replace Newton's second law of motion in the submicroscopic domain of nature. At the end we have a remarkably complete theory, so far as the prediction of experiments are concerned.

Conceptually, the quantum mechanical theory introduces very large departures from the classical view of nature. Matter and light both are recognized as dual—as both particle and wave. But matter waves are not like ordinary waves; they must be interpreted as probability waves. Thus the statistical concept of probability enters physics, and physics no longer is expressible in completely deterministic terms. The indeterministic character of the submicroscopic world is summarized by the Heisenberg uncertainty relation:

$$\Delta p\ \Delta x \gtrsim h$$

The duality of matter can be understood from the principle of complementarity. A complete view of nature can be obtained only as separate polar opposites, like matter and waves; the reality is transcendental, it is beyond duality. Although quantum mechanical theory always explains experimental data, some physicists find the underlying framework of uncertainty and the probabilistic aspects unsatisfactory.

Review and reason

1. Explain the concept of a quantum of energy. Give an example of quantization from everyday life. What is the difference between such "quanta" in everyday life and nature's quanta?

2. Explain the concept of blackbody radiation. Draw a picture of the blackbody radiation curve.

3. What are energy levels? Draw the energy level diagram of the quantum oscillator. How does a quantum oscillator emit or absorb energy? Compare its performance with a classical oscillator.

4. What is a photon? Explain how photons produce the photoelectric effect.

5. Explain the concept of the "stationary states" of an atom. How does an atom emit or absorb radiation?

6. What is meant by matter waves? How is the wavelength of matter waves related to the momentum? Why isn't the wave nature of ordinary matter obvious to us?

7. How did de Broglie combine the stationary wave idea with the Bohr orbits?

8. What is a Schrödinger equation?

9. Explain the concept of duality of matter and light.

10. What is a probability wave? How do we know that matter waves are probability waves? Discuss briefly the gedanken double-slit experiment with electrons cited in the text.

11. An electron is both a wave and a particle. Does this mean that the electron is something intermediate between a particle and a wave? Perhaps we should concoct a name like "wavicle" in describing such things as electrons. Give your own understanding of this.

12. State the uncertainty principle in regard to the uncertainties in the position and the momentum of an object.

13. Is it easier to predict the position of a more massive particle than the position of one of lesser mass? Explain.

14. Much of the atom is empty space. Is it possible to squeeze the atoms and make use of the empty space? Why or why not?

15. Explain how the uncertainty principle affects our ideas about causality and determinism.

16. Explain the complementarity principle. Find an everyday life example of complementarity.

17. Describe the paradox known as "Schrödinger's cat." Whether the paradox has been resolved or not is somewhat a matter of opinion. What is your opinion?

18. If Planck's constant were zero, would there be any necessity for involving quantum mechanics? Explain your answer.

Arithmetic

1. The energy of oscillation of a pendulum clock is of the order of 10^{-3} J. From its frequency of oscillation of 0.5 Hz, calculate the energy spacing of the permitted quantized levels of the pendulum. You should be able to see that this is extremely small compared to the value of 10^{-3} J, which can be regarded as an "average" energy. Since the quantum of energy is very small compared to the average energy, we can neglect quantum effects entirely for motion of macroscopic objects like the pendulum.

2. A photon has an energy of 10^{-19} J. What is its frequency? What is its wavelength?

3. Your eye can respond to a single photon. How much energy does your eye respond to when it absorbs a photon of frequency 6×10^{14} Hz?

4. The frequency of an ultraviolet photon is of the order of 10^{16} Hz. Calculate its energy. Given 1 electronvolt (eV) $= 1.6 \times 10^{-19}$ J, express the energy of the given photon in electronvolts. Note that this is of the same order as the energy required to ionize an atom (in other words, the binding energy of an electron to the atom). This is the reason that photons of ultraviolet and higher frequencies are so damaging to living tissue.

5. The velocity of a certain electron is $0.01c$. Given the mass of the electron as 9.1×10^{-31} kg, what is the momentum of the electron. What is the electron's de Broglie wavelength?

6. The electron of a hydrogen atom can be anywhere from zero to roughly 5.3×10^{-11} m away from the nucleus. Therefore, we can say that the uncertainty of the electron's position is 5.3×10^{-11} m. Using the uncertainty relation, calculate the electron's momentum uncertainty. What does the value of the momentum uncertainty mean?

22 Atoms and Solids

Hush—the electrons sing
Dropping
Down
Through their orbits
From ring
To glowing ring
Lower
And lower
Hush—*

LOUISE McNEILL

■ 22.1 Atomic Physics

With the advent of quantum mechanics, some of the mysteries associated with atomic phenomena began to have explanations. When light from an electrical discharge tube (a partially evacuated tube through which an electric current flows) containing a little of the element hydrogen is passed through a prism, what we see are a few narrow lines of light of several colors. This is very different from the **continuous spectrum** of color that we get when we decompose white light, either from the sun or from an incandescent lamp. This kind of **line spectrum** is a common thing for chemical elements; in fact, each element gives its own characteristic line spectrum. We can actually identify a chemical element from the observation of its line spectrum. The line spectra of atoms

---

*"The Tuning Fork," by Louise McNeill, in *The Christian Science Monitor*, July 25, 1977. Reprinted by permission from The Christian Science Monitor. © 1977 The Christian Science Publishing Society. All rights reserved.

were some of the first things to be explained by the new theory. We will examine this in some detail in this section. And, of course, most physical theories find important technological applications eventually. The atomic theory has given us the laser, which we will also discuss.

One of the biggest surprises of nineteenth-century chemistry was a piece of work by a Russian chemist named Dmitri Mendeleev, who found some strange but systematic repeated behavior among the elements. He arranged the elements in a table called the **periodic table** (shown in Fig. 22.10) to exhibit this order. Elements in the same column of the table are remarkably similar in their physical and chemical properties. An explanation of this has been given by the quantum theory.

Atoms and light

The first triumph of the new quantum theory of the atom came from the analysis of the light emitted by an atom as one of its electrons jumps from a level of higher energy to one of lower energy. How can we study the light from an individual atom? As you know, a solid when hot emits continuous radiation. The atoms in a solid are packed so closely together that they affect each other too much. The "music" each makes gets mixed with its neighbors' and all that comes out is noise, optical noise, which is what the continuous radiation is. So in order to study the "music" of each atom, we have to get it in a relatively free state of existence.

Fortunately, this is not very hard to do since it is known that in the gaseous state the atoms are practically free of interaction. So all we have to do is look at the light emitted from a heated dilute gas to observe the spectra of the atoms.

Historically, physicists studied the light from a heated gas without having a clear notion of why they were doing it. They found that it is easier to get light from a gas by exciting it electrically. To this end they put a little of the gas in an evacuated tube and passed an electrical spark through the gas. Then they studied the light that came out with a simple prism arrangement (Fig. 22.1). To their surprise they found that the light consisted of individual lines. Since light of different wavelengths was dispersed by different amounts, by noting the place on the screen where a particular line appeared, the wavelength of that line could be determined.

Thus these spectroscopists, as these experimenters of optical spectra are called, amassed a huge amount of data on the lines of light (more formally these

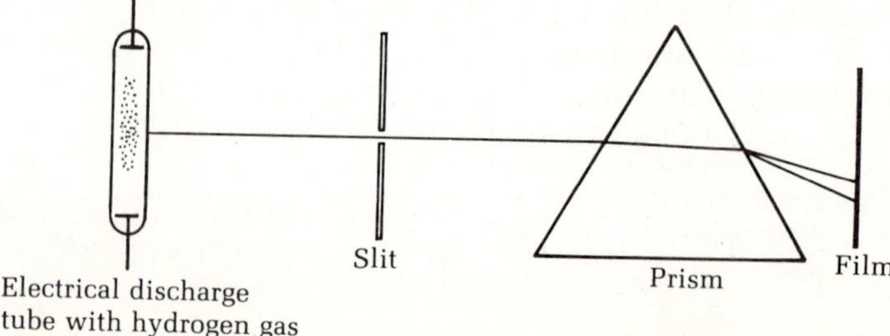

FIGURE 22.1   An arrangement for the study of the emission spectra of gases. The spectrum consists of individual lines.

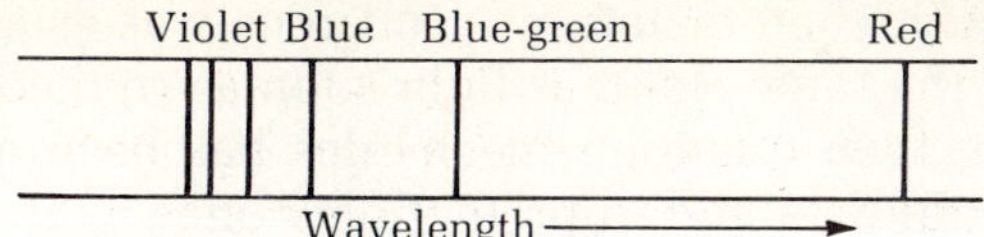

FIGURE 22.2 The Balmer lines of the hydrogen atom. There are other lines in the hydrogen spectrum, but these played a very special role in the development of the Bohr atomic theory.

are called line spectra) emitted by various gases. The most famous of all these spectral data came from a German school teacher named J. J. Balmer in his work on the hydrogen atom (Fig. 22.2).

When Niels Bohr was working out the details of his theory of the hydrogen atom around 1913, one of the first things he did was to apply his theory to the calculation of its line spectra. In this theory the line spectra of the hydrogen atom were due to the transition of the atomic electron from various excited higher energy levels to lower levels. Bohr could determine the energy of the light of the lines using Planck's formula $E = h\nu$, except E now was the difference of energy between the two levels involved in the transition. Calling E_i and E_f the energies of the initial and final energy levels, respectively, the frequency of the photon emitted was given as

$$h\nu = E_i - E_f$$

or

$$\nu = \frac{E_i - E_f}{h} \tag{22.1}$$

Since his theory had only a few discrete energy levels for the hydrogen electrons, he predicted only a few discrete frequencies, or lines. For each final level f, corresponding to the various initial states i (that are higher in energy), he predicted a series of spectral lines related by very simple formulas.

Imagine Bohr's surprise, then, when somebody told him that such correlation of the spectral data of hydrogen into different series already existed. A few hours of intense excitement, and he had it. His theory explained all the known spectral data for the hydrogen atom, including those of Balmer. Each of the Balmer lines were due to the transition of an electron from one of the higher levels to the first excited state (Fig. 22.3).

The excellent agreement of Bohr's theory with the spectroscopic data was what made people interested in the quantum theory of the atom. When Einstein heard about this, it is said that his eyes became big, and he said softly something to the effect that this will be considered one of the best pieces of physics ever done.

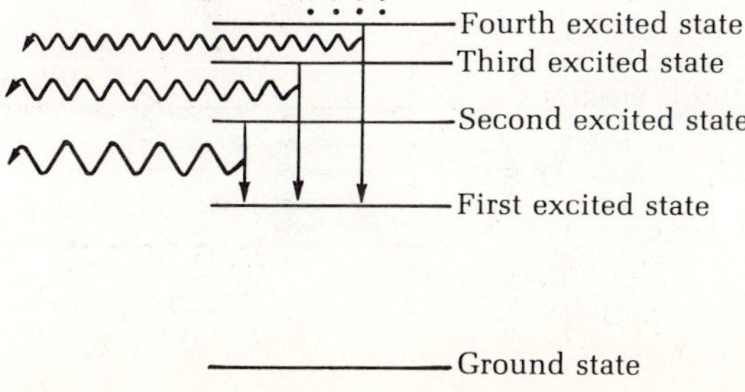

FIGURE 22.3 How Bohr explained the Balmer lines. Each line is due to a transition from a higher excited state to the first excited state of the hydrogen atom. The calculated frequencies [using Eq. (22.1)] agree with the experimentally observed frequencies of the lines.

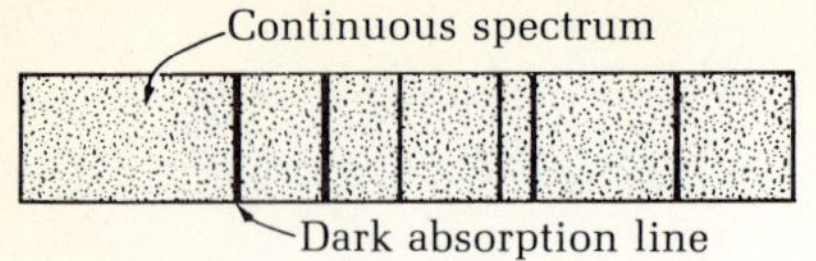

FIGURE 22.4 The absorption lines in the background of a continuous spectrum.

Later when quantum mechanics was established, the now complete theory gave the same result as Bohr's tentative theory in the simple case of hydrogen. Since then quantum mechanics has been able to explain many more of the spectral data and to predict new ones, which have been verified later.

Absorption spectra

When light falls on an atom, if the energy is right—that is, if $h\nu$ for the photon matches the difference of energy between two levels of the atom—the atom will absorb the light. The absorption gives rise to dark lines in the otherwise continuous spectrum of a solid or of some other source that has been contaminated by a few gaseous atoms (Fig. 22.4). The sun's own spectrum showed quite a few of these dark lines under the careful scrutiny first carried out by German physicist Joseph Fraunhofer (hence the lines are called Fraunhofer lines). The dark **absorption lines** match exactly the frequencies of the emission lines from the same gas. Thus looking at the absorption lines of the spectrum of a star is one way to determine the chemical composition of that star. The element helium was discovered through the observation of its absorption lines in the solar spectrum. The name helium (from the Greek "helios," meaning "sun") is a reminder of its discovery story.

Fluorescence

The formula $E = h\nu$ makes clear that the energy of a photon increases in direct proportion to its frequency. Ultraviolet light, with a greater frequency, has more energy than visible light. In fact, an ultraviolet photon has just enough energy to disrupt molecules of the human body, knock electrons out of it, and so on. This is the reason that you get a sunburn; the ultraviolet component of the sunlight does the trick. The visible parts of sunlight are quite harmless.

There is a phenomenon called **fluorescence** in which ultraviolet photons are absorbed by some atoms, which then emit low-energy photons; this is the basis of the fluorescent lamp. A fluorescent lamp in its essential features is shown in Fig. 22.5. Basically it is a glass tube filled with mercury vapor at low pressure, containing a hot filament which emits electrons near the base of the tube, and with a coating on the inside surface of the tube of special fluorescent atoms. The electrons from the filament collide with the mercury atoms and excite their electrons to high energy levels. When the electrons de-excite to lower levels,

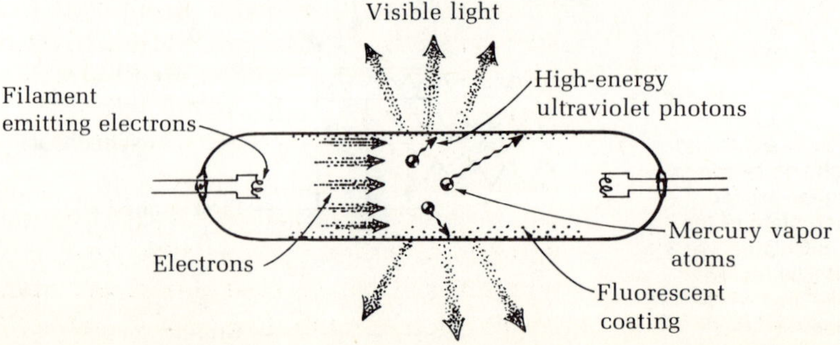

FIGURE 22.5 A fluorescent lamp.

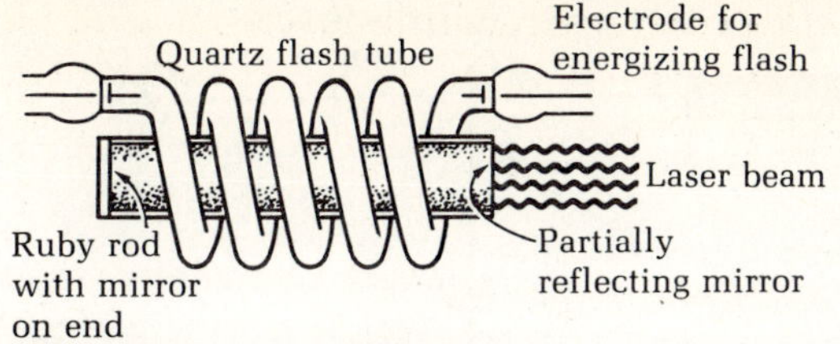

high-frequency ultraviolet light is emitted. The energy of the photon is equal to the difference of energy of the two electronic levels. The atoms of the coating now absorb some of these ultraviolet photons and transfer some of their electrons to higher energy levels. But the energy level structure of these atoms is such that when their electrons de-excite, only photons of lower energy in the visible spectrum are emitted.

The transition of an electron from each higher energy level to a lower one gives a photon of a characteristic frequency or color a sharp line in the spectrum Usually there can be several such lines, and their mixture determines the color or light that we get from the lamp.

The laser

We can now discuss another aspect of the laser. We have previously shown how the stationary wave principle is used to get a unidirectional beam of light in the laser. The other feature of the laser is that it gives a very intense beam of a single wavelength. We can now understand how it does that.

Fig. 22.6 shows a ruby laser, which uses a pink ruby rod with a partially reflecting mirror at one end. Ruby is an oxide of aluminum (a chemical compound of aluminum and oxygen atoms) with just about five parts in ten thousand of an oxide of another metal, chromium. Very intense ultraviolet photons from a flashtube filled with the gas xenon are directed into the ruby rod. The electrons of the chromium atoms are lifted to a specific higher energy level by the incident photons. A few of these electrons come down to the ground level spontaneously and make red light in the process. And now a most interesting thing happens. The red photons, upon being reflected from the mirror at the end, fall back on the excited chromium atoms. This causes a **stimulated emission** of radiation from the chromium atoms, which are de-exciting in unison, all giving radiation of the same color. Such an "in phase" pulse of single-frequency photons may be very intense (Fig. 22.7). So the key to a laser beam's intensity is really stimulated emission. In fact, the name laser is an acronym for "light amplification by stimulated emission of radiation." The laser can be such a concentrated beam of energy that it can burn a hole in a steel plate, as you may have heard.

The fact that laser light is single-frequency light, referred to as "frequency coherence of the light," gives us another technological possibility. A laser wave can be modulated just as a radio-frequency wave can. Thus a laser wave can carry information such as speech or music and can be used in long-distance communication. The use of laser in communication (called light-wave communication) is a brand new field of research.

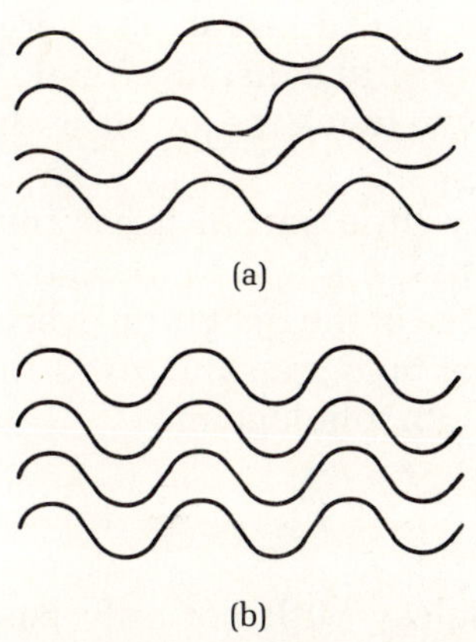

(a)

(b)

FIGURE 22.7   (a) Waves out of phase; an ordinary light beam is like this. (b) Waves in phase; laser light.

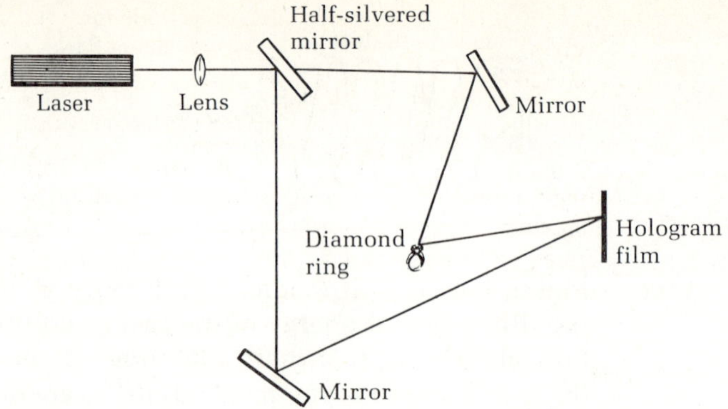

Holography

No discussion of lasers is complete without the mention of **holography,** or "holistic photography." This is the use of a laser beam to produce a three-dimensional picture of an object by using a method first suggested by physicist Dennis Gabor. The arrangement is shown in Fig. 22.8. The laser beam is spread out by means of a lens in order to illuminate the whole body of the object to be holographed. After the spreading is accomplished, the beam is split in two by means of a half-silvered mirror. One part of the split beam is directed by means of a second mirror onto the hologram film directly: this is called the reference beam. The other part is directed onto the object and is reflected by it. The reflected beam from the object forms an interference pattern with the reference beam at the location of the film, which records the interference pattern. This is the hologram: it is an interference pattern. If you look at a hologram, all you see is a pattern of whirling lines in which, however, is recorded the whole picture of the original object (Fig. 22.9).

In order to retrieve the picture, you have to shine a monochromatic laser beam on the hologram and then look through the hologram. The picture is three-dimensional, as you can verify by shifting the angle of your vision. The image is virtual; you can't touch it, but it looks like the real object in just about every respect. Tiffany's store in New York City has a window with a displayed hologram of its jewelry, and people have been known to try to touch the jewels with their hands, they appear so real!

One important aspect of the hologram is that every little part of it has the complete picture. The record was made without the use of a lens to focus the beam on the film. Thus every point of the film had a piece of the light from every part of the object. So you can tear off a little part of the hologram and you still have the whole picture. So a part is also the whole with a hologram.

The explanation of the periodic table

The table of the atomic elements that Dmitri Mendeleev built not only has survived with very few changes but also has become one of the fundamental aspects of atomic physics and chemistry. One of the most important achievements of the new physics was the explanation of the periodic table.

The table is shown in Fig. 22.10. As we have mentioned before, the table exhibits a spectacular periodicity (or repetition) of the properties of atomic elements when arranged in order of increasing **atomic number,** which is the number of electrons in the neutral atom. All elements having similar properties are put in the same column. The rows of the table define the periods. Thus the first period has 2 elements, the second and third have 8, the fourth and fifth have 18, the sixth has 32. It is believed that the seventh period is not complete. This is due to the fact that the atomic nuclei of elements heavier than 92 become more and more unstable, so much so that it is increasingly difficult for nature to produce them. Beyond element No. 106, it has not been possible to produce any element, even in the laboratory.

If you are interested in numerology, you can notice one interesting thing about these numbers right away. If you divide each number by 2, you get 1, 4, 4, 9, 9, 16. These are the squares of the first four integers, 1, 2, 3, and 4.

Ancient Greeks used to be interested in such numerology. In particular, the philosophers led by Pythagoras were fascinated by numbers. Pythagoras very probably discovered the musical octave and knew that other musical harmony—for example, the major chords—is based on ratios of small integers. It was but one step more for him to suggest that the motion of the entire world system should be based on numbers. (One person influenced by Pythagorean thought was Johannes Kepler, whose work on the five perfect solids has already been mentioned.) This numerological approach does not work, not as a world system in the macroworld of planets and such. But strangely, in the microworld of the atoms, the Pythagorean idea of the importance of certain numbers seems to have validity.

The stationary quantum states of the atomic electrons are characterized by a set of four numbers, which we call the quantum numbers. The first of these, called the **principal quantum number,** takes on the successive values of the integers starting with 1 (i.e., 1, 2, 3, 4, and so forth). It is usually designated by the letter n and is really a measure of the discrete value of the electron's energy in the relevant stationary state.* $n = 1$ is the lowest stationary state, or the ground state, $n = 2$ is next, and so forth.

For each value of n, a second quantum number, l, can be assigned, which takes on the values starting from zero and going all the way to $n - 1$. Thus for $n = 5$, l can assume values 0, 1, 2, 3, and 4. For multielectron atoms, the states corresponding to the same n but different l have different energy. The quantum number l actually measures the discrete value of the angular momentum of the electron associated with its motion around the nucleus. The third quantum number, designated by the letter m_l, denotes the direction of the angular momentum vector. It takes on the value from $-l$ all the way to $+l$, increasing in steps of 1. Thus, for $l = 3$, m_l can be -3, -2, -1, 0, $+1$, $+2$, and $+3$. The quantum states that have the same value of n and l but different values of m_l are found to have the same energy, a situation often referred to by saying that these states are **degenerate.**

---

*n also indicates the average distance of the electron from the nucleus; the larger n is, the larger is the average distance of the electron from the atomic nucleus.

PERIODIC TABLE OF ELEMENTS

1	2	3	4	5	6	7	8	9	10	11	12	13	14	15	16	17	18
H 1 1.01 Hydrogen																	**He** 2 4.00 Helium
Li 3 6.94 Lithium	**Be** 4 9.01 Beryllium											**B** 5 10.81 Boron	**C** 6 12.01 Carbon	**N** 7 14.01 Nitrogen	**O** 8 16.00 Oxygen	**F** 9 19.00 Fluorine	**Ne** 10 20.18 Neon
Na 11 22.99 Sodium	**Mg** 12 24.31 Magnesium											**Al** 13 26.98 Aluminum	**Si** 14 28.09 Silicon	**P** 15 30.97 Phosphorus	**S** 16 32.06 Sulfur	**Cl** 17 35.45 Chlorine	**Ar** 18 39.95 Argon
K 19 39.10 Potassium	**Ca** 20 40.08 Calcium	**Sc** 21 44.96 Scandium	**Ti** 22 47.90 Titanium	**V** 23 50.94 Vanadium	**Cr** 24 52.00 Chromium	**Mn** 25 54.94 Manganese	**Fe** 26 55.85 Iron	**Co** 27 58.93 Cobalt	**Ni** 28 58.71 Nickel	**Cu** 29 63.55 Copper	**Zn** 30 65.37 Zinc	**Ga** 31 69.72 Gallium	**Ge** 32 72.59 Germanium	**As** 33 74.92 Arsenic	**Se** 34 78.96 Selenium	**Br** 35 79.90 Bromine	**Kr** 36 83.80 Krypton
Rb 37 85.47 Rubidium	**Sr** 38 87.62 Strontium	**Y** 39 88.91 Yttrium	**Zr** 40 91.22 Zirconium	**Nb** 41 92.91 Niobium	**Mo** 42 95.94 Molybdenum	**Tc** 43 (99) Technetium	**Ru** 44 101.07 Ruthenium	**Rh** 45 102.91 Rhodium	**Pd** 46 106.4 Palladium	**Ag** 47 107.87 Silver	**Cd** 48 112.40 Cadmium	**In** 49 114.82 Indium	**Sn** 50 118.69 Tin	**Sb** 51 121.75 Antimony	**Te** 52 127.60 Tellurium	**I** 53 126.90 Iodine	**Xe** 54 131.30 Xenon
Cs 55 132.91 Cesium	**Ba** 56 137.34 Barium	**La** 57 138.91 Lanthanum	**Hf** 72 178.49 Hafnium	**Ta** 73 180.95 Tantalum	**W** 74 183.85 Wolfram	**Re** 75 186.2 Rhenium	**Os** 76 190.2 Osmium	**Ir** 77 192.2 Iridium	**Pt** 78 195.09 Platinum	**Au** 79 196.97 Gold	**Hg** 80 200.59 Mercury	**Tl** 81 204.37 Thallium	**Pb** 82 207.19 Lead	**Bi** 83 208.98 Bismuth	**Po** 84 (210) Polonium	**At** 85 (210) Astatine	**Rn** 86 (222) Radon
Fr 87 (223) Francium	**Ra** 88 (226) Radium	**Ac** 89 (227) Actinium	**Rf** 104 (261) Rutherfordium	**Ha** 105 (260) Hahnium	106												

58	59	60	61	62	63	64	65	66	67	68	69	70	71
Ce 140.12 Cerium	**Pr** 140.91 Praseodymium	**Nd** 144.24 Neodymium	**Pm** (147) Promethium	**Sm** 150.35 Samarium	**Eu** 151.96 Europium	**Gd** 157.25 Gadolinium	**Tb** (158.92) Terbium	**Dy** 162.50 Dysprosium	**Ho** 164.93 Holmium	**Er** 167.26 Erbium	**Tm** 168.93 Thulium	**Yb** 173.04 Ytterbium	**Lu** 174.97 Lutetium

90	91	92	93	94	95	96	97	98	99	100	101	102	103
Th 232.04 Thorium	**Pa** (231) Protactinium	**U** 238.03 Uranium	**Np** (237) Neptunium	**Pu** (242) Plutonium	**Am** (243) Americium	**Cm** (247) Curium	**Bk** (247) Berkelium	**Cf** (251) Californium	**Es** (254) Einsteinium	**Fm** (253) Fermium	**Md** (256) Mendelevium	**No** (254) Nobelium	**Lw** (257) Lawrencium

FIGURE 22.10 The periodic table. Each row of elements defines a period. For example, the first period consists of the two elements hydrogen and helium. The number at the top left of each element symbol indicates the atomic number Z. The number below the element symbol denotes the atomic weight.

The final quantum number, m_s, represents a two-valuedness (m_s can be only $+\frac{1}{2}$ or $-\frac{1}{2}$) the electron acquires, which is unique to some particles in the quantum world. What this means is that for each state characterized by a unique set of n, l, and m_l, the electron has two degenerate states represented by another quantum number m_s, which takes on two values. We often express this by saying that in one state the electron's spin is "up" and in the other state the electron's spin is "down." This designation implies that the electron has a spinning motion about its own axis. This is true in some sense, but the spinning motion of a quantum particle is something very special, certainly not like the spinning motion of daily rotation of the earth around its axis. So it is better to think that the electron has this extra quantum number taking on two values $+\frac{1}{2}$ and $-\frac{1}{2}$ for each state with a designated n, l, and m_l.

For each value of n, when we add up all the possible values that l, m_l, and m_s can take, we get $2n^2$ individual quantum states, some of them degenerate. These energy states of the electrons are shown in Fig. 22.11. We already have pointed out that the periods also are given as $2n^2$. This must be something more than a coincidence.

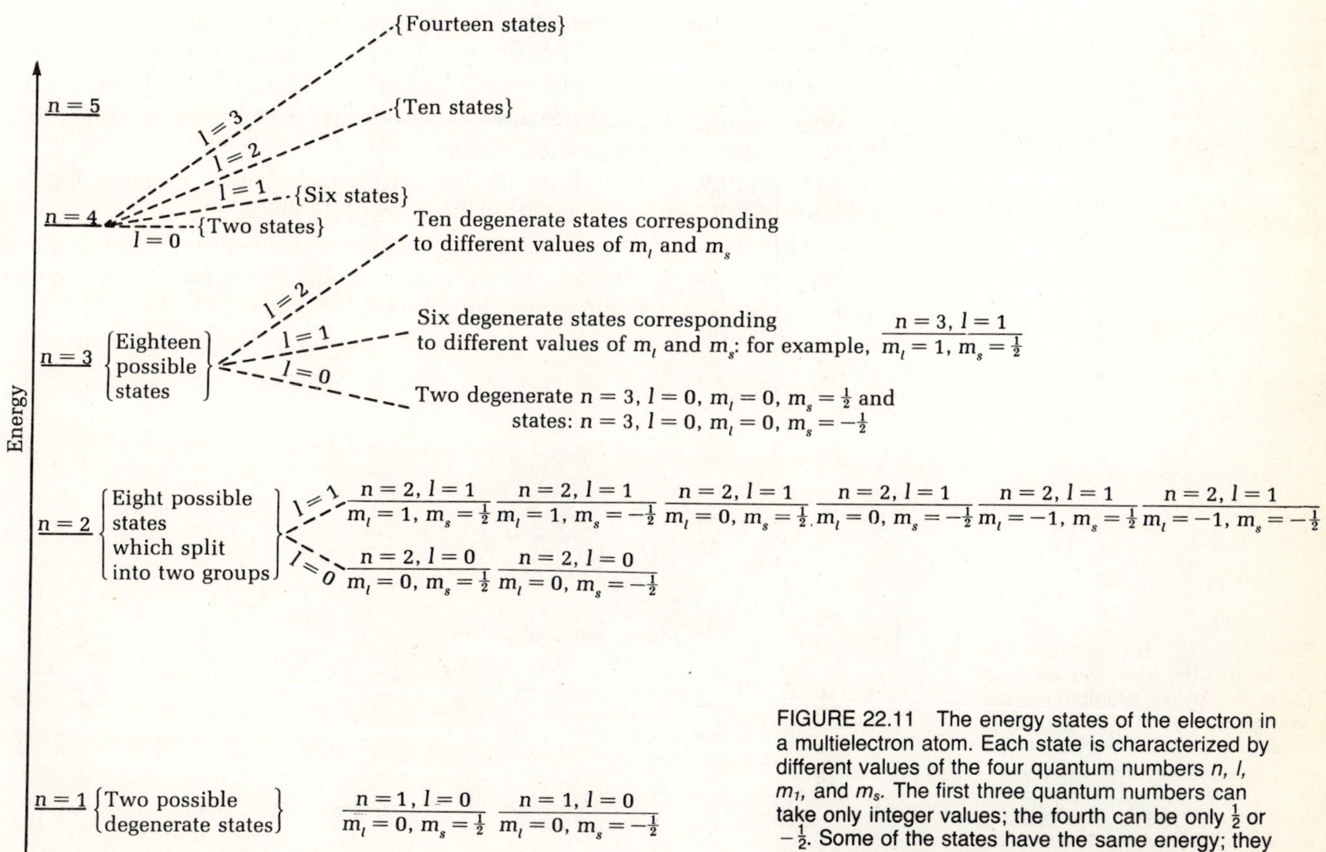

FIGURE 22.11 The energy states of the electron in a multielectron atom. Each state is characterized by different values of the four quantum numbers n, l, m_l, and m_s. The first three quantum numbers can take only integer values; the fourth can be only $\frac{1}{2}$ or $-\frac{1}{2}$. Some of the states have the same energy; they are called degenerate. Notice the large "energy gap" separation between groups of levels of similar energy. This is called a "shell structure."

And indeed it is so. It was German physicist Wolfgang Pauli who proposed that only one electron can occupy a single quantum state characterized by the four numbers—n, l, m_l, and m_s—that we have mentioned. This is called the **Pauli exclusion principle,** exclusion because the occupation of one quantum state by an electron excludes any other from occupying it. The atomic electrons obey the rules of a game of musical chairs: they do not sit on each other's lap; only one electron per chair is allowed, so to speak.

So the first period is explained simply by realizing that for $n = 1$, we have only two quantum states [$2n^2 = 2(1)^2 = 2$]. If we have only one electron, then that electron will occupy only one of the states. But if there is a second electron, it will take up the second state. And we have both states belonging to $n = 1$ occupied.

Such a configuration is specially stable, the reason being that the energy of the quantum states of $n = 2$ and higher is much larger than that of $n = 1$. This is an extremely interesting fact of the electronic energy levels: the levels cluster in energy around some values. This is called **shell structure.** There are large "energy gaps" between the shells.

Fig. 22.11 shows this shell structure of the energy levels of the electrons. Notice that for the higher shells, the clustering is not strictly according to the value of n, but nevertheless, the energy gaps between the shells persist.

This then is the explanation of the periodic table of the atomic elements: the Pauli principle, shell structure, and the assertion that for a nonexcited multi-electronic atom, its electrons must be placed in the electronic states of lowest available energy. This assertion makes sense since we are constructing the state of lowest energy for the atom: its "ground" state. When an atom has just the right number of electrons to fill up the complete shell, the atom is particularly stable. This is the case with helium, neon, argon, krypton, xenon, and radon; they are all noble gases, very noninteracting (see Fig. 22.12 for an example).

We can characterize any atom, then, by the number of electrons it has outside the closed shells, since the closed-shell electrons do not take part in the chem-

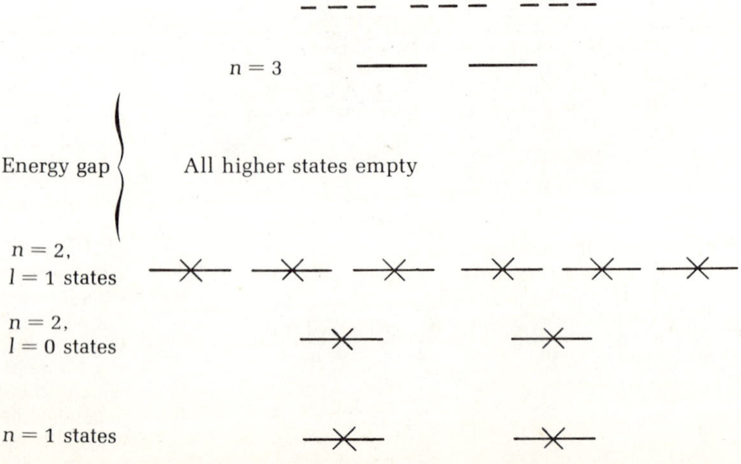

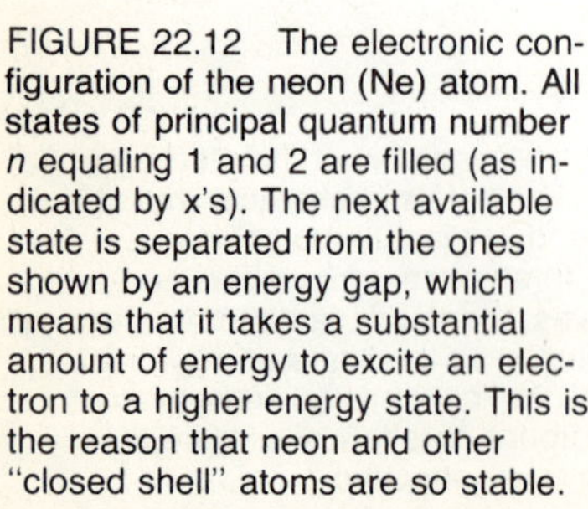

FIGURE 22.12 The electronic configuration of the neon (Ne) atom. All states of principal quantum number n equaling 1 and 2 are filled (as indicated by x's). The next available state is separated from the ones shown by an energy gap, which means that it takes a substantial amount of energy to excite an electron to a higher energy state. This is the reason that neon and other "closed shell" atoms are so stable.

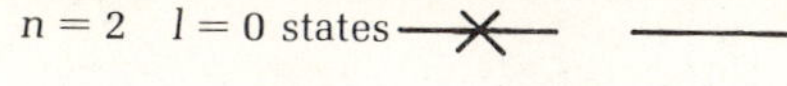

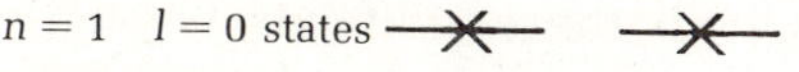

FIGURE 22.13   The electronic configuration of lithium (Li). It has one electron outside the closed-shell configuration of helium. The electron in the open shell can easily jump into another state without needing much energy

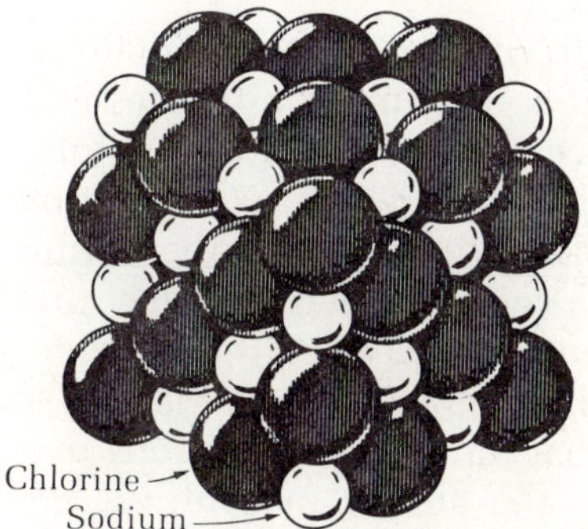

FIGURE 22.14   The sodium chloride crystal, showing the ionic bond between the sodium and chlorine atoms. The interionic bond between the sodium chloride molecules is large, explaining why molecules with this kind of bonding are ordinarily found in large aggregates like a crystal.

ical reactions. These extra electrons outside the closed shells are called **valence electrons.** Atoms that have the same number of valence electrons are put in the same column of the periodic table and have similar chemical properties. For example, the "alkali" metals—lithium, sodium, potassium, cesium, and so on—all have one electron outside the closed-shell atomic configuration of helium, neon, argon, and krypton, respectively (see Fig. 22.13 for an example). And it has been known for a long time that these metals have similar chemical properties.

Since the closed shells are very stable structures, it is understandable that in chemical reactions atoms that get together will try to make closed shells with each other's help. An atom like chlorine (Cl), which lacks one electron from a closed shell, has a particular affinity for an atom that has just the right excess of electrons, such as the sodium (Na). Together they make up for each other's lack of closed-shellness. The compound sodium chloride (NaCl), which is common table salt, is the result. As the sodium atom donates an electron to the chlorine, it becomes a positive ion, whereas the chlorine atom becomes a negative ion. The molecule NaCl is held together by the electrical attraction of the two ions (Fig. 22.14). Such a chemical bond in a molecule is called the **ionic bond.** There is also a tight bond when two atoms "share" a few of their electrons to achieve closed-shellness. This is the **covalent bond.** An example is the hydrogen molecule, H_2. By sharing each other's electrons, both H atoms seem like closed-shell atoms (Fig. 22.15).

Carbon is a very special atom because it combines with a variety of elements, and it can make very large chains of atoms forming very big molecules (see Fig. 22.16 for an example). Such "macromolecules" are the essential ingredient of life on earth. The special ability of carbon atoms to form a variety of compounds comes from the fact that its electronic structure is very special. It has four valence electrons over the closed-shell configuration of helium. Viewed in another way, it lacks exactly the same four electrons from the closed-shell atom of neon. It turns out that this is almost unique to carbon; only one other element has this property, silicon. This is the reason some people have suggested that if ever we encounter life based on any element other than carbon, that element would be silicon. As you may know, silicon compounds make up our rocks and sands. Can you imagine a rocky creature? Perhaps, you have seen the "Star Trek" episode that featured such silicon creatures (imagined, of course).

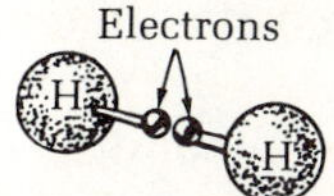

FIGURE 22.15   The covalent, or electron-sharing, bond. The two hydrogen atoms share their electrons to make the stable molecule H_2. The molecule itself is stable and exists quite independently of others. In fact, the intermolecular force between two hydrogen molecules is small.

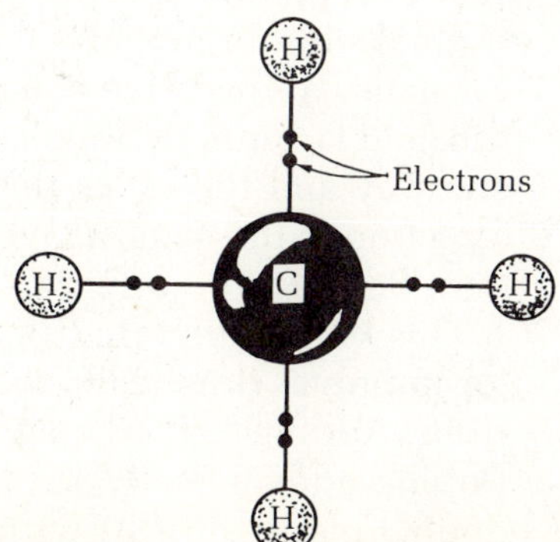

FIGURE 22.16   The covalent bonding in methane, CH_4, which is also called natural gas.

Solid-state physics is the largest discipline of physics practiced today. Solid-state physics deals with solids and materials, and in general, with their thermal, optical, electrical, and magnetic properties. Of all these subfields we will discuss only one, the electrical properties. In particular, we will ask the question: What is the difference between conductors and insulators?

It has been well known for sometime that there are also substances that are intermediate between a conductor and an insulator in their electrical attributes. Such substances are called **semiconductors.** Recently, solid-state physics has initiated a revolution in technology, the most well known symbol of which is the transistor. As we will see, the transistor is based on the physics of semiconductors. And so are the solar cells, which we mentioned previously. Solar cells are devices that convert solar energy directly into electricity.

The carriers of electricity

In Chapter 16 a picture was given of an electric current through a conductor in an electrical circuit. The basic idea of the picture is that in a conductor there exists an ample supply of current carriers—the free electrons—which are relatively free to wander around inside the body of the solid. These free electrons constitute the basic difference between conductors and insulators. In the atomic picture we can say that in a conductor some of the valence electrons (electrons that lie outside closed shells) are available in the entire solid as current carriers. The electron wave of these electrons spreads itself through the entire lattice of the crystal. This is what we mean when we say that the valence electrons in a metal are bound to the entire metal, not just to the atoms. In contrast, the valence electrons of a nonconductor are cemented to nearby atoms; they are not available as current carriers. In an insulator this cement, or chemical bond, is so strong that an electron cannot be freed, short of melting down or evaporating the crystal altogether.

The semiconductors, as we said earlier, are intermediate between these two. The valence electrons are bonded to atoms but not very strongly. Energy in the form of heat or light can free a few electrons from the atomic bonds and make a small supply of free-electron current carriers. Actually, it works even better than this picture implies. It turns out that the electrons leave holes at their original sites upon thermal excitation to higher energy levels. These "holes," or vacancies, of negative charge, act as positive charge carriers. So when an electric field is applied to such a crystal with electrons and "holes," the electrons go one way and the holes the other. But positive charge giving a current in one direction is the same as negative charge giving a current in the opposite direction. So the two currents add together to increase the net current.

This kind of picture can also explain why the performance of a semiconductor improves drastically with the addition of a small amount of specific impurities. Adding a small amount of impurity to a semiconductor is called *doping*. Doping adds a ready supply of free electrons. As an example, consider the doping of silicon (Si) with phosphorus (P). Silicon has 4 electrons over and above the closed-shell atom neon, so it has 4 valence electrons. Phosphorus, if you look at the periodic table, has five electrons outside the closed-shell configuration of neon. So the number of valence electrons of phosphorus is five. Now if a phosphorus atom replaces an atom of silicon at one of the lattice sites (Fig. 22.17), four of its valence electrons will bond with four neighboring silicon atoms, but the fifth one has none to make a bond with and will be quite free. Atoms that donate free electrons to a semiconductor are called donors of negative charge, and the semiconductor with such doping is called an **n-type semiconductor.**

We also can think of donors of holes, or, which is behaviorally the same thing, of positive charges. Consider silicon again but this time doped with boron (B) which has three valence electrons (in the same way phosphorus has five). In this case three silicon atoms can fulfill their bonds, but the fourth bond is incomplete; it has a hole in it (Fig. 22.18). Thus a boron atom impurity donates holes or positive charges to silicon, and the semiconductor is called a **p-type semiconductor** in this case.

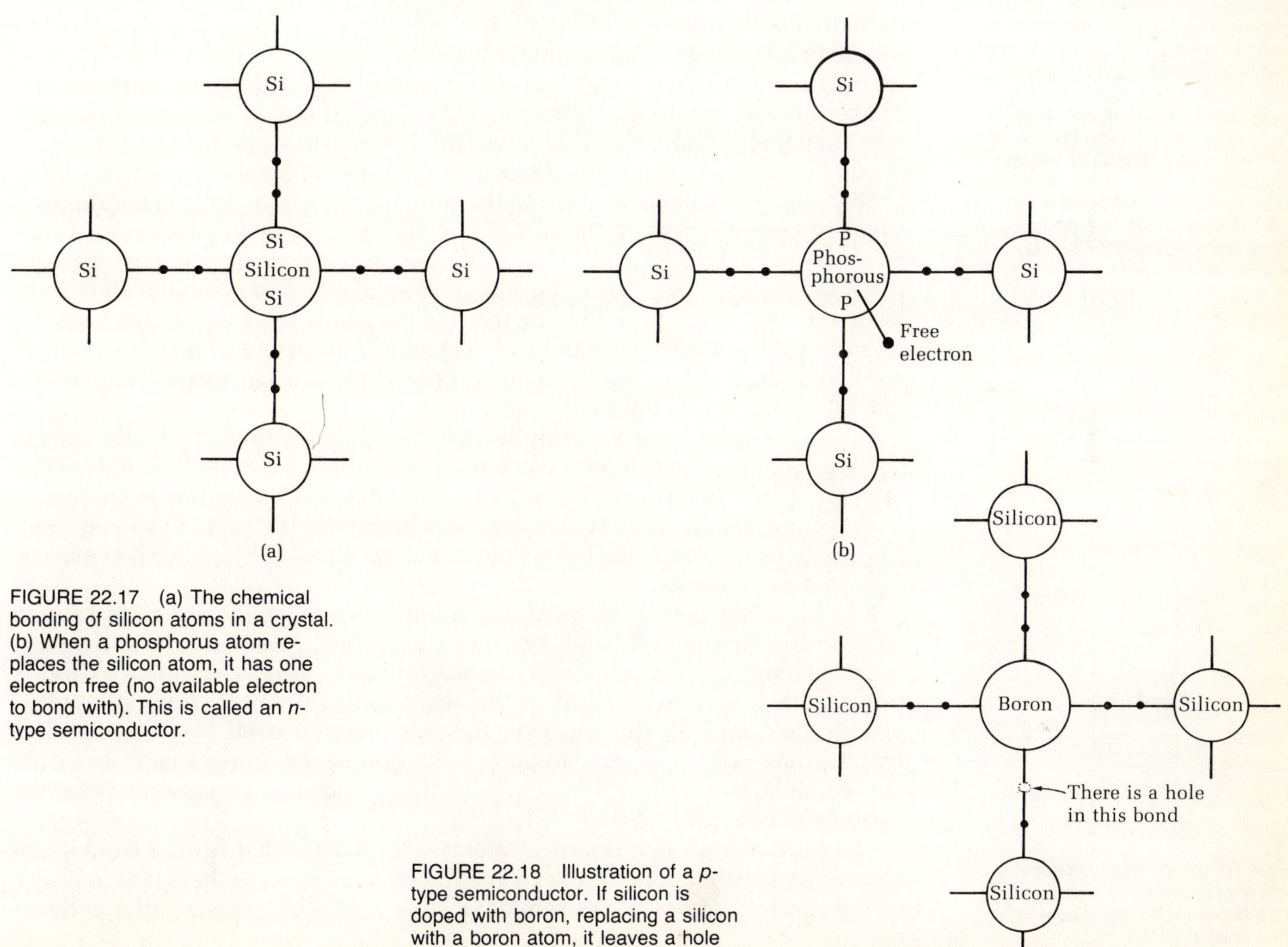

FIGURE 22.17   (a) The chemical bonding of silicon atoms in a crystal. (b) When a phosphorus atom replaces the silicon atom, it has one electron free (no available electron to bond with). This is called an *n*-type semiconductor.

FIGURE 22.18   Illustration of a *p*-type semiconductor. If silicon is doped with boron, replacing a silicon with a boron atom, it leaves a hole in one of the bonds.

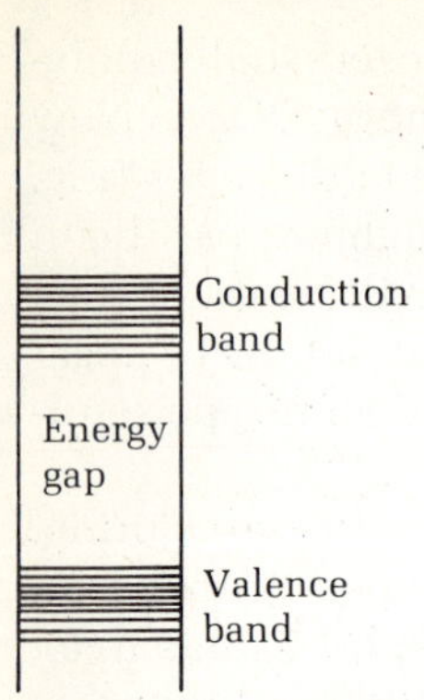

FIGURE 22.19 Energy bands in solids. Usually there is a large energy gap between the conduction band and the valence band. Insulators have the valence band completely filled, but the conduction band is empty. Thus electrons in an insulator cannot accept a small amount of energy and then flow when a voltage is applied: there is no energy level the electron can go into with just a little bit of energy. Conductors, on the other hand, have their conduction bands partially filled, and therefore the electrons have many energy levels to go into when they are given a small amount of energy by an applied voltage.

Energy bands in solids

An even better understanding of the differences in the electrical properties of conductors, insulators, and semiconductors is provided by the energy band theory, a detailed theory of the electronic energy levels in a solid. You know that isolated free atoms have energy levels often separated by large gaps (Fig. 22.11). How does this picture change due to the presence of the other atoms in a closely packed material such as a solid? What results from the interactions is a transformation of each atomic energy level into a band of closely spaced levels (Fig. 22.19). These are the **energy bands.** The number of levels in a band is as large as the number of atoms present in the crystal. Also, since the bands arise from the energy levels of the free atoms, if a certain atomic energy level is filled, so will be the corresponding band. The electrons obey the Pauli principle and fill successively the levels of lowest available energy, as in the case of the isolated atom.

For a metal the upper band, called the **conduction band** (Fig. 22.19), is partially filled. Since the difference in energy between successive levels within a band is very small, when a voltage is applied to a metal, the conduction electrons can accept some energy and get excited to a higher level within the band. They will now drift in the direction of the electric field. In an insulator, on the other hand, there is no electron in the conduction band, and the bands below it are all filled. Since the energy difference between the highest completely filled band (called the **valence band**) and the conduction band is large, if we supply a small amount of energy by means of a regular-sized voltage, the energy is not going to be enough for individual electrons to overcome the energy gap and reach the empty band. Since the energy of an electron in a solid is allowed to have only the values permitted by those of the levels lying within the bands, it follows that the electrons in an insulator cannot accept energy and flow with an applied voltage. Thus the insulator property really originates from quantum effects, from the graininess of energy.

The conduction band is empty in semiconductors also, but now the energy gap between the filled valence band and the unfilled conduction band is not so large as in the insulators (Fig. 22.20). Even thermal excitation is enough to enable a few electrons in these substances to go over the energy gap and reach the empty band. These electrons now can be accelerated by an applied electric field to give a current.

It is also clear that, as thermal excitation lifts an electron from the valence band to the conduction band, there is a hole left in the latter. The hole can propagate and conduct electricity in the following way. The valence band is now unfilled, and an electron in this band can gain energy from an applied electric field and fill the vacancy. But this electron itself leaves a vacancy, which some other electron, upon acceleration and gaining energy from the applied field, can occupy, creating in its turn a vacancy. Thus we find that the vacancy is now propagating.

The introduction of "impurity" atoms introduces either further conduction electrons or further holes in the valence band. In either case there is an increase of the current carriers of the semiconductor, and it becomes a better conductor.

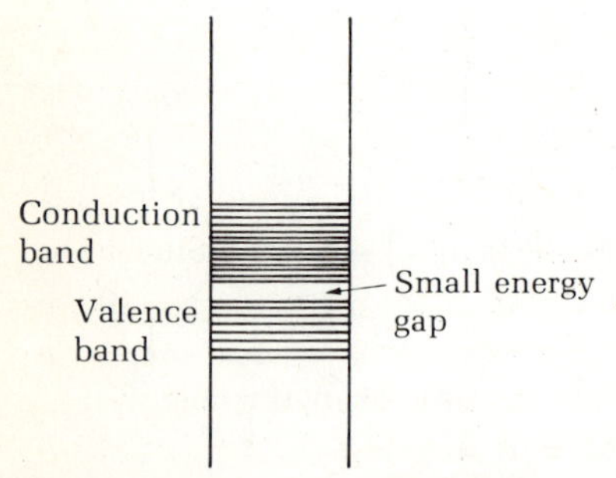

FIGURE 22.20 Semiconductors are characterized by a small energy gap between the last filled and the first unfilled bands.

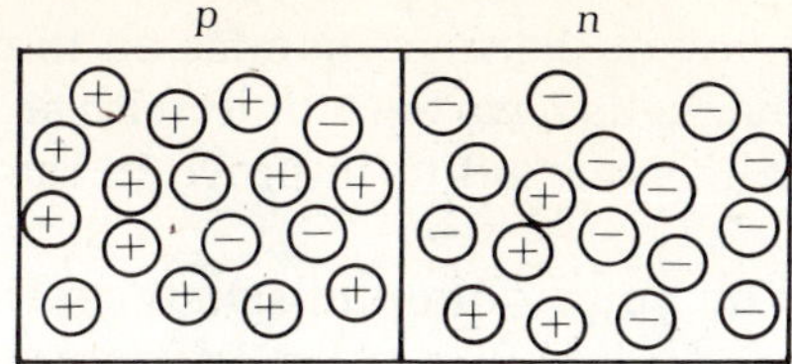

FIGURE 22.21 A *p-n* junction. In the absence of an applied voltage, some electrons (⊖) find their way into the *p*-region and some holes (⊕) into the *n*-region.

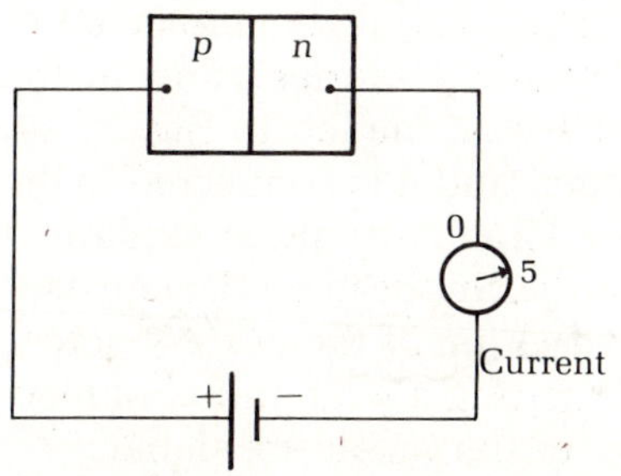

FIGURE 22.22 The *p-n* junction connected to a battery the "right" way. The current can continue. See text for explanation.

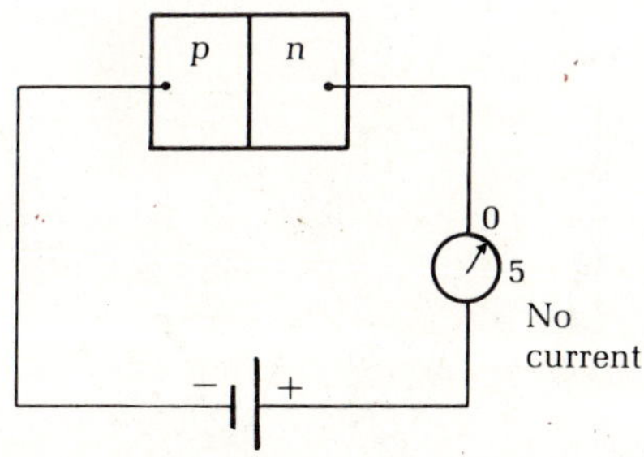

FIGURE 22.23 A *p-n* junction connected to a battery the "wrong" way. The current stops. See text for explanation.

The solid-state rectifier

Now we are ready to consider a *p-n* junction, a junction of an *n*-type semiconductor with a *p*-type one. The *n*-type has free elctrons, and the *p*-type has free holes. (Fig. 22.21). Some of the free electrons from the *n*-region will flow into the *p*-region by ordinary diffusion, the same process by which a vial of open perfume mixes with the adjacent air. (Free electrons in a solid behave in some ways the same as the free molecules in a gas. This is the reason that a collection of free electrons in a solid is called an electron gas.) This is accompanied by the diffusion of holes from the *p*-region into the *n*-region. The result of both motions is to make the *n*-side somewhat positively charged and the *p*-side simultaneously negatively charged by the same amount. The opposite charge on the two sides will provide an attractive force, which will make further migration difficult. Thus an equilibrium situation will be established in which there will be a few electrons among the holes on the *p*-side and vice versa for the *n*-side. Perhaps you are thinking that electrons and electron holes existing side by side on both sides should neutralize each other (the electron can jump into the hole). Such neutralization does occur (but at a slow rate), and so some diffusion must continue to make up for the neutralization.

Now imagine that we apply an electric voltage to the two sides of the *p-n* junction (Fig. 22.22). If we connect the negative terminal of the battery to the *n*-type crystal and the postive terminal to the *p*-type crystal, there will be a force driving the electrons to the left and holes to the right (remember, electrons flow from the negative to the positive electrode; since holes are equivalent to positive charge, they flow from the positive to the negative terminal). The net direction of both currents is the same; they augment each other. As there are a larger number of electrons moving into the *p*-region, more of the holes will be neutralized, and the same is true for the *n*-region due to the invasion of the holes. Thus both sides need a fresh supply of electrons or holes, as the case may be, which is precisely what the job of the battery is. So the current will continue.

If, on the other hand, we connect the positive terminal to the *n*-side and the negative terminal to the *p*-side, the situation is messed up. The electrons are still pulled to the positive terminal and the holes to the negative terminal, but this just pulls them both away from the junction, creating an empty region in between (Fig. 22.23). Thus there cannot be any current in this case; the current stops. This is the same kind of affair that is accomplished by the vacuum tube diode (Chapter 18). Both of these devices allow current in one direction only and are therefore able to "rectify," or make a dc current out of an ac current.

The solar cell

We have seen how an applied electric voltage can affect the equilibrium situation existing in a *p-n* junction. The equilibrium can also be affected by radiation; if sunlight falls on one side, it will release electrons from that crystal by knocking them out. Just as a voltage offsets the equilibrium of the junction, offsetting the equilibrium of the junction with sunlight will generate a voltage. This is the principle of the **solar cell.**

The big advantage of the solar cell is that we don't have to go through the intermediary of thermal energy in converting solar energy to electrical energy.

Thus there is no limit imposed by the second law of thermodynamics on the efficiency of conversion. At the present time, solar cells have been built with an efficiency of only about 15%, but the situation is expected to improve in the future.

The major disadvantage of using a solar cell on your rooftop to satisfy your electrical need is an economic one. The material needed for solar cells, pure silicon, is very costly. Currently electricity generated by a solar cell is 50 to 100 times costlier than that obtained from conventional methods.

The transistor

One of the most significant changes to occur in modern technology has been the replacement of the vacuum tube triode (Chapter 18) with *transistors*, which are made out of semiconductor *p-n* junctions sandwiched back to back (Fig. 22.24). We can easily understand how the transistor works from what we have said before about the *p-n* junction.

The transisitor has three distinct regions, which are variously connected to the input signal to be amplified, to sources of power (batteries), and to the output circuit accepting the input signal. The battery on the left in Fig. 22.24, connected to the n-p junction, is a low-voltage battery and it is connected to the n-p junction in the right way to give a relatively large current, as explained before. The incoming signal voltage is, of course, superimposed on the constant voltage of the battery, and this produces a large variation of the current across the junction. The net result is that the p-region is getting a large influx of electrons from the left, which varies with the voltage of the incoming signal.

Now look at the p-n junction on the right. This one is connected to a high-voltage battery but in the wrong way; the negative pole of the battery is connected to the p-region and the positive pole to the n-region. As we saw earlier, when the p-n junction is connected to a voltage source this way, it hardly transmits any current at all. However, there is a difference in this case; there is a large influx of electrons coming in from the left. These will diffuse to the right, giving a current that can be as large as the current in the input circuit if the design is right.

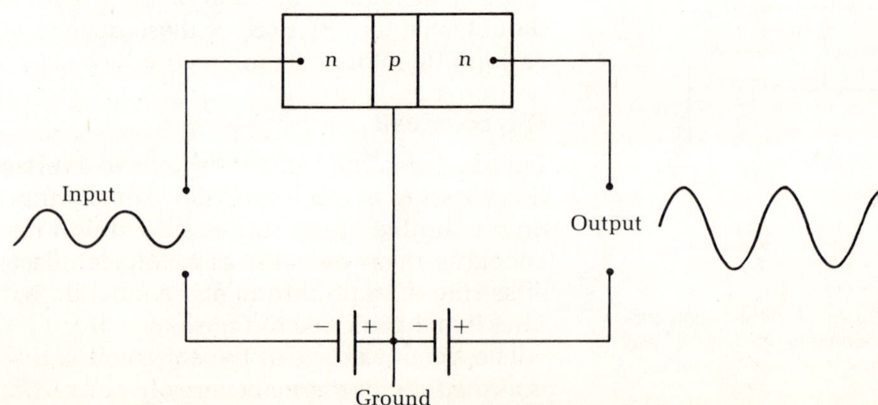

FIGURE 22.24   A transistor, or *n-p-n*, junction, as an amplifier.

This does not sound so spectacular. Actually, though, something great has been accomplished. The voltage fluctuation in the input signal is small, and it produces a certain current fluctuation. But now the same current fluctuation, as it passes through the junction on the right, is accompanied by a large voltage fluctuation. In other words, the signal in the output circuit is an amplified one. Thus the transistor serves the same role—an amplifier—as does the triode. Its advantages over the triode are the tremendous savings in space and the greater reliability in performance.

In today's microelectronic circuits, a 1-inch square of silicon may hold some 100,000 transistors of this kind. As a result, devices like the pocket calculator have been possible to manufacture.

SUMMARY

The development of quantum mechanics mostly occurred in conjunction with the physics of the atom, which is the subject of the first section of this chapter. In previous chapters we have talked only about the continuous spectrum of light emitted by incandescent solids. However, with gases we can get two other kinds of spectra: the emission line spectrum and the absorption line spectrum. These last types of spectra are directly connected with the energy levels of the electron within the atom. The emission spectrum occurs when an electron of an atom makes a transition to a lower state from an excited energy state. The opposite case—where an electron absorbs incident light energy to make a transition from a lower energy level to a higher one—gives us the absorption lines. A landmark in the history of science was established when Niels Bohr, with his atomic theory, explained the Balmer series of emission lines of the hydrogen atom. One of the spectacular applications of the atomic theory of light is the laser, which uses "stimulated emission."

The quantum theory of the atom also explains some puzzling things that the chemists discovered while classifying the chemical elements in the so-called periodic table. The physicist's explanation of the periodic table rested on the following ideas. The first is that of quantum numbers; each of the stationary states of the atomic electron is characterized by a set of four discrete numbers called the quantum numbers. The second idea is the Pauli exclusion principle: no two electrons can have an identical set of the four quantum numbers. And the third idea is simply that, in order to form the ground state, the electrons occupy in order the states of lowest available energy.

In the second section of the chapter we discussed the electrical properties of solids, in particular, the differences among conductors, insulators, and semiconductors. The quantum theoretic explanation of the differences rests on the structure of the energy bands of the particular solid. In conductors there is a partially filled band (called the conduction band). In insulators there is a large energy gap between the highest completely filled band (the valence band) and the lowest empty band (the conduction band). But in semiconductors this energy gap between the conduction and valence bands is small, making possible the special properties of a semiconductor. Two important applications of semiconductors also were discussed: solar cells and transistors.

Review and reason

1. Discuss how, when a gas is hot, the collision among its atoms can cause the gas to emit light. How does such emitted light verify the Bohr theory of the atom?
2. What is the difference between the emission and the absorption spectra of a substance? What is the difference between these kinds of spectra and continuous spectra?
3. Why do incandescent solids emit a continuous spectrum?
4. Describe how a laser works. What are the special properties of a laser beam that an ordinary light beam does not have? What are some of the applications of the laser?
5. What is meant by holography? Is the hologram a photograph of the same kind as an ordinary photograph? Explain.
6. State the Pauli exclusion principle. Why is it called the exclusion principle? What does it exclude? Give an example of how the Pauli principle works in atoms.
7. What is a quantum number?
8. Discuss the physical principles that are instrumental in the explanation of the periodic table.
9. Among all the elements, carbon and silicon have a very high affinity in making compounds, whereas the noble gases (for example, helium) have the lowest such affinity. Explain this in terms of the electronic structure of the atoms of these elements.
10. What is the difference between the ionic and the covalent bonds?
11. Explain how the difference in the electrical properties among conductors, insulators, and semiconductors comes about.
12. What is accomplished by the "doping" of a semiconductor?
13. What is an energy band? Draw the energy band diagram for an insulator, a semiconductor, and a conductor.
14. Discuss how a semiconductor p-n junction can act as a rectifier of ac current.
15. What is a solar cell and how does it work?
16. Discuss how a transistor amplifies an incoming signal.
17. The following statements have mistakes. Correct them.
    (a) The periodic table is an arrangement of the elements in order of increasing atomic weights.
    (b) The larger the energy gap between the valence band and the conduction band of a semiconducting solid, the better a semiconductor it is.
    (c) What is the difference between the n-type and the p-type semiconductors? The n stands for negative and p stands for positive; so the n-type semiconductors must be negatively charged and the p-type positively charged.

Arithmetic

1. Fig. 22.25 shows the energy levels of an atomic electron; the number shown beside each level corresponds to the energy (in electronvolts) of the level relative to the ground state. [Note: 1 electronvolt (eV) = 1.6×10^{-19} J.] Calculate the frequency of light emitted for transitions of an electron starting from any of the excited energy levels to the ground level.
2. An atom emits visible light of frequency 7×10^{14} Hz when an electron jumps from one of its excited states to another excited state. Calculate the energy of the initial excited state of the atom relative to the final state. In joules? In electron volts?
3. What are the number of valence electrons for each of these elements: aluminum, calcium, chlorine, and strontium? To answer this, consult the periodic table.

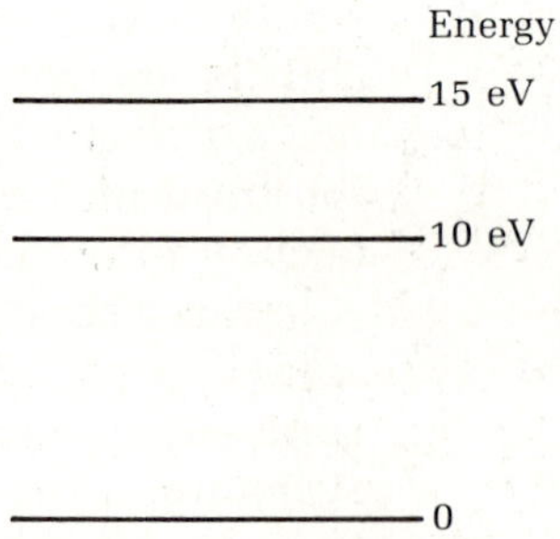

Energy

———————— 15 eV

———————— 10 eV

———————— 0

FIGURE 22.25   See text for explanation.

23

Nuclei, Plasma, and Elementary Particles

The research of the physics in this chapter is expensive. Mostly it needs expensive devices called **accelerators,** machines that speed up submicroscopic particles to enormous energy, which then are used to shoot at matter. When the latest and the most sophisticated of the accelerators now built (at Batavia, Illinois) was proposed, its value was immediately questioned. The following is an excerpt from congressional testimony given by physicist Robert R. Wilson, who later became the director of the accelerator center at Batavia.

Senator Pastore: Is the accelerator connected in any way with the security of our country?

Wilson: No, sir, I do not believe so.

Senator Pastore: It has no value in this respect?

Wilson: It only has to do with the respect with which we regard one another, the dignity of men, our love of culture. It has to do with those things. It has nothing to do with the military, I am sorry.

Senator Pastore: Don't be sorry for it.

Wilson: I am not but I cannot in honesty say it has such applications; but it has to do with whether we are good painters, good sculptors, great poets, I mean all the things that we really venerate and are patriotic about in our country. In that sense, this new knowledge has everything to do with honor and country but it has nothing to do with defending our country except to help make it worth defending.*

■ 23.1 The Story of the Nucleus: Its Physics and its Technology

The nuclear story began to unfold with the coming of the present century. It stayed dormant for a while during a glorious period of physics that saw the rise of quantum mechanics. It was revived in the thirties and enjoyed a spectacular period of startling, epoch-making discoveries. Toward the end of World War II,

---

*Quoted in D. Giancoli, *The Ideas of Physics* (New York: Harcourt Brace Jovanovich, 1974), p. 437.

a controversy was stirred by the development of the atomic bomb. The controversy has continued although, at least in part, nuclear technology has shifted to peaceful uses, namely, generation of power in an energy-hungry world. But peaceful uses may have their own perils, and thus the debate continues.

A chance discovery by French physicist Henri Becquerel marked the very beginning. Becquerel discovered that compounds of uranium emitted a strange invisible radiation, strange because he could not do a thing to change the behavior or emission of this radiation. It just kept right on.

Of course, this was in 1896, and physicists at that time were pretty much convinced that radiation that represented energy could not be created from nothing. Where was it coming from? Could it be from the uranium atom itself?

The phenomenon was called **radioactivity.** Uranium and another naturally occurring element, thorium, were found to be radioactive. Were there other elements that shared this strange property of uranium and thorium? The famous Curies, Marie and her husband Pierre, discovered two more, radium and polonium; and another gaseous element called radon was discovered among the residue of radium and was found to be radioactive. Soon it was established that all these radioactive elements actually form sequences: parent, child, grandchild, and so on—a sort of dynasty. Radioactivity is a kind of atomic transmutation; by emitting the radiation, uranium becomes other elements (Fig. 23.1).

Meanwhile, a New Zealander, Ernest Rutherford, who subsequently settled in England, proved a very important thing. He showed that the radiation from the radioactive elements really consists of three different kinds of radiation. The nature of these radiations was to be clarified later, so he tentatively named them **alpha, beta,** and **gamma** rays.

Soon it was established that the alpha rays are really doubly ionized helium atoms, the beta rays are electrons, and the gamma rays are high-frequency electromagnetic radiation. The alpha particle has two electronic units of positive charge and a mass about four times that of the hydrogen atom.* Most importantly, often the alpha particles are emitted from the radioactive substance at high speed. Thus they gave Rutherford a kind of high-speed atomic bullet. Rutherford decided that his chances of discovering new things about the still mysterious atom were best if he probed them with these alpha particle bullets.

At the time, English physicist J. J. Thompson, who discovered the electron back in 1897, had a model for the atom that proposed that atoms are like lumps of positive electric charge in which are immersed the negatively charged electrons, very similar to little plums in a pudding. When Rutherford directed his alpha bullets at a gold foil, he found no evidence in favor of this positively charged lump model of the atom. Most of the time the alphas passed through the atom undisturbed, suggesting that most of the atom consisted of empty space. Most strangely, there were a few alphas that were deflected at very sharp angles, even coming back out where they entered.

---

*It is convenient to express charges of nuclei and elementary particles in units of electronic charge, $e = 1.6 \times 10^{-19}$ C. Thus the magnitude of the charge of the alpha particle is 2e, or $2 \times 1.6 \times 10^{-19}$ C $= 3.2 \times 10^{-19}$ C.

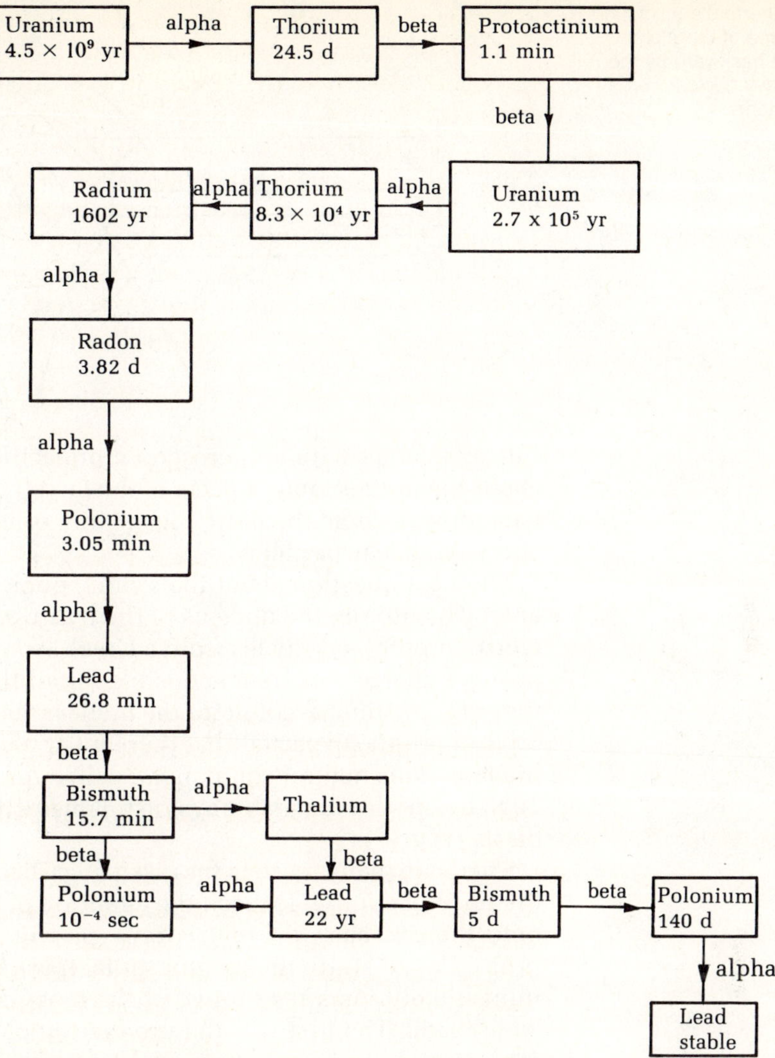

FIGURE 23.1 The uranium dynasty. By emitting alpha and beta radiations, radioactive nuclei transform into other nuclei. In each decay the original nucleus is called the "parent" and the product nucleus is the "child." Note that some elements exist in different forms with different radioactive properties. Such different forms of the same element are called isotopes. See p. 484.

This last fact, argued Rutherford mathematically, can be explained only if the positive charge of the atom resides in a very tiny "nucleus" that also contains most of the mass of the atom. The positively charged alpha particles were strongly repelled by these positively charged atomic nuclei of the gold foil whenever they came close, hence the large-angle scattering (Fig. 23.2). Thus the atomic electrons were not at all like plums in a pudding, said Rutherford. Rather, they formed a tiny solar system, with the nucleus as the central sun and the electrons orbiting around them.

As you know already, this picture is not quite right, and quantum ideas are necessary to get a complete picture of the atom. We have discussed quantum mechanics in earlier chapters and will not repeat the arguments here. The purpose of mentioning some detail of Rutherford's experiment is twofold. First, of course, it is a landmark in the history of the nucleus; this experiment in fact established the nucleus. Second, it showed that a scattering experiment—

FIGURE 23.2 Rutherford's scattering experiment. Some of the alpha particles are turned backward by the repulsion of the heavy nuclear core of the gold target atoms.

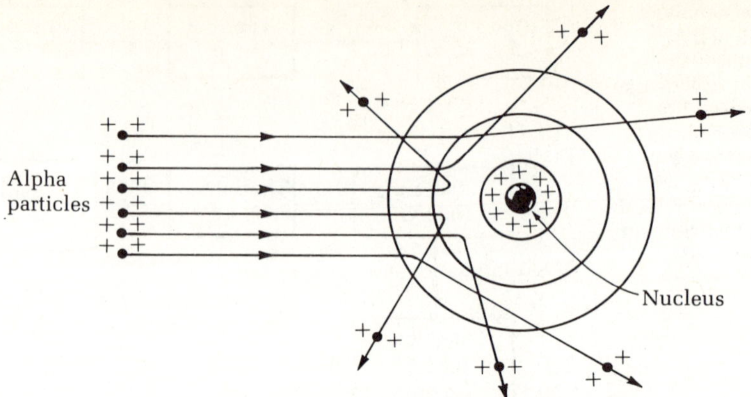

hitting a target with a microscopic projectile—is a very effective way to learn about the microscopic aspects of the target. Indeed, such collision experiments have now become the most cogent way of investigating the physics of nuclear and subnuclear particles.

The next question about the atomic nucleus was this: What are its constituents? The proton, the nucleus of the hydrogen atom, is the first constituent (the word "proton" originates from a Greek word meaning first). The protons carry positive charge and are responsible for all the charge of the nucleus. However, they alone cannot account for the nuclear mass of nuclei heavier than hydrogen. So Rutherford suspected that there must be a neutral component of the atomic nucleus. This neutral component, called the neutron, was finally discovered in 1932 by another English physicist, James Chadwick. Now nuclear physics was on its way.

There are many reasons for regarding Chadwick's discovery of the neutron as the turning point: it focused the attention of physicists away from the atom and onto the nucleus. But that is only part of the story. The use of neutrons as projectiles to study nuclei and, in fact, transform them was thought of almost immediately. As a result, two of the most far-reaching discoveries in the field were made. The first was the production of new radioactive substances in the laboratory by the French physicist couple Irene Curie and Fredric Joliot, and the second was the discovery of nuclear fission.

An atom is characterized by the number of protons in its nucleus, which is the same as the number of electrons in the neutral atom. This is the **atomic number** Z, sometimes called the proton number, introduced earlier. Since the nucleus also contains the neutrons, it is obviously important to keep track of the number of neutrons as well. This is done by a statement of the total number of neutrons plus protons inside the nucleus, a number called the **mass number,** designated by A. The number of neutrons in the nucleus is then $A - Z$, and this is called the neutron number N of the nucleus.

It was known for some time that the same element could exist in forms having somewhat different atomic weights. Such forms were called **isotopes** of each other. The isotopes now found a natural explanation with the realization that two isotopes of the same element had the same number of protons inside their atomic nuclei but a different number of neutrons. The atomic or chemical prop-

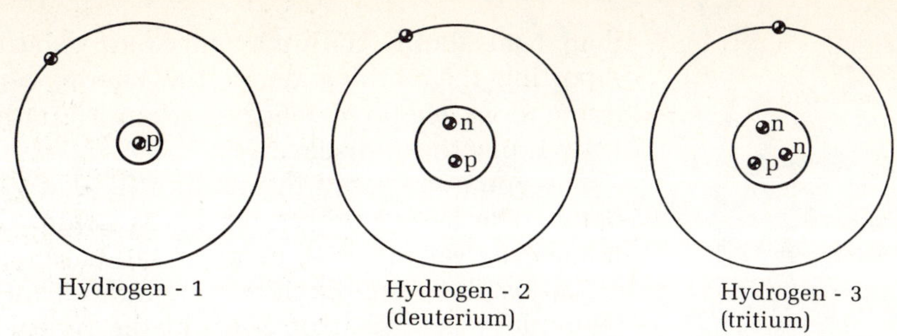

FIGURE 23.3 The three isotopes of hydrogen. They belong to the same element, having the same number of protons and electrons, and have the same chemistry which depends only on the electronic structure. But the nuclei of the isotopes contain different numbers of neutrons. The hydrogen-1 nucleus is just a proton (p) with no neutrons. The hydrogen-2 or deuterium, nucleus contains 1 proton (p) and 1 neutron (n). And hydrogen-3 nucleus has one proton and two neutrons.

erties of two isotopes are practically the same, having the same number of electrons in the neutral atomic state. But the different number of neutrons suggested that their nuclei were different, and therefore they could have different nuclear properties. The most spectacular demonstration of this was the discovery of radioactive isotopes of naturally occurring stable elements produced in the laboratory with the help of nuclear reactions. Clearly, radioactivity also had to be a property of the atomic nucleus, a transmutation of the nucleus with the emission of radiation.

The existence of isotopes also makes it imperative that when referring to an atomic nucleus, we not only use the chemical symbol but also the mass number. For example, ^{14}C or ^{235}U. The superscript on the left of the symbol denotes the mass number.

The discovery story of **nuclear fission** is both interesting and dramatic. It started in 1935 with the Italian-born American physicist Enrico Fermi, in Italy at the time, who had the idea to produce transuranic elements (elements beyond uranium in Z number) by bombarding uranium with slow neutrons. The term "slow neutron" may be confusing; how slow is slow? Slow means roughly having energy comparable to that of a molecule at room temperature, which is about 10^{-21} J. Fermi and his collaborators did their experiment and found something new, but they misinterpreted the results, thinking they had discovered transuranic material. Actually, what they had seen was the breakup of an isotope of uranium of mass number 235 into two lighter nuclei, a process now called fission (Fig. 23.4). The mistake was not cleared up until later by two German chemists Otto Hahn and Fritz Strassmann. Being chemists they were able to separate and identify the products of uranium bombardment with slow neutrons. Soon afterwards physicist Lise Meitner and her nephew, Otto Frisch, suggested that tremendous energy must be liberated in the fission process.

Meanwhile, Fermi had moved to America in order to avoid the Fascist regime of Mussolini in Italy. The news of the work of Hahn and Strassmann and of the suggestion of Meitner and Frisch reached him through the mouth of no other

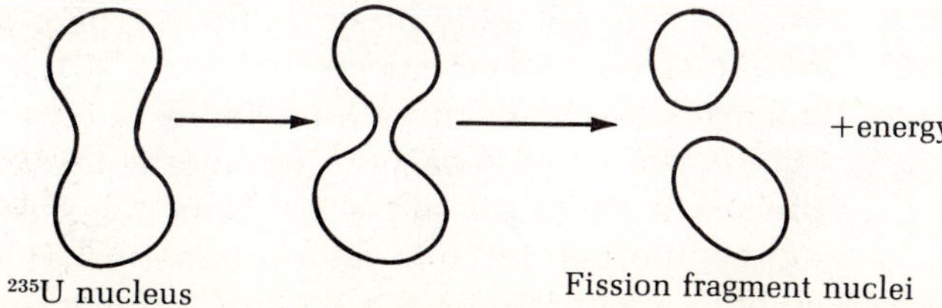

FIGURE 23.4 The fission of a uranium nucleus. Energy is released.

than Niels Bohr. Within twenty-four exciting hours, the newspapers were reporting the verification of the energy aspect of nuclear fission. Indeed, tremendous energy comes out when a uranium nucleus divides into two. A truly Promethean discovery.

The energy released in uranium fission was soon to be used in the atomic bomb. The fission process also was harnessed as a source of electricity. The nuclear power plants of today are all based on fission power.

Radioactivity and nuclear fission: these are the two fronts of today's nuclear technology (see Chapter 24). But the future may hold another very exciting nuclear process: thermonuclear reactions.

It was a major discovery in the 1930s that **thermonuclear fusion reactions,** fusion of light nuclei into heavier ones, is responsible for the heat generation in stars. The same energy is used in the hydrogen bomb, and now research continues to develop fusion power plants.

Nuclear structure

The discovery of the neutron paved the path to the beginning of a theoretical understanding of nuclear physics. Since nuclei were now perceived clearly as tight packets of neutrons and protons, the theorists became busy in their attempt to understand the force that holds the nucleons together (the word "nucleon" refers collectively to both the neutron and the proton, the inhabitants of the nucleus). The nuclear force was found to be the strongest of all known forces, but restricted to a very short region of space not extending beyond the boundaries of the nucleus itself. Strong, attractive, and short range: these are the characteristics of the forces that bind the nucleons. And only nucleons and a family of particles that we call **hadrons** possess this special strong way of interacting. Other particles—electrons, for example—do not interact with this force.

The properties of the nuclear force are very different from the two other fundamental forces that you have encountered so far: electromagnetic and gravitational. The last two forces are of infinite range. The short range of the nuclear force is an essential ingredient of nature's order, because if such a strong force were longer range, it certainly would have disturbing effects on things such as chemical bonds, molecular structure, and so on. We have no idea how nature would have adjusted to that.

The same short range, however, prevents our using the nuclear force on the macroscopic scale. Because of their long range, both electromagnetism and gravity allow for superposition. We can add little charges together and get a big effect from their sum. The same goes for masses, of course. No such thing exists for nuclear force. Because of the short range, the forces of an assembly of nucleons do not add up as they do for charges. So we cannot make large macroscopic-sized nuclear objects that would attract other nuclear objects outside the nuclear distances with a huge force.

Now we come to a very important idea. So far we have talked about the nuclear force of the nucleons. This is their binding force, and it is by far the most important force in their interaction. However, they also have other interactions. The proton has electrical interactions with other charged protons and other charged particles by virtue of their electrical charge. Actually, the electrical interactions of protons play quite a significant role in nuclear dynamics, as we saw in the alpha particle scattering experiment described earlier. The positively charged alphas are repelled electrically by the positive charge of the nuclear protons. The weak interaction of nucleons is responsible for radioactivity with beta emission. The nucleons also have gravitational interaction by virtue of their masses, but gravity is just too small to have any significant effects, and we ignore it when talking about nucleons and nuclei.

The nuclear physicists don't stop here. They are particularly interested in the associations of the nucleons inside the nuclei—how they cooperate, what rules determine the greater stability of some nuclei over others, the structure of the average distribution of the nucleons inside their residence, the rules of their reactions and scattering when two nuclei hit each other or a nucleus is hit by a nucleon or some other projectile. And on and on goes the list.

The strangest thing that emerges in the theorist's picture is the fact that inside the nucleus the nucleons behave as basically independent objects, although bound inside the overall nuclear box. This is surprising, because nucleons are such strongly interacting objects that we would expect them to interact violently with one another all the time. Instead, they seem to settle down in a fairly ho-hum existence. One consequence is that the energy levels of nucleons inside nuclei also exhibit a shell structure like the atomic energy levels of the electron. Nuclei with filled shells are especially stable and are called "magic nuclei." This independent particle picture is usually called the **nuclear shell model.**

Within the shell model, nucleons seem to have some weaker, residual interactions among themselves, which give rise to additional interesting cooperative phenomena. For example, if the independent particle picture were totally true, all nuclei would have a spherical shape. The nucleons would distribute themselves in the nuclear well in a spherically symmetric manner. This turns out to be not true. Because of the residual interactions, the nucleons often manage to distribute themselves in the shape of an ellipsoid (such as a football); such nuclei are called deformed. And there are nuclei that have shapes somewhat intermediate; their deformation is kind of "soft." But now we are really getting into the intimate workings of nuclear theory.

Are you beginning to get the picture? Admittedly these things are hopelessly remote from your reality. There is a story about a physicist who actually calculated the distribution of the nucleons inside a nucleus of deformed shape (Fig. 23.6). He was rightly very excited. When he came home from work, he took this drawing out of his briefcase and said to his wife, "I finally know what the distribution of the nucleons inside a complicated nucleus really looks like."

She took the picture, looked at it for a second, and said, "You are out of your mind."

Nevertheless, the models of the theoretical nuclear physicist and their experimental study are very important because of the predictive power of these models. This is the point where nuclear physics is at today. As one example, one of the models has predicted the existence of long-lived superheavy nuclei

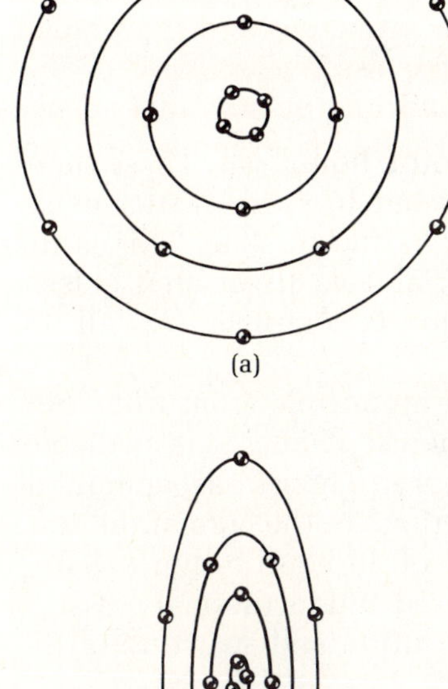

FIGURE 23.5   (a) A spherical nucleus: the nucleons are distributed in a spherically, symmetric way. (b) A deformed nucleus: the nucleons are distributed in an ellipsoidally symmetric way.

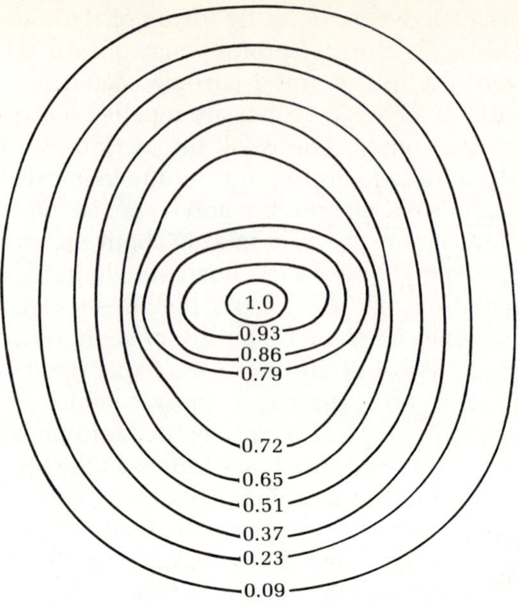

FIGURE 23.6 The detailed shape of a deformed nucleus. The lines join points of equal density of the nucleons inside the nucleus. The numbers are relative to the central density, which is assumed to be 1. [From A. L. Goodman et al., *Physical Review C*, vol. 2 (1970), p. 380.]

with a Z of around 114 (the heaviest nucleus produced so far is one of $Z = 106$). Such superheavy nuclei could influence nuclear technology in a radical manner, and they are being searched for by physicists as one of their priority items.

■ 23.2 Plasma in the Heavens and on Earth: Astrophysics and Plasma Physics

In Chapter 8 we mentioned three states of matter: solid, liquid, gas. These states of matter certainly dominate the earth scene, but out in space matter exists mostly in a form that many physicists refer to as its fourth state. This is the **plasma state,** a state in which the atoms of the material have dissociated, at least in part, into nuclei and electrons. If the dissociation is complete, we call it a perfect plasma.

It takes a great amount of energy to dissociate the atomic electrons from their nuclear core, so a plasma state is possible only when such energy is available. An obvious place for this to occur is in a stellar interior. There a vast amount of heat energy is provided by the thermonuclear reactions in the core of the star. The confinement of the stellar plasma is a result of an equilibrium existing between the vast pressure of gravity pushing it inward and its thermal pressure directed outward (Fig. 10.9). The study of the evolution and structure of the stars is a branch of physics called astrophysics.

The problem of using thermonuclear reactions for the purpose of generation of electricity on earth has gotten attention from physicists ever since the thermonuclear mechanism of the heat of the stars was discovered. The problem here is to confine a plasma; very special problems are involved in this, as we will see. The branch of modern physics dealing with these problems is called plasma physics.

By far the most consequential development in the history of nuclear physics was the making of the atomic bomb. The atomic bomb was first invented in America as a result of the celebrated Manhattan Project at Los Alamos, New Mexico. Since then five other nations have joined the nuclear bomb club.

We will discuss neither the history of the development of the bomb nor its technology. Rather, we will consider this question: What is the responsibility of a scientist in matters like this?

Before I can handle such a difficult question, perhaps it is worthwhile to start with some simpler ones. Why did the scientists involved with the Manhattan Project develop such a vehicle of ultimate destruction? Did they know about its destructive power before the bomb was dropped on Hiroshima? Did they consent?

In answer to the last question first, there is definite evidence that the majority of scientists in the project were against dropping the bomb. Yes, most of them knew about the destructive power before the actual drop, because there was a test run made some time before the big event.

Why did they develop the atomic bomb? Well, initially it was for fear that Nazi Germany might do it before they could. This was the stage at which we found humanitarian scientists like Einstein or American physicist Leo Szilard actively convincing the politicians to go ahead with the project, and another great humanist, physicist J. Robert Oppenheimer, accepting the directorship of the project. Their motives at later stages of the development are less clear. Germany had been defeated already and Japan was not doing well in the war when the final test on the bomb was carried out. What kept them going?

Partly, I suppose, it was curiosity. There is a Sanskrit story that four disciples of a guru once learned from him the art of restoring life to the remains of animals, any animal. The art consisted of saying out loud four mantras in succession.* So the eager students found a few bones. The first mantra was uttered and, lo and behold, the various parts of the animal body assembled rapidly. With the second mantra the bone structure took shape, and with the third one were added flesh, blood, and skin. The animal now could be recognized as a tiger. It was the turn of the fourth student to say his mantra. He looked at everybody else, looked at the tiger, and finally his curiosity won. He just had to find out if his mantra really worked. It worked—and the tiger killed all of them. The story has an unnerving similarity to that of the atomic bomb; we can only hope that it is not prophetic.

Yet curiosity as the only motivating force is too simplistic. There are nationalism, pride, fear of others being first, and many other contributing motives. And then, to top it off, there is this attitude: scientists have a responsibility to develop what is possible. And this involves inventing the blessings as well as the curses, electricity as well as the bomb. It is the people who decide what to do with scientific inventions. The politicians drop the bomb, not the scientist.

There is some truth to this attitude and some reasonableness. I personally have always felt uncomfortable with this view. But one more thing is clear. If the people are to be ultimately responsible, then they should directly participate in the decision as to whether to begin such projects as well as in the ex post facto decisions, particularly if public funds are used.

———
*A mantra is an invocation sometimes assumed to have magical power.

Astrophysics: thermonuclear reaction in stars and stellar evolution

> Twinkle, twinkle, little star
> How I wonder what you are.
> Are you a giant, are you a dwarf?
> Or are you a main sequence star on the wharf?

When we photograph stars through a telescope, especially those powerful telescopes that are available in our national observatories, we find stars of a variety of color and brightness. Of course, the brightness depends on the distance, but if the distance is known, we can figure out the intrinsic brightness, or the **luminosity,** of a star, which is the amount of energy radiated per second.

TABLE 23.1 Spectral classification of stars.

Spectral Class	Color	Approximate Surface Temperature (°K)
O	Blue	over 25,000
B	Blue	11,000–25,000
A	Blue	7,500–11,000
F	Blue to white	6,000–7,500
G	White to yellow	5,000–6,000
K	Orange to red	3,500–5,000
M	Red	Less than 3,500

The color, or, in general, the characteristics of the spectral emission of the star, tells us about its temperature. The stars are put into different **spectral classes** according to the light they emit (Table 23.1).

The blue stars are of the highest temperature; they are classified as O, B, and A in order of decreasing temperature. Then come the blue-white (F) stars and the white-yellow (G) stars. Our own sun belongs to this last class. Still lower in temperature are the orange-red stars (K) and finally the red stars (M). If you have trouble remembering the letters corresponding to the different spectral classes, the following sentence may be helpful (the first letter of each word is a spectral class): Oh, be a fine girl (or guy), kiss me.

When we plot the luminosity of the stars against their spectral classes, the resulting graph shows an extremely interesting feature (Fig. 23.7). Notice that the temperature increases from right to left in the diagram. The majority of the stars lie on a slightly curved line running diagonally across the diagram. These stars are called the **main sequence stars.** Then there is a group of stars in the upper right-hand corner; their luminosity is high, but their temperature is low. It can be shown that their size is huge compared to the sun. These are called the **red giants.** There is a second cluster of stars in the lower left-hand corner; these have high temperature and low luminosity and are considerably smaller in size than our sun. These are called the **white dwarfs.**

This diagram, which provides a simple division of stars into three major groups, is called the **Hertzprung-Russell diagram** (abbreviated H-R diagram). It

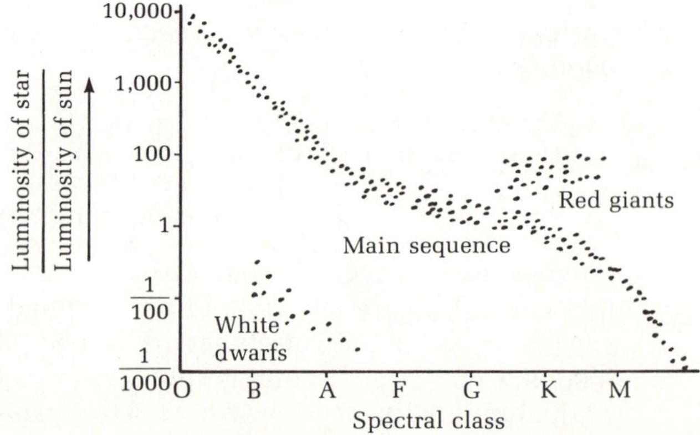

FIGURE 23.7 The Hertzprung-Russell diagram. The luminosity of stars relative to our sun is plotted against the spectral classes of the stars. The majority of the stars lie along the main sequence.

provides us with the basis for an understanding of the evolution of stars. The picture the astrophysicists have put together suggests that the main sequence is the mainstay of stars in their evolution. Every star takes a place on the diagonal line of the main sequence in the H-R diagram for the major part of its life, the exact position depending on the mass and composition of the star. The stars of the heaviest masses lie high on the curve toward the upper left-hand corner; the less massive stars occupy positions near the lower right-hand corner. After the main sequence phase, the stars expand and become red giants. The red color of the giants is due to the cooling of their surfaces because of expansion. The final phase for many stars is as a white dwarf; the white dwarfs are one of the end products of stellar evolution.

A star is born when interstellar matter condenses under the influence of mutual gravity of the dust and debris that make up such matter. As the matter comes together, gravitational potential energy is converted into kinetic energy of its particles, producing heat. Part of the heat energy is radiated away, but the contraction continues, providing even more heating of the matter, which now has become a perfect plasma in the core region. When the temperature of the plasma is hot enough, the energy of the individual nuclei (of hydrogen) is enough (this corresponds to a temperature of 10 million degrees Kelvin or more at the core) for them to overcome the mutual electrical repulsion and fuse. The fusion releases a high amount of heat energy, which compensates for the radiation loss and enables the star to maintain a state of equilibrium at a constant size. No further gravitational contraction is necessary at this stage. The thermal pressure of the ionized gas at the prevailing temperature of the star balances the inward pressure of gravity.

We can estimate the time a star like the sun spends on the main sequence in the following manner. The nuclear fuel of main sequence stars is hydrogen. By means of a series of reactions, the hydrogen nuclei fuse to make helium nuclei. Since the energy release can be regarded as a conversion of mass energy into heat, we have to know the masses of the reactants and the products. It is convenient to express nuclear masses in a new unit of **atomic mass unit** (abbreviated amu): 1 amu = 1.66×10^{-27} kg. In this unit the mass of a hydrogen nucleus is 1.00813 amu and that of a helium nucleus is 4.00389 amu. Thus the total initial mass per conversion, that of four hydrogen nuclei, is given as

$$4 \times 1.00813 = 4.03252 \text{ amu}$$

The final mass is 4.00389 amu, that of a single helium nucleus. The difference between the initial and the final mass is

$$4.03252 - 4.00389 = 0.02863 \text{ amu}$$

This may be a small mass (the reason we had to keep all the decimal figures in order to calculate it), but its energy equivalent is huge, as we know from the mass-energy relation.

The 0.02863 amu that is converted into heat, we notice, constitutes $0.02863/4.03252 = 0.0071 = 0.71\%$ of the initial hydrogen mass. Thus if 1 kg of hydrogen burns this way, 0.0071 kg of mass is converted into energy. We can calculate the amount of energy from $E = mc^2$. Since $m = 0.0071$ kg and $c = 3 \times 10^8$ m/

sec, the energy is

$$0.0071 \times (3 \times 10^8)^2 = (7.1 \times 10^{-3}) \times (9 \times 10^{16})$$
$$= 64 \times 10^{13} \text{ J}$$

Even at the rate our sun radiates, 4×10^{26} J/sec, it takes the burning of

$$\frac{4 \times 10^{26}}{64 \times 10^{13}} \text{ kg}$$

of hydrogen per second. This comes out approximately as 6×10^{11} kg of hydrogen per second, or $6 \times 10^{11} \times (3.2 \times 10^7)$ kg per year (the last factor comes in because 1 year $= 3.2 \times 10^7$ sec). So the sun burns 1.92×10^{19} kg of hydrogen in just 1 year. Fortunately, the sun has a huge supply of its fuel, roughly 10^{30} kg of hydrogen. Even at this huge rate it will take the sun

$$\frac{10^{30}}{1.92 \times 10^{19}} = 5 \times 10^{10} \text{ years}$$

to exhaust its fuel. Of course, the sun will become a red giant long before its hydrogen is completely exhausted, but a lifetime of $\sim 10^{10}$ years is expected for it in its present condition.

The calculation above is designed to give you some idea of how thermonuclear reactions generate the huge amount of energy that a star like our sun radiates for such a long time. The length of time is due to the slow rate of the thermonuclear reactions. Previously we called this the nuclear reaction hang-up of a star.

Until recently, this picture seemed to be firmly established; the theoretical calculations based on this model were able to fit the experimental data rather well. But recently an ominous hitch has developed. The reactions in our sun should produce energetic neutrinos, according to the stellar model. These neutrinos have been looked for quite thoroughly, but they have not been found in sufficient numbers. Until this case of the missing solar neutrino is solved, many physicists will hesitate to accept the theory above as the final one.

Pauli principle in the stars

We mentioned white dwarf stars as an end product of stellar evolution. The explanation of the stability of white dwarf stage of the evolution comes from the Pauli principle. These stars are in a state in which their atoms are all more or less dissociated into nuclei and electrons. All these electrons make up a gas usually referred to as an electron gas. But in the larger sense, all these electrons within the confines of the star define a sort of atom, a giant star-size atom. As such they share one very important property of atomic electrons, the Pauli principle.

According to the Pauli principle, no two electrons can occupy the same quantum state. So in the white dwarf star, since each electron must occupy one and only one quantum state, a minimum volume is required to accommodate all the electrons. In other words, there is a resistance on the part of the electrons if nature makes any attempt to compress them any further. In fact, nature cannot do that without getting rid of the electrons in some way. The resistive pressure

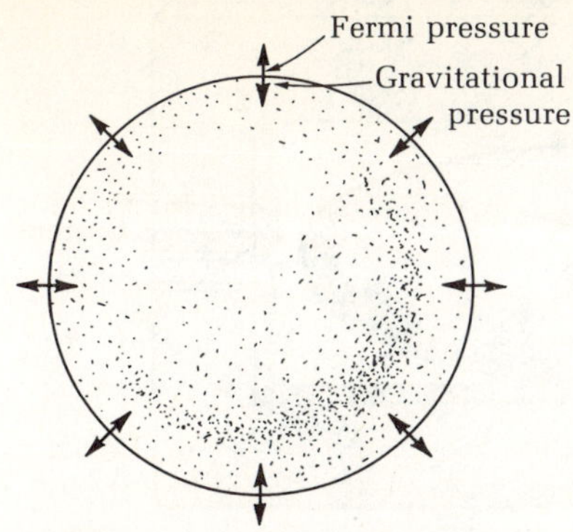

FIGURE 23.8 The equilibrium of a white dwarf star. The electron Fermi pressure cancels the inward pressure of gravity and prevents collapse of the star.

of the electrons against further compression is called the **Fermi pressure.** If the Fermi pressure is just right to balance the inward gravity pressure, we have an equilibrium situation, which is the white dwarf star (Fig. 23.8).

The same Fermi pressure explains the equilibrium that exists in the neutron stars, which are made up almost entirely of neutrons. This time it is the Fermi pressure of neutrons that balances the gravity pressure. Neutrons also obey the Pauli principle.

The matter in white dwarf stars exists with a density of some million times that of the sun. The density of the neutron star is even more spectacular—it is 100 trillion times more dense than the sun. Outside of black holes, the neutron star matter is the most dense encountered in the universe.

White dwarfs, neutron stars, and black holes: these are the three end products of stellar evolution, according to the present theory. A particular star will end up as any one of these, mainly depending on its mass. If the mass is less than or equal to 1.4 times the solar mass, the white dwarf is the most likely state of the star's finale. For a mass bigger than 1.4 but smaller than 2 or so times that of solar mass, the star probably will become a neutron star at the end of its life. And if it's more massive than that, nothing will stop its gravitational contraction from becoming a black hole. Perhaps.

Plasma physics

The possibility of using thermonuclear reactions for producing electricity here on earth is a fantastic idea. We cannot use the most abundant fusion fuel—namely, ^{1}H—on earth, because nuclear burning of hydrogen is too slow. We have to use one or both of the heavier isotopes of hydrogen, deuterium (^{2}H) and tritium (^{3}H). But deuterium also is so abundant that we could have all our energy needs met for the entire lifetime of our planet. And even if we can manage to burn profitably only a deuterium-tritium mixture, we can get enough tritium (tritium is radioactive and does not occur naturally; it has to be manufactured) for the next million years.

Energetically, the burning of deuterium and tritium together is more favorable than that of deuterium alone. Thus most of the work today focuses on the combination fuel. In order for a thermonuclear reaction between the tritium and deuterium nuclei to occur, we must have an assembly of these gases heated and stripped of electrons so that they are in their plasma state. In addition, the nuclei of the two elements must have sufficient energy to overcome their Coulomb repulsion. A plasma reactor performs these two tasks.

In terms of temperature, the plasma has to be as hot as 100 million degrees Kelvin or so. Even the temperature at the core of the sun (20,000,000°K) is puny compared to this. In addition, we have to make a profit on energy; that is, we must get more energy out of the system than we spend in heating the plasma. Then it turns out that we must satisfy a most difficult criterion. The product of the particle density of the plasma, the number of nuclei per cubic meter, and the particle confinement time (in seconds) must be greater than or equal to 6×10^{16} for deuterium nuclei and tritium nuclei to combine in a thermonuclear way. This is known as the **Lawson criterion.**

The containment of the plasma is a very difficult problem, because if you put

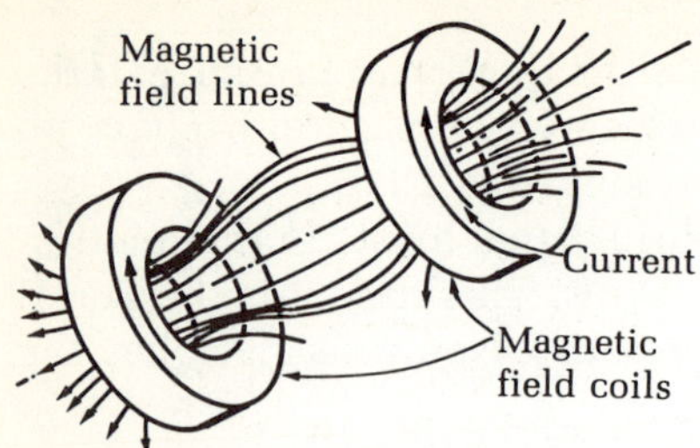

FIGURE 23.9 A magnetic mirror confinement system.

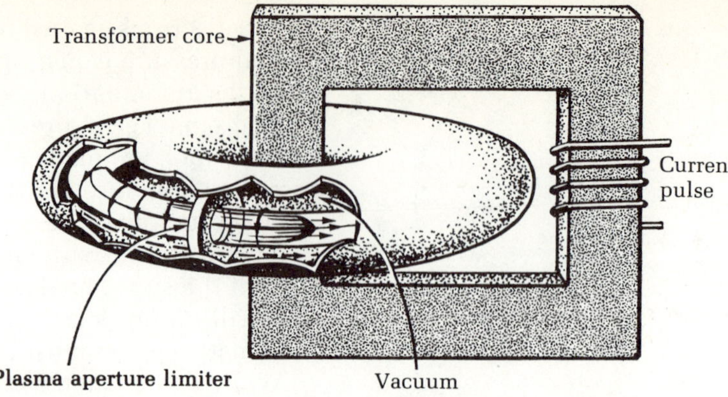

FIGURE 23.10   The Tokamak configuration for confining plasma.

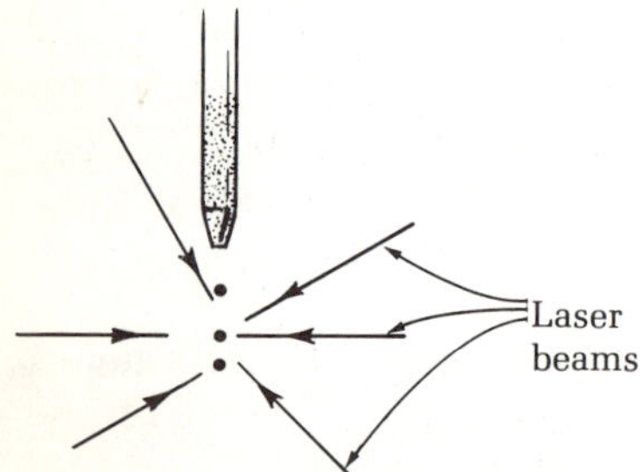

FIGURE 23.11   Laser-induced fusion. Laser beams are directed in short bursts on the pellet of deuterium and tritium from all sides. The pellet gets hot, and perhaps the nuclei will fuse before they can escape.

it in an ordinary vessel, the collision with the walls will cool the plasma very quickly and eventually neutralize it (actually the walls will get holes, too). Thus we must confine the plasma while keeping it away from the walls.

We can learn a few tricks from nature. We have mentioned the Van Allen radiation belts, which are really a plasma confined by the magnetic field of the earth. Thus the first attempts to build a plasma confinement system used the model of the earth (Fig. 23.9). Called a magnetic mirror scheme, the lines of force of the magnetic field here are pretty much like the dipole field of the earth. And the particles of the plasma can indeed be made to go back and forth from one end to the other in this system, just as do the trapped particles of the Van Allen belts. Unfortunately, they also follow one other tendency of the Van Allen belts: they leak at the necks of the bottle at the ends.

Right now the most promising confinement system is the Tokamak, first developed by the Russians (Fig. 23.10). Hundreds of plasma physicists spend literally all their time and energy in trying to figure out an improved Tokamak system that will give a plasma satisfying the Lawson criterion. This has not been accomplished yet, but optimists say that perhaps within the next decade we will have such a system.

Another possibility being actively pursued by plasma physicists is to hit a pellet of liquid deuterium and tritium with high-power laser light from all directions (Fig. 23.11). The laser is expected to heat up the pellet to the desired temperature so that we get an instant plasma with thermonuclear reactions going on inside. The practical problems with such a scheme are also very difficult to solve, the major problem being that of developing a laser of sufficiently high power.

■ 23.3   Elementary Particle Physics: The Quest for Elementarity

When we look at a forest from a distance, it looks like a unified whole. But as you approach it, you find that it is made up of individual units. The philosophers of ancient Greece found this intriguing. Is it possible, they argued, that matter is also like this? When we look close at an object, perhaps its wholeness will disappear and the object will be found to be made up of elementary particles, just as the forest is made up of trees. Among the Greek philosophers, the most outspoken proponent of this "atomic" view of matter, was Democritus: matter is composed of "indestructible" units called "atomos," he said.*

Now, of course, more than two thousand years after Democritus, the atomistic constitution of matter is well established. We know that a solid object is not solid after all; looked at closely, it is seen as a conglomerate of atoms separated by empty space. But the modern atom is not the indestructible elementary constituent of matter that Democritus envisioned; it also has a structure. Not even the atomic nucleus, the core of the atom, is elementary. It also is a conglomerate of other particles called neutrons and protons. Are neutrons and protons elementary in the sense intended by Democritus? Or are there elementary particles from which protons and neutrons are made? This is one of the motivating quests of elementary particle physics.

Another motivation comes from a desire to understand the nature of the nuclear force, the strong interaction and forces between subatomic particles in general. Even in quantum physics we know that we really ought to talk about a force field rather than a force. But in the quantum world of subatomic particles, the field itself acquires a graininess. This is the idea of the quantum field—a grainy field rather than a continuous one. The quantum field theory, although by no means a completely understood theory, has added new dimensions to the concept of a force between two particles and to the subject of the interactions.

But before we discuss either of these two topics, let's discuss how we study our subjects (which are so tiny) at a close level. We can think of a light by which to see, such as small wavelength gamma rays as small as the size of the proton or smaller. The problem is that these gamma photons have such a high energy that they disrupt the particle they hit in a major way, so the result is not unambiguously clear. But this also tells us that it is not necessary to use light alone. We can hit the particles with other projectiles, such as protons, electrons or even neutrinos, and study the result. In order to get close to the target subatomic particles, our projectiles must, like the gamma ray, have high energy. Thus physicists have built machines known as particle accelerators, which create beams of electrons or protons of extremely high energy. For this reason elementary particle physics is also called high-energy physics.

When we hit our particles with these high-energy projectiles, the outcomes of the collision is usually a whole collection of particles. But we must not jump to the conclusion that the particles that result are the constituents of the target particle. This is not true because of a strange aspect of the relativistic equivalence of mass and energy. Whenever sufficient energy is available, some of the energy manifests itself as mass, which takes the form of these particles that come out—of course, subject to all the other laws of physics. For example, the conservation law of number of baryons, which we referred to earlier, must hold in determining which particles can emerge.

The size of the particles that result from a collision process between subatomic particles is found to be the same as the colliding particles themselves,

---

*The word "atom" originated from this Greek word.

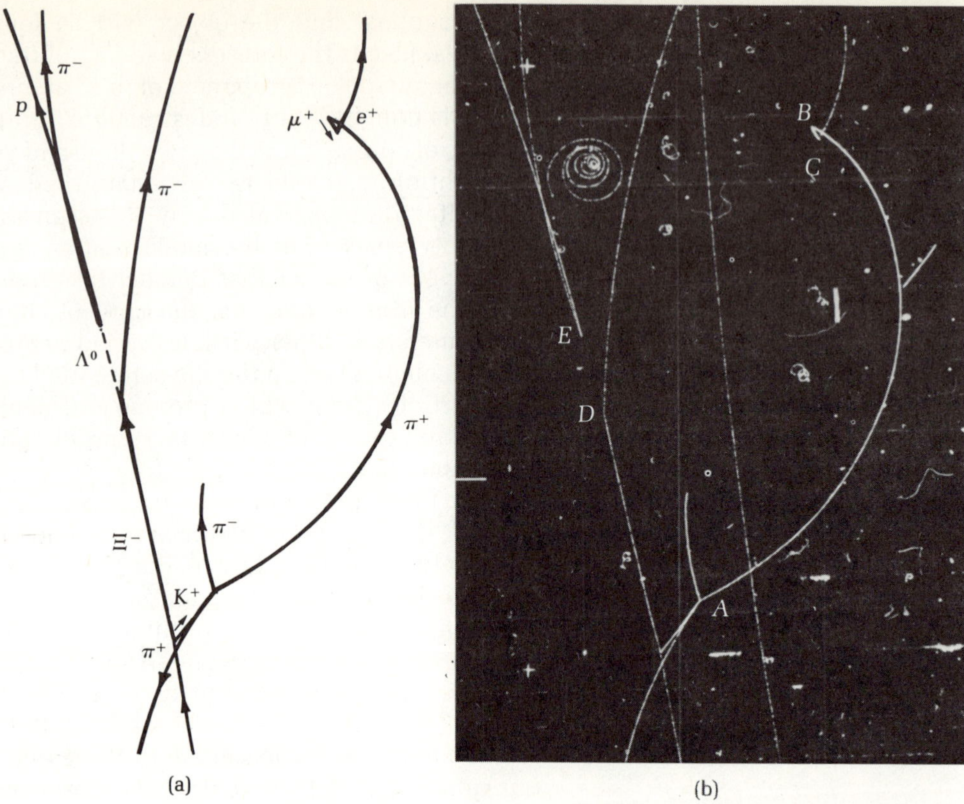

the exceptions being the photon, the neutrino, and the electron. These last three seem to be geometrical point particles, and perhaps they can be regarded as "elementary" but they are not the constituents of the protons and neutrons of which the bulk of matter is made.

The outcoming particles from a collision event are detected by the tracks they create in a **bubble chamber,** a device containing a liquid that bubbles as a charged particle passes through it. The line of the bubbles gives the track of the particles (Fig. 23.12). And the elementary particle physicist tries to find answers in these tracks.

Perhaps you can appreciate the plight of the particle physicist after you've read the following episode in *Winnie-the-Pooh:*

> "Hallo!" said Piglet, "what are you doing?"
> "Hunting," said Pooh.
> "Hunting what?"
> "Tracking something," said Winnie-the-Pooh very mysteriously.
> "Tracking what?" said Piglet, coming closer.
> "That's just what I ask myself. I ask myself, what?"
> "What do you think you'll answer?"
> "I shall have to wait until I catch up with it," said *Winnie-the-Pooh.* "Now

*By A. A. Milne (New York: Dutton, 1926).

there." He pointed to the ground in front of him. "What do you see there?"

"Tracks," said Piglet. "Pawprints." He gave a little squeak of excitement. "Oh, Pooh! Do you think it's a—a—a Woozle?"

"It may be," said Pooh. "Sometimes it is, and sometimes it isn't. You never can tell with pawmarks."

"Wait a moment," said Winnie-the-Pooh, holding up his paw. He sat down and thought, in the most thoughtful way he could think. Then he fitted his paw into one of the tracks . . . and then he scratched his nose twice, and stood up.

"Yes," said Winnie-the-Pooh. "I see it now. I have been Foolish and Deluded," said he, "and I am a Bear of No Brain at All."

"You're the Best Bear in All the World," said Christopher Robin soothingly.

Like Pooh, the particle physicist looks for his "Woozle," the mysterious elementary constituent of nature, if such exists. And like Pooh, he is constantly plagued by the possibility of making a fool of himself, because deciphering these "Woozle" tracks of the bubble chamber is an extremely tricky operation. In spite of this, the knowledge amassed from such collision experiments is amazing.

The particles and their interaction

Experimental particle physicists have discovered with their accelerators and bubble chambers hundreds of subatomic particles. Table 23.2 shows some representative ones. There are so many of them that the designation of "elementary particle" is now considered meaningless for most of them.

The first two decades of particle physics were spent classifying the "zoo" of particles. The result has shed considerable light on the affair. One very broad classification results from consideration of the Pauli exclusion principle: if a particle obeys it, it is called a **fermion;** if it does not, it is called a **boson.** Two fermions cannot occupy the same space at the same time, but two bosons can.

The fermions are further classified into two major groups: the **baryons** (meaning heavy) and the **leptons** (meaning light). The baryons consist of the proton, the neutron, a group of particles under the generic name of **hyperon,** and another group of particles that are considered to be excited states of baryons, called **baryon resonances.** All hyperons are heavier than the nucleons (neutron and proton) and, if in the free state, they always decay into a proton plus some other particles. All baryons interact via both strong and weak nuclear interactions.

The leptons are the group of particles consisting of the electron, the muon, and the tau (this is a very recent addition to the lepton family), and their individual neutrinos. Here is another interesting thing: each lepton seems to be associated with its own brand of neutrino; the number of each lepton plus its neutrino is a conserved quantity. For example, the total number of electrons and electron neutrinos always remains the same in all reactions or particle decay processes. If an electron is emitted in a process, then we know from this that an anti(electron)-neutrino must also be emitted in order that the number of electrons and electron neutrinos remain the same. The emission of the electron increases the number of electrons by one, but the emission of the antineutrino

TABLE 23.2 The particle zoo (a selected sample).

Name			
Proton Neutron }	Nucleon		
Lambda Sigmas Cascades Omega }	Hyperons (they are strange)	Baryons	Fermions (they obey Pauli principle)
Delta and others }	Baryon Resonances (they are excited states of baryons)		
Electron Electron neutrino Muon Muon neutrino Tau Tau neutrino*	Leptons (they interact weakly)		
Pions (π mesons) K mesons Eta meson and others	Mesons		Bosons (they do not obey Pauli principle)
Photon W boson* Graviton* and others			

*Undiscovered so far.

decreases the number of electron neutrinos by one, so the total number of electrons and electron neutrinos remains the same. The same applies to the muon and, presumably, to the tau.

The bosons that have been discovered so far are found to be of two categories: the photon, which is massless in the sense that its rest mass is zero, and the *mesons*, which do have masses ("meson" means "intermediate particle"; most mesons have a mass intermediate between the leptons and the baryons). There is one very interesting thing about these bosons. Whereas the fermions are the sources of force fields, the bosons seem to be the mediators of the interaction between two fermions.

This last statement needs clarification. Consider the interaction of two charged particles—for example, two electrons. In the field description of Chapter 15, we think of them as continuously interacting with each other's fields. But in quantum theory, the field (the electromagnetic field, in this case) assumes a graininess. The grains of the electromagnetic field are the photons. The interaction of two electrons now is pictured to arise from the exchange of photons.

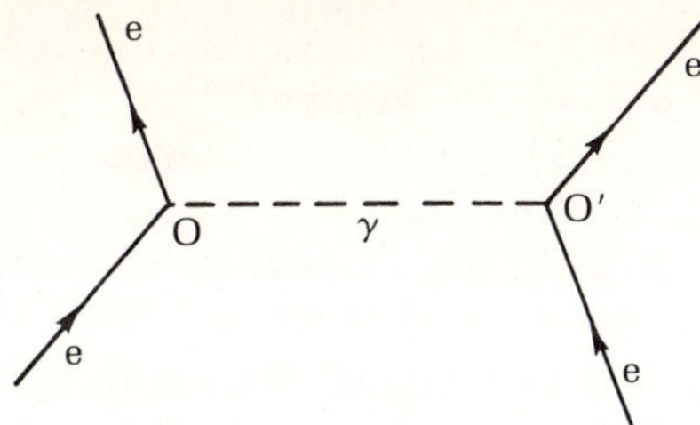

FIGURE 23.13 A Feynman diagram for the interaction of two electrons with the exchange of a photon. One electron emits a photon (γ) at O which the other electron absorbs at O'.

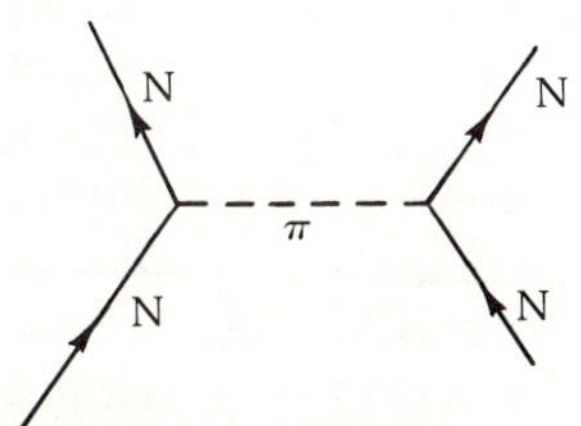

FIGURE 23.14 Feynman diagram for the interaction of two nucleons, denoted as N. Nucleons interact by exchanging a π meson, according to the theory of Hideki Yukawa.

Is this picture too difficult to grasp? Perhaps it will be a little easier to understand if we use a technique proposed by Richard Feynman to picture this way of viewing the interaction of two electrons. First, we draw the world lines (see Chapter 20) in space-time for the two interacting electrons in their initial state, before the interaction takes place (Fig. 23.13). Then one of them emits a photon (the dotted line denoted by γ) at O, and the other one absorbs the photon at O'. The emission of the photon causes a change in the direction and speed of the first electron, so the direction of its world line changes. The same thing happens to the absorbing electron; the process of absorption changes its velocity.

The kind of diagram in Fig. 23.13 is called a **Feynman diagram** and is really a pictorial representation of a complicated mathematical procedure. A complete interaction between two electrons is given as a sum of several such photon-exchange diagrams. The repulsion between the two electrons in the figure looks jerky, but when we sum up all the diagrams contributing to the process, the electrons are found to repel each other following smooth trajectories, as given by the familiar Coulomb force.

In quantum field theory all interactions are mediated by the exchange of particles. The photons, as we saw above, mediate the electromagnetic interaction of all charged particles. Other bosons, called the mesons, are now known to mediate the nuclear strong interaction (Fig. 23.14). Thus a neutron and a proton may interact by exchanging a π meson, or pion (see Table 23.2). The man who first proposed this in 1935, Japanese physicist Hideki Yukawa, did so without any knowledge of the mesons whatsoever. The discovery of the π meson is a triumph of theoretical physics at its very best. The baryons and the group of mesons that take part in strong interaction are now collectively referred to as hadrons (strongly interacting particles).

Along this line of thought, it is believed that the weak interaction (common to baryons, mesons, and leptons) is also mediated by bosons (named W bosons, also first suggested by Yukawa), and the discovery of these bosons is one of the primary objectives of many particle experiments today. Similarly, it is expected that the gravitational interaction also occurs via the exchange of a boson; this has not been discovered either, but few physicists doubt its existence.

The latest work in this area is an attempt to give a unified description of the weak, the electromagnetic and, perhaps, even the strong interaction in terms of a quantum field theory. The mediators of the interactions are the quanta of this unified quantum field; some of them are the old ones mentioned above, like the photon and the W boson, but now looked at as a class of similar particles. One cannot claim this theory to have provided the "final" answer, but it is generally considered to be one with good possibilities.

The quarks: the end of the quest?

Physicist Murray Gell-Mann of the California Institute of Technology, one of the most successful builders of models in elementary particle theory, has a weakness for literary expressions. Whenever he gets a chance, he injects some literary flavor in the otherwise abstract mathematical land of theoretical particle physics. He first came to prominence by showing that, in order to understand the previously mentioned hyperons, we must assign them a new quantum number that protons and neutrons do not have. He called this quantum number "*strangeness*," which is a rather nice name for a quantum number.*

It was also Gell-Mann who discovered that a class of baryons, which includes the neutron, the proton, and some hyperons, can be described by a set of eight quantum states, together making up a family. He called his finding **the eightfold way.**† Whereas Buddha's eightfold way was designed to relieve the suffering of all humanity, but never quite achieved its goal, Gell-Mann's eightfold way did relieve the sufferings of many particle physicists who were groping in the dark trying to find ways of understanding the mysteries of the baryons.

Then in 1964 Gell-Mann made his most daring proposal ever. He suggested that the neutrons and protons (and all other hadrons) are made up of even more elementary constituents, the ultimate elementary particles, perhaps. To name them he got help from the literary master, James Joyce. Joyce, in his novel *Finnegan's Wake*, has a phrase "three quarks for muster mark." Since Gell-Mann needed three of these elementary particles to make up a neutron or a proton, and since he is devoted to James Joyce, he called these particles **quarks.**‡ Incidentally, if you look it up in a dictionary, you will see that the word "quark," has many meanings, one of which is "rubbish." It is not known if Gell-Mann knew of this when he named his new elementary particles quarks.

The quarks are truly amazing particles. For one thing, they have a charge of one-third or two-thirds that of the electronic charge. This goes against a trusted axiom of physics—that charges occur only in quanta of one electronic charge. For another, when everything is said and done about the quarks, it may very well be (at least many physicists belive so) that no quark will ever see daylight. Quarks are destined to be forever confined within the baryons (and mesons) they compose.

Let's go into a little more detail of how the quarks are supposed to make up the various baryons and mesons. The proton and the neutron are each made from three of two kinds of quark; the up or u quark and the down or d quark. (Don't pay any attention to the designations; they are arbitrary names given to distinguish the two quarks.) The u quark has a charge of $+\frac{2}{3}e$ (e is the electronic charge) and the d quark has a charge of $\frac{1}{3}e$. Thus the proton is imagined to be made up of two u's and a d; then its total charge is $\frac{2}{3}e + \frac{2}{3}e - \frac{1}{3}e = e$, which is correct. Similarly, if you picture the neutron to be made up of two d's and one u, you get the correct charge of the neutron, namely, zero.

One thing was immediately clear to Gell-Mann. Because of the eightfold way, he knew that any such scheme of elementary particles must also explain the other particles that fall in the same class. So he introduced a strange quark, called s, in order to explain the strange baryons, baryons that possess a strangeness quantum number.

For a time it was thought that these three types of quarks, u, d, and s, were all you needed to understand all the strongly interacting particles. For example, the mesons are seen to be made up of quark-antiquark pairs (as you recall,

---

*This idea was independently proposed by Japanese physicist K. Nishijima.
†This idea was independently proposed also by Israeli physicist Y. Ne'eman.
‡The idea of quarks was independently arrived at by G. Zweig.

particles have antiparticles, according to relativity; antiparticles of quarks are called antiquarks). Then in 1974 a new class of particles was discovered in the newest particle accelerators. The first one of these made headlines worldwide and was called the J on the East Coast and ψ on the West Coast of the United States (the particle was discovered independently by a group on the West Coast and another on the East Coast.) It was soon established that these particles are like the mesons and can be understood as consisting a quark-antiquark pair, except that their constitution demanded a new kind of quark. This quark was given a new quantum number, **charm,** and the new quark was called a **charmed quark.** So now there are four quarks.

There is now some indication that there may be another quark, and still another quantum number to designate it. And there is no indication that the list will stop here.

It has been known for some time that each of the quarks proposed above comes in three individual quantum states, an affair that particle physicists, in the tradition of Gell-Mann, describe by saying that each quark comes with three **colors.** This increases the number of quarks by another factor of three. So even if the quarks are elementary, there seem to be quite a large number of them.

We expect that elementary particles are point particles, geometrical points of mass or energy. Are the quarks point particles? This question has been answered affirmatively through ingenious analysis of the collision of high-energy electron beams with a target containing protons. In general, the results of such collision experiments depend on both the energy and the angle at which the collision occurs. But if a particle is composed of pointlike constituents, then at the very high energies of the electron projectiles, the results of the collision can be shown to depend on just one quantity. This one quantity is a complicated combination of the two quantities mentioned above, the energy and the angle. Richard Feynman himself pointed this out when he showed that the experimental results tend to give a clear indication that the protons are made up of point particles (Feynman called these partons). Thus the elementary constituents of the protons, the quarks, or partons, whichever you call them, must be point particles. In this sense the quarks can be regarded as elementary, just as the electrons are.

Still many physicists are dissatisfied with the quarks. One main reason is that nobody has yet "seen" a quark, not even in the sense that one has "seen" an electron. That is, we have no evidence as yet that quarks can exist in an isolated state. But the proponents of quarks actually like this. If quarks are forever confined within the particles they make up, then perhaps they are the end of the quest for elementary constituents of matter. If we cannot isolate quarks, we can never ask the question: what is a quark made of?

One final historical note. We mentioned at the outset that the idea of elementary indivisible constituents of matter (the "atomos") was due to Democritus. It turns out that other Greek philosophers also contributed to this idea. One of them, Epicurus (supposedly the founding father of Epicurean philosophy: eat, drink, and be merry), talked about *minimae parte* (minimal parts) of the indivisible atoms that were destined to stay forever confined within the atoms. Thus it has been conjectured that the idea of the confined quarks really originated in the Greek atomic theory.*

SUMMARY

This chapter consists of a discussion of four important divisions of modern physics: nuclear physics, plasma physics, astrophysics, and elementary particle physics. Nuclear physics started with Rutherford's discovery of the atomic nucleus, but it did not catch on until the 1930s when the basic structure of nuclei as a composite of neutrons and protons was established. Soon scientists were producing radioactive isotopes of elements in the laboratory, a trend of research that culminated in the discovery of nuclear fission. More recently, the study of nuclear physics has centered around models of nuclear structure. These models portray the very complicated motion of the nucleons inside the nucleus in terms of simple pictures, to various degrees of approximation. For example, it is found that in the simplest approximation, nucleons in a nucleus, can be assumed to move independently of each other.

Plasma is the state of matter in which the atoms are partially or completely ionized, that is, separated into positively charged ions and negatively charged electrons. Astrophysics, the study of stellar evolution and structure, involves the physics of the plasma inside the stars. Thermonuclear reactions in stars are responsible for the different stages of stellar evolution, which are usually studied in the context of the Hertzsprung-Russell diagram. In particular, the main sequence stars are hydrogen-burning stars. When the hydrogen is exhausted, stars become red giants, which burn more complex nuclei like helium for their energy. When all nuclear fuel is gone, stars go into one of their final stages of evolution. For less massive stars of one solar mass or so, the final stage is that of the white dwarfs, whose stability is a manifestation of the Pauli exclusion prin-

---

*Read an article entitled "Partons in Antiquity," by J. H. Gaisser and T. K. Gaisser, in the *American Journal of Physics*, vol. 45, 1977, p. 439.

ciple applied to the electrons of the star. If the star is about two or three solar masses, it may end up as a neutron star, a star consisting almost entirely of neutrons. For more massive stars the end product is theorized to be the black holes (which are further discussed in Chapter 25).The study of plasma physics on earth centers around the problem of building power plants based on thermonuclear fusion. The problem is one of confining the plasma; magnetic confinement is being tried.

Elementary particle physics is in the forefront of physics today. The main concern of this branch of physics is to find the ultimate constituents of matter. A partial list of subatomic particles is given in Table 23.2. There has been considerable excitement recently about quarks, which have been proposed as the ultimate elementary particles of nature.

QUESTIONS

Review and reason

1. A king hides a gold vessel in a bale of hay and puts the bale with the gold among other bales of hay. Now he wants his men to find which bale contains the gold without ransacking them. His men are allowed to use probes like arrows from a distance. Discuss the scattering of fast arrows from the bales and how this will give away the one containing the gold vessel.

2. Discuss the significance of Rutherford's experiment on the scattering of alpha particles from an aluminum target.

3. The alpha, beta, and gamma rays emitted by radioactive substances were first distinguished by observing their behavior in a magnetic field. Discuss, from what you know about these rays, why their behavior in a magnetic field would be different and in what ways.

4. Can the radioactivity of a radioactive substance be changed by subjecting it to very intense pressure or very low temperature? Explain.

5. What is an isotope? Can there be a radioactive isotope of an element when the most commonly found isotope of that element is stable?

6. What is meant by nuclear shell model? What is a deformed nucleus?

7. How do you feel about the atomic bomb and the issues that go with it—for example, the disarmament treaty and test bans? Express your views concisely.

8. Give the spectral classification of stars.

9. Describe the major steps of the evolution of the stars. Draw a Hertzprung-Russell diagram and indicate on it the probable course of the evolution of a star like our sun.

10. What does the Pauli exclusion principle have to do with the equilibrium of white dwarf stars? Explain.

11. What are neutron stars? How are they explained?

12. Why can't we confine plasma in a regular bottle? How does a magnetic field help in plasma confinement?

13. Discuss two methods that are being studied for the generation of electrical power from fusion reactions.

14. Explain the following terms in connection with the classification of elementary particles: fermion, boson, baryon, hyperon, meson, and lepton.

15. All processes that we know of, including reactions involving nuclei and subatomic particles, conserve baryon number and charge. On this basis, figure out which of these reactions are not permitted by nature.

$$p + \pi^- \rightarrow n + \pi^0$$
$$^2H + {}^2H \rightarrow {}^3He + n + p$$
$$p + p \rightarrow p + \bar{p}$$

(π^- is a pion of negative charge and π^0 is a neutral pi meson. $\bar{p}$ denotes antiproton.)

16. How do two particles interact in the picture of quantum field theory? Draw a Feynman diagram.

17. Up quark, down quark
Strangeness and charm:
Quarks are bottled up
They can't do any harm.
Elucidate the first two lines in connection with the proposed quarks. The idea of the last two lines, perhaps, is that if we can't get hold of free quarks, we can't make a bomb with them. Can you explain how, if we had free quarks, we could make a quark bomb?

1. Calculate the number of neutrons in each of the following isotopes: ^{14}C, ^{28}Si, ^{238}U, ^{235}U, ^{239}Pu, ^{232}Th, ^{40}K. You may need the periodic table to help you with this problem.

2. When U-238 emits an alpha particle, to what element does the child nucleus belong? Use the periodic table, if necessary

3. A nucleus of cobalt-60 decays, emitting a beta particle. What is the child nucleus in this case? Use the periodic table, if necessary.

4. An excited state of the nucleus of oxygen-16 decays, emitting gamma radiation. What is the child nucleus in this case? Use the periodic table, if necessary.

Radioactivity and Nuclear Fission Power

I am an advocate of nuclear energy. I believe that the technology of nuclear power reactors offers great benefits to mankind, without commensurate costs, and that we can reduce (and have reduced) the risks associated with it to an acceptable, and indeed very low level.

BERNARD I. SPINRAD

Fission energy is safe only if a number of critical devices work as they should, if a number of people in key positions follow all of their instructions, if there is no sabotage, no hijacking of transports, if no reactor fuel processing plant or waste repository anywhere in the world is situated in a region of riots or guerilla activity, and no revolution or war—even a "conventional one"—takes place in these regions....

HANNES ALFVEN

■ 24.1 Radioactivity: Applications and Threats

Radioactivity is the process by which unstable nuclei emit radiations of different varieties. Often in the process the nuclear species itself undergoes a transmutation (e.g., alpha and beta decay). Sometimes the nucleus just de-excites from an excited energy level to another lower energy level, giving off gamma radiation in the process.

Before we go into the applications and threats of radioactivity, it is necessary to answer a few questions. Why are some nuclei unstable while others are not? A decaying nucleus must be characterized by a rate of decay. In other words, we can ask: How long does it take a given radioactive sample to decay away so that no further activity is left?

Nuclear stability

There are several and sometimes conflicting tendencies of nuclear interactions that determine the stability of nuclei. First of all, two protons, in addition to their attractive nuclear interaction, have the repulsive electrical interaction to reckon with. When there are a large number of protons and each feels the repulsion due to all the other protons, the magnitude of the repulsion starts to be competitive with that of the nuclear attraction. Since repulsive forces tend to take the protons apart from the nucleus, a large number of protons do not contribute to stability. This is why heavy stable nuclei are usually found to contain a hefty excess of neutrons. Since neutrons only attract (via the nuclear force), a neutron excess gives adequate "glue" in the stabilization process.

But we may wonder why nuclei are not just bundles of neutrons alone. Why have the proton at all with its problem of electrical repulsion? The answer is that the attractive nuclear force between neutrons and protons is the strongest (stronger than the nuclear attraction between a pair of protons or a pair of neutrons). Because of this there is a tendency for nuclei to have an equal number of protons and neutrons. The nuclei we find in the universe occur as a result of the competition between these two tendencies: (1) to avoid having a large number of protons and (2) to have an equal number of protons and neutrons. Thus we get the chart shown in Fig. 24.1.

Let's discuss the chart, which is a plot of the proton number Z of the nucleus against the neutron number N. The stability region is shown by the boxes in the figure. Unstable nuclei that decay by alpha emission are shown by x's. Notice that alpha instability occurs only for heavy nuclei.

In Fig. 24.1, below the region of stability, we are dealing with unstable nuclei that have too many neutrons relative to what is optimal for stability. These

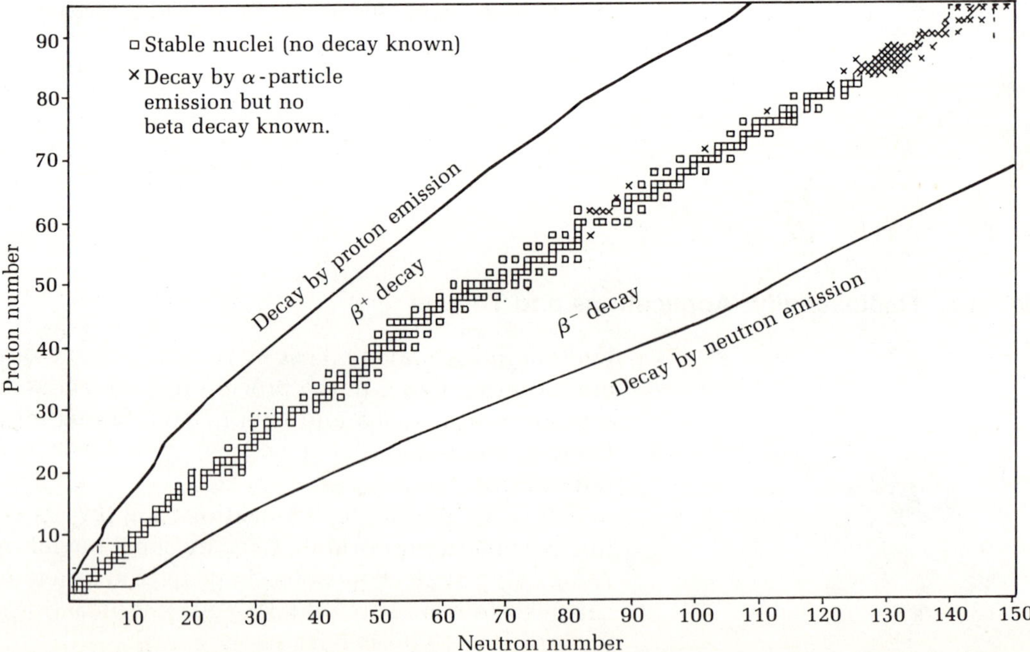

FIGURE 24.1 Plot of number of protons versus number of neutrons for all possible nuclei, showing stable nuclei and showing how unstable nuclei decay. (Excerpted from *Nuclear Science and Society* by Bernard L. Cohen. Copyright © 1974 by Bernard L. Cohen. Reprinted by permission of Doubleday & Company, Inc.)

nuclei tend to decay by (beta) emission of electrons. The basic process contributing to the decay is the beta decay of the neutron converting itself to a proton:

$$n \rightarrow p + e^- + \bar{\nu}$$

where n denotes a neutron, e^- an electron, p a proton, and $\bar{\nu}$ an antineutrino. The child nucleus from such a decay has one more proton and one less neutron than its parent and is therefore relatively more stable. An example is the beta decay of strontium (^{90}Sr) into yttrium (^{90}Y):

$$^{90}\text{Sr} \rightarrow {}^{90}\text{Y} + e^- + \bar{\nu}$$

Likewise, nuclei lying above the region of stability are unstable because they contain too many protons. These nuclei therefore exhibit positron beta (β^+) decay. The basic process here is the conversion of a proton into a positron, a neutron, and a neutrino.

$$p \rightarrow n + e^+ + \nu$$

An example of this kind of decay is that of cobalt-56 into iron-56:

$$^{56}\text{Co} \rightarrow {}^{56}\text{Fe} + e^+ + \nu$$

It is extremely important to note that, whereas a free neutron does show the beta decay process, a free proton never decays: it is stable. Only a proton inside a nucleus can undergo the positron decay process.

Nuclei in Fig. 24.1 that lie far below the stability region are unstable against neutron emission. Such nuclei decay very fast and therefore are difficult to produce even for a very short time in the laboratory. Nevertheless, such nuclei are sometimes encountered in nuclear fallout or similar situations. The nuclei far above the stability region are unstable against proton emission, and the same comments apply to them.

Nuclei with a Z significantly greater than 92 often decay by fission, spontaneously. The nuclei that result from the fission of the heavier nucleus are usually radioactive, and neutrons are emitted in the fission process itself. So spontaneous fission is also often classified as a radioactive decay.

Half-life

The number of radioactive decays depends on the number of nuclei present at any moment. Such a quantity is exponential, and the decay of radioactive nuclei is **exponential decay** (Fig. 24.2). As with the doubling time for exponential growth (Chapter 17), exponential decay can be characterized by the concept of **half-life**: the time it takes a given radioactive sample to become half the original amount.

Each species of unstable nuclei is characterized by its own half-life of decay, which is determined by nuclear dynamics. The half-life easily (in most cases) can be determined experimentally.

As an example, take the case of radium. It decays by alpha emission with a half-life of 1600 years. So a given sample of radium, say a gram, becomes half a gram of radium after 1600 years, 1/4 of a gram in 3200 years, and so on.

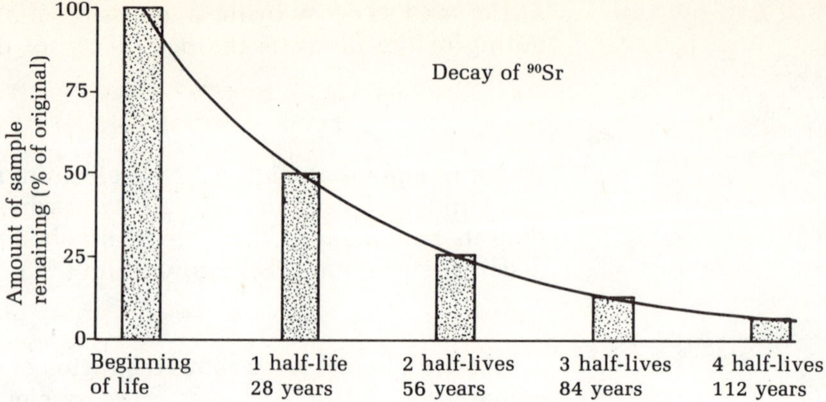

Tritium, the heaviest isotope of hydrogen, decays by electron emission with a half-life of 12 years. Suppose we start with 2 grams of tritium. How much will be left after 60 years? 60 years is 5 half-lives of tritium. After each half-life, the amount of tritium is halved. After five half-lives, the amount of tritium will be reduced by a factor of 2^5. So the amount of tritium left will be $2/2^5$ gram, or $2/32$ gram, or $1/16$ gram.

The half-life of different radioactive elements varies widely, from billions of years to minute fractions of seconds. Table 24.1 gives a chart of some of the important radioactive nuclei.

There are two other important things to consider. First, there is a direct connection between the half-life and the **specific activity** (the number of disintegrations per second per unit mass of the sample) of a radioactive sample. The shorter the half-life, the stronger is the specific activity; half-life and specific activity are inversely proportional to each other. Thus, kilogram per kilogram, the radioactivity from uranium (half-life of 4.5 billion years) is relatively harmless compared to that of ^{239}Pu (plutonium), with a half-life of 25,000 years. And plutonium's specific activity is small compared to that of radium, and so on it goes. Second, for a given nuclear sample, the radioactivity decreases with time since the number of decaying nuclei decreases with time.

TABLE 24.1 Some important radioactive elements.

Name and Symbol	Radiation Emitted	Half-Life(years)
Tritium, ^{3}H	Beta (electrons)	12
Carbon, ^{14}C	Beta (electrons)	5,730
Potassium, ^{40}K	Beta (electrons)	1.3×10^9
Strontium, ^{90}Sr	Beta (electrons)	28
Technetium, ^{99}Tc	Beta (electrons)	2.1×10^5
Cesium, ^{137}Cs	Beta (electrons)	30
Radium, ^{226}Ra	Alpha	1,602
Thorium, ^{232}Th	Alpha	1.4×10^{10}
Uranium, ^{238}U	Alpha	4.5×10^9
Plutonium, ^{239}Pu	Alpha	24,390
Sodium, ^{22}Na	Positrons	2.60

Applications of radioactive decay: a revolution in geology and archeology

One of the amazing things about radioactivity is that it is not affected by any of the usual physical and chemical means, such as heat, cold, pressure, or chemical reactions. The radioactivity of an element stays just the same when it is present in a chemical compound. For example, oxides of uranium are as radioactive as pure uranium. Radioactivity is a nuclear process, and nothing short of nuclear processes can change it. This particular property is unfortunate in many ways—we cannot get rid of radioactive substances easily—but it also is a boon in some instances. For example, because of this nonchangeability, radioactivity gives us a very reliable time clock. We can identify material from ancient times by finding out how much of a particular radioactive nuclear sample is left in the object. This has revolutionized archeology and geology. This field is called **radioactive dating**.

An example of dating that is very useful in archeology is **radiocarbon dating**. Radiocarbon refers to an isotope of carbon, ^{14}C. Ordinary carbon is mostly ^{12}C, but trace amounts of radiocarbon are contained in all living objects.

Several things conspire to create the radiocarbon in living objects. The cosmic rays that bombard the earth constantly contain neutrons as one ingredient. These neutrons are energetic enough to initiate a nuclear reaction, involving nitrogen-14 nuclei (of atmospheric nitrogen), that converts them into carbon-14:

$$^{14}N + n \rightarrow {}^{14}C + p$$

Here n denotes a neutron and p denotes a proton. The radiocarbon atoms produced in this way chemically react with oxygen of the atmosphere making carbon dioxide. Thus some of the carbon nuclei in the carbon dioxide in our atmosphere are really radiocarbon, about one part in 10^8. When plants absorb the atmospheric carbon dioxide by photosynthesis, they absorb a trace of the radiocarbon as well. Animals eat plants, and other animals eat animals: as a result every living things acquires some amount of radiocarbon (Fig. 24.3).

In the body of a plant or any other living object, the C-14 nuclei continue their radioactive (beta) decay:

$$^{14}C \rightarrow {}^{14}N + e^- + \bar{\nu}$$

But the supply is continually replenished by new intakes, so what results is a stable situation, a fixed ratio of radiocarbon nuclei and regular ^{12}C nuclei. Thus all living objects have this property: their carbon atoms include a fixed proportion of the radioactive variety. And this is responsible for a fixed amount of radioactivity per unit mass of carbon in a living sample: about 15,000 beta emissions occur every minute for a kilogram of carbon in living tissue.

Once the living object dies, there is no further replenishment of radiocarbon. The radiocarbon nuclei, however, continue to decay, and as a result their proportion to regular carbon nuclei decreases with time for a dead object. The radioactivity of a dead sample will decrease as time goes on.

So if we find a few bones from prehistoric time, or the remnants of a coffin, we can determine the percentage of radiocarbon in the object, which determines

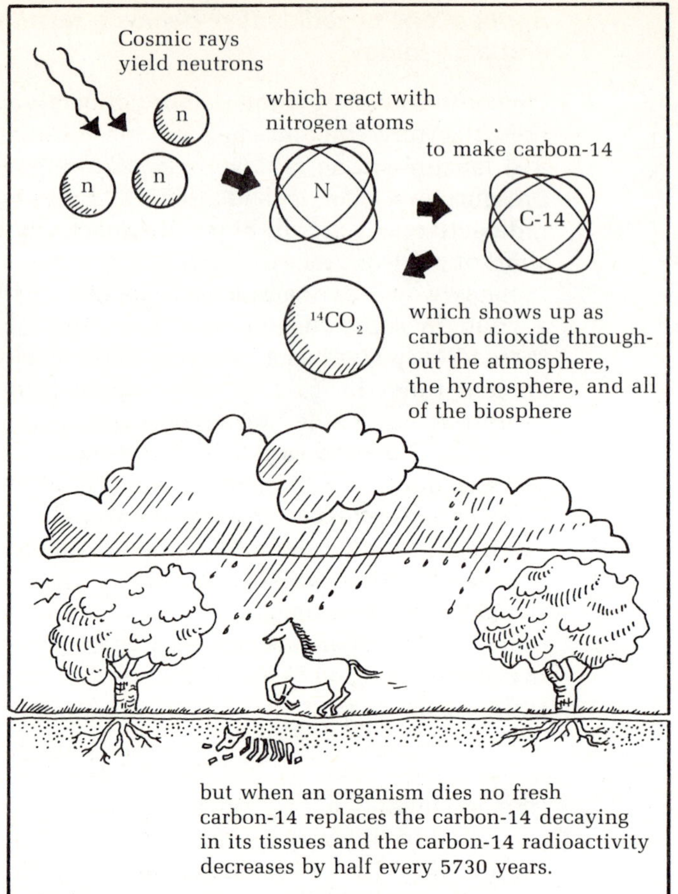

FIGURE 24.3   How carbon-14 is formed and how it enters into the biosphere. (Courtesy: Energy Research and Development Adminstration.)

how much time has elapsed since the skeleton belonged to a living creature or the wood of the coffin was part of live trees. Since ^{14}C has a half-life of 5,730 years, after 5,730 years the activity will be only 7,500 emissions (1/2 of 15,000) per minute per kilogram; we will get 3,750 emissions per minute per kilogram after 11,460 years; and so forth. Conversely, if a sample shows 3,750 emissions per minute per kilogram, we know that its age must be about 11,460 years.

How far back in time can we go with radiocarbon dating? Because the number of emissions decreases by a factor of 2 every 5,730 years, after about ten half-lives (that is, about 57,000 years), the activity will have dwindled to a meager 15 ($15,000/2^{10} = 15,000/1,024 \simeq 15$) emissions per minute per kilogram of carbon. This is pretty close to the limits of experimental accuracy. Although ages as old as 100,000 years have been estimated by this method, the accuracy is questionable.

Thus radiocarbon dating has its maximum usefulness in the age determination of archeological findings. For most geological dating we have to use other radionuclei.

The uranium-lead method

When the first lunar rocks were brought to earth, geologists immediately measured and declared the ages of these samples. The oldest one so far has been found to be some 4.5 billion years old. How can we be so sure?

If you read an article on evolution, geological or biological, you frequently come across time scales of the order of 10^8 years. Even the fastest geological processes occur with a time scale of a million years. How do we determine such old ages?

There are several radioactive dating methods available to the geologist, and usually one of these is more suitable than another in a given situation. In each case we look for the remnants of the decay of a long-lived radioactive isotope; by finding out how much is left we can tell how long the sample has been around.

We will examine one of these techniques, **uranium-lead dating**. We previously have displayed the decay chain of uranium (Fig. 23.1), in which the final child nuclei, the end of the line, are two isotopes of lead, lead-206 and lead-207. Naturally occurring lead is mostly lead-208. So by measuring how much lead-206 or lead-207 is present, we can tell how long the uranium has been around, since we know that the half-life of uranium is 4.5 billion years. And there is one more thing that is important. The amount of lead *really* tells us the time that has elapsed after the rock sample settled down to cool. In a hot environment, like that in the interior of the earth, the lead melts and separates out as it is formed. When cold, however, the lead stays within the sample. Thus the dating tells us exactly the time we want to know: when the rock sample was deposited from the interior to the surface.

Dangers from radioactivity

How dangerous is radioactivity? What kind of danger does it represent to living organisms? Recently there has developed the real threat of environmental pollution from radioactive substances, both from the fallout of nuclear bomb tests and from nuclear power plants. In order to evaluate how dangerous these pollutions are, you ought to be aware of some rudimentary facts.

The first thing to consider is what happens when the energetic radiation from the radioactive nuclei—the alpha, beta, and gamma rays and neutrons—pass through a material and impart its energy to it. Usually the radiation imparts its energy to the atomic electrons of the material, often even knocking the electrons out of their atoms. That is, the radiations are ionizing; by knocking out the electrons, they make the atoms into ions. The charged particle radiations, like the alpha and the beta, do this directly by virtue of their electrical forces. Neutral radiation like the gamma photons and the neutrons do it in two steps. They first give their energy to a charged particle in the body of the material, an electron or a proton, which then causes the ionization.

The properties of materials strongly depend on their molecules. If the molecules are disrupted, the properties may change, sometimes quite drastically. If the material receiving the ionizing radiation is biological tissue or a single cell, it may even die as a result of the change in its atoms.

The reason radiation is so effective in killing cells is an interesting thing to consider. The way a cell functions depends on the instructions written in code form in its large molecules that we call DNA (DNA is an abbreviation for deoxyribonucleic acid). The DNA's are the blueprints for the cell. If incident radiation on a living cell damages even one-thousandth of 1 percent of the DNA molecules, this will either kill the cell or severely alter its performance. Thus very serious damage results when only one-thousandth of 1 percent of all the molecules of the cell have been affected by the radiation. Compared to this, burning, or the direct application of heat to a cell, also kills it, but only by destroying nearly all the molecules of the cell. Thus the process of radiation can do the same damage as burning but with practically one-thousandth of 1 percent of the energy of burning. Radiation is extremely effective.

Actually, this is not all we should consider. The effects of radiation on living cells can be divided into several categories:

1. The cell dies outright; this is the process we mentioned above.
2. The reproductive or dividing capability of the cell is damaged, and the cell dies without leaving replacements.
3. The cell may cease to function, a characteristic of cells in a benign tumor.
4. The cell may start to divide in an uncontrollable manner, as is found in cancer.
5. In a sex cell the change in a DNA can lead to mutation, a change in the characteristics of the offspring from those of the parents.

On the bright side, the cells do have a certain amount of capability to repair the damage. So a small exposure distributed over a long period of time does not produce much damage. This is good to know since in the course of our lives we receive a certain amount of natural background radiation (for example, from cosmic rays) about which we can do very little.

Before we go into some quantitative details of radiation exposure, we must mention the units of measurement of radiation. The unit of dosage is devised by keeping in mind the amount of energy the radiation can deliver to the absorbing material. One **rad** of radiation, by definition, imparts 10^{-5} joules of energy to 1 gram of water or living tissue when it is absorbed. In dealing with the biological effects of radiation, another unit, called the **rem** (an abbreviation for radiation equivalent *man*), is more frequently used. One rem is the same as 1 rad for radiation exposure with gamma rays or X rays. But for other kinds of radiation, a rad and a rem are not the same; we have to multiply the dosage in rad by a factor called the **relative biological effectiveness (RBE)** to determine the dosage in rems:

$$\text{dosage in rems} = \text{dosage in rads} \times \text{RBE}$$

Table 24.2 gives the RBE's for a few common types of radiation. Note that the RBE for X rays and gamma rays is 1, consistent with what we have said earlier.

Let's consider an example. A person receives an exposure of 2 millirads [1 millirad (mrad) = 10^{-3} rad] of low-energy neutrons from a nuclear bomb test. What is the dosage in rems? From the table the RBE of the neutrons is 5. So the dosage in rems is equal to 5 times the dosage in rads, which is $5 \times 2 \times 10^{-3} =$

TABLE 24.2 **Relative biological effectiveness (RBE) for various kinds of radiation.**

Name of radiation	RBE
X rays, gamma rays	1
Alpha rays	10
Beta rays	1
Slow neutrons	5

10×10^{-3} rem. It is convenient here to introduce the millirem (abbreviated mrem), which is 10^{-3} rem. Thus the dosage is 10 mrem.

We now can talk about the effect on human subjects for various amounts of radiation exposure. If the received dosage is larger than 750 rems, death from radiation sickness is almost certain within a few weeks. For a dose lower than this lethal one, the repair mechanisms of the cells get a chance to operate. For a dosage of about 450 rems, the chance of imminent death is only fifty–fifty. And for dosages of less than 200 rems, recovery from radiation sickness is almost a certainty. For a dosage of 25 rems or less, there is no observable effect on the body in the short-term sense.

Unfortunately there is a long-term effect. Radiation exposure increases people's susceptibility to cancer many years later (there seems to be about a 15-year lag). From what we said before about the effect of radiation on a single cell, this cancer-producing feature is not surprising. However, the mechanism of this process is not totally understood, although the fact is well established. And here much smaller radiation doses may have a considerable effect.

So when we talk about biological effects of radiation, radiation sickness is only a minor threat, because only a minor portion of the population is exposed to that kind of danger (unless we get into an atomic war or there is a severe accident involving nuclear bomb testing or a nuclear power plant). But the second threat, cancer, occurs with relatively small doses of radiation, and this is where the real danger of radioactive pollution lies. We will return to this subject again toward the end of the next section.

Radiation therapy and other applications

There is a Chinese ideogram (for the word "crisis," actually) that means both danger and opportunity. It is important to note that the biological effects of radiation are somewhat like this ideogram. Although the dangers are undeniably potent, radiation from radioactive substances does present some opportunities.

Since radiation can destroy cells, we can use it in cancer therapy to destroy cancerous cells; 4 million cancer patients are treated this way each year. Mostly gamma rays are used, but treatment with neutrons and even pi mesons is available also.

The killing property of radiation applies, of course, to microorganisms like bacteria. Radiation is now widely used for sterilizing medical supplies and for preserving some types of food.

In 1942, while World War II raged, Italian-American physicist Enrico Fermi and his collaborators made a secret demonstration of the controlled use of nuclear fission on the campus of the University of Chicago. One witness was American physicist Arthur Compton, who later telephoned another interested party.

"The Italian navigator has crossed the Atlantic," Compton said.
"How were the natives?" the man at the other end asked.
"Very friendly," answered Compton, and the conversation ended.

Thus began the stormy story of nuclear fission power. What Fermi built was a nuclear reactor, a device that converts nuclear energy from the fission of an isotope of uranium, ^{235}U, into heat; this device forms the core of the nuclear power plants of today. The war itself had produced nuclear weapons, the threat of which still hangs over the entire human race. Nuclear power plants are a peaceful use of atomic energy, yet they also embody some real threats to the public. Consequently, public reaction to nuclear power has been anything but friendly.

The following are perhaps the most disturbing questions about nuclear power:

1. Nuclear power is accompanied by environmental pollution from radioactive material. How serious is the threat of such pollution?
2. Are nuclear plants really safe against accidents or even sabotage?
3. Nuclear power plants also produce fissionable material, which, if isolated in bulk, can be used to make a nuclear bomb by unfriendly parties. Is this a serious threat?

Before we can answer these questions, we have to discuss some of the physics of the fission process.

The physics of nuclear fission reactors

Fission is the process by which a very heavy (relative to other elements) atomic nucleus splits into two fragments of smaller nuclei and a few neutrons, releasing a large amount of energy. The energy release can be understood on the basis of the binding energy. We previously have stated, in connection with chemical reactions, that energy is given off in a reaction (i.e., the reaction is exothermic) if the product is a system of greater binding energy than that with which we started (see Section 13.1). So uranium splits up because the product nuclei have a combined binding energy that exceeds that of the uranium nucleus.

Nevertheless, this fact may come as a surprise, because you may remember the nuclear fusion process, where we found that lighter nuclei are less bound than the heavier nuclei they fuse into. In fact, this is the reason that thermonuclear fusion works. So you may ask: How can both of these be true?

The answer lies in a truly interesting behavior of the binding energy of atomic nuclei.* The behavior is best depicted in a graph of the binding energy per

____
*Recall that the binding energy of the nucleus is the absolute value of the negative total energy (kinetic plus potential) of its constituent nucleons.

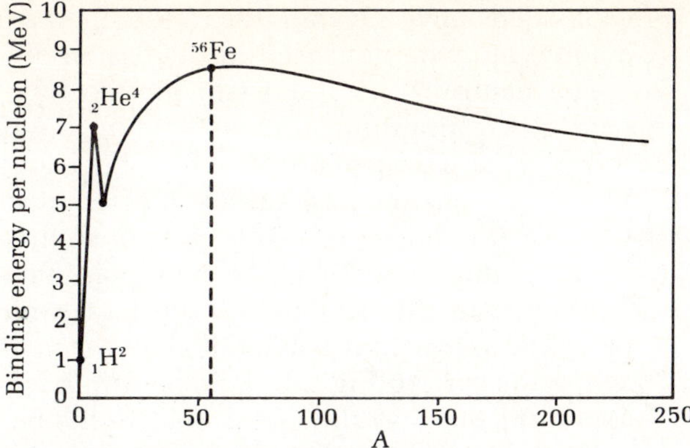

FIGURE 24.4 The graph of nuclear binding energy per nucleon versus mass number A. The binding energy per nucleon increases only until iron-56 (^{56}Fe) is reached.

nucleon (the total binding energy of each nucleus divided by its mass number A gives the binding energy per nucleon) against the mass number (Fig. 24.4). Notice that energy is plotted in terms of a new unit, the MeV, which is an abbreviation for megaelectronvolt (1 MeV = 1.6×10^{-13} J). As you can see, this graph shows an increase, initially, for low mass numbers, or light nuclei. So for light nuclei it is energetically more favorable to fuse nuclei together to make bigger nuclei. But the graph also makes it clear that this state of affairs remains true only until we reach ^{56}Fe. Soon after this, the binding energy per particle actually decreases as we go to heavier and heavier nuclei. Thus from the graph we can see that a heavy nucleus like uranium has less binding energy than do two fragments it may split into.

We can discuss this quantitatively. For ^{235}U, the binding energy per nucleon is 7.6 MeV. The total binding energy of the uranium nucleus is then

$$7.6 \times 235 = 1786 \text{ MeV}$$

In a typical fission event, uranium splits into one fragment that has a mass number of around 100 and another that has a mass number of about 135. The binding energy per nucleon for the first fragment is about 8.6 MeV, giving a total of

$$8.6 \times 100 = 860 \text{ MeV}$$

For the other heavier fragment, the binding energy per nucleon is 8.4 MeV, and this gives a total of

$$8.4 \times 135 = 1136 \text{ MeV}$$

The total combined binding energy of the two fragments is then

$$860 + 1136 = 1996 \text{ MeV}$$

This exceeds the binding energy of 1786 MeV of the uranium nucleus by more than 200 MeV. This excess is released as heat when a uranium-235 nucleus undergoes fission.

Actually, one of the important aspects of the fission event is that, in addition to the two fragments, a few neutrons also are emitted. Since the number of nucleons cannot change, this means that one of the two fragments will have to be slightly lighter, leading to slightly less binding energy for the fragments than

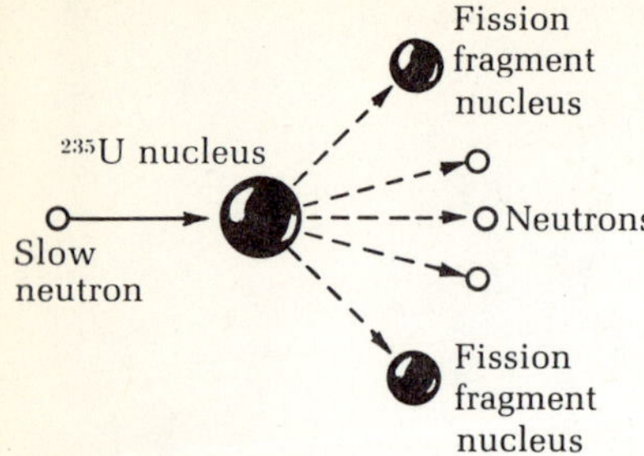

FIGURE 24.5 When a slow neutron is absorbed by the uranium-235 nucleus, it readily fissions. Usually the products are a couple of fragment nuclei and a few (two or three) neutrons.

calculated above. Overall, the round number of 200 MeV represents the average amount of energy released in the fission of an atomic nucleus.

The neutrons are important for the following reason. Without the aid of neutrons, a uranium nucleus seldom divides. The nucleus of uranium-235 behaves like a drop of liquid, which has a lot of resistance to splitting because of its surface tension. The surface tension erects a barrier against fragmentation. Left by itself, once every 10^{17} years or so the nucleus may split and separate by virtue of the probabilistic laws of quantum mechanics. But a neutron (such a neutron is readily available from the cosmic rays) when absorbed by the nucleus takes it over the barrier; it gives the nucleus a piggyback on which to jump across the barrier (Fig. 24.5). So neutrons are essential for the fission process. It also turns out that slow neutrons are the best.

You can see now the advantage of the emission of neutrons in the fission process itself. These neutrons can keep the reaction going if we confine them in some way. This is the idea of a chain reaction. Once the reaction is started, it goes on by itself. The products of the reaction process itself guarantee that the reaction will continue once started.

So the basic problem of the reactor assembly is to have enough size to confine the product neutrons from the fission. Since slow neutrons are the most effective, the reactor should also have some means to slow down the relatively fast neutrons emitted in uranium-235 fission. Since collision with hydrogen is very effective in doing just that, water (which contains hydrogen) is often used to do the slowing down. The water is called a *moderator* in performing this function.

The energy release in fission is in the form of kinetic energy of the fragments, which are radioactive, and the neutrons. The fragments start to decay immediately, producing radiation. All this generates a lot of heat. The heat can be extracted by a coolant and used to generate electricity, using the usual components of a regular power plant.

In the **boiling water reactor** (BWR), which is a commonly used type in this country, the same water is used as the moderator and as the coolant. Moreover, the water is allowed to boil inside the reactor itself and the steam so generated is used to turn the turbine [Fig. 24.6(a)]. In this case one has to be extra careful to avoid contamination, since this water contains radioactive material to some extent.

In the **pressurized water reactor** (PWR), water again is used as both coolant and moderator, but by means of extremely high pressure (as you know, pressure increases the boiling point of water), the water is not allowed to become steam inside the reactor. The hot pressurized water later is allowed to exchange heat with other water in a steam chamber [Fig. 24.6(b)]. This other water becomes steam and is used to turn the turbine.

Collectively, the two forms of reactors are called **light-water reactors** (LWR). Most of the reactors in America are one of these two types. The efficiency of power plants using steam generated by reactors is about 32%, quite comparable to the average of 33% for coal power plants. This means that both types of power plants produce roughly equal amounts of thermal pollution.

The nuclear power assembly needs one basic safety feature (other safety features will be mentioned later): a means of shutting down the reactor in case

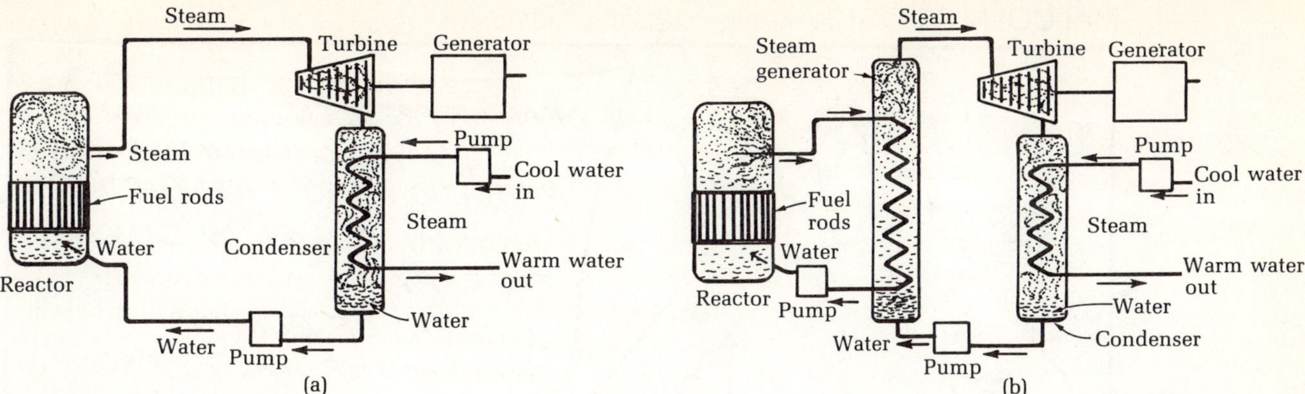

FIGURE 24.6 The components of nuclear power plants. (a) Boiling water reactor (BWR). (b) Pressurized water reactor (PWR).

the chain reactions go out of control. This is accomplished by having a set of cadmium or boron rods (called the **control rods**), which are very efficient in absorbing neutrons and thus stop the reactions when necessary.

Finally, let's discuss one feature of the fuel rods. The concentration of the fissionable isotope uranium-235 in the rods is only about 3%; the rest is the nonfissionable uranium-238. This is way below the concentration one needs in order to make a bomb. Natural uranium contains only 0.7% of ^{235}U; its enrichment to the 3% or higher level requires highly sophisticated technology. And this is why there are only a few nations with nuclear know-how.

Pollution from nuclear power plants

Nuclear power plants do not give us air pollution, but they are not pollution free; they give us radioactive pollution. The fission products are radioactive, some with short and some with long half-lives. Some of the radioactive substances involved are in gaseous form and therefore leak more easily. The remnants of the fuel rods after the fuel is exhausted constitute the main radioactive waste from a nuclear power plant.

There now seems to be some consensus about the radioactive leakage from a nuclear plant: most experts agree that this adds about 1 to 5 mrem per year to the adjacent environment. The problem is that we don't know how serious this amount is. From natural sources alone (such as cosmic rays, radioactivity within the human body, etc.) we receive about 100 mrem per year. Moreover, things like X-ray diagnosis, to which we routinely subject ourselves, add about 50 mrem on the average to this dosage. The Energy Research and Development Administration (ERDA) has recommended a total of 170 mrem of annual exposure as the maximum allowable exposure for a person, suggesting that even an exposure of 170 mrem is probably not dangerous. Looked at from this point of view, even 5 mrem a year exposure from a local nuclear power plant seems to be safe.

On the other side, there is some concern that even the 100-mrem-per-year exposure we receive from nature is causing some extra cancers. If this is true, then a 5 percent increase of the radioactive dosage certainly will lead to more cancer deaths. This in itself, however, should not be all that alarming. We can

also measure the pollution from coal power plants this way: the number of deaths per year caused by air pollution. It is almost certain that coal power will fare no better than nuclear power in terms of the number of extra deaths per year, and it probably will prove much worse.

There is also leakage of radioactive substances in transporting and handling the fuel and the reactor wastes. It is not known how dangerous this type of leakage is.

The most publicized nuclear pollution problem is the disposal of wastes. The latest attitude on this question seems to be that if the waste is put into solid form (the technology for this is already available) and then buried in small bundles (so that they do not melt down from getting the benefit of each other's heat) in salt mines where there is no underground seepage of water, it should be quite safe environmentally. There is no fault to be found in the technical argument. Whether it can be carried out without mishap remains to be seen.

The question of safety

There are several kinds of safety questions that can be asked about nuclear power. The first one we will consider is the safety against an accident. A reactor ordinarily works in a condition that is called *critical*; just enough fission neutrons are produced to continue the chain reaction. If the available number of neutrons exceeds the critical condition, the reactor becomes supercritical. Under this condition the rate of the generation of heat by the fission process increases rapidly until everything is melted down. This is the most dangerous reactor accident.

The control rods mentioned previously are an excellent safeguard against such accidents. In the case of any indication of supercriticality, they are immediately inserted into the reactor, stopping the chain reaction. There are backup safeguards, in addition, and many checks and balances in the system to prevent this kind of an accident.

A second kind of accident can occur if there is a large leak in the circulation pipes for the coolants of the reactor. In the absence of cooling, it takes a very short time for the reactor core to heat up to a temperature that is beyond the critical temperature, and this can lead to a meltdown. To prevent this, aside from safety checks and inspections, all nuclear reactor systems have an emergency core cooling system, which comes into operation almost immediately if the main cooling system fails.

Suppose that, in spite of the backup precautions, an accident happens anyway. In this event, reactor engineers arrange a series of barriers to prevent the escape of the molten material. The last of these barriers is the large steel-lined reinforced concrete structure that the whole system is put into. It is called the **containment** and is expected to contain the radioactive material from an accident even under the worst of circumstances.

How about sabotage? Saboteurs want maximum effect for their efforts. A reactor system with all its complications is an unlikely system for saboteurs' actions. With all these safety features, their chances of getting any results are negligible.

A third kind of safety problem comes in when we consider nuclear reprocessing plants and breeder reactors. The spent material from a reactor, besides containing radioactive wastes, also contains valuable amounts of both old fuel residue and a new fuel: plutonium-239. The plutonium-239 is routinely manufactured in a reactor by a set of reactions in which uranium-238 (which is the bulk of the fuel rods) absorbs some of the neutrons and eventually changes into plutonium. This series of reactions that makes plutonium, which is fissionable, can be used more profitably in a specially designed reactor assembly called the **breeder reactor;** it breeds its own fuel. The reprocessing plant extracts the fuel from the reactor trash.

Both the reprocessing plants and the breeder reactors must handle large amounts of already concentrated fissionable material. If some of this material were diverted and used for nonpeaceful purposes, the world could become a more dangerous place to live.

Advantages and disadvantages of nuclear power

The major advantage of nuclear power is the energy storage efficiency of uranium as a fuel. One kilogram of uranium-235 is able to generate as much heat by fission as does the chemical burning of 1 kiloton of coal (discussed later). This means for the power company a reduction in the cost of handling the fuel: the day-to-day operating costs are reduced as a result.

Another advantage of nuclear power over conventional coal power is the almost total absence of air pollution. However, this must be balanced against the radioactive pollution that nuclear power creates. In general, it can be said also that mining for uranium is not as environmentally destructive as for coal; again the difference comes from the high storage efficiency of uranium.

One disadvantage of nuclear power plants is the large capital cost necessary initially. Also, from the consumer's point of view, there remain questions about the pollution from nuclear power plants and the safe handling of the nuclear fuel. That such questions cannot be answered at present to the satisfaction of most people must be counted against nuclear power.

Nuclear power has brought to light several issues that previously have not been discussed in the public arena in this country. Only some of these issues are peculiar to the nuclear power industry (like the disposal of radioactive waste); there are other issues that apply to energy use in general—for example, the whole area of ecology. Thus this is a good place not only to try to assess the situation for nuclear power but also to discuss the entire energy question.

It is important to understand why there is so much push for the public acceptance of nuclear power. First, nuclear power does provide, at face value, an easy solution to our energy problem. It is a fact that, in spite of a lot of empty talk about conservation, the American public has not yet opted in its favor. The consumption of energy is still on the upswing. Under this circumstance, people who feel they have the responsibility to provide the public the power it wants naturally opt for the only new solution, which is commensurate with the profit motive, that technology has come up with. After all, nuclear fission power is here, whereas economical and profit-saving solar power or fusion power is still many years away.

Second, when everything is said and done, it is probably correct to say that the pollution danger from nuclear power is no greater than that from fossil fuel power, which the public has long accepted.

Third, it is very likely that other nations are going to go into nuclear power in a big way. For example, they may develop and use breeder reactors, which will solve their power problems perhaps for hundreds of years. Can we risk getting behind in this important technology?

On the other hand, these same reasons also make it clear why the opposition to nuclear power has been so vehement. Consider the matter of pollution. It is true that ecology has long been a very neglected issue in the industrial nations of the world. In fact, the word "ecology" appeared in the English dictionary no earlier than 1879. Therefore, in the earlier days of industrialization, there was very little awareness of its consequences to the environment, and therefore very little protest. Since we have begun to understand the nature of pollution and its effect on our environment, it is difficult not to be concerned.

It is true that, to date, the American public has not voted in favor of the conservation of energy. But part of the reason for this is that power comes to the people at a disguised price. The cost of pollution is never included. If the correct higher price were charged, serious conservation of energy would very likely begin.

And as for what other countries are doing, their actions certainly do not make it right for us to develop such potentially dangerous means for electric power as the breeder reactor. If plutonium is in the hands of too many countries, there is very little doubt that it will be used some day in nuclear warfare. If we join the crowd instead of urging other nations to stop these programs, it will only add to this dreadful probability.

And so the debate continues. There are many other facets of the debate; we have only mentioned a few. Who is right? What should be done?

It is most important to see and agree on the following two aspects of the problem. First, the demand for useful energy produces the demand for resources. The energy must come from some resource, and the choice may be very limited. At the moment, with our present technology, the choice really boils down to that between coal and nuclear fission fuels.

Second, burning any fuel produces pollution. All pollutions are potentially very dangerous and come with both short-term and long-term effects. For example, some radioactive substances last for a long time, making our environment more cancer producing in the future than it is today. But air pollution in the form of carbon dioxide also stays in the atmosphere for a long time, and eventually could cause immense environmental changes, such as the melting of the ice caps.

I believe that once we have some kind of agreement on these two issues, perhaps it will be easier to develop a national energy plan that takes into account both of the problems. The priority of such an energy plan should be conservation and developmental research on solar and nuclear fusion power. In addition, the use of small-scale resources, more efficient generation of power, and cheaper methods of transportation of electricity should get appropriate attention in this plan. And finally, I feel that the use of both coal and nuclear fission power should be held to a minimum, just enough to meet the demands of our essential needs.

We should note that, by the best estimates, nuclear power is like coal power—no more than a temporary solution to the energy problem. Without the use of breeders, if we were to get all our electricity from nuclear fission, the fuels would not last more than 50 years or so. With breeding, the lifetime of the fission fuel is extended considerably, perhaps to a few hundred years, which is quite comparable to that of coal.

A comparison of energy storage efficiency of uranium and coal

Let's estimate the energy release per kilogram of ^{235}U when it splits. The splitting of one ^{235}U nucleus produces 200 MeV of energy. A kilogram-mole of uranium-235, which has a mass of 235 kg, contains 6.02×10^{26} uranium nuclei. So for every 235 kg of ^{235}U, we get, for the amount of energy release,

$$6.02 \times 10^{26} \times 200 \text{ MeV} = 12.04 \times 10^{28} \text{ MeV}$$

Since 1 MeV $= 1.6 \times 10^{-13}$ J, the energy released in joules is

$$12.04 \times 10^{28} \times 1.6 \times 10^{-13} = 1.92 \times 10^{16} \text{ J}$$

This is for every 235 kg of ^{235}U; the energy release per kilogram is obtained by dividing this number by 235. This gives us

$$\frac{1.92 \times 10^{16}}{235} = 8 \times 10^{13} \text{ J/kg}$$

for the energy release in uranium fission.

Compare this to the corresponding figure for the chemical burning of coal. From Table 13.2, 1 kg of coal gives us the energy of

$$5600 \text{ Cal} = 5600 \times 4200 \text{ J} = 5.6 \times 10^3 \times 4.2 \times 10^3 \text{ J}$$
$$= 2.35 \times 10^7 \text{ J}$$

Thus, kilogram per kilogram, the fission of uranium gives us

$$\frac{8 \times 10^{13}}{2.35 \times 10^7} = 3.4 \times 10^6$$

or 3.4 million times more energy than the burning of coal.

SUMMARY

This chapter deals with nuclear fission power and radioactivity. In connection with radioactivity, an important concept is that of half-life, the time it takes for a given radioactive sample to decay to half the original amount. The half-life phenomenon can be used for the purpose of dating ancient material. Two such dating procedures, radiocarbon dating and uranium dating, are described.

Recently the prospects of radioactive material as pollutants of our environment have increased due both to weapons testing and to the advent of nuclear fission power. The biological effects of radiation from radioactive material are discussed in some detail.

The section on nuclear fission power discusses the advantages and disadvantages of fission as a power source. The mechanism of nuclear fission as a physical process is also discussed.

Review and reason

1. Explain why some nuclei are stable and some are unstable.
2. Explain the concept of half-life of a radioactive element. Give an example.
3. Discuss the carbon-14 method of dating archeological objects.
4. Discuss the uranium-lead method of dating geological deposits.
5. Explain the difference between a rad and a rem of radiation.
6. What are the major aspects of the biological effects of radiation from radioactive substances? Be concise.
7. Why is a weak dosage of radiation effective in food preservation?
8. Is the mass of an atomic nucleus greater than, smaller than, or the same as the sum of the masses of its constituent subnuclear particles? Why?
9. Why don't natural deposits of uranium undergo chain reaction?
10. What are the major components of a nuclear power plant? What are the safety features of such a system?
11. What is the purpose of a breeder reactor?
12. Compare coal and nuclear fission power. Consider the following aspects: the availability of the resources, the operating costs of the power plants, the capital costs, pollution problems, safety problems, public opinion problems, and any other issues you think are important. You may have to do some outside reading to answer this question.
13. Compare nuclear fission and nuclear fusion power (assuming that nuclear fusion power plants can be built). Again consider all the aspects mentioned in question 12. Do some outside reading, if necessary.

Arithmetic

1. Consider a sample with a mass of 1 gram, of a certain radioactive substance with a half-life of 2 days. How much of the sample will be left after 16 days?
2. Some human bones are found in an ancient cave. The radioactivity from 1 kg of carbon of the sample is found to be 1875 beta emissions per minute. How old are the bones?
*3. In the text we showed that ^{235}U gives an amount of energy of 8×10^{13} J/kg. Suppose we have a 1000-MW nuclear power plant running at an efficiency of 30%. How much U-235 per second do we need to run the power plant? How much per year? U-235 occurs in nature in the proportion of 1 kg for every 140 kg of natural uranium. How much natural uranium do we have to mine annually to run the power plant?

---

*Optional.

General Relativity: Einstein's Theory of Gravitation

A philosopher, a physicist, and a mathematician were looking intently at an animal in the distance. First the philosopher spoke up.

"That's very interesting," he said. "I didn't know there were black sheep in this area."

"Wait a minute," retorted the physicist. "You still don't know there are black sheep in this area; all you know is that there is *a* black sheep."

"How do you prove that?" The mathematician sounded irritated. "All we see, and therefore all we have proof for, is that there is a sheep in this area, *one side of which is black.*"

The story points out this one important thing about the mathematician: he doesn't like to accept anything on faith; he wants proof. Many mathematicians therefore were very unhappy about one aspect of Euclid's geometry: the parallel postulate. From time to time many famous mathematicians devoted many years of their lives to trying to find a "proof" of the parallel postulate. In their attempts to find a "proof," new notions were conceived about the geometry of space. These notions brought a revolution in mathematics and later formed the basis of Einstein's theory of general relativity—Einstein's theory of space and gravitation. So this is where we begin, with the work of the mathematicians.

■ 25.1 The Geometry of Space

A very long time ago, Greek mathematician Euclid developed the geometry that has since become gospel to billions of minds. We learn Euclid's geometry in school and accept its validity for physical space without much scrutiny. And why not? Euclid's theorems make sense when we apply them to the physical world around us; his postulates do seem "self-evident" to a large extent. So when a poet writes, "Euclid alone has looked at beauty bare,"* we accept the statement about not only Euclid, who certainly was one of the greatest mathe-

---

*Edna St. Vincent Millay wrote this in a poem, which was quoted in *Mathematics in Western Culture,* by Morris Kline (New York: Oxford, 1953), p. 40.

FIGURE 25.1 Gauss's experiment. Do three angles of a space triangle add up to 180°?

matical minds who ever lived, but also his geometry, attributing to it a sense of infallibility that we like to associate with beauty. We become so awestruck by the monumental nature of Euclid's work that we forget it is just a model for our physical space, a figment of human-created reality.

Look back into your childhood memories. Did it bother you that a straight line in physical space could be extended indefinitely? Parallel lines do not meet in your limited experience, but perhaps you wondered if that remains true even when the lines are extended without bounds.

Fortunately for us, some people get concerned with questions like this as children and decide to look for answers when they grow up. Mathematician Karl Friedrich Gauss of Germany, regarded by many as *the* greatest mathematician who ever lived, was one such person. Gauss asked a simple question: What is the experimental evidence for the validity of Euclid's geometry for physical space, beyond that of everyday experience? Science has taught us that everyday experience is not all that reliable a guide to truth. Gauss found that there weren't any experiments conducted to verify Euclid's theorems about physical space. So he decided to carry out an experiment himself.

At the time, Gauss was working as a surveyor. Taking advantage of his official capacity, Gauss set up survey apparatus in three places in Germany, Brocken, Hohegen, and Inselburg. His idea was to make up a space triangle, with apexes at these three places of different altitudes, and test the Euclidean theorem that says that the sum of the angles of a triangle add to 180° (Fig. 25.1).

Gauss's measurements uncovered no deviation from Euclidean rules of geometry outside experimental errors. Did everything return to the status quo after this? No; all Gauss's experiment proved was that Euclid's geometry is valid within the limitations of our experimental apparatus for the physical space in a small scale. The large-scale question remained open. And the most important point Gauss made through his work was that geometry is a physical subject: it must be settled by experiment. Euclid's geometry is one of many mathematically possible geometries of space. There are other geometries, and only experiments can settle which of the geometries applies to real space.

The men who first developed non-Euclidean geometries were contemporaries of Gauss, John Bolyai from Hungary and Nicholas Lobachevsky from Russia. "Out of nothing I have created a new and another world," declared Bolyai after successfully formulating a geometry that could prove to be an alternative to Euclid's. What caused this exultation? Let's find out.

The creation of the new world starts with a different parallel postulate. Suppose there is a point P and a line L [Fig. 25.2(a)]. According to Euclid, one and only one line can be drawn through P that is parallel to L. Defying Euclid,

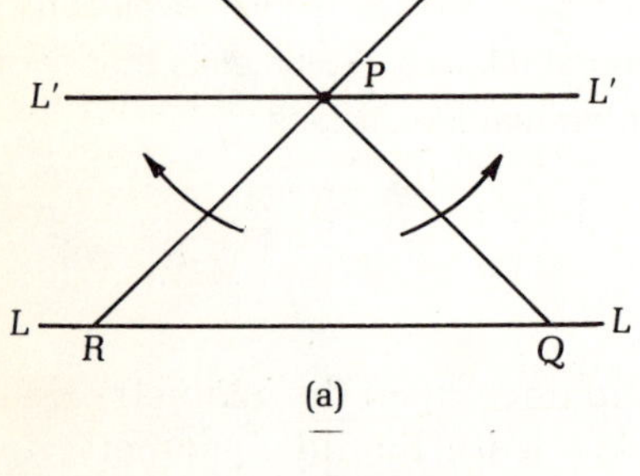

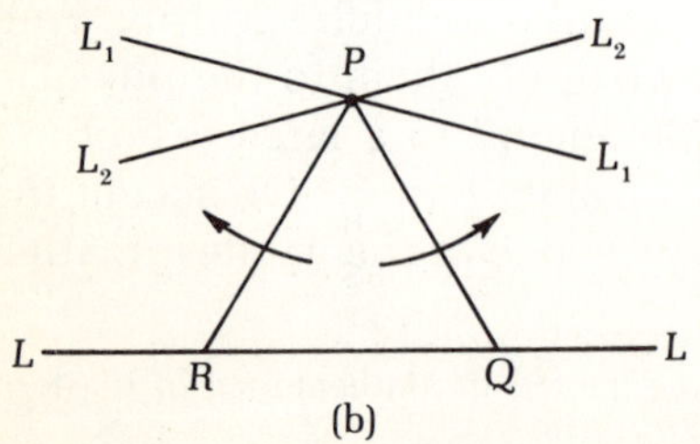

FIGURE 25.2 (a) Parallel lines in Euclid's geometry. As the lines PR and PQ are rotated, as shown, both approach the limiting parallel line L'. There is only one line parallel to a given line L through a point P in Euclid's geometry: this is the parallel postulate. (b) In the Bolyai-Lobachevsky geometry, the line PQ when rotated, as shown, approaches the parallel line L_1. The line PR, on the other hand, approaches the parallel line L_2 when rotated, as shown. The two limits are not the same. In this geometry we can draw an infinite number of lines through a point, all of which are parallel to a given line. See text for further clarification.

FIGURE 25.3   The surface of a saddle as an example of the Bolyai-Lobachevsky geometry in two dimensions. Through a point P on the saddle, we can indeed draw an infinite number of lines parallel to a given line L and lying between L_1 and L_2. None of these lines can ever meet L when extended; hence they are parallel to L. Note that if we construct a triangle made of the "straight lines" of the saddle surface, the three angles of the triangle add up to a total that is less than 180°.

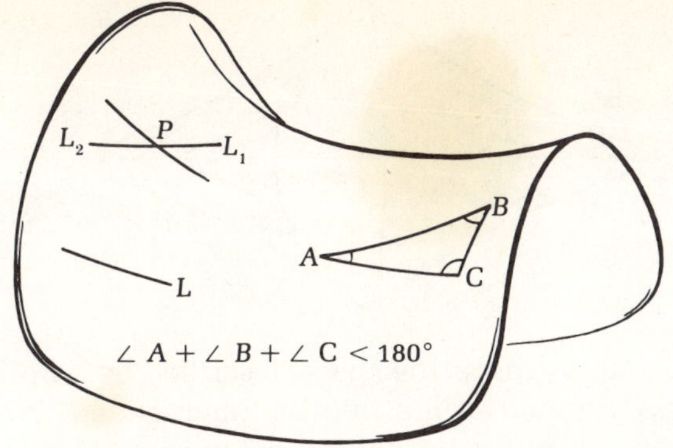

Bolyai and Lobachevsky asserted that not one but an infinite number of parallel lines could be drawn through P. How they arrived at such a seemingly preposterous axiom can be appreciated from a look at Figs. 25.2(a) and 25.2(b). In Fig. 25.2(a), as the point Q on L moves to the right, the line PQ moves counterclockwise and approaches the parallel line L'; and as R moves to the left, the line PR rotates clockwise and again approaches the limiting parallel line L'. Thus in both cases we get the same limit L'. In the geometry of Bolyai and Lobachevsky, the two limits are not the same [Fig. 25.2(b)]. Here the right limit is the line L_1 and the left is the line L_2; both lines are different and neither meets L when extended to infinity. Most importantly, now there are an infinite number of parallel lines passing through P lying between L_1 and L_2.

On the face of the diagram we have drawn, this seems absurd. Surely one look shows that the lines L_1 and L_2 are going to intersect with L eventually, at least in one direction. But maybe what we are calling straight lines in the diagram are not what such a geometry implies by a straight line. Why? Because the rudimentary concepts of geometry, such as points and straight lines, are basically undefined and easily may be adjusted to fit the new postulate of this geometry.

Consider the surface of a saddle, a two-dimensional example of the Bolyai-Lobachevsky geometry (Fig. 25.3). A "straight" line defined as the shortest distance between two points on this kind of a surface is curved. Such a curve is called a **geodesic.** On the surface of a saddle, with the geodesic defined as the straight line, it is easy to verify that the parallel postulate of Euclid doesn't hold; instead of one, we can draw an infinite number of "lines" through a point, all of which are parallel to a given line.

A more unfamiliar but appropriate two-dimensional model for physical space is the surface of the so-called **pseudosphere,** or opened-up sphere of sorts (Fig. 25.4). Again the geodesics of this surface have the properties demanded by the postulates of the Bolyai-Lobachevsky geometry. Notice that on such a surface (as with the saddle) the angles of a triangle total less than 180°. The ratio of the circumference to the diameter of a circle, which is π in Euclidean geometry, is greater than π in this geometry. Thus having a different "parallel" postulate leads to other pronounced differences in the theorems or rules of these geometries. One can now use these differences in an experimental fashion to decide which geometry holds for physical space.

FIGURE 25.4 Introducing the mathematician's pseudosphere, of which a three-dimensional model is impossible to construct. The two-dimensional surface of the pseudosphere shown here is an example of the Bolyai-Lobachevsky geometry.

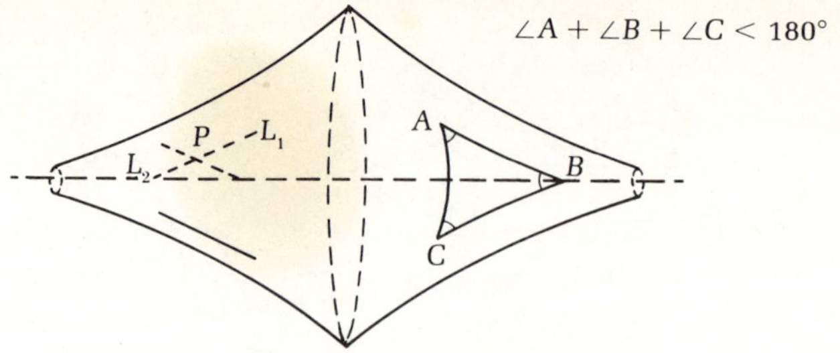

Interestingly, the space described by both Euclid's and the Bolyai-Lobachevsky geometries are infinite, open-ended. We don't know how you feel about infinity, but many people feel uneasy about it. This was particularly true for the ancient Greeks, who could not, for example, conceive of natural motion as motion in ordinary straight-line paths, perhaps because such motion could proceed without bounds. Euclid himself regarded his geometry as quite finite: the straight lines could be extended indefinitely in either direction, but he begrudged that anybody would want to make such extensions. However, a finite Euclidean geometry as envisaged by him has a severe problem. Witness this anecdote by author Anderas de Rhoda:

> One clear night before the advent of kindergarten science books, my sister Ursula-Marie stood in our parent's garden in Germany gazing at the stars.
>
> "Do you think," she whispered, "there's a place where the world comes to an end?"
>
> "Of course, silly."
>
> "Like a fence? Like a wall?" There was disappointment in her voice.
>
> "Well, it's got to be something."
>
> She fell silent. Suddenly she cried: "But what about the other side of the fence? What about the other side?"

It is difficult for a child, or an adult, to imagine physical space having a boundary. A finite Euclidean space that necessitates the concept of a boundary is impossible. But can we have a finite space without a boundary? Can a geometry be defined this way?

The man who figured out such a geometry was a brilliant student of Gauss named Bernhard Riemann. Like Bolyai and Lobachevsky, Riemann started with a different approach to Euclid's parallel postulate. In the geometry of Riemann, all "straight lines" are finite. Let's see where this leads us in the consideration of parallel lines. Look at Fig. 25.5. As before, the point Q moves to the right and the point R to the left; but L is finite, so the two points Q and R must eventually meet. The line PR thus rotates around P and merges with PQ, but there is no loss of contact with L. So the line thus generated is not parallel to L at all: there are no parallel lines in this geometry. Again, from Fig. 25.5 it may be hard to visualize this because we have been using Euclidean straight lines.

Let's look at a two-dimensional analogue of Riemann's geometry, which is provided by the surface of a sphere. Like Bolyai's, this is also a "curved" geometry, the "straight lines" (shortest distance between two points) are arcs of curves called **great circles.** Great circles are circles concentric with the sphere itself (Fig. 25.6). The great circles are clearly finite; they do not extend to infin-

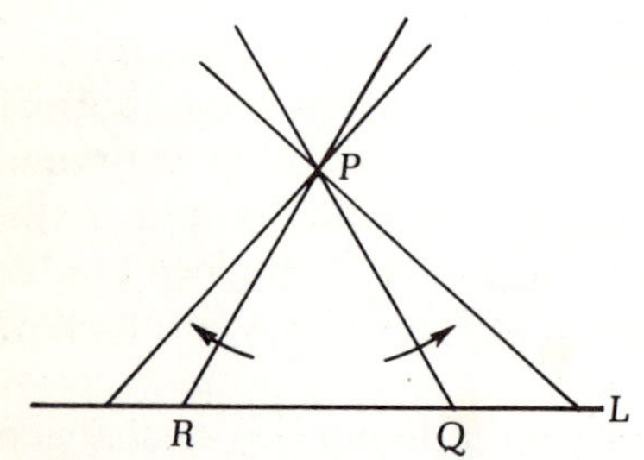

FIGURE 25.5 Riemannian geometry. Suppose the straight line L is finite. Then as we rotate PR and PQ in the ways indicated, the points R and Q must meet since L is finite. Thus PR rotates about P and becomes one with PQ, while still maintaining contact with L. So the resulting line is not parallel to L; there are no parallel lines in this geometry. See text.

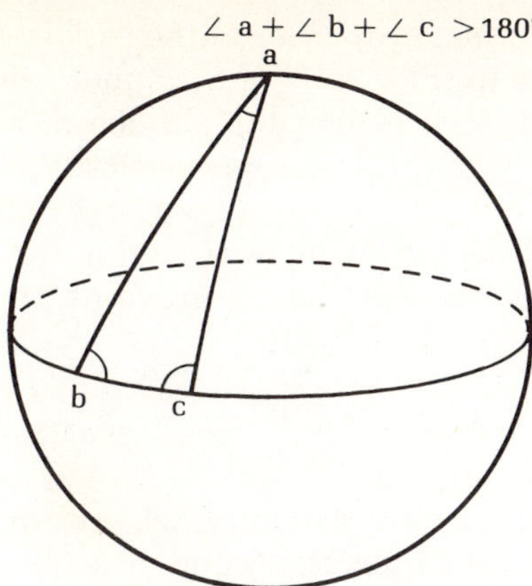

FIGURE 25.6　The surface of a sphere: a two-dimensional example of Riemann's geometry. The "straight lines" of this geometry are great circles, circles that are concentric with the sphere itself. All great circles intersect; there are no parallel lines. If we construct a triangle, using arcs of great circles, the three angles of such a triangle add up to a total greater than 180°.

ity, yet they have no ends. There are no parallel lines; any two great circles intersect. Again we get some new rules. Three angles of a triangle now total greater than 180°; the ratio of the circumference to the diameter of a circle turns out to be less than π. Riemann's geometry gives us a model of space that is finite yet without a boundary.

Mathematicians have hailed the development of non-Euclidean geometry as a revolution: mathematics could be regarded no longer as ultimate truth but, like physical theories, only as human-created models. And only experiments can decide which geometry is the right one for our physical universe.

Let's clarify a point of terminology. Mathematicians refer to Euclidean space as **flat space,** as opposed to the **curved space** of the geometries of Bolyai-Lobachevsky and Riemann. To distinguish between the two curved spaces, we note that the "straight lines" of Bolyai-Lobachevsky geometry bend up one way and down the other way (see Figs. 25.3 or 25.4). This is called **space of negative curvature.** In contrast, the "lines" of the Riemann geometry always bend the same way and define a **space of positive curvature.** Flat space has **zero curvature.**

One problem may be bothering you. Is there a way to think about the curvature of space without taking a look from outside? In a story named *Sphereland*, author Dionys Burger writes about imaginary two-dimensional people inhabiting the surface of a sphere, only they think they live on a flatland. Then one day a creature from three dimensions appears before them and points out that their world is curved.* Do we have to wait for the visit of a creature from a four-dimensional world (if it exists) to find out whether our space is really curved and, if curved, which way?

Fortunately, the answer is no. Every geometry is characterized by two kinds of features, extrinsic and intrinsic. For inhabitants of a certain space obeying a certain geometry, the extrinsic properties are meaningless because to see them we do need to look from the outside. Thus the sphere analogy may let you imagine that Riemannian space somehow curves around in the fourth dimension, but then you get stuck. You don't know how the fourth dimension would behave; nobody in our three-dimensional universe knows.

So in making a picture of the curvature of space, we must depend on the intrinsic properties of the relevant geometry. How do we do this? Try an experiment. Cut out two pieces of material, one from a soccerball and another from a saddle. Now try to straighten them out on a flat surface, say on a table. The idea is to get a representation of a three-dimensional object in two dimensions. What do you find? First of all, you won't be able to straighten them out without distorting them by stretching or shrinking. Before you can lay them down in two dimensions, the soccerball piece will need stretching and the saddle piece will need shrinking at the periphery.

Actually, this distortion of the scale in the representation of a three-dimensional object in two dimensions is quite familiar to you from maps or atlases of the spherical earth. In such a map the distances are distorted around the edge; thus, for example, Canada looks much bigger than it really is. On the other hand, by studying the distortion, we can find out everything we need to know

---

*Dionys Burger, *Sphereland* (New York: Crowell, 1965).

FIGURE 25.7   The map of the spherical earth is a better analogy of space of positive curvature than the curved surface of a sphere. The map zeros in on the intrinsic properties alone. From the nature of the distortions of distances on the map, the geometry can be figured out.

about the sphericity of the earth. That is, all intrinsic properties of the "spherical" geometry are contained in the map. So a map (Fig. 25.7) of the spherical earth is a much better analogy of a Riemannian space than the surface of a sphere, because for a physical space we see only the intrinsic properties; the extrinsic properties are not accessible to us.

Fig. 25.8 is an artist's concept of the map of a pseudosphere showing the distortion of the scale for a space of negative curvature. The genius of M. C. Escher created this particular picture.

Kant's antinomy*

While the mathematicians were busy developing the new geometry, what were the physicists doing? Very surprisingly, few people in physics before Einstein were aware of curved geometry. Most physicists who wondered about space did so within the context of Euclid's geometry. And then space can be viewed in one of two interesting ways.

One of these ways was that of Newton. Newton, being deeply religious, believed that only God could be infinite, not His creation. Thus Newton's universe was finite. But because of the boundary, such a universe must be inhomogeneous; there is matter inside it and nothing outside it. The underlying geometry of Newton's universe is flat, Euclidean; with such a geometry it is impossible to think of a finite space with a homogeneous distribution of matter. Thus Newton was forced to give up the very attractive idea of homogeneity of space in order to hold on to his belief of a finite universe.

Newton's contemporary, Leibniz, whom we have mentioned before, also was influenced by theology, but his God was different than Newton's apparently. Leibniz believed that God would never create an inhomogeneous universe,

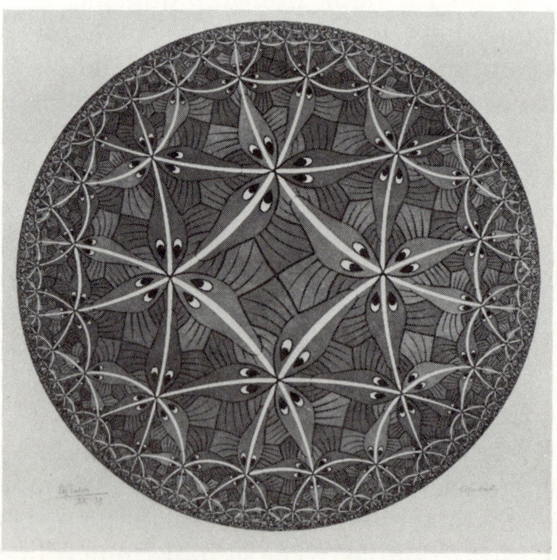

FIGURE 25.8   The "map" of a pseudosphere, produced by the genius of Escher. Called *Circle limit III*, this drawing properly represents the distortion of scale that is inherent in the two-dimensional map of a three-dimensional object of negative curvature. (M. C. Escher, "Circle Limit III," Escher Foundation—Haags gemeente museum—The Hague.)

---

*An antinomy is a conflict between equally valid ideas, which cannot be resolved with currently available knowledge.

since then He would have to be partial to one part of space over another in choosing where to put the stars. The problem does not arise if the universe is infinite, because then there is no boundary to deal with and space can be homogeneous. To Leibniz homogeneity was most important; the infiniteness of space did not bother him.

So Newton and Leibniz were at an impasse. One way to resolve such an impasse is to say they both were right. The German philosopher Immanuel Kant did just that; he developed an antinomy that space is neither finite nor infinite; to him it was a meaningless question. Kant's argument says that this is a question we can't possibly settle by experimentation, so it is not in the realm of objective properties of space. The antinomy thus insists that we have a basic limitation in our perception of space.

A recent detailed analysis of Kant's antinomy has been made.* It points out that there are three ways out of the antinomy. First, the universe could be finite, but then it has to be inhomogeneous because of the boundary, if one still insists on space being flat. This is the solution that Newton accepted. Secondly, as Leibniz insisted, space could be homogeneous, that is, one part of it would look and behave the same as any other part. Therefore, he was forced to the infinite model. Leibniz's predicament also was due to his insistent use of flat-space geometry. The third way is to attribute a different geometry to space. Today most physicists believe in the homogeneity of space but they are not forced thereby to accept an infinite model of space. Space can be finite yet homogeneous and unbounded in a curved-space geometry. Thus the use of curved space geometry gave a resolution to Kant's antinomy.

■ 25.2 General Relativity

Special relativity shows us how to construct physical laws so that they look the same from all inertial reference frames. It abolishes the idea of a preferred frame among all the inertial reference frames, a very democratic situation. Can this democracy be further extended?

Remember the equivalence principle? This principle tells us that, at least locally, the effect of gravity can be completely simulated by a pseudoforce that appears in an accelerated reference frame. But acceleration with respect to what? The inertial reference frames, of course. Are the inertial reference frames then to be regarded as preferred?

Nature may not intend it that way. One of the consequences of the equivalence principle is that we can think of the following situation. Suppose there is gravity in some reference frame—for example, take a small elevator on earth. If such an elevator is freely falling under gravity, the pseudoforce of acceleration completely cancels the gravity force locally within the confinement of the elevator. So the accelerated reference frame of the elevator behaves exactly as if it were an inertial reference frame. How can we regard the inertial reference frame as special and preferred if it is so easily simulated?

---

*J. J. Callahan, "The Curvature of Space in a Finite Universe," *Scientific American*, August 1976, p. 90.

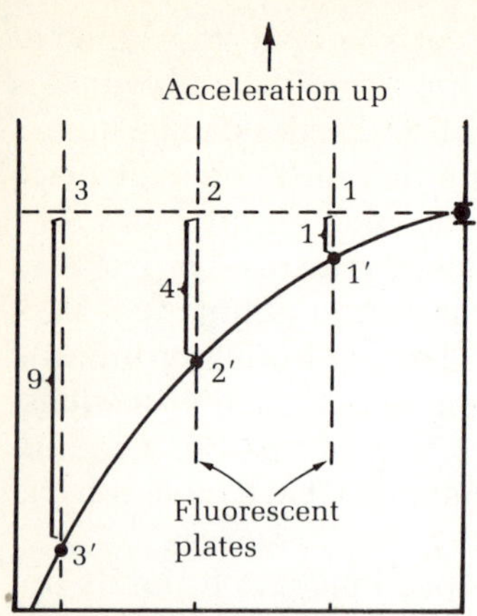

FIGURE 25.9   From an accelerating elevator, the path of a light beam appears to be bent like a parabola. See text for explanation.

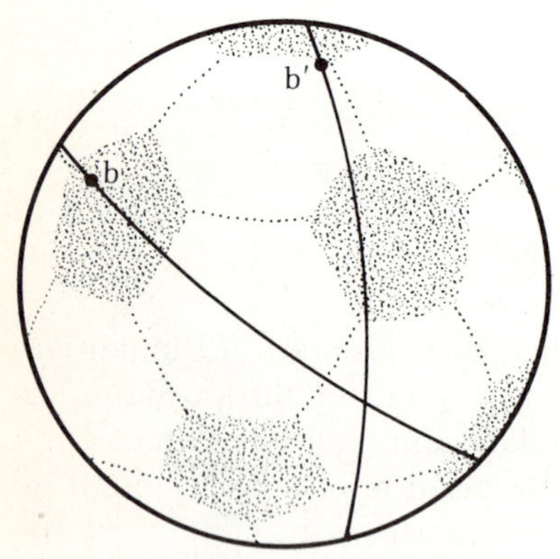

FIGURE 25.10   The motion of bodies in "straight lines" on the surface of a soccerball. Do they collide because of a force between them or because the underlying geometry is curved? Both interpretations are valid.

The objective of general relativity is to establish a democracy among all reference frames, including those that are accelerated. Physical laws should look the same by construction from all such reference frames. From what we have said above, it should be clear that the key to establishing this democracy among reference frames is gravity. Thus the theory of general relativity is also a theory of gravitation.

The equivalence principle is not a complete theory of gravitation; an accelerating elevator in space cannot reproduce all the effects of gravity (for example, the tidal effects). The equivalence principle is a beginning, a very auspicious one. The question is, how do we go from here to find a complete theory of gravity?

The equivalence principle has a consequence that gave Einstein the right idea for the path to progress. Suppose we trace the path of a light beam through an elevator accelerating upward while in space. What should we see? Specifically, we imagine that light enters the elevator through a window (Fig. 25.9); we put a few transparent plates of fluorescent glass in its path at equal distances from the window and each other. The straight line from the window intersects the plates at the points 1, 2, and 3, giving the path of the light beam when the elevator is stationary. When the elevator is accelerating, light will take equal intervals of time to travel from the window to point 1, from point 1 to point 2, and from 2 to 3. But now the plates themselves will undergo displacement during the successive periods in proportion to the square of the time that is, in the ratio 1:4:9. Thus the spots 1′, 2′, and 3′ that light makes this time will appear to be on a parabola and not a straight line. If you were an observer inside the elevator—in which case you consider yourself to be under gravity (equivalence principle)—you must conclude that a light ray is bent by gravity, just as is a bullet or a baseball.

How do we interpret such a bending of a light ray near a source of gravity? We can quite legitimately take the point of view that light is deflected by a source of gravity because it has a nonzero mass (arising from its energy according to special relativity). Such a deflection can be calculated by using the Newtonian theory of gravitation. However, an alternative point of view is possible.

The path of light rays gives us the shortest path between two points of space, which is also the definition of a "straight line" for any geometry, except that the "straight" lines are not really straight in some geometries. Thus the bending of the light ray near a source of gravity perhaps means that the "straight lines " of the geometry of space near a source of gravity are curved. This gave Einstein the idea that the correct theory of gravitation must use a curved geometry instead of the flat Euclidean geometry.

In the theory that Einstein developed, gravity does not exist as a force at all but only as a source of curvature of space. To see how this works, let's consider motion on the curved surface of a soccerball as an example (Fig. 25.10).

We have already seen that the "straight lines" of a spherical surface, the geodesics, are the great circles. Now imagine two bodies b and b′ moving in two such "straight lines," which appear parallel in a certain region of the sphere. However, we know that on a spherical surface all such "straight lines" eventually intersect, so clearly the two objects will collide in time before they sepa-

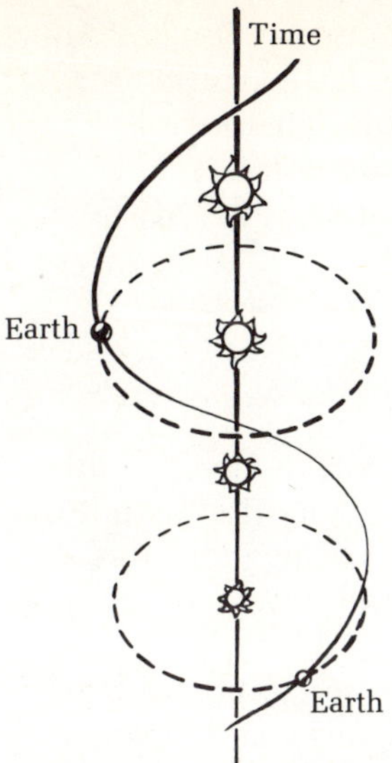

FIGURE 25.11   The world line of the earth moving around the sun. The world line of the sun is along the time axis.

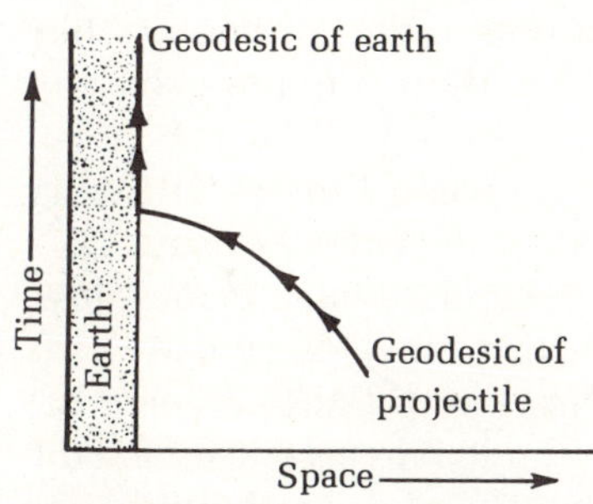

FIGURE 25.12   The geodesic, or curved world line, of a projectile on earth. The collision of the projectile with the earth is an intersection of its geodesic with that of the earth in curved space-time.

rate again. How do inhabitants of the two-dimensional soccerball (for example, as imagined in *Sphereland*) interpret the collision?

In their eyes a collision such as this can only mean that the bodies came together because of a mutual force of attraction, which they may decide to call gravity. This conclusion is the only one that is consistent with their belief that their world is flat. If they had good mathematicians like Gauss or Einstein among them, eventually they might realize that the space in question is not flat, and the two bodies came together not because of the "gravity force" but because of the curvature of the underlying space. The apparent gravity force is one way to interpret collision, but it is really a consequence of the curvature of space. Thus if we consider the motion in curved geometry, even motion under gravity can be regarded as natural motion, motion along the "straight lines" of the appropriate geometry.

The soccerball analogy, however, goes only so far. On a soccerball, neither the apparent gravity force nor the underlying curvature of space for the path of b' depends on the other body b. In our real universe the gravity force that b' experiences is determined by b and, generally, also by the distribution of matter in the vicinity of b' and, ultimately speaking, on the mass distribution of the entire universe. Thus the curvature of space that determines the geodesic of travel for b' must depend on the same factors.

Another important generalization is needed. Special relativity has taught us that space and time are wedded and cannot be considered separately. So the geodesics of natural motion must be those of space-time. Actually, time adds a glorious new visual element to the picture. This is because the curvature in time is more noticeable near the gravitating object. Whereas the curvature of space is certainly present but is too small to visualize, the curvature in time gains by the factor of the speed of light and becomes noticeable. An important analogy is a small bump on a road; it is hardly noticeable to a pedestrian, but it is a menace to a high-speed automobile because it is magnified by the automobile's speed.

As an example, consider the motion of the earth around the sun; customarily we exhibit this as a nearly circular orbit. However, time moves on during earth's revolution, so in a space-time diagram, the path will look like a helix, as indicated in Fig. 25.11. In the picture each plane perpendicular to the time axis shows the position of the earth at the time corresponding to the position of the particular plane. Then we join all these consecutive positions, which gives us the helix. This is the earth's world line; the sun's world line is the straight line going up along the time axis.

In the flat-space point of view, the earth's world line is curved, indicating the presence of the gravity force on earth due to the sun. In the new view it is space-time that is curved; the helix is the geodesic in space-time that earth follows—it is the shortest path in space-time. And earth's motion is now viewed as natural motion. A parabola of a projectile (Fig. 25.12) and the path of light near the sun (Fig. 25.13) are both examples of the appropriate geodesic in space-time.

Ironically, we discover that falling motion is natural motion after all, whether it be falling motion of an object on earth or the falling motion of the planets around the sun. A long time ago the immortal Aristotle thought in this fashion. Does this mean that Aristotle was right?

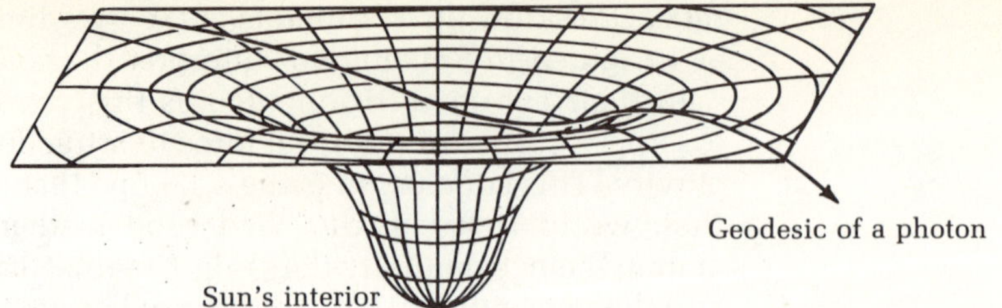

FIGURE 25.13 The geodesic of light near the sun. "Warped space-time," from *Relativity and Cosmology*, 2nd ed., by William J. Kaufmann, III. Copyright © 1977 by Kaufmann Industries, Inc. By permission of Harper & Row, Publishers, Inc.

Not quite. Aristotle was thinking in terms of Euclidean space. If you hold on to flat space, then there is no way you can perceive falling motion as natural motion. But it also becomes clear that Aristotle was not all wrong in his description of falling bodies. His idea was right, but the background in which he stated the idea was wrong.

Almost the same statement can be made about Newton's perception of gravity and space in comparison to Einstein's. Newton held that space was Euclidean—in fact, he made it absolute—and then he corrected Aristotle the only way he could within the context of flat absolute space. Was Newton wrong?

Definitely not; be very clear on this. Einstein himself repeatedly tried to reassert the status of Newton's work in spite of the new insights yielded by general relativity. In his autobiography, while he was writing about the difficulties faced by Newton's theory, we find the following outburst of emotion toward Newton:

> Newton, forgive me. You found the only way that, in your day, was at all possible for a man of the highest powers of intellect and creativity. The concepts that you created still dominate the way we think in physics, although we now know that they must be replaced by others further removed from the sphere of immediate experience if we want to try for a more profound understanding of the way things are interrelated.*

Actually, Einstein's work has established the evolutionary chain of physical thought regarding falling motion: from Aristotle, through Newton, and on to his own work. Physical theories seldom replace one another; more commonly they display an evolutionary pattern. The theories of gravitation provide excellent examples of this.

There are, of course, very important differences between Einstein's theory and Newton's. Whereas Newton's was an action-at-a-distance theory, Einstein's is a field theory, strictly local. Einstein recognized the equivalence principle—the fact that gravity can be eliminated in a freely falling elevator on earth—as fundamental and so incorporated it in his theory; Newton did not perceive the equivalence principle. Einstein realized that true gravity reveals itself in the tidal effects; Newton explained the tides with his theory but did not appreciate the fundamental significance of the tidal phenomenon. It is the local changes in the gravity force that cause the tides. These local changes are described natually in terms of a local curvature of space-time in Einstein's theory. When we add up the local changes of dimension, we get the global structure that we associate

---

*Quoted in B. Hoffmann, *Albert Einstein: Creator and Rebel* (New York: Viking Press, 1972), p. 247.

with the gravity force, which is described to an excellent approximation by Newton's inverse square law.

Einstein summarized his theory in terms of a very elegant set of equations called the **gravitational field equations.** The solutions of these equations determine the curvature of space-time and the geodesics of objects under the influence of the field. Previously we have encountered fields in electromagnetic theory. The gravitational fields, however, are more complicated than the electromagnetic fields in their mathematical structure. Electric and magnetic fields are vector fields; a statement of three quantities (the three components of the field) is necessary at every point of space for the electric or the magnetic field. In contrast, the gravitational field is a **symmetrical tensor field;** a knowledge of ten components is necessary for each point of space-time for the complete determination of the field. Fortunately, tensor mathematics was already known from the work of mathematician Riemann, so this particular aspect of the work did not present any difficulty for Einstein. The physical interpretation of the field, however, is astounding and totally new. The field in space-time determines the structure of the latter. The field is the panorama of space-time; the painting cannot be separated from the canvas, and together they comprise a unified whole. It is this unification of geometry and physics that is the major accomplishment of Einstein's general relativity.

Equality of the inertial and gravitational mass

In the next section we will discuss the experimental verification of Einstein's general relativity. Since the general relativity is founded on the principle of equivalence, it is natural to ask: Is there a solid experimental basis for believing the equivalence principle?

As you know, Newton's second law of motion introduces us to the concept of mass through the equation

$$F = ma$$

We can label the mass that appears here as the **inertial mass** m_I for the purpose of the present discussion. Mass also makes an appearance in Newton's law of gravitation: the gravitational force on an object due to the earth is proportional to the object's mass. Let us label this mass as the **gravitational mass** m_{grav}:

$$F_{grav} = \frac{GM_e m_{grav}}{R_e{}^2}$$

M_e and R_e are the mass and radius, respectively, of the earth. For an object falling in earth's gravity, we can substitute F_{grav} for F in the first of the equations above. Also, recall that the acceleration due to earth's gravity is called g. Thus we get

$$\frac{GM_e m_{grav}}{R_e{}^2} = m_I g$$

Rearranging slightly, we get the following expression for g:

$$g = \left(\frac{GM_e}{R_e{}^2}\right)\left(\frac{M_{grav}}{m_I}\right)$$

Galileo's observations tell us that g is independent of the inertial mass of a falling object. In order for this to be true, we must have

$$\frac{m_{\text{grav}}}{m_{\text{I}}} = 1 \qquad (25.1)$$

Only if g is independent of the mass of an object does the elevator example discussed in Chapter 7 work out right. That is, the statement that in a freely falling elevator on earth, all objects become weightless, which is the basis of the equivalence principle, is strictly correct only if Eq. (25.1) is strictly valid.

So it is not surprising that scientists have made a few renewed and vigorous attempts of verifying the validity of Eq. (25.1). The most famous of these experiments is the one carried out by physicist E. B. Eötvös over a period of 32 years, from 1890 to 1922. Eötvös's experiment has been repeated with improved techniques, and the latest result is that the equality of inertial and gravitational mass holds to within one part in 10^{12}.

These experiments then establish the validity of the equivalence principle.

■ 25.3   The Three Classic Tests of General Relativity

Now comes the most important issue: testing the theory. Einstein himself suggested what are now known as the three classic tests of general relativity. These are (1) the correct description of the orbit of the planet Mercury, (2) the bending of starlight near the sun, which is the most famous of the tests, and (3) the phenomena of gravitational blue and red shifts.

Newton's theory, as presented in Chapter 6, tells us that the orbits of planets around the sun are ellipses fixed in space, which is more or less borne out by observations, but there are some discrepancies. The orbits do not remain fixed in space but actually rotate very slowly, a phenomenon known as the **precession of the perihelion** (Fig. 25.14). The bulk of the precessions are caused by the perturbing effect of the gravity force of other planets. However, even when these effects are accounted for, a slight discrepancy remains between the observed value of the precession and the Newtonian estimates, including the planetary perturbations. The discrepancy is particularly large for the planet Mercury, about 42 seconds of arc per century.

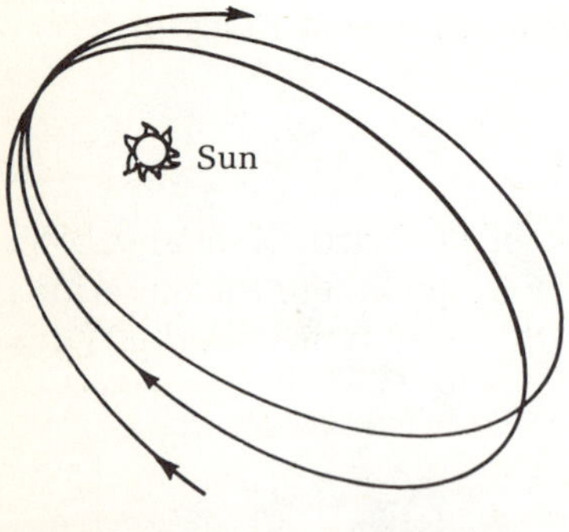

When Einstein formulated the field equation of his theory, the first application he thought of was the case of the precession of the perihelion of Mercury. It was already encouraging to him that Mercury had the largest discrepancy with Newton's theory, because the orbit of Mercury is the closest to the sun, where the predictions of the new theory differed most from Newton's. He did the calculation and found the advance of the perihelion of Mercury to be 43 seconds per arc per century, an excellent agreement with the experimental data. Jubilant, he wrote to a friend: "Imagine my joy . . . at the result that the equations yield the correct perihelion motion for Mercury. I was beside myself with ecstasy for days."*

The biggest test came when a group of British astronomers and physicists decided to measure the deflection of starlight while passing near the sun during

FIGURE 25.14   Precession, or the rotation of the axis, of the elliptical orbit of Mercury in space (highly exaggerated).

*Quoted in Hoffmann, *Albert Einstein*, p. 124.

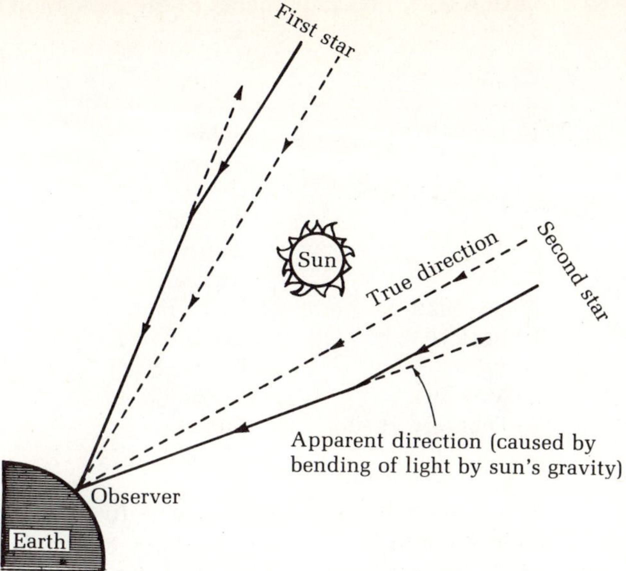

the solar eclipse of 1919 (Fig. 25.15). This famous experiment was carried out at Principe, on the west coast of Africa.

The deflection of light near the sun, it can be argued, also may be predicted just from the Newtonian theory and the equivalence principle. Actually, it turned out, as Einstein himself was able to show, that when the deflection is calculated using the complete machinery of general relativity, one gets an answer that is double that predicted from the equivalence principle alone. The old theory would predict a deflection of 0.87 seconds of arc, whereas general relativity gave the deflection as 1.74 seconds of arc. These numbers are for the deflection of a light ray that just grazes the sun's surface.

The experiment at Principe conclusively proved the validity of the equivalence principle, that light has weight. For the numerical value of the deflection, it gave 1.61 seconds of arc with an error margin of 0.40 seconds. This means that any number between 2.01 seconds and 1.21 seconds was consistent with experiment. Obviously, the number favored the complete theory of general relativity.

Since the days of Principe many more starlight deflection experiments have been carried out. The results are summarized in Table 25.1. It should be amply clear from this table that light deflection experiments generally have provided excellent confirmation of the general relativity theory.

A third test involves the change of frequency of a light ray in a gravitational field. Actually, this is only a test of the equivalence principle and not of the complete machinery of general relativity. Even so, the effect is extremely interesting and well verified and, therefore, we will give you some of the details.

From the equivalence of mass and energy, a photon of energy E has an inertial mass of E/c^2; from the equivalence principle the photon must also have the same amount of gravitational mass. Let us then consider a photon emitted by a source at a height H above the surface of the earth. Let the photon be of frequency ν, in

TABLE 25.1.   Measurements of the deflection of starlight by the sun.

Date of Eclipse	Site	Number of Stars Observed	Angle of Deflection (seconds of arc)
29 May 1919	Sobral	7	1.98 ± 0.16
	Principe	5	1.61 ± 0.40
21 September 1922	Australia	11–14	1.77 ± 0.40
	Australia	18	1.42 to 2.16
	Australia	62–85	1.72 ± 0.15
	Australia	145	1.82 ± 0.20
9 May 1929	Sumatra	17–18	2.24 ± 0.10
19 June 1936	U.S.S.R.	16–29	2.73 ± 0.31
	Japan	8	1.28 ± 2.13
20 May 1947	Brazil	51	2.01 ± 0.27
25 February 1952	Sudan	9–11	1.70 ± 0.10

which case it has kinetic energy $h\nu$ and an amount of gravitational potential energy given by the product mass $\times$ g $\times$ H $= (h\nu/c^2)gH$. As it falls toward the earth, its gravitational potential energy changes to kinetic energy, and its energy as it is received on the surface is given as

$$h\nu' \approx h\nu + \left(\frac{h\nu}{c^2}\right)gH \tag{25.2}$$

The expression is one of approximate equality because we have assumed that the mass $m = h\nu/c^2$ of the photon remains a constant during the process, which is not exactly true (however, the error does not affect the result much). Canceling h from both sides, we get the following equation:

$$\nu' = \nu + \left(\frac{\nu}{c^2}\right)gH = \nu\left(1 + \frac{gH}{c^2}\right) \tag{25.3}$$

It is a blue shift. The frequency is increased [Fig. 25.16(a)]. Now light emitted by an atom can be used as a timekeeping device called the atomic clock. So the expression above really tells us that the clock of an observer sitting on the "top" of the gravity field of the earth seems to have a smaller period (remember the inverse relationship between frequency and period; if the frequency is increased, as shown above, the period must decrease). That is, the clock will go faster in the reckoning of the earth observer.

But now we have an important question. How does the observer at the top see the performance of an atomic clock on earth? Since the photon now loses kinetic energy as it recedes from the gravity field of the earth, the frequency will decrease, or be red-shifted. [Fig. 25.16(b)]. Such a red shift is called a **gravitational red shift** to distinguish it from the Doppler red shift we discussed in an earlier chapter. The red shift means in clock terms that the period will be longer, or the earth observer's clock will appear to be slow to the observer at the top. Interestingly, there is no reciprocal symmetry, as in the case of special relativistic time dilation.

The increase in frequency, or the blue shift of light as it falls in the field of earth's gravity, extremely small as it is, has nevertheless been measured by two

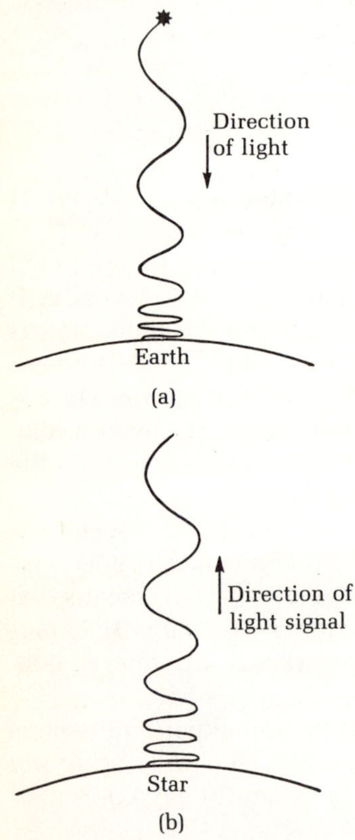

FIGURE 25.16   The gravitational frequency shift. (a) If light falls from a source at the top of earth's gravity field, its gravitational potential energy is converted to kinetic energy, leading to an increase of frequency. Thus the light is blue-shifted. (b) When starlight leaves the gravity field of the star, its frequency is shifted toward the red.

American experimentalists, R. B. Pound and G. A. Rebka, using a gamma ray source. Their results are in complete accord with the theoretical formula of Eq. (25.3).

The gravitational red shift of starlight (unrelated to the red shift arising from Doppler recession) has also been verified to a reasonable degree of accuracy for the case of a white dwarf star named Sirius B.

■ 25.4 New Frontiers

If gravitational phenomena are described in terms of a field, shouldn't we expect **gravitational waves,** propagating disturbances of the gravity field? The tests of general relativity we described in the last section all pertained to the case of rather weak gravity fields producing rather small curvatures of space. What happens to the curvature of space when the field is very strong? What is the nature of the curvature of space when the field is very strong? What is the nature of the curvature of space on the grand scale of the entire universe?

Interestingly, when some of these puzzling questions were solved, the results were new triumphs of the theory and the establishment of entirely new frontiers of knowledge. These frontiers are the most important contributions of general relativity to physical thought.

And then there are other questions that have not been answered to everybody's satisfaction. One such question, which plagues not only the science fiction writer but also the serious scientist, is this: Does antigravity exist? General relativity gives us a pessimistic answer, which nevertheless can be questioned with some legitimacy. If antigravity is discovered experimentally, the whole theory may have to be revised.

There are also efforts at further modification (evolution?) of general relativity to bring it even closer to being a complete theory of space-time. Without describing these in detail, the following ideas are some of the ones currently being explored. There have been attempts to combine general relativity with quantum mechanics; John Wheeler has developed a dynamical theory of the geometry of space (geometrodynamics), which is an extension of general relativity along such lines. There also have been some attempts to incorporate Mach's principle, the principle that says that the inertial mass of an object really derives from the distant matter of the universe. Einstein himself spent the last thirty years or so of his life pondering another generalization of his theory. His quest was to find an ultimate unified field that would encompass all physical phenomena. He never did succeed.

Gravitational waves

An accelerated charge radiates electromagnetic waves; it creates disturbances of the electromagnetic field, which propagate in space. It is easy to imagine that there should be similar effects in the case of gravity. A shaking mass should generate gravity waves. Yet there are two problems associated with the waves of gravity.

The first problem is one of conceptualizing them. The gravity field is the curvature of space-time itself. Therefore, disturbances of the gravity field are

also those of the curved space itself; they are like ripples of curved space. Measurement of space must involve some kind of measuring rods, a system of coordinates. If the measuring rods are themselves wavy, a certain phenomenon may be found to have wavelike properties arising entirely from the waviness of the measuring rods. For example, imagine yourself jumping rope, up and down; in your view a ball moving perfectly straight will appear to go up and down, but the result is entirely due to your use of a peculiar coordinate system. Of course, in this case there is no problem, because you know that you are using a special coordinate system. However, in general relativity we don't know that, since there is no special coordinate system; all coordinate systems are equivalent.

This riddle is indeed a very formidable one, and to this day it remains difficult to prove the existence of gravity waves in the case of strong fields that produce pronounced curvatures of space-time. Fortunately, for weak gravity fields the problem has been untangled, and it has been possible to predict the existence of gravitational waves. For example, a double star system should emit gravity waves. The two components of the double star system accelerate each other; accelerated masses radiate gravitational waves just as accelerated charges radiate electromagnetic waves.

The second problem arises from the extreme smallness of the effect of the gravity waves. This makes it very hard to detect them. The waves are believed to travel with velocity c, just as do light waves. Their wavelengths are predicted to be huge, some one hundred miles or more, which is quite unique and can be used advantageously in their detection.

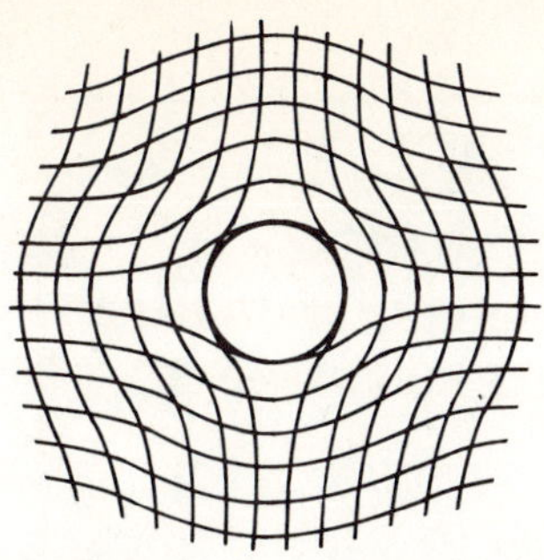

FIGURE 25.17 The geodesics of light near a black hole. Inside light cannot connect to these geodesics and therefore cannot escape.

A University of Maryland physicist, Joseph Weber, claimed in 1969 that he had detected gravity waves. Since then, however, other experimentalists have repeated Weber's experiment with even more sensitive apparatus and have not found anything. There now seems to be a general consensus that Weber was perhaps too hasty in his claim.

In spite of the setback, it remains very important to observe these waves, and the attempts to do so, no doubt, will continue.

Gravitational collapse*

German astrophysicist Karl Schwarzschild discovered a very strange consequence of general relativity when applied to the case of immensely strong fields. A spherical object of enormous mass could produce such a strong curvature of space that even light would be turned back, unable to find a way to escape such an object. If we drew the geodesics of light in the space of such pronounced curvature, they would look like Fig. 25.17. Notice the big hole in the middle where the geodesics from the outside does not penetrate. Consequently, the geodesics of light from inside (not shown in the figure) cannot connect to the outside. Such a hole is called a black hole (you may remember it from Chapter 10), and its radius is the Schwarzschild radius. The value of the Schwarzschild radius for an object of mass M is given by the expression

$$R = \frac{2GM}{c^2} \tag{25.4}$$

where G is the constant of gravity and c is the velocity of light [refer to Eq. (10.3)].

Our discussion in Chapter 10 has given you some ideas about these strange objects in connection with the consideration of the escape velocity for a gravity field, but now we will say a little more about them. One of the strange things that follow from general relativity is the behavior of time clocks in such highly curved space. A distant outside observer will see the time clock near a black hole slow down so much that a particle on its way to being captured by the black hole will seem to take forever to do so (Fig. 25.18). By contrast, if an observer is brave enough to venture within the black hole himself, he will notice no such slowing down of time; the time clocks inside the black hole will look entirely normal to him. Unfortunately, such an observer can never come out of the hole; his future is sealed.

General relativity theorizes convincingly that if there is enough mass, a star will collapse to become a black hole. Such a collapse is called a **gravitational collapse.** The general theory also tells us some very new and dramatic things about the nature of the collapse. The curvature of space-time increases without limit as we approach (figuratively speaking) the center of the hole in space-

---

*I would like to share a personal anecdote. My first in-depth exposure to the ideas of gravitational collapse was in 1964 through two lectures American physicist John Wheeler gave at Case-Western Reserve (then Western Reserve) University. The title of the first lecture was "Gravitational Collapse, From What?" and the second was "Gravitational Collapse, To What?" One of my colleagues, physicist Joe Weinberg, who was introducing Professor Wheeler, gravely suggested the following title for still another talk: "Gravitational Collapse: So What?"

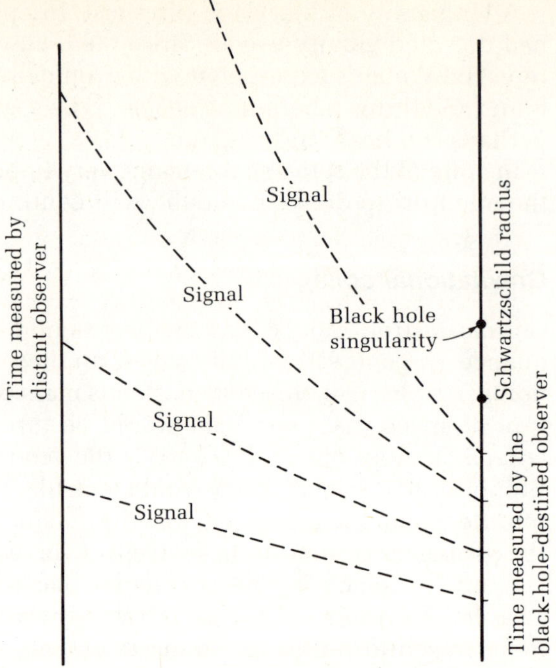

FIGURE 25.18   A signal sent by the black hole adventurer will take longer and longer to reach a distant outside observer as the black hole explorer approaches closer and closer to the Schwarzschild radius. The outside observer interprets this as a slowing down of clocks near the black hole. The black hole explorer does not notice any change in his time clock. Once the black hole explorer has ventured inside the Schwarzschild radius, his signals will not reach the outside at all.

time.* As a consequence, the matter that collapses is squeezed to infinite density. Thus the center of a black hole in space-time is what mathematicians abhor, because infinity defies mathematics; they describe it by the word "singularity." At the singularity, the physical laws as we know them break down.

What, then, is the meaning of the size of a black hole? All the matter literally is squeezed out of existence; in this sense there is no size. On the other hand, the gravitational effect of the object remains intact as far as the outside world is concerned, following the usual rules,† and thus we still must regard the black hole as an object of the same mass as before and radius R, the Schwarzschild radius. The size of the black hole can be determined experimentally, at least in principle, by throwing objects at it sideways with sufficient energy. If the particle misses the Schwarzschild radius—that is, gets no closer than that—it comes back, making a parabolic or hyperbolic trajectory. Otherwise, it is absorbed. In either case, the result can give us an estimate of the black hole's size.

There are many speculations and models, some of them quite interesting, about black holes, which we will not go into in detail. Until a black hole is found and can be studied experimentally, such detail makes little difference. We already have discussed the present status of the observational situation and will not repeat it here (see Chapter 10).

---

*You must not think of the center of the black hole as a center in space, it's more like a center in time. Within the black hole, space becomes time and time space. Very peculiar.

†John Wheeler has compared this state of affairs with the Cheshire cat; the matter has disappeared but the gravitational field lingers on.

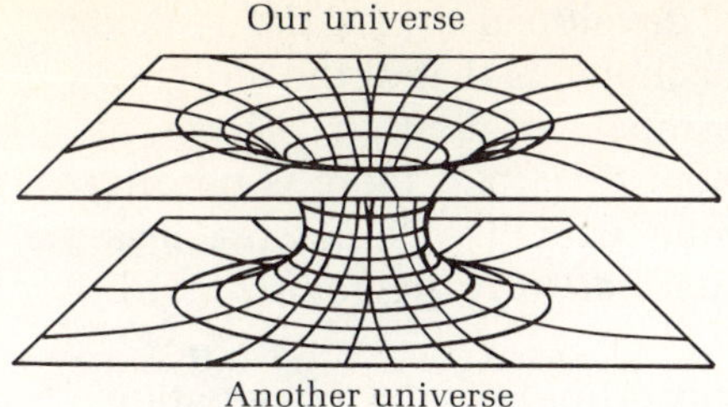

FIGURE 25.19 The Einstein-Rosen bridge connecting two universes. "The wormhole," from *Relativity and Cosmology,* 2nd ed., by William J. Kaufmann, III. Copyright © 1977 by Kaufmann Industries, Inc. By permission of Harper & Row, Publishers, Inc.

However, we will briefly consider the following question: Where does the matter go after reaching the singularity? We do not know for sure. It is conjectured that the singularity is a theoretical one, predicted by the imperfections of our model; for example, quantum effects may be important at such a scale of space-time and save us from the singularity—maybe. Another interesting speculation is that the matter of the black hole passes through the so-called Einstein-Rosen bridge (perhaps "tunnel" is a better word) into a parallel universe, creating a fountain of matter called a white hole in that universe (Fig. 25.19). Some have suggested that the quasars of our universe are such white holes, arising from the black holes of another universe. Anyway, the important thing is to realize that in black holes there is a chance to encounter a new domain of physical reality. Therefore, it is of the utmost importance to find them.

General relativity and the expanding universe

General relativity gives us a natural place to begin a study of the structure of the universe. This is the subject of the next chapter.

Russian physicist Alexander Friedmann made a most novel theoretical prediction based on general relativity: that our universe, provided it is closed and roughly of uniform density, must expand until it reaches a maximum dimension and then contract, ending in a collapse. This was around 1922, several years before the galactic red shifts were discovered, making the idea of an expanding universe a respectable experimental fact. Curiously, when Friedmann's expansion model was first published, Einstein did not like it at all, although later it turned out to be a monumental consequence of general relativity. At that time Einstein believed the universe to be an unchanging and static one. Mathematically, of course, there was no escape from Friedman's solution if general relativity was kept unchanged. So convinced was Einstein of his belief in a static universe that he did not hesitate to modify his field equations by introducing an extra repulsive force (he called it the "cosmological term") between matter in the cosmological scale. With the introduction of the cosmological constant, he got his wish; a static model of the universe now could be arrived at from the theory. After Hubble discovered the red shift phenomenon in the late twenties which obviously leads to interpretation in terms of an expanding universe, Einstein gave up his idea of the cosmological term, calling it "the worst blunder" of his life.

The story of the cosmological constant illustrates how easy it is to fool around with cosmological models, introducing quantities that do not affect our observations in our limited experience yet in the cosmological scale make very large contributions. This does not mean that no such thing exists, only that we have to be extra careful about accepting it.

Is antigravity possible?

A gravity shield that can block off the gravitational force of an object, as is possible with an electric force, is a concept that is used over and over again in science fiction literature. H. G. Wells used the idea for going to the moon; the hero in his book, *The First Men in the Moon,* developed a material named "tectight," which was a gravity shield. A similar theme is that of the antigravitating (or negatively gravitating) object, which repels a gravitating object, as two like charges do. Arthur C. Clarke writes about this: ". . . there is nothing inherently absurd in the idea that there may be substances which possess negative gravity, so that by some treatment we may permanently 'degravitize' ordinary substances."*

Clearly the idea of negative gravity or antigravity stimulates our imagination. Why is it that it has proven impossible (at least so far) to find objects possessing negative gravity or to build a gravity shield? Are these theoretically viable concepts?

One is inclined to think that an antigravity object may easily be constructed theoretically by assigning it a negative mass. In Newton's law of gravitation [Eq. (6.4)], if one of the masses is negative, the force is negative or the opposite of attraction, which is repulsion. So an object of negative mass repels an object of regular positive mass; we now have both attractions and repulsions in gravitation, just as in electricity. It is very tempting to assume that all the rest of the electrical properties will also follow—for example, the shielding property.

Unfortunately, it is not so for the following reason. The second law of motion of Newton tells us that the acceleration of an object due to a force is given by the equation

$$F = ma$$

If m is negative in this equation, the direction of the acceleration is the negative or opposite of the direction of the force. So if we push a negative mass one way, it accelerates the opposite way.

If we look at the mutual effect of a positive and negative mass, keeping in mind this perverse behavior of the acceleration of the latter, we find the following result. Look at Fig. 25.20; the gravitational repulsion from the negative mass accelerates the positive mass to the right. The force on the negative mass is toward the left; but from above it accelerates in the direction opposite to the force and therefore to the right also, just as does the positive mass. So instead of moving away from the positive mass, as you may have intuitively expected, it chases it. This is very different from the behavior of two charges that repel each other; in that case we really see repelling motion between the two.

So the negative mass is not the object that we hoped it is; it does not accomplish what it was supposed to. It is not an antigravity object; it does not repel ordinary matter. For this and other reasons most physicists consider it unlikely to exist.

Nevertheless, experimental physicists have an interest in looking for a genuine antigravity object. If it exists, it will provide a most important challenge to the equivalence principle. The perverse behavior that we found for the negative mass (Fig. 25.20) is based on one tacit assumption, that the object has the same inertial mass and gravitational mass; that is, it obeys the equivalence principle. But suppose it didn't. If its inertial mass were positive and its gravitational mass negative, then, as you can easily verify, a negative mass would be

____
*Arthur C. Clark, *Profiles of the Future* (New York: Bantam Books, 1958), p. 46.

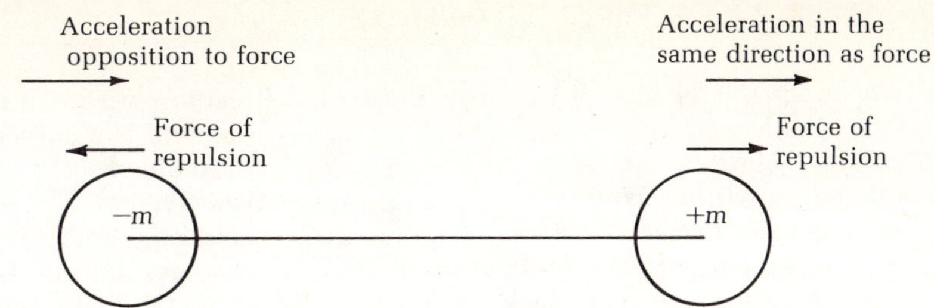

FIGURE 25.20 The negative mass follows the positive mass.

repelled from a positive mass, and we would have a truly negative gravity object. So the existence of a truly negative gravity object is tied down to the validity of the equivalence principle. Arthur Clarke is right, of course; "there is nothing inherently absurd in the idea" of the negative gravity substance, except that if it is found, then we probably would have to discard the equivalence principle and the theory of general relativity that rests on it.

One natural question to ask is this: Are antiparticles antigravity objects? If they are—that is, if they violate the equivalence principle—they should "fall" away (or "up") from the earth. They should be repelled from the earth's gravity field. On the other hand, if they obey the equivalence principle, then they should fall "down" toward the earth, irrespective of the sign of their mass. There is some indirect evidence that antiparticles fall downward toward the earth, and hence there is no violation of the equivalence principle. However, we do not know the sign of their mass. We assume that it is positive because, as explained above, there is no inherent interest in a negative mass object unless it exists in violation of the principle of equivalence.

SUMMARY

This chapter puts forth the basic ideas of general relativity, which is really Einstein's theory of gravitation. General relativity looks upon gravitation as a manifestation of the curvature of space-time. We started with the work of mathematicians on the possible geometries of space. We find that space can have negative or zero curvature, in which case space is open. On the other hand, space of positive curvature is closed space—finite and unbounded.

Einstein's theory of gravitation rests on the idea of the equivalence principle—one cannot distinguish the gravity force from pseudoforces arising in accelerated reference frames. Another way of stating the equivalence principle is to say that the inertial mass and the gravitational mass of an object are equal. This aspect of the equivalence principle has been verified experimentally to a great degree of accuracy.

In Einstein's theory, the gravitational field is the curved space-time, the curvature being determined from the matter density. Material bodies follow the geodesics, the "straight lines" of the curved space-time. Matter acts on space, and space acts on matter, and the idea of absolute space is abandoned forever. Also, the force of gravity becomes a consequence of the motion of objects along the geodesics of the underlying curved space.

The theory of general relativity has been experimentally verified by a series of three classic tests, the most famous of which is the Principe experiment on the

For some two hundred and fifty years, until Einstein, Newton was unchallenged in this proclamation that space is absolute: "Absolute space, in its own nature, without relation to anything external, remains always similar and immovable." General relativity shattered the concept of absolute space. In Newton's view space acted on matter, but matter did not react back. In Einstein's view space affects matter, but matter affects space, too. The two are equitably connected. Matter follows the geodesics or the natural path of curved space, but the curvature of space is given by the solutions to the equations of the gravity field, which is intimately associated with the matter density in the neighborhood.

This democracy between space (emptiness) and matter (form) is also an interesting aspect of the philosophy of some of the Eastern schools. It is abundant in the Upanishad of the Hindus and in the writings of the Buddhist traditions, where emptiness affects form and form affects emptiness, and the two must be regarded as two parts of a dynamical whole. Consider the following quotation from Lama Govinda: "The relationship of form and emptiness cannot be conceived as a state of mutually exclusive opposites, but only as two aspects of the same reality, which coexist and are in continual cooperation."*

But the parallel with general relativity seems to end here. Going on one step further, the Eastern philosophers conceive of an ultimate reality from which both emptiness and form originate. The Hindus call this ultimate reality "Brahman" and the Buddhists, "Shunyata," but the idea is the same—the idea of one absolute from which springs everything in the universe. Incidentally, the Sanskrit word "Shunyata" literally means "void," but the word is nevertheless used to denote the ultimate reality beyond both void and matter. "Brahman" and "Shunyata" are conceived to be beyond our ordinary perception; in order for us to grasp the ultimate reality, the Eastern tradition demands that we be able to transcend duality, the dual existence of emptiness and form, space and matter.

Curiously, Einstein himself spent the last years of his life in an attempt to generalize general relativity partway along this line of thought, very possibly without being aware of the similarity of his idea to the Eastern concept. In Einstein's work the role of the absolute was played by the "unified field."

Let me discuss the idea in some detail. General relativity, which is a field theory of gravity, never got rid of matter entirely. It got rid of a separate concept of space; the field is a curved space. But matter density had to be introduced into the field equations externally. Conversely, matter appeared as the "singular solution" of the field equations, singular meaning that the field energy becomes infinite at these points. Infinity is not mathematically tractable. (A similar situation exists for the case of the electromagnetic field, where the charge also has to be introduced into the theory from outside, and again the field is singular at the position of the charges.)

Einstein did not like this situation at all. He believed that the ultimate physical theory should be a "true" continuous field theory and there should be no necessity of the concept of matter at the outset. Properties of both matter and space should follow from the field. "We may therefore regard matter as being constituted by the regions of space in which the field is extremely intense.... There is no place in this new kind of physics both for the field and matter, for the field is the only reality," Einstein used to say.* In this view, matter represents those regions of space where the field is immensely stronger than in the surrounding space, yet the field is continuous; the same field equations should hold even inside matter. Matter is no longer a singularity in the field. Einstein also believed that in the final theory the concept of a unified field would encompass not only gravitational but also other phenomena (he himself concentrated on unifying electromagnetism and gravity as a first step).

We see in this work a clear attempt to replace one absolute by another, that of absolute space by one of absolute field. Human beings seem to have an immense need for an absolute, and even Einstein was no exception to this.

Einstein never found a satisfactory unified field theory, although he kept at it until the very end. In the process he became almost completely isolated from the mainstream of physics and from most active physicists of the time. No doubt this is what inspired the following lines of a Bob Dylan song about Einstein:

> But he was famous long ago
> For playing the electric violin
> On the desolation row.†

Perhaps there are absolutes as the ancients believed and as all theologians lead us to believe, even today. Perhaps some day a human will discover a unified field.

---

*Lama Govinda, *Foundations of Tibetan Mysticism* (New York: Weiser, 1974), p. 223.

*"On the Generalized Theory of Gravitation," *Scientific American*, April 1950, p. 13.

†From "Desolation Row," by Bob Dylan. © 1965 Warner Brothers, Inc. All rights reserved. Used by permission.

Perhaps. But if not, we must conclude that there is no absolute, there is no ultimate field that is responsible for every phenomena of the universe.

I will tell you a story, not a true one, by an anonymous writer, which is almost made to order for the point that I want to make. The hero of the story is a brilliant twentieth-century physicist of the quantum era, the late Wolfgang Pauli. Pauli was very famous ("notorious" might be a better adjective) for his critical ability, which he often expressed very caustically in "put downs." According to the story, when Pauli died and went to heaven (of course, all famous physicists go to heaven), his friends set up an appointment with God for him almost immediately upon his arrival. When Pauli arrived at His office, God was working at a desk. Pauli wasted no time. He asked *the* question and held his breath: "Tell me God, how did you do it? What's the ultimate theory behind the workings of the universe?" God went to the blackboard immediately and wrote down an equation. (Don't be surprised to hear

that God has a blackboard in His office. He is the best mathematical physicist around, so He has to have a blackboard.) It was a complicated equation and involved only a field with no reference to matter. Pauli studied it for a moment and then blurted out: "Sorry, but I have tried it. It doesn't work."

Most physicists today agree with Pauli on this one. Few believe that there is such an absolute as an ultimate, unified field from which everything else follows.

Yet, a third alternative is also possible. Einstein's failure may simply indicate that the ultimate reality is transcendental: we cannot catch it and analyze it with mathematics; it is beyond mathematics. The concepts of matter and the fields are dual, contrapuntal aspects of the ultimate reality. In any mathematical theory, both will remain. If this view turns out to be right, then modern physics will indeed have converged to a position that is strikingly similar to that of the Eastern philosophers.

bending of starlight near the sun as observed during a solar eclipse. Other important verifications are offered by the explanations of the phenomenon of the precession of the perihelion of the orbit of Mercury, and the gravitational red shift (and blue shift) phenomenon.

This phenomenon of the gravitational red shift implies a slowing down of clocks near a source of intense gravity field. An observer sitting at the "top" of a gravity field sees all clocks at the "bottom" to have slowed down. Also, the observers at the bottom see the clock of the observer on top to have speeded up.

Two of the predicted consequences of the general theory of relativity are now being intensively sought: the gravitational waves and the black holes. In addition, general relativity seems to exclude the idea of antigravity. Finally, we have briefly discussed Einstein's unified field theory, on which he spent the last thirty years or so of his life, in vain.

QUESTIONS

Review and reason

1. What are some of the differences among the Euclidean, Bolyai-Lobachevsky, and Riemannian geometries?
2. Three angles of a triangle do not necessarily add up to 180°. Give two examples of surfaces (or two-dimensional spaces) on which three angles of a triangle add up to something less than 180° and something greater than 180°, respectively.
3. What do we have to do when we try to map a spherical surface on a plane? Why does Canada look so much bigger than it actually is on a map of our earth?

4. Why is Fig. 25.8 a good example of the two-dimensional map of a pseudosphere? It has been described as being able "to catch infinity in a finite space." Explain.

5. State and explain Kant's antinomy. How is it resolved?

6. What is meant by general relativity as opposed to special relativity? General relativity is considered a theory of gravitation. Explain.

7. The theory of gravitation has evolved from Aristotle's, through Newton's, and on to Einstein's theory. Discuss the evolutionary chains of the theory.

8. Distinguish between inertial mass and gravitational mass of an object. What if they were not equal? Discuss.

9. What is meant by the precession of the perihelion of Mercury? How does it constitute a test of the theory of general relativity?

10. Why do we need a solar eclipse to observe the bending of a light ray near the sun? What do such experiments prove about the geometry of space near the sun?

11. What is meant by the gravitational red shift? How is this related to the slowing down of the time clocks near a source of intense gravity as seen by a distant observer?

12. What is meant by a gravitational wave? How are these proposed waves different from, say, a sound wave?

13. Define the Schwarzschild radius of a black hole. What happens to time clocks near a black hole?

14. A negative mass obeying the principle of equivalence will have a gravitational interaction of repulsion with a positive mass, yet it will follow the positive mass. How is this possible?

15. Suppose there were a negative mass particle that did not obey the equivalence principle. In which direction will this particle fall, up or down, when under earth's gravitational field? Explain.

26 The Structure of the Universe

I have listened,
And I have looked
  with open eyes.
I've poured my soul
  into this world
Seeking the unknown
  within the known.
And I sing out loud
  in amazement.

■ 26.1 Cosmology

Cosmology is the study of the origin, evolution, and structure of the entire universe. As a scientific discipline, modern cosmology is a relatively young field. Even the rudimentary fact that the universe consists of conglomerates of stars or island universes (called galaxies) separated by vast domains of almost empty space was established only about fifty years ago. It is also a very unusual scientific discipline, since many of the astronomical events are not reproducible or controllable at will and correspond to a very different epoch, even billions of years ago; the observations are limited by our observing apparatus and by the fact that we are restricted to a small place and a small epoch of time. With all these limitations, it is a marvel that we seem to know so much about the universe, enough to make very plausible models for its structure and its future. Einstein once said, "The most incomprehensible thing about the universe is its comprehensibility." Some of the successes of cosmology support his view. On the other hand, we must remember that much of cosmology is based on premises that are hard to prove and therefore could be wrong.

As an example, one of our basic assumptions is that physical laws, as we know them, remain valid in the cosmological scale. This assumption is very hard to verify to the degree of accuracy needed. Yet without such an assumption, it is almost impossible to make any kind of progress. By contrast, consider a somewhat different assumption: all the physical laws in the cosmological scale (in both space and time) are those that also play a role in our earth-fixed laboratory and are discoverable from our studies therein; that is, there are no new physical laws needed. Is this a reasonable assumption? We don't know. If the physical laws can undergo basic changes in the microdomain, as we have discovered, are basic changes not likely also when it comes to the superlarge? The problem is that from our experience we can guess that if there are such changes, they are going to be rather subtle; with our limited observational capability, our chances of discovering them are not high. Yet one of the lofty motives of studying cosmology must be to discover such new laws, if they exist.

How do we proceed, then? We proceed very cautiously on the basis of what is known, but we are prepared to confront the unforeseen.

The observational basis of cosmology

One of the most important starting points of cosmology is the observation that matter in the universe seems to exist primarily in the form of clusters of stars known as the galaxies. The story of the discovery of the galaxies is interesting. William Herschel, the eighteenth-century astronomer who discovered the planet Neptune, published an astronomical catalogue of the sky in which there was this catchy phrase, "island universe." But it was not clear whether these "island universes," or galaxies, were vast conglomerates of stars or, like the nebulas, glowing clouds of gas. The confusion existed for quite a while until American astronomer Edwin Hubble resolved the controversy in 1924. With the advanced observational apparatus of the Mt. Wilson Observatory, Hubble was able to show that galaxies are indeed conglomerates of stars (Fig. 26.1).

FIGURE 26.1   The Andromeda galaxy. (Lick Observatory Photographs)

Since Hubble's research, extensive studies have established that there are perhaps some hundred billion galaxies in our universe, each with that many stars as well. The average diameter of a galaxy is 100,000 light-years, and neighboring galaxies are separated on the average by distances of the order of a million light-years.

Hubble himself made yet another great discovery. Earlier an astronomer named V. M. Slipher at the Lowell Observatory, making some observations on what he thought were nebulas, found these objects to recede from us with large velocities. Hubble identified these receding objects as galaxies. He then made some new measurements on both the distances and the velocity of recession of these galaxies. The result of Hubble's measurements can be stated in the form of a law, now called **Hubble's law:**

> *The velocity v of recession of the galaxies is proportional to their distance r:*

$$v \propto r$$

Introducing a constant of proportionality H, we can rewrite Hubble's law in the form

$$v = Hr \tag{26.1}$$

This means that the further a galaxy is from us, the faster it recedes in direct proportion to its distance. If we consider two galaxies, one twice as far from us as the other, we find that the more distant galaxy moves away from us at twice the speed of the closer one.

How do we interpret the galactic recession? It is tempting to consider ourselves the center of the universe, and claim that the recession implies an expansion of the universe with us (our galaxy) at the center. Such an interpretation is not consistent with a basic symmetry of the physical laws, which says that physical laws do not depend on the choice of the origin of the coordinate system used to interpret the results of measurements (Chapter 7). Even more importantly, the interpretation above is in conflict with another basic axiom of cosmology, **the cosmological principle:**

> *On a sufficiently large scale, the distribution of matter in the universe is the same everywhere in all directions, no matter from where you look.*

If you imagine yourself to be the hero of Arthur Clarke's *2001*, after you go through an interspace tunnel, you wake up in another part of the universe. However, looking at the sky, you would not be able to tell whether you are in a different part of the universe or not. You could tell small-scale differences— that you are not in the same solar system, even that you are not in the same galaxy—but beyond that it would be very difficult for you to tell where you were.

The correct interpretation of the expansion can be visualized with the help of the following analogy. Imagine that you are baking raisin bread in the form of a spherical ball. As the bread rises the raisins on the surface become more and more separated (Fig. 26.2). Also, you see the same kind of expansion from the vantage point of any and all raisins. No one raisin is special; all raisins are moving away from each other. The recession of the raisins is due to an expansion of the fabric of the bread itself. Our picture of the expanding universe is

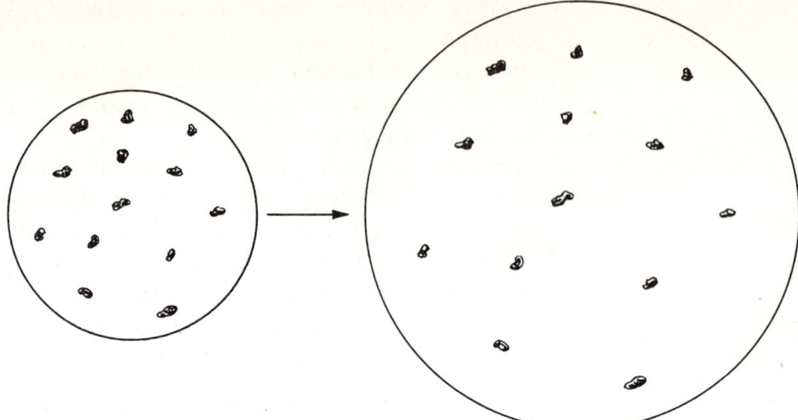

very much like this expanding surface of raisin bread. The expansion looks the same from one and all points of the universe; no point is the center.

The discovery of the expansion of the universe also resolves a paradox that carries the name of eighteenth-century German astronomer Heinrich Olbers. **Olbers' paradox** calls attention to a familiar property of the night sky, which is truly astonishing when you stop to think about it—the night sky is dark. If the universe were infinite and static and the stars uniformly dispersed in it, then the night sky should not be dark because of the following argument. Consider yourself at the center, with stars distributed uniformly all around you in all directions. In such a case, how many stars are there at a distance r from you? Those that are on the surface of a sphere of radius r with you as the center. Since the surface of a sphere of radius r is proportional to the square of r, r^2, and since we have assumed a uniform density of distribution of the stars, the number of stars at distance r also goes up in proportion to r^2.

In Fig. 26.3, if you take the two spherical surfaces, one twice as distant from the center as the other, this argument tells us that the number of stars on the surface of the more distant sphere is larger by a factor of four. Now consider the light reaching you from the stars on the surfaces of these same two spheres. If the stars have the same intrinsic brightness, a very reasonable assumption, then the light reaching us from each of the spherical surfaces can be figured out by using the inverse square law of light intensities [Eq. (12.1)]. According to this equation, the intensity falls as $1/r^2$, so we receive one-fourth the light from the more distant stars as from the ones on the nearby sphere. Thus the decrease in intensity suffered by light in traveling a greater distance is exactly canceled out by the greater number of stars at the greater distance. The net result is that we receive the same amount of light from all such spherical surfaces. So if we can draw an infinite number of these spheres—which is the case if the universe is infinite and static—then the total light coming to us from all the stars is infinite.

The point is that our common sense assumption that the distant stars contribute less light to the night sky is not true, because the number of contributing stars goes up with distance. Even if we allow for the blocking of distant starlight by other nearby stars, we find that the night sky still should be very bright, which it is not. This is the paradox.

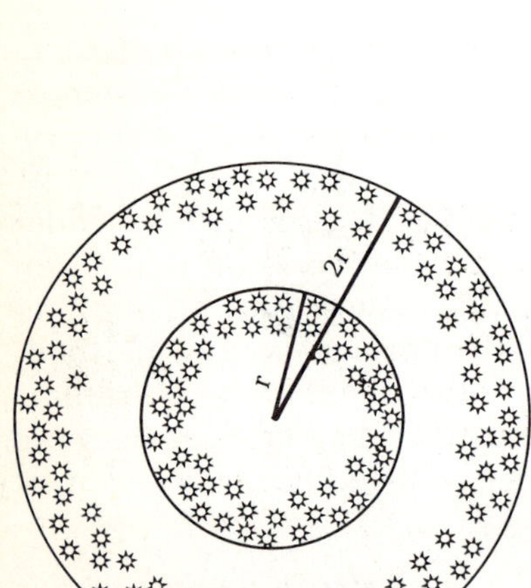

The expansion resolves the paradox to a large extent because if the galaxies are receding, we receive less and less light from the faraway galaxies. Furthermore, because of the Doppler shift, the light from the receding galaxies is redshifted (this red shift is what tells us about the expansion in the first place; see Section 12.3); it has smaller frequency and hence less energy [see Eq. (21.1)]. The net effect is that the intensity of radiation from the faraway stars and galaxies is less than that from the nearby stars after all. The contributions of light do decrease with distance, as we intuitively expect, and therefore the total of all the light reaching us is not large. So the darkness of the night sky prevails.

■ 26.2   The Big Bang Theory

The big bang theory starts with the idea that the universe was created in a huge explosion. We will have more to say about the act of creation itself, but first let's see how the big bang idea simply explains Hubble's law. In any explosion—of a bomb, for example—the various fragments are imparted vastly different velocities. If we make the simple assumption that the velocity of a fragment remains constant in its entire journey, then the distance traveled by a fragment relative to another is its relative velocity times the time of travel (Fig. 26.4). The time of travel is, of course, the age of the universe in this case. Thus, we get the simple relation

$$\text{distance of a galaxy } r = \text{age of the universe} \times \text{velocity of recession } v \qquad (26.2)$$

FIGURE 26.4   The big bang. The distance traveled by one fragment relative to another should be equal to its velocity (relative to the other) times the time of travel.

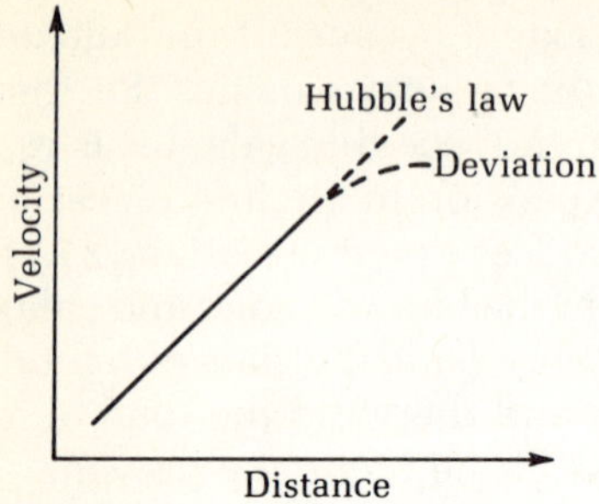

FIGURE 26.5 If we plot velocity versus distance for galaxies, we should get a straight line, according to Hubble's law. Any deviation from this straight line gives us valuable information about the nature of the expansion.

We can write Hubble's law, Eq. (26.1), in the form

$$\text{distance } r = \frac{1}{H} \times \text{velocity } v$$

Comparing this equation with Eq. (26.2), we find that the two are identical if the age of the universe is identified with $1/H$, the reciprocal of Hubble's constant.

Hubble's constant H is an experimentally established quantity, although its value is not known to much accuracy. Most recent measurements give us an approximate value of

$$H = 1.58 \times 10^{-18} \text{ per sec}$$

Thus the age of the universe is given as

$$\frac{1}{H} = \frac{1}{1.58 \times 10^{-18}} \text{ sec}$$

$$= \frac{10^{18}}{1.58} \times \frac{1}{3.2 \times 10^{7}} \text{ years} = 1.9 \times 10^{10} \text{ years}$$

The age of the universe determined this way is called the **Hubble time.** It is really an upper limit, for the following reason. The velocity of recession of the galaxies in all probability does not remain constant because of the action of the gravity forces between the galaxies. It is to be expected that the galaxies have been slowing down since the time of creation—that their initial velocities were larger than those we measure for them. If this is the case, then the distances at which we see them must have been reached in a time less than the Hubble time.

Of course, the last argument also tells us that we should expect some deviation from Hubble's law (Fig. 26.5). In fact, the amount of this deviation should tell us the value of the deceleration of the galaxies due to the braking effect of gravity.

Anyway, to finish the story of the age of the universe, we can also determine a lower limit for its age from another independent source. Astrophysical models of the evolution of stars give us estimates of the ages of stars. Since the universe must be at least as old as the oldest stars, their age gives a lower limit for the age of the universe. Unfortunately, the uncertainty of the stellar models precludes us from getting a very accurate value of the lower limit, so we get a number between 10 and 16 billion years. The important thing to notice is that this is quite consistent with the upper limit given above.

In the Prologue we mentioned the steady state theory, which maintains that the universe remains the same at all times and does not evolve with time, as depicted by the big bang theory. In the steady state theory, matter is continuously created to offset the effect of the expansion; because of the creation of new matter, any part of the universe looks like any other part at all times in spite of the expansion. Also, if you are worried about how matter can be created without violating the law of conservation of energy, the rate of creation of matter is very small. In fact, it is so small that we cannot possibly find any deviation from the energy conservation law within the limited accuracy of our experiments.

So which theory is right? An idea from geology suggested a solution to the controversy. One way to resolve the problem is to find some kind of remnant from the big bang, an analogue of the geological fossils which give us the clue to the existence of animals and things in the past that do not exist today. Can we find a "fossil" from the big bang?

Recently it has been discovered that we are bathed by a cosmic background radiation (background because it seems to be everywhere) coming from outer space. The intensity of the radiation is the same from all directions. Presumably the whole universe is full of this radiation. The frequency of the radiation is in the microwave region: between 5×10^8 and 5×10^9 Hz, which is typical of the radiation from a blackbody at 3°K.

The proponents of the big bang theory have backtracked in time and shown that this 3°K microwave background is the same light that was emitted at some early stage of the big bang when the temperature of the primordial matter was something like 5000°C (about the same as the surface temperature of our sun). The Doppler shift due to the cosmic expansion has red-shifted the original light (which was very much like sunlight) to the microwave frequency that we observe.

To a majority of physicists, the evidence is conclusively in favor of the big bang theory.

The major questions now remaining concern the act of creation itself and, perhaps even more importantly, the future of the universe. Is the expansion going to stop or not?

The theoretical framework of general relativity is natural for dealing with the last question. This brings us to the subject of cosmological models of a class known as the Friedmann models.

The Friedmann models

The Friedmann models are based on the solution of the Einstein equation of general relativity applied to the entire universe. The solutions are obtained with two simplifying assumptions. The first one we have already stated: the cosmological principle. The second one defines the important concept of **cosmic time.**

A cosmic time can be defined if we assume that the galaxies follow simple nonintersecting (except at an "origin") world lines, as shown in Fig. 26.6. The significance of the origin relates to the event of creation, as we will see later.

Since the galaxies are all moving at great speeds, they must have their own time clock, according to special relativity. If the world lines along which they travel were haphazard and intersecting (Fig. 26.7), we could find no way of synchronizing their clocks in a globally useful manner. But with the regular world lines of Fig. 26.6, we can always imagine a surface that is perpendicular to all the world lines. The different galactic clocks can now be synchronized at the points of intersection of this surface with the galactic world lines. The time read by the synchronized clocks is the cosmic time. Since the earth has very little relative motion with respect to the local galaxies, earth's time clock can be assumed to be synchronized to the cosmic time.

Now we need a parameter to describe the expansion. Suppose we focus on

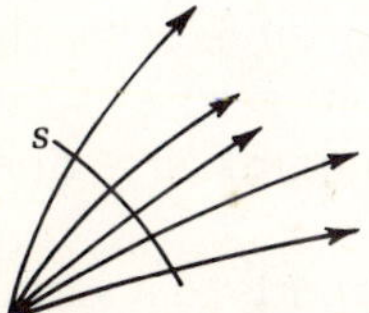

FIGURE 26.6 If the galactic world lines are regular, as shown here, we can think of cosmic time clocks that are synchronized at the points of intersection of the world lines with any surface s that is perpendicular to all of them. The time read by these synchronized time clocks is cosmic time.

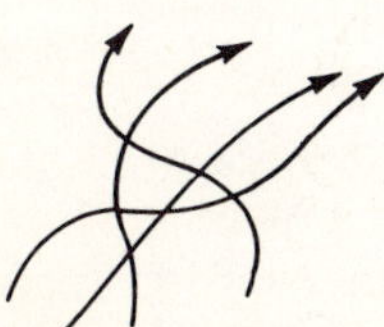

FIGURE 26.7 It would be impossible to define cosmic time if galactic world lines were jumbled like this.

two galaxies and suppose they are separated by a distance R at the present moment. Before the present moment, their separation must have been less, and at some later time the separation will be greater, all as a result of the expansion. Thus R depends on time. We can display this dependence on a graph, plotting R along the ordinate and time along the abscissa. If we are careful to use the cosmic time as the time variable in this graph, then the graph of R against time will not depend on which two galaxies we use for our reference. Even if we use two different galaxies, the dependence of R on time will come out to be the same.

We previously have referred to the solution that Friedmann himself obtained of the Einstein equation. Other theoretical physicists have since expanded on Friedmann's work. Friedmann's equations give three kinds of solutions, each associated with one of the three kinds of geometry of space (of negative, zero, and positive curvature) that we discussed in the previous chapter. These solutions are shown as three different curves in the plot of R versus t in Fig. 26.8.

All graphs start at the origin of time, t = 0, with the value R = 0. But then the curves for the three curvatures differ, sometimes quite drastically, from one another. The curve for negative curvature space is a continuously increasing curve; thus in this case R increases forever with time. The universe is forever expanding.

The curve for zero curvature, or flat space, is similar to the one for negative curvature space. Here also R increases without bounds with the passage of time, although not as fast as in the former case.

But the curve for positive curvature is a drastically different one. Here the value of R increases to a maximum and then decreases back to zero at some future time. Thus for a space geometry of positive curvature, the universe does not expand forever; it stops at some point and then turns back.

Let's discuss the meaning of R = 0. In mathematics it corresponds to a singularity of the same kind that we mentioned in connection with the black holes. Thus for geometry of positive curvature, the universe ends as a mathematical impossibility, where our equations break down, the density of energy or matter becomes infinite, and physical laws as we know them cease to be meaningful.

Of course, the same things can be said for the point R = 0 at the beginning of time, the big bang; it is the same mathematical catastrophe only going the other way. So space then explodes and matter flies apart.

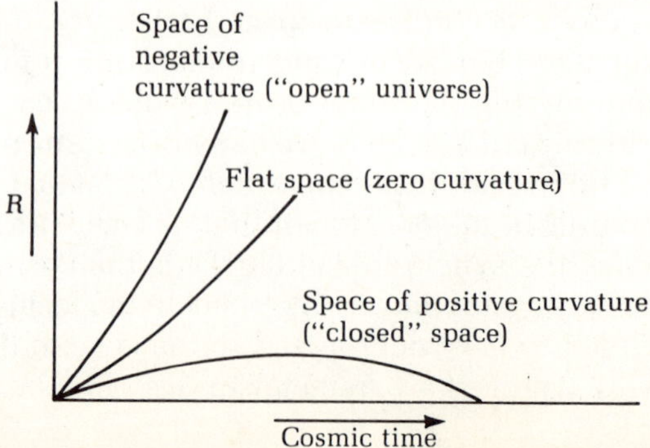

FIGURE 26.8 The graph of the intergalactic separation R versus cosmic time t. The three curves correspond to three different geometries of space.

We do not get an oscillating or cyclical model solution of the universe (as shown in Fig. P.4) from this theory. Physicists of John Wheeler's school of thought believe that just before the catastrophe, other physical considerations, perhaps those of quantum mechanics, enter and change things around to start another expansion phase. Or, perhaps, there is some new physics here that we do not know about.

But forgetting about the oscillating model, this question becomes paramount: Which one of these curves are we on? Ultimately speaking, of course, this is an observational question. Have the observational astronomers found the answer?

What clues should the experimenter look for in an attempt to answer this question? This is one of the greatest puzzles human beings have ever faced. After all, this is a cosmic puzzle; we are talking about the structure and future of the entire universe. We must be humble enough to admit that we may not know how to recognize the right clues.

Even so, within the context of the Friedmann models, we know what to look for. Which solution is the right solution depends, in theory, on the average density of the universe. If the observed average density happens to be larger than a calculated critical value, then the mutual gravitational forces between the galaxies will be strong enough to stop the expansion. In such a case the gravitational forces are strong enough to close space. Thus for high density, a closed and finite universe of positive curvature is predicted. In this case the expansion stops and the universe subsequently goes through a contraction phase.

On the other hand, if the observed average density of matter is smaller than the critical value, then the galaxies have sufficient energy to escape from the gravitational bondage of the universe. If such is the case, the universe will keep expanding forever, and the universe is then an open one of negative curvature.

For the special case of the average density of the universe being equal to the critical density, we get the flat space universe of zero curvature.

Unfortunately, the theoretical value of the critical density depends crucially on the age of the universe. If we assume the value of the Hubble time to be the 19 billion years mentioned previously, the critical density is estimated to be 5×10^{-27} kg/m³. But there is a lot of uncertainty here since we do not know the exact value of the Hubble time.

What is the actual matter density of the universe? We can determine the average density by determining the masses of the galaxies in a given volume of space, add these masses, hope that there is not much missing mass, and divide by the volume of space taken. This gives us a value for the average density. So the determination of the average density is really a matter of measuring, or rather estimating, the mass of the galaxies.

In 1976, astronomers were ready to conclude that their estimates of the galactic masses give a value of the average density that is much too small compared to the critical density, perhaps a few orders of magnitude smaller than the critical density needed to stop the expansion.* But there is another issue.

It has been suspected for some time that the clustering of matter that we call

---

*Gott, Gunn, Schramm, and Tinsley, "Will the Universe Expand Forever?", *Scientific American*, March 1976, p. 62.

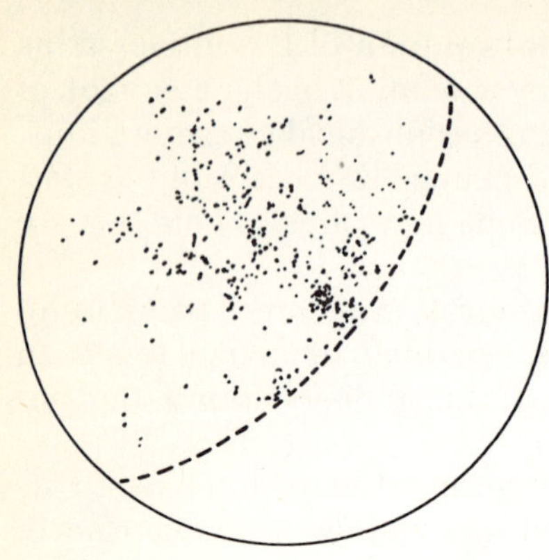

FIGURE 26.9 Clustering of galaxies. The distribution of galaxies (the dots) in the northern celestial hemisphere. The cluster near the center is called the Virgo cluster. The band of galaxies extending from the Virgo cluster toward the 10-o'clock position forms a supercluster of galaxies. (From Groth, Peebles, Seldner, and Soneira, "The Clustering of Galaxies," *Scientific American*, November 1977, p. 76.

galaxies is not the end of the clustering phenomenon. Galaxies have been known to form clusters of galaxies. And now some researchers have discovered that these local clusters form even larger clusters.* In fact, there is a whole hierarchy of such clustering (Fig. 26.9).

This tendency of the galaxies to form the hierarchy of superclusters cannot be explained with present knowledge, unless we assume a large enough mass of the galaxies. But this changes the observational estimate of the average density to just about the critical value. Thus if this argument is right, then the expansion will stop and we will have to conclude that we have a closed universe.

You may be curious about how the previous observational estimates could have gone wrong. It is quite possible that they did not account for all the mass of a galaxy. But where exactly this missing mass lies will be an interesting thing to find out.

The uncertainties of both observational estimates are so large, however, that nobody is betting on either of them at this time. So we really don't know for sure, not yet anyway, which of the Friedmann models is right for our universe. Not knowing has one advantage—we can speculate.

■ 26.3 The Beginning, the End, and Time's Arrow

> Who knows for certain? Who shall here declare it?
> Whence was it born, whence came creation?
> The gods are later than this world's formation;
> Who then can know the origins of the world?
>
> None knows when creation arose;
> And whether he has or has not made it;
> He who surveys it from the lofty skies,
> Only he knows—or perhaps he knows not.†
>
> *The Vedas*

A colleague of Einstein, quite a celebrity in his field, once conjured up a model for the creation of the universe. As is often the case with a creator, the more he looked at his model, the surer he became of its validity. Finally, he became so convinced that he felt an irresistible urge to communicate his finding

---

*Groth, Peebles, Seldner, and Soneira, "The Clustering of Galaxies," *Scientific American*, November 1977, p. 76.

†Quoted in *Intelligent Life in the Universe*, by I. S. Shklovski and Carl Sagan (San Francisco: Holden Day, 1968).

to others. And he wanted Einstein to be the first to know. "Blah, blah, blah, . . ."
he went on; Einstein listened patiently, as was usual with him.

"Well, what do you say? Do you think that the universe could be created this
way?" Einstein was asked at the end of the story.

"I don't know," Einstein said. "I was not there."

This story goes to the heart of the problem of making theories about the creation, the beginning of the universe. Nobody was there at the time of creation.
Yet on the basis of the cosmological theories of the last section, we can speculate about the beginning and even about the end. And speculate we will do in
this section.

So far we have looked at the structure of the universe more or less from the
point of view of space. We will begin the discussion here with the consideration
of time, in particular, with time's arrow. Where does it fit in the overall nature of
the universe?

In the Prologue we talked about processes in our environment that occur in
the direction of increasing entropy, and we concluded that an arrow of time can
be associated with the direction of increasing entropy. In Chapter 10 we saw
that the statement, "for an isolated system, entropy always increases," is a
physical law, which we call the second law of thermodynamics. If entropy
always increases, then time must always progress forward in one direction,
from the past through the present into the future.

This arrow of time based on the thermodynamic concept of entropy is often
called the **thermodynamic arrow of time.** It rests on the observation that the
universe is in a state of large amount of order today containing low entropy, and
that it is running downhill in entropy toward increasing disorder.

But this way of looking at things also presupposes that the entropy was even
lower in the past. Was the universe created with a very high amount of order, or
did it get this way as an accident, a big "statistical fluctuation" that shows up
once in a while, creating order in even the most disorderly system?

Of course, it is possible that the universe was created with high order. But it is
also interesting to speculate, what if it was not? After all, it is more likely for a
system to be in equilibrium; this is the state of higher probability. There is also
some indirect evidence that the universe was in a state of thermal equilibrium
sometime after its creation, at which time it emitted the light that we see today
as the microwave background. This light, all indications suggest, came from an
object in thermal equilibrium.

So now the important question is this: Once in equilibrium, how could the
universe get itself into its present state of very high order? How could a stable
universe represented by the state of thermal equilibrium become unstable, as it
is today? Does this not violate the second law of thermodynamics, which says
that order must give way to disorder, not the other way around?

To understand this we have to consider the material of which the primordial
universe was composed: hydrogen nuclei, protons. This primordial material
would have existed in a state called "the fireball," a state of extremely high
density at immense temperature. The key part of our answer is that this fireball
was in a state of local equilibrium only, not in one of stablest equilibrium. To
understand this, compare the two pictures in Fig. 26.10 and 26.11. In the first
picture the ball is totally unstable. The slightest disturbance will bring it to the

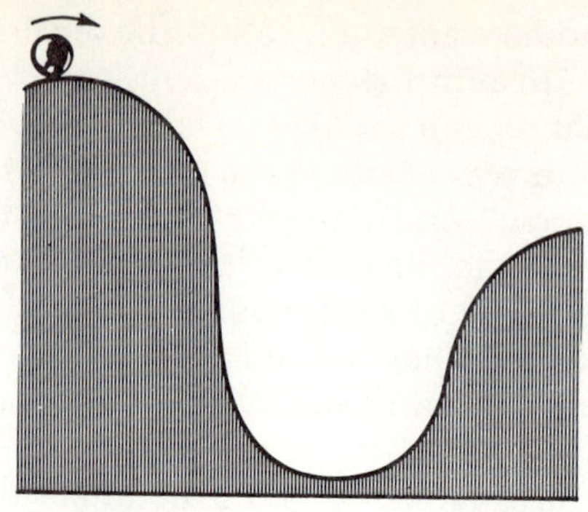
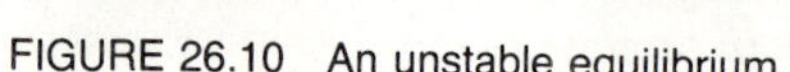

FIGURE 26.10   An unstable equilibrium.

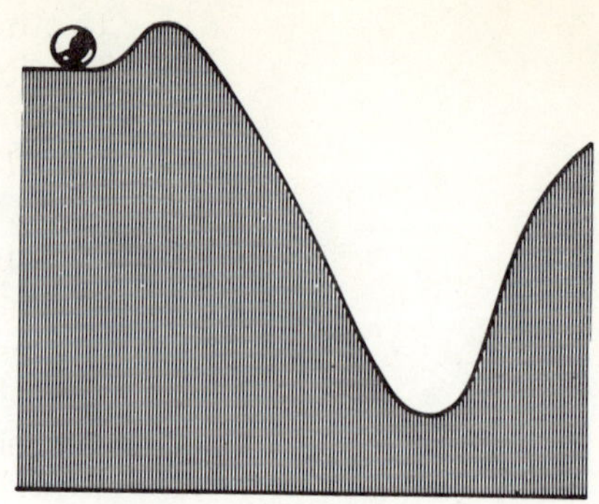

FIGURE 26.11   A metastable equilibrium.

valley of stability where, after a few oscillations, it would acquire total equilibrium at the bottom of the valley. In Fig. 26.11 the situation is different. The ball is in a local equilibrium, and it needs a big push to go over the hill to the valley of more stable equilibrium. What we are saying is that the primitive fireball may have been like this ball, not unstable nor stable, but metastable, stable against small pushes.

The hydrogen nuclei are the basic ingredients of this metastability. These particles can fuse to make helium (hydrogen fusion), but the electrical repulsion prevents them from doing so unless they come together with high energy. The order of the universe we see today is due to the large amount of hydrogen that escaped intact from that primordial fireball. Every once in a while, some of the hydrogen nuclei come together through gravity to make a star. The star's primary process is the burning of the hydrogen nuclei, which is the process that brings the hydrogen down to a lower valley of equilibrium.

The local equilibrium that the fireball had can be explained in the following way. In the intense temperature of the fireball, the hydrogen could fuse, of course, but the temperature and density were so huge that the helium that resulted had to disintegrate right back. The two competing processes created the metastable equilibrium. The radiation from this state is what we see today as the 3°K microwave radiation.

And now we come to the most important question: What enabled the metastable equilibrium to become a state of disequilibrium? Here the answer is, the expansion of the universe. Due to the expansion, and also due to the cooling by radiation, the temperature of the fireball was decreasing. Eventually the temperature became so low that the hydrogen could not fuse together; the nuclei did not have enough energy to go over the reaction barrier.

If this picture is correct, there must have been a stage when the temperature was just right for the hydrogen to fuse into helium but not high enough for the helium to be broken down again. Actually, some deuterium is also created this way. There is good evidence now that the amount of helium (and deuterium) we find in the universe could not have come just from the fusion in the stars. Some of the helium came from the big bang itself.

To summarize, the universe very likely was in a metastable state shortly after the creation. Further expansion, however, destroyed this local equilibrium and gave us the state of disequilibrium that we see today. From this state it is a long way to the lowest equilibrium state, mainly because of the peculiar nature of hydrogen nuclei.*

Thus the expansion is the key to the relatively low entropy state of today and hence to the (thermodynamic) time's arrow. This is interesting, because it is known that the expansion of the universe can be used also as the direction finder for the arrow of time in the following way. In the past the galaxies were closer together than they are now. And in the future they will be further apart. Both conclusions follow from the cosmic expansion. Thus we can tell the arrow of time just by keeping track of the density of the galaxies, perhaps taking a photograph of them every now and then. The arrow of time determined this way is called the **cosmic arrow of time.** Our discussion above makes it clear that in the present phase of the universe, the two arrows work the same way; they are coupled.

If the universe were open and continued to expand forever, then this coupling would never be broken. The thermodynamic arrow for such a universe predicts the heat death (see Section 10.3), which is an end of time in a way. The cosmic arrow says just about the same thing: one by one the galaxies move so far away from ours that we won't be able to see any of them some day in the far future, and this arrow also stops.† Thus there is an end of time of sorts in this kind of a model. It takes a very long time to achieve this, of course.

If the universe is closed, then in the expansion phase, the cosmic arrow and the thermodynamic arrow are in the same direction, but how about the contraction phase? Does the entropy go on increasing while the universe contracts? This represents a decoupling of the two arrows, which some physicists do not accept. Instead, they would like to have the following picture. The cosmic arrow is the dominant arrow, and therefore it just turns the thermodynamic arrow around. So now the entropy actually decreases instead of increasing (as reckoned by us). You can speculate about the strange ways of such a contracting universe: human beings would be born from the grave as old folks and become younger and younger, ending up in the mother's womb. But we really do not know if this happens, because nobody has been able to prove that the cosmic arrow indeed turns the entropy arrow around. So such a time-reversed world is no more than fiction.

The last speculation we will mention is the nature of time in the cyclical model of the universe. Here time is destroyed at the end of the contraction cycle along with everything else, only to be created again, starting a new expansion-contraction cycle in a new universe.

EPILOGUE

So we end this chapter without giving you a final answer. In fact, many a time we have seen this in physics—that we do not have the final answer, not yet,

---

*For further details, read: P. C. W. Davies, *Space and Time in the Modern Universe* (Cambridge, England: Cambridge Univ. Press, 1977), p. 172.

†This is a consequence of the Doppler shift of the light coming from the galaxies. When the Doppler shifted wavelength approaches infinity, no energy reaches us and we cannot see the galaxy.

anyway. We have not found the ultimate elementary particles of nature. We do not know if quantum mechanics is the ultimate description of the reality of atomic particles. These are just a couple of unresolved issues. In more practical matters, we do not know if physics will bring a solution to today's energy crisis. And, of course, in the most abstract matters, we do not know how our universe is going to end.

If you are dazzled by all these nonanswers, consider these lines

> Yours is no unique condition,
> Your type of search and conflict and construction,
> Don't worry if you have no answer ready to the lasting question.
> Hold out, meditate, listen.
> Explore, explore. . . .*

---

*From "Zima Junction," by Yevgeny Yevtushenko, in *Yevtushenko: Selected Poems*, translated by Robin Milner-Gulland and Peter Levi, S.J. (Penguin Modern European Poets, 1962), p. 51. Copyright © Robin Milner-Gulland and Peter Levi, 1962. Reprinted by permission of Penguin Books Ltd.

QUESTIONS

Review and reason

1. What is the arena of the subject of cosmology? In what sense are some of the queries of cosmology not completely scientific?
2. State Hubble's law. How does the big bang theory offer an explanation for Hubble's law? Should we expect any deviation from Hubble's law? What should such deviation tell us?
3. State the cosmological principle.
4. What is the paradox in Olbers' paradox? How is the paradox resolved?
5. What is meant by cosmic time?
6. Briefly discuss the three types of cosmological models of the Friedmann type corresponding to the three types of geometries of space.
7. What do you make of the conjectures presented in the text for the beginning of the universe?
8. How do we define the cosmological arrow of time? How is it reconciled with the thermodynamic arrow of time?
9. How would the universe end in the open-universe versus the closed-universe Friedmann models?

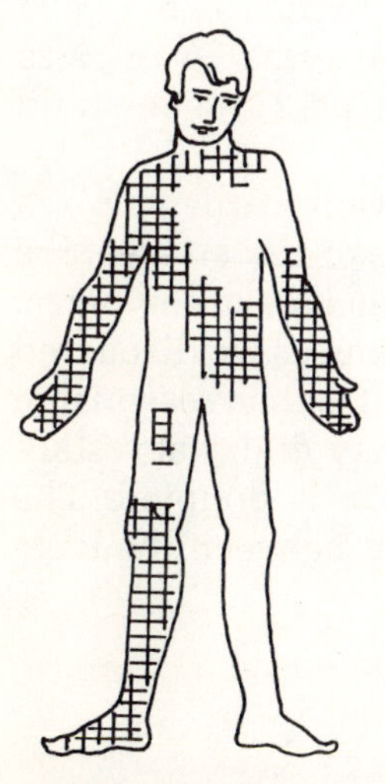

FIGURE A.1   One way to determine the area of a human body is to cover it with 1-inch-square papers and count the squares.

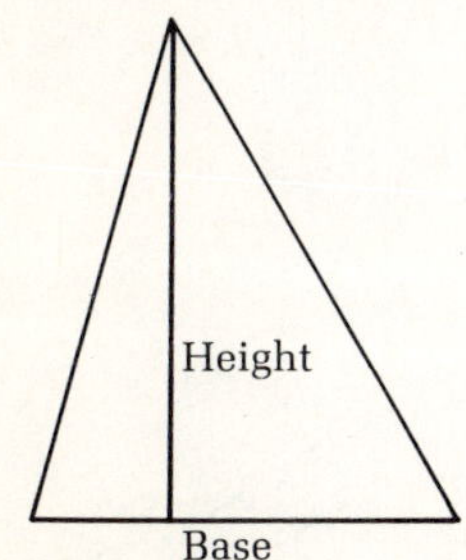

Area of a triangle = $\frac{1}{2}$ base $\times$ height

FIGURE A.2

If something is one-dimensional, such as a line, we can only measure distances or lengths along that line. But in the case of a two-dimensional surface, such as the walls of a room, not only can we measure lengths along a line on the surface, we can also think of the amount of surface. For example, if we were to paint those walls, the relevant question for determining how much paint we need is; what's the surface area of the walls? Thus area refers to the amount of surface.

How do we measure an area? It is advantageous to take a regular-shaped surface for the purpose, and we choose a 1-foot square for making the measurements. The unit of area in this case is the square foot.

One important thing. If you have it programmed in you that area is length times breadth, think for a minute. For an irregular surface—for example, a human body—how do we find *a* length and *a* breadth? They are not the same everywhere. But we still can take pieces of 1-inch-square paper and paste them all over the body; it may look funny but the area of the body will be given correctly in square inches, the number of pieces of paper pasted on the body (Fig. A.1). Once we know the area in terms of one unit, finding it in another unit is just a matter of using a conversion table.

For a rectangular surface, the area indeed is given by the product of the length and the breadth. For other kinds of regular surfaces, geometrical rules are available for determining the surface area in terms of some of the linear dimensions. For example, for a triangular surface, the area is $\frac{1}{2}$ the length of the base $\times$ the height (Fig. A.2).

The metric unit of area used in this book is the square meter, which is the area of a square of side 1 meter. All conversion factors are given in Appendix 4.

Likewise, when we consider three-dimensional space, we can talk about the amount of space, or the volume. In measuring the volume, again we must choose a unit. A convenient one is a cubic foot, a cube of side 1 foot. In the metric system we choose the cubic meter. The volume of a given space is then given as the number of these unit cubes that can fit into that space. Alternatively, of course, you can use geometry to figure out the volume in terms of other given dimensions for regular-shaped space. If the space is that inside a room of 10 ft by 15 ft and height 8 ft, the volume is the product of the length, width, and height, which in this case comes out as 1200 ft^3.

■ **Scaling**

Scaling is the uniform expansion or reduction of a geometric shape so that none of the internal geometric relationships undergoes any change. For example, a scale model is one that scales a shape up or down while leaving the relationships of the linear dimensions to one another unchanged.

For a surface of a given shape, the area increases in proportion to the square of the linear dimension. For example, if the linear dimension increases by a factor of 3, the area increases by a factor of the square of 3, or 3^2, which is 9.

Let's consider an example. In pizza places they often make a small 6-in pizza and a large 12-in pizza. Suppose the smaller pizza is priced at $2. What would be a reasonable price for the large size?

If you think it should be twice the $2, think again if you want to open a pizza place. It will not take you long to go broke. Since pizza is basically an area, the raw material required for the large size is proportional to the area of the pizza. How does the area scale? As the square of the linear dimension, as we discussed above. So the area goes up by a factor of the square of 2, or 4. It is then reasonable to price the large pizza at $8. Actually, if you check, you may find that restaurants don't charge quite 4 times more when the linear size is doubled. The reason is that the labor part of the price is not very different between the large and the small sizes.

Appendix 2
The Power of Mathematics

When talking about mathematics to nonscientists, I cannot help thinking of the reaction that Linus had after receiving a calendar from Charlie Brown as a present. "It's got numbers on it. I can't understand something with numbers all over it," he says. "Why can't people give simple things? Why does everything have to be so complicated?"

I know you will sympathize with Linus, because you also may not take kindly to numbers or, in general, to mathematics. For one thing, mathematics is probably not your mode of expression; if it were, it is likely that you would be doing something related to it.

Why then talk about mathematics at all? Can't I just describe the various physical theories to you in words and be satisfied with that? I could and I will for most things. I will also, however, try to give you some flavor of the intimate relation of mathematics and physics.

James Jeans, a notable scientist of the early twentieth century, said, "the Great architect seems to be a mathematician." Rene Descartes has described a dream, a mystical experience as he put it, which revealed to him that mathematics was the key to opening the doors of nature, the "open sesame." If you close your mind and heart completely to all aspects of mathematics, you will probably miss the experience of understanding the simplicity of the way nature behaves. I cannot ask you to learn mathematics at this stage in order to appreciate what goes on in the universe, but I do expect you to be willing to have a look at the other culture. Instead of saying "the twain shall never meet," let's see what happens when the twain meet, at least cursorily, and proceed from there.

From this perspective I will review some basic principles of arithmetic and algebra for your benefit. The following topics are discussed: fractions, equations, exponentiation, and ratio and proportion.

In talking about a fraction like x/y, we refer to the top part x as the numerator and the bottom part y as the denominator. We can multiply or divide both the denominator and the numerator by the same factor without changing the value of the fraction. This is often very useful. As an example, suppose we get the following end product as a result of some algebraic manipulations:

$$\frac{8x}{12y}$$

We can divide the numerator and the denominator by the factor of 4 without changing anything according to the rule above. This gives us the fraction

$$\frac{2x}{3y}$$

which is equivalent to the previous fraction. This process of dividing by a common factor is also called canceling a common factor. In the example, we could say that we have canceled the common factor of 4.

In arithmetic computations with fractions, there are two things to note. First, it is often simpler to use decimals in the manipulations. For example, this enables us to use the power-of-ten notations for large or small numbers. Examples are

$$\tfrac{1}{2} = 0.5, \ \tfrac{1}{3} = 0.33, \ \tfrac{1}{300} = \tfrac{1}{3} \times 10^{-2} = 0.33 \times 10^{-2}$$

The other helpful thing to remember is the right way to manipulate the denominators of fractions that are themselves fractional. For example, how do we handle $5/(\tfrac{7}{8})$? In a fraction like this, the denominator can be inverted and multiplied like so:

$$\frac{5}{7/8} = \frac{5 \times 8}{7} = \frac{40}{7} = 5.71$$

◼ Equations

Suppose we have an equation involving an unknown quantity x,

$$4x - 3 = 5$$

It is usual to ask, what is x? The important rules for the solution of equations are the following:

1. Transposition: an isolated quantity can be transposed from one side of an equation to the other by adding or subtracting the quantity to both sides of the equation. In the equation above, in effect, 3 can be transposed from the left-hand to the right-hand side, provided we change its sign; the equation now becomes

$$4x - 3 + 3 = 5 + 3 \qquad \text{or} \qquad 4x = 8$$

2. Transposition of factors: A common factor of all the terms on one side can be transposed to the other side of an equation if we multiply or divide both

sides of the equation by the factor. For the equation above,

$$4x = 8$$

this procedure gives us

$$\frac{4x}{4} = \frac{8}{4} \quad \text{or} \quad x = 2$$

Cross multiplication is a special case of this rule. If we have an equation

$$\frac{a}{b} = \frac{c}{d}$$

we can write it as

$$ad = bc$$

3. Both sides of an equation can be squared. For example, the equations

$$3x = 5 \quad \text{and} \quad 9x^2 = 25$$

are equivalent. (Note that each factor, 3 and x, is squared.)

4. Substitution: suppose the equation for x involves another quantity y, for which we have a second equation. We can then substitute for y in the equation for x. Suppose

$$4x = 3y$$

and

$$y = 3$$

We can substitute for y from the second equation into the first equation. This gives

$$4x = 3 \times 3 = 9$$

or

$$x = \tfrac{9}{4} = 2.25$$

■ **Exponents**

You know that the square of 2 is 4, the square of 4 is 16, and so on. We write these results as

$$2^2 = 4 \quad \text{and} \quad 4^2 = 16$$

We say that 2 raised to the power, or exponent, 2 gives 4, and so on. The way we get the result of such an exponentiation is by multiplication. Thus

$$2^2 = 2 \times 2 = 4, \quad 4^2 = 4 \times 4 = 16$$

What is 10^6? It is $10 \times 10 \times 10 \times 10 \times 10 \times 10 = 1,000,000$. In general, then, if

n is an integer exponent, a^n can be obtained by multiplying n number of a's together.

When one quantity raised to a certain exponent is multiplied by the *same* quantity raised to another exponent, the exponents add:

$$2^3 \times 2^4 = 2^7$$
$$10^3 \times 10^8 = 10^{11}$$

On the other hand, division is achieved by subtracting the exponents. We recall that the notation a^{-n} stands for $1/a^n$. Then

$$\frac{a^n}{a^m} = a^n \times a^{-m} = a^{n-m}$$

For example, $10^9/10^5 = 10^{9-5} = 10^4$.

One important corollary is that any quantity raised to the exponent 0 gives 1.

$$a^0 = 1, \qquad \text{for all values of } a \text{ except } a = 0$$

Also,

$$(a^n)^2 = a^{n+n} = a^{2n}, \qquad \sqrt{a^n} = a^{n/2}$$

■ Ratio and Proportion

How do we compare things? Say we want to compare the sizes of the earth and the moon. What do we do? The radius of the earth is 6400 km, and the radius of the moon is about 1900 km. To compare the two values we can subtract one from the other, $6400 - 1900 = 4500$, and say that the earth is 4500 km larger than the moon in radius. So comparing can be done by subtraction.

But there is another way. We could also do the comparison by division. We could say, since

$$\frac{6400}{1900} = 3.37$$

that the earth is 3.37 times larger than the moon in radius. Such a comparison by division is called taking a ratio. The ratio of earth's radius to the moon's is 3.37, or

$$\frac{\text{earth's radius}}{\text{moon's radius}} = 3.37$$

This is a very common practice in physics, comparison by taking a ratio, and it means this simple thing, division.

What is proportionality? If two quantities are proportional, then their values always bear the same ratio. So proportionality is really a statement of equality of ratios. If one of the quantities increases by a factor of 3, the other one must increase by the same factor.

Suppose $x \propto y$. The symbol $\propto$ stands for "is proportional to." When x assumes a value x_1, suppose y has the value y_1; and when x has the value x_2, y assumes the value y_2. From what we have said in the paragraph above, the ratio of x_1 and

x_2 must be equal to that of y_1 and y_2:

$$\frac{x_1}{x_2} = \frac{y_1}{y_2}$$

There is still another way of looking at this. When two quantities are proportional, they bear to each other a constant ratio, a number whose value does not change. Thus if distance d is proportional to time t,

$$d \propto t$$

we can write

$$\frac{d}{t} = C$$

where C is called a constant of proportionality. Once C is determined numerically, given either one of the quantities d or t, we can always determine the other.

Appendix 3
Graphs

■ **How to Draw Graphs**

There is more than one way to express and communicate our understanding of a phenomenon. Words, of course, and mathematical equations, are two ways. There is a third very effective means of communication, graphs.

Exactly what is a graph? It is a picture of a mathematical relationship. For example, consider the relation of distance traveled with time for an automobile moving at a constant velocity of 10 m/sec:

$$d \text{ (in meters)} = 10 \times t \text{ (in seconds)}$$

Now let's see how to express this as a picture.

We start with drawing two lines perpendicular to each other from an origin O, as in Fig. A.3. The two lines are called the axes: the horizontal one is the

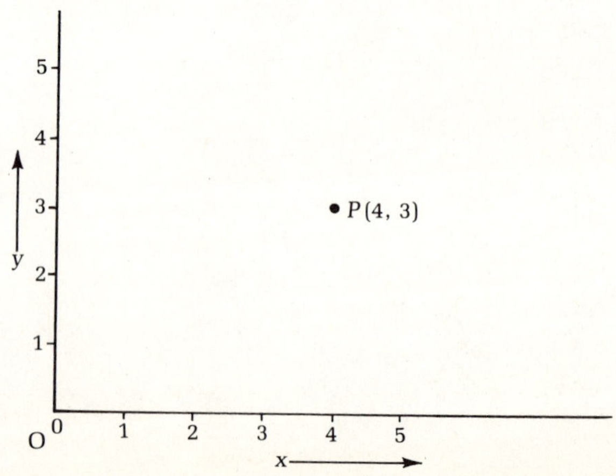

FIGURE A.3

abscissa or x-axis, the vertical one the ordinate or y-axis. Any point on the graph can be described by two numbers (x, y), which specify the distances from the origin along each of the two axes. These two numbers are called the coordinates of the point. The point P in the graph has the coordinates $(4, 3)$, meaning that to locate P we have to travel 4 units along the x-axis and 3 units along the y-axis: 4 is the x-coordinate and 3 is the y-coordinate. The coordinates are always written this way: x-coordinate first, that is, (x, y).

Now from an equation like $d = 10t$, we can make a table in the following form:

time t (in seconds)	0	1	2	3	4
distance d (in meters)	0	10	20	30	40

Thus we now have pairs of numbers, $(0, 0)$, $(1, 10)$, $(2, 20)$, $(3, 30)$, and $(4, 40)$. We can think of these pairs of numbers as coordinates of points on a graph, (t, d), and we get our desired picture by joining these points with a curve, in this case a straight line (Fig. A.4).

That's all there is to it to get a graphical display of a relationship between two quantities. Incidentally, it is customary to use a certain terminology. Notice that given the time and an equation like $d = 10t$, we can always calculate the distance traveled. Time in this case is called the independent variable and distance is called a **function** of time. The value of time is chosen first, which automatically determines the distance through the functional relationship $d = 10t$. In a graph it is customary to plot the independent variable along the x-axis and the function along the y-axis.

Two important uses of the graphical technique are interpolation and extrapolation. **Interpolation** is guessing the value of one of the quantities in the graph for a given value of the other, between those values that we know. In our case, from the graph in Fig. A.4, we can tell that a distance of 25 meters is traveled in 2.5 seconds. In contrast, **extrapolation** is guessing a value beyond what we

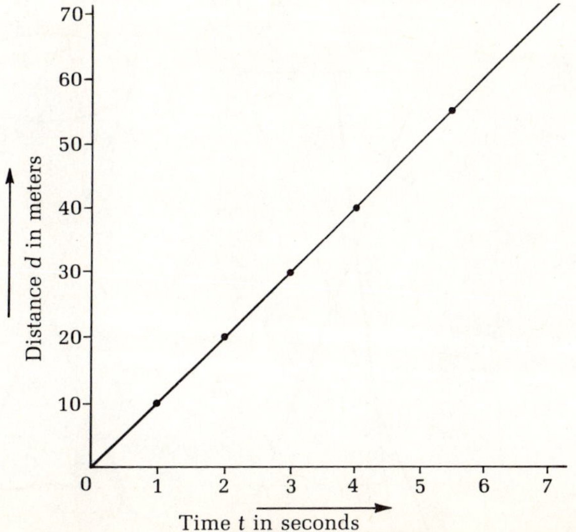

FIGURE A.4   The graph of distance d versus time t according to the relation $d = 10t$.

know. This is where the power of a graph lies. We can continue our straight line and find the value(s) of one of the quantities for any other given value of the other, provided we know that their functional relationship continues to hold. For example, we can predict that the car will travel 70 meters in the time of 7 seconds from the graph.

The graphical technique can be used even when we do not have a definite mathematical equation describing a function in terms of the independent variable. Remember that all we need for drawing a graph is a table such as the one above. Actually, we often work it this way. Once the graph is drawn, we can now guess at a mathematical relationship from the shape of the curve.

The Conic Curves

The ancient Greeks did not have ice cream cones, but that didn't prevent them from studying a class of mathematical curves that are associated with the cone. I am talking about the circle, ellipse, parabola, and hyperbola, the family of curves known as **conic sections** because they can be obtained from the intersection of the so-called conic surface with a plane (Fig. A.5). As you can see in the figure, the conical surface consists of two cones extending on opposite sides from an origin O.

The different conic curves result from different positions of the plane in relation to the cones. Thus if the plane intersects through only one part of the cone, a circle $CC'C''$ or an ellipse $EE'E''$ is the result. If the cutting plane is parallel to one of the lines of the cone, like LOL', we get the parabola $PP'P''$ as a result of the intersection. Finally, if the intersecting plane is inclined in such a way that it cuts through both parts of the conical surface, we get a pair of opposite curves called the two "branches" of a hyperbola (Fig. A.6).

I hope you have noticed how important the role of these curves has been in the development of physical theories. Imagine what would have happened if Galileo hadn't known about the parabola or Kepler about ellipses.

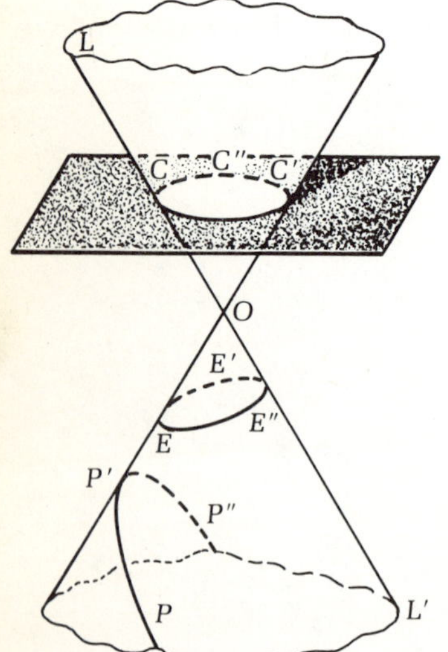

FIGURE A.5 The conic curves. See text for description.

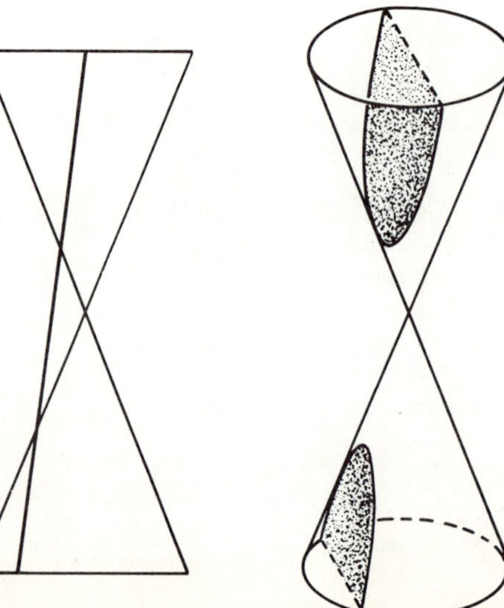

FIGURE A.6 The hyperbola.

Appendix 4
Physical Constants and Conversion Factors

TABLE 1. Physical constants.

Acceleration due to gravity	g	$9.8 \text{ m/sec}^2 = 32.2 \text{ ft/sec}^2$
Universal constant of gravity	G	$6.67 \times 10^{-11} \text{ N·m}^2/\text{kg}^2$
Speed of light	c	$3 \times 10^8 \text{ m/sec}$
Boltzmann's constant	k_B	$1.38 \times 10^{-23} \text{ J/°K}$
Avogadro's number	N_o	$6.02 \times 10^{26}/\text{kg-mole} = 6.02 \times 10^{23}/\text{gram mole}$
Electronic charge	e	$1.6 \times 10^{-19} \text{ C}$
Coulomb's law constant	k	$9 \times 10^9 \text{ N·m}^2/\text{C}^2$

TABLE 2. Conversion factors.

Mass	$1 \text{ kg} = 10^3 \text{ g} = 6.85 \times 10^{-2} \text{ slug}$
	$1 \text{ g} = 10^{-3} \text{ kg} = 6.85 \times 10^{-5} \text{ slug}$
	$1 \text{ slug} = 14.6 \text{ kg} = 1.46 \times 10^4 \text{ g}$
	$1 \text{ amu} = 1.66 \times 10^{-27} \text{ kg}$
Length	$1 \text{ m} = 10^2 \text{ cm} = 3.28 \text{ ft}$
	$1 \text{ ft} = 12 \text{ in} = 0.3048 \text{ m}$
	$1 \text{ mi} = 5280 \text{ ft} = 1609 \text{ m} = 1.609 \text{ km}$
	$1 \text{ km} = 10^3 \text{ m} = 0.62 \text{ mi}$
Area	$1 \text{ m}^2 = 10^4 \text{ cm}^2 = 10.76 \text{ ft}^2$
	$1 \text{ cm}^2 = 10^{-4} \text{ m}^2 = 1.08 \times 10^{-3} \text{ ft}^2$
	$1 \text{ ft}^2 = 144 \text{ in}^2 = 9.29 \times 10^{-2} \text{ m}^2$
Volume	$1 \text{ m}^3 = 10^6 \text{ cm}^3 = 10^3 \text{ liter (l)} = 35.3 \text{ ft}^3 = 264 \text{ gal}$
	$1 \text{ cm}^3 = 10^{-6} \text{ m}^3 = 3.53 \times 10^{-5} \text{ ft}^3$
	$1 \text{ l} = 10^3 \text{ cm}^3 = 10^{-3} \text{ m}^3 = 0.264 \text{ gal}$
	$1 \text{ gal} = 4 \text{ qt} = 231 \text{ cm}^3 = 3.785 \text{ l}$

Time	1 h = 60 min = 3600 sec
	1 day = 24 h = 8.64×10^4 sec
	1 year = 365 days = 3.16×10^7 sec
Force	1 N = 0.225 lb
	1 lb = 4.45 N
Pressure	1 atm = 14.7 lb/in^2 = 1.013×10^5 N/m^2 = 30 in of Hg = 76 cm of Hg
	1 bar = 10^5 N/m^2 = 10^3 millibar
	1 millibar = 10^{-3} bar
Energy	1 J = 10^7 erg = 0.738 ft·lb = 2.39×10^{-4} Cal = 0.239 cal = 9.48×10^{-4} Btu
	1 Cal = 1 kcal = 10^3 cal = 4186 J = 3.968 Btu
	1 Btu = 1055 J = 778 ft·lb = 0.252 Cal
	1 ft·lb = 1.36 J = 1.29×10^{-3} Btu
	1 kWh = 3.6×10^6 J
	1 eV = 1.6×10^{-19} J
Power	1 W = 0.738 ft·lb/sec = 1.34×10^{-3} hp
	1 ft·lb/sec = 1.36 W = 1.82×10^{-3} hp
	1 hp = 550 ft·lb/sec = 746 W

Answers to Arithmetic Exercises

Prologue

1. 100,000,000,000   2. 1.4×10^{-15} m   3. 30,000 light years; $\approx 3 \times 10^{20}$ m

Chapter 1

1. $\frac{1}{5}$ mi/h   2. 45 mi/h   3. -5 mi/h   4. 40 mi/h   5. 5 mi/h   6. 5 ft/sec; 25 ft/sec   7. 1 sec

Chapter 2

1. 5 m/sec² 2. 9 N 3. 50 kg 5. 2040.82 kg; 15918.36 N 6. 2000 lb 8. 4.8 in.

Chapter 3

1. (c)   2. Your velocity is half of hers   3. 0   4. 60 m/sec 5. 30.1 m/sec   6. 2 ft from the heavier weight

Chapter 4

1. 5 ft/sec²   2. 408.16 sec   3. 4096 ft   5. 90 million square miles 6. 8.

Chapter 5

1. 1 meter-newton   2. 2.5 ft from center   3. 2.57 ft from center 4. 1.96×10^{-7} radian/sec

Chapter 6

1. 26.67 m/sec^2 2. 2.67 $\times$ 10^{-8} N   3. 3.56 $\times$ 10^{22} N   4. $\frac{1}{8}$; $\frac{1}{2}$; $\frac{1}{2}$
5. 10 lb; increases   6. 12766 mi/h   7. 42 min   8. 7.75 $\times$ 10^8 m from
the center of the sun on the line joining the sun and Jupiter.

Chapter 7

1. Weight reduced by 3 $\times$ (your mass in slugs) lb   3. 5 $\times$ (your mass in kg) N
4. 99 m/sec

Chapter 8

1. 144   2. Pressure = 1248 lb/ft^2 3. 62.4 lb/ft^2 4. 3.03 $\times$ 10^5 N/m^2
5. 860 mm of mercury   6. 57%   7. 2 kg-moles; 12.04 $\times$ 10^{26}

Chapter 9

1. 3500 ft·lb; 175,000 ft·lb; 700,000 ft·lb; 194.44 ft·lb/sec   2. (a) 7350 J (b)
2880 J (c) 200,000 J   3. (c)   4. The 1-lb hammer   5. 1.21 hp
6. 1152 W   7. $\approx$ 2 $\times$ 10^{12} W; 6.4 $\times$ 10^{19} J/year   8. 9.8 $\times$ 10^4 J; 4.9 $\times$ 10^4 J;
4.9 $\times$ 10^4 J; 9.8 $\times$ 10^4 J; 23.33 Cal   9. 13,000 Cal   10. 60 lb

Chapter 10

1. 30%; 70%   2. geothermal   3. 5.05 km/sec

Chapter 11

1. $\frac{1}{4}$ ft; 4400 Hz   2. $\frac{1}{2}$ Hz   3. $\frac{1}{3}$ Hz; 3 sec; 1.5 in   4. 9 m/sec   6. 165 m
7. 2.26 $\times$ 10^8 m/sec

Chapter 12

1. 1   2. 1000   3. 500 Hz; 750 Hz   5. Intensity is quartered when
observed from the larger distance   6. 0.5c

Chapter 13

1. 550 Cal   2. 720 Cal   3. $\frac{1}{2}$ 4. 68%   5. 4.8°C   6. Heat pump;
direct electrical heating   7. 6.67%.

Chapter 15

1. 3.6 $\times$ 10^{-4} N   2. +4 μC   3. 1.6 $\times$ 10^{-15} N   4. 10^4 V/m; zero.

Chapter 16

1. 12.48 $\times$ 10^{18} 2. 806.67 Ω 3. 0.5 A; 55 V   4. 8.18 A   5. (a) 27.5 V
(b) 55 V   (c) 5.5 A   (d) 16.5 A   7. 200 ampere-meter

Chapter 17

1. $\frac{1}{11}$ 2. (a) 182.5 kWh (b) 328.5 kWh   4. 3 days   5. (a) 15% (b) 28% (c) 25.2%

Chapter 19

1. 29.9 min   2. 11.74 light year   3. (a) 1.67 (b) 1.25   4. 67.5 years; 40.42 years.

Chapter 20

1. 9.282×10^{-31} kg   2. 1.74×10^{-27} kg   3. 1.8×10^{17} J   4. 3.6×10^{17} J   5. $\cdot 10^{13}$ J; 1.11×10^{-4} kg; 0.74 kg

Chapter 21

1. 3.31×10^{-34} J   2. 1.5×10^{14} Hz; 2×10^{-6} m   3. 3.97×10^{-19} J   4. 6.62×10^{-18} J; 41.38 eV   5. 2.73×10^{-24} kg·m/sec; 2.42×10^{-10} m   6. 1.25×10^{-23} kg·m/sec

Chapter 22

2. 4.63×10^{-19} J; 2.9 eV   3. 3; 2; 7; 2

Chapter 23

1. 8; 14; 146; 143; 145; 142; 21   2. ^{234}Th   3. ^{60}Ni   4. ^{16}O

Chapter 24

1. $\frac{1}{256}$ g   2. 17,190 years   3. 1.25×10^{-5} kg/sec; 400 kg; 56,000 kg

Index

1 2 3 4 5 6 7 8 9 0

3 DC Theory   AC Theory
1 DC Lab.   DC Lab. 1
3 Math   Math

Solid State
Lab
Math

MIN

flat tire
Chicago   MAX } pressure
color of sky or moon   MIN }   pgd
fish (scaling)
brush   MAX

$$t = \frac{1}{f}$$

$$f = \frac{1}{t}$$